New Foundations for Classical Mechanics

Fundamental Theories of Physics

A New International Book Series on The Fundamental Theories of Physics: Their Clarification, Development and Application

Editor: ALWYN VAN DER MERWE
University of Denver, U.S.A.

Editorial Advisory Board:

ASIM BARUT, *University of Colorado, U.S.A.*
HERMANN BONDI, *University of Cambridge, U.K.*
BRIAN D. JOSEPHSON, *University of Cambridge, U.K.*
CLIVE KILMISTER, *University of London, U.K.*
GÜNTER LUDWIG, *Philipps-Universität, Marburg, F.R.G.*
NATHAN ROSEN, *Israel Institute of Technology, Israel*
MENDEL SACHS, *State University of New York at Buffalo, U.S.A.*
ABDUS SALAM, *International Centre for Theoretical Physics, Trieste, Italy*
HANS-JÜRGEN TREDER, *Zentralinstitut für Astrophysik der Akademie der Wissenschaften, G.D.R.*

New Foundations for Classical Mechanics

by

David Hestenes
Department of Physics, Arizona State University, Tempe, Arizona, U.S.A.

Kluwer Academic Publishers
Dordrecht / Boston / Lancaster / Tokyo

Library of Congrèss Cataloging in Publication Data

CIP

Hestenes, David, 1933–
New foundations for classical mechanics.

(Fundamental theories of physics)
Bibliography: p.
Includes index.
1. Mechanics. I. Title. II. Series.
QA805.H58 1986 531 86–620
ISBN 90–277–2090–8
ISBN 90–277–2526–8 (pbk)

Published by Kluwer Academic Publishers,
P.O. Box 17, 3300 AA Dordrecht, The Netherlands.

Kluwer Academic Publishers incorporates
the publishing programmes of
D. Reidel, Martinus Nijhoff, Dr W. Junk and MTP Press.

Sold and distributed in the U.S.A. and Canada
by Kluwer Academic Publishers,
101 Philip Drive, Norwell, MA 02061, U.S.A.

In all other countries, sold and distributed
by Kluwer Academic Publishers,
P.O. Box 322, 3300 AH Dordrecht, The Netherlands.

First published in 1986 in hardbound edition.
Reprinted with corrections in 1987 in hardbound and paperback editions.
Reprinted in 1990 with corrections.

Printed in The Netherlands

Table of Contents

Preface ix

Chapter 1: Origins of Geometric Algebra 1
1-1. Geometry as Physics 1
1-2. Number and Magnitude 5
1-3. Directed Numbers 11
1-4. The Inner Product 16
1-5. The Outer Product 20
1-6. Synthesis and Simplification 30
1-7. Axioms for Geometric Algebra 34

Chapter 2: Developments in Geometric Algebra 39
2-1. Basic Identities and Definitions 39
2-2. The Algebra of a Euclidean Plane 48
2-3. The Algebra of Euclidean 3-Space 54
2-4. Directions, Projections and Angles 64
2-5. The Exponential Function 73
2-6. Analytic Geometry 78
2-7. Functions of a Scalar Variable 96
2-8. Directional Derivatives and Line Integrals 104

Chapter 3: Mechanics of a Single Particle 120
3-1. Newton's Program 120
3-2. Constant Force 126
3-3. Constant Force with Linear Drag 134
3-4. Constant Force with Quadratic Drag 140
3-5. Fluid Resistance 146
3-6. Constant Magnetic Field 151
3-7. Uniform Electric and Magnetic Fields 155
3-8. Linear Binding Force 164
3-9. Forced Oscillations 174
3-10. Conservative Forces and Constraints 181

Chapter 4: Central Forces and Two-Particle Systems 195
4-1. Angular Momentum 195
4-2. Dynamics from Kinematics 198
4-3. The Kepler Problem 204
4-4. The Orbit in Time 216
4-5. Conservative Central Forces 219
4-6. Two-particle Systems 230
4-7. Elastic Collisions 236
4-8. Scattering Cross Sections 243

Chapter 5: Operators and Transformations 252
5-1. Linear Operators and Matrices 252
5-2. Symmetric and Skewsymmetric Operators 263
5-3. The Arithmetic of Reflections and Rotations 277
5-4. Transformation Groups 295
5-5. Rigid Motions and Frames of Reference 306
5-6. Motion in Rotating Systems 317

Chapter 6: Many-Particle Systems 334
6-1. General Properties of Many-Particle Systems 334
6-2. The Method of Lagrange 350
6-3. Coupled Oscillations and Waves 360
6-4. Theory of Small Oscillations 378
6-5. The Newtonian Many Body Problem 398

Chapter 7: Rigid Body Mechanics 419
7-1. Rigid Body Modeling 419
7-2. Rigid Body Structure 434
7-3. The Symmetrical Top 454
7-4. Integrable Cases of Rotational Motion 476
7-5. Rolling Motion 492
7-6. Impulsive Motion 501

Chapter 8: Celestial Mechanics 512
8-1. Gravitational Forces, Fields and Torques 513
8-2. Perturbations of Kepler Motion 527
8-3. Perturbations in the Solar System 541
8-4. Spinor Mechanics and Perturbation Theory 564

Chapter 9: Foundations of Mechanics 574
9-1. Models and Theories 575
9-2. The Zeroth Law of Physics 582
9-3. Generic Laws and Principles of Particle Mechanics 588
9-4. Modeling Processes 595

Appendixes 603
A Spherical Trigonometry 603
B Elliptic Functions 610
C Units, Constants and Data 614

Hints and Solutions for Selected Exercises 616

References 632

Index 636

Preface

This is a textbook on classical mechanics at the intermediate level, but its main purpose is to serve as an introduction to a new mathematical language for physics called *geometric algebra*. Mechanics is most commonly formulated today in terms of the vector algebra developed by the American physicist J. Willard Gibbs, but for some applications of mechanics the algebra of complex numbers is more efficient than vector algebra, while in other applications matrix algebra works better. Geometric algebra integrates all these algebraic systems into a coherent mathematical language which not only retains the advantages of each special algebra but possesses powerful new capabilities.

This book covers the fairly standard material for a course on the mechanics of particles and rigid bodies. However, it will be seen that geometric algebra brings new insights into the treatment of nearly every topic and produces simplifications that move the subject quickly to advanced levels. That has made it possible in this book to carry the treatment of two major topics in mechanics well beyond the level of other textbooks. A few words are in order about the unique treatment of these two topics, namely, rotational dynamics and celestial mechanics.

The spinor theory of rotations and rotational dynamics developed in this book cannot be formulated without geometric algebra, so a comparable treatment is not to be found in any other book at this time. The relation of the spinor theory to the matrix theory of rotations developed in conventional textbooks is completely worked out, so one can readily translate from one to the other. However, the spinor theory is so superior that the matrix theory is hardly needed except to translate from books that use it. In the first place, calculations with spinors are demonstrably more efficient than calculations with matrices. This has practical as well as theoretical importance. For example, the control of artificial satellites requires continual rotational computations that soon number in the millions. In the second place, spinors are essential in advanced quantum mechanics. So the utilization of spinors in the classical theory narrows the gap between the mathematical formulations of classical and quantum mechanics, making it possible for students to proceed more rapidly to advanced topics.

Celestial mechanics, along with its modern relative astromechanics, is essential for understanding space flight and the dynamics of the solar system. Thus, it is essential knowledge for the informed physicist of the space age. Yet celestial mechanics is scarcely mentioned in the typical undergraduate physics curriculum. One reason for this neglect is the belief that the subject is too advanced, requiring a complex formulation in terms of Hamilton-Jacobi theory. However, this book uses geometric algebra to develop a new formulation of perturbation theory in celestial mechanics which is well within the reach of undergraduates. The major gravitational perturbations in the solar system are discussed to bring students up to date in space age mechanics. The new mathematical techniques developed in this book should be of interest to anyone concerned with the mechanics of space flight.

The last chapter of this book presents a new analysis of the foundations of mechanics. The main objective is a formulation of mechanics which is *complete*, in the sense that the essential premises of the theory are explicitly formulated, and *externally coherent*, in the sense that it articulates smoothly with neighboring branches of physics, principally electromagnetic theory and special relativity. The entire analysis is carried out from the perspective of *Modeling Theory*, a general theory about the development and deployment of mathematical models proposed as a definite philosophy of science.

To provide an introduction to geometric algebra suitable for the entire physics curriculum, the mathematics developed in this book exceeds what is strictly necessary for a mechanics course, including a substantial treatment of linear algebra and transformation groups with the techniques of geometric algebra. Since linear algebra and group theory are standard tools in modern physics, it is important for students to become familiar with them as soon as possible. There are good reasons for integrating instruction in mathematics and physics. It assures that the mathematical background will be sufficient for the needs of physics, and the physics provides nontrivial applications of the mathematics as it develops. But most important, it affords an opportunity to teach students that the design and development of an efficient mathematical language for representing physical facts and concepts is the business of theoretical physics. That is one of the objectives of this book.

In a sequel to this book called *New Foundations for Mathematical Physics* (NFII), the geometric algebra developed here will be extended to a complete mathematical language for electrodynamics, relativity and quantum theory, in short, a unified language for physics. The chapter on relativity in NFII is a smooth continuation of the present book, so it could easily be included at the end of a two semester course on mechanics.

The most complete available treatment of geometric algebra and calculus is given in *Clifford Algebra to Geometric Calculus*, published by Reidel in the same series as the present book. That book is written at an advanced mathematical level and contains no direct applications to physics, so it is not recommended for beginners. However, it should be useful to mathematicians and theoretical physicists, and it will be much more accessible to readers who have mastered the mathematics in the present book.

The making of this book turned out to be much more difficult than I had anticipated, and could not have been completed without help from many sources. I am indebted to my NASA colleagues for educating me on the vicissitudes of celestial mechanics; in particular, Phil Roberts on orbital mechanics, Neal Hulkower on the three body problem, and, especially, Leon Blitzer for permission to draw freely on his lectures. I am indebted to Patrick Reany, Anthony Delugt and John Bergman for improving the accuracy of the text, and to Carmen Mendez and Denise Jackson for their skill and patience in typing a difficult manuscript. Most of all I am indebted to my wife Nancy for her unflagging support and meticulous care in preparing every one of the diagrams. Numerous corrections have been made in the second and third printings. I am grateful to the many students and colleagues who have helped root out the errors.

DAVID HESTENES

Chapter 1

Origins of Geometric Algebra

There is a tendency among physicists to take mathematics for granted, to regard the development of mathematics as the business of mathematicians. However, history shows that most mathematics of use in physics has origins in successful attacks on physical problems. The advance of physics has gone hand in hand with the development of a mathematical language to express and exploit the theory. Mathematics today is an immense and imposing subject, but there is no reason to suppose that the evolution of a mathematical language for physics is complete. The task of improving the language of physics requires intimate knowledge of how the language is to be used and how it refers to the physical world, so it involves more than mathematics. It is one of the fundamental tasks of theoretical physics.

This chapter sketches some historical high points in the evolution of geometric algebra, the mathematical language developed and applied in this book. It is not supposed to be a balanced historical account. Rather, the aim is to identify explicit principles for constructing symbolic representations of geometrical relations. Then we can see how to *design* a compact and efficient geometrical language tailored to meet the needs of theoretical physics.

1-1. Geometry as Physics

Euclid's systematic formulation of Greek geometry (in 300 BC) was the first comprehensive theory of the physical world. Earlier attempts to describe the physical world were hardly more than a jumble of facts and speculations. But Euclid showed that from a mere handful of simple assumptions about the nature of physical objects a great variety of remarkable relations can be deduced. So incisive were the insights of Greek geometry that it provided a foundation for all subsequent advances in physics. Over the years it has been extended and reformulated but not changed in any fundamental way.

The next comparable advance in theoretical physics was not consummated until the publication of Isaac Newton's *Principia* in 1687. Newton was fully aware that geometry is an indispensible component of physics; asserting,

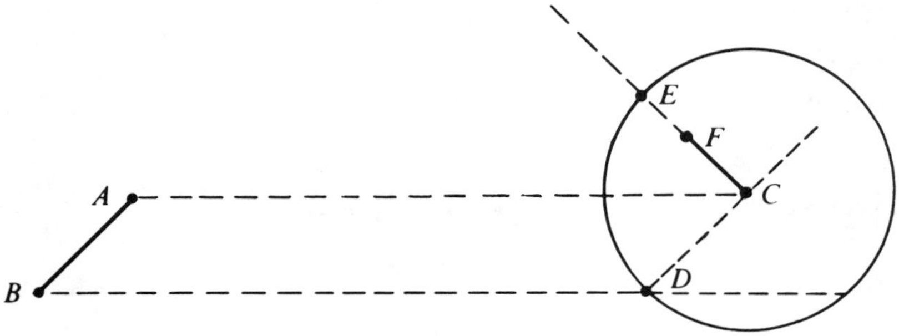

Fig. 1.1. Congruence of Line Segments.

Euclid's axioms provide rules which enable one to compare any pair of line segments. Segments AB and CF can be compared as follows.

First, a line parallel to AB can be drawn through C. And a line parallel to AC can be drawn through B. The two lines intersect at a point D. The line segment CD is *congruent* to AB.

Second, a circle with center C can be drawn through D. It intersects the line CF at a point E. The segment CE is congruent to CD and, by the assumed transitivity of the relation, congruent to AB.

Third, the point F is either inside, on, or outside the circle, in which cases we say that the *magnitude* of CF is respectively, less than, equal to, or greater than the magnitude of AB.

The procedure just outlined can, of course, be more precisely characterized by a formal deductive argument. But the point to be made here is that this procedure can be regarded as a theoretical formulation of basic physical operations involved in measurement.

If AB is regarded as the idealization of a standard stick called a "ruler", the first step above may be regarded as a description of the *translation* of the stick to the place CD without changing its magnitude. Then the second step idealizes the reorientation of the ruler to place it contiguous to an idealized body CF so that a comparison (third step) can be made. Further assumptions are needed to supply the ruler with a "graduated scale" and so assign a unique magnitude to CF.

". . . the description of right lines and circles, upon which geometry is founded, belongs to mechanics. Geometry does not teach us to draw these lines, but requires them to be drawn . . . To describe right lines and circles are problems, but not geometrical problems. The solution of these problems is required from mechanics and by geometry the use of them, when so solved, is shown; and it is the glory of geometry that from those few principles, brought from without, it is able to produce so many things. *Therefore geometry is founded in mechanical practice, and is nothing but that part of universal mechanics which accurately proposes and demonstrates the art of measuring* . . ." (italics added)

As Newton avers, geometry is the theory on which the practice of measurement is based. Geometrical figures can be regarded as idealizations of physical bodies. The theory of congruent figures is the central theme of geometry, and it provides a theoretical basis for measurement when it is regarded as an idealized description of the physical operations involved in classifying physical bodies according to size and shape (Figure 1.1). To put it

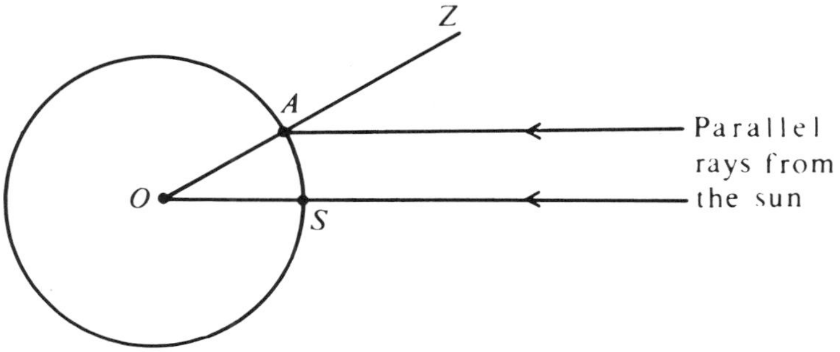

Fig. 1.2. Measurement of the Earth.

The most accurate of the early measurements of the earth's circumference was made by Eratosthenes (~200 BC). He observed that at noon on the day of the summer solstice the sun shone directly down a deep well at Syene. At the same time at Alexandria, taken to be due north and 5000 stadia (≈500 miles) away, the sun cast a shadow indicating it was 1/50 of a circle from zenith. By the equality of corresponding angles in the diagram this gives $50 \times 500 = 25\,000$ miles for the circumference of the earth.

another way, the theory of congruence specifies a set of rules to be used for classifying bodies. Apart from such rules the notions of size and shape have no meaning.

Greek geometry was certainly not developed with the problem of measurement in mind. Indeed, even the idea of measurement could not be conceived until geometry had been created. But already in Euclid's day the Greeks had carried out an impressive series of applications of geometry, especially to optics and astronomy (Figure 1.2), and this established a pattern to be followed in the subsequent development of trigonometry and the practical art of measurement. With these efforts the notion of an experimental science began to take shape.

Today, "to measure" means to assign a number. But it was not always so. Euclid sharply distinguished "number" from "magnitude". He associated the notion of number strictly with the operation of counting, so he recognized only integers as numbers; even the notion of fractions as numbers had not yet been invented. For Euclid a magnitude was a line segment. He frequently represented a whole number n by a line segment which is n times as long as some other line segment chosen to represent the number 1. But he knew that the opposite procedure is impossible, namely, that it is impossible to distinguish all line segments of different length by labeling them with numerals representing the counting numbers. He was able to prove this by showing the side and the diagonal of a square cannot both be whole multiples of a single unit (Figure 1.3).

The "one way" correspondence of counting numbers with magnitudes shows that the latter concept is the more general of the two. With admirable consistency, Euclid carefully distinguished between the two concepts. This is born out by the fact that he proves many theorems twice, once for numbers and once for magnitudes. This rigid distinction between number and magnitude proved to be an impetus to progress in some directions, but an impediment to progress in others.

As is well known, even quite elementary problems lead to quadratic equations with solutions which are not integers or even rational numbers. Such problems have no solutions at all if only integers are recognized as numbers. The Hindus and the Arabs resolved this difficulty directly by generalizing their notion of number, but Euclid sidestepped it cleverly by reexpressing problems in arithmetic and algebra as problems in geometry. Then he solved for line segments instead of for numbers. Thus, he represented the product x^2 as a square with a side of magnitude x. In fact, that is why we use the name "x squared" today. The product xy was represented by a rectangle and called the "rectangle" of the two sides. The term "x cubed" used even today originates from the representation of x^3 by a cube with side of magnitude x. But there are no corresponding representations of x^4 and higher powers of x in Greek geometry, so the Greek correspondence between algebra and geometry broke down. This "breakdown" impeded mathematical progress from antiquity until the seventeenth century, and its import is seldom recognized even today.

Commentators sometimes smugly dismiss Euclid's practice of turning every

Fig. 1.3. The diagonal of a square is incommensurable with its side.

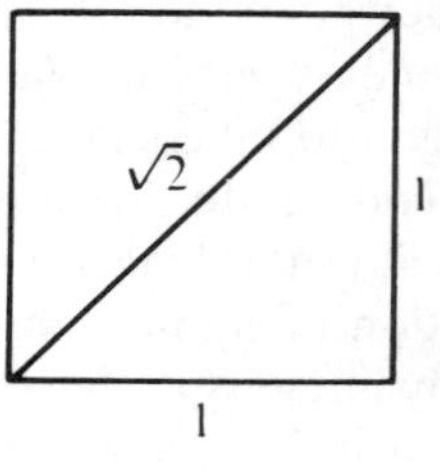

This can be proved by showing that its contrary leads to a contradiction. Supposing, then, that a diagonal is an m-fold multiple of some basic unit while a side is an n-fold multiple of the same unit, the Pythagorean Theorem implies that $m^2 = 2n^2$. This equation shows that the integers m and n can be assumed to have no common factor, and also that m^2 is even. But if m^2 is even, then m is even, and m^2 has 4 as a factor. Since $n^2 = 1/2m^2$, n^2 and so n is also even. But the conclusion that m and n are both even contradicts the assumption that they do not have a common factor.

Euclid gave an equivalent proof using geometric methods. The proof shows that $\sqrt{2}$ is not a rational number, that is, not expressible as a ratio of two integers. The Greeks could represent $\sqrt{2}$ by a line segment, the diagonal of a unit square. But they had no numeral to represent it.

algebra problem into an equivalent geometry problem as an inferior alternative to modern algebraic methods. But we shall find good reasons to conclude that, on the contrary, they have failed to grasp a subtlety of far-reaching significance in Euclid's work. The real limitations on Greek mathematics were set by the failure of the Greeks to develop a simple symbolic language to express their profound ideas.

1-2. Number and Magnitude

The brilliant flowering of science and mathematics in ancient Greece was followed by a long period of scientific stagnation until an explosion of scientific knowledge in the seventeenth century gave birth to the modern world. To account for this explosion and its long delay after the impressive beginnings of science in Greece is one of the great problems of history. The "great man" theory implicit in so many textbooks would have us believe that the explosion resulted from the accidental birth of a cluster of geniuses like Kepler, Galileo and Newton. "Humanistic theories" attribute it to the social, political and intellectual climate of the Renaissance, stimulated by a rediscovery of the long lost culture of Greece. The invention and exploitation of the experimental method is a favorite explanation among philosophers and historians of science. No doubt all these factors are important, but the most critical factor is often overlooked. The advances we know as modern science were not possible until an adequate number system had been created to express the results of measurement, and until a simple algebraic language had been invented to express relations among these results. While social and political disorders undoubtedly contributed to the decline of Greek culture, deficiencies in the mathematical formalism of the Greek science must have been an increasingly powerful deterrent to every scientific advance and to the transmission of what had already been learned. The long hiatus between Greek and Renaissance science is better regarded as a period of incubation instead of stagnation. For in this period the decimal system of arabic numerals was invented and algebra slowly developed. It can hardly be an accident that an explosion of scientific knowledge was ignited just as a comprehensive algebraic system began to take shape in the sixteenth and seventeenth centuries.

Though algebra was associated with geometry from its beginnings, René Descartes was the first to develop it systematically into a geometrical language. His first published work on the subject (in 1637) shows how clearly he had this objective in mind:

"Any problem in geometry can easily be reduced to such terms that a knowledge of the lengths of certain straight lines is sufficient for its construction. Just as arithmetic consists of only four or five operations, namely, addition, subtraction, multiplication, division and the extraction of roots, which may be considered a kind of division, so in geometry, to find required lines it is merely

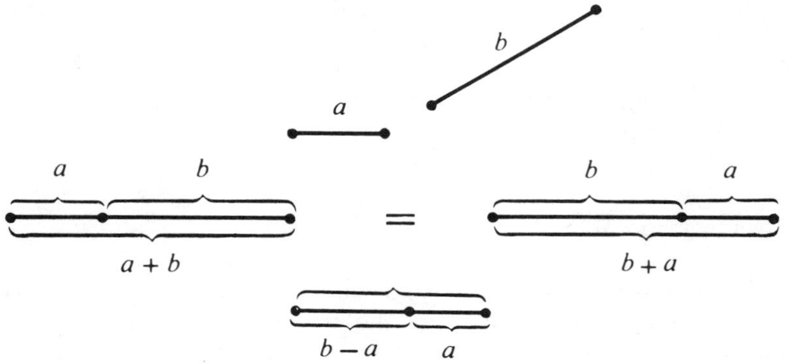

Fig. 2.1a. Addition and Subtraction of Line Segments.

The geometrical theory of congruence (illustrated in Figure 1) gives a precise mathematical expression of the idea that a line segment can be moved around without changing its length. Taking congruence for granted, two line segments labeled by their lengths a and b can be joined end to end to create a new line segment with length $a + b$. This is the geometrical equivalent of the addition of numbers a and b. The commutative law of addition is reflected in the fact that the line segments can be joined at either end with the same result. The geometrical analog of subtraction is obtained as illustrated by joining the line segments to create a new line segment of length $b-a$. Since the length of a line segment is a positive number, negative numbers cannot be represented geometrically by labeling line segments by length alone.

necessary to add or subtract lines; or else, taking one line which I shall call unity in order to relate it as closely as possible to numbers, and which can in general be chosen arbitrarily, and having given two other lines, to find a fourth line which shall be to one of the given lines as the other is to unity (which is the same as multiplication); or, again, to find a fourth line which is to one of the given lines as unity is to the other (which is equivalent to division); or, finally to find one, two, or several mean proportionals between unity and some other line (which is the same as extracting the square root, cube root, etc., of the given line). And I shall not hesitate to introduce these arithmetical terms into geometry, for the sake of greater clearness . . ."

Descartes gave the Greek notion of magnitude a happy symbolic form by assuming that every line segment can be uniquely represented by a number. He was the first person to label line segments by letters representing their numerical lengths. As he demonstrated, the aptness of this procedure resides in the fact that the basic arithmetic operations such as addition and subtraction can be supplied with exact analogs in geometrical operations on line segments (Figures 2.1a, 2.1b). One of his most significant innovations was to discard the Greek idea of representing the "product" of two line segments by a rectangle. In its stead he gave a rule for "multiplying" line segments which yielded another line segment in exact correspondence with the rule for multiplying numbers (Figure 2.2). This enabled him to avoid the apparent

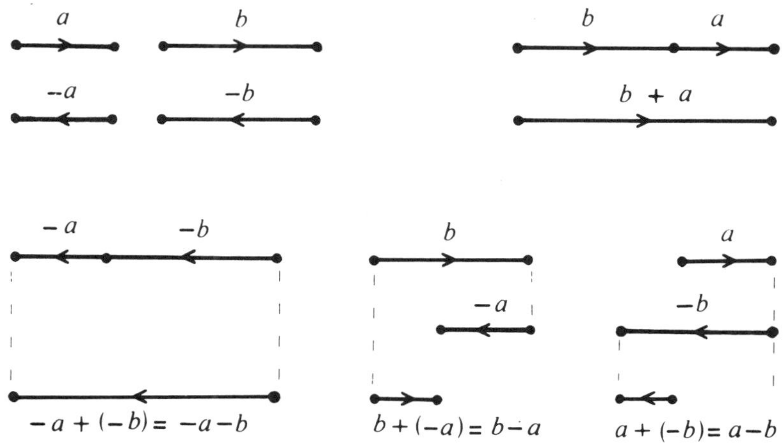

Fig. 2.1b. Oriented Line Segments.

Line segments can be assigned an *orientation* or *sense* as well as a length. There are exactly two different orientations called positive and negative; they are denoted respectively, by the signs + and − in arithmetic and represented by arrow heads in the above diagram. If a line segment is labeled by a, then $-a$ denotes a line segment of the same length but opposite orientation.

As indicated in the diagram, oriented line segments are added by joining them end to end to produce a new line segment with a unique orientation. Subtraction is reduced to addition; subtraction by a is defined as addition of $-a$.

Orientation is a geometric notion which has been given a symbolic rendering in algebra by the signs + and −.

Descartes had not grasped the notion of orientation. This accounts for the fact that he was prone to error when a problem called for the geometric representation of a negative number.

limitations of the Greek rule for "geometrical multiplication". Descartes could handle geometrical products of any order and he put this new ability to good use by showing how to use algebraic equations to describe geometric curves. This was the beginning of analytic geometry and a crucial step in the development of the mathematical language that makes modern physics possible. Finally, Descartes made significant improvements in algebraic notations, putting algebra in a form close to the one we use today.

It has been said that the things a man takes for granted is a measure of his debt to his culture. The assumption of a complete correspondence between numbers and line segments was the foundation of Descartes' union of geometry and algebra. A careful Greek logician like Eudoxus, would have demanded some justification for such a farreaching assumption. Yet, Descartes' contemporaries accepted it without so much as a raised eyebrow. It did not seem revolutionary to them, because they were accustomed to it. In fact,

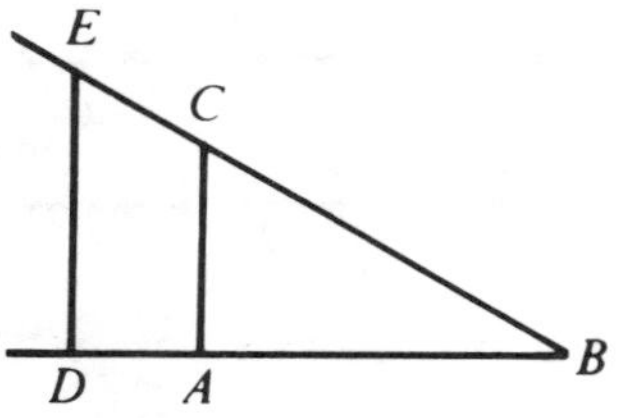

Fig. 2.2. Multiplication of Line Segments.

Given line segments BD of length a and BC of length b, Descartes constructed a new line segment BE of length ab by the following procedure: First lay off the segment AB of unit length. Then draw the line through D parallel to AC; it intersects the line BC at the point E.

To make the correspondence between algebra and geometry more direct, Descartes introduced the notation

$$a = BD, \; b = BC,$$
$$AB = 1, \; BE = ab,$$

(except he used the symbol α instead of =).

The fact that BE has length ab justifies calling the geometrical construction of BE "multiplication". A geometrical proof of this fact had already been given by Euclid in his account of the Greek "theory of proportions". Descartes' rule gave the Greek theory at last a proper symbolic expression in algebraic language.

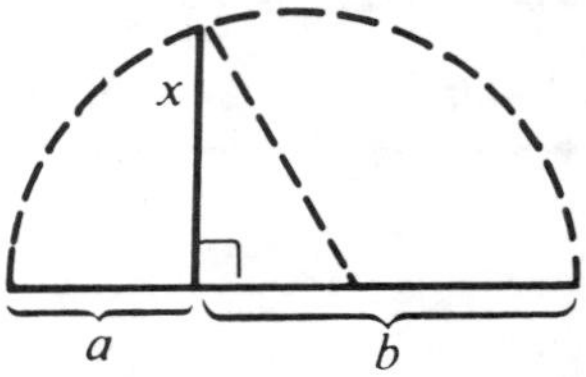

Fig. 2.3. Extraction of Roots.

Given line segments a and b, construct a half circle with diameter $a + b$. Construct the half chord x intersecting the diameter a distance a from the circle. The Pythagorean theorem applied to the triangle in the above diagram gives $x = \sqrt{ab}$, or $x = \sqrt{a}$ if $b = 1$. So the construction of x is a geometric analog of the arithmetic computation of a square root.

This construction appears in Book II of Euclid's *Elements*, from whence, no doubt, Descartes obtained it. But Descartes expresses it in the language of algebra.

algebra and arithmetic had never been free of some admixture of geometry. But a union could not be consummated until the notion of number and the symbolism of algebra had been developed to a degree commensurate with Greek geometry. That state of affairs had just been reached when Descartes arrived on the historical scene.

Descartes stated explicitly what everyone had taken for granted. If Descartes had not done it someone else would have. Indeed, Fermat independently achieved quite similar results. But Descartes penetrated closer to the heart of the matter. His explicit union of the notion of number with the Greek geometric notion of magnitude sparked an intellectual explosion unequalled in all history.

Descartes was not in the habit of acknowledging his debt to others, but in a letter to Mersenne in 1637 he writes,

> "As to the suggestion that what I have written could easily have been gotten from Vieta, the very fact that my treatise is hard to understand is due to my attempt to put nothing in it that I believed to be known either by him or by any one else. . . . I begin the rules of my algebra with what Vieta wrote at the very end of his book, . . . Thus, I begin where he left off."

The contribution of Vieta has been too frequently undervalued. He is the one who explicitly introduced the idea of using letters to represent constants as well as unknowns in algebraic equations. This act lifted algebra out of its infancy by separating the study of special properties of individual numbers from the abstract study of the general properties of all numbers. It revealed the dependence of the number concept on the nature of algebraic operations. Vieta used letters to denote numbers, and Descartes followed him by using letters to denote line segments. Vieta began the abstract study of rules for manipulating numbers, and Descartes pointed out the existence of similar rules for manipulating line segments. Descartes gives some improvements on the symbolism and algebraic technique of Vieta, but it is hard to say how much of this comes unacknowledged from the work of others. Before Vieta's innovations, the union of geometry and algebra could not have been effected.

The correspondence between numbers and line segments presumed by Descartes can be most simply expressed as the idea that numbers can be put into one to one correspondence with the points on a geometrical line (Figure 2.4). This idea seems to be nearly as old as the idea of a geometrical line itself. The Greeks may have believed it at first, but they firmly rejected it when incommensurables were discovered (Figure 1.3). Yet Descartes and his contemporaries evidently regarded it as obvious. Such a significant change in attitude must have an interesting history! Of course, such a change was possible only because the notion of number underwent a profound evolution.

Diophantes (250 AD), the last of the great Greek mathematicians, was probably the first to regard fractions as numbers. But the development most pertinent to the present discussion was the invention of algebraic numbers. This came about by presuming the existence of solutions to algebraic equations and devising symbols to represent them. Thus the symbol was invented to designate a solution of the equation $x^2 = 2$. Once the symbol $\sqrt{2}$ had been invented, it was hard to deny the reality of the number it names, and this number takes on a more concrete appearance when identified as the diagonal length of a unit square. In this way the incommensurables of the Greeks received number names at last, and with no reason to the contrary, it must have seemed natural to assume that all points on a line can be named by numbers. So it seemed to Descartes. Perhaps it is a good thing that there was no latter-day Eudoxus to dampen Descartes' ardor by proving that it is

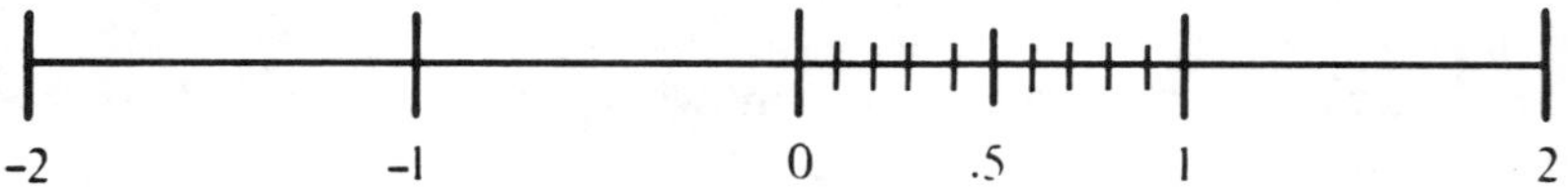

Fig. 2.4. The number line.

The points on a line can be put into one to one correspondence with numbers by labeling (naming) them with decimal numerals. More specifically, every point can be uniquely labeled by an (infinite) decimal numeral with the form

$$\pm\, a_1a_1 \ldots a_m \cdot b_1b_2 \ldots b_n \ldots .$$

Here, of course the a's and b's can have only integer values from zero to nine, and m and n represent natural numbers.

The "real number line" can be defined "arithmetically" simply as the set of all decimal numerals. This definition may seem to be devoid of any geometrical content. However, the familiar arithmetic rules for adding and multiplying numbers in decimal form correspond exactly to the rules needed to define a geometrical construction. Moreover, a decimal numeral can be interpreted as a set of instructions for the unique determination of a geometrical point (or, equivalently, for the construction of a line segment) by elementary geometrical operations. Only a few simple conventions are required:

(1) On the given line a point must be chosen and labeled zero.
(2) The two orientations of the line must be labeled positive and negative.
(3) A convenient line segment must be chosen as a unit.

Then the "integer part" of the numeral can be interpreted as an instruction to begin at zero and lay off $a_1a_2 \ldots a_m$ units in the direction designated by the sign of the numeral. The "decimal part" of the numeral $b_1b_2 \ldots b_n \ldots$ then "says" to divide the next consecutive unit segment into ten equal parts and "move forward" b_1 of these units, etc. If the decimal is infinite, an infinite sequence of geometrical operations will be needed to determine the point.

It should be noted that geometry requires that lines, points and units be *given*, but the nature of these entities is actually determined only by the geometrical relations they enter into. Geometry only specifies rules. Arithmetic may be regarded as a formulation of geometrical rules without reference to undefined entities. This does not mean that undefined entities can be dispensed with. They are essential when arithmetic and geometry are applied to the physical world; then, a unit is typically *given* in the form of a physical object, and a line may be *given* by a ray of light; then a number typically specifies operations which have been or are to be carried out on physical objects. In modern physics the relations of mathematical entities to physical objects and operations are extremely intricate.

impossible to name every point on a line by an algebraic number. Descartes did not even suspect that the circumference of a unit circle is not an algebraic number, but then, that was not proved until 1882.

Deficiencies in the notion of number were not felt until the invention of calculus called for a clear idea of the "infinitely small". A clear notion of "infinity" and with it a clear notion of the "continuum of real numbers" was not achieved until the latter part of the nineteenth century, when the real number system was "arithmeticized" by Weierstrass, Cantor and Dedekind. "Arithmeticize" means define the real numbers in terms of the natural numbers and their arithmetic, without appeal to any geometric intuition of "the continuum". Some say that this development separated the notion of number from geometry. Rather the opposite is true. It consumated the union of number and geometry by establishing at last that the real numbers can be put into one to one correspondence with the points on a geometrical line. The arithmetical definition of the "real numbers" gave a precise symbolic expression to the intuitive notion of a continuous line (Figure 2.4).

Descartes began the explicit cultivation of algebra as a symbolic system for representing geometric notions. The idea of number has accordingly been generalized to make this possible. But the evolution of the number concept does not end with the invention of the real number system, because there is more to geometry than the linear continuum. In particular, the notions of direction and dimension cry out for a proper symbolic expression. The cry has been heard and answered.

1-3. Directed Numbers

After Descartes, the use of algebra as a geometric language expanded with ever mounting speed. So rapidly did success follow success in mathematics and in physics, so great was the algebraic skill that developed that for sometime no one noticed the serious limitations of this mathematical language.

Descartes expressed the geometry of his day in the algebra of his day. It did not occur to him that algebra could be modified to achieve a fuller symbolic expression of geometry. The algebra of Descartes could be used to classify line segments by length. But there is more to a line segment than length; it has direction as well. Yet the fundamental geometric notion of direction finds no expression in ordinary algebra. Descartes and his followers made up for this deficiency by augmenting algebra with the ever ready natural language. Expressions such as "the x-direction" and "the y-direction" are widely used even today. They are not part of algebra, yet ordinary algebra cannot be applied to geometry without them.

Mathematics has steadily progressed by fashioning special symbolic systems to express ideas originally expressed in the natural language. The first mathematical system, Greek geometry, was formulated entirely in the natural language. How else was mathematics to start? But, to use the words of Descartes, algebra makes it possible to go "as far beyond the treatment of ordinary geometry, as the rhetoric of Cicero is beyond the *a*, *b*, *c*, of children". How much more can be expected from further refinements of the geometrical language?

The generalization of number to incorporate the geometrical notion of direction as well as magnitude was not carried out until some two hundred years after Descartes. Though several people might be credited with conceiving the idea of "*directed number*", Herman Grassmann, in his book of 1844, developed the idea with precision and completeness that far surpassed the work of anyone else at the time. Grassmann discovered a rule for relating line segments to numbers that differed slightly from the rule adopted by Descartes, and this led to a more general notion of number.

Before formulating the notion of a directed number, it is advantageous to substitute the short and suggestive word "*scalar*" for the more common but clumsy expression "*real number*", and to recall the key idea of Descartes' approach. Descartes united algebra and geometry by corresponding the arithmetic of scalars with a kind of arithmetic of line segments. More specifically, if two line segments are congruent, that is, if one segment can be obtained from the other by a translation and rotation, then Descartes would designate them both by the same positive scalar. Conversely, every positive scalar designates a "line segment" which possesses neither a place nor a direction, all congruent line segments being regarded as one and the same. Or, to put it in modern mathematical terminology, every positive scalar designates an *equivalence class* of congruent line segments. This is the rule used by Descartes to relate numbers to line segments.

Alternatively, Grassmann chose to regard two line segments as "*equivalent*" *if and only if* one can be obtained from the other by a translation; only then would he designate them by the same symbol. If a rotation was required to obtain one line segment from another, he regarded the line segments as "possessing different directions" and so designated them by different symbols. These conventions lead to the idea of a "directed line segment" or *vector* as a line segment which can be moved freely from place to place without changing either its magnitude or its direction. To achieve a simple symbolic expression of this idea and yet distinguish vectors from scalars, vectors will be represented by letters in bold face type. If two line segments, designated by vectors **a** and **b** respectively, have the same magnitude and direction, then the vectors are said to be *equal*, and, as in scalar algebra, one writes

$$\mathbf{a} = \mathbf{b}. \tag{3.1}$$

Of course the use of vectors to express the geometrical fact that line segments may differ in direction does not obviate the value of classifying line segments by length. But a simple formulation of the relation between scalars and vectors is called for. It can be achieved by observing that to every vector **a** there corresponds a unique positive scalar, here denoted by $|\mathbf{a}|$ and called the "magnitude" or the "length" of **a**. This follows from the correspondence between scalars and line segments which has already been discussed. Suppose, now, that a vector **b** has the same direction as **a**, but $|\mathbf{b}| = \lambda |\mathbf{a}|$, where λ is a positive scalar. This can be expressed simply by writing

$$\mathbf{b} = \lambda \mathbf{a}. \tag{3.2}$$

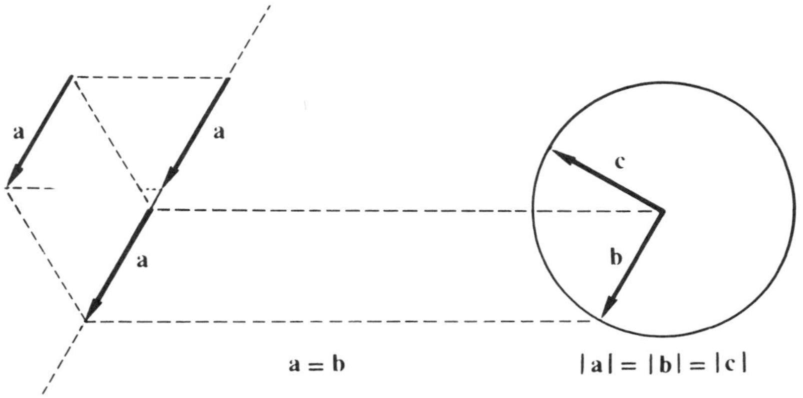

Fig. 3.1. Vectors and Arrows.

A vector **a** can be pictured as a "directed line segment" or arrow. The length of the arrow corresponds to the magnitude of the vector. Arrows representing vectors with the same direction lie on parallel lines. Arrows with the same length and direction can be regarded as different representations of one and the same vector no matter where they are located; so they can be labeled by one and the same vector symbol.

Arrows that do not lie on parallel lines must be labeled by different vector symbols, even if they have the same length.

But this can be interpreted as an equation defining the multiplication of a vector by a scalar. Thus, multiplication by a positive scalar changes the magnitude of a vector but not its direction (Figure 3.2). This operation is commonly called a *dilation*. If $\lambda > 1$, it is an *expansion*, since then $|\mathbf{b}| > |\mathbf{a}|$. But if $\lambda < 1$, it is a *contraction*, since then $|\mathbf{b}| < |\mathbf{a}|$. Descartes' geometrical construction for "multiplying" two line segments (Figure 2.2) is a dilation of one line segment by the magnitude of the other.

Equation (3.2) allows one to write

$$\mathbf{a} = |\mathbf{a}|\,\hat{\mathbf{a}}\,,\ \text{where}\ |\hat{\mathbf{a}}| = 1. \qquad (3.3)$$

This expresses the vector **a** as the product of its magnitude $|\mathbf{a}|$ with a "unit vector" $\hat{\mathbf{a}}$. The "unit" $\hat{\mathbf{a}}$ uniquely specifies the direction

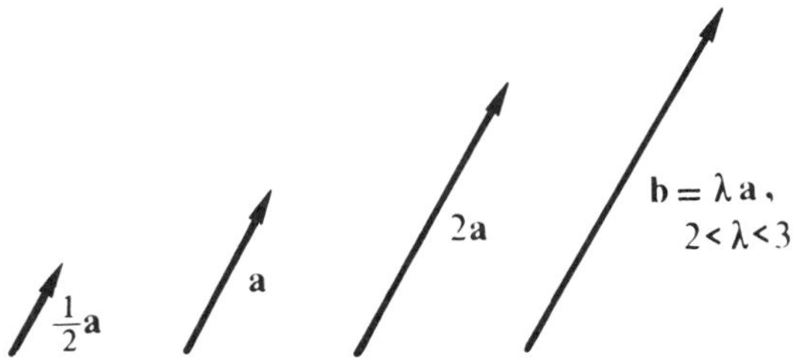

Fig. 3.2. Illustrating dilation.

Multiplication by a positive scalar changes the length but not the direction of a vector.

of **a**, so Equation (3.3) can be regarded as a decomposition of **a** into magnitude and direction.

If Equation (3.2) is supposed to hold, then multiplication of a vector **a** by zero results in a vector with zero magnitude. Express this by writing

$$(0)\mathbf{a} = 0. \tag{3.4}$$

Since the direction associated with a line segment of zero length seems to be of no consequence, it is natural to assume that the zero vector on the right side of (3.4) is a unique number no matter what the direction of **a**. Moreover, it will be seen later that there is good reason to regard the zero vector as one and the same number as the zero scalar. So the zero on the right side of (3.4) is not written in bold face type.

Grassmann may have been the first person to clearly understand that the significance of a number lies not in itself but solely in its relation to other numbers. The notion of number resides in the rules for combining two numbers to get a third. Grassmann looked for rules for combining vectors which would fully describe the geometrical properties of directed line segments. He noticed that two directed line segments connected end to end determine a third, which may be regarded as their sum. This "geometrical sum" of directed line segments can be simply represented by an equation for corresponding vectors **a**, **b**, **s**:

$$\mathbf{a} + \mathbf{b} = \mathbf{s}. \tag{3.5}$$

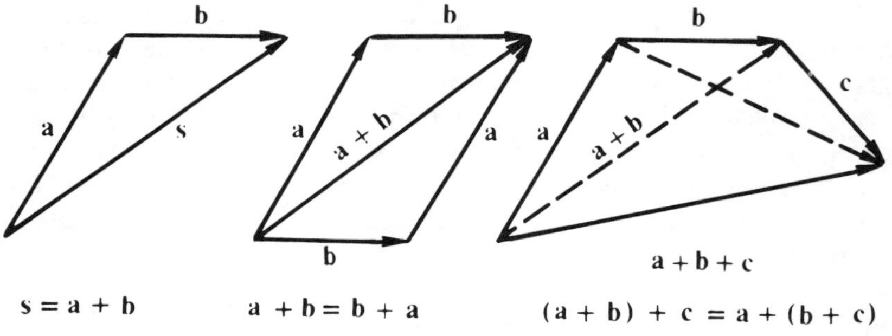

Fig. 3.3. Addition of Arrows.

Two arrows, labeled respectively by vectors **a** and **b** above can be "added" geometrically by joining the tip of one with the tail of the other and drawing in the arrow labeled **s** that connects the remaining tip and tail. The properties of vector addition are determined by assuming that addition of vectors **a** and **b** to get **s** corresponds exactly to this geometrical construction.

Since the same arrow **s** is obtained whether the tip of **a** is joined to the tail of **b** or the tip of **b** is joined to the tail of **a**, vector addition must be commutative.

Since, as the figure shows, the result of adding arrows **a** + **b** to **c** is the same as the result of adding **a** to **b** + **c**, vector addition must be associative.

This procedure is like the one used by Descartes to relate "geometrical addition of line segments" to addition of scalars, except that the definition of "geometrically equivalent" line segments is different.

The rules for adding vectors are determined by the assumed correspondence with directed line segments. As shown in Figure 3.3, vector addition, like scalar addition, must obey the commutative rule,

$$\mathbf{a} + \mathbf{b} = \mathbf{b} + \mathbf{a}, \tag{3.6}$$

and the associative rule,

$$(\mathbf{a} + \mathbf{b}) + \mathbf{c} = \mathbf{a} + (\mathbf{b} + \mathbf{c}). \tag{3.7}$$

As in scalar algebra, the number zero plays a special role in vector addition. Thus,

$$\mathbf{a} + 0 = \mathbf{a}. \tag{3.8}$$

Moreover, to every vector there corresponds one and only one vector **b** which satisfies the equation

$$\mathbf{a} + \mathbf{b} = 0. \tag{3.9}$$

This unique vector is called the *negative* of **a** and denoted by $-\mathbf{a}$ (Figure 3.4).

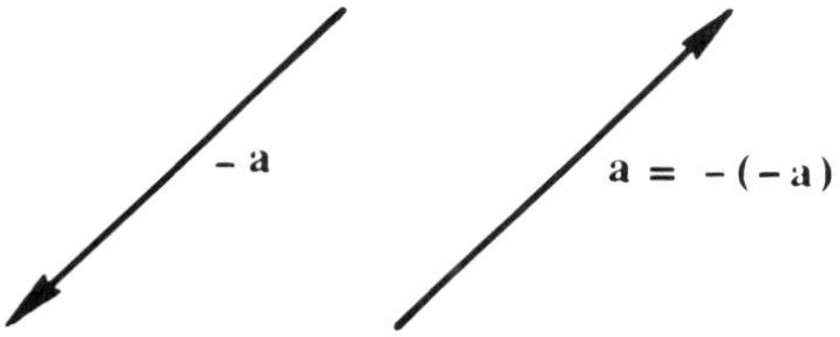

Fig. 3.4. Negative vectors.

An arrow representing $-\mathbf{a}$ differs from one representing **a** only in having its tip point in the opposite direction. We say that **a** and $-\mathbf{a}$ have *opposite orientation*.

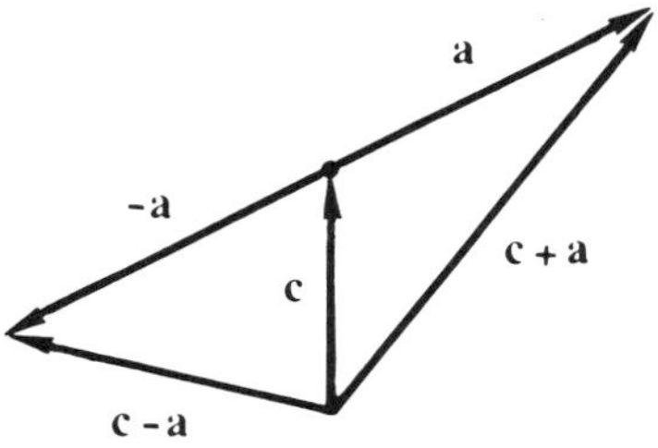

Fig. 3.5. Comparing addition and subtraction of arrows and vectors.

The existence of *negatives* makes it possible to define subtraction as addition of a negative. Thus

$$\mathbf{c} - \mathbf{a} = \mathbf{c} + (-\mathbf{a}). \tag{3.10}$$

Subtraction and addition are compared in Figure 3.5.

The existence of negatives also makes it possible to define multiplication by the scalar -1 by the equation

$$(-1)\mathbf{a} = -\mathbf{a}. \tag{3.11}$$

This equation justifies interpreting -1 as a representation of the operation of reversing direction (i.e. orientation, as explained in Figure 2.1b). Then the equation $(-1)^2 = 1$ simply expresses the obvious geometrical fact that by reversing direction twice one reproduces the original direction. In this way the concept of directed numbers leads to an operational interpretation of negative numbers.

Now that the geometrical meaning of multiplication by minus one is understood, it is obvious that Equation (3.2) is meaningful even if λ is a negative scalar. Vectors which are scalar multiples of one another are said to be *codirectional* or *collinear*.

1-4. The Inner Product

A great many significant geometrical theorems can be simply expressed and proved with the algebraic rules for vector addition and scalar multiplication which have just been set down. However, the algebraic system as it stands cannot be regarded as a complete symbolic expression of the geometric notions of magnitude and direction, because it fails to fully indicate the difference between scalars and vectors. This difference is certainly not reflected in the rules for addition, which are the same for both scalars and vectors. In fact, the distinction between scalars and vectors still resides only in their geometric interpretations, that is, in the different rules used to correspond them with line segments.

The opportunity to give the notion of direction a full algebraic expression arises when the natural question of how to multiply vectors is entertained. Descartes gave a rule for "multiplying" line segments, but his rule does not depend on the direction of the line segments, and it already has an algebraic expression as a dilation. Yet the general approach of Descartes can be followed to a different end. One can look for a significant geometrical construction based on two line segments that *does* depend on direction; then, by correspondence, use this construction to define the product of two vectors.

One need not look far, for one of the most familiar constructions of ordinary geometry is readily seen to meet the desired specifications, namely, the *perpendicular projection* of one line segment on another (Figure 4.1). A study of this construction reveals that, though it depends on the relative directions of the line segments to be "multiplied", the result depends on the magnitude of only one of them. This result can be multiplied by the magnitude of the other to get a more symmetrical relation. In this way one is led to the following rule for multiplying vectors:

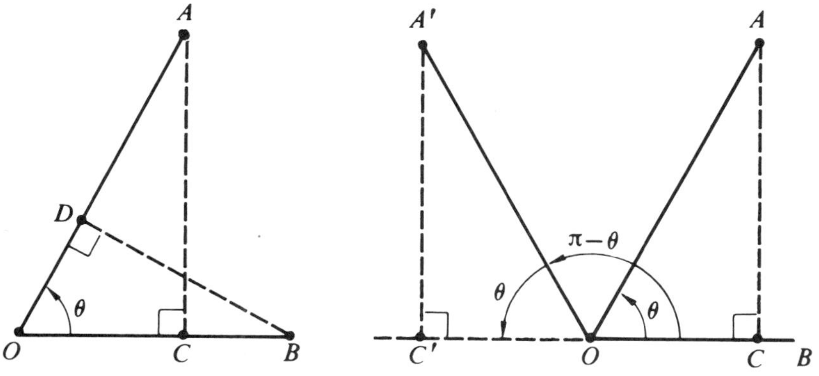

Fig. 4.1. Perpendicular projection.

The perpendicular projection of OA on the line OB is the line segment OC. The adjective "perpendicular" expresses the stipulation that the line AC be perpendicular to OC. If θ is the radian measure of the angle AOB and if $a = OA$ is the length of OA, then, one can write

$$OC = a \cos \theta.$$

Likewise, if $b = OB$, then

$$OD = b \cos \theta.$$

In the second diagram, to distinguish the projection of OA' on OB from the projection of OA on the same line where $OA' = a = OA$, it is convenient to regard OC' and OC as the line segments with the same magnitude but opposite orientation. This can be expressed by writing

$$OC' = a \cos(\pi - \theta) = -a \cos \theta = -OC.$$

This scalar quantity clearly depends on the orientation of OB, because OB determines the line from which the angle is measured.

By taking orientation into account, we go slightly beyond the Greek idea of perpendicular projection.

Define the "*inner product*" of two *directed* line segments, denoted by vectors **a** and **b** respectively, to be the *oriented* line segment obtained by dilating the projection of **a** on **b** by the magnitude of **b**. The magnitude and the orientation of the resulting line segment is a scalar; denote this scalar by **a·b** and call it the *inner product* of vectors **a** and **b**.

This definition of **a·b** implies the following relation to the angle θ between **a** and **b**:

$$\mathbf{a}\cdot\mathbf{b} = |\mathbf{a}|\,|\mathbf{b}| \cos \theta. \qquad (4.1)$$

This expression is commonly taken as the definition of **a·b**, but that calls for an independent definition of cos θ, which would be out of place here.

It should be noted that the geometrical construction on which the definition of $\mathbf{a}\cdot\mathbf{b}$ is based actually gives a line segment directed along the same line as $\mathbf{a}$ or $\mathbf{b}$; the magnitude and relative orientation of this line segment were used in the definition of $\mathbf{a}\cdot\mathbf{b}$, but its direction was not. There is good reason for this. It is necessary if the algebraic rule for multiplication is to depend only on the *relative* directions of $\mathbf{a}$ and $\mathbf{b}$. Thus, as defined, the numerical value of $\mathbf{a}\cdot\mathbf{b}$ is unaffected by any change in the directions of $\mathbf{a}$ and $\mathbf{b}$ which leaves the angle between $\mathbf{a}$ and $\mathbf{b}$ fixed. Moreover, $\mathbf{a}\cdot\mathbf{b}$ has the important symmetry property

$$\mathbf{a}\cdot\mathbf{b} = \mathbf{b}\cdot\mathbf{a}. \tag{4.2}$$

This expresses the fact that the projection of $\mathbf{a}$ on $\mathbf{b}$ dilated by $|\mathbf{b}|$ gives the same result as the projection of $\mathbf{b}$ on $\mathbf{a}$ dilated by $|\mathbf{a}|$ (Figure 4.2).

The inner product has, besides (4.2), several basic algebraic properties which can easily be deduced from its definition by correspondence with perpendicular projection. Its relation to scalar multiplication of vectors is expressed by the rule

$$(\lambda\mathbf{a})\cdot\mathbf{b} = \lambda(\mathbf{a}\cdot\mathbf{b}) = \mathbf{a}\cdot(\lambda\mathbf{b}). \tag{4.3}$$

Here λ can be any scalar — positive, negative or zero. Its relation to vector addition is expressed by the distributive rule

$$\mathbf{a}\cdot(\mathbf{b} + \mathbf{c}) = \mathbf{a}\cdot\mathbf{b} + \mathbf{a}\cdot\mathbf{c}. \tag{4.4}$$

(Figure 4.3). The magnitude of a vector is related to the inner product by

$$\mathbf{a}\cdot\mathbf{a} = |\mathbf{a}|^2 \geqslant 0. \tag{4.5}$$

Of course, $\mathbf{a}\cdot\mathbf{a} = 0$ if and only if $\mathbf{a} = 0$.

The inner product greatly increases the usefulness of vectors, for it can be used to compute angles and the lengths of line segments. Important theorems of geometry and trigonometry can be proved easily by the methods of "vector algebra", so easily, in fact, that it is hardly necessary to single them out by calling them theorems. Results which men once went to great pains to prove have been worked into the algebraic rules where they can be exploited routinely. For example, everyone knows that a great many theorems about triangles are proved in trigonometry and geometry. But, such theorems seem superfluous when it is realized that a triangle can be completely characterized by the simple vector equation

$$\mathbf{a} + \mathbf{b} = \mathbf{c}. \tag{4.6}$$

From this equation various properties of a triangle can be derived by simple steps. For instance, by "squaring" and using the distributive rule (4.4) one gets an equation relating sides and angles:

$$\begin{aligned}\mathbf{c}\cdot\mathbf{c} &= (\mathbf{a} + \mathbf{b})\cdot(\mathbf{a} + \mathbf{b})\\ &= \mathbf{a}\cdot(\mathbf{a} + \mathbf{b}) + \mathbf{b}\cdot(\mathbf{a} + \mathbf{b})\\ &= \mathbf{a}\cdot\mathbf{a} + \mathbf{b}\cdot\mathbf{b} + \mathbf{a}\cdot\mathbf{b} + \mathbf{b}\cdot\mathbf{a}.\end{aligned}$$

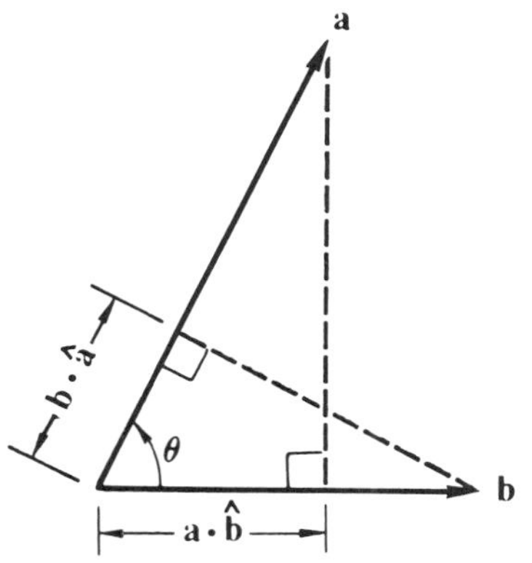

Fig. 4.2. Symmetry of the Scalar Product.

The perpendicular projections of **a** on **b**, and **b** on **a** give respectively

$$\mathbf{a} \cdot \hat{\mathbf{b}} = |\mathbf{a}| \cos \theta,$$

and $$\mathbf{b} \cdot \hat{\mathbf{a}} = |\mathbf{b}| \cos \theta,$$

which after dilation results in the symmetrical form

$$\mathbf{a} \cdot \mathbf{b} = |\mathbf{b}|\, \mathbf{a} \cdot \hat{\mathbf{b}} = |\mathbf{a}|\, \mathbf{b} \cdot \hat{\mathbf{a}} = \mathbf{b} \cdot \mathbf{a}.$$

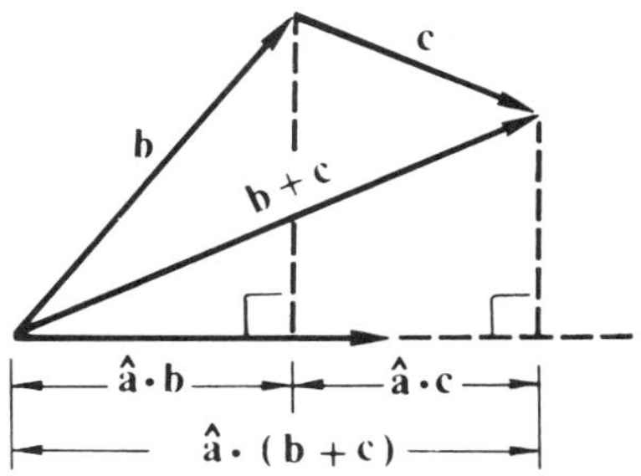

Fig. 4.3. Distributive rule.

The inner product is distributive with respect to vector addition. Projection gives

$$\hat{\mathbf{a}} \cdot (\mathbf{b} + \mathbf{c}) = \hat{\mathbf{a}} \cdot \mathbf{b} + \hat{\mathbf{a}} \cdot \mathbf{c}.$$

Multiply this by $|\mathbf{a}|$, and use the distributive rule for scalars. Finally use the relation (4.3) and $|\mathbf{a}|\, \hat{\mathbf{a}} = \mathbf{a}$ to get the general distributive (4·4).

Or, using (4.2) and (4.5),

$$|\mathbf{c}|^2 = |\mathbf{a}|^2 + |\mathbf{b}|^2 + 2\mathbf{a}\cdot\mathbf{b}. \tag{4.7}$$

This equation can be reexpressed in terms of scalar labels still commonly used in trigonometry. Figure 4.4 indicates the relations

$$a = |\mathbf{a}|, \; b = |\mathbf{b}|, \; c = |\mathbf{c}|, \; \mathbf{a}\cdot\mathbf{b} = -ab \cos C.$$

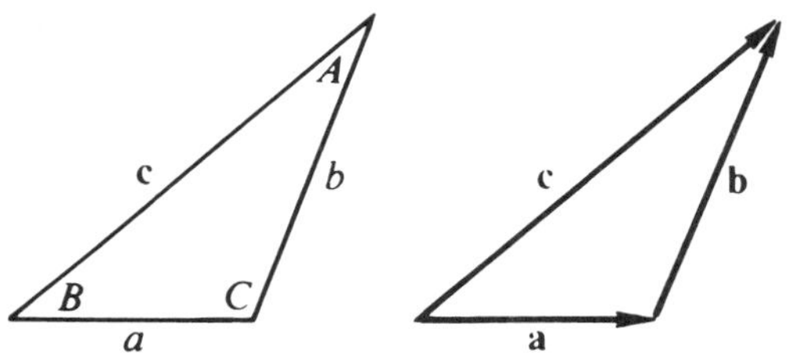

Fig. 4.4. Scalar and vector labels for a triangle.

So (4.7) can be written in the form

$$c^2 = a^2 + b^2 - 2ab \cos C. \tag{4.8}$$

This formula is called the "law of cosines" in trigonometry. If C is a right angle, then $\cos C = 0$, and Equation (4.8) reduces to the Pythagorean Theorem.

By rewriting (4.6) in the form $\mathbf{a} = \mathbf{c} - \mathbf{b}$ and squaring one gets a formula similar to (4.8) involving the angle A. Similarly, an equation involving angle **B** can be obtained. In this way one gets three equations relating the scalars a, b, c, A, B, C. These equations show that given the magnitude of three sides, or of two sides and the angle between, the remaining three scalars can be computed. This result may be recognized as encompassing several theorems of geometry. The point to be made here is that these results of geometry and trigonometry need not be remembered as theorems, since they can be obtained so easily by the "algebra" of scalars and vectors.

Trigonometry is founded on the Greek theories of proportion and perpendicular projection. But the principle ideas of trigonometry did not find their simplest symbolic expression until the invention of vectors and the inner product by Grassmann. Grassmann originally defined the inner product just as we did by correspondence with a perpendicular projection. But he also realized that once the basic algebraic properties have been determined by correspondence, no further reference to the idea of projection is necessary. Thus, the "inner product" can be fully defined abstractly as a rule relating scalars to vectors which has the properties specified by Equations (4.2) through (4.5).

With the abstract definition of $\mathbf{a}\cdot\mathbf{b}$ the intuitive notion of relative direction at last receives a precise symbolic formulation. The notion of number is thereby nearly developed to the point where the principles and theorems of geometry can be completely expressed by algebraic equations without the need to use natural language. For example, the statement "lines OA and OB are perpendicular" can now be better expressed by the equation

$$\mathbf{a}\cdot\mathbf{b} = 0. \tag{4.9}$$

Trigonometry can now be regarded as a system of algebraic equations and relations without any mention of triangles and projections. However, it is precisely the relation of vectors to triangles and of the inner product to projections that makes the algebra of scalars and vectors a useful language for describing the real world. And that, after all, is what the whole scheme was designed for.

1-5. The Outer Product

The algebra of scalars and vectors based on the rules just mentioned has been so widely accepted as to be routinely employed by mathematicians and physicists today. As it stands, however, this algebra is still incapable of providing a full expression of geometrical ideas. Yet there is nothing close to a consensus on how to overcome this limitation. Rather there is a great proliferation of different mathematical systems designed to express geometrical ideas – tensor algebra, matrix algebra, spinor algebra – to name just a few of the most common. It might be thought that this profusion of systems reveals the richness of mathematics. On the contrary, it reveals a widespread confusion – confusion about the *aims and principles of geometric algebra*. The intent here is to clarify these aims and principles by showing that the preceding arguments leading to the invention of scalars and vectors can be continued in a natural way, culminating in a single mathematical system which facilitates a simple expression of the full range of geometrical ideas.

The principle that the product of two vectors ought to describe their relative directions presided over the definition of the inner product. But the inner product falls short of a complete fulfillment of that principle, because it fails to express the fundamental geometrical fact that two non-parallel lines determine a plane, or, better, that two non-collinear directed line segments determine a parallelogram. The possibility of giving this important feature of geometry a direct algebraic expression becomes apparent when the parallelogram is regarded as a kind of "geometrical product" of its sides. But to make this possibility a reality, the notion of number must again be generalized.

A parallelogram can be regarded as a directed plane segment. Just as vectors were invented to characterize the notion of a directed line segment, so a new kind of directed number, called a *bivector* or *2-vector*, can be introduced to characterize the notion of directed plane segment (Figure 5.1). Like

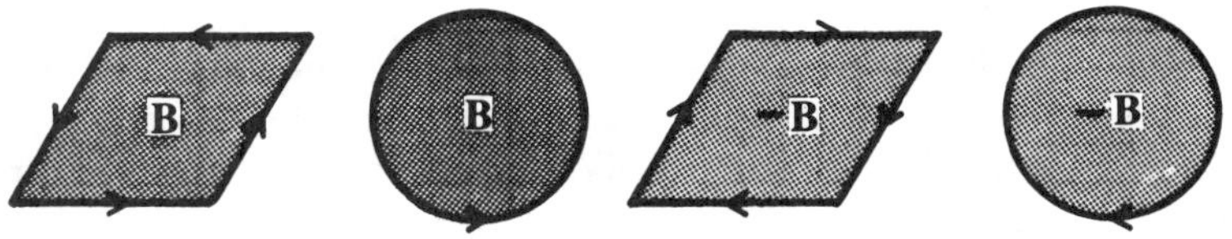

Fig. 5.1. Bivectors and Plane Segments.

A bivector **B** can be pictured as a plane segment. Just as vectors with the same direction can be represented by line segments on parallel lines, so bivectors with the same direction can be represented by plane segments in parallel planes.

The *magnitude* of **B** is a scalar denoted by | **B** |. The magnitude of **B** is equal to the area of the corresponding plane segment. The shape of the plane segment, is irrelevant, or rather, is not associated with any property of **B**. However, a circular shape suggests the fact that the **B** does not distinguish any one direction in the plane from any other, while a parallelogram indicates a relation of the plane segment to line segments.

The orientation of the bivector (and the corresponding plane) can be indicated by an arrowhead assigning a "sense" to the curve bounding the plane segment. A bivector **B** and its *negative*, denoted by −**B**, can be pictured as the same figure but with opposite orientations. The two orientations of a plane (or bivector) are commonly distinguished by the words "clockwise" and "counter clockwise".

Like a vector, a bivector should not be regarded as having a place. Plane segments with the same magnitude and direction can be regarded as different representations of one and the same bivector no matter where they are located; so they can be labeled by one and the same bivector symbol. Plane segments that do not lie in parallel planes must be labeled with different bivector symbols even if they have the same magnitudes.

a vector, a bivector has magnitude, direction and orientation, and only these properties. But here the word "direction" must be understood in a sense more general than is usual. Just as the direction of a vector corresponds to an (oriented straight) line, so the direction of a bivector corresponds to an (oriented flat) plane. The distinction between these two kinds of direction involves the geometrical notion of *dimension* or *grade*. Accordingly, the direction of a bivector is said to be 2-*dimensional* to distinguish it from the *1-dimensional* direction of a vector. And it is sometimes convenient to call a *vector* a *1-vector* to emphasize its dimension. Also, a scalar can be regarded as a 0-*vector* to indicate that it is a 0-dimensional number. Since, as already shown, the only directional property of a scalar is its orientation, orientation can be regarded as a 0-dimensional direction. Thus the idea of numbers with different geometrical dimension begins to take shape.

In ordinary geometry the concepts of line and plane play roles of comparable significance. Indeed, the one concept can hardly be said to have any significance at all apart from the other, and the mathematical meanings of "line" and "plane" are determined solely by specifying relations between them. To give "planes" and "lines" *equal* algebraic representation, the notion of directed number must be enlarged to include the notion of bivector as well as vector, and the relations of lines to planes must be reflected in the relations of vectors to bivectors. It may be a good idea to point out that both line and plane, as commonly conceived, consist of a set of points in definite relation to one another. It is the nature of this relation that distinguishes line from plane. A single vector completely characterizes the directional relation of points in a given line. A single bivector completely characterizes the directional relation of points in a given plane. In other words, a bivector does not describe a set of points in a plane; rather it describes the directional property of such a set, which, so to speak, specifies the plane the points are "in". Thus, the notion of a plane as a relation can be separated from the notion of a plane as a point set. After the directional properties of planes and lines have been fully incorporated into an algebra of directed numbers, the geometrical properties of point sets can be more easily and completely described than ever before, as we shall see.

Now return to the problem of giving algebraic expression to the relation of line segments to plane segments. Note that a point moving a distance and direction specified by a vector **a** sweeps out a directed line segment. And the points on this line segment, each moving a distance and direction specified by a vector **b**, sweep out a parallelogram, (Figure 5.2). Since the bivector **B** corresponding to this parallelogram is clearly uniquely determined by this geometrical construction, it may be regarded as a kind of "product" of the vectors **a** and **b**. So write

$$\mathbf{a}\wedge\mathbf{b} = \mathbf{B}. \tag{5.1}$$

A "wedge" is used to denote this new kind of multiplication to distinguish it

Fig. 5.2. The "parallelogram rule" for outer multiplication.

Note that the order of arrows on the boundary determines an orientation for the parallelogram. The arrows indicate the path of a point which first sweeps out a line segment and then, as the line segment moves, an edge of the parallelogram.

from the "dot" denoting the inner product of vectors. The bivector $\mathbf{a}\wedge\mathbf{b}$ is said to be the *outer product* of vectors **a** and **b**.

Now note that the parallelogram obtained by "sweeping **b** along **a**" differs only in orientation from the parallelogram obtained by "sweeping **a** along **b**" (Figure 5.2). This can be simply expressed by writing

$$\mathbf{b}\wedge\mathbf{a} = -\mathbf{a}\wedge\mathbf{b} = -\mathbf{B}. \tag{5.2}$$

Thus, reversing the order of vectors in an outer product "reverses" the orientation of the resulting bivector. This is expressed by saying that the outer product is *anticommutative*.

The relation of vector orientation to bivector orientation is fixed by the rule

$$\mathbf{b}\wedge\mathbf{a} = \mathbf{a}\wedge(-\mathbf{b}) = (-\mathbf{b})\wedge(-\mathbf{a}) = (-\mathbf{a})\wedge\mathbf{b}. \tag{5.3}$$

This rule, like the others, follows from the correspondence of vectors and bivectors with oriented line segments and plane segments. It can be simply "read off" from Figure 5.3.

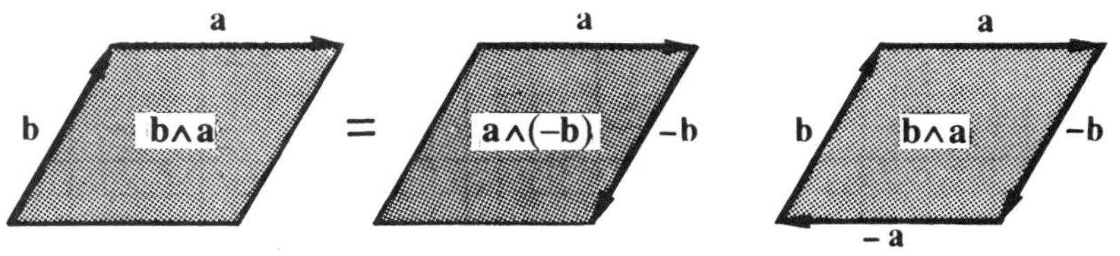

Fig. 5.3. Relative orientations of vectors and bivectors.

The same bivector is obtained from the outer product of any pair of vectors labeling consecutive oriented line segments bounding an oriented parallelogram. Note that directed line segments on opposite sides of an oriented parallelogram correspond to vectors of opposite orientation.

Since the magnitude of the bivector $\mathbf{a}\wedge\mathbf{b}$ is just the area of the corresponding parallelogram,

$$|\mathbf{B}| = |\mathbf{a}\wedge\mathbf{b}| = |\mathbf{b}\wedge\mathbf{a}| = |\mathbf{a}||\mathbf{b}| \sin \theta, \tag{5.4}$$

where θ is the angle between vectors **a** and **b**. This formula expresses the relation between vector magnitudes and bivector magnitudes. The relation to $\sin \theta$ is given in (5.4) for comparison with trigonometry; it is not part of the definition.

Scalar multiplication can be defined for bivectors in the same way as it was for vectors. For bivectors **C** and **B** and scalar λ, the equation

$$\mathbf{C} = \lambda\mathbf{B} \tag{5.5}$$

means that the magnitude of **B** is dilated by the magnitude of λ, that is,

$$|\mathbf{C}| = |\lambda|\,|\mathbf{B}|\,, \tag{5.6}$$

and the direction of **C** is the same as that of **B** if λ is positive, or opposite to it if λ is negative. This last stipulation can be expressed by equations for multiplication by the unit scalars one and minus one:

$$(1)\mathbf{B} = \mathbf{B}, \qquad (-1)\mathbf{B} = -\mathbf{B}. \tag{5.7}$$

Bivectors which are scalar multiples of one another are said to be *codirectional*.

Scalar multiplications of vectors and bivectors are related by the equation

$$\lambda(\mathbf{a}\wedge\mathbf{b}) = (\lambda\mathbf{a})\wedge\mathbf{b} = \mathbf{a}\wedge(\lambda\mathbf{b}). \tag{5.8}$$

For $\lambda = -1$, this is equivalent to Equation (5.3). For positive λ, Equation (5.8) merely expresses the fact that dilation of one side of a parallelogram dilatates its area by the same amount.

Note that, by (5.4), $|\mathbf{a}\wedge\mathbf{b}| = 0$ for nonzero **a** and **b** if and only if $\sin \theta = 0$, which is a way of saying that **a** and **b** are collinear. Adopting the principle, already applied to vectors, that a directed number is zero if and only if its magnitude is zero, it follows that $|\mathbf{a}\wedge\mathbf{b}| = 0$ if and only if $\mathbf{a}\wedge\mathbf{b} = 0$. Hence, the outer product of nonzero vectors is zero if and only if they are collinear, that is,

$$\mathbf{a}\wedge\mathbf{b} = 0 \tag{5.9}$$

if and only if $\mathbf{b} = \lambda\mathbf{a}$. Note that if $\lambda \neq 0$, (5.9) together with (5.8) implies that

$$\mathbf{a}\wedge\mathbf{a} = 0. \tag{5.10}$$

This is as it should be, for the anticommutation rule (5.2) implies that $\mathbf{a}\wedge\mathbf{a} = -\mathbf{a}\wedge\mathbf{a}$, and only zero is equal to its own negative. All of this is in complete accord with the geometric interpretation of outer multiplication, for if **a** and **b** are collinear, then "sweeping **a** along **b**" produces no parallelogram at all.

The relation of addition to outer multiplications is determined by the *distributive rule*:

$$\mathbf{a}\wedge(\mathbf{b} + \mathbf{c}) = \mathbf{a}\wedge\mathbf{b} + \mathbf{a}\wedge\mathbf{c}. \tag{5.11}$$

The corresponding geometrical construction is illustrated in Figure 5.4. Note that (5.11) relates addition of vectors on the left to addition of bivectors on the right. So the algebraic properties and the geometrical interpretation of bivector addition are completely determined by the properties and interpretation already accorded to vector addition. For example, the sum of two bivectors is a unique bivector, and again, bivector addition is associative. For

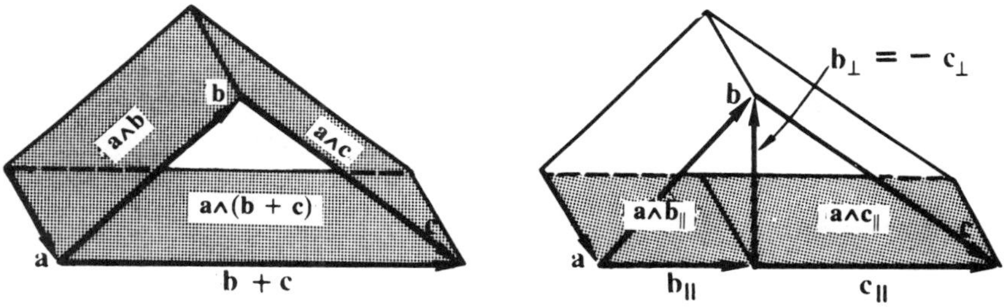

$$\mathbf{a}\wedge(\mathbf{b} + \mathbf{c}) = \mathbf{a}\wedge\mathbf{b} + \mathbf{a}\wedge\mathbf{c} = \mathbf{a}\wedge(\mathbf{b}_\parallel + \mathbf{c}_\parallel) = \mathbf{a}\wedge\mathbf{b}_\parallel + \mathbf{a}\wedge\mathbf{c}_\parallel$$

Fig. 5.4. Distributive rule for outer multiplication.

To prove the distributive rule, express $\mathbf{b}$ as the sum of a part $\mathbf{b}_\parallel$ collinear with $\mathbf{b} + \mathbf{c}$ and a part $\mathbf{b}_\perp$ orthogonal to $\mathbf{b} + \mathbf{c}$. Do the same for $\mathbf{c}$ and observe that $\mathbf{b}_\perp = -\mathbf{c}_\perp$, so $\mathbf{b} + \mathbf{c} = (\mathbf{b}_\parallel + \mathbf{b}_\perp) + (\mathbf{c}_\parallel + \mathbf{c}_\perp) = \mathbf{b}_\parallel + \mathbf{c}_\parallel$ and

$$\mathbf{a}\wedge(\mathbf{b} + \mathbf{c}) = \mathbf{a}\wedge(\mathbf{b}_\parallel + \mathbf{c}_\parallel) = \mathbf{a}\wedge\mathbf{b}_\parallel + \mathbf{a}\wedge\mathbf{c}_\parallel .$$

This reduces the distributive to the usual rule for adding areas, since all the bivectors in the equation are codirectional. Thus, if $\mathbf{b}_\parallel$ and $\mathbf{c}_\parallel$ have the same orientation as $\mathbf{a} + \mathbf{b}$, which is the case in the above diagram, then

$$|\mathbf{a}\wedge(\mathbf{b} + \mathbf{c})| = |\mathbf{a}\wedge(\mathbf{b}_\parallel + \mathbf{c}_\parallel)| = |\mathbf{a}\wedge\mathbf{b}_\parallel| + |\mathbf{a}\wedge\mathbf{c}_\parallel|.$$

It may happen, however, that the orientation of $\mathbf{c}_\parallel$ is opposite to that of $\mathbf{a} + \mathbf{b}$, in which case the orientation of $\mathbf{a}\wedge(\mathbf{b} + \mathbf{c})$ is opposite to that of $\mathbf{a}\wedge\mathbf{c}_\parallel$,

$$|\mathbf{a}\wedge(\mathbf{b} + \mathbf{c})| = |\mathbf{a}\wedge(\mathbf{b}_\parallel + \mathbf{c}_\parallel)| = |\mathbf{a}\wedge\mathbf{b}_\parallel| - |\mathbf{a}\wedge\mathbf{c}_\parallel|.$$

Construction of a diagram corresponding to this case is left to the student.

It should be evident from the diagram that quite generally

$$|\mathbf{a}\wedge(\mathbf{b} + \mathbf{c})| \leqslant |\mathbf{a}\wedge\mathbf{b}| + |\mathbf{a}\wedge\mathbf{c}| ,$$

with equality possible only if $\mathbf{a}$, $\mathbf{b}$ and $\mathbf{c}$ are coplanar. The quantities $|\mathbf{a}\wedge\mathbf{b}|$ and $|\mathbf{a}\wedge\mathbf{c}|$ are ordinary areas and can be added like any other scalars. But $\mathbf{a}\wedge\mathbf{b}$ and $\mathbf{a}\wedge\mathbf{c}$ are "directed areas" and add "like vectors".

bivectors with the same direction, it is easily seen that the distributive rule (5.11) reduces to the usual rule for adding areas.

Both the inner and outer products are measures of relative direction, but they *complement* one another. Relations which are difficult or impossible to obtain with one may be easy to obtain with the other. Whereas the equation $\mathbf{a}\cdot\mathbf{b} = 0$ provides a simple expression of "perpendicular", $\mathbf{a}\wedge\mathbf{b} = 0$ provides a simple expression of "parallel". To illustrate the point, reconsider the vector equation for a triangle, which was analyzed above with the help of the inner product. Take the outer product of $\mathbf{a} + \mathbf{b} = \mathbf{c}$ successively with vectors $\mathbf{a}$, $\mathbf{b}$, $\mathbf{c}$, and use the rules (5.10) and (5.11) to obtain the three equations

$$\begin{aligned}&\mathbf{a}\wedge\mathbf{b} = \mathbf{a}\wedge\mathbf{c},\\ &\mathbf{b}\wedge\mathbf{a} = \mathbf{b}\wedge\mathbf{c},\\ &\mathbf{c}\wedge\mathbf{a} + \mathbf{c}\wedge\mathbf{b} = 0.\end{aligned}$$

Only two of these equations are independent; the third, for instance, is the sum of the first two. It is convenient to write the first two equations on a single line, like so:

$$\mathbf{a}\wedge\mathbf{c} = \mathbf{a}\wedge\mathbf{b} = \mathbf{c}\wedge\mathbf{b}. \tag{5.12}$$

Here are three different ways of expressing the same bivector as a product of vectors. This gives three different ways of expressing its magnitude:

$$|\mathbf{a}\wedge\mathbf{c}| = |\mathbf{c}\wedge\mathbf{b}| = |\mathbf{a}\wedge\mathbf{b}|\ . \tag{5.13}$$

Using (5.4) and the scalar labels for a triangle indicated in Figure 4.4, one gets, after dividing by abc,

$$\frac{\sin A}{a} = \frac{\sin B}{b} = \frac{\sin C}{c}. \tag{5.14}$$

This formula is called the "law of sines" in trigonometry. We shall see in Chapter 2 that all the formulas of plane and spherical trigonometry can be easily derived and compactly expressed by using inner and outer products.

The theory of the outer product as described so far calls for an obvious generalization. Just as a plane segment is swept out by a moving line segment, a "space segment" is swept out by a moving plane segment. Thus, the points on an oriented parallelogram specified by the bivector $\mathbf{a}\wedge\mathbf{b}$ moving a distance and direction specified by a vector $\mathbf{c}$ sweep out an oriented parallelepiped (Figure 5.5), which may be characterized by a new kind of directed number $\mathbf{T}$ called a *trivector* or *3-vector*. The properties of $\mathbf{T}$ are fixed by regarding it as equal to the *outer product* of the bivector $\mathbf{a}\wedge\mathbf{b}$ with the vector $\mathbf{c}$. So write

$$(\mathbf{a}\wedge\mathbf{b})\wedge\mathbf{c} = \mathbf{T}. \tag{5.15}$$

The study of trivectors leads to results quite analogous to those obtained above for bivectors, so the analysis need not be carried out in detail. But one

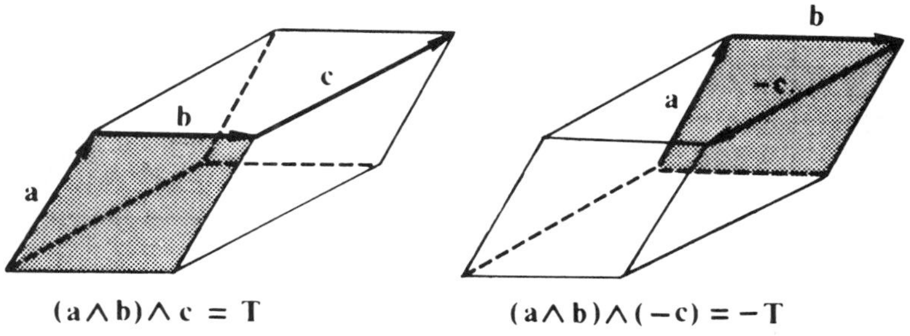

Fig. 5.5. The "parallelepiped rule" for outer multiplication.

Displacement of an oriented parallelogram sweeps out an oriented parallelepiped. Displacement in the opposite direction sweeps out a parallelepiped with opposite orientation.

new result obtains, namely, the conclusion that outer multiplication should obey the *associative rule*:

$$(\mathbf{a}\wedge\mathbf{b})\wedge\mathbf{c} = \mathbf{a}\wedge(\mathbf{b}\wedge\mathbf{c}). \tag{5.16}$$

The geometric meaning of associativity can be ascertained with the help of the following rule:

$$(\mathbf{b}\wedge\mathbf{a})\wedge\mathbf{c} = (-\mathbf{a}\wedge\mathbf{b})\wedge\mathbf{c} = -\mathbf{T}. \tag{5.17}$$

This is an instance of the general rule that the orientation of a product is reversed by reversing the orientation of one of its factors. Repeated applications of (5.16) and (5.17) makes it possible to rearrange the vectors in a product to get

$$(\mathbf{a}\wedge\mathbf{b})\wedge\mathbf{c} = (\mathbf{b}\wedge\mathbf{c})\wedge\mathbf{a} = (\mathbf{c}\wedge\mathbf{a})\wedge\mathbf{b}.$$

This says the same oriented parallelepiped is obtained by sweeping "$\mathbf{a}\wedge\mathbf{b}$ along **c**", "$\mathbf{b}\wedge\mathbf{c}$ along **a**" or "$\mathbf{c}\wedge\mathbf{a}$ along **b**". So the associative rule is needed to express the equivalence of different ways of "building up" a space segment out of line segments.

Of course, if **c** "lies in the plane of $\mathbf{a}\wedge\mathbf{b}$", then "sweeping $\mathbf{a}\wedge\mathbf{b}$ along **c**" does not produce a 3-dimensional object. Accordingly, write

$$(\mathbf{a}\wedge\mathbf{b})\wedge\mathbf{c} = \mathbf{a}\wedge\mathbf{b}\wedge\mathbf{c} = 0. \tag{5.18}$$

This equation provides a simple algebraic way of saying that 3 lines (with directions denoted by vectors **a**, **b**, **c**) lie in the same plane, just as Equation (5.9) provides a simple way of saying that 2 lines are parallel.

Like any other directed number, a 3-vector has magnitude, direction and orientation, and only these properties. The dimensionality of a 3-vector is expressed by the fact that it can be factored into an outer product of three vectors, though this can be done in an unlimited number of ways. The magnitude of $\mathbf{a}\wedge\mathbf{b}\wedge\mathbf{c}$ is denoted by $|\mathbf{a}\wedge\mathbf{b}\wedge\mathbf{c}|$ and is equal to the volume of the parallelepiped determined by the vectors **a**, **b** and **c**.

The orientation of a trivector depends on the order of its factors. The anticommutation rule together with the associative rule imply that exchange of any pair of factors in a product reverses the orientation of the result. For instance, $\mathbf{c}\wedge\mathbf{b}\wedge\mathbf{a} = -\mathbf{a}\wedge\mathbf{b}\wedge\mathbf{c}$. Thus the idea of relative orientation is very easily expressed with the help of the outer product. Without such algebraic apparatus the geometrical idea of orientation is quite difficult to express, and, not surprisingly, was only dimly understood before the invention of vectors and the outer product.

The essential aspects of outer multiplication and the generalized notions of number and direction it entails have now been set down. No fundamentally new insights into the relations between algebra and geometry are achieved by considering the outer product of four or more vectors. But it should be mentioned that if vectors are used to describe the 3-dimensional space of ordinary geometry, then displacement of the trivector $\mathbf{a}\wedge\mathbf{b}\wedge\mathbf{c}$ in a direction specified by **d** fails to sweep out a 4-dimensional space segment. So write

$$(\mathbf{a}\wedge\mathbf{b}\wedge\mathbf{c})\wedge\mathbf{d} = \mathbf{a}\wedge\mathbf{b}\wedge\mathbf{c}\wedge\mathbf{d} = 0. \tag{5.19}$$

The parenthesis is unnecessary because of the associative rule (5.16). Equation (5.19) must hold for any four vectors **a**, **b**, **c**, **d**. This is a simple way of saying that space is 3-dimensional. Note the similarity in form and meaning of Equations (5.19), (5.18) and (5.9). It should be clear that (5.19) does not follow from any ideas or rules previously considered. By supposing that the outer product of four vectors is not zero, one is led to an algebraic description of spaces and geometries of four or more dimensions, but we already have what we need to describe the geometrical properties of physical space.

The outer product was invented by Herman Grassmann, and, following a line of thought similar to the one above, developed into a complete mathematical theory before the middle of the nineteenth century. His theory has been accorded a prominent place in mathematics only in the last forty years, and it is hardly known at all to physicists. Grassmann himself was the only one to use it during the first two decades after it was published. There are several reasons for this. The most important one arises from the fact that Grassmann's understanding of the abstract nature of mathematics was far ahead of his time. He was the first person to arrive at the modern conception of algebra as a system of rules relating otherwise undefined entities. He realized that the nature of the outer product could be defined by specifying the rules it obeys, especially the distributive, associative, and anticommutive rules given above. He rightly expounded this momentous insight in great detail. And he proved its significance by showing, for the first time, how abstract algebra can take us beyond the 3-dimensional space of experience to a conception of space with any number of dimensions. Unfortunately, in his enthusiasm for abstract developments, Grassmann deemphasized the geometric origin and interpretation of his rules. No doubt many potential readers would have appreciated the geometrical applications of Grassmann's system, but most were simply confounded by the profusion of unfamiliar abstract ideas in Grassmann's long books.

The seeds of Herman Grassmann's great invention were sown by his father, Günther, who in 1824, when Herman was 15, published these words in a book intended for elementary instruction:

"the rectangle itself is the *true geometrical product*, and the construction of it is really *geometrical multiplication. . . . A rectangle is the geometric product of its base and height, and this product behaves in the same way as the arithmetic product.*" (italics added)

The elder Grassmann elaborated this idea at some length and must have advocated it with considerable enthusiasm to his young son. As it stands, however, Günther's idea is hardly more than a novel way of expressing the central idea of Book II of Euclid's *Elements*. The Greeks made frequent use of the correspondence between the product of numbers and the construction of a parallelogram from its base and height. For example, Euclid represented the distributive rule of algebra as addition of areas and proved it as a geometrical theorem. This correspondence between arithmetic and geometry was rejected by Descartes and duly ignored by the mathematicians that followed him. However, as already explained, Descartes merely associated arithmetic multiplication with a different geometric construction. The old Greek idea lay dormant until it was reexpressed in strong arithmetic terms by Günther Grassmann. But the truly significant advance, from the idea of a geometrical product to its full algebraic expression by outer multiplication, was made by his son.

Herman Grassmann completed the algebraic formulation of basic ideas in Greek geometry begun by Descartes. The Greek theory of ratio and proportion is now incorporated in the properties of scalars and scalar multiplication. The Greek idea of projection is incorporated in the inner product. And the Greek geometrical product is expressed by outer multiplication. The invention of a system of directed numbers to express Greek geometrical notions makes it possible, as Descartes had already said, to go far beyond the geometry of the Greeks. It also leads to a deeper appreciation of the Greek accomplishments. Only in the light of Grassmann's outer product is it possible to understand that the careful Greek distinction between number and magnitude has real geometrical significance. It corresponds roughly to the distinction between scalar and vector. Actually the Greek magnitudes added like scalars but multiplied like vectors, so multiplication of Greek magnitudes involves the notions of direction and dimension, and Euclid was quite right in distinguishing it from multiplication of "Greek numbers" (our scalars). Only in the work of Grassmann are the notions of direction, dimension, orientation and scalar magnitude finally disentangled. But his great accomplishment would have been impossible without the earlier vague distinction of the Greeks, and perhaps without its reformulation in quasi-arithmetic terms by his father.

1-6. Synthesis and Simplification

Grassmann was the first person to define multiplication simply by specifying a set of algebraic rules. By systematically surveying various possible rules, he discovered several other kinds of multiplication besides his inner and outer products. Nevertheless, he overlooked the most important possibility until late in his life, when he was unable to follow up on its implications. There is one fundamental kind of geometrical product from which all other significant geometrical products can be obtained. All the geometrical facts needed to discover such a product have been mentioned above.

It has already been noted that the inner and outer products seem to complement one another by describing independent geometrical relations. This circumstance deserves the most careful study. The simplest approach is to entertain the possibility of introducing a new kind of product **ab** by the equation

$$\mathbf{ab} = \mathbf{a}\cdot\mathbf{b} + \mathbf{a}\wedge\mathbf{b}. \tag{6.1}$$

Here the scalar $\mathbf{a}\cdot\mathbf{b}$ has been added to the bivector $\mathbf{a}\wedge\mathbf{b}$. At first sight it may seem absurd to add two directed numbers with different grades. That may have delayed Grassmann from considering it. For centuries the notion that you can only add "like things" has been relentlessly impressed on the mind of every schoolboy. It is a kind of mathematical taboo – its real justification unknown or forgotten. It is supposedly obvious that you cannot add apples and oranges or feet and square feet. On the contrary, it is only obvious that addition of apples and oranges is not usually a practical thing to do – unless you are making a salad.

Absurdity disappears when it is realized that (6.1) can be justified in the abstract "Grassmannian" fashion which has become standard mathematical procedure today. All that mathematics really requires is that the indicated relations and operations be well defined and consistently employed. The mathematical meaning of adding scalars and bivectors is determined by specifying that such addition satisfy the usual commutative and associative rules. Use of the "equal sign" in (6.1) is justified by assuming that it obeys the same rules as those governing equality in ordinary scalar algebra. With this understood, it now can be shown that the properties of the new product are almost completely determined by the obvious requirement that they be consistent with the properties already accorded to the inner and outer products.

The commutative rule $\mathbf{b}\cdot\mathbf{a} = \mathbf{a}\cdot\mathbf{b}$ together with the anticommutative rule $\mathbf{b}\wedge\mathbf{a} = -\mathbf{a}\wedge\mathbf{b}$ imply a relation between **ab** and **ba**. Thus,

$$\mathbf{ba} = \mathbf{b}\cdot\mathbf{a} + \mathbf{b}\wedge\mathbf{a} = \mathbf{a}\cdot\mathbf{b} - \mathbf{a}\wedge\mathbf{b}. \tag{6.2}$$

Comparison of (6.1) with (6.2) shows that, in general, **ab** is not equal to **ba** because, though their scalar parts are equal, their bivector parts are not. However, if $\mathbf{a}\wedge\mathbf{b} = 0$, then

$$\mathbf{ab} = \mathbf{a}\cdot\mathbf{b} = \mathbf{ba}. \tag{6.3}$$

And if $\mathbf{a}\cdot\mathbf{b} = 0$, then

$$\mathbf{ab} = \mathbf{a}\wedge\mathbf{b} = -\mathbf{b}\wedge\mathbf{a} = -\mathbf{ba}. \tag{6.4}$$

It should not escape notice that to get (6.3) and (6.4) from (6.1) the usual "additive property of zero" is needed, and no distinction between a scalar zero and a bivector zero is called for.

The product **ab** inherits a geometrical interpretation from the interpretations already accorded to the inner and outer products. It is an algebraic measure of the relative direction of vectors **a** and **b**. Thus, from (6.3) and (6.4) it should be clear that vectors are *collinear* if and only if their product is commutative, and they are *orthogonal* if and only if their product is anticommutative. But more properties of the product are required to understand its significance when the relative direction of two vectors is somewhere between the extremes of collinearity and orthogonality.

To give due recognition to its geometric significance **ab** will henceforth be called the **geometric product** of vectors **a** and **b**.

From the distributive rules (4.4) and (5.11) for inner and outer products, it follows that the geometric product must obey the left and right *distributive rules*

$$\mathbf{a}(\mathbf{b} + \mathbf{c}) = \mathbf{ab} + \mathbf{ac}, \tag{6.5}$$

$$(\mathbf{b} + \mathbf{c})\mathbf{a} = \mathbf{ba} + \mathbf{ca}. \tag{6.6}$$

Equation (6.5) can be derived from (6.1) by the following steps

$$\begin{aligned} \mathbf{a}(\mathbf{b} + \mathbf{c}) &= \mathbf{a}\cdot(\mathbf{b} + \mathbf{c}) + \mathbf{a}\wedge(\mathbf{b} + \mathbf{c}) \\ &= (\mathbf{a}\cdot\mathbf{b} + \mathbf{a}\cdot\mathbf{c}) + (\mathbf{a}\wedge\mathbf{b} + \mathbf{a}\wedge\mathbf{c}) \\ &= (\mathbf{a}\cdot\mathbf{b} + \mathbf{a}\wedge\mathbf{b}) + (\mathbf{a}\cdot\mathbf{c} + \mathbf{a}\wedge\mathbf{c}) \\ &= \mathbf{ab} + \mathbf{ac}. \end{aligned}$$

Note that the usual properties of equality and the commutative and associative rules of addition have been employed. Equation (6.6) can be derived from (6.2) in the same way. The distributive rules (6.5) and (6.6) are independent of one another, because multiplication is not commutative. To derive them, the distributive rules for both the inner and outer products were needed.

The relation of scalar multiplication to the geometric product is described by the equations

$$\lambda(\mathbf{ab}) = (\lambda\mathbf{a})\mathbf{b} = \mathbf{a}(\lambda\mathbf{b}). \tag{6.7}$$

This is easily derived from (4.3) and (5.8) with the help of the definition (6.1). It says that scalar and vector multiplication are mutually commutative and associative. If the commutative rule is separated from the associative rule, it takes the simple form

$$\lambda\mathbf{a} = \mathbf{a}\lambda. \tag{6.8}$$

Now observe that by taking the sum and difference of equations (6.1) and (6.2), one gets

$$\mathbf{a}\cdot\mathbf{b} = \tfrac{1}{2}(\mathbf{ab} + \mathbf{ba}), \tag{6.9}$$

and

$$\mathbf{a}\wedge\mathbf{b} = \tfrac{1}{2}(\mathbf{ab} - \mathbf{ba}). \tag{6.10}$$

This points the way to a great simplification. Instead of regarding (6.1) as a definition of **ab**, consider **ab** as fundamental and regard (6.9) and (6.10) as definitions of $\mathbf{a}\cdot\mathbf{b}$ and $\mathbf{a}\wedge\mathbf{b}$. This reduces two kinds of vector multiplication to one. It is curious, then, to note that by (6.7) the commutativity of the inner product arises from the commutativity of addition, and by (6.10) the anticommutativity of the outer product arises from the anticommutativity of subtraction.

The algebraic properties of the geometric product of two vectors have already been ascertained. It should be evident that the corresponding properties of the inner and outer products can be derived from the definitions (6.9) and (6.10) simply by reversing the arguments already given.

The next task is to examine the geometric product of three vectors **a**, **b**, **c**. It is certainly desirable that this product satisfy the *associative rule*

$$\mathbf{a}(\mathbf{bc}) = (\mathbf{ab})\mathbf{c} = \mathbf{abc}, \tag{6.11}$$

for that greatly simplifies algebraic manipulations. But it must be shown that this rule reproduces established properties of the inner and outer products. This can be done by examining the product of a vector with a bivector.

The product **aB** of a vector **a** with a bivector **B** can be expressed as a sum of "symmetric" and "antisymmetric" parts in the following way

$$\begin{aligned}\mathbf{aB} &= \tfrac{1}{2}(\mathbf{aB} + \mathbf{aB}) + \tfrac{1}{2}(\mathbf{Ba} - \mathbf{Ba}) \\ &= \tfrac{1}{2}(\mathbf{aB} - \mathbf{Ba}) + \tfrac{1}{2}(\mathbf{aB} + \mathbf{Ba}).\end{aligned}$$

Anticipating results to be obtained, introduce the notations

$$\mathbf{a}\cdot\mathbf{B} = \tfrac{1}{2}(\mathbf{aB} - \mathbf{Ba}) = -\mathbf{B}\cdot\mathbf{a} \tag{6.12}$$

$$\mathbf{a}\wedge\mathbf{B} = \tfrac{1}{2}(\mathbf{aB} + \mathbf{Ba}) = \mathbf{B}\wedge\mathbf{a}, \tag{6.13}$$

So

$$\mathbf{aB} = \mathbf{a}\cdot\mathbf{B} + \mathbf{a}\wedge\mathbf{B}. \tag{6.14}$$

As the notation indicates, $\mathbf{a}\wedge\mathbf{B}$ is to be regarded as identical to the *outer product* of vector and bivector which has already been introduced for geometric reasons. The quantity $\mathbf{a}\cdot\mathbf{B}$ is something new; as the notation suggests, it is to be regarded as a generalization of the *inner product* of vectors.

Note that (6.13) differs from (6.10) by a sign, because (6.13) has a bivector where (6.10) has a vector. The sign in (6.13) is justified by showing that (6.13) yields the properties already ascribed to the outer product. To this end, it is sufficient to show that (6.13) implies the associative rule $(\mathbf{a}\wedge\mathbf{b})\wedge\mathbf{c} = \mathbf{a}\wedge(\mathbf{b}\wedge\mathbf{c})$. Of course the properties of the geometric product, including the associative rule (6.11) must be freely used in the proof. Utilizing the definitions (6.10) and (6.13) for the outer product.

$$\begin{aligned}(\mathbf{a}\wedge\mathbf{b})\wedge\mathbf{c} &= \tfrac{1}{2}[\tfrac{1}{2}(\mathbf{ab} - \mathbf{ba})\mathbf{c} + \mathbf{c}\tfrac{1}{2}(\mathbf{ab} - \mathbf{ba})]\\ &= \tfrac{1}{4}[\mathbf{abc} - \mathbf{bac} + \mathbf{cab} - \mathbf{cba}].\end{aligned}$$

Similarly,

$$\begin{aligned}\mathbf{a}\wedge(\mathbf{b}\wedge\mathbf{c}) &= \tfrac{1}{2}[\mathbf{a}\tfrac{1}{2}(\mathbf{bc} - \mathbf{cb}) + \tfrac{1}{2}(\mathbf{bc} - \mathbf{cb})\mathbf{a}]\\ &= \tfrac{1}{4}[\mathbf{abc} - \mathbf{acb} + \mathbf{bca} - \mathbf{cba}].\end{aligned}$$

On taking the difference of these expressions several terms cancel and the remaining terms can be arranged to give

$$\begin{aligned}(\mathbf{a}\wedge\mathbf{b})\wedge\mathbf{c} - \mathbf{a}\wedge(\mathbf{b}\wedge\mathbf{c}) &= -\tfrac{1}{4}\,\mathbf{b}(\mathbf{ac} + \mathbf{ca}) + \tfrac{1}{4}(\mathbf{ca} + \mathbf{ac})\mathbf{b}\\ &= -\tfrac{1}{2}\,\mathbf{b}(\mathbf{a}\cdot\mathbf{c}) + \tfrac{1}{2}(\mathbf{a}\cdot\mathbf{c})\mathbf{b} = 0.\end{aligned}$$

Note that the fact that the vector inner product is a scalar is needed in the last step of the proof.

Now to understand the significance of $\mathbf{a}\cdot\mathbf{B}$, let $\mathbf{B} = \mathbf{b}\wedge\mathbf{c}$. Use the definitions as before to eliminate the dot and wedge:

$$\begin{aligned}\mathbf{a}\cdot(\mathbf{b}\wedge\mathbf{c}) &= \tfrac{1}{2}[\mathbf{a}\,\tfrac{1}{2}(\mathbf{bc} - \mathbf{cb}) - \tfrac{1}{2}(\mathbf{bc} - \mathbf{cb})\mathbf{a}]\\ &= \tfrac{1}{4}[\mathbf{abc} - \mathbf{acb} - \mathbf{bca} + \mathbf{cba}].\end{aligned}$$

To this, add

$$0 = \tfrac{1}{4}[\mathbf{bac} - \mathbf{cab} - \mathbf{bac} + \mathbf{cab}],$$

and collect terms to get

$$\begin{aligned}\mathbf{a}\cdot(\mathbf{b}\wedge\mathbf{c}) &= \tfrac{1}{4}[(\mathbf{ab} + \mathbf{ba})\mathbf{c} - (\mathbf{ac} + \mathbf{ca})\mathbf{b} - \mathbf{b}(\mathbf{ca} + \mathbf{ac}) + \mathbf{c}(\mathbf{ba} + \mathbf{ab})]\\ &= \tfrac{1}{4}[2(\mathbf{a}\cdot\mathbf{b})\mathbf{c} - 2(\mathbf{a}\cdot\mathbf{c})\mathbf{b} - \mathbf{b}2(\mathbf{a}\cdot\mathbf{c}) + \mathbf{c}2(\mathbf{b}\cdot\mathbf{a})]\\ &= (\mathbf{a}\cdot\mathbf{b})\mathbf{c} - (\mathbf{a}\cdot\mathbf{c})\mathbf{b}.\end{aligned}$$

Thus

$$\mathbf{a}\cdot(\mathbf{b}\wedge\mathbf{c}) = (\mathbf{a}\cdot\mathbf{b})\mathbf{c} - (\mathbf{a}\cdot\mathbf{c})\mathbf{b}. \tag{6.15}$$

This shows that inner multiplication of a vector with a bivector results in a vector. So Equation (6.14) expresses the fact that the geometric product $\mathbf{aB}$ of a vector with a bivector results in the sum of a vector $\mathbf{a}\cdot\mathbf{B}$ and a trivector $\mathbf{a}\wedge\mathbf{B}$. The similarity between (6.1) and (6.14) should be noted. They illustrate the general rule that outer multiplication by a vector "raises the dimension" of any directed number by one, whereas inner multiplication "lowers" it by one. Clearly, the generalized inner and outer products provide an algebraic vehicle for expressing geometric notions about "increasing or decreasing the dimension of space".

1-7. Axioms for Geometric Algebra

Let us examine now what we have learned about building a geometric algebra. To begin with, the algebra should include the graded elements 0-vector, 1-vector, 2-vector and 3-vector to represent the directional properties of points, lines, planes and space. We introduced three kinds of multiplication, the scalar, inner and outer products, to express relations among the elements. But we saw that inner and outer products can be reduced to a single geometric product if we allow elements of different grade (or dimension) to be added. For this reason, we conclude that the algebra should include elements of "mixed grade", such as

$$A = A_0 + \mathbf{A}_1 + \mathbf{A}_2 + \mathbf{A}_3, \tag{7.1}$$

Before continuing, a note about nomenclature is in order. The term "dimension" has two distinct but closely related mathematical meanings. To separate them we will henceforth use the term "grade" exclusively to mean "dimension" in the sense that we have used the term up to this point. We have preferred the term "dimension" in our introductory discussion, because it is likely to have familiar and helpful connotations for the reader. However, now we aim to improve the precision of our language with an axiomatic formulation of the basic concepts. The alternative meaning for "dimension" will be explained in Section 2.2, where for $k = 0, 1, 2, 3$, $\mathbf{A}_k$ is an element of grade k called *k-vector part* of A. Thus, (7.1) presents A as the sum of a scalar A_0, a vector $\mathbf{A}_1$, a bivector $\mathbf{A}_2$, and a trivector $\mathbf{A}_3$. We refer to A as a *multivector*, a (*directed*) *number* or a *quantity*. Any element of the geometric algebra can be called a multivector, because it can be represented in the form (7.1). For example, a vector $\mathbf{a}$ can be expressed trivially in the form (7.1) by writing $A = \mathbf{A}_1 = \mathbf{a}$, $A_0 = \mathbf{A}_2 = \mathbf{A}_3 = 0$ and using the property $\mathbf{a} + 0 = \mathbf{a}$. Note the k-vectors which are not scalars are denoted by symbols in boldface type. Such k-vectors are sometimes called *k-blades* or, simply, *blades* to emphasize the fact that, in contrast to 0-vectors (scalars), they have "directional properties".

Now another simplification becomes possible. It will be noted that the geometric product of vectors which we have just considered has, except for commutivity, the same algebraic properties as scalar multiplication of vectors and bivectors. In particular, both products are associative and distributive with respect to addition. Rather than regard them as *two different kinds of multiplication*, we can regard them as *instances of a single geometric product among different kinds of multivector*. Thus, scalar multiplication is the geometric product of any multivector by a special kind of multivector called a scalar. The special geometric nature of the scalars is expressed algebraically by the fact that they commute with every other multivector.

Thus, we *define* addition and multiplication of multivectors by the following familiar rules: For multivectors A, B, C, . . ., addition is *commutative*,

$$A + B = B + A; \tag{7.2}$$

addition and multiplication are *associative*,

$$(A + B) + C = A + (B + C), \tag{7.3}$$

$$(AB)C = A(BC); \tag{7.4}$$

multiplication is *distributive* with respect to addition,

$$A(B + C) = AB + AC, \tag{7.5}$$

$$(B + C)A = BA + CA. \tag{7.6}$$

There exist unique multivectors 0 and 1 such that

$$A + 0 = A, \tag{7.7}$$

$$1A = A. \tag{7.8}$$

Every multivector A has a unique *additive inverse* $-A$, that is,

$$A + (-A) = 0. \tag{7.9}$$

Of course, the whole algebra is assumed to be *algebraically closed*, that is, the sum or product of any two multivectors is itself a unique multivector.

It is hardly necessary to discuss the significance of the above axioms, since they are familiar from the elementary algebra of scalars. They can be used to manipulate multivectors in exactly the same way that numbers are manipulated in arithmetic. For example, axiom (7.9) is used to define subtraction of arbitrary multivectors in the same way that subtraction of vectors was defined by Equation (3.10).

To complete our system of axioms for geometric algebra, we need some axioms that characterize the various kinds of k-vectors. First of all, we assume that the set of all scalars in the algebra can be identified with the real numbers, and we express the commutivity of scalar multiplication by the axiom

$$\lambda A = A\lambda \tag{7.10}$$

for every scalar λ and multivector A. Vectors are characterized by the following axiom. The "square" of any nonzero *vector* $\mathbf{a}$ is a unique positive scalar $|\mathbf{a}|^2$, that is,

$$\mathbf{a}^2 = |\,\mathbf{a}\,|^2 > 0. \tag{7.11}$$

We characterize k-vectors of higher grade by relating them to vectors. It will be most convenient to do that after we introduce a couple of definitions.

For a vector $\mathbf{a}$ and any k-blade $\mathbf{A}_k$ we define the *inner product* by

$$\mathbf{a}\cdot\mathbf{A}_k = \tfrac{1}{2}(\mathbf{a}\mathbf{A}_k - (-1)^k\mathbf{A}_k\mathbf{a}), \tag{7.12a}$$

and the *outer product* by

$$\mathbf{a}\wedge\mathbf{A}_k = \tfrac{1}{2}(\mathbf{a}\mathbf{A}_k + (-1)^k\,\mathbf{A}_k\mathbf{a}), \tag{7.12b}$$

adding these equations, we get

$$\mathbf{a}\mathbf{A}_k = \mathbf{a}\cdot\mathbf{A}_k + \mathbf{a}\wedge\mathbf{A}_k. \tag{7.12c}$$

Note that (7.12a) includes (6.9) and (6.12) as special cases, while (7.12b) includes (6.10) and (6.13), and (7.12c) includes (6.1) and (6.4).

Using the definitions (7.12a) and (7.12b), we adopt the following propositions as axioms:

$$\mathbf{a}\cdot\mathbf{A}_k \text{ is a } (k\text{-}1)\text{-vector} \tag{7.13a}$$

$$\mathbf{a}\wedge\mathbf{A}_k \text{ is a } (k+1)\text{-vector.} \tag{7.13b}$$

Thus, the inner product lowers the grade of a k-vector, while the outer product raises the grade. According to (7.1), however, all k-vectors with grade $k \geqslant 4$ must vanish. To assure this, we need one more axiom: For every vector $\mathbf{a}$ and 3-vector $\mathbf{A}_3$,

$$\mathbf{a}\wedge\mathbf{A}_3 = 0 \tag{7.14a}$$

By virtue of the definition (7.12b), this can alternatively be written

$$\mathbf{a}\mathbf{A}_3 = \mathbf{A}_3\mathbf{a}. \tag{7.14b}$$

In other words, vectors always commute with trivectors.

Finally, to assure that the whole algebraic system is not vacuous, we must assume that nonzero multivectors with all grades $k \leqslant 3$ actually exist.

This completes our formulation of the axioms for geometric algebra. We have neglected some logical fine points (e.g., Exercise 7.1), but our axioms suffice to show exactly how geometric algebra generalizes the familiar algebra of scalars. We have chosen a notation for geometric algebra that is as similar as possible to the notation of scalar algebra. This is a point of great importance, for it facilitates the transfer of skills in manipulations with scalar algebra to manipulations with geometric algebra. Let us note exactly how the basic operations of scalar algebra transfer to geometric algebra.

Axioms (7.2) to (7.9) implicitly define the operations of addition, subtraction and multiplication. Except for the absence of a general commutative law for multiplication, they are identical to the axioms of scalar algebra. Therefore, multivectors can be equated, added, subtracted and multiplied in exactly the same way as scalar quantities, provided one does not reorder multiplicative factors which do not commute. Division by multivectors can be defined in terms of multiplication, just as in scalar algebra. But we need to pay special attention to notation on account of noncommutivity, so let us consider the matter explicitly.

In geometric algebra, as in elementary algebra, the solution of equations is greatly facilitated by the possibility of division. We can divide by a multivector A if it has a *multiplicative inverse*. The *inverse* of A, if it exists, is denoted by A^{-1} or $1/A$ and defined by the equation

$$A^{-1}A = 1. \tag{7.15}$$

We can divide any multivector B by A in two ways, by multiplying it by A^{-1} on the left,

$$A^{-1}B = \frac{1}{A}B,$$

or on the right,

$$BA^{-1} = B\frac{1}{A} = B/A.$$

Obviously, the "left division" is not equivalent to the "right division" unless B commutes with A^{-1}, in which case the division can be denoted unambiguously by

$$\frac{B}{A}.$$

Every nonzero vector **a** has a multiplicative inverse. To determine the inverse of **a**, we multiply the equation $\mathbf{a}^{-1}\mathbf{a} = 1$ on the right by **a** and divide by the scalar $\mathbf{a}^2 = |\,\mathbf{a}\,|^2$; thus

$$\mathbf{a}^{-1} = \frac{\mathbf{a}}{|\,\mathbf{a}\,|^2}. \tag{7.16}$$

With due regard for the order of factors, many tricks of elementary algebra, such as "rationalizing the denominator", are equally useful in geometric algebra. It should be noted, however, that some multivectors do not have multiplicative inverses (see Exercise 7.2), so it is impossible to divide by them.

7-1. Exercises

Hints and solutions for selected exercises are given at the back of the book.

(7.1) The axioms given in this chapter do not suffice to prove the elementary

(a) *Addition Property of Equality*:

If $B = C$, then $A + B = A + C$.

(b) *Multiplication Property of Equality*:

$$\text{If } B = C, \text{ then } AB = AC.$$

Add these properties to our list of axioms and prove the converses
(c) *Cancellation Principle of Addition*:

$$\text{If } A + B = A + C, \text{ then } B = C.$$

(d) *Cancellation Principle of Multiplication*:

$$\text{If } AB = AC \text{ and } A^{-1} \text{ exists, then } B = C.$$

Specify the justification for each step in the proof. The proofs in geometric algebra are identical to those in elementary algebra.

(7.2) Let $A = \alpha + \mathbf{a}$ where α is a scalar and $\mathbf{a}$ is a nonzero vector.

(a) Find A^{-1} as a function of α and $\mathbf{a}$. What conditions on α and $\mathbf{a}$ imply that A^{-1} does not exist?

(b) Show that if A^{-1} does not exist, then A can be normalized so that

$$A^2 = A.$$

A quantity with this property is said to be *idempotent*.

(c) Show that if $A \neq 1$ is idempotent, its product with any other multivector is not invertible. It can be proved that every multivector which does not have an inverse has an idempotent for a factor.

(d) Find an idempotent which does have an inverse.

(7.3) Prove that every left inverse is also a right inverse and that this inverse is unique.

Chapter 2

Developments in Geometric Algebra

In Chapter 1 we developed geometric algebra as a symbolic system for representing the basic geometrical concepts of direction, magnitude, orientation and dimension. In this chapter we continue the development of geometric algebra into a full-blown mathematical language. The basic grammar of this language is completely specified by the axioms set down at the end of Chapter 1. But there is much more to a language than its grammar!

To develop geometric algebra to the point where we can express and explore the ideas of mechanics with fluency, in this chapter we introduce auxiliary concepts and definitions, derive useful algebraic relations, describe simple curves and surfaces with algebraic equations, and formulate the fundamentals of differentiation and integration with respect to scalar variables. Further mathematical developments are given in Chapter 5.

2-1. Basic Identities and Definitions

In Chapter 1 we were led to the geometric product for vectors by combining inner and outer products according to the equation

$$\mathbf{ab} = \mathbf{a}\cdot\mathbf{b} + \mathbf{a}\wedge\mathbf{b}. \tag{1.1}$$

Then we reversed the procedure, defining the inner and outer products in terms of the geometric product by the equations

$$\mathbf{a}\cdot\mathbf{b} = \tfrac{1}{2}(\mathbf{ab} + \mathbf{ba}) \tag{1.2}$$

$$\mathbf{a}\wedge\mathbf{b} = \tfrac{1}{2}(\mathbf{ab} - \mathbf{ba}). \tag{1.3}$$

This did more than reduce two different kinds of multiplication to one. It made possible the formulation of a simple axiom system from which an unlimited number of geometrical relations can be deduced by algebraic manipulation. In this section we aim to improve our skills at carrying out such deductions and establish some widely useful results.

The inner and outer products appear frequently in applications, because they have straightforward geometrical interpretations, as we saw in Chapter

1. For this reason, it is often desirable to operate directly with inner and outer products, even though we regard the geometric product as more fundamental. To make this possible, we need a system of algebraic identities relating inner and outer products. We derive these identities, of course, by using the geometric product and the axioms of geometric algebra set down in Section 1-7.

The different products are most easily related by the equation

$$\mathbf{a}\mathbf{A}_r = \mathbf{a}\cdot\mathbf{A}_r + \mathbf{a}\wedge\mathbf{A}_r, \tag{1.4}$$

which generalizes (1.1) to apply to any r-blade $\mathbf{A}_r$, that is, any r-vector with grade $r > 0$. Recall that the corresponding definitions of inner and outer products are given by

$$\mathbf{a}\cdot\mathbf{A}_r = \tfrac{1}{2}(\mathbf{a}\mathbf{A}_r - (-1)^r\mathbf{A}_r\mathbf{a}) = (-1)^{r+1}\mathbf{A}_r\cdot\mathbf{a} \tag{1.5}$$

$$\mathbf{a}\wedge\mathbf{A}_r = \tfrac{1}{2}(\mathbf{a}\mathbf{A}_r + (-1)^r\mathbf{A}_r\mathbf{a}) = (-1)^r\mathbf{A}_r\wedge\mathbf{a}. \tag{1.6}$$

We will make frequent use of the fact that $\mathbf{a}\cdot\mathbf{A}_r$ is an $(r-1)$-vector (a scalar if $r = 1$), while $\mathbf{a}\wedge\mathbf{A}_r$ is an $(r+1)$-vector.

To illustrate the use of (1.4) and its special case (1.1), let us derive the associative rule for the outer product. Beginning with the associative rule for the geometric product,

$$\mathbf{a}(\mathbf{b}\mathbf{c}) = (\mathbf{a}\mathbf{b})\mathbf{c},$$

we use (1.1) to get

$$\mathbf{a}(\mathbf{b}\cdot\mathbf{c} + \mathbf{b}\wedge\mathbf{c}) = (\mathbf{a}\cdot\mathbf{b} + \mathbf{a}\wedge\mathbf{b})\mathbf{c}.$$

Applying the distributive rule and (1.4), we get

$$\mathbf{a}(\mathbf{b}\cdot\mathbf{c}) + \mathbf{a}\cdot(\mathbf{b}\wedge\mathbf{c}) + \mathbf{a}\wedge(\mathbf{b}\wedge\mathbf{c}) = (\mathbf{a}\cdot\mathbf{b})\mathbf{c} + (\mathbf{a}\wedge\mathbf{b})\cdot\mathbf{c} + (\mathbf{a}\wedge\mathbf{b})\wedge\mathbf{c}.$$

Now we identify the terms $\mathbf{a}(\mathbf{b}\cdot\mathbf{c})$ and $\mathbf{a}\cdot(\mathbf{b}\wedge\mathbf{c})$ as vectors, and the term $\mathbf{a}\wedge(\mathbf{b}\wedge\mathbf{c})$ as a trivector. Since vectors are distinct from trivectors, we can separately equate vector and trivector parts on each side of the equation. By equating trivector parts, we get the associative rule

$$\mathbf{a}\wedge(\mathbf{b}\wedge\mathbf{c}) = (\mathbf{a}\wedge\mathbf{b})\wedge\mathbf{c}. \tag{1.7}$$

And by equating the vector parts we find an algebraic identity which we have not seen before,

$$\mathbf{a}(\mathbf{b}\cdot\mathbf{c}) + \mathbf{a}\cdot(\mathbf{b}\wedge\mathbf{c}) = (\mathbf{a}\cdot\mathbf{b})\mathbf{c} + (\mathbf{a}\wedge\mathbf{b})\cdot\mathbf{c}. \tag{1.8}$$

For more about this identity, see Exercise 1.11.

This derivation of the associative rule (1.9) should be compared with our previous derivation of the same rule in Section 1-6. That derivation was considerably more complicated, because it employed a direct reduction of the outer product to the geometric product. Moreover, the indirect method employed here gives us the additional "vector identity" (1.8) at no extra cost.

Note the general structure of the method: an identity involving geometric products alone is expanded into inner and outer products by using (1.4) and then parts of the same grade are separately equated. This will be our principal method for establishing identities involving inner and outer products. As another example, note that the method immediately gives the distributive rules for inner and outer products. Thus, if $\mathbf{a}$ is a vector and $\mathbf{B}_r$ and $\mathbf{C}_r$ are r-blades, then by applying (1.4) to

$$\mathbf{a}(\mathbf{B}_r + \mathbf{C}_r) = \mathbf{a}\mathbf{B}_r + \mathbf{a}\mathbf{C}_r,$$

and separating parts of different grade, we get

$$\mathbf{a}\cdot(\mathbf{B}_r + \mathbf{C}_r) = \mathbf{a}\cdot\mathbf{B}_r + \mathbf{a}\cdot\mathbf{C}_r, \tag{1.9a}$$

and

$$\mathbf{a}\wedge(\mathbf{B}_r + \mathbf{C}_r) = \mathbf{a}\wedge\mathbf{B}_r + \mathbf{a}\wedge\mathbf{C}_r. \tag{1.9b}$$

Here we have the distributive rules in a somewhat more general (hence more useful) form than they were presented in Chapter 1.

These examples show the importance of separating a multivector or a multivector equation into parts of different grade. So it will be useful to introduce a special notation to express such a separation. Accordingly, we write $\langle A\rangle_r$ to denote the *r-vector part* of a multivector A. For example, if $A = \mathbf{abc}$, this notation enables us to write

$$\langle\mathbf{abc}\rangle_3 = \mathbf{a}\wedge\mathbf{b}\wedge\mathbf{c}$$

for the trivector part,

$$\langle\mathbf{abc}\rangle_1 = \mathbf{a}(\mathbf{b}\cdot\mathbf{c}) + \mathbf{a}\cdot(\mathbf{b}\wedge\mathbf{c})$$

for the vector part, while the vanishing of scalar and bivector parts is described by the equation

$$\langle\mathbf{abc}\rangle_0 = 0 = \langle\mathbf{abc}\rangle_2.$$

According to axiom (7.1) of Chapter 1, every multivector can be decomposed into a sum of its r-vector parts, as expressed by writing

$$A = \sum_r \langle A\rangle_r = \langle A\rangle_0 + \langle A\rangle_1 + \langle A\rangle_2 + \langle A\rangle_3. \tag{1.10}$$

If $A = \langle A\rangle_r$, then A is said to be *homogeneous* of grade r, that is, A is an r-vector. A multivector A is said to be *even* (*odd*) if $\langle A\rangle_r = 0$ when r is an odd (even) integer. Obviously every multivector A can be expressed as a sum of an *even part* $\langle A\rangle_+$, and an *odd part* $\langle A\rangle_-$. Thus

$$A = \langle A\rangle_+ + \langle A\rangle_- \tag{1.11a}$$

where

$$\langle A\rangle_+ = \langle A\rangle_0 + \langle A\rangle_2, \tag{1.11b}$$

$$\langle A\rangle_- = \langle A\rangle_1 + \langle A\rangle_3. \tag{1.11c}$$

We shall see later that the distinction between even and odd multivectors is important, because the even multivectors form an algebra by themselves but the odd multivectors do not.

According to (1.10), we have $\langle A \rangle_k = 0$ for all $k > 3$, that is, every blade with grade $k > 3$ must vanish. We adopted this condition in Chapter 1 to express the fact that physical space is three dimensional, so we will be assuming it in our treatment of mechanics throughout this book. However, such a condition is not essential for mathematical reasons, and there are other applications of geometric algebra to physics where it is not appropriate. For the sake of mathematical generality, therefore, all results and definitions in this section are formulated without limitations on grade, with the exception, of course, of (1.10) and (1.11b, c). This generality is achieved at very little extra cost, and it has the advantage of revealing precisely what features of geometric algebra are peculiar to three dimensions.

Before continuing, it will be worthwhile to discuss the use of parentheses in algebraic expressions. Note that the expression $\mathbf{a}\cdot\mathbf{bc}$ is ambiguous. It could mean $(\mathbf{a}\cdot\mathbf{b})\mathbf{c}$, which is to say that the inner product $\mathbf{a}\cdot\mathbf{b}$ is performed first and the resulting scalar multiplies the vector $\mathbf{c}$. On the other hand, it could mean $\mathbf{a}\cdot(\mathbf{bc})$, which is to say that the geometric product $\mathbf{bc}$ is performed before the inner product. The two interpretations give completely different algebraic results. To remove such ambiguities without using parentheses, we introduce the following *precedence convention*: If there is ambiguity, indicated inner and outer products should be performed before an adjacent geometric product. Thus

$$(\mathbf{A}\wedge\mathbf{B})\mathbf{C} = \mathbf{A}\wedge\mathbf{BC} \neq \mathbf{A}\wedge(\mathbf{BC}), \tag{1.12a}$$

$$(\mathbf{A}\cdot\mathbf{B})\mathbf{C} = \mathbf{A}\cdot\mathbf{BC} \neq \mathbf{A}\cdot(\mathbf{BC}). \tag{1.12b}$$

This convention eliminates an appreciable number of parentheses, especially in complicated expressions. Other parentheses can be eliminated by the convention that outer products have "preference" over inner products, so

$$\mathbf{A}\cdot(\mathbf{B}\wedge\mathbf{C}) = \mathbf{A}\cdot\mathbf{B}\wedge\mathbf{C} \neq (\mathbf{A}\cdot\mathbf{B})\wedge\mathbf{C}, \tag{1.13}$$

but we use this convention much less often than the preceding one.

The most useful identity relating inner and outer products is, of course, its simplest one:

$$\mathbf{a}\cdot(\mathbf{b}\wedge\mathbf{c}) = \mathbf{a}\cdot\mathbf{bc} - \mathbf{a}\cdot\mathbf{cb}. \tag{1.14}$$

We derived this in Section 1-6 before we had established our axiom system for geometric algebra. Now we can derive it by a simpler method. First we use (1.2) in the form $\mathbf{ab} = -\mathbf{ba} + 2\mathbf{a}\cdot\mathbf{b}$ to reorder multiplicative factors as follows:

$$\mathbf{abc} = -\mathbf{bac} + 2\mathbf{a}\cdot\mathbf{bc} = -\mathbf{b}(-\mathbf{ca} + 2\mathbf{a}\cdot\mathbf{c}) + 2\mathbf{a}\cdot\mathbf{bc}.$$

Rearranging terms and using (1.1), we obtain

$$\mathbf{a}\cdot\mathbf{bc} - \mathbf{a}\cdot\mathbf{cb} = \tfrac{1}{2}(\mathbf{abc} - \mathbf{bca}) = \tfrac{1}{2}(\mathbf{ab}\wedge\mathbf{c} - \mathbf{b}\wedge\mathbf{ca}),$$

which, by (1.5), gives us (1.14) as desired.

By the same method we can derive the more general *reduction formula*

$$\mathbf{a}\cdot(\mathbf{b}\wedge\mathbf{C}_r) = \mathbf{a}\cdot\mathbf{b}\mathbf{C}_r - \mathbf{b}\wedge(\mathbf{a}\cdot\mathbf{C}_r), \tag{1.15}$$

where $\mathbf{a}$ and $\mathbf{b}$ are vectors while $\mathbf{C}_r$ is an r-blade. We use (1.2) and (1.5) to reorder multiplicative factors as follows:

$$\mathbf{abC}_r = -\mathbf{baC}_r + 2\mathbf{a}\cdot\mathbf{bC}_r = -\mathbf{b}((-1)^r\,\mathbf{C}_r\,\mathbf{a} + 2\mathbf{a}\cdot\mathbf{C}_r) + 2\mathbf{a}\cdot\mathbf{bC}_r.$$

Rearranging terms and using (1.4) and (1.5), we obtain

$$\mathbf{a}\cdot\mathbf{bC}_r - \mathbf{ba}\cdot\mathbf{C}_r = \tfrac{1}{2}[\mathbf{a}(\mathbf{bC}_r) + (-1)^r\,(\mathbf{bC}_r)\mathbf{a}] = \mathbf{a}\cdot(\mathbf{b}\cdot\mathbf{C}_r) + \mathbf{a}\cdot(\mathbf{b}\wedge\mathbf{C}_r).$$

The r-vector part of this equation gives us (1.15) as desired.

By iterating (1.15), we obtain the expanded reduction formula

$$\begin{aligned}\mathbf{a}\cdot(\mathbf{a}_1\wedge\mathbf{a}_2\wedge\ldots\wedge\mathbf{a}_r) &= \sum_{k=1}^{r}(-1)^{k+1}\,\mathbf{a}\cdot\mathbf{a}_k\mathbf{a}_1\wedge\ldots\check{\mathbf{a}}_k\ldots\wedge\mathbf{a}_r\\ &= \mathbf{a}\cdot\mathbf{a}_1\mathbf{a}_2\wedge\mathbf{a}_3\wedge\ldots\wedge\mathbf{a}_r - \mathbf{a}\cdot\mathbf{a}_2\mathbf{a}_1\wedge\mathbf{a}_3\wedge\ldots\wedge\mathbf{a}_r +\\ &\quad+\ldots+(-1)^{r+1}\,\mathbf{a}\cdot\mathbf{a}_r\mathbf{a}_1\wedge\mathbf{a}_2\wedge\ldots\wedge\mathbf{a}_{r-1}.\end{aligned} \tag{1.16}$$

The inverted circumflex in the product $\mathbf{a}_1\wedge\ldots\check{\mathbf{a}}_k\ldots\wedge\mathbf{a}_r$ means that the kth factor $\mathbf{a}_k$ is to be omitted. Equation (1.16) determines the inner product of the r-vector $\mathbf{A}_r = \mathbf{a}_1\wedge\ldots\wedge\mathbf{a}_r$ with $\mathbf{a}$ in terms of its vector factors $\mathbf{a}_k$ and their inner products $\mathbf{a}\cdot\mathbf{a}_k$. It would be quite appropriate to refer to (1.15) as the *Laplace expansion* of the inner product, because of its relation to the expansion of a determinant (see Chapter 5).

Our definitions (1.5) and (1.6) for inner and outer products require that one of the factors in the products be a vector. It will be useful to generalize these definitions to apply to blades of any grade. For any r-blade $\mathbf{A}_r$ and s-blade $\mathbf{B}_s$, the *inner product* $\mathbf{A}_r\cdot\mathbf{B}_s$ is defined by

$$\mathbf{A}_r\cdot\mathbf{B}_s \equiv \langle\mathbf{A}_r\mathbf{B}_s\rangle_{|r-s|}, \tag{1.17a}$$

and the *outer product* $\mathbf{A}_r\wedge\mathbf{B}_s$ is defined by

$$\mathbf{A}_r\wedge\mathbf{B}_s \equiv \langle\mathbf{A}_r\mathbf{B}_s\rangle_{r+s}. \tag{1.17b}$$

Thus, the inner product produces an $|r-s|$–vector, while the outer product produces an $(r+s)$–vector. The symbol $\equiv$ denotes a definition or identity.

The reduction of an inner product between two blades can be accomplished by using the formula

$$(\mathbf{A}_r\wedge\mathbf{b})\cdot\mathbf{C}_s = \mathbf{A}_r\cdot(\mathbf{b}\cdot\mathbf{C}_s), \tag{1.18}$$

which holds for $0 < r < s$, with $\mathbf{A}_r = \langle\mathbf{A}_r\rangle_r$, $\mathbf{b} = \langle\mathbf{b}\rangle_1$ and $\mathbf{C}_s = \langle\mathbf{C}_s\rangle_s$. Note that the factor $\mathbf{b}\cdot\mathbf{C}_s$ on the right side of (1.18) can be further reduced by (1.15) or (1.16) if $\mathbf{C}_s$ is expressed as a product of vectors. Equation (1.18) can be proved in the same way as (1.8). Use (1.4) to expand $(\mathbf{A}_r\mathbf{b})\mathbf{C}_s = \mathbf{A}_r(\mathbf{b}\mathbf{C}_s)$ and ascertain that the $(s - r - 1)$-vector part is equivalent to (1.18).

The student should beware that the geometric product of blades **A** and **B** is *not* generally related to inner and outer products by the formula

$$\mathbf{AB} = \mathbf{A}\cdot\mathbf{B} + \mathbf{A}\wedge\mathbf{B}$$

unless one of the factors is a vector, as in (1.4). In particular, this formula does not hold if both **A** and **B** are bivectors. To prove that, express **A** as a product of orthogonal vectors by writing $\mathbf{A} = \mathbf{a}\wedge\mathbf{b} = \mathbf{ab}$.

Then

$$\begin{aligned}\mathbf{AB} &= \mathbf{a}(\mathbf{b}\cdot\mathbf{B} + \mathbf{b}\wedge\mathbf{B})\\ &= \mathbf{a}\cdot(\mathbf{b}\cdot\mathbf{B}) + \mathbf{a}\wedge(\mathbf{b}\cdot\mathbf{B}) + \mathbf{a}\cdot(\mathbf{b}\wedge\mathbf{B}) + \mathbf{a}\wedge\mathbf{b}\wedge\mathbf{B}.\end{aligned}$$

Hence

$$\mathbf{AB} = \mathbf{A}\cdot\mathbf{B} + \langle\mathbf{AB}\rangle_2 + \mathbf{A}\wedge\mathbf{B}, \tag{1.19}$$

where

$$\begin{aligned}\mathbf{A}\cdot\mathbf{B} &= \langle\mathbf{AB}\rangle_0 = \mathbf{a}\cdot(\mathbf{b}\cdot\mathbf{B}),\\ \langle\mathbf{AB}\rangle_2 &= \mathbf{a}\wedge(\mathbf{b}\cdot\mathbf{B}) + \mathbf{a}\cdot(\mathbf{b}\wedge\mathbf{B}),\\ \mathbf{A}\wedge\mathbf{B} &= \langle\mathbf{AB}\rangle_4 = \mathbf{a}\wedge\mathbf{b}\wedge\mathbf{B}.\end{aligned}$$

Note that we have 3 terms in (1.19) in contrast to the two terms in (1.4). To learn more about the product between bivectors, we use the trick that any geometric product can be decomposed into symmetric and antisymmetric parts by writing

$$\mathbf{AB} = \tfrac{1}{2}(\mathbf{AB} + \mathbf{BA}) + \tfrac{1}{2}(\mathbf{AB} - \mathbf{BA}).$$

Comparing this with (1.19), it is not difficult to establish that, for bivectors **A** and **B**,

$$\mathbf{A}\cdot\mathbf{B} + \mathbf{A}\wedge\mathbf{B} = \tfrac{1}{2}(\mathbf{AB} + \mathbf{BA}) = \mathbf{B}\cdot\mathbf{A} + \mathbf{B}\wedge\mathbf{A} \tag{1.20a}$$

$$\langle\mathbf{AB}\rangle_2 = \tfrac{1}{2}(\mathbf{AB} - \mathbf{BA}) = -\langle\mathbf{BA}\rangle_2. \tag{1.20b}$$

The expression $\tfrac{1}{2}(\mathbf{AB} - \mathbf{BA})$ is sometimes called the *commutator* or *commutator product* of **A** and **B**, because it vanishes if **A** and **B** commute. Equation (1.20b) tells us that the commutator product of bivectors produces another bivector. Equation (1.20a) tells us that the *symmetric product* of bivectors $\tfrac{1}{2}(\mathbf{AB} + \mathbf{BA})$ produces a scalar $\mathbf{A}\cdot\mathbf{B}$ and a 4-vector $\mathbf{A}\wedge\mathbf{B}$. Of course, we can take $\mathbf{A}\wedge\mathbf{B} = 0$ when we employ our grade restriction axiom as in (1.10).

The procedure which gave us the expansion (1.19) for the product of bivectors can be applied to the product of blades of any grade. Note, that for the product $\mathbf{A}_r\mathbf{B}_s = \mathbf{a}_1\mathbf{a}_2 \ldots \mathbf{a}_r\mathbf{B}_s$, if $r \leqslant s$, we reduce the initial grade of the factor $\mathbf{B}_s$ on the right by successive inner products with vectors, and the term of lowest grade will be

$$\mathbf{a}_1\cdot(\mathbf{a}_2\cdot \ldots (\mathbf{a}_r\cdot\mathbf{B}_s)) = \langle \mathbf{A}_r\mathbf{b}_s\rangle_{s-r} = \mathbf{A}_r\cdot\mathbf{B}_s .$$

Thus, we conclude that $\mathbf{A}_r\cdot\mathbf{B}_s$ is the term of lowest grade in the product $\mathbf{A}_r\mathbf{B}_s$. And as a useful corollary, we note that the product $\mathbf{A}_r\mathbf{B}_s$ can have a nonzero scalar part $\mathbf{A}_r\cdot\mathbf{B}_s$ only if $r = s$.

Factorization

We know the great importance of factoring in scalar algebra. In geometric algebra we have a new kind of factoring which is equally important, namely, the factoring of a k-blade into a product of vectors. Consider, for example, a unit vector **a** and a nonzero bivector **B** such that

$$\mathbf{a}\wedge\mathbf{B} = 0. \tag{1.21}$$

By virtue of (1.4), this implies that

$$\mathbf{aB} = \mathbf{a}\cdot\mathbf{B} \equiv \mathbf{b}, \tag{1.22}$$

which defines a vector **b**. We can solve this equation for **B** by division, that is, by multiplying it by $\mathbf{a}^{-1} = \mathbf{a}$. Thus, we obtain

$$\mathbf{B} = \mathbf{ab} = \mathbf{a}\wedge\mathbf{b}. \tag{1.23}$$

The last equality follows from (1.1), since (1.22) implies $\mathbf{a}\cdot\mathbf{b} = 0$. Equation (1.23) is a factorization of the bivector **B** into a product of orthogonal vectors, so (1.21) is a condition that **a** be a factor of **B**. Of course, **b** is also a factor of **B** and $\mathbf{b}\wedge\mathbf{B} = 0$. Equation (1.22) shows that **b** is a unique factor of **B** orthogonal to **a**. In Section 2-2 it will become obvious that **B** can be factored into orthogonal vector pairs in an infinite number of ways. Blades of higher grade can be factored in a similar way.

Reversion

In algebraic computations, it is often desirable to reorder the factors in a product. For this reason, it is convenient to introduce the operation of *reversion* defined by the equations

$$(AB)^\dagger = B^\dagger A^\dagger , \tag{1.24a}$$

$$(A + B)^\dagger = A^\dagger + B^\dagger , \tag{1.24b}$$

$$\langle A^\dagger\rangle_0 = \langle A \rangle_0, \tag{1.24c}$$

$$\mathbf{a}^\dagger = \mathbf{a} \quad \text{if} \quad \mathbf{a} = \langle\mathbf{a}\rangle_1 . \tag{1.24d}$$

We say that $A^\dagger$ is the *reverse* of the multivector A. It follows easily from (1.24a) and (1.24d) that the reverse of a product of vectors is

$$(\mathbf{a}_1\mathbf{a}_2 \ldots \mathbf{a}_r)^\dagger = \mathbf{a}_r \ldots \mathbf{a}_2\mathbf{a}_1 . \tag{1.25}$$

This justifies our choice of the name "reverse".

The reverse of a bivector $\mathbf{B} = \mathbf{a}\wedge\mathbf{b}$ is given by

$$\mathbf{B}^\dagger = (\mathbf{a}\wedge\mathbf{b})^\dagger = \mathbf{b}\wedge\mathbf{a} = -\mathbf{a}\wedge\mathbf{b} = -\mathbf{B}, \tag{1.26}$$

where the anticommutivity of the outer product was used to reorder vector factors. In a similar way we can reorder vector factors in a trivector one at a time.

$$\mathbf{c}\wedge\mathbf{b}\wedge\mathbf{a} = -\mathbf{b}\wedge\mathbf{c}\wedge\mathbf{a} = \mathbf{b}\wedge\mathbf{a}\wedge\mathbf{c} = -\mathbf{a}\wedge\mathbf{b}\wedge\mathbf{c}.$$

Hence,

$$(\mathbf{a}\wedge\mathbf{b}\wedge\mathbf{c})^\dagger = \mathbf{c}\wedge\mathbf{b}\wedge\mathbf{a} = -\mathbf{a}\wedge\mathbf{b}\wedge\mathbf{c}. \tag{1.27}$$

Thus, reversion changes the signs of bivectors and trivectors, while, according to (1.24c) and (1.24d), it does not affect scalars and vectors. All this is summed up by applying reversion to a general multivector in the expanded form (1.10), with the result

$$A^\dagger = \langle A\rangle_0 + \langle A\rangle_1 - \langle A\rangle_2 - \langle A\rangle_3 . \tag{1.28}$$

Magnitude

To every multivector A there corresponds a unique scalar $|A|$, called the *magnitude* or *modulus* of A, defined by the equation

$$|A| = \langle A^\dagger A\rangle_0^{1/2}. \tag{1.29}$$

Existence of the square root is assured by the fact that

$$|A|^2 = \langle A^\dagger A\rangle_0 \geqslant 0, \tag{1.30}$$

where $|A| = 0$ if and only if $A = 0$. To prove this, first observe that

$$|\mathbf{a}_1 \ldots \mathbf{a}_r|^2 = (\mathbf{a}_1 \ldots \mathbf{a}_r)^\dagger (\mathbf{a}_1 \ldots \mathbf{a}_r) = |\mathbf{a}_1|^2 \ldots |\mathbf{a}_r|^2 \geqslant 0. \tag{1.31}$$

If the vectors here are orthogonal, then they are factors of an r-blade $\mathbf{a}_1\mathbf{a}_2 \ldots \mathbf{a}_r = \mathbf{a}_1\wedge\mathbf{a}_2\wedge \ldots \wedge\mathbf{a}_r$. It follows that the squared modulus of any non-zero r-blade is positive, that is,

$$|\langle \mathrm{A}\rangle_r|^2 \geqslant 0 \text{ for any grade } r. \tag{1.32}$$

Now, when we expand the product $A^\dagger A$ in terms of r-vector parts, the "cross terms" multiplying blades of different grades have no scalar parts, so they can be ignored when we take scalar parts. Thus from (1.10) we obtain the expansion

$$|A|^2 = \langle A^\dagger A\rangle_0 = |\langle A\rangle_0|^2 + |\langle A\rangle_1|^2 + |\langle A\rangle_2|^2 + |\langle A\rangle_3|^2. \tag{1.33}$$

None of the terms in this expansion can be negative according to (1.32), so (1.30) is proved.

2-1. Exercises

(1.1) Establish the following "vector identities":

(a) $(\mathbf{a}\wedge\mathbf{b})\cdot(\mathbf{c}\wedge\mathbf{d}) = \mathbf{b}\cdot\mathbf{c}\mathbf{a}\cdot\mathbf{d} - \mathbf{b}\cdot\mathbf{d}\mathbf{a}\cdot\mathbf{c} = \mathbf{b}\cdot(\mathbf{c}\wedge\mathbf{d})\cdot\mathbf{a}$.

(b) $\mathbf{a}\cdot(\mathbf{b}\wedge\mathbf{c}\wedge\mathbf{d}) = \mathbf{a}\cdot\mathbf{b}\mathbf{c}\wedge\mathbf{d} - \mathbf{a}\cdot\mathbf{c}\mathbf{b}\wedge\mathbf{d} + \mathbf{a}\cdot\mathbf{d}\mathbf{b}\wedge\mathbf{c}$.

(c) $$(\mathbf{u}\wedge\mathbf{v})\cdot(\mathbf{a}\wedge\mathbf{b}\wedge\mathbf{c}) = (\mathbf{u}\wedge\mathbf{v})\cdot(\mathbf{a}\wedge\mathbf{b})\mathbf{c} - (\mathbf{u}\wedge\mathbf{v})\cdot(\mathbf{a}\wedge\mathbf{c})\mathbf{b} + (\mathbf{u}\wedge\mathbf{v})\cdot(\mathbf{b}\wedge\mathbf{c})\mathbf{a}.$$

(1.2) Vectors **a**, **b**, **c** are said to be *linearly dependent* if there exists scalars α, β, γ (not all zero) such that $\alpha\mathbf{a} + \beta\mathbf{b} + \gamma\mathbf{c} = 0$. Prove that $\mathbf{a}\wedge\mathbf{b}\wedge\mathbf{c} = 0$ if and only if **a**, **b**, **c** are linearly dependent. Express the coefficients α, β, γ for linearly dependent vectors in terms of the inner products of the vectors.

(1.3) Solve the following vector equation for the vector **x**:

$$\alpha\mathbf{x} + \mathbf{a}\mathbf{x}\cdot\mathbf{b} = \mathbf{c}.$$

(1.4) In the following vector equation **B** is a 2-blade,

$$\alpha\mathbf{x} + \mathbf{x}\cdot\mathbf{B} = \mathbf{a}.$$

Solve for the vector **x**. A good plan of attack in this kind of problem is to eliminate inner and/or outer products in favor of geometric products, so one can "divide out" multiplicative factors.

(1.5) Solve the following simultaneous equations for the vector **x** under the assumption that $\mathbf{c}\cdot\mathbf{a} \neq 0$: $\mathbf{a}\wedge\mathbf{x} = \mathbf{B}$, $\mathbf{c}\cdot\mathbf{x} = \alpha$.

(1.6) Prove the related vector identities

$$\mathbf{b}^2\,\mathbf{a}\cdot\mathbf{c} = \mathbf{a}\cdot\mathbf{b}\mathbf{b}\cdot\mathbf{c} + (\mathbf{a}\wedge\mathbf{b})\cdot(\mathbf{b}\wedge\mathbf{c}),$$

$$\langle\mathbf{a}\wedge\mathbf{b}\mathbf{c}\wedge\mathbf{b}\rangle_2 = (\mathbf{a}\wedge\mathbf{b}\wedge\mathbf{c})\cdot\mathbf{b}.$$

(1.7) Reduce $(\mathbf{a}\wedge\mathbf{b}\wedge\mathbf{c})\cdot(\mathbf{u}\wedge\mathbf{v}\wedge\mathbf{w})$, and $\langle\mathbf{abuv}\rangle_0$ to inner products of vectors.

(1.8) The identity $\mathbf{ab} = -\mathbf{ba} + 2\mathbf{a}\cdot\mathbf{b}$ can be used to reorder vectors in a product. Use it to establish the expansion formula

$$\langle\mathbf{abcuvw}\rangle_0 = \mathbf{a}\cdot\mathbf{b}\langle\mathbf{cuvw}\rangle_0 - \mathbf{a}\cdot\mathbf{c}\langle\mathbf{buvw}\rangle_0 + \mathbf{a}\cdot\mathbf{u}\langle\mathbf{bcvw}\rangle_0 - - \mathbf{a}\cdot\mathbf{v}\langle\mathbf{bcuw}\rangle_0 + \mathbf{a}\cdot\mathbf{w}\langle\mathbf{bcuv}\rangle_0.$$

(1.9) Prove that for vectors $\mathbf{a}_k$,

$$\langle\mathbf{a}_1\mathbf{a}_2\ldots\mathbf{a}_r\rangle_s = 0 \text{ if } r + s \text{ is an odd integer.}$$

(1.10) Prove the identity (1.15).

(1.11) Establish the "Jacobi identity" for vectors:

$$\mathbf{a}\cdot(\mathbf{b}\wedge\mathbf{c}) + \mathbf{b}\cdot(\mathbf{c}\wedge\mathbf{a}) + \mathbf{c}\cdot(\mathbf{a}\wedge\mathbf{b}) = 0$$

(1.12) Prove that if $\mathbf{A}_r$ is an r-blade, then $\mathbf{A}_r^{-1} = \dfrac{\mathbf{A}_r^\dagger}{|\mathbf{A}_r|^2}$.

(1.13) Prove $\langle A^\dagger\rangle_r = \langle A\rangle_r^\dagger = (-1)^{r(r-1)/2}\langle A\rangle_r$
and $\langle AB\rangle_r = (-1)^{r(r-1)/2}\langle B^\dagger A^\dagger\rangle_r$.

(1.14) Prove that $\langle AB\rangle_0 = \langle BA\rangle_0$.

(1.15) Prove that the bivector $\mathbf{a}\wedge\mathbf{b} + \mathbf{c}\wedge\mathbf{d}$ can be factored into a product of vectors if and only if $\mathbf{a}\wedge\mathbf{b}\wedge\mathbf{c}\wedge\mathbf{d} = 0$.

(1.16) Prove that

$$\langle AB\rangle_+ = \langle A\rangle_+\langle B\rangle_+ + \langle A\rangle_-\langle B\rangle_-$$
$$\langle AB\rangle_- = \langle A\rangle_+\langle B\rangle_- + \langle A\rangle_-\langle B\rangle_+ .$$

Use this to prove that the product of even (or odd) multivectors is always even.

(1.17) Define $A\cdot B$ and $A\wedge B$ for arbitrary multivectors A and B so that the distributive properties of inner and outer products are preserved. Show that if Equation (1.4) were to be generalized to $\mathbf{a}A = \mathbf{a}\cdot A + \mathbf{a}\wedge A$ for arbitrary A, then it would be necessary to require that $\mathbf{a}\cdot\lambda = 0$ for scalar λ, which is inconsistent with the definition (1.17a).

2-2. The Algebra of a Euclidean Plane

Every vector $\mathbf{a}$ determines an *oriented line*, namely, the set of all vectors which are scalar multiples of $\mathbf{a}$. Thus, every vector $\mathbf{x}$ on the line is related to $\mathbf{a}$ by the equation

$$\mathbf{x} = \alpha\mathbf{a}. \tag{2.1}$$

This is said to be a *parametric equation* for the $\mathbf{a}$-line. Each value of the *scalar parameter* α determines a unique point $\mathbf{x}$ on the line. A vector $\mathbf{x}$ is said to be *positively directed* (relative to $\mathbf{a}$) if $\mathbf{x}\cdot\mathbf{a} > 0$, or *negatively directed* if $\mathbf{x}\cdot\mathbf{a} < 0$. This distinction between positive and negative vectors is called an *orientation* (or *sense*) of the line. The unit vector $\hat{\mathbf{a}} = \mathbf{a}\,|\,\mathbf{a}\,|^{-1}$ is called the *direction* of the oriented line. The *opposite orientation* (or *sense*) for the line is obtained by reversing the assignments of positive and negative to vectors, that is, by designating $-\hat{\mathbf{a}}$ as the direction of the line. If a distinction between the two possible orientations is not made, the line is said to be *unoriented*.

Outer multiplication of (2.1) by $\mathbf{a}$ gives the equation

$$\mathbf{x}\wedge\mathbf{a} = 0. \tag{2.2}$$

This is a *nonparametric* equation for the $\mathbf{a}$-line. The $\mathbf{a}$-line is the *solution set* $\{\mathbf{x}\}$ of this equation. Note that we use curly brackets $\{\quad\}$ to indicate a set. To show that, indeed, every solution of (2.2) has the form (2.1), use (1.1) to obtain from (2.2), the equation

$$\mathbf{x}\mathbf{a} = \mathbf{x}\cdot\mathbf{a}.$$

Multiplying this equation on the right by $\mathbf{a}^{-1} = \mathbf{a}\mathbf{a}^{-2}$ and writing $\alpha = \mathbf{x}\cdot\mathbf{a}\mathbf{a}^{-2} = \mathbf{x}\cdot\mathbf{a}^{-1}$, one gets (2.1) as promised.

Two Dimensional Vector Space

There is an algebraic description of a plane which is quite analogous to that of a line. Given a (nonzero) bivector **B**, the set of all vectors **x** which satisfy the equation

$$\mathbf{x}\wedge\mathbf{B} = 0 \tag{2.3}$$

is said to be a 2-*dimensional vector space* and may be referred to as the **B**-*plane*. An *orientation* for the **B**-plane is determined by designating a unit bivector **i** proportional to **B** as the *direction* of the plane. Obviously, if the relation

$$\mathbf{B} = B\mathbf{i} \tag{2.4}$$

is substituted in (2.3), the scalar B can be divided out to get the equivalent equation

$$\mathbf{x}\wedge\mathbf{i} = 0. \tag{2.5}$$

So every bivector which is a non-zero scalar multiple of **i** determines the same plane as **i**. Such a bivector is called a *pseudoscalar* of the plane.

A parametric equation for the **i**-plane can be derived from (2.5) by factoring **i** into the product

$$\mathbf{i} = \boldsymbol{\sigma}_1\boldsymbol{\sigma}_2 = \boldsymbol{\sigma}_1\wedge\boldsymbol{\sigma}_2 = -\boldsymbol{\sigma}_2\boldsymbol{\sigma}_1\,, \tag{2.6}$$

where $\boldsymbol{\sigma}_1$ and $\boldsymbol{\sigma}_2$ are *orthogonal unit vectors*, that is, $\boldsymbol{\sigma}_1\cdot\boldsymbol{\sigma}_2 = 0$ and $\boldsymbol{\sigma}_1^2 = \boldsymbol{\sigma}_2^2 = 1$. Using (1.4), we obtain from (2.5),

$$\mathbf{x}\mathbf{i} = \mathbf{x}\cdot\mathbf{i}, \tag{2.7}$$

or, by (2.6) and (1.16),

$$\mathbf{x}\boldsymbol{\sigma}_1\boldsymbol{\sigma}_2 = \mathbf{x}\cdot(\boldsymbol{\sigma}_1\wedge\boldsymbol{\sigma}_2) = \mathbf{x}\cdot\boldsymbol{\sigma}_1\boldsymbol{\sigma}_2 - \mathbf{x}\cdot\boldsymbol{\sigma}_2\boldsymbol{\sigma}_1\,.$$

Multiplying this on the right by $\mathbf{i}^\dagger = \boldsymbol{\sigma}_2\boldsymbol{\sigma}_1$, we obtain

$$\mathbf{x} = x_1\boldsymbol{\sigma}_1 + x_2\boldsymbol{\sigma}_2\,, \tag{2.8}$$

where $x_1 = \mathbf{x}\cdot\boldsymbol{\sigma}_1$ and $x_2 = \mathbf{x}\cdot\boldsymbol{\sigma}_2$. Equation (2.8) is a parametric equation for the **i**-plane. Scalars x_1 and x_2 are called *rectangular components* of the vector **x** with respect to the *basis* $\{\boldsymbol{\sigma}_1, \boldsymbol{\sigma}_2\}$. Equation (2.8) determines a distinct vector **x** for each distinct pair of values of the components. A typical vector **x** is represented by a directed line segment in Figure 2.1. Orthogonal vectors like $\boldsymbol{\sigma}_1$ and $\boldsymbol{\sigma}_2$ are represented by perpendicular line segments in the figure, and the unit pseudoscalar **i** is represented by a plane segment. To be precise, **i** is the directed area of the plane segment, and the directed area of every plane segment in the **i**-plane is proportional to **i**.

Geometric Interpretations of a Bivector

The unit bivector **i** has two distinct geometric interpretations corresponding to two basic properties of the plane. First, as already mentioned, it is the *unit of directed area*, or simply, the *direction* of the plane. Second, as we shall see, it is the *generator of rotations* in the plane. The first interpretation is exemplified by Equation (2.6), which expresses the fact that the unit of area **i** can be obtained from the product of two orthogonal units of length $\boldsymbol{\sigma}_1$ and $\boldsymbol{\sigma}_2$. Equations exemplifying the second interpretation can be obtained by multiplying (2.6) on the left by $\boldsymbol{\sigma}_1$ and $\boldsymbol{\sigma}_2$ to get

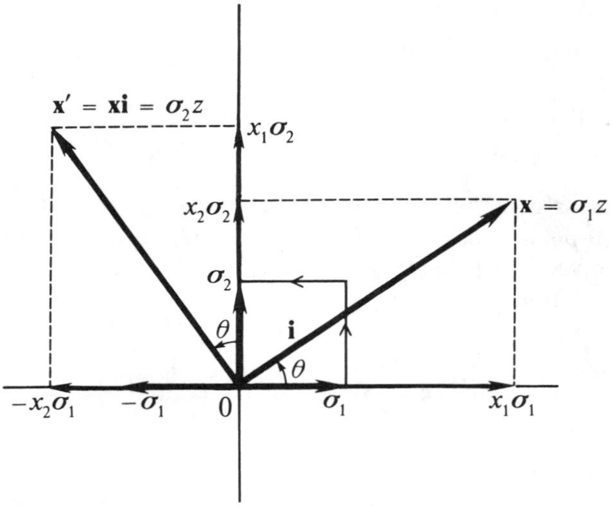

Fig. 2.1. Diagram of the **i**-*plane* of vectors, commonly known as the "real plane".

$$\boldsymbol{\sigma}_1\mathbf{i} = \boldsymbol{\sigma}_2\,, \tag{2.9a}$$

$$\boldsymbol{\sigma}_2\mathbf{i} = -\boldsymbol{\sigma}_1\,. \tag{2.9b}$$

According to (2.9a), multiplication of $\boldsymbol{\sigma}_1$ on the right by **i** transforms $\boldsymbol{\sigma}_1$ into $\boldsymbol{\sigma}_2$. Since $\boldsymbol{\sigma}_2$ has the same magnitude as $\boldsymbol{\sigma}_1$ but is orthogonal to it, this transformation is a *rotation* of $\boldsymbol{\sigma}_1$ *through a right angle*. Similarly, Equation (2.9b) expresses the rotation of $\boldsymbol{\sigma}_2$ through a right angle into $-\boldsymbol{\sigma}_1$.

Substitution of (2.9a) into (2.9b) gives

$$(\boldsymbol{\sigma}_1\mathbf{i})\mathbf{i} = \boldsymbol{\sigma}_2\mathbf{i} = -\boldsymbol{\sigma}_1\,,$$

which expresses the fact *two consecutive rotations through right angles reverses the direction of a vector*. This provides a geometric interpretation for the equation

$$\mathbf{i}^2 = -1, \tag{2.10}$$

when **i** and -1 are both regarded as *operators* (by multiplication) on vectors.

Right multiplication by **i** of any vector **x** in the **i**-plane rotates **x** by a right angle into a vector $\mathbf{x}' = \mathbf{x}\mathbf{i}$. From (2.8) and (2.9) we can get the components of $\mathbf{x}'$ from the components of **x**; thus,

$$\mathbf{x}' = \mathbf{x}\mathbf{i} = x_1\boldsymbol{\sigma}_2 - x_2\boldsymbol{\sigma}_1\,. \tag{2.11}$$

The relation of **x** to $\mathbf{x}'$ is represented in Figure 2.1. Notice that multiplication by **i** rotates vectors *counterclockwise* by a right angle. It is conventional to

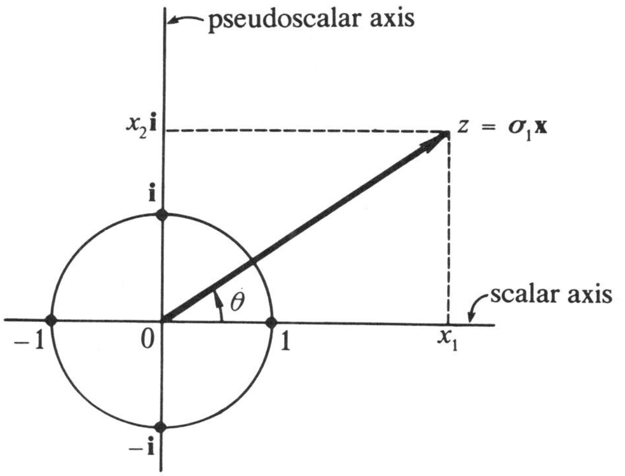

Fig. 2.2. Diagram of the *Spinor* **i**-*plane* commonly known as an "Argand diagram of the complex plane". Each point in the spinor plane represents a rotation-dilation. Points on the unit circle represent pure rotations, while points on the positive scalar axis represent pure dilations.

correspond a *positive orientation* (or *sense*) of a plane (and its unit pseudoscalar) with a *counterclockwise* rotation, as indicated in Figure 2.1. A *negative orientation* (or *sense*) then corresponds to a *clockwise* rotation. Another common way to distinguish between the two orientations of a plane is with the terms "left turn" and "right turn".

Plane Spinors

By multiplying two vectors in the **i**-plane, we get a quantity called a *spinor* of the **i**-plane. For example, by (2.6) and (2.8), from the product of vectors $\boldsymbol{\sigma}_1$ and $\mathbf{x}$, we get a spinor z in the form

$$z = \boldsymbol{\sigma}_1\mathbf{x} = x_1 + \mathbf{i}x_2 . \tag{2.12}$$

Quantities of the form $x_1 + \mathbf{i}x_2$ are commonly called *complex numbers*. It will be convenient for us to adopt that terminology when we wish to emphasize some relation to traditional concepts. However, it must be remembered that besides the property $\mathbf{i}^2 = -1$ ascribed to the traditional *unit imaginary*, our $\mathbf{i}$ is a bivector, so it has geometric and algebraic properties beyond those traditionally accorded to "*imaginary numbers*". The real and imaginary parts of a complex number z are commonly denoted by $\mathcal{R}e\{z\}$ and $\mathcal{I}m\{z\}$. Separation of a complex number into real and imaginary parts is equivalent to separating a spinor into scalar and pseudoscalar (bivector) parts, that is,

$$x_1 = \mathcal{R}e\{z\} = \langle z\rangle_0 = \frac{z+z^\dagger}{2} , \tag{2.13a}$$

$$x_2 = \mathcal{I}m\{z\} = \frac{\langle z\rangle_2}{\mathbf{i}} = \frac{z-z^\dagger}{2\mathbf{i}} . \tag{2.13b}$$

Note that reversion of a spinor corresponds exactly to conventional *complex conjugation*, that is,

$$z^\dagger = \mathbf{x}\boldsymbol{\sigma}_1 = x_1 - \mathbf{i}x_2 . \tag{2.14}$$

Also, our notation $|\, z\, |$ agrees with the conventional notation for the *modulus* of a complex number, thus,

$$|\, z\, |^2 = zz^\dagger = z^\dagger z = \mathbf{x}^2 = x_1^2 + x_2^2 . \tag{2.15}$$

The set of all spinors of the form (2.12) is a 2-dimensional space, which is aptly called the *spinor plane* (or *spinor* **i**-*plane*, to emphasize the special role of **i**). The elements of the spinor plane can be represented by directed line segments or points in a diagram such as Figure 2.2. Comparison of Figures 2.1a and 2.1b shows that the points of the **i**-plane of vectors can be put into one-to-one correspondence with points of the spinor plane. A correspondence is determined by an arbitrary choice of some vector in the vector plane, say $\boldsymbol{\sigma}_1$; then the product of $\boldsymbol{\sigma}_1$ with each vector $\mathbf{x}$ is a unique spinor z, as is expressed by equation (2.12). Conversely, each spinor z determines a unique vector $\mathbf{x}$ according to the equation

$$\mathbf{x} = \boldsymbol{\sigma}_1 z , \tag{2.16}$$

derived from (2.12). Note that $\boldsymbol{\sigma}_1$ distinguishes a line in the vector plane to be associated with the scalar axis in the spinor plane.

In spite of the correspondence between the vector and spinor planes, each plane has a different geometric significance, just as their elements have different algebraic properties. The distinction between the two planes corresponds to the two distinct interpretations of **i**. The interpretation of **i** as a directed area is indicated in the Figure 2.1 for the vector plane. On the other hand, the operator interpretation of **i** as a rotation (of vectors) through a right angle is indicated in the Figure 2.2 by the right angle that the **i**-axis makes with the scalar axis. This observation leads to an operator interpretation for all the spinors, and justifies calling **i** the generator of rotations. Consider, in particular, the interpretation of Equation (2.16) and its representation in Figures 2.1 and 2.2. Operating on $\boldsymbol{\sigma}_1$ (by right multiplication), the spinor z transforms $\boldsymbol{\sigma}_1$ into a vector $\mathbf{x}$. As indicated in the figures, this transformation is a *rotation* of $\boldsymbol{\sigma}_1$ through some angle θ combined with a *dilation* of $\boldsymbol{\sigma}_1$ by an amount $|\, z\, |$. Our choice of $\boldsymbol{\sigma}_1$ was arbitrary, so z evidently has the same effect on every vector in the **i**-plane. Thus, each spinor can be regarded as an algebraic representation of a *rotation-dilation*. This connection of spinors with rotations provides some justification for the terminology "spinor". Further justification comes from the fact that our use of the term here and in a more general sense later on is consistent with established use of the term "spinor" in advanced quantum mechanics.

The Algebra of a Plane

Some of our language from this point on will be simpler and clearer if we take a moment to introduce a few general concepts and definitions. Any expression of the form $\alpha_1 A_1 + \alpha_2 A_2 + \ldots + \alpha_n A_n$ with scalar coefficients α_k is called a *linear combination* of the quantities $A_1, A_2, \ldots, A_n$. The A_k are *linearly dependent* if some linear combination of the A_k with at least one nonzero coefficient is identically zero. Otherwise, they are *linearly independent*. If the A_k $(k=1, \ldots, n)$, are linearly independent, then the set of all linear combinations of the A_k is said to be an *n-dimensional linear space*, and the set $\{A_k\}$ is said to be a *basis* for that space.

Returning to the particular case at hand, we recall that in connection with Equation (2.8) the vectors $\boldsymbol{\sigma}_1$ and $\boldsymbol{\sigma}_2$ have already been identified as comprising a basis for the **i**-plane. Now, when we multiply vectors $\boldsymbol{\sigma}_1$ and $\boldsymbol{\sigma}_2$, we generate the *unit scalar* $1 = \boldsymbol{\sigma}_1^2 = \boldsymbol{\sigma}_2^2$ and the *unit pseudoscalar* $\mathbf{i} = \boldsymbol{\sigma}_1\boldsymbol{\sigma}_2$ but no other new quantities. It follows that every multivector A which can be obtained from vectors of the **i**-plane by addition and multiplication can be written as a linear combination

$$A = \alpha_0 + \alpha_1\boldsymbol{\sigma}_1 + \alpha_2\boldsymbol{\sigma}_2 + \alpha_3\mathbf{i}\,, \tag{2.17}$$

with scalar coefficients α_0, α_1, α_2 and α_3. We call the set of all such multivectors the *(Geometric) Algebra of the* **i**-*plane* or *simply the* **i**-*algebra*, and we denote it by $\mathcal{G}_2(\mathbf{i})$. We suppress the **i** and write $\mathcal{G}_2$ when we do not wish to refer to a particular plane. The subscript 2 here refers both to the grade of the pseudoscalar and the dimension of the plane.

It is clear from (2.17) that any multivector A in $\mathcal{G}_2(\mathbf{i})$ can be expressed as the sum of a vector $\mathbf{a} = \alpha_1\boldsymbol{\sigma}_1 + \alpha_2\boldsymbol{\sigma}_2$ and a spinor $z = \alpha_0 + \alpha_3\mathbf{i}$, that is

$$A = \mathbf{a} + z\,. \tag{2.18}$$

The vectors are odd, while the spinors are even multivectors. Accordingly, we can express $\mathcal{G}_2$ as the sum of two linear spaces,

$$\mathcal{G}_2 = \mathcal{G}_2^- + \mathcal{G}_2^+\,, \tag{2.19}$$

where $\mathcal{G}_2^-$ is the 2-dimensional space of vectors and $\mathcal{G}_2^+$ is the 2-dimensional space of spinors. Being the sum of two 2-dimensional spaces, the algebra $\mathcal{G}_2$ is itself a 4-dimensional linear space. The four unit multivectors 1, $\boldsymbol{\sigma}_1$, $\boldsymbol{\sigma}_2$ and $\mathbf{i} = \boldsymbol{\sigma}_1\boldsymbol{\sigma}_2$ make up a basis for this space.

A Distinction between Linear Spaces and Vector Spaces

A comment on nomenclature is in order here. In most mathematical literature the term "vector space" is synonymous with "linear space". This is because any quantities that can be added and multiplied by scalars are commonly called vectors. However, geometric algebra ascribes other properties to vectors, in particular, that they can be multiplied in a definite way. So we restrict our use of the term "vector" to the precise sense we have given it in geometric algebra. Accordingly, we restrict our use of the term *vector space* to refer to a linear space of vectors. We continue to use the term "linear space" in its usual more general sense. Thus both $\mathcal{G}_2^-$ (**i**) and $\mathcal{G}_2^+$ (**i**) can be called linear spaces of two dimensions, but of the two, only $\mathcal{G}_2^-$ (**i**) can be called a vector space, and $\mathcal{G}_2$ (**i**) is a 4-dimensional linear space but not a vector space.

The 2-dimensional vector space $\mathcal{G}_2^-$ (**i**) is, of course, the **i**-plane itself under another name. To describe the fact that every nonzero vector **a** in this space has a positive square $\mathbf{a}^2 = \mathbf{a}\cdot\mathbf{a} = |\,\mathbf{a}\,|^2$, this vector space is said to be *Euclidean*. For, as we have seen in Chapter 1, Euclidean geometry can be represented algebraically when the magnitude of a vector is interpreted as the length of a line segment. Accordingly, we may write

$$\mathcal{G}_2^-(\mathbf{i}) = \mathcal{E}_2 \tag{2.20}$$

to express the fact that the **i**-plane is a *Euclidean plane*, that is, a *2-dimension Euclidean vector space* $\mathcal{E}_2$. And we may refer to the **i**-algebra $\mathcal{G}_2$ *as the algebra of a Euclidean plane*.

One other comment about nomenclature is in order here. In Chapter 1 it was mentioned that a distinction between the concepts of dimension and grade must be made. That distinction should be clear by now. The *dimension* of a linear space is the number of linearly independent elements in the space. This definition of dimension is well established in mathematics, so we have adopted it. On the other hand, the concept of grade derives from a concept of vector multiplication producing new entities distinguished by grade. For a vector space and its algebra, the concepts of dimension and grade are closely related. We have seen that a vector space of dimension 2 is determined by a pseudoscalar of grade 2 and *vice-versa*. It is not difficult to show that a vector space of any finite dimension n is similarly related to a pseudoscalar of grade n.

2-3. The Algebra of Euclidean 3-Space

The concept of a *3-dimensional Euclidean space* $\mathcal{E}_3$ is fundamental to physics, because it provides the mathematical structure for the concept of physical space. Moreover, the properties of physical space are presupposed in every aspect of mechanics, not to mention the rest of physics. For this reason we cultivate the geometric algebra of $\mathcal{E}_3$ as the basic conceptual tool for representing and analyzing geometrical relations in physics.

We can analyze the algebra of $\mathcal{E}_3$ in the same way that we analyzed the algebra of $\mathcal{E}_2$. Let i be a unit 3-blade. The set of all vectors **x** which satisfy the equation

$$\mathbf{x}\wedge i = 0 \tag{3.1}$$

is the Euclidean 3-dimensional vector space $\mathcal{E}_3$. Scalar multiples of i are called *pseudoscalars* of this vector space, and we refer to i as the *unit-pseudoscalar*.

Note that the symbol "i" is an exception to our convention that k-blades be represented in boldface type. We make this exception to emphasize the singular importance of i and to distinguish it from unit 2-blades which we have represented by $\mathbf{i}$. The set of all multivectors generated from the vectors of $\mathcal{E}_3$ by addition and multiplication is the *(geometric) algebra* of $\mathcal{E}_3$, and it will be denoted by $\mathcal{G}_3$ or $\mathcal{G}_3(i)$. One way to study the structure of $\mathcal{G}_3$ is by constructing a basis for the algebra.

Because of (3.1), we can factor i into a product of three orthonormal vectors: thus,

$$i = \boldsymbol{\sigma}_1\boldsymbol{\sigma}_2\boldsymbol{\sigma}_3 = \boldsymbol{\sigma}_1\wedge\boldsymbol{\sigma}_2\wedge\boldsymbol{\sigma}_3 . \tag{3.2}$$

The term "orthonormal" means orthogonal and normalized to unity. The normalization of vectors $\boldsymbol{\sigma}_1$, $\boldsymbol{\sigma}_2$, $\boldsymbol{\sigma}_3$ is expressed by

$$\boldsymbol{\sigma}_1^2 = \boldsymbol{\sigma}_2^2 = \boldsymbol{\sigma}_3^2 = 1. \tag{3.3a}$$

The orthogonality of the vectors is expressed by the equations

$$\boldsymbol{\sigma}_i\cdot\boldsymbol{\sigma}_j = 0 \quad \text{if} \quad i \neq j, \tag{3.3b}$$

or equivalently, by

$$\boldsymbol{\sigma}_i\boldsymbol{\sigma}_j = \boldsymbol{\sigma}_i\wedge\boldsymbol{\sigma}_j = -\boldsymbol{\sigma}_j\boldsymbol{\sigma}_i \quad \text{if} \quad i \neq j, \tag{3.4}$$

with $i, j = 1, 2, 3$ understood.

We further assume that $\boldsymbol{\sigma}_1$, $\boldsymbol{\sigma}_2$, $\boldsymbol{\sigma}_3$ make up a *righthanded* or *dextral* set of vectors. The term "righthanded" actually concerns the interpretation of $\mathcal{G}_3$, rather than some intrinsic property of the algebra. This interpretation arises from the correspondence of vectors with directions in physical space as indicated in Figure 3.1. As the figure shows, a righthanded screw pointing in the $\boldsymbol{\sigma}_3$ direction will advance in that direction when given a *counterclockwise* rotation in the $(\boldsymbol{\sigma}_1\boldsymbol{\sigma}_2)$-plane. Equation (3.2) specifies a definite relation of the pseudoscalar i to the righthanded set of vectors, which we express in words by saying the i is the *dextral* or *righthanded unit pseudoscalar*. By reversing the directions of the $\boldsymbol{\sigma}_k$, we get a *lefthanded* set of vectors, $\{-\boldsymbol{\sigma}_k\}$ and the lefthanded unit pseudoscalar $(-\boldsymbol{\sigma}_1)(-\boldsymbol{\sigma}_2)(-\boldsymbol{\sigma}_3) = -i$. The terms "righthanded" and "lefthanded" distinguish the two possible *orientations* of a 3-dimensional vector space, just as the terms "counterclockwise" and "clock-

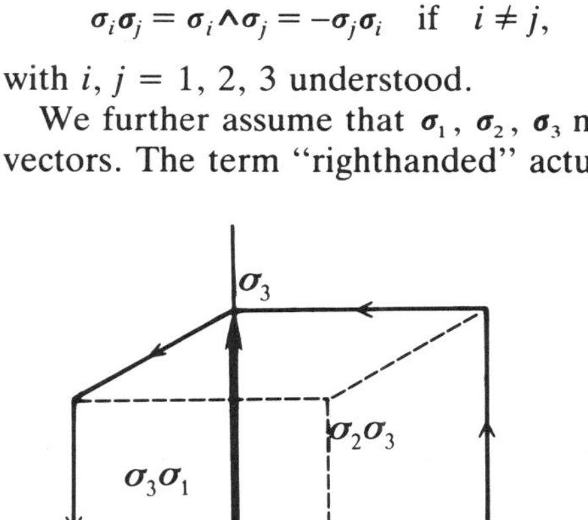

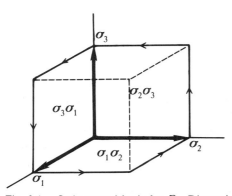

Fig. 3.1. Orthonormal basis for $\mathcal{G}_3$. Directed line segments represent unit vectors. Directed plane segments represent unit bivectors. The oriented cube represents the unit pseudoscalar i.

wise" distinguish the two possible orientations of a plane.

Every vector **x** in $\mathcal{G}_3$ is related to the $\boldsymbol{\sigma}_k$ by an equation of the form

$$\mathbf{x} = x_1\boldsymbol{\sigma}_1 + x_2\boldsymbol{\sigma}_2 + x_3\boldsymbol{\sigma}_3, \tag{3.5}$$

as represented in Figure 3.2. The scalars $x_k = \mathbf{x}\cdot\boldsymbol{\sigma}_k$ are called *rectangular components* of the vector **x** with respect to the basis $\{\boldsymbol{\sigma}_1, \boldsymbol{\sigma}_2, \boldsymbol{\sigma}_3\}$. Equation (3.5) can be derived as follows; from (3.1), we deduce

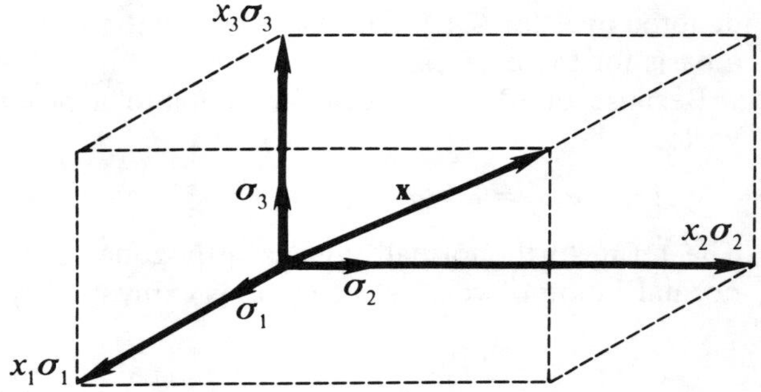

Fig. 3.2. Rectangular components of a vector.

$$\mathbf{x}i = \mathbf{x}\cdot i,$$

but by (3.2) and (1.16),

$$\mathbf{x}i = \mathbf{x}\boldsymbol{\sigma}_1\boldsymbol{\sigma}_2\boldsymbol{\sigma}_3 = \mathbf{x}\cdot(\boldsymbol{\sigma}_1\wedge\boldsymbol{\sigma}_2\wedge\boldsymbol{\sigma}_3) = \mathbf{x}\cdot\boldsymbol{\sigma}_1\boldsymbol{\sigma}_2\wedge\boldsymbol{\sigma}_3 - \mathbf{x}\cdot\boldsymbol{\sigma}_2\boldsymbol{\sigma}_1\wedge\boldsymbol{\sigma}_3 + \mathbf{x}\cdot\boldsymbol{\sigma}_3\boldsymbol{\sigma}_1\wedge\boldsymbol{\sigma}_2.$$

Multiplying this on the right by $i^\dagger = \boldsymbol{\sigma}_3\boldsymbol{\sigma}_2\boldsymbol{\sigma}_1$ and using (3.3a) and (3.4) we get (3.5) as expected.

Bivectors of $\mathcal{E}_3$

Inspection of (3.4) shows that by multiplication of the $\boldsymbol{\sigma}_i$ we obtain exactly three linearly independent bivectors, namely

$$\begin{aligned} \mathbf{i}_1 &= \boldsymbol{\sigma}_2\boldsymbol{\sigma}_3 = i\boldsymbol{\sigma}_1, \\ \mathbf{i}_2 &= \boldsymbol{\sigma}_3\boldsymbol{\sigma}_1 = i\boldsymbol{\sigma}_2, \\ \mathbf{i}_3 &= \boldsymbol{\sigma}_1\boldsymbol{\sigma}_2 = i\boldsymbol{\sigma}_3. \end{aligned} \tag{3.6}$$

The last equalities in (3.6) were obtained by multiplying (3.2) successively by $\boldsymbol{\sigma}_1$, $\boldsymbol{\sigma}_2$ and $\boldsymbol{\sigma}_3$. Note that the three equations (3.6) differ only by a cyclic permutation of the indices 1, 2, 3.

Since the $\mathbf{i}_k$ are the only bivectors which can be obtained from the $\boldsymbol{\sigma}_k$ by multiplication, any bivector **B** in $\mathcal{G}_3$ can be expressed as the linear combination

$$\mathbf{B} = B_1\mathbf{i}_1 + B_2\mathbf{i}_2 + B_3\mathbf{i}_3 \tag{3.7}$$

with scalar coefficients B_k. Thus, the set of all bivectors in $\mathcal{G}_3$ is a 3-dimensional linear space with a basis $\{\mathbf{i}_1, \mathbf{i}_2, \mathbf{i}_3\}$. Now, by substituting (3.6) into (3.7), we find that every bivector **B** is uniquely related to a vector $\mathbf{b} = B_1\boldsymbol{\sigma}_1 + B_2\boldsymbol{\sigma}_2 + B_3\boldsymbol{\sigma}_3$ by the equation

$$\mathbf{B} = i\mathbf{b}. \tag{3.8}$$

This relation is expressed in words by saying that the bivector **B** is the *dual* of the vector **b**. In general, we define the *dual of any multivector A* in $\mathcal{G}_3$ to be its product iA with the dextral unit pseudoscalar.

From (3.2) and (3.5) it should be clear that every trivector in $\mathcal{G}_3$ is proportional to i. It follows, then, from (3.1) that $\mathcal{G}_3$ contains no nonzero k-vectors with $k \geqslant 4$. Therefore, by axiom (1.11), every multivector A in $\mathcal{G}_3$ can be expressed in the expanded form

$$A = \langle A \rangle_0 + \langle A \rangle_1 + \langle A \rangle_2 + \langle A \rangle_3 . \tag{3.9}$$

Introducing the notations $\alpha = \langle A \rangle_0$ and $\mathbf{a} = \langle A \rangle_1$ for the scalar and vector parts of A, and expressing the bivector and pseudoscalar parts of A as duals of a vector and a pseudoscalar by writing $\langle A \rangle_2 = i\mathbf{b}$ and $\langle A \rangle_3 = i\beta$, we can put (3.9) in the form

$$A = \alpha + i\beta + \mathbf{a} + i\mathbf{b}. \tag{3.10}$$

This multivector has one scalar component, 3 vector components, 3 bivector components and one pseudoscalar component. Thus, $\mathcal{G}_3$ is a linear space with $1 + 3 + 3 + 1 = 8$ dimensions. As a basis for that space we may use the 8 unit multivectors $\{1, \boldsymbol{\sigma}_k, i\boldsymbol{\sigma}_k, i\}$, with $k = 1, 2, 3$ understood. The subspace of k-vectors in $\mathcal{G}_3$ can be denoted by $\langle \mathcal{G}_3 \rangle_k$; thus, $\langle \mathcal{G}_3 \rangle_1$ is a 3-dimensional space of vectors, $\langle \mathcal{G}_3 \rangle_2$ is a 3-dimensional space of bivectors and $\langle \mathcal{G}_3 \rangle_3$ is a 1-dimensional space of trivectors.

The Pseudoscalar of $\mathcal{E}_3$

Although we established some important properties of $\mathcal{G}_3$ in the course of determining a basis for the algebra, reference to a basis was not at all necessary, and we shall avoid it in the future except when it is an essential part of the problem at hand. In computations with $\mathcal{G}_3$ the pseudoscalar i plays a crucial role, so we list now its basic properties:

$$i^\dagger = -i, \tag{3.11a}$$

$$i^2 = -1, \tag{3.11b}$$

$$iA = Ai \text{ for every } A \text{ in } \mathcal{G}_3, \tag{3.11c}$$

$$\mathbf{a} \wedge \mathbf{b} \wedge \mathbf{c} = \lambda i \tag{3.11d}$$

for any vectors $\mathbf{a}$, $\mathbf{b}$, $\mathbf{c}$ in $\mathcal{G}_3$; the scalar λ is positive if and only if the vectors make up a righthanded set in the order given. Properties (3.11a, b) follow from the fact that i is a 3-blade normalized to unity, Property (3.11d) obviously generalizes (3.2). Properties (3.11a, c) are both consequences of Equation (3.1), but, conversely, Equation (3.1) can easily be derived from them.

Complex Numbers

The symbol i has been chosen for the unit pseudoscalar, because the properties (3.11a, b, c) are similar to those usually attributed to "the square root of minus one" in mathematics. We have seen, however, that there is not just one root of minus one in geometric algebra, but many. By inspection of the elements in a basis of the algebra, we see that in $\mathcal{G}_3$ there are two distinct kinds of solutions to the equation

$$x^2 = -1;$$

either x is a pseudoscalar, whence $x = \pm i$, or x is a bivector such as $\mathbf{i}_3 = \sigma_1\sigma_2$. To get a unique bivector solution of the equation we must have some information which determines the plane of the bivector, such as the directions of two noncollinear vectors in the plane.

Complex numbers are widely used in mathematical physics. To translate a specific application of complex numbers into geometric algebra, it is necessary to identify the geometrical role that the $\sqrt{-1}$ *tacitly* plays in the application. Usually, it will be found that the $\sqrt{-1}$ can be associated with some plane in physical space, so it should be interpreted as a bivector. Whenever this is done, the physical significance of the mathematical apparatus becomes more transparent, and the power of the theory is enhanced. This will be demonstrated by many examples in the rest of the book. In some applications of complex number to physics it is not so easy to attribute some physicogeometrical significance to the $\sqrt{-1}$. Our experience with geometric algebra then suggests that there must be a better way to formulate a problem.

Quaternions are Spinors

We have seen that the algebra of complex numbers appears with a geometric interpretation as the subalgebra $\mathcal{G}_2^+$ of even multivectors in $\mathcal{G}_2$. Similarly, we can express $\mathcal{G}_3$ as the sum of an odd part $\mathcal{G}_3^-$ and even part $\mathcal{G}_3^+$, that is,

$$\mathcal{G}_3 = \mathcal{G}_3^- + \mathcal{G}_3^+ .$$

According to (3.10) then, we can write a multivector A in the form

$$A = \langle A \rangle_- + \langle A \rangle_+ , \qquad (3.12a)$$

where

$$\langle A \rangle_- = \mathbf{a} + i\beta, \qquad (3.12b)$$

$$\langle A \rangle_+ = \alpha + i\mathbf{b}, \qquad (3.12c)$$

As is easily verified, $\mathcal{G}_3^+$ is closed under multiplication, so it is a subalgebra of $\mathcal{G}_3$, though $\mathcal{G}_3^-$ is not. For this reason $\mathcal{G}_3^+$ is sometimes called the *even subalgebra* of $\mathcal{G}_3$. But it may be better to refer to $\mathcal{G}_3^+$ as the *spinor algebra or subalgebra*, to emphasize the geometric significance of its elements. Just as every spinor in $\mathcal{G}_2^+$ represents a rotation-dilation in 2-dimensions, so every spinor in $\mathcal{G}_3^+$ represents a rotation-dilation in 3-dimensions. The representation of rotations by spinors is discussed fully in Chapter 5.

Equation (3.12c) shows that each spinor can be expressed as the sum of a scalar and a bivector. In view of (3.7), then, the four quantities 1, $\mathbf{i}_1$, $\mathbf{i}_2$, $\mathbf{i}_3$ make up a basis for $\mathcal{G}_3^+$. Thus $\mathcal{G}_3^+$ is a linear space of 4 dimensions. For this reason, the elements of $\mathcal{G}_3^+$ were called *quaternions* by William Rowan Hamilton, who invented them in 1843 independently of the full geometric algebra from which they arise here. Following Hamilton, we may also use the name *Quaternion Algebra* for $\mathcal{G}_3^+$.

Quaternions are well known to mathematicians today as the largest possible *associative division* algebra. But few are aware of how quaternions fit in the more general system of geometric algebra. Some might say that the quaternion algebra is actually distinct from the spinor algebra $\mathcal{G}_3^+$, that these algebras are not identical but only isomorphic. But such a distinction only serves to complicate mathematics unnecessarily. The identification of quaternions with spinors is fully justified not only because they have equivalent algebraic properties, but more important, because they have the same geometric significance. Hamilton's choice of the name quaternion is unfortunate, for the name merely refers to the comparatively insignificant fact that the quaternions compose a linear space of four dimensions. The name quaternion diverts attention from the key fact that Hamilton had invented a geometric algebra. Hamilton's work itself shows clearly the crucial role of geometry in his invention. Hamilton was consciously looking for a system of numbers to represent rotations in three dimensions. He was looking for a way to describe geometry by algebra, so he found a geometric algebra.

Hamilton developed his quaternion algebra at about the same time that Herman Grassmann developed his "algebra of extension" based on the inner and outer products. In spite of the fact that both Hamilton and Grassmann eventually came to know and admire one another's work, for several decades neither of them could see how their respective geometric algebras were related. It was only late in his life that Grassmann realized that Hamilton's quaternions can be derived simply by adding his inner and outer products to get the geometric product $\mathbf{ab} = \mathbf{a}\cdot\mathbf{b} + \mathbf{a}\wedge\mathbf{b}$, but it was too late for him to pursue the implications of this insight very far. At about the same time, the English mathematician W. K. Clifford independently realized that Hamilton and Grassmann were approaching one and the same subject from different points of view. By combining their algebraic ideas, he was led, in 1876, to the geometric product. Unfortunately, death claimed him before he was able to fully delineate the rich mixture of geometric and algebraic ideas he discovered, and no successor appeared to continue his work with the same depth of geometric insight. Consequently, the mathematical world continued to regard Grassmann's and Hamilton's algebras as independent systems. Divided, they fell into relative disuse.

Quaternions today reside in a kind of mathematical limbo, because their place in a more general geometric algebra is not recognized. The prevailing attitude toward quaternions is exhibited in a biographical sketch of Hamilton by the late mathematician E. T. Bell. The sketch is titled "An Irish Tragedy", because for the last twenty years of his life, Hamilton concentrated all his enormous mathematical powers on the study of quaternions in, as Bell would have it, the quixotic belief that quaternions would play a central role in the mathematics of the future. Hamilton's judgement was based on a new and profound insight into the relation between algebra and geometry. Bell's evaluation was made by surveying the mathematical literature nearly a century later. But union with Grassmann's algebra puts quaternions in a different perspective. It may yet prove true that Hamilton looking ahead saw further than Bell looking back.

Clifford may have been the first person to find significance in the fact that two different interpretations of number can be distinguished, the *quantitative* and the *operational*. On the first interpretation, number is a measure of "how much" or "how many" of something. On the second, number describes a relation between different quantities. The distinction is nicely illustrated by recalling the interpretations already given to a unit bivector **i**. Interpreted quantitatively, **i** is a measure of directed area. Operationally interpreted, **i** specifies a rotation in the **i**-plane. Clifford observed that Grassmann developed the idea of directed number from the quantitative point of view, while Hamilton emphasized the operational interpretation. The two approaches are brought together by the geometric product. Either a quantitative or an operational interpretation can be given to any number, yet one or the other may be more important in most applications. Thus, vectors are usually interpreted quantitatively, while spinors are usually interpreted operationally. Of course the algebraic properties of vectors and spinors can be studied abstractly with no reference whatsoever to interpretation. But interpretation is crucial when algebra functions as a language.

The Vector Cross Product

Vector algebra, as conceived by J. Willard Gibbs in 1884, is widely used as the basic mathematical language in physics textbooks today, so it is important to show that this system fits naturally into $\mathcal{G}_3$. The demonstration is easy. We need only introduce the *vector cross product* $\mathbf{a} \times \mathbf{b}$ defined by the equation

$$\mathbf{a} \times \mathbf{b} = -i\mathbf{a}\wedge\mathbf{b},$$

or, equivalently,

$$\mathbf{a}\wedge\mathbf{b} = i\mathbf{a} \times \mathbf{b}. \tag{3.13}$$

Thus, $\mathbf{a} \times \mathbf{b}$ is the vector dual to the bivector $\mathbf{a}\wedge\mathbf{b}$. As shown in Figure 3.3, the sign of the duality is chosen so that the vectors $\mathbf{a}$, $\mathbf{b}$, $\mathbf{a} \times \mathbf{b}$, in that order, form a righthanded set. This agrees with our convention for the handedness of the pseudoscalar i, for by comparing (3.13) with (3.6), we see that

$$\boldsymbol{\sigma}_1 \times \boldsymbol{\sigma}_2 = \boldsymbol{\sigma}_3. \tag{3.14}$$

To remember the correct sign in the duality relation (3.13), it is helpful to note that the geometric product of vectors in $\mathcal{G}_3$ can be written

$$\mathbf{ab} = \mathbf{a}\cdot\mathbf{b} + \mathbf{a}\wedge\mathbf{b} = \mathbf{a}\cdot\mathbf{b} + i\mathbf{a} \times \mathbf{b}, \tag{3.15}$$

and (3.13) can be obtained from this by separately equating bivector parts. Finally, note that by squaring (3.13) we deduce

$$(\mathbf{a} \times \mathbf{b})^2 = -(\mathbf{a}\wedge\mathbf{b})^2 = |\,\mathbf{a}\wedge\mathbf{b}\,|^2. \tag{3.16}$$

Hence, the magnitude $(\mathbf{a} \times \mathbf{b})$ is equal to the area of the parallelogram in Figure 3.3, which was identified as $|\,\mathbf{a}\wedge\mathbf{b}\,|$ in Section 1-5.

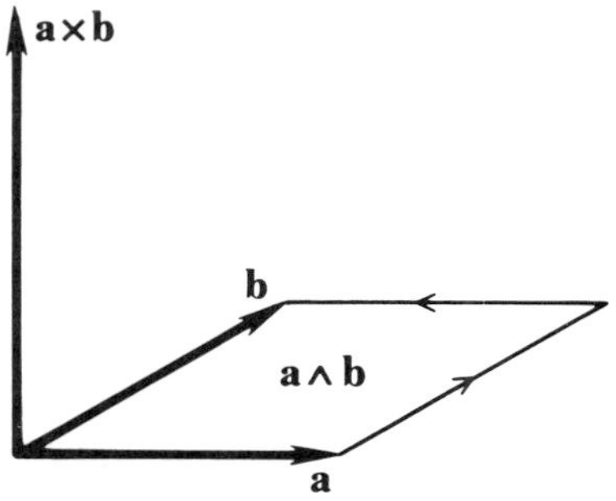

Fig. 3.3. Duality of the cross product and the outer product.

At this point, a caveat is in order. Books on vector algebra commonly make a distinction between *polar* vectors and *axial* vectors, with $\mathbf{a} \times \mathbf{b}$ identified as an axial vector if $\mathbf{a}$ and $\mathbf{b}$ are polar vectors. This confusing practice of admitting two kinds of vectors is wholly unnecessary. An "axial vector" is nothing more than a bivector disguised as a vector. So with bivectors at our disposal, we can do without axial vectors. As we have defined it in (3.13), the quantity $\mathbf{a} \times \mathbf{b}$ is a vector in exactly the same sense that $\mathbf{a}$ and $\mathbf{b}$ are vectors.

The ease with which conventional vector algebra fits into $\mathcal{G}_3$ is no accident. Gibbs constructed his system from the same ideas of Grassmann and Hamilton that have gone into geometric algebra. By the end of the 19th century a lively controversy had developed as to which system was more suitable for the work of theoretical physics, the quaternions or vector algebra. A glance at modern textbooks shows that the votaries of vectors were victorious. However, quaternions have reappeared disguised as matrices and proved to be essential in modern quantum mechanics. The ironic thing about the vector-quaternion controversy is that there was nothing substantial to dispute. Far from being in opposition, the two systems complement each other and, as we have seen, are perfectly united in the geometric algebra $\mathcal{G}_3$. The whole controversy was founded on the failure of everyone involved to appreciate the distinction between vectors and bivectors. Indeed, the word "vector" was originally coined by Hamilton for what we now call a bivector. Gibbs changed the meaning of the word to its present sense, but no one at the time understood the real significance of the change he had made.

2-3. Exercises

In the following exercises and throughout the book, the symbols σ_i and i always have the meanings assigned to them in this chapter. Also, unless otherwise indicated, we will assume that every multivector is an element of $\mathcal{G}_3$. The perceptive reader will be able to identify those instances where such an assumption is unnecessary.

(3.1) (a) Prove that a vector $\mathbf{x}$ is in the $\mathbf{B}$-plane if and only if $\mathbf{xB} = -\mathbf{Bx}$

(b) Prove then that $\mathbf{x}' = \mathbf{xB}$ is a vector with the properties $|\,\mathbf{x}'\,| = |\,\mathbf{B}\,||\,\mathbf{x}\,|$ and $\mathbf{x}\cdot\mathbf{x}' = 0$.

(3.2) If $\mathbf{a}$ and $\mathbf{b}$ are vectors in the plane of $\mathbf{i} = \boldsymbol{\sigma}_1\boldsymbol{\sigma}_2$, show that the ratio of $\mathbf{a}\wedge\mathbf{b}$ to $\mathbf{i}$ is equal to the determinant

$$\begin{vmatrix} \mathbf{a}\cdot\boldsymbol{\sigma}_1 & \mathbf{b}\cdot\boldsymbol{\sigma}_1 \\ \mathbf{a}\cdot\boldsymbol{\sigma}_2 & \mathbf{b}\cdot\boldsymbol{\sigma}_2 \end{vmatrix} = \mathbf{a}\cdot\boldsymbol{\sigma}_1\,\mathbf{b}\cdot\boldsymbol{\sigma}_2 - \mathbf{b}\cdot\boldsymbol{\sigma}_1\,\mathbf{a}\cdot\boldsymbol{\sigma}_2$$

(3.3) Prove the following important identities:

$$\mathbf{a}\cdot\mathbf{b} = -i[\mathbf{a}\wedge(i\mathbf{b})].$$

$$\mathbf{b}\times\mathbf{a} \equiv i(\mathbf{a}\wedge\mathbf{b}) = \mathbf{a}\cdot(i\mathbf{b}) = -(i\mathbf{b})\cdot\mathbf{a}.$$

$$\mathbf{a}\cdot(\mathbf{b}\wedge\mathbf{c}) = -\mathbf{a}\times(\mathbf{b}\times\mathbf{c}) = \mathbf{a}\cdot\mathbf{b}\mathbf{c} - \mathbf{a}\cdot\mathbf{c}\mathbf{b}.$$

$$\mathbf{a}\wedge(\mathbf{b}\wedge\mathbf{c}) = i\mathbf{a}\cdot(\mathbf{b}\times\mathbf{c}) = \tfrac{1}{2}(\mathbf{abc} - \mathbf{cba}).$$

Note that the first identity expresses the inner product in terms of the outer product and two duality operations. With the help of the remaining identities, any result of conventional vector algebra can easily be derived from the more powerful results for inner and outer products established in Section 2.1.

(3.4) Reexpress the identities of Exercise (1.1) in terms of the dot and cross products alone.

(3.5) Use an identity in Exercise (1.1) to prove that

$$\mathbf{a}\times\mathbf{b} = \begin{vmatrix} a_2 & b_2 \\ a_3 & b_3 \end{vmatrix}\boldsymbol{\sigma}_1 - \begin{vmatrix} a_1 & b_1 \\ a_3 & b_3 \end{vmatrix}\boldsymbol{\sigma}_2 + \begin{vmatrix} a_1 & b_1 \\ a_2 & b_2 \end{vmatrix}\boldsymbol{\sigma}_3,$$

where $a_k \equiv \mathbf{a}\cdot\boldsymbol{\sigma}_k$, $b_k \equiv \mathbf{b}\cdot\boldsymbol{\sigma}_k$.

(3.6) From Equation (3.10), show that

$$A^\dagger = \alpha - i\beta + \mathbf{a} - i\mathbf{b},$$

$$|\,A\,|^2 = \alpha^2 + \beta^2 + \mathbf{a}^2 + \mathbf{b}^2.$$

(3.7) The quaternions can be defined as the set of quantities Q of the form

$$Q = Q_0 + Q_1\mathbf{i}_1 + Q_2\mathbf{i}_2 + Q_3\mathbf{i}_3,$$

where the Q_k $(k = 0, 1, 2, 3)$ are scalar coefficients and the $\mathbf{i}_k$ satisfy the equations

$$\mathbf{i}_1^2 = \mathbf{i}_2^2 = \mathbf{i}_3^2 = -1,$$

$$\mathbf{i}_1\mathbf{i}_2\mathbf{i}_3 = 1.$$

Show that the bivector basis given by Equations (3.6) has these properties.

Hamilton used the symbols $\mathbf{i}$, $\mathbf{j}$, $\mathbf{k}$ instead of $\mathbf{i}_1$, $\mathbf{i}_2$, $\mathbf{i}_3$ and wrote down the famous equations

$$\mathbf{i}^2 = \mathbf{j}^2 = \mathbf{k}^2 = -1,$$

$$\mathbf{ijk} = -1.$$

Of what geometrical significance is the difference in sign between this last equation and the corresponding equation above?

(3.8) The expansion of a vector **b** in terms of its components $b_k = \mathbf{b}\cdot\boldsymbol{\sigma}_k$ is commonly expressed by any of the notations

$$\mathbf{b} = b_k\boldsymbol{\sigma}_k = \sum_k b_k\boldsymbol{\sigma}_k = \sum_{k=1}^{3} b_k\boldsymbol{\sigma}_k = b_1\boldsymbol{\sigma}_1 + b_2\boldsymbol{\sigma}_2 + b_3\boldsymbol{\sigma}_3.$$

The most abbreviated form $b_k\boldsymbol{\sigma}_k$ employs the so called *summation convention*, which calls for summation over all allowed values of a repeated pair of indicies in a single term. By this convention, the expansion of a bivector **B** in terms of components

$$B_{ij} = \boldsymbol{\sigma}_i\cdot\mathbf{B}\cdot\boldsymbol{\sigma}_j = (\boldsymbol{\sigma}_j\wedge\boldsymbol{\sigma}_i)\cdot\mathbf{B} = -B_{ji}$$

can be written

$$\mathbf{B} = \tfrac{1}{2}B_{ij}\boldsymbol{\sigma}_j\wedge\boldsymbol{\sigma}_i = B_{12}\boldsymbol{\sigma}_2\wedge\boldsymbol{\sigma}_1 + B_{31}\,\boldsymbol{\sigma}_1\wedge\boldsymbol{\sigma}_3 + B_{23}\boldsymbol{\sigma}_3\wedge\boldsymbol{\sigma}_2.$$

Show that the duality relation

$$\mathbf{B} = i\mathbf{b}$$

can be expressed in terms of components by the equations

$$B_{ij} = \varepsilon_{ijk}b_k,$$

where ε_{ijk} is defined by

$$\varepsilon_{ijk} = \frac{\boldsymbol{\sigma}_i\wedge\boldsymbol{\sigma}_j\wedge\boldsymbol{\sigma}_k}{i} = i^\dagger\boldsymbol{\sigma}_i\wedge\boldsymbol{\sigma}_j\wedge\boldsymbol{\sigma}_k.$$

Note that $\varepsilon_{ijk} = 0$ if any pair of indicies have the same value, then

$$\varepsilon_{ijk} = \varepsilon_{123} = 1 \text{ if } \{i, j, k\} \text{ is an even permutation of } \{1, 2, 3\},$$
$$\varepsilon_{ijk} = \varepsilon_{321} = -1 \text{ if } \{i, j, k\} \text{ is an odd permutation of } \{1, 2, 3\}.$$

Also prove that

$$\mathbf{a}\times\mathbf{b} = a_ib_j\varepsilon_{ijk}\boldsymbol{\sigma}_k,$$

$$\frac{\mathbf{a}\wedge\mathbf{b}\wedge\mathbf{c}}{i} = \varepsilon_{ijk}a_ib_jc_k.$$

(3.9) Prove

$$\boldsymbol{\sigma}_k\mathbf{a}\boldsymbol{\sigma}_k = -\mathbf{a},$$
$$\boldsymbol{\sigma}_k\mathbf{a}\wedge\mathbf{b}\boldsymbol{\sigma}_k = \mathbf{b}\wedge\mathbf{a},$$
$$\boldsymbol{\sigma}_k\mathbf{a}\wedge\mathbf{b}\wedge\mathbf{c}\boldsymbol{\sigma}_k = 3\mathbf{a}\wedge\mathbf{b}\wedge\mathbf{c}. \qquad (\text{sum over } k)$$

(3.10) Solve the following vector equation for $\mathbf{x}$:

$$\alpha\mathbf{x} + \mathbf{b} \times \mathbf{x} = \mathbf{a}.$$

2-4. Directions, Projections and Angles

In Chapter 1 we saw how first the inner and outer products and finally the geometric product were invented to give algebraic expression to the geometric concept of direction. We are now prepared to express the primitive relations among directions in the simplest possible algebraic terms. Having done so, we will be able to analyze complex relations among directions by straightforward computation.

The geometric product provides us with an algebraic measure of relative direction. Indeed, the geometric product $\mathbf{ab}$ was specifically designed to contain *all* the information about the *relative* directions of vectors $\mathbf{a}$ and $\mathbf{b}$. Part of this information can be extracted by decomposing $\mathbf{ab}$ into symmetric (scalar) and antisymmetric (bivector) parts according to the fundamental formula

$$\mathbf{ab} = \mathbf{a}\cdot\mathbf{b} + \mathbf{a}\wedge\mathbf{b}. \tag{4.1}$$

The fact that the product $\mathbf{ab}$ is indeed a direct measure of the relative directions of vectors $\mathbf{a}$ and $\mathbf{b}$ follows from the interpretations associated with $\mathbf{a}\cdot\mathbf{b}$ and $\mathbf{a}\wedge\mathbf{b}$ in chapter 1. Accordingly, vectors $\mathbf{a}$ and $\mathbf{b}$ are *collinear* if and only if $\mathbf{ab} = \mathbf{ba}$, and they are *orthogonal* if and only if $\mathbf{ab} = -\mathbf{ba}$. In general, $\mathbf{ab}$ has an intermediate "degree of commutativity" and, hence, describes a relative direction somewhere between these two extremes.

Given a vector $\mathbf{b}$, any vector $\mathbf{a}$ can be resolved into a vector $\mathbf{a}_\parallel$ collinear with $\mathbf{b}$ and a vector $\mathbf{a}_\perp$ orthogonal to $\mathbf{b}$. Explicit algebraic expressions for this resolution are obtained from Equation (4.1) by dividing by the vector $\mathbf{b}$; thus,

$$\mathbf{a} = \mathbf{a}_\parallel + \mathbf{a}_\perp, \tag{4.2a}$$

where

$$\mathbf{a}_\parallel \equiv \mathbf{a}\cdot\mathbf{b}\mathbf{b}^{-1}, \tag{4.2b}$$

$$\mathbf{a}_\perp \equiv \mathbf{a}\wedge\mathbf{b}\mathbf{b}^{-1} = (\mathbf{a}\wedge\mathbf{b})\cdot\mathbf{b}^{-1}. \tag{4.2c}$$

These relations are represented in slightly different ways in Figures (4.1a) and (4.1b). The collinearity and orthogonality properties are expressed by the equations

$$\mathbf{a}_\parallel\mathbf{b} = \mathbf{a}\cdot\mathbf{b} = \mathbf{b}\mathbf{a}_\parallel, \tag{4.3a}$$

$$\mathbf{a}_\perp\mathbf{b} = \mathbf{a}\wedge\mathbf{b} = -\mathbf{b}\mathbf{a}_\perp. \tag{4.3b}$$

Note that (4.3b) expresses the directed area $\mathbf{a}\wedge\mathbf{b}$ as the product of the "altitude" $\mathbf{a}_\perp$ and "base" $\mathbf{b}$ of the $(\mathbf{a}, \mathbf{b})$-parallelogram.

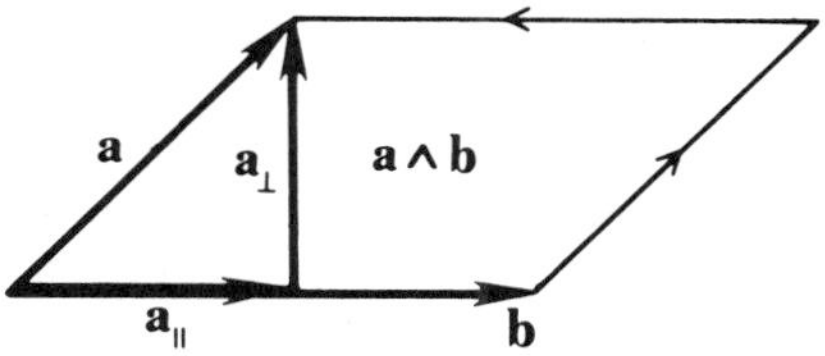

Fig. 4.1. (a)

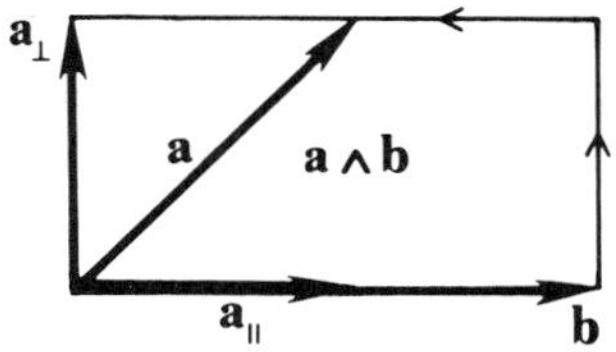

Fig. 4.1. (b)

Our considerations are easily generalized as follows. In preceding sections, we have seen that a k-blade $\mathbf{B}$ determines a k-dimensional vector space called $\mathbf{B}$-space. The relative direction of $\mathbf{B}$ and some vector $\mathbf{a}$ is completely characterized by the geometric product

$$\mathbf{aB} = \mathbf{a}\cdot\mathbf{B} + \mathbf{a}\wedge\mathbf{B}. \tag{4.4}$$

The vector $\mathbf{a}$ is uniquely resolved into a vector $\mathbf{a}_\parallel$ in $\mathbf{B}$-space and a vector $\mathbf{a}_\perp$ orthogonal to $\mathbf{B}$-space by the equations

$$\mathbf{a} = \mathbf{a}_\parallel + \mathbf{a}_\perp, \tag{4.5a}$$

where

$$\mathbf{a}_\parallel = P_{\mathbf{B}}(\mathbf{a}) \equiv \mathbf{a}\cdot\mathbf{B}\mathbf{B}^{-1}, \tag{4.5b}$$

$$\mathbf{a}_\perp = P_{\mathbf{B}}^{\perp}(\mathbf{a}) \equiv \mathbf{a}\wedge\mathbf{B}\mathbf{B}^{-1}. \tag{4.5c}$$

Besides the case $k = 1$ which we have already considered, we are most interested in the case $k = 2$, when $\mathbf{B}$ is a bivector. The latter case is depicted in Figure 4.2.

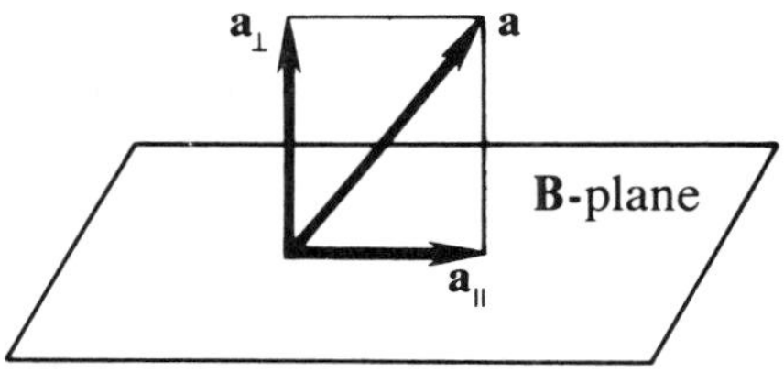

Fig. 4.2. Projection and Rejection of a vector $\mathbf{a}$ by a bivector $\mathbf{B}$.

The vector $\mathbf{a}_\parallel$ determined by Equation (4.5b) is called the *projection* of $\mathbf{a}$ into $\mathbf{B}$-space, while $\mathbf{a}_\perp$ determined by (4.5c) is called the *rejection* of $\mathbf{a}$ from $\mathbf{B}$-space. The new term "rejection" has been introduced here in the absence of a satisfactory standard name for this important concept. Although we will not make much use of it, the notation $P_{\mathbf{B}}(\mathbf{a})$ has been introduced in (4.5b) to emphasize that the projection is a function (or operator) $P_{\mathbf{B}}$ depending on the blade $\mathbf{B}$ with the value $\mathbf{a}_\parallel$ when operating on $\mathbf{a}$. This function is explicitly defined in terms of the geometric product by Equation (4.5b). Similarly, the rejection $P_{\mathbf{B}}^{\perp}$ is a function determined by Equation (4.5c).

From Equations (1.5) we get the generalization of Equations (4.3a, b):

$$\mathbf{a}_\parallel\mathbf{B} = \mathbf{a}\cdot\mathbf{B} = (-1)^{k+1}\mathbf{B}\mathbf{a}_\parallel, \tag{4.6a}$$

$$\mathbf{a}_\perp\mathbf{B} = \mathbf{a}\wedge\mathbf{B} = (-1)^{k}\mathbf{B}\mathbf{a}_\perp. \tag{4.6b}$$

For the important case $k = 2$, these equations imply that a vector is in the **B**-plane if and only if it anticommutes with **B**, and it is orthogonal to the **B**-plane if and only if it commutes with **B**.

The conventional notion of "a direction" is given a precise mathematical representation by "a unit vector", so it is often convenient to refer to the unit vectors themselves as *directions*. An *angle* is a relation between two directions. To give this relation a precise mathematical expression, let θ be the angle between directions **a** and **b**. The sine and cosine of the angle are defined, respectively, as the components of the rejection and projection of one direction by the other, as indicated in Figures (4.3a, b). These relations can be expressed by the equations

$$\mathbf{b}_{\|} = \mathbf{a}\mathbf{a}\cdot\mathbf{b} = \mathbf{a}\cos\theta, \tag{4.7a}$$

$$\mathbf{b}_{\perp} = \mathbf{a}\mathbf{a}\wedge\mathbf{b} = \mathbf{a}\mathbf{i}\sin\theta, \tag{4.7b}$$

or, more simply, by the equations

$$\mathbf{a}\cdot\mathbf{b} = \cos\theta, \tag{4.8a}$$

$$\mathbf{a}\wedge\mathbf{b} = \mathbf{i}\sin\theta, \tag{4.8b}$$

where **i** is the unit blade of the $\mathbf{a}\wedge\mathbf{b}$-plane. Equations (4.8a, b) are just parts of the single fundamental equation

$$\mathbf{a}\mathbf{b} = e^{\mathbf{i}\theta} \tag{4.9}$$

where

$$e^{\mathbf{i}\theta} = \cos\theta + \mathbf{i}\sin\theta \tag{4.10}$$

For the time being, Equation (4.10) can be regarded as a definition of the exponential function $e^{\mathbf{i}\theta}$.

So far, $\cos\theta$ and $\sin\theta$ are not definite functions of the angle θ, because we have not specified a definite measure for the angle. Two measures of angle are in common use, "degree" and "radian". We will employ the radian measure almost exclusively, because, as will be seen, the degree measure is not compatible with the fundamental definition of the exponential function.

In Equation (4.9), we interpret θ as the *radian measure* of the angle from **a** to **b**, that is, the numerical magnitude of θ is equal to the length of arc on the unit circle from **a** to **b**, as indicated in Figures (4.3a, b). The common convention of representing angles by scalars like θ fails to represent the fact that angles refer to planes, in the present case, the plane containing vectors **a** and **b**. This deficiency is remedied by representing angles by bivectors, so the angle from **a** to **b** is represented by the bivector

$$\boldsymbol{\theta} = \mathbf{i}\theta. \tag{4.11}$$

Here, **i** specifies the plane of the angle, while $|\,\theta\,|$ specifies the magnitude of the angle. Note that the sign of θ in (4.11) depends on the orientation assigned to the unit blade **i**. In Figures (4.3a, b) the orientation was chosen so that θ is positive.

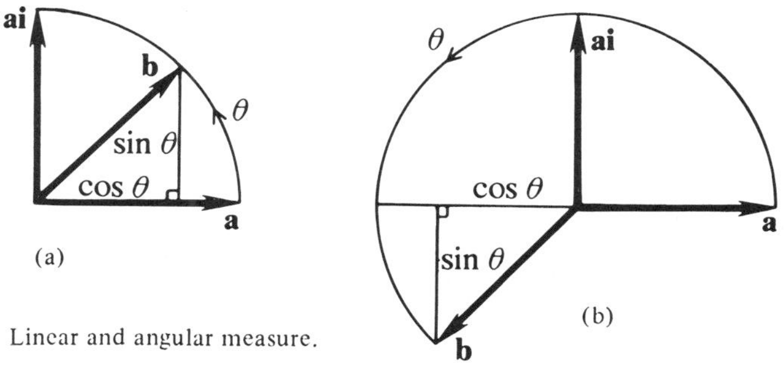

Fig. 4.3a, b. Linear and angular measure.

The fact that angles are best represented by bivectors suggests that the magnitude of an angle would be better interpreted as an area than as an arc length. As shown in Figure 4.4, the angle θ is just twice the directed area of the circular sector between **a** and **b**. This can be ascertained from the simple proportion

$$\frac{\text{area of sector}}{\text{arc length}} = \frac{\text{area of circle}}{\text{circumference}} = \frac{\pi\mathbf{i}}{2\pi}.$$

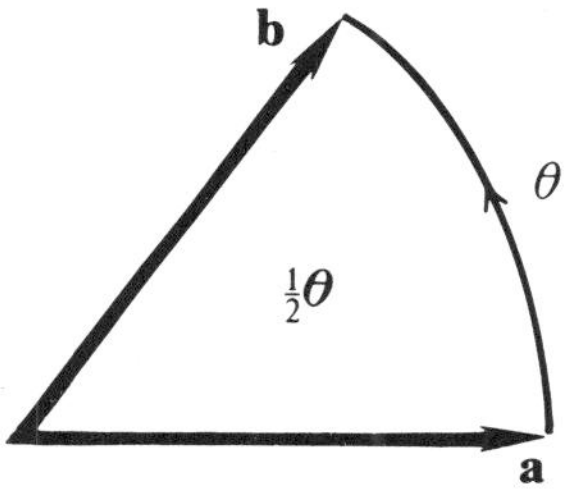

Fig. 4.4. Angle and area of a circular sector.

The radian (arc length) measure of angle is so well established that it is hardly worth changing, especially since it is related to the area of a circular sector by a mere factor of two. But it will be seen that the "areal measure" plays a more direct role in applications to geometry and physics.

Quite apart from the interpretation of $\boldsymbol{\theta} = \mathbf{i}\theta$ as the angle (in radian measure) from **a** to **b**, Equation (4.9) should be regarded as a functional relation of the bivector $\boldsymbol{\theta}$ to the vectors **a** and **b**. In Section 2.2 we called the quantity

$$z = \mathbf{ab} = e^{\mathbf{i}\theta} \tag{4.12}$$

a *spinor* of the **i**-plane, and noted that each such spinor (with $|z| = 1$) determines a rotation in the plane. The spinor z rotates each vector **a** in the **i**-plane into a vector **b** according to the equation

$$\mathbf{b} = \mathbf{a}z = \mathbf{a}e^{\mathbf{i}\theta}. \tag{4.13}$$

Thus, the exponential function $e^{\mathbf{i}\theta}$ represents a rotation in **i**-plane as a function of the angle of rotation.

The operational interpretation of $e^{\mathbf{i}\theta}$ as a rotation enables us to write down several important properties of the exponential function immediately, without appeal to the algebraic definition to be given in the next section. To begin with, we saw in Section 2-2 that a rotation through a right angle (with radian measure $2\pi/4 = \pi/2$) is represented by the unit bivector **i**. Hence,

$$e^{\mathbf{i}\pi/2} = \mathbf{i}. \tag{4.14a}$$

Similarly, a rotation through two right angles (measure π) reverses direction, hence we have the famous formula

$$e^{\mathbf{i}\pi} = -1, \tag{4.14b}$$

which relates several remarkable constants of elementary mathematics. Rotation through four right angles is equal to the *identity transformation* represented by the "multiplicative identity" 1. Hence,

$$e^{2\pi\mathbf{i}} = 1. \tag{4.14c}$$

The fact that a rotation through an angle $\mathbf{i}\theta$ followed or preceded by a rotation through an angle $\mathbf{i}\phi$ is equivalent to a rotation through an angle $\mathbf{i}(\theta + \phi)$ is expressed by the equation

$$e^{\mathbf{i}\theta}e^{\mathbf{i}\phi} = e^{\mathbf{i}(\theta + \phi)} = e^{\mathbf{i}\phi}e^{\mathbf{i}\theta}. \tag{4.15}$$

Consequently, for n rotations through an angle $\mathbf{i}\theta$ we get *de Moivre's theorem*:

$$(e^{\mathbf{i}\theta})^n = e^{\mathbf{i}n\theta}. \tag{4.16}$$

Clearly the exponential function is a great aid to the *arithmetic of rotations*.

Plane Trigonometry

Now that the basic relations of vectors to angles and rotations in a plane have been established, *all* the standard results of plane trigonometry follow by simple algebraic manipulations. Trigonometry can therefore be regarded as an elementary part of geometric algebra.

A central problem of trigonometry is the determination of all numerical relations among the sides and angles of an arbitrary triangle. Consider a triangle with sides of length a, b, c and angles of measure α, β, γ as shown in Figure 4.5a. The relations among the sides and angles are completely determined by the algebraic representation of the triangle as a vector equation

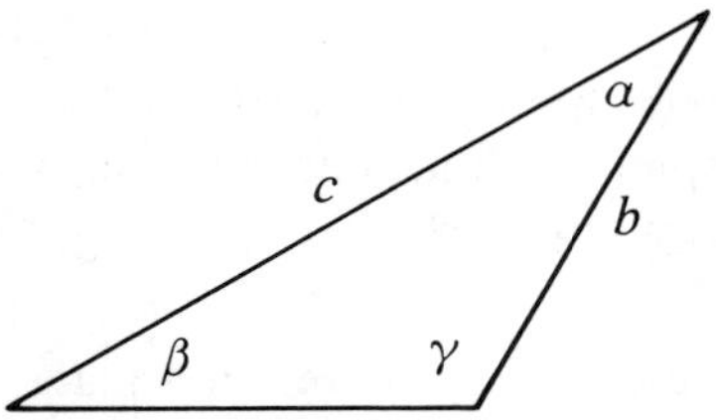

Fig. 4.5a. Scalar labels for a triangle.

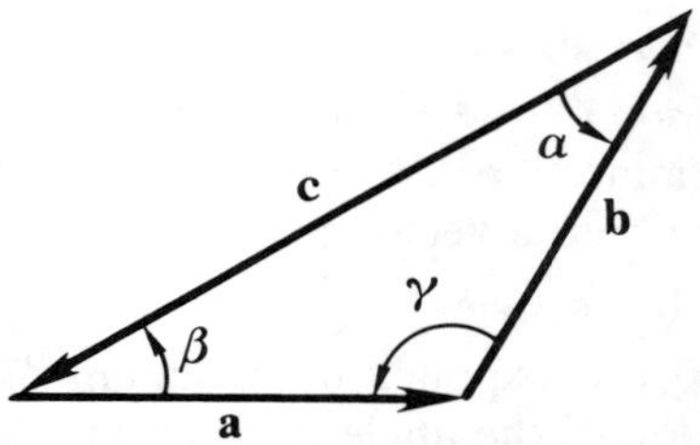

Fig. 4.5b. Vector labels for a triangle.

$$\mathbf{a} + \mathbf{b} + \mathbf{c} = 0, \tag{4.17}$$

with $|\mathbf{a}| = a$, $|\mathbf{b}| = b$ and $|\mathbf{c}| = c$ (Figure 4.5b). According to (4.12), the angles are related to the vectors by the equations

$$-\mathbf{ba} = bae^{\mathbf{i}\gamma}, \tag{4.18a}$$

$$-\mathbf{cb} = cbe^{\mathbf{i}\alpha}, \tag{4.18b}$$

$$-\mathbf{ac} = ace^{\mathbf{i}\beta}. \tag{4.18c}$$

Of course, each of these three equations supplies us with two separate equations when separated into scalar and bivector parts; for example, from (4.18a) we get

$$-\mathbf{b}\cdot\mathbf{a} = ba\cos\gamma, \tag{4.19a}$$

and

$$\mathbf{a}\wedge\mathbf{b} = ba\mathbf{i}\sin\gamma. \tag{4.19b}$$

The minus sign appears in (4.18a) and (4.19a) because, as Figure 4.5b shows, the angle $\boldsymbol{\gamma} = \mathbf{i}\gamma$ is from $\mathbf{b}$ to $-\mathbf{a}$, and not from $\mathbf{b}$ to $\mathbf{a}$ or $\mathbf{a}$ to $\mathbf{b}$.

A word about the logical status of Equation (4.19a) is in order, because many books on vector analysis use such an equation to define the inner product of vectors in terms of the cosine of the angle between the vectors; thus, they take trigonometry as an established subject whose content is merely to be reexpressed in vector language. On the contrary, we have defined inner and outer products with no reference whatever to trigonometric functions, and most of our applications of geometric algebra, including some to trigonometry, require no mention of angles. We regard (4.19a) and the more general Equation (4.18a) as *functional relations* between angles and vectors, rather than primary definitions of any sort. Using these relations the basic trigonometric identities can be derived from the simpler and more general identities of geometric algebra almost as easily as they can be written down from memory. Of course, the main reason for regarding geometric algebra as logically prior to trigonometry is the fact that its scope is so much greater.

Now let us derive the trigonometric formulas for a triangle (4.17) by using inner and outer products. We already did this in Chapter 1, but without justifying the relation to angles established above. Solving (4.17) for $\mathbf{c}$ and squaring we get

$$\mathbf{c}^2 = (\mathbf{a} + \mathbf{b})^2 = \mathbf{a}^2 + \mathbf{b}^2 + \mathbf{ab} + \mathbf{ba},$$

or

$$c^2 = a^2 + b^2 + 2\mathbf{a}\cdot\mathbf{b} \tag{4.20}$$

and, by using (4.19a) to express $\mathbf{a}\cdot\mathbf{b}$ as a function of angle, we get

$$c^2 = a^2 + b^2 - 2ab\cos\gamma. \tag{4.21}$$

This is the *law of cosines* in trigonometry. The same name may be given to the equivalent equation (4.20), though it does not explicitly refer to a cosine and has many applications which require no such reference.

By taking the outer product of (4.17) first by $\mathbf{a}$ and then by $\mathbf{b}$ or $\mathbf{c}$, we get the equations

$$\mathbf{a}\wedge\mathbf{b} = \mathbf{b}\wedge\mathbf{c} = \mathbf{c}\wedge\mathbf{a}. \tag{4.22}$$

From these equations we easily get relations among the angles of the triangle by using (4.19b) and the corresponding relations from (4.18b) and (4.18c); thus

$$\frac{\sin\alpha}{a} = \frac{\sin\beta}{b} = \frac{\sin\gamma}{c}\,. \tag{4.23}$$

This set of equations is the *law of sines* in trigonometry. It should be noted that, although (4.23) is an immediate consequence of (4.22), the former is somewhat more general than the latter because it includes the factor $\mathbf{i}$ representing the direction of the plane. The full generality is helpful when trigonometric relations in a plane are to be related to 3-dimensional space, as in the laws of reflection and refraction.

Equation (4.22) can be regarded as giving three equivalent ways of determining the area of the triangle. We have discussed the interpretation of $\mathbf{a}\wedge\mathbf{b}$ as the directed area of a parallelogram; our triangle has only half that area. Hence, the directed area $\mathbf{A}$ of the triangle is given by

$$\mathbf{A} = \tfrac{1}{2}\mathbf{a}\wedge\mathbf{b} = \tfrac{1}{2}\mathbf{b}\wedge\mathbf{c} = \tfrac{1}{2}\mathbf{c}\wedge\mathbf{a}. \tag{4.24}$$

Using (4.19b), we get the more conventional expression for the area of a triangle,

$$|\mathbf{A}| = \frac{\mathbf{a}\wedge\mathbf{b}}{2\mathbf{i}} = \frac{ab}{2}\sin\gamma. \tag{4.25}$$

or, one half the base a times the altitude $b\sin\gamma$.

The laws of sines and cosines refer to the scalar and bivector parts of Equations (4.18a, b, c) separately. A property of the triangle which makes more direct use of these equations is derived by multiplying the three equations together and dividing by $a^2b^2c^2$ to get

$$e^{\mathbf{i}\alpha}e^{\mathbf{i}\beta}e^{\mathbf{i}\gamma} = -1. \tag{4.26}$$

This says that successive rotations through the three interior angles of a triangle is equivalent to rotation through a straight angle. By virtue of (4.14b) and (4.15), then, we can conclude that

$$\alpha + \beta + \gamma = \pi. \tag{4.27}$$

This familiar result is traditionally regarded as a theorem of geometry rather than of trigonometry. But we use exactly the same techniques of geometric algebra to prove the theorems of both subjects.

Geometric algebra is just as effective for formulating and deriving the results of spherical trigonometry, as shown in Appendix A.

2-4. Exercises

(4.1) Deduce Equations (4.14b, c) from (4.14a) by (4.16).

(4.2) Prove and interpret the identities

$$e^{\mathbf{i}(\theta+2\pi n)} = e^{\mathbf{i}\theta}, \quad e^{\mathbf{i}\theta}e^{-\mathbf{i}\theta} = 1,$$

where n is an integer.

(4.3) From Equation (4.15) derive the trigonometric identities

$$\cos(\theta + \phi) = \cos\theta\cos\phi - \sin\theta\sin\phi$$
$$\sin(\theta + \phi) = \cos\theta\sin\phi + \sin\theta\cos\phi$$

(4.4) Prove that

$$\cos\theta = \frac{e^{\mathbf{i}\theta} + e^{-\mathbf{i}\theta}}{2}, \qquad \sin\theta = \frac{e^{\mathbf{i}\theta} - e^{-\mathbf{i}\theta}}{2\mathbf{i}},$$

$$\mathbf{i}\tan\theta = \frac{e^{2\mathbf{i}\theta} - 1}{e^{2\mathbf{i}\theta} + 1}.$$

(4.5) Prove that $\mathbf{ab} = e^{\mathbf{i}\theta}$ and $\mathbf{a}^2 = \mathbf{b}^2 = 1$ imply

(a) $(\mathbf{ab})^\dagger = \mathbf{ba} = e^{-\mathbf{i}\theta}$
(b) $(\mathbf{a} - \mathbf{b})^2 = 4\sin^2\frac{1}{2}\theta$
(c) $(\mathbf{a} + \mathbf{b})^2 = 4\cos^2\frac{1}{2}\theta$.

(4.6) Locate the following points on an Argand diagram (Figure 2.2):

$$e^{\mathbf{i}\pi/4}, \qquad e^{\mathbf{i}3\pi/2}, \qquad \frac{3}{2}e^{\mathbf{i}\pi/6}, \qquad \frac{1}{\sqrt{2}}(1 + \mathbf{i}).$$

(4.7) Solve the equation

$$1 + e^{\mathbf{i}\theta} + e^{2\mathbf{i}\theta} + e^{3\mathbf{i}\theta} + e^{4\mathbf{i}\theta} + e^{5\mathbf{i}\theta} = 0$$

by interpreting the terms as operators on a vector and identifying the geometrical figure generated.

(4.8) Prove the following identities, and identify trigonometric identities to which they reduce when $\mathbf{a}\wedge\mathbf{b}\wedge\mathbf{c} = 0$.

(a) $(\mathbf{a}\cdot\mathbf{b})^2 - (\mathbf{a}\wedge\mathbf{b})^2 = \mathbf{a}^2\mathbf{b}^2$.
(b) $\mathbf{b}\cdot(\mathbf{a}\wedge\mathbf{b}\wedge\mathbf{c}) = \mathbf{b}\cdot\mathbf{a}\mathbf{b}\wedge\mathbf{c} - \mathbf{b}^2\mathbf{a}\wedge\mathbf{c} + \mathbf{b}\cdot\mathbf{c}\mathbf{a}\wedge\mathbf{b}$,
(c) $(\mathbf{a}\wedge\mathbf{b})\cdot(\mathbf{b}\wedge\mathbf{c}) = \mathbf{b}^2\mathbf{a}\cdot\mathbf{c} - \mathbf{a}\cdot\mathbf{b}\mathbf{b}\cdot\mathbf{c}$.
(d) $2(\mathbf{a}\wedge\mathbf{b})\cdot(\mathbf{b}\wedge\mathbf{c}) = \mathbf{b}^2\mathbf{a}\cdot\mathbf{c} - \mathbf{a}\cdot(\mathbf{bcb}) = -2(\mathbf{a}\times\mathbf{b})\cdot(\mathbf{b}\times\mathbf{c})$
(e) $2\mathbf{a}\cdot\mathbf{b}\mathbf{b}\cdot\mathbf{c} = \mathbf{b}^2\mathbf{a}\cdot\mathbf{c} + \mathbf{a}\cdot(\mathbf{bcb})$.

(4.9) For the triangle in Figure 4.5a, establish the results

(a) $|\,\mathbf{A}\,| = \frac{1}{2}a^2\dfrac{\sin\beta\sin\gamma}{\sin\alpha}$, $\alpha + \beta + \gamma = \pi$.

(b) Hero's formula:

$$|\mathbf{A}|^2 = s(s-a)(s-b)(s-c) = s^2r^2,$$

where s is half the perimeter of the triangle and r is the radius of the inscribed circle.

(c) Half-angle formulas:

$$\tan\frac{\alpha}{2} = \frac{r}{s-a}, \ \tan\frac{\beta}{2} = \frac{r}{s-b}, \ \tan\frac{\gamma}{2} = \frac{r}{s-c}.$$

(d) The *Law of tangents*:

$$\frac{a-b}{a+b} = \frac{\tan\frac{1}{2}(\alpha-\beta)}{\tan\frac{1}{2}(\alpha+\beta)}, \text{ etc.}$$

(4.10) Prove that the angle inscribed in a semicircle is a right angle, by using the vectors indicated in Figure 4.6.

(4.11) Let **a**, **b**, **c** be the verticies of a triangle, as shown in Figure 4.7. Note that vectors designating points in the figure are not represented by arrows; this is because we are not interested in the relation of these points to some arbitrarily designated origin. Prove the following general theorems about triangles:

(a) The altitudes intersect at a point. This point **p** is called the *orthocenter* of the triangle.

(b) The perpendicular bisectors of the sides intersect at a point. Why do you think that this point **q** is called the *circumcenter*?

(c) The medians intersect at a point $\mathbf{r} = \frac{1}{3}(\mathbf{a} + \mathbf{b} + \mathbf{c})$ which lies on the line segment joining the orthocenter to the circumcenter and divides it in the ratio 2/1.

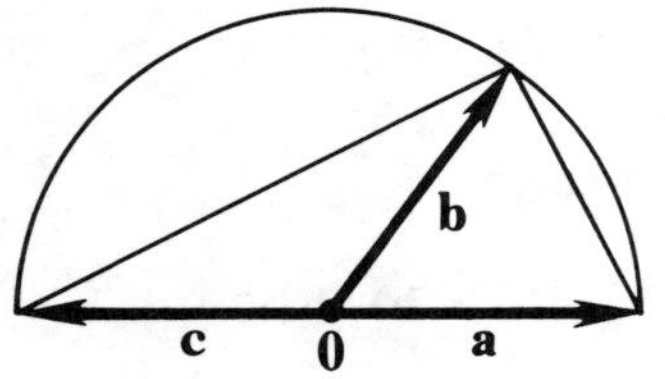

Fig. 4.6.

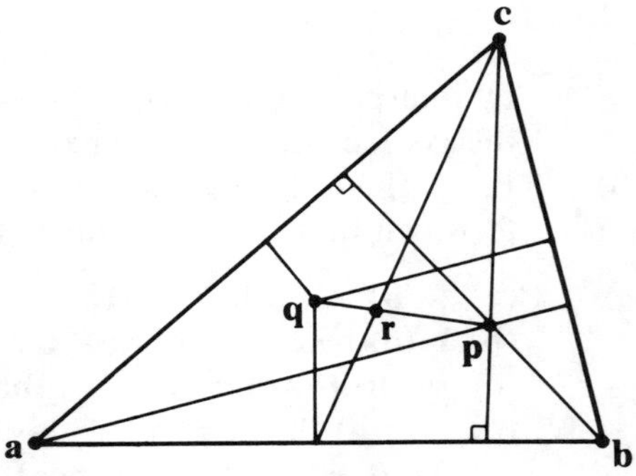

Fig. 4.7.

2-5. The Exponential Function

We have seen the great value of the exponential function for expressing the relation of angles to directions. It first appeared as a combination of sines and cosines. Our aim now is to define the exponential function and ėstablish its properties from first principles. This will simplify some of our calculations and extend the range of applications for the function.

The *exponential function* of a multivector A is denoted by $\exp A$ or e^A and defined by

$$\exp A = e^A \equiv \sum_{k=0}^{\infty} \frac{A^k}{k!}$$

$$= 1 + \frac{A}{1!} + \frac{A^2}{2!} + \ldots + \frac{A^k}{k!} + \ldots . \tag{5.1}$$

This series can be shown to be absolutely convergent for all values of A by standard mathematical arguments. Standard texts give the proof assuming that A is a real or complex number, but the proof actually requires only that A have a definite magnitude $|A|$, so it is easily extended to general multivectors, and we shall take it for granted.

Definition (5.1) is an *algebraic definition* of the exponential function. This is to say that the function is completely defined in terms of the basic operations of addition and multiplication, so all its properties can be determined by using these operations. The most immediate and obvious property of the exponential function is that the value of e^A is a definite multivector. This follows from (5.1) by the closure of geometric algebra under the operations of addition and multiplication. More particularly, it follows that if A is an element of any subalgebra, such as $\mathcal{G}_3$, $\mathcal{G}_2^+$ or $\mathcal{G}_2$, then the value of e^A is an element of the same algebra.

The most important property of the exponential function is the "additivity rule"

$$e^A e^B = e^{A+B}, \tag{5.2}$$

which holds if and only if $AB = BA$, although there are some trivial exceptions to the "only if" condition (Exercise (5.9)). Indeed, if A and B commute, then

$$e^A e^B = \sum_{m=0}^{\infty} \frac{A^m}{m!} \sum_{n=0}^{\infty} \frac{B^n}{n!} = \sum_{n=0}^{\infty} \sum_{k=0}^{n} \frac{A^{n-k}}{(n-k)!} \frac{B^k}{k!} .$$

But, by the binomial expansion,

$$(A+B)^n = \sum_{k=0}^{n} \frac{n!}{(n-k)!k!} A^{n-k}B^k ,$$

hence,

$$e^A e^B = \sum_{n=0}^{\infty} \frac{(A+B)^n}{n!} = e^{A+B}.$$

Notice that the commutativity of A with B is needed to apply the binomial expansion, so if it is lacking the result (5.2) cannot be obtained.

We will always be able to write

$$e^A e^B = e^C,$$

but when $C \neq A + B$, the problem of finding C from A and B does not have a general solution, so different cases must be considered separately. In Chapter 5, when we study rotations we will solve the problem when A and B are bivectors.

The *hyperbolic cosine* and *sine* functions are defined by the usual series expansions

$$\cosh A \equiv \sum_{k=0}^{\infty} \frac{A^{2k}}{(2k)!} = 1 + \frac{A^2}{2!} + \frac{A^4}{4!} + \ldots, \tag{5.3a}$$

$$\sinh A \equiv \sum_{k=0} \frac{A^{2k+1}}{(2k+1)!} = A + \frac{A^3}{3!} + \frac{A^5}{5!} + \ldots. \tag{5.3b}$$

These are just the *even* and *odd* parts of the exponential series (5.1), thus,

$$e^A = \cosh A + \sinh A. \tag{5.3c}$$

The multivector A is called the *argument* of each of the functions in (5.3c).

The *cosine* and *sine* functions are defined by the usual series expansions

$$\cos A \equiv \sum_{k=0}^{\infty} (-1)^k \frac{A^{2k}}{(2k)!} = 1 - \frac{A^2}{2!} + \frac{A^4}{4!} - \frac{A^6}{6!} + \ldots, \tag{5.4a}$$

$$\sin A \equiv \sum_{k=0}^{\infty} (-1)^k \frac{A^{2k+1}}{(2k+1)!} = A - \frac{A^3}{3!} + \frac{A^5}{5!} - \frac{A^7}{7!} + \ldots. \tag{5.4b}$$

If I is a multivector with the properties $I^2 = -1$ and $IA = AI$, then it is easily shown by substitution in the series (5.3a, b) that

$$\cosh IA = \cos A, \tag{5.5a}$$

$$\sinh IA = I \sin A, \tag{5.5b}$$

and

$$e^{IA} = \cos A + I \sin A. \tag{5.5c}$$

It will be noticed that this last equation generalizes (4.10).

When A is a scalar, the definitions (5.4a, b) for the trigonometric functions reduce to the well-known series expansions for sines and cosines established in elementary calculus. Therefore, we are assured that they apply to the trigonometric functions employed in Section 2.4. So, considering (5.5c), we have from (5.1) and (4.9) the following expansion of the product of two unit vectors in terms of their relative angle:

$$\mathbf{ab} = e^{\mathbf{i}\theta} = \sum_{k=0}^{\infty} \frac{(\mathbf{i}\theta)^k}{k!} = 1 + \mathbf{i}\theta - \frac{\theta^2}{2!} - \mathbf{i}\frac{\theta^3}{3!} + \ldots . \tag{5.6}$$

As emphasized before, the exponential function and its series expansion require that the angle θ be measured in radians, though of course it can be expressed in degrees by inserting a conversion constant.

As indicated by (5.3c) and (5.5c) the hyperbolic and trigonometric functions are best regarded as parts of the more fundamental exponential function. We have considered only the basic algebraic properties of the exponential function in this section. Differentiation of the exponential function will be considered when the need arises, and we will learn more about this remarkable function as we encounter it in physical applications.

Logarithms

The following discussion of the natural log function can be omitted by readers just beginning the study of geometric algebra, as it involves some subtle points which can be ignored in elementary applications. However, it will be needed as background for our discussion of rotations in Chapter 5.

The exponential function associates a unique multivector B with every multivector A according to the formula

$$B = e^{A} . \tag{5.7}$$

We know that if A is any scalar, then B is a positive scalar, and further, that for positive scalars the exponential function has an inverse called the *logarithmic function*, for which we write

$$A = \log B . \tag{5.8}$$

This brings up the question: Since we have extended the domain of the exponential function to all multivectors, can we not do the same for the logarithmic function? The answer is "not quite", as long as the logarithm is required to be the inverse of the exponential. Recall from Exercise 1-7.2 that geometric algebra contains idempotents other than the scalars zero and one. It can be shown that, like zero, these idempotents do not have finite logarithms.

Although we cannot define the logarithmic function on all multivectors, evidently we can define it for any multivector B which can be expressed as an exponential of some other multivector as in (5.7). But here we meet another difficulty, for the exponential function is many-to-one, that is, there exist many different multivectors A_k (k=1, 2, . . .) such that

$$B = e^{A_1} = e^{A_2} = \ldots = e^{A_k} = \ldots . \tag{5.9}$$

Therefore, the inverse function must be many-valued, and for any of the A_k we can write

$$A_k = \log B. \tag{5.10}$$

We can make the logarithmic function single-valued by introducing a rule to pick out just one of the A_k as its value. It is most convenient to choose the one with the smallest magnitude. If $|A_1| < |A_k|$ for $k > 1$, we write

$$A_1 = \operatorname{Log} B, \tag{5.11}$$

and call A_1 the *principal part* of the logarithm. The restriction to the principal part is indicated in (5.11) by the capital L. Ambiguity can still arise, however, because for some B there is more than one A_k with smallest magnitude; to choose between them a new rule must be introduced, such as one considered below.

To be more specific, let us consider the logarithmic function on spinors. Any spinor z in $\mathcal{G}_3$ can be written in the equivalent forms

$$z = \lambda e^{\mathbf{i}\theta_k} = e^{\alpha} e^{\mathbf{i}\theta_k} = e^{\alpha + \mathbf{i}\theta_k}, \tag{5.12}$$

where $\mathbf{i}$ is a unit bivector, θ_k is a scalar and α is the logarithm of the positive scalar λ. Evidently,

$$\log z = \alpha + \mathbf{i}\theta_k . \tag{5.13}$$

Thus, every spinor has a logarithm. The logarithm is not unique, however, because of the multiplicity of possible values for the angle θ_k. Let θ be the smallest of these angles. Noting that $e^{\mathbf{i}2\pi k} = 1$ for any integer k, we verify that

$$e^{\mathbf{i}\theta} = e^{\mathbf{i}\theta} e^{\mathbf{i}2\pi k} = e^{\mathbf{i}(\theta + 2\pi k)} . \tag{5.14}$$

Therefore, any of the angles

$$\theta_k = \theta + 2\pi k \tag{5.15}$$

will satisfy (5.12), and evidently no other angles will. Since the possible angles differ by any positive or negative multiple of 2π, the smallest of them, θ, must be confined to the interval $-\pi \leqslant \theta \leqslant \pi$. Having determined the allowed range of its bivector part, the principle part of the logarithm is well defined, and we write

$$\operatorname{Log} z = \alpha + \mathbf{i}\theta \tag{5.16}$$

Ambiguity arises when $|\theta| = \pi$, for then both $\mathbf{i}\pi$ and $-\mathbf{i}\pi$ might be allowed in (5.16). Choice of one of them, say $\mathbf{i}\pi$, amounts to a choice of orientation for the $\mathbf{i}$–plane. Accordingly, we write

$$\operatorname{Log}(-1) = \mathbf{i}\pi. \tag{5.17}$$

Note also that $e^{\pm i\pi} = -1$, where i is the unit pseudoscalar for $\mathcal{G}_3$. So we have

$$\log(-1) = \pm i\pi \tag{5.18}$$

as well. This possibility is eliminated if we are interested only in spinor-valued logarithms. In applications the appropriate value for a logarithm will be determined by the problem at hand.

Finally, we note that, although vectors do not have logarithms, the product of vectors **a** and **b** is a spinor, so we can write

$$\mathrm{Log}(\mathbf{ab}) = \log|\mathbf{a}| + \log|\mathbf{b}| + \mathbf{i}\theta, \tag{5.19}$$

where $\mathbf{i}\theta$ is the directed angle from **a** to **b**. The exponential being the inverse of the logarithm, we have

$$\mathbf{ab} = e^{\mathrm{Log}(\mathbf{ab})} = e^{\log(\mathbf{ab})}. \tag{5.20}$$

2-5. Exercises

(5.1) Establish the following general properties of the exponential function

(a) The "algebraic inverse" of e^A is $(e^A)^{-1} = e^{-A}$.

(b) $(e^A)^n = e^{nA}$ for scalar values of n.

(c) $(e^A)^\dagger = e^{A^\dagger}$.

(d) If $AB = BA$, then $e^A B = Be^A$.

(e) If $AB = -BA$, then $e^A B = Be^{-A}$.

(5.2) The hyperbolic functions are strictly even or odd functions in the following sense

$$\cosh A = \frac{e^A + e^{-A}}{2} = \cosh(-A),$$

$$\sinh A = \frac{e^A - e^{-A}}{2} = -\sinh(-A).$$

Similarly,

$$\cos(-A) = \cos A,$$
$$\sin(-A) = -\sin A.$$

Justify these relations.

(5.3) Show that if $A^2 = |A|^2$, then

$$\cosh A = \cosh |A|,$$
$$\sinh A = \hat{A} \sinh |A|.$$

and if $A^2 = -|A|^2$, then

$$\cosh A = \cos|A|,$$
$$\sinh A = \hat{A}\sin|A|.$$

(5.4) Let $\mathbf{a}$ be a vector in $\mathcal{G}_3(i)$, and write $\mathbf{a} = a\mathbf{n}$, where $\mathbf{n}$ is a unit vector, and $a = \mathbf{na} = \mathbf{n}\cdot\mathbf{a}$ is possibly negative. Prove that

$$\cos\mathbf{a} = \cos a, \qquad \cosh\mathbf{a} = \cosh a,$$
$$\sin\mathbf{a} = \mathbf{n}\sin a, \qquad \sinh\mathbf{a} = \mathbf{n}\sinh a,$$
$$e^{i\mathbf{a}} = \cos a + i\mathbf{n}\sin a, \qquad e^{\mathbf{a}} = \cosh a + \mathbf{n}\sinh a.$$

(5.5) Prove that if $J^2 = 1$ and $AJ = JA$, then

$$\cosh JA = \cosh A, \qquad \sinh JA = J\sinh A,$$
$$e^{JA} = \cosh A + J\sinh A.$$

(5.6) Prove that if $I^2 = -1$ and A, B, I mutually commute, then

$$\cos(A + IB) = \cos A\cosh B - I\sinh B\sin A,$$
$$\sin(A + IB) = \sin A\cosh B + I\sinh B\cos A.$$

(5.7) Evaluate $\sin(\pi/4 + \mathrm{I})$ and $\cos(\pi/3 + \mathrm{I})$ when $I^2 = -1$.

(5.8) For vectors $\mathbf{a}$ and $\mathbf{b}$, show that if $|\mathbf{a}| > |\mathbf{b}|$, then

$$\frac{1}{\mathbf{a}-\mathbf{b}} = \frac{1}{\mathbf{a}} + \frac{1}{\mathbf{a}}\mathbf{b}\frac{1}{\mathbf{a}} + \frac{1}{\mathbf{a}}\mathbf{b}\frac{1}{\mathbf{a}}\mathbf{b}\frac{1}{\mathbf{a}} + \ldots.$$

(5.9) Let the angle between unit vectors $\mathbf{a}$ and $\mathbf{b}$ be $2\pi/3$. Show $\mathbf{c} = \mathbf{a} + \mathbf{b}$ is also a unit vector. Define $A = \alpha i\mathbf{a}$, $B = \alpha i\mathbf{b}$ and show that A and B do not commute. Show that if α is an integer multiple of 2π, then Equation (5.2) is satisfied. However, Equation (5.2) is not satisfied for any other value of α. Show that for $\alpha = \pi$, $e^A e^B = -e^{A+B}$.

2-6. Analytic Geometry

This section can be skipped or lightly perused by readers who are in a hurry to get on with mechanics. It is included here as a reference on elementary concepts and results of Analytic Geometry expressed in terms of geometric algebra.

Analytic Geometry is concerned with the description or, if you will, the representation of geometric curves and surfaces by algebraic equations. The traditional approach to Analytic Geometry is accurately called *Coordinate Geometry*, because it represents each geometrical point by a set of scalars called its coordinates. Curves and surfaces are then represented by algebraic equations for the coordinates of their points. A major drawback of Coordinate Geometry is the fact that coordinates carry superfluous information which often entails unnecessary complications. Thus, rectangular coordinates (x, y, z) of a point specify the distances of the point from the three coordinate planes, the coordinate z, for example, specifying the distance from the (xy)–plane. Therefore, equations for a geometric figure in rectangular coordinates describe the relation of that figure to three arbitrarily chosen planes. Obviously, it would be more efficient to describe the figure in terms of its intrinsic properties alone, without introducing extrinsic relations to lines or planes which are frequently of no interest. Geometric algebra makes this possible.

In the language of geometric algebra, each geometrical point is represented or labelled by a vector. Indeed, for mathematical purposes it is often simplest to regard the point and the vector that labels it as one and the same. Of course, we can label a given point by any vector we please, and problems can often be simplified by a judicious selection of the point to be labelled by the zero vector. But, as a rule, once a labelling has been selected, it is unnecessary to change it.

The distinction between a point and its vector label becomes important when geometric algebra is used as a language, for then the point, which is undefined as a mathematical entity, might be identified with a mark on a piece of paper or a "place" among physical objects. However, the vector label retains its status as a purely mathematical entity, and geometric algebra precisely describes its geometric properties. It will be noticed also in the following that, although some vectors designate (or are designated as) points, other vectors describe relations between points or have some other geometrical significance.

The simplest relation between two points $\mathbf{a}$ and $\mathbf{b}$ is the vector $\mathbf{a} - \mathbf{b}$, which, for want of standard terminology, we propose to call the *chord* from $\mathbf{b}$ to $\mathbf{a}$. The magnitude of the chord $|\mathbf{a} - \mathbf{b}|$ is called the (Euclidean) *distance* between $\mathbf{b}$ and $\mathbf{a}$. The zero vector designates a point called the *origin*. Since $\mathbf{a} - \mathbf{0} = \mathbf{a}$, the vector $\mathbf{a}$ specifies both the point and the chord from the origin to the point $\mathbf{a}$, and $|\mathbf{a}| = |\mathbf{a} - \mathbf{0}|$ is the distance between the point $\mathbf{a}$ and the origin.

Geometric spaces and figures are sets of points. *Euclidean Geometry* is concerned with distance relations of the form $|\mathbf{a} - \mathbf{b}|$ among pairs of points in such spaces and figures. *Non-Euclidean* geometries are based on alternative definitions of the distance between points, such as $\log |\mathbf{a} - \mathbf{b}|$. However interesting it is to explore the implications of alternative definitions of distance, we want our definition to correspond to the relations among physical objects determined by the operational rules for *measuring distance*, and it is a physical fact that, at least to a high order of approximation, such relations conform to the Euclidean definition of distance. For this reason, we will be concerned with the Euclidean concept of distance only, and the adjective "Euclidean" will be unnecessary.

A set $\mathcal{E}^n$ with elements called *points* is said to be an *n-dimensional Euclidean Space* if it has the following properties:

(1) The points in $\mathcal{E}^n$ can be put in one-to-one correspondence with the vectors in an n-dimensional vector space. (Each vector then is said to *designate* or *label* the corresponding point.)

(2) There is a rule for assigning a positive number to every pair of points called the *distance* between the points, and the points can be labelled in such a way that the distance between any pair of points $\mathbf{a}$ and $\mathbf{b}$ is given by $|\mathbf{a} - \mathbf{b}| = [(\mathbf{a} - \mathbf{b})^2]^{1/2}$.

Obviously, any vector space can be regarded as a Euclidean space simply by regarding each vector as identical with the point it labels. For most mathematical purposes it is quite sufficient to regard each Euclidean space as a vector space. However, in physical applications it is essential to distinguish between each point and the vector which labels it. This is apparent in the most fundamental application of all, the application of geometry to measurement.

In Chapter 1 we saw that a complex system of operational rules is needed to determine *physical points* (i.e. *positions* or *places*) and *measure distances* between them. The set of all physical points determined by these rules is called *Physical Space*. We saw that Euclidean geometry has certain physical implications when interpreted as a physical theory. We can now *completely* formulate the physical implications of geometry in the single proposition: *Physical Space is a 3-dimensional Euclidean Space*. This proposition could be called the *Zeroth Law* of physics, because it is presumed in the theory of measurement and so in every branch of physics, although the Law must be modified or reinterpreted somewhat to conform to Einstein's theory of relativity and gravitation. Chapter 9 gives a more complete formulation and discussion of the Zeroth Law in relation to the other laws of mechanics.

We label the points of Physical Space by vectors in the geometric algebra $\mathcal{G}_3$. These vectors compose a 3-dimensional Euclidean space $\mathcal{E}_3 = \langle \mathcal{G}_3 \rangle_1$ which can be regarded as a mathematical *model* of Physical Space. The properties of points in $\mathcal{E}_3$, such as their relations to other points, to lines and to planes, require the complete algebra $\mathcal{G}_3$ for their description. Since $\mathcal{G}_3$ thereby provides us with the necessary language to describe relations among points in Physical Space, it is appropriate to call $\mathcal{G}_3$ *the geometric algebra of Physical Space*.

The study of curves and surfaces in $\mathcal{E}_3$ is a purely mathematical enterprise, but its relevance to physics is assured by the correspondence of $\mathcal{E}_3$ with Physical Space. By appropriate semantic assumptions a curve in $\mathcal{E}_3$ can be variously interpreted as the path of a particle, a boundary on a surface or the edge of a solid body. But in the rest of this section, all such interpretations are deliberately ignored, as we learn to describe the *form* of curves and surfaces with geometric algebra. Our results can then be used in a variety of physical contexts when we introduce semantic assumptions later on.

Straight Lines

The most basic equations in analytic geometry are those for lines and planes. In Section 2-2 we saw that the equation

$$\mathbf{x}\wedge\mathbf{u} = 0 \tag{6.1}$$

determines a line through the origin when $\mathbf{u}$ is a fixed nonzero vector. The substitution $\mathbf{x} \rightarrow \mathbf{x} - \mathbf{a}$ in (6.1) has the effect of rigidly displacing each point on the line by the same amount $\mathbf{a}$. From this we conclude that the line $\mathscr{L} = \{\mathbf{x}\}$ with *direction* $\hat{\mathbf{u}}$ passing through the point $\mathbf{a}$ is determined by the equation

$$(\mathbf{x} - \mathbf{a})\wedge\mathbf{u} = 0. \tag{6.2}$$

It should be noted that this equation determines the line without reference to any space in which the line might be imbedded, although, of course, we are most interested in lines in $\mathscr{E}_3$.

Moment and Directance of a Line

Equation (6.2) is a necessary and sufficient condition for a point $\mathbf{x}$ to lie on the line $\mathscr{L}$. All the properties of a line, such as its relations to specified points, lines and planes can be derived from the defining equation (6.2) by geometric algebra. To see how this can best be done, we derive and study various alternative forms of the defining equation; each reveals a different property of the line. On writing $\mathbf{M}$ for the bivector $\mathbf{a}\wedge\mathbf{u}$, Equation (6.2) takes the form

$$\mathbf{x}\wedge\mathbf{u} = \mathbf{M}. \tag{6.3}$$

Since $\mathbf{x}\wedge\mathbf{u}\wedge\mathbf{u} = 0$, multiplication of (6.3) by $\mathbf{u}^{-1}$ and use of (1.4) as well as (1.14) yields

$$(\mathbf{x}\wedge\mathbf{u})\cdot\mathbf{u}^{-1} = \mathbf{x} - \mathbf{x}\cdot\mathbf{u}\mathbf{u}^{-1} = \mathbf{M}\mathbf{u}^{-1}.$$

Hence, for fixed $\mathbf{M}$ and $\mathbf{u}$,

$$\mathbf{x} = (\mathbf{M} + \alpha)\mathbf{u}^{-1} \tag{6.4}$$

is a parametric equation for the line $\mathscr{L}$, each point $\mathbf{x} = \mathbf{x}(\alpha)$ being determined by a value of the parameter α.

Introducing the vector

$$\mathbf{d} = \mathbf{M}\mathbf{u}^{-1} = \mathbf{M}\cdot\mathbf{u}^{-1}, \tag{6.5}$$

Equation (6.4) takes the form

$$\mathbf{x} = \mathbf{d} + \alpha\mathbf{u}^{-1}. \tag{6.6}$$

Note that $\mathbf{d}$ is orthogonal to $\mathbf{u}$, since, by (6.5),

$$\mathbf{d}\cdot\mathbf{u} = \langle\mathbf{d}\mathbf{u}\rangle_0 = \langle\mathbf{M}\rangle_0 = 0.$$

So, by squaring (6.6), one obtains the following expression for the distance $|\mathbf{x}| = |\mathbf{x} - \mathbf{0}|$ between the origin and a point $\mathbf{x}$ on $\mathscr{L}$:

$$|\mathbf{x}|^2 = \mathbf{x}^2 = \mathbf{d}^2 + \alpha^2\mathbf{u}^{-2}.$$

This has its minimum value when $\mathbf{x} = \mathbf{d}$. Thus, $\mathbf{d}$ is that point on the line $\mathscr{L}$ which is "closest" to the origin. We call $\mathbf{d}$ the *directance* (= directed distance) from the point $\mathbf{0}$ to the line $\mathscr{L}$. The magnitude $|\mathbf{d}|$ is called the *distance* from the point $\mathbf{0}$ to the line $\mathscr{L}$ (Figure 6.1).

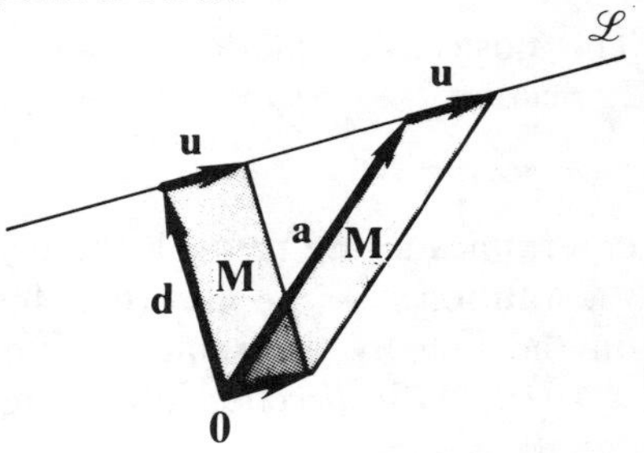

Fig. 6.1. For a line $\mathscr{L}$ with direction $\mathbf{u}$ and directance $\mathbf{d}$ from the origin. Note the two different representations of the moment $\mathbf{M} = \mathbf{du} = \mathbf{a}\wedge\mathbf{u}$.

By substituting (6.3) into (6.5) we get the useful expression

$$\mathbf{d} = \mathbf{x}\wedge\mathbf{u}\mathbf{u}^{-1} \equiv P_{\mathbf{u}}^{\perp}(\mathbf{x}), \tag{6.7}$$

where $P_{\mathbf{u}}^{\perp}$ is the rejection operator defined in Section 2-4. This tells us how to find the directance of a line from any point of the line.

The bivector $\mathbf{M}$ is called the *moment* of the line $\mathscr{L}$. From (6.5) one finds that $|\mathbf{d}| = |\mathbf{M}|\,|\mathbf{u}|^{-1}$, showing, in particular, that if $|\mathbf{u}| = 1$, the magnitude of the moment is equal the distance from the origin to $\mathscr{L}$. From Equation (6.3), it is clear that any oriented line $\mathscr{L}$ is uniquely determined by specifying its direction $\mathbf{u}$ and its moment $\mathbf{M}$, or equivalently, the single quantity $L \equiv \mathbf{u} + \mathbf{M} = (1 + \mathbf{d})\mathbf{u}$. We shall see that the last way of characterizing a line is useful in rigid body mechanics.

Points on a Line

A line determines relations among pairs of points on the line. To analyze such relations, we put the defining equation (6.2) in a different form. Equation (6.2) is equivalent to the statement that the chord $\mathbf{x} - \mathbf{a}$ is collinear with the vector $\mathbf{u}$. Since $\mathbf{x}$ and $\mathbf{a}$ are any pair of points on the lines, it follows (by transitivity) that all chords of the line are collinear. If $\mathbf{x}$, $\mathbf{a}$, $\mathbf{b}$ are any three points on the line, the collinearity of chords $\mathbf{x} - \mathbf{a}$ and $\mathbf{b} - \mathbf{a}$ is expressed by the equation

$$(\mathbf{x} - \mathbf{a})\wedge(\mathbf{b} - \mathbf{a}) = 0. \tag{6.8}$$

This differs from (6.2) only in the replacement of $\mathbf{u}$ by the chord $\mathbf{b} - \mathbf{a}$ which is proportional to it. So (6.8) is equivalent to (6.2) if $\mathbf{b}$ and $\mathbf{a}$ are distinct points. Thus, we have shown that *two distinct points determine the equation for a line.*

Barycentric Coordinates

By expanding (6.8) with the distributive rule and introducing a factor of $\frac{1}{2}$, we get the equivalent equation

$$\tfrac{1}{2}\mathbf{a}\wedge\mathbf{b} = \tfrac{1}{2}\mathbf{a}\wedge\mathbf{x} + \tfrac{1}{2}\mathbf{x}\wedge\mathbf{b}. \tag{6.9}$$

Now $\frac{1}{2}\mathbf{a}\wedge\mathbf{b}$ is the directed area of a triangle with vertices $\mathbf{a}$, $\mathbf{0}$, $\mathbf{b}$ and "sides" (or chords) $\mathbf{a}$, $\mathbf{b}$, $\mathbf{b} - \mathbf{a}$. The other two terms in (6.9) can be interpreted similarly, and it will be noted that any two of the three triangles have one side in common. So (6.9) merely expresses the area of a triangles as the sum of areas of two triangles into which it can be decomposed; this is depicted in Figure 6.2a when $\mathbf{x}$ is between $\mathbf{a}$ and $\mathbf{b}$ and in Figure 6.2b when it is not. From (6.9) it follows that

$$\mathbf{a}\wedge\mathbf{b}\wedge\mathbf{x} = 0, \tag{6.10}$$

so all three vectors and the three triangles they determine are in the same plane. Denoting the unit bivector for this plane by $\mathbf{i}$, we introduce the notation for directed areas

$$\begin{aligned} \mathbf{B} &\equiv \tfrac{1}{2}\,\mathbf{a}\wedge\mathbf{x} = B\mathbf{i}, \\ \mathbf{A} &\equiv \tfrac{1}{2}\,\mathbf{x}\wedge\mathbf{b} = A\mathbf{i}. \end{aligned} \tag{6.11}$$

This notation is used to denote areas in Figures (6.2a, b). Note that the orientation of $\mathbf{A}$ and hence the sign of A is opposite in the two figures.

Let us regard $\mathbf{a}$ and $\mathbf{b}$ as fixed and let $\mathbf{x}$ be any point on the line they determine, as indicated in Figures (6.2a, b). We now show that the bivectors

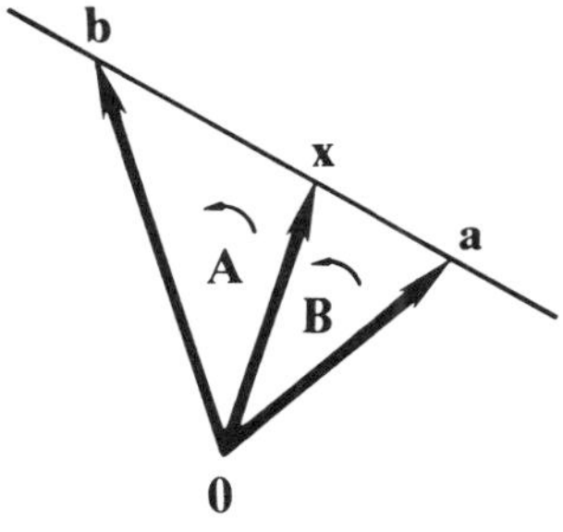

Fig. 6.2a. $\mathbf{A} = |\,\mathbf{A}\,|\,\mathbf{i}$, $\mathbf{B} = |\,\mathbf{B}\,|\,\mathbf{i}$, $\mathbf{A} + \mathbf{B} = (|\,\mathbf{A}\,| + |\,\mathbf{B}\,|)\mathbf{i}$.

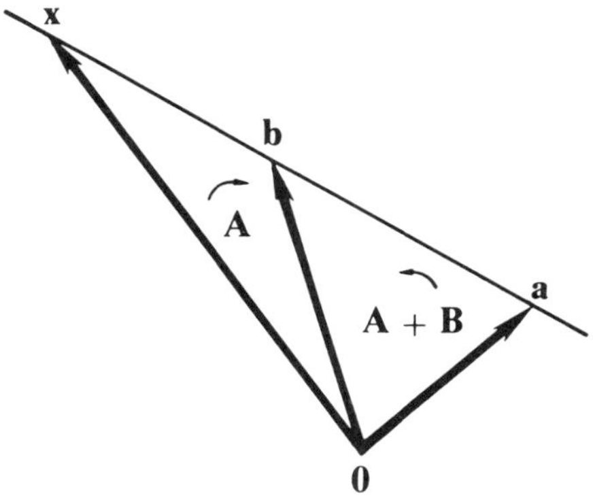

Fig. 6.2b. $\mathbf{A} = -|\,\mathbf{A}\,|\,\mathbf{i}$, $\mathbf{B} = |\,\mathbf{B}\,|\,\mathbf{i}$, $\mathbf{A} + \mathbf{B} = (|\,\mathbf{B}\,| - |\,\mathbf{A}\,|)\mathbf{i}$.

$\mathbf{A}$ and $\mathbf{B}$ or the determinants A and B can be used as coordinates for the point $\mathbf{x}$. Recall the Jacobi identity for vectors proved in Exercise (1.11).

$$(\mathbf{a}\wedge\mathbf{b})\cdot\mathbf{x} + (\mathbf{b}\wedge\mathbf{x})\cdot\mathbf{a} + (\mathbf{x}\wedge\mathbf{a})\cdot\mathbf{b} = 0.$$

Now, because of (6.10) the dots in the equation can be dropped, so, if we introduce the notation (6.11) and use (6.9), we get

$$(\mathbf{A} + \mathbf{B})\mathbf{x} - \mathbf{A}\mathbf{a} - \mathbf{B}\mathbf{b} = 0.$$

If the origin is not on the line, then (6.11) implies **A** and **B** cannot both vanish, so we can solve for

$$\mathbf{x} = \left(\frac{\mathbf{A}}{\mathbf{A}+\mathbf{B}}\right)\mathbf{a} + \left(\frac{\mathbf{B}}{\mathbf{A}+\mathbf{B}}\right)\mathbf{b}. \tag{6.12}$$

Since **A** and **B** are codirectional, we can express this in terms of the scalars A and B defined by (6.11); thus,

$$\mathbf{x} = \frac{A\mathbf{a} + B\mathbf{b}}{A + B} . \tag{6.13}$$

In the mathematical literature, the scalars A and B in (6.13) are called *homogeneous* (line) *coordinates* for the point **x**. They are also called *barycentric coordinates*, because of a similarity of (6.13) to the formula for center of mass (defined in Chapter 6). But unlike masses, the scalars A and B can be negative and, as we have seen, they can be interpreted geometrically as oriented areas. They have another geometrical interpretation which we now determine.

Division and Intersection of Lines

Since all chords of a line are collinear, we can write

$$\mathbf{a} - \mathbf{x} = \lambda(\mathbf{x} - \mathbf{b}), \tag{6.14}$$

where λ is a scalar. The outer product of this with **x** gives

$$\mathbf{x}\wedge\mathbf{a} = -\lambda\mathbf{x}\wedge\mathbf{b}.$$

Solving both these equations for λ, and using (6.11) we find

$$\lambda = \frac{\mathbf{a}-\mathbf{x}}{\mathbf{x}-\mathbf{b}} = \frac{\mathbf{a}\wedge\mathbf{x}}{\mathbf{x}\wedge\mathbf{b}} = \frac{\mathbf{B}}{\mathbf{A}} = \frac{B}{A} , \tag{6.15a}$$

or

$$\lambda = \pm\frac{|\mathbf{a}-\mathbf{x}|}{|\mathbf{x}-\mathbf{b}|} = \pm\frac{|\mathbf{B}|}{|\mathbf{A}|} = \frac{B}{A} , \tag{6.15b}$$

where the positive sign applies if **x** is between **a** and **b** and the negative sign applies if it is not. The point **x** is sometimes called a *point of division* for the oriented line segment [**a**, **b**], and because of (6.15), **x** is said to *divide* [**a**, **b**] *in ratio B*/*A*. The *division ratio* λ can be used as a coordinate for points on the line through **a** and **b** simply by solving (6.14) to get

$$\mathbf{x} = \frac{\mathbf{a} + \lambda\mathbf{b}}{1 + \lambda} , \tag{6.16}$$

which, of course, is equivalent to (6.13). Thus the *midpoint* of [**a**, **b**] is defined by the condition $\lambda = 1$ and is given by $\frac{1}{2}(\mathbf{a} + \mathbf{b})$.

It should be noted that the division ratio (6.15) is not only independent of an orientation for the line or for the $(\mathbf{a}\wedge\mathbf{b})$-plane, it is also, independent of the location of the origin (Even the ratio $B/A = 0/0$, which occurs when the origin is on the line, is determined by (6.15)). This helps us deduce a large number of geometrical facts, for by displacing the origin by an arbitrary vector $\mathbf{c}$, we get from (6.15) the following general relations among the quantities indicated by Figure 6.3:

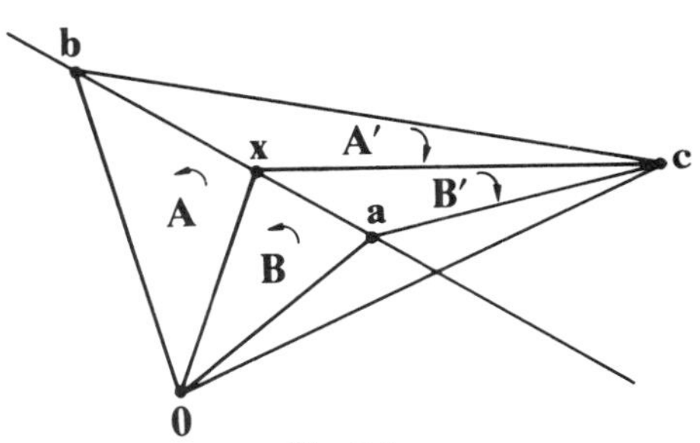

Fig. 6.3.

$$\lambda = \frac{\mathbf{a}-\mathbf{x}}{\mathbf{x}-\mathbf{b}} = \frac{\mathbf{B}}{\mathbf{A}} = \frac{\mathbf{B}'}{\mathbf{A}'} = \frac{\mathbf{B} \pm \mathbf{B}'}{\mathbf{A} \pm \mathbf{A}'}, \tag{6.17a}$$

or, in terms of determinants for the directed areas,

$$\lambda = \frac{B}{A} = \frac{B'}{A'} = \frac{B \pm B'}{A \pm A'}, \tag{6.17b}$$

or, in terms of vectors,

$$\lambda = \frac{\mathbf{a}\wedge\mathbf{x}}{\mathbf{x}\wedge\mathbf{b}} = \frac{(\mathbf{a}-\mathbf{c})\wedge(\mathbf{x}-\mathbf{c})}{(\mathbf{x}-\mathbf{c})\wedge(\mathbf{b}-\mathbf{c})} = \frac{\mathbf{a}\wedge\mathbf{x} + \mathbf{c}\wedge\mathbf{a} + \mathbf{x}\wedge\mathbf{c}}{\mathbf{x}\wedge\mathbf{b} + \mathbf{b}\wedge\mathbf{c} + \mathbf{c}\wedge\mathbf{x}}$$

$$= \frac{\mathbf{c}\wedge\mathbf{a} + \mathbf{x}\wedge\mathbf{c}}{\mathbf{b}\wedge\mathbf{c} + \mathbf{c}\wedge\mathbf{x}} = \frac{(\mathbf{x}-\mathbf{a})\wedge\mathbf{c}}{\mathbf{c}\wedge(\mathbf{x}-\mathbf{b})}. \tag{6.17c}$$

These relations hold even if $\mathbf{c}$ is not in the $(\mathbf{a}\wedge\mathbf{b})$-plane.

For the special case when $\mathbf{c}$ is collinear with $\mathbf{x}$, we can have one of the three cases depicted by Figures (6.4a, b, c). All three cases are governed by (6.17). However, the point $\mathbf{x}$ also divides the

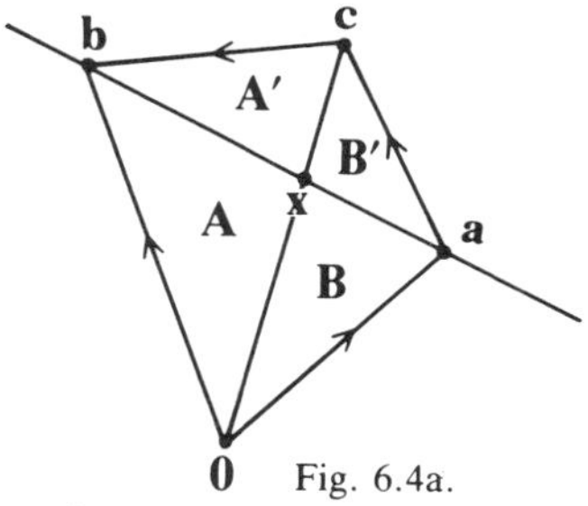

Fig. 6.4a.

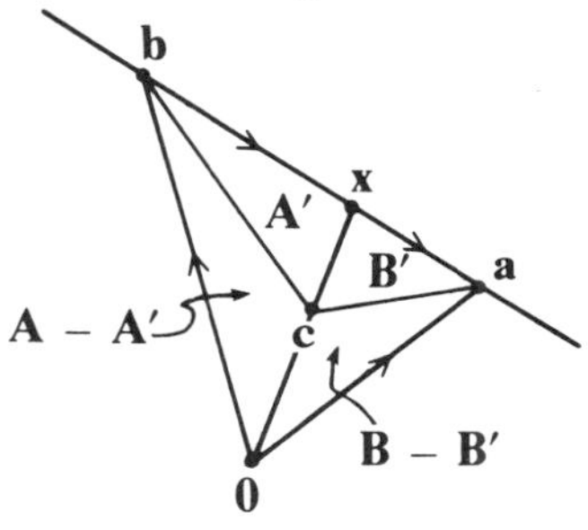

Fig. 6.4b.

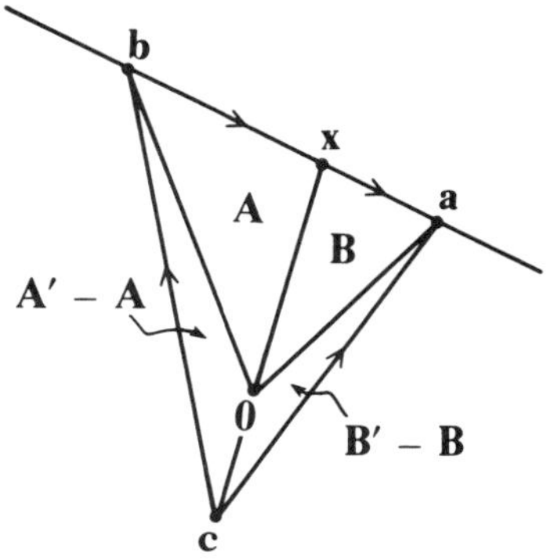

Fig. 6.4c.

line segment [**c**, **0**] in some ratio λ' given by

$$\lambda' = \frac{\mathbf{c}-\mathbf{x}}{\mathbf{x}-\mathbf{0}} = \frac{\mathbf{B}}{\mathbf{B}'} = \frac{\mathbf{A}}{\mathbf{A}'} = \frac{\mathbf{A} \pm \mathbf{B}}{\mathbf{A}' \pm \mathbf{B}'} . \tag{6.18}$$

The point **x** is, in fact, the *point of intersection* of the line through points **0**, **c** with the line through points **a**, **b**. From any one of the figures we see that

$$\mathbf{A} + \mathbf{B} = \tfrac{1}{2}\mathbf{a}\wedge\mathbf{b}, \tag{6.19a}$$

and

$$\mathbf{A}' + \mathbf{B}' = \tfrac{1}{2}(\mathbf{a}-\mathbf{c})\wedge(\mathbf{b}-\mathbf{c}). \tag{6.19b}$$

So (6.18) gives us

$$\lambda' = \frac{\mathbf{a}\wedge\mathbf{b}}{(\mathbf{a}-\mathbf{c})\wedge(\mathbf{b}-\mathbf{c})} , \tag{6.20a}$$

and

$$\mathbf{x} = \frac{\mathbf{c}}{1+\lambda'} . \tag{6.20b}$$

These equations give us the point of intersection in terms of the vectors **a**, **b**, **c**. They also determine the point **x** in Figure 6.5, and, by interchanging **a** and **c**, they determine the point **y** in the same figure. Thus, a small number of algebraic formulas describe a wide variety of geometric relations.

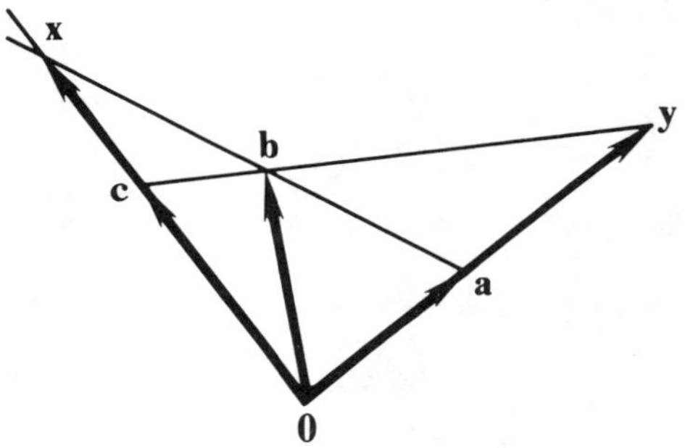

Fig. 6.5.

Planes

The algebraic description of a plane is similar to that for a line, so we consider planes only briefly to make this fact clear. The plane with *direction* **U** passing through a given point **a** is determined by the equation

$$(\mathbf{x}-\mathbf{a})\wedge\mathbf{U} = 0, \tag{6.21}$$

where **U** is a non-zero 2-blade. Planes with equal or opposite directions are said to be *parallel* to one another. Every plane $\mathcal{P} = \{\mathbf{x}\}$ with direction **U** is the solution set of an equation with the form

$$\mathbf{x}\wedge\mathbf{U} = \mathbf{T}. \tag{6.22}$$

This equation is analogous to (6.21) when $\mathbf{T} = \mathbf{a}\wedge\mathbf{U}$. The trivector **T** is called the moment of the plane. The *directance* **d** from the origin to $\mathcal{P}$ is given by the vector

$$\mathbf{d} = \mathbf{T}\mathbf{U}^{-1} = \mathbf{T}\cdot\mathbf{U}^{-1}. \tag{6.23}$$

The magnitude $|\mathbf{d}| = |\mathbf{T}||\mathbf{U}|^{-1}$ is the distance from the origin to the plane. Finally, using (6.22) in (6.23) we get

$$\mathbf{d} = \mathbf{x}\wedge\mathbf{U}\mathbf{U}^{-1} \equiv P_{\mathbf{U}}^{\perp}(\mathbf{x}). \tag{6.24}$$

Thus the directance can be obtained by "rejecting" any point of the plane.

Spheres and Circles

Besides lines and planes, the most elementary geometrical figures are circles and spheres. A *sphere* with radius r and center **c** is defined as the set of all points **x** in $\mathcal{E}_3$ satisfying the equation

$$|\mathbf{x}-\mathbf{c}| = r, \tag{6.25a}$$

or, equivalently,

$$(\mathbf{x}-\mathbf{c})^2 = r^2. \tag{6.25b}$$

Besides (6.25), the points of the sphere satisfy the equation $(\mathbf{x} - \mathbf{c})\wedge i = 0$, which is an algebraic formulation of the condition that each point **x** belongs to $\mathcal{E}_3$.

The intersection of the sphere with a plane through its center is a *circle*. Thus, a circle with center **c** and radius **a** in the **i**-plane is determined by supplementing (6.25) with the condition

$$(\mathbf{x}-\mathbf{c})\wedge\mathbf{i} = 0. \tag{6.26}$$

Equation (6.25) can be regarded as the equation for a circle without explicitly writing (6.26) if it is understood that the points are in $\mathcal{E}_2$.

The general solution of the simultaneous Equations (6.25) and (6.26) has the convenient parametric form

$$\mathbf{x}-\mathbf{c} = \mathbf{r}_0 e^{\mathbf{i}\theta}, \tag{6.27}$$

where $\mathbf{r}_0$ is a fixed vector in the **i**-plane ($\mathbf{r}_0\wedge\mathbf{i} = 0$) with magnitude $|\mathbf{r}_0| = r$. The fact that (6.27) solves (6.25) and (6.26) is readily verified by substitution. With the bivector **i** normalized to unity, (6.27) associates exactly one value of θ with each point **x** on the circle if the values of θ are restricted by the condition $0 \leqslant \theta < 2\pi$. The generalization of (6.27) to a parametric equation for a sphere will be made later in the chapter on rotations. We concentrate here on general properties of circles. Equation (6.27) is only one of many useful parametric equations for a circle. Equation (6.27) is not very helpful when one is concerned only with points on the circle and not with the center of the circle. A more useful parametric equation can be derived from the *constant angle theorem* for a circle. This theorem states that a given arc of a circle subtends the same angle ϕ at every point $\mathbf{x}$ on the circle outside that arc. We prove the theorem by showing that $\phi = \frac{1}{2}\theta$ for the angles indicated in Figure 6.6. According to the figure,

$$\mathbf{a}^{-1}\mathbf{b} = e^{\mathrm{i}\theta}.$$

Our arguments will apply to all circles through the points $\mathbf{a}$, $\mathbf{b}$ if we allow θ to have any value in the interval $[-\pi, \pi]$. The angle of interest ϕ is defined by the equation

$$(\mathbf{a}-\mathbf{x})^{-1}(\mathbf{b}-\mathbf{x}) = \lambda e^{\mathrm{i}\phi}. \tag{6.28}$$

Writing $\mathbf{a}^{-1}\mathbf{x} = e^{\mathrm{i}\alpha}$ for convenience, we observe that

$$\mathbf{a}^{-1}(\mathbf{b}-\mathbf{x}) = e^{\mathrm{i}\theta} - e^{\mathrm{i}\alpha}$$

and

$$(\mathbf{a}-\mathbf{x})^{-1}\mathbf{a} = [\mathbf{a}^{-1}(\mathbf{a}-\mathbf{x})]^{-1} = (1-e^{\mathrm{i}\alpha})^{-1}.$$

Hence

$$(\mathbf{a}-\mathbf{x})^{-1}(\mathbf{b}-\mathbf{x}) = (\mathbf{a}-\mathbf{x})^{-1}\mathbf{a}\mathbf{a}^{-1}(\mathbf{b}-\mathbf{x}) = \frac{e^{\mathrm{i}\theta}-e^{\mathrm{i}\alpha}}{1-e^{\mathrm{i}\alpha}}.$$

Inserting this in (6.28), we get

$$\frac{e^{\mathrm{i}\theta}-e^{\mathrm{i}\alpha}}{1-e^{\mathrm{i}\alpha}} = \lambda e^{\mathrm{i}\phi}.$$

We can eliminate λ by dividing this equation by its reverse, and we find that α disappears as well; thus,

$$e^{\mathrm{i}2\phi} = \frac{e^{\mathrm{i}\theta}-e^{\mathrm{i}\alpha}}{1-e^{\mathrm{i}\alpha}}\,\frac{1-e^{-\mathrm{i}\alpha}}{e^{-\mathrm{i}\theta}-e^{-\mathrm{i}\alpha}} = e^{\mathrm{i}\theta}.$$

To solve for ϕ, we must consider the two square roots

$$e^{\mathrm{i}\phi} = \pm\, e^{\mathrm{i}\theta/2}. \tag{6.29}$$

This gives us two different values for ϕ, which we denote by ϕ and ϕ' respectively.

The positive root from (6.29) gives $\phi = \frac{1}{2}\theta$, as claimed earlier. Since this result is independent of α, it holds for every point $\mathbf{x}$ on the circle outside the arc. This completes the proof of the constant angle theorem, but we can deduce more. The negative root from (6.29) gives $\phi' = \frac{1}{2}\theta - \pi$ if $\theta > 0$ and $\phi' = \frac{1}{2}\theta + \pi$ if $\theta < 0$. This relation holds for every point $\mathbf{x}'$ *on* the given arc of the circle, as indicated in Figure 6.6. Angles for the two cases are obviously related by $\phi' = \phi \pm \pi$, or equivalently by

$$e^{i\phi'} = \pm e^{i\phi}. \tag{6.30}$$

From this we can conclude that (6.28) applies with fixed ϕ to *all* points on the circle if the parameter λ is allowed to have negative values.

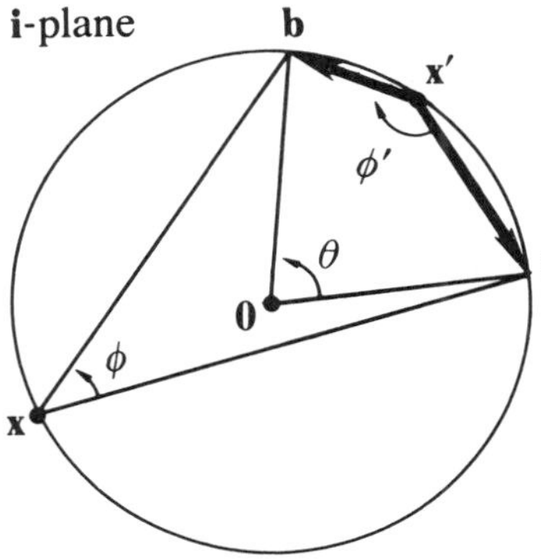

Fig. 6.6.

We have, in fact, proved that (6.28) can be regarded as a parametric equation for a circle in the **i**-plane passing through the points **a** and **b**. Points on the circle are distinguished by the signed ratio of their distances from **a** and **b**,

$$\lambda = \pm \frac{|\mathbf{x} - \mathbf{b}|}{|\mathbf{x} - \mathbf{a}|}, \tag{6.31}$$

The two signs of λ corresponding to the two arcs into which the circle is cut by **a** and **b**. Each finite value of λ determines a distinct point of the circle, while the singular values $\lambda = \pm\infty$ determine the point $\mathbf{x} = \mathbf{a}$.

Equation (6.28) helps us answer many questions about circles. For example, to find an equation for the circle passing through distinct points **a**, **b**, **d**, we write

$$(\mathbf{a} - \mathbf{d})^{-1}(\mathbf{b} - \mathbf{d}) = \delta e^{i\phi}. \tag{6.32}$$

This determines the angle ϕ in (6.28). Taking the ratio of (6.28) to (6.32) we have the desired parametric equation,

$$\frac{(\mathbf{a} - \mathbf{x})^{-1}(\mathbf{b} - \mathbf{x})}{(\mathbf{a} - \mathbf{d})^{-1}(\mathbf{b} - \mathbf{d})} = \frac{\lambda}{\delta} \equiv \lambda'. \tag{6.33}$$

The parameter λ' has the values $\pm\infty$, 0, 1 at the points $\mathbf{x} = \mathbf{a}, \mathbf{b}, \mathbf{d}$ respectively.

The quantity on the left of (6.33) is called the *cross ratio* of the points **a**, **b**, **x**, **d**. It is well-defined for any four distinct points. From our derivation of (6.33) we can conclude that four distinct points lie on a circle if and only if their cross ratio is a scalar.

Returning to (6.28), we observe that each value of ϕ in the interval $(-\pi/2, \pi/2)$ determines a distinct circle, with the value $\phi = 0$ determining a straight line, which may be regarded as a circle passing through infinity. Thus, (6.28) describes the 1-parameter family of circles in the **i**-plane which pass through the points **a** and **b**, as shown in Figure 6.7. On the other hand, if λ is fixed and positive while ϕ varies, then (6.28) describes the set of all points in the **i**-plane whose distances from **a** and **b** have the fixed ratio λ. This set is also a circle, called the *circle of Appolonius*. By varying λ we get the 1-parameter family of all such circles (Figure 6.8). As shown in Figure 6.9, the circle with constant ϕ intersects the circle with constant λ in two points **x** and **x**$'$ distinguished by the

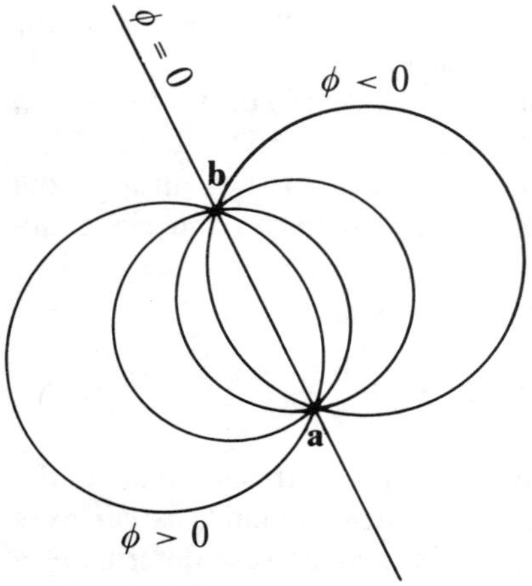

Fig. 6.7. The 1-parameter family of circles through points **a** and **b**.

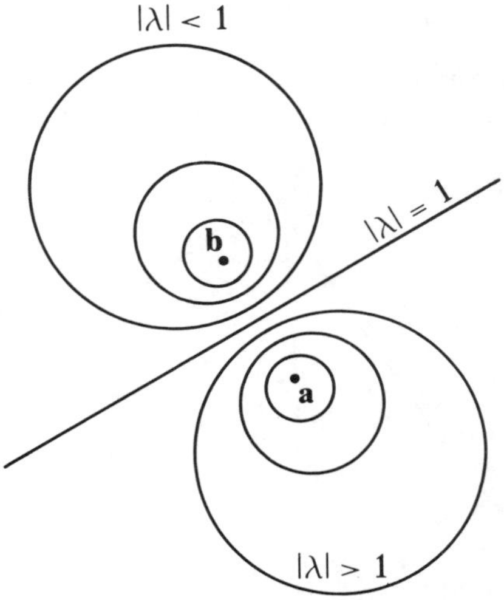

Fig. 6.8. Circles of Appolonius.

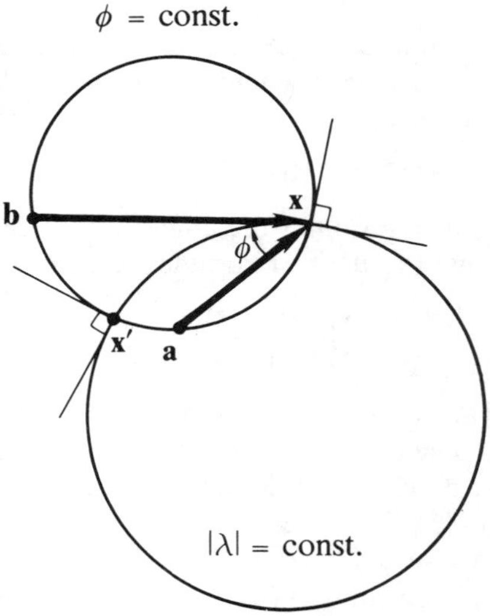

Fig. 6.9.

values $+|\lambda|$ and $-|\lambda|$ respectively. Each point in the **i**-plane can be designated in this way, so (6.28) can be regarded as a parametric equation for the **i**-plane. The parameters λ and ϕ are then called *bipolar coordinates* for the plane.

Conic Sections

Next to straight lines and circles, the simplest curves are the *conic sections*, so-called because each can be defined as the intersection of a cone with a plane. We shall prefer the following alternative definition, because it leads directly to a most valuable parametric equation: A *conic* is the set of all points in the Euclidean plane $\mathscr{E}_2$ with the property that the distance of each point from a fixed point (the *focus*) is in fixed ratio (the *eccentricity*) to the distance of that point from a fixed line (the *directrix*). To express this as an equation, we denote the eccentricity by ε, the directance from the focus to the directrix by $\mathbf{d} = d\hat{\boldsymbol{\varepsilon}}$ with $\hat{\boldsymbol{\varepsilon}}^2 = 1$, and the directance from the focus to any point on the conic by $\mathbf{r}$ (see Figure 6.10). The defining condition for a conic can then be written

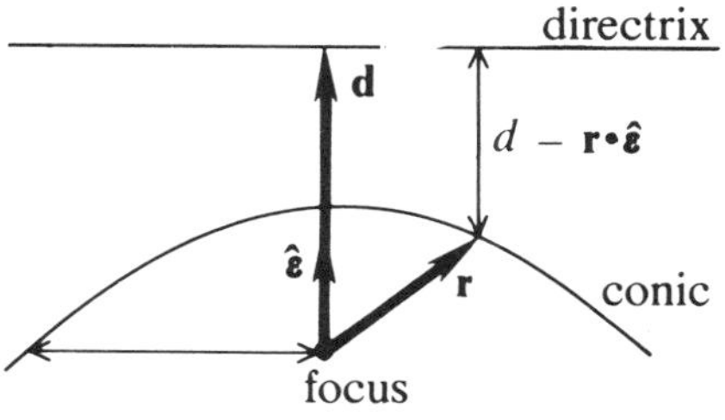

Fig. 6.10.

$$\frac{|\mathbf{r}|}{d - \mathbf{r}\cdot\hat{\boldsymbol{\varepsilon}}} = \varepsilon. \qquad (6.34)$$

Solving this for $r = |\mathbf{r}|$ and introducing the *eccentricity vector* $\boldsymbol{\varepsilon} = \varepsilon\hat{\boldsymbol{\varepsilon}}$ along with the so-called *semi-latus rectum* $\ell = \varepsilon d$, we get the more convenient equation

$$r = \frac{\ell}{1 + \boldsymbol{\varepsilon}\cdot\hat{\mathbf{r}}}\,. \qquad (6.35)$$

This expresses the distance r from the focus to a point on the conic as a function of the direction $\hat{\mathbf{r}}$ to the point. Alternatively, the same condition can be expressed as a parametric equation for r as a function of the angle θ between $\boldsymbol{\varepsilon}$ and $\hat{\mathbf{r}}$. Thus, substituting $\boldsymbol{\varepsilon}\cdot\hat{\mathbf{r}} = \varepsilon\cos\theta$ into (6.35), we get

$$r = \frac{\ell}{1 + \varepsilon\cos\theta}\,. \qquad (6.36)$$

This is a standard equation for conics, but we usually prefer (6.35), because it shows the dependence of r on the directions $\hat{\boldsymbol{\varepsilon}}$ and $\hat{\mathbf{r}}$ explicitly, while this dependence in (6.36) is expressed only indirectly through the definition of θ.

Equation (6.35) determines a curve when r is restricted to directions in a plane, but if r is allowed to range over all directions in $\mathcal{E}_3$, then (6.35) describes a 2-dimensional surface called a *conicoid*. Our definition of a conic can be used for a conicoid simply by interpreting the directrix as a plane instead of a line. Both the conics and the conicoids are classified according to the values of the eccentricity as shown in Table 6.1.

TABLE 6.1. Classification of Conics and Conicoids.

Eccentricity	Conic	Conicoid
$\varepsilon > 1$	hyperbola	hyperboloid
$\varepsilon = 1$	parabola	paraboloid
$0 < \varepsilon < 1$	ellipse	ellipsoid
$\varepsilon = 0$	circle	sphere

The 1-parameter family of conics with a common focus and pericenter is illustrated in Figure 6.11. The *pericenter* is the point on the conic at which r has a minimum value. In the hyperbolic case there are actually two pericenters, one on each branch of the hyperbola. Only one of these is shown in Figure 6.11. If the conics in Figure 6.11 are rotated about the *axis* through the focus and pericenter, they "sweep out" corresponding conicoids.

The conics and conicoids have quite a remarkable variety of properties, which is related to the fact that they can be described by many different equations besides (6.35). Rather than undertake a systematic study of those properties, we shall wait for them to arise in the context of physical problems, and we will be better prepared for this when we have the tools of differential calculus at our disposal.

Our study of analytic geometry has just begun. The study of particle trajectories, which we undertake in the next chapter, is largely analytic geometry in $\mathcal{E}_3$. For those who wish to study the classical analytic geometry in $\mathcal{E}_2$ in more detail, the book of Zwikker (1963) is recommended. He formulates analytic geometry in terms of complex numbers and shows how much this improves on the traditional methods of coordinate geometry. Of course, everything he does is easily reexpressed in the language of geometric algebra, which has all the advantages of complex numbers and more. Indeed, geometric algebra brings further improvements to Zwikker's treatment by enlarging the algebraic system from $\mathcal{G}_2^+$ to $\mathcal{G}_2$, and so introducing the fundamental distinction between vectors and spinors and along with it the concepts of inner and outer products. Most important, geometric algebra provides for the generalization of the geometry in $\mathcal{E}_2$ to $\mathcal{E}_3$. The present book develops all the principles and techniques needed for analytic geometry, but Zwikker's book is a valuable storehouse of particular facts about curves in $\mathcal{E}_2$. Among other things, it includes the remarkable proof that conic sections as defined by (6.34) really are sections of a cone.

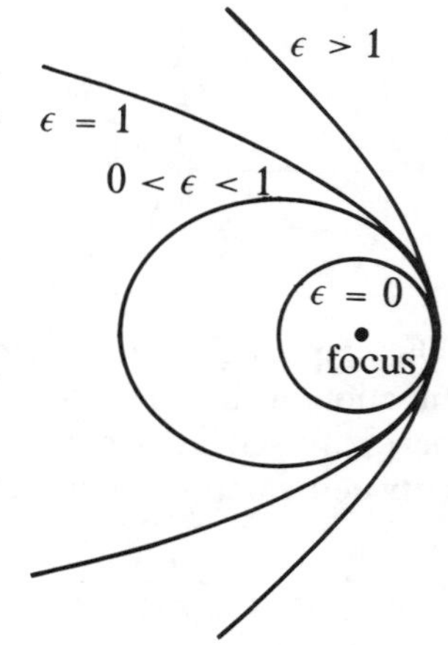

Fig. 6.11. Conics with a common focus and pericenter.

2-6. Exercises

(6.1) From Equation (6.2) derive the following equations for the line $\mathcal{L}$ in terms of rectangular coordinates in $\mathcal{E}_3$:

$$\frac{x_1 - a_1}{u_1} = \frac{x_2 - a_2}{u_2} = \frac{x_3 - a_3}{u_3},$$

where $x_k = \mathbf{x}\cdot\boldsymbol{\sigma}_k$, $a_k = \mathbf{a}\cdot\boldsymbol{\sigma}_k$, $u_k = \mathbf{u}\cdot\boldsymbol{\sigma}_k$.

(6.2) (a) Show that Equation (6.2) is equivalent to the parametric equation

$$\mathbf{x} = \mathbf{a} + \lambda\mathbf{u}^{-1}.$$

(b) Describe the solution set $\{\mathbf{x} = \mathbf{x}(t)\}$ of the parametric equation

$$\mathbf{x} = \mathbf{a} + t^2\mathbf{u}$$

for all scalar values of the parameter t.

(6.3) (a) Compute the directance to the line through points **a** and **b** from the origin.
(b) Compute the directance to this line from an arbitrary point **c**.

(6.4) Prove the theorem "Three points not on a line determine a plane" by using geometric algebra to derive an equation for the plane from three points **a**, **b**, **c**.

(6.5) Describe the solution set $\{x\}$ of the simultaneous equations

$$\mathbf{x}\wedge\mathbf{A} = 0, \qquad \mathbf{x}\wedge\mathbf{B} = 0$$

if **A** and **B** are noncommuting blades of grade 2.

(6.6) Find the point of intersection of the line $\{\mathbf{x}\}$ determined by the equation $(\mathbf{x} - \mathbf{a})\wedge\mathbf{u} = 0$ with the plane $\{\mathbf{y}\}$ determined by the equation $(\mathbf{y} - \mathbf{b})\wedge\mathbf{B} = 0$. What are the conditions on **a**, **b**, **u**, **B** that this point exists and is unique?

(6.7) The *directance* from one point set to another can be defined quite generally as the chord of minimum length between points in the two sets, provided there is only one such chord.

Determine the directance **d** from a line with direction **u** through a point **a** to a line with direction **v** through a point **b**. Show that the lines intersect only if $(\mathbf{a} - \mathbf{b})\wedge\mathbf{u}\wedge\mathbf{v} = 0$.

(6.8) Compute the directance from a point **b** to the plane $\{\mathbf{x}\colon (\mathbf{x} - \mathbf{a})\wedge\mathbf{U} = 0\}$.

(6.9) Show that the equation

$$\mathbf{x} = \alpha\mathbf{a} + \beta\mathbf{b} + \gamma\mathbf{c}$$

subject to the conditions

$$\alpha + \beta + \gamma = 1 \quad \text{and} \quad \mathbf{a}\wedge\mathbf{b}\wedge\mathbf{c} \neq 0$$

can be regarded as a parametric equation for a plane. Find a nonparametric equation for this plane.

(6.10) *Ceva's Theorem*: Suppose that concurrent lines from the verticies **a**, **b**, **c** of a triangle divide the opposing sides at **a′**, **b′**, **c′** (Figure 6.12). Then the division ratios satisfy

$$\left(\frac{\mathbf{a} - \mathbf{c}'}{\mathbf{c}' - \mathbf{b}}\right)\left(\frac{\mathbf{b} - \mathbf{a}'}{\mathbf{a}' - \mathbf{c}}\right)\left(\frac{\mathbf{c} - \mathbf{b}'}{\mathbf{b}' - \mathbf{a}}\right) = 1.$$

Prove by showing that the areas indicated in the figure satisfy

$$\frac{\mathbf{A}_1}{\mathbf{A}_2}\frac{\mathbf{B}_1}{\mathbf{B}_2}\frac{\mathbf{C}_1}{\mathbf{C}_2} = 1.$$

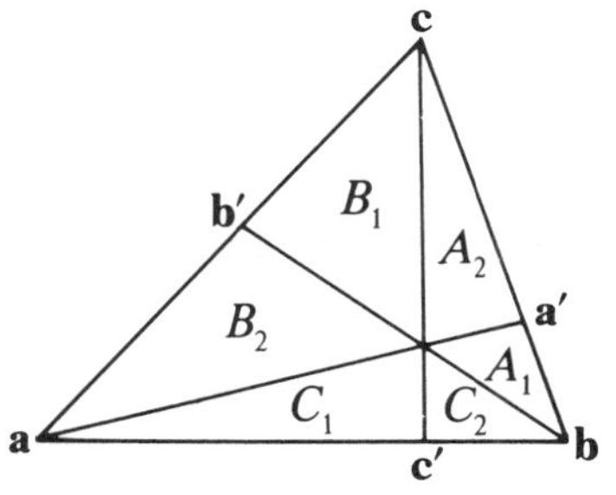

Fig. 6.12. *Ceva's Theorem.*

(6.11) In Figure 6.5 we have the division ratios

$$\lambda = \frac{\mathbf{x} - \mathbf{b}}{\mathbf{b} - \mathbf{a}}, \qquad \mu = \frac{\mathbf{c} - \mathbf{0}}{\mathbf{x} - \mathbf{c}}, \qquad \nu = \frac{\mathbf{a} - \mathbf{y}}{\mathbf{y} - \mathbf{0}}.$$

Prove the *theorem of Menelaus*: $\lambda\mu\nu = -1$. Note that the theorem can be interpreted as expressing a relation among intersecting sides of a quadrilateral with verticies **0**, **a**, **b**, **c** or as a property of a transversal cutting the triangle with verticies **0**, **a**, **x**.

(6.12) Prove that three points **a**, **b**, **c** lie on a line iff there exist nonzero scalars α, β, γ such that $\alpha\mathbf{a} + \beta\mathbf{b} + \gamma\mathbf{c} = 0$ and $\alpha + \beta + \gamma = 0$.

(6.13) *Desargues' Theorem*. Given two triangles **a**, **b**, **c** and **a**′, **b**′, **c**′. Then lines through corresponding verticies are concurrent (at a point **s**) iff lines along corresponding sides intersect at collinear points (**p**, **q**, **r**). (Figure 6.13) Note that the triangles need not lie in the same plane.

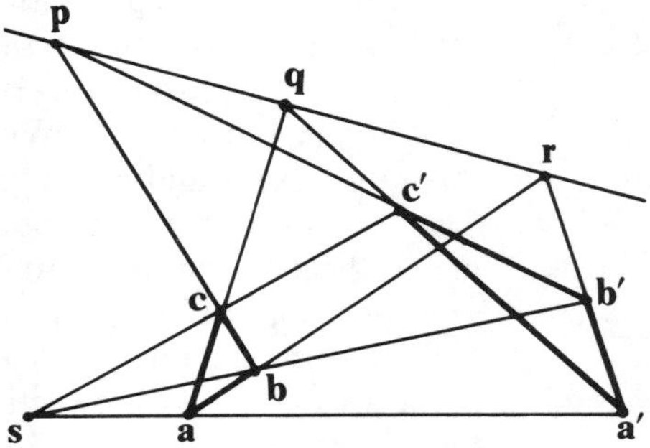

Fig. 6.13. *Desargue's Theorem.*

(6.14) The equation $(\mathbf{x} - \mathbf{b})\cdot\mathbf{u} = 0$ describes a plane in $\mathcal{E}_3$ with normal **u**. Derive this equation from Equation (6.21).

(6.15) Four points **a**, **b**, **c**, **d** determine a tetrahedron with directed volume

$$\begin{aligned} V &= \tfrac{1}{6}\,(\mathbf{b} - \mathbf{a})\wedge(\mathbf{c} - \mathbf{a})\wedge(\mathbf{d} - \mathbf{a}) \\ &= \tfrac{1}{6}\,(\mathbf{b}\wedge\mathbf{c}\wedge\mathbf{d} - \mathbf{c}\wedge\mathbf{d}\wedge\mathbf{a} + \mathbf{d}\wedge\mathbf{a}\wedge\mathbf{b} - \mathbf{a}\wedge\mathbf{b}\wedge\mathbf{c}). \end{aligned}$$

Use this to determine the equation for a plane through three distinct points **a**, **b**, **c**.

(6.16) Let **a**, **b**, **c** be the directions of three coplanar lines. The relative directions of the lines are then specified by $\alpha = \mathbf{b}\cdot\mathbf{c}$, $\beta = \mathbf{a}\cdot\mathbf{c}$, $\gamma = \mathbf{a}\cdot\mathbf{b}$. Prove that

$$2\alpha\beta\gamma = \alpha^2 + \beta^2 + \gamma^2 - 1.$$

(6.17) Determine the parametric values λ_1, λ_2 for which the line $\mathbf{x} = \mathbf{x}(\lambda) = \mathbf{a} + \lambda\mathbf{u}$ ($\mathbf{u}^2 = 1$) intersects the circle with equation $\mathbf{x}^2 = r^2$, and show that $\lambda_1\lambda_2 = a^2 - r^2$ for every line through **a** which intersects the circle.

(6.18) Show that tangents to the circle of radius r and center at the origin in $\mathcal{E}_2$ which pass through a given point **a** intersect the circle at the points.

$$\mathbf{d}_\pm = \left(1 + \frac{\alpha_\pm}{r}\mathbf{i}\right)^{-1}\mathbf{a} = (r - \alpha_\pm\mathbf{i})r\mathbf{a}^{-1},$$

where $\alpha_{\pm} = \pm(a^2 - r^2)^{1/2}$ and $\mathbf{i}$ is the unit bivector.

(6.19) Find the radius r and center $\mathbf{c}$ of the circle determined by Equation (6.28).

(6.20) Let x and y be rectangular coordinates of the point $\mathbf{x}$. Show that the defining equation (6.35) of a conic is equivalent to the equations

$$\frac{x^2}{a^2} + \frac{y^2}{b^2} = 1\,, \quad \frac{x^2}{a^2} - \frac{y^2}{b^2} = 1\,,$$

for an ellipse and a hyperbola respectively, where

$$a = \frac{\ell}{|1 - \varepsilon^2|}\,, \qquad b^2 = a\ell, \qquad \mathbf{x} = \mathbf{r} + a\boldsymbol{\varepsilon}.$$

The curves and related parameters are shown in Figures 6.14a, b.

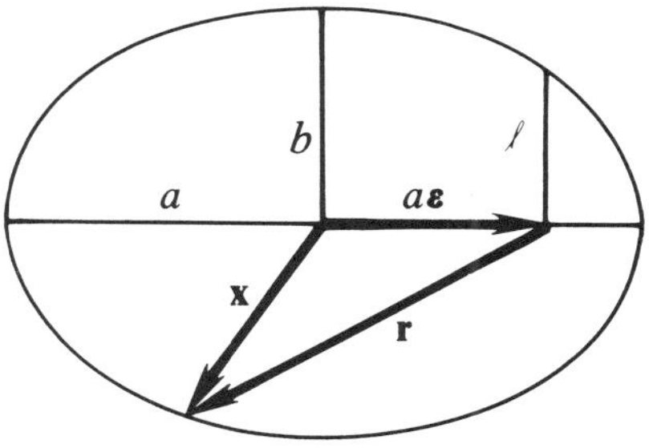

Fig. 6.14a. Ellipse.

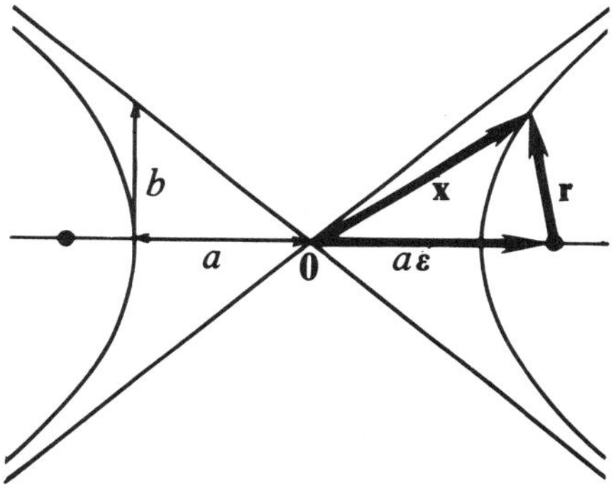

Fig. 6.14b. Hyperbola.

Use the above equations to show that an ellipse has a parametric equation $\mathbf{x} = \mathbf{x}(\phi)$ with the explicit form

$$\mathbf{x} = \mathbf{a} \cos \phi + \mathbf{b} \sin \phi,$$

while a hyperbola has the parametric equation

$$\mathbf{x} = \mathbf{a} \cosh \phi + \mathbf{b} \sinh \phi,$$

where $\mathbf{a}^2 = a^2$, $\mathbf{b}^2 = b^2$ and $\mathbf{a}\cdot\mathbf{b} = 0$.

(6.21) Parametric curves $\mathbf{x} = \mathbf{x}(\lambda)$ of the second order are defined by equation

$$\mathbf{x} = \frac{\mathbf{a}_0 + \mathbf{a}_1\lambda + \mathbf{a}_2\lambda^2}{\alpha_0 + \alpha_1\lambda + \alpha_2\lambda^2}\,.$$

Note that this generalizes the Equation (6.16) for a line. By the change of parameters $\lambda \to \lambda - \alpha_1/2\alpha_2$, this can be reduced to the form

$$\mathbf{x} = \frac{\mathbf{a}_0 + \mathbf{a}_1\lambda + \mathbf{a}_2\lambda^2}{\alpha + \lambda^2} .$$

We now aim to show that this equation describes an *ellipse* iff $\alpha > 0$, a *parabola* iff $\alpha = 0$, and a *hyperbola* iff $\alpha < 0$, where iff means "if and only if". Thus, all conics are second order curves and conversely. Show that

(a) For $\alpha = 1$, the change of parameters $\lambda = \tan \frac{1}{2}\phi$ enables us to put the equation in the form

$$\mathbf{x} = \mathbf{a} \cos \phi + \mathbf{b} \sin \phi + \mathbf{c},$$

which we recognize as a general equation for an ellipse.

(b) For $\alpha = -1$, $\lambda = \tanh \frac{1}{2}\phi$ gives

$$\mathbf{x} = \mathbf{a} \cosh \phi + \mathbf{b} \sinh \phi + \mathbf{c}.$$

(6.22) Solve Equation (6.28) for $\mathbf{x}$ and put it in the general form given in exercise (6.21).

(6.23) Let $\mathbf{x} = \mathbf{x}(\lambda) = \boldsymbol{\sigma}_1 z(\lambda)$ describe a curve in $\mathcal{E}_2$. Identify and draw diagrams of the curves determined by the following specific forms for the spinor z.

(a) $z = (1 + \mathbf{i}\lambda)^{1/2}$

(b) $z = \dfrac{1}{1 - \mathbf{i}\lambda}$

(c) $z = (1 - \mathbf{i}\lambda)\, e^{\mathbf{i}\lambda}$

(d) $z = (1 - \mathbf{i}\lambda)^{-1/2}$.

(6.24) Describe the solution set $\{\mathbf{x}\}$ in $\mathcal{E}_3$ determined by the following equations. Comment, especially, on the dependence of the solution set on vector parameters $\mathbf{a}$, $\mathbf{b}$, $\mathbf{c}$.

(a) $(\mathbf{a}\cdot\mathbf{x})^2 = \mathbf{x}^2$.

(b) $\mathbf{a}\cdot\mathbf{x} \geqslant |\,\mathbf{x}\,|$.

(c) $\langle(\mathbf{ax})^2\rangle_0 = 0$; $\qquad \langle(\mathbf{ax})^2\rangle_2 = 0$.

(d) $\langle(\mathbf{ax})^2\rangle_0 = 1$.

(e) $\mathbf{x}\cdot\mathbf{a} + (\mathbf{x}\wedge\mathbf{a})^2 = 0$.

(f) $(\mathbf{a}\cdot\mathbf{x})^2 = \mathbf{x}^2$ and $\quad (\mathbf{x} - \mathbf{c})\cdot\mathbf{b} = 0$.

2-7. Functions of a Scalar Variable

In this section we review some basic concepts of differential and integral calculus to show how they apply to multivector-valued functions.

If to each value of a scalar variable t there corresponds a multivector $F(t)$, then $F = F(t)$ is said to be a (multivector-valued) *function* of t. It is important to distinguish between the function F and the functional value $F(t_0)$, the particular multivector which corresponds to the particular scalar t_0. However, it is often inconvenient to make that distinction explicit in the mathematical notation, so the reader will be left to infer it from the context. Thus, $F(t)$ will denote a functional value if t is understood to be a specific real number, but $F(t)$ will denote a function if no specific value is attributed to t.* Similarly, when the variable t is suppressed, F may indicate a value of the function instead of the function itself. So $F = F(t)$ may refer either to a function or a functional value. It should be understood, also, that the function $F(t)$ is not completely defined until the values of the variable t for which it is defined have been specified. However, the reader will usually be left to infer the allowed values of a variable from the context.

Continuity

The function $F(t)$ is said to be *continuous* at t_0 if

$$\lim_{t \to t_0} | F(t) - F(t_0) | = 0. \tag{7.1a}$$

We write

$$\lim_{t \to t_0} [F(t) + G(t)] \quad \text{or} \quad F(t) \to F(t_0). \tag{7.1b}$$

The definition of "limit" presumed in (7.1a) is the same as the one introduced in elementary calculus. It applies to multivectors, because we have already introduced an appropriate definition of the "absolute value" $| F(t) |$, and, in spite of the fact that the geometric product is not commutative, it can be proved that for multivector-valued functions $F(t)$ and $G(t)$ we have the elementary results

$$\lim_{t \to t_0} [F(t) + G(t)] = \lim_{t \to t_0} F(t) + \lim_{t \to t_0} G(t), \tag{7.2a}$$

$$\lim_{t \to t_0} F(t)G(t) = \lim_{t \to t_0} F(t) \lim_{t \to t_0} G(t). \tag{7.2b}$$

To this we can add the (almost trivial) result

$$\langle \lim_{t \to t_0} F(t)\rangle_k = \lim_{t \to t_0} \langle F(t)\rangle_k . \tag{7.2c}$$

It follows, then, that the function F is continuous if its k-vector parts $\langle F\rangle_k$ are continuous functions.

*It may be noted that a variable is itself a function. Given a set, the *variable* on that set is just the identity function, namely that function which associates each element of the set with itself. So $F(t)$ can be interpreted as the composite of two functions F and t.

Scalar differentiation

The *derivative* of the function $F = F(t)$ at the point t_0 is denoted by $\dot{F} = \dot{F}(t_0)$ or $\mathrm{d}F/\mathrm{d}t = \mathrm{d}F/\mathrm{d}t\ (t_0)$ and defined by

$$\dot{F}(t_0) = \frac{\mathrm{d}F(t_0)}{\mathrm{d}t} \equiv \lim_{\Delta t \to 0} \frac{F(t_0 + \Delta t) - F(t_0)}{\Delta t}\,. \tag{7.3}$$

This will sometimes be referred to as a *scalar derivative* to emphasize that the variable is a scalar and to distinguish it from the vector derivative to be defined in a subsequent volume, NFII. Unless there is some reason to believe otherwise, it will usually be convenient to make the tacit assumption that functions we deal with have derivatives that "exist" in the mathematical sense. Such functions are said to be *differentiable*.

The derivative of $F = F(t)$ is itself a function $\dot{F} = \dot{F}(t)$, so we can contemplate its derivative. This is called the second derivative and denoted by $\ddot{F} = \mathrm{d}^2F/\mathrm{d}t^2$. Similarly, derivatives of higher order are defined as in elementary calculus.

We will be particularly interested in curves representing paths (trajectories) of physical particles. Such a curve is described by a parametric equation $\mathbf{x} = \mathbf{x}(t)$, a vector-valued function of the time t. The derivative $\dot{\mathbf{x}} = \dot{\mathbf{x}}(t)$ is called the *velocity* of the particle; it is, of course, defined by (7.3), which we can put in the abbreviated form

$$\dot{\mathbf{x}} = \frac{\mathrm{d}\mathbf{x}}{\mathrm{d}t} = \lim_{\Delta t \to 0} \frac{\Delta \mathbf{x}}{\Delta t}\,. \tag{7.4}$$

The curve and vectors involved in the derivative are shown in Figure 7.1. The derivative of the velocity,

$$\ddot{\mathbf{x}} = \frac{\mathrm{d}^2\mathbf{x}}{\mathrm{d}t^2} = \lim_{\Delta t \to 0} \frac{\Delta \dot{\mathbf{x}}}{\Delta t}\,, \tag{7.5}$$

is called the *acceleration* of the particle.

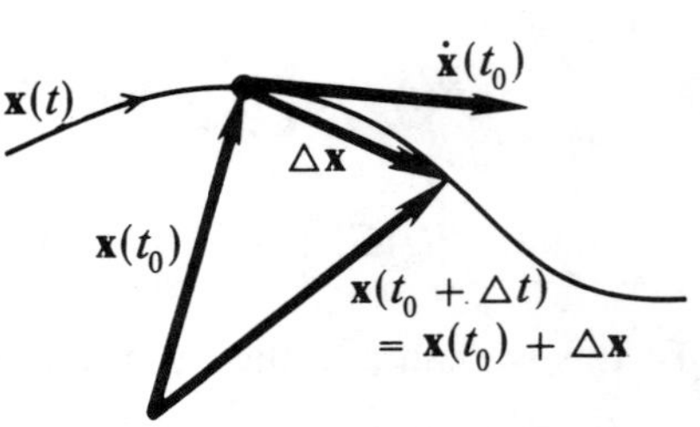

Fig. 7.1.

From multivector-valued functions $F = F(t)$ and $G = G(t)$ we can form new functions by addition and multiplication. Their derivatives are subject to the rules

$$\frac{\mathrm{d}}{\mathrm{d}t}(F + G) = \dot{F} + \dot{G}, \tag{7.6a}$$

$$\frac{\mathrm{d}}{\mathrm{d}t}(FG) = \dot{F}G + F\dot{G}, \tag{7.6b}$$

$$\frac{\mathrm{d}}{\mathrm{d}t}\langle F\rangle_k = \langle \dot{F}\rangle_k\,. \tag{7.6c}$$

These rules can be derived from (7.2a, b, c) by arguments of elementary calculus. The proof of (7.6c) depends in addition on the axioms of geometric algebra. It must be realized that (7.6b) differs from the result in elementary calculus by requiring that the order of factors in the product, be retained. In particular, (7.6b) implies

$$\frac{\mathrm{d}F^2}{\mathrm{d}t} = \frac{\mathrm{d}}{\mathrm{d}t}(FF) = \dot{F}F + F\dot{F}\,. \tag{7.7}$$

The right side of (7.7) is equal to the elementary result $2F\dot{F}$ only if F commutes with $\dot{F}$. To understand the significance of this deviation from the elementary result, consider the special case of a vector function $\mathbf{v} = \mathbf{v}(t)$ which might be, for example, the velocity of a particle. Now $\mathbf{v}^2 = v^2$, where $v = |\,\mathbf{v}\,|$, so from (7.7) we get

$$\frac{\mathrm{d}\mathbf{v}^2}{\mathrm{d}t} = 2v\dot{v} = \dot{\mathbf{v}}\mathbf{v} + \mathbf{v}\dot{\mathbf{v}} = 2\dot{\mathbf{v}}\cdot\mathbf{v}\,. \tag{7.8}$$

With a change in t, the vector $\mathbf{v} = v\hat{\mathbf{v}}$ can undergo changes in magnitude v and in direction $\hat{\mathbf{v}}$. If v is constant (i.e. constant speed), then (7.8) implies

$$\dot{\mathbf{v}}\cdot\mathbf{v} = 0\,, \tag{7.9a}$$

or equivalently,

$$\dot{\mathbf{v}}\mathbf{v} = -\mathbf{v}\dot{\mathbf{v}}\,. \tag{7.9b}$$

Thus a vector which undergoes changes in direction only is orthogonal to its derivative. On the other hand, if the direction $\hat{\mathbf{v}}$ is constant, then $\dot{\mathbf{v}} = \dot{v}\hat{\mathbf{v}}$, and (7.8) reduces to

$$\frac{\mathrm{d}\mathbf{v}^2}{\mathrm{d}t} = 2v\dot{v} = 2\mathbf{v}\dot{\mathbf{v}} = 2\dot{\mathbf{v}}\cdot\mathbf{v}\,. \tag{7.10}$$

We see, then, that (7.7) need not reduce to the equation $\mathrm{d}F^2/\mathrm{d}t = 2F\dot{F}$ if F changes in direction, but it will if F changes in magnitude only. Continuous changes in direction are not considered in elementary calculus, which deals with scalar-valued functions only; changes in direction of scalar-valued functions are limited to changes in sign.

Constant Magnitude

It is now merely an algebraic exercise to show that the vector-valued function $\mathbf{v} = \mathbf{v}(t)$ has constant magnitude if and only if there exists a bivector-valued function $\boldsymbol{\Omega} = \boldsymbol{\Omega}(t)$ such that

$$\dot{\mathbf{v}} = \mathbf{v}\cdot\boldsymbol{\Omega}\,. \tag{7.11}$$

Expressing $\boldsymbol{\Omega}$ as the dual of a vector $\boldsymbol{\omega}$ by writing $\boldsymbol{\Omega} = i\boldsymbol{\omega}$, we have

$$\mathbf{v}\cdot\boldsymbol{\Omega} = \mathbf{v}\cdot(i\omega) = i\mathbf{v}\wedge\omega = -\mathbf{v}\times\omega = \omega\times\mathbf{v};$$

hence (7.11) is equivalent to the equation

$$\dot{\mathbf{v}} = \omega\times\mathbf{v}\,. \tag{7.12}$$

To show that (7.11) implies that $|\,\mathbf{v}\,|$ is constant, we use the algebraic identity

$$\mathbf{v}\cdot(\mathbf{v}\cdot\boldsymbol{\Omega}) = (\mathbf{v}\wedge\mathbf{v})\cdot\boldsymbol{\Omega} = 0$$

to prove that $\mathbf{v}\cdot\dot{\mathbf{v}} = 0$, so the result follows from our previous considerations. To prove the converse, suppose that $\boldsymbol{\Omega}$ exists and we introduce a bivector $\mathbf{B}$ to write

$$\boldsymbol{\Omega} = \mathbf{v}^{-1}\wedge\dot{\mathbf{v}} + \mathbf{B}\,. \tag{7.13}$$

If $|\,\mathbf{v}\,|$ is constant, then $\mathbf{v}\cdot\dot{\mathbf{v}} = 0$, which implies that $\mathbf{v}^{-1}\wedge\dot{\mathbf{v}} = \mathbf{v}^{-1}\dot{\mathbf{v}}$ and $\mathbf{v}\cdot(\mathbf{v}^{-1}\wedge\dot{\mathbf{v}}) = \dot{\mathbf{v}}$. Therefore, (7.13) satisfies (7.11) if $\mathbf{v}\cdot\mathbf{B} = 0$, but this condition implies that the dual of $\mathbf{B}$ is collinear with $\mathbf{v}$, so there exists a scalar λ such that $\mathbf{B} = \lambda i\mathbf{v}^{-1}$, and

$$\boldsymbol{\Omega} = \mathbf{v}^{-1}\wedge\dot{\mathbf{v}} + \lambda i\mathbf{v}^{-1} = \mathbf{v}^{-1}(\dot{\mathbf{v}} + i\lambda)\,. \tag{7.14}$$

This shows not only that (7.11) has a bivector solution as required for our proof, but that the solution is not unique without some condition to determine the scalar-valued function $\lambda = \lambda(t)$.

Chain Rule

To complete our review of scalar differentiation, we consider the effect of changing variables. When $F = F(s)$ is a function of a scalar variable s, and $s = s(t)$ is in turn a function of a scalar variable t, then by substitution one has $F = F(s(t)) = F(t)$. Now $F(s)$ is not generally the same function as $F(t)$, but the values of both functions are identical for corresponding values of s and t, and this is emphasized by using the same symbol F for both functions as well as functional values. The derivatives of F with respect to both variables are related by the familiar *chain rule* of elementary calculus:

$$\frac{\mathrm{d}F}{\mathrm{d}t} = \frac{\mathrm{d}F}{\mathrm{d}s}\frac{\mathrm{d}s}{\mathrm{d}t} = \frac{\mathrm{d}s}{\mathrm{d}t}\frac{\mathrm{d}F}{\mathrm{d}s}\,. \tag{7.15}$$

Scalar Integration

Like the rules for differentiation, the rules for integration of multivector-valued functions of a scalar variable can be taken over directly from elementary calculus as long as the order of noncommuting factors is retained in any products. Accordingly, we have the familiar formulas

$$\int_a^b F(t)\,\mathrm{d}t = -\int_b^a F(t)\,\mathrm{d}t, \tag{7.16a}$$

$$\int_a^b [F(t) + G(t)]\,\mathrm{d}t = \int_a^b F(t)\,\mathrm{d}t + \int_a^b G(t)\,\mathrm{d}t, \tag{7.16b}$$

and for $a < c < b$,

$$\int_a^b F(t)\,\mathrm{d}t = \int_a^c F(t)\,\mathrm{d}t + \int_c^b F(t)\,\mathrm{d}t. \tag{7.16c}$$

If A is a constant multivector, then

$$\int_a^b AF(t)\,\mathrm{d}t = A\int_a^b F(t)\,\mathrm{d}t \tag{7.17a}$$

$$\int_a^b F(t)A\,\mathrm{d}t = \left[\int_a^b F(t)\,\mathrm{d}t\right]A. \tag{7.17b}$$

These two formulas are not equivalent unless A commutes with all values of $F(t)$ over the interval $[a, b]$. Integrals can be separated into k-vector parts according to the formula

$$\left\langle\int_a^b F(t)\,\mathrm{d}t\right\rangle_k = \int_a^b \langle F(t)\rangle_k\,\mathrm{d}t. \tag{7.18}$$

This follows from (1.10), for, being the limit of a sum, the integral has the algebraic properties of a sum.

Scalar differentiation and integration are related by two basic formulas; first, the "*fundamental formula of integral calculus*",

$$\int_a^b \dot{F}(t)\,\mathrm{d}t = F(t)\Big|_a^b = F(b) - F(a), \tag{7.19}$$

evaluating the integral of a derivative; second, the formula for the derivative of an integral:

$$\frac{\mathrm{d}}{\mathrm{d}t}\int_a^t F(s)\,\mathrm{d}s = F(t). \tag{7.20}$$

To simplify the notation, the latter formula is sometimes written in the ambiguous form

$$\frac{\mathrm{d}}{\mathrm{d}t}\int^t F(t)\,\mathrm{d}t = F(t),$$

wherein t has two different meanings.

Taylor Expansion

With the fundamental formula (7.19), one can generate *Taylor's formula*

$$F(t+s) = F(t) + s\dot{F}(t) + \frac{s^2}{2}\ddot{F}(t) + \ldots$$

$$= \sum_{k=0}^{\infty} \frac{s^k}{k!} \frac{\mathrm{d}^k}{\mathrm{d}t^k} F(t). \tag{7.21}$$

This power series expansion applies to any function F possessing derivatives of all orders in an interval of the independent variable containing t and $t + s$. It is of great utility in mathematical physics, because it often enables one to approximate a complicated function F by a tractable polynomial function of the independent variable.

The derivation of Taylor's formula is worth reviewing to recall how the expansion is generated by the fundamental formula (7.19). To begin with the fundamental formula allows us to write

$$I \equiv \int_t^{t+s} \dot{F}(v)\,\mathrm{d}v = F(t+s) - F(t).$$

With the change of variables $v = t + s - u$, the integral can be written

$$I = \int_0^s \dot{F}(t+s-u)\,\mathrm{d}u.$$

Next, the fundamental formula (7.19) and the product formula (7.6b) enable us to "integrate by parts" to get

$$I = u\dot{F}(t+s-u)\Big|_{u=0}^{s} + \int_0^s u\ddot{F}(t+s-u)\,\mathrm{d}u$$

$$= s\dot{F}(t) + \int_0^s u\ddot{F}(t+s-u)\,\mathrm{d}u.$$

Again, integration of the second term by parts yields

$$I = s\dot{F}(t) + \frac{s^2}{2!}\ddot{F}(t) + \int_0^s \frac{u^2}{2!}\dddot{F}(t+s-u)\,\mathrm{d}u.$$

Thus we have generated the first three terms in the series (7.21). Moreover, if the series is terminated at this point, we have an exact integral expression for the remainder, namely $\int_0^s (u^2/2!)\,\dddot{F}(t+s-u)\,\mathrm{d}u$, which is sometimes useful for estimating the error incurred in the approximation.

2-7. Exercises

(7.1) Let $\mathbf{u} = \mathbf{u}(t)$ be a vector-valued function, and write $u = |\,\mathbf{u}\,|$. Show that

$$\frac{\mathrm{d}}{\mathrm{d}t}\left(\frac{\mathbf{u}}{u}\right) = \frac{\mathbf{u}(\mathbf{u}\wedge\dot{\mathbf{u}})}{u^3} = \frac{(\mathbf{u}\times\dot{\mathbf{u}})\times\mathbf{u}}{u^3} .$$

(7.2) For a multivector-valued function $F = F(t)$, show that:
(a) For any integer $k > 1$,

$$\frac{\mathrm{d}F^k}{dt} = \dot{F}F^{k-1} + F\dot{F}F^{k-2} + \ldots + F^{k-1}\dot{F} .$$

(b) If for each t the value of F has an algebraic inverse F^{-1}, then

$$\frac{\mathrm{d}F^{-1}}{\mathrm{d}t} = -F^{-1}\dot{F}F^{-1} .$$

(c) If $|F|^2 = F^\dagger F$, then

$$\frac{\mathrm{d}\,|F|^2}{\mathrm{d}t} = 2\langle F^\dagger\dot{F}\rangle_0 .$$

(d) If $|F|^2$ is constant, then

$$\langle F^\dagger\dot{F}\rangle_0 = 0, \quad\text{and}\quad \langle F\dot{F}\rangle_0 = 0 .$$

(7.3) Prove that $F(t) = e^{At}B$ for any constant multivectors A and B if and only if

$$\dot{F}(t) = AF(t) .$$

This relation between F and $\dot{F}$, the simplest possible relation between a multivector-valued function and its scalar derivative, could be taken as the defining property of the exponential function, since it can be used to generate the infinite series (5.1) representing the exponential function. Note that $F(1) = e^A B$.

(7.4) For $F = F(t) = |F|\,\hat{F}$ show that $F\dot{F} = \dot{F}F$ if

$$\dot{F} = \hat{F}\frac{\mathrm{d}\,|F|}{\mathrm{d}t} \quad\text{or}\quad \frac{\mathrm{d}\hat{F}}{\mathrm{d}t} = 0 .$$

Show that the converse is true if F is a k-blade; find a counter-example to prove that the converse is not true more generally.

(7.5) Use Equation (7.6b) to prove

$$\frac{\mathrm{d}}{\mathrm{d}t}\mathbf{r}\wedge\mathbf{p} = \dot{\mathbf{r}}\wedge\mathbf{p} + \mathbf{r}\wedge\dot{\mathbf{p}},$$

$$\frac{\mathrm{d}}{\mathrm{d}t}\mathbf{r}\cdot\mathbf{p} = \dot{\mathbf{r}}\cdot\mathbf{p} + \mathbf{r}\cdot\dot{\mathbf{p}},$$

where $\mathbf{r} = \mathbf{r}(t)$ and $\mathbf{p} = \mathbf{p}(t)$ are vector-valued functions.

It should be clear that as long as the order of factors is retained, the rule for differentiating all products is essentially the same, whether applied to inner or outer products as above or to the geometric product as in (7.6b). Express the derivatives of $\mathbf{p}\cdot(\mathbf{q}\wedge\mathbf{r}) = \mathbf{p}\times(\mathbf{r}\times\mathbf{q})$ and $\mathbf{p}\wedge\mathbf{q}\wedge\mathbf{r} = i\mathbf{p}\cdot(\mathbf{q}\times\mathbf{r})$ in terms of $\dot{\mathbf{p}}$, $\dot{\mathbf{q}}$ and $\dot{\mathbf{r}}$.

(7.6) From Exercise (7.3) we have,

$$\frac{\mathrm{d}}{\mathrm{d}t}e^{At} = Ae^{At} = e^{At}A\,.$$

for any constant multivector A. Use this to prove

$$\frac{\mathrm{d}}{\mathrm{d}t}\cosh(At) = A\sinh(At)\,,$$

$$\frac{\mathrm{d}}{\mathrm{d}t}\sinh(At) = A\cosh(At)\,,$$

$$\frac{\mathrm{d}}{\mathrm{d}t}\cos(At) = -A\sin(At)\,,$$

$$\frac{\mathrm{d}}{\mathrm{d}t}\sin(At) = A\cos(At)\,.$$

(7.7) Show that the derivative of any vector-valued function $\mathbf{v} = \mathbf{v}(t)$ can be expressed in the form

$$\dot{\mathbf{v}} = \mathbf{v}\cdot\Omega + \tfrac{1}{2}\mathbf{v}\,\frac{\mathrm{d}}{\mathrm{d}t}\log\mathbf{v}^2\,.$$

The first term on the right describes the rate of direction change, while the second term describes the rate of magnitude change.

2-8. Directional Derivatives and Line Integrals

The laws of physics are expressed as mathematical functions of position as well as time. To deal with such functions, we must extend the differential and integral calculus of functions with scalar variables to a calculus of functions with vector variables. In a subsequent volume, NFII, the general concept of differentiation with respect to a vector variable will be developed, but here we restrict our considerations to a special case which is closely related to scalar differentiation. The main results of this section will first be used in Sections 3-8 and 3-10, so study of this section can be deferred to that point.

Let $F = F(\mathbf{x})$ be a multivector-valued function of a vector variable $\mathbf{x}$ defined on some region of the Euclidean space $\mathcal{E}_3$. In physical applications the symbol $\mathbf{x}$ will denote a place in Physical Space, in which case F is said to be a "function of position". Such a function is also called a *field*, a *vector field* if it is vector-valued, a *scalar field* if it is scalar-valued, a *spinor field* if it is spinor-valued, etc.

Directional Derivatives

The *directional derivative* of the function $F = F(\mathbf{x})$ is denoted by $\mathbf{a}\cdot\nabla F$ or $\mathbf{a}\cdot\nabla F(\mathbf{x})$ and can be defined in terms of the scalar derivative by

$$\mathbf{a}\cdot\nabla F \equiv \left.\frac{\mathrm{d}F(\mathbf{x}+\mathbf{a}\tau)}{\mathrm{d}\tau}\right|_{\tau=0} = \lim_{\tau\to 0} \frac{F(\mathbf{x}+\tau\mathbf{a}) - F(\mathbf{x})}{\tau} . \tag{8.1}$$

Many authors call $\mathbf{a}\cdot\nabla F$ a directional derivative only if $\mathbf{a}$ is a unit vector. In NFII it will be seen that the dot in $\mathbf{a}\cdot\nabla$ can actually be interpreted as an inner product, but for the time being, it can be regarded as a special notation. However, it is important to note that, just as the distributive property of the inner product would require $(\mathbf{a}+\mathbf{b})\cdot\nabla = \mathbf{a}\cdot\nabla + \mathbf{b}\cdot\nabla$, so, by an elementary mathematical exercise with limits, it can be proved from (8.1) that

$$(\mathbf{a}+\mathbf{b})\cdot\nabla F = \mathbf{a}\cdot\nabla F + \mathbf{b}\cdot\nabla F. \tag{8.2a}$$

Moreover, for any scalar λ,

$$(\lambda\mathbf{a})\cdot\nabla F = \lambda(\mathbf{a}\cdot\nabla F). \tag{8.2b}$$

Besides this, the directional derivative obviously has all the general properties of the scalar derivative which were mentioned in the preceding section; thus, for multivector-valued functions $F = F(\mathbf{x})$ and $G = G(\mathbf{x})$, we have

$$\mathbf{a}\cdot\nabla(F+G) = \mathbf{a}\cdot\nabla F + \mathbf{a}\cdot\nabla G, \tag{8.3a}$$

$$\mathbf{a}\cdot\nabla(FG) = (\mathbf{a}\cdot\nabla F)G + F(\mathbf{a}\cdot\nabla G), \tag{8.3b}$$

$$\mathbf{a}\cdot\nabla\langle F\rangle_k = \langle\mathbf{a}\cdot\nabla F\rangle_k . \tag{8.3c}$$

Also, if $F = F(\lambda(\mathbf{x}))$ where $\lambda = \lambda(\mathbf{x})$ is a scalar-valued function, then we have the *chain rule*

$$\mathbf{a}\cdot\nabla F = (\mathbf{a}\cdot\nabla\lambda)\,\frac{\mathrm{d}F}{\mathrm{d}\lambda} . \tag{8.4}$$

The most basic function of a vector variable is the "identity function" $F(\mathbf{x}) = \mathbf{x}$. To determine its directional derivative, we observe that $\mathrm{d}/\mathrm{d}\tau\,(\mathbf{x}+\mathbf{a}\tau) = \mathbf{a}$ so according to the definition (8.1)

$$\mathbf{a}\cdot\nabla\mathbf{x} = \mathbf{a}. \tag{8.5}$$

Obviously, the "constant function" $F(\mathbf{x}) = A$ has the trivial derivative

$$\mathbf{a}\cdot\nabla A = 0. \tag{8.6}$$

From these basic derivatives, the derivatives of more complicated functions can be determined by using the general rules (8.3) and (8.4) without further appeal to the definition (8.1). In particular, the derivative of any algebraic function of $\mathbf{x}$ can be determined in this way. For example, to differentiate the function $|\,\mathbf{x}\,|$, we note that it is related to $\mathbf{x}$ by the algebraic equation

$|\mathbf{x}| = (\mathbf{x}^2)^{1/2}$, or, more simply, by $|\mathbf{x}|^2 = \mathbf{x}^2$. Using the product rule (8.3b), we have

$$\mathbf{a}\cdot\nabla\mathbf{x}^2 = \mathbf{ax} + \mathbf{xa} = 2\mathbf{a}\cdot\mathbf{x}. \tag{8.7}$$

On the other hand, the chain rule gives

$$\mathbf{a}\cdot\nabla\,|\mathbf{x}|^2 = 2\,|\mathbf{x}|\,\mathbf{a}\cdot\nabla\,|\mathbf{x}|.$$

So, equating this with (8.7), we get the desired result

$$\mathbf{a}\cdot\nabla\,|\mathbf{x}| = \frac{\mathbf{a}\cdot\mathbf{x}}{|\mathbf{x}|} = \mathbf{a}\cdot\hat{\mathbf{x}}. \tag{8.8}$$

It is helpful to know the derivative of the "direction function" $\hat{\mathbf{x}}$ as well as the "magnitude function" $|\mathbf{x}|$. We can find it by using the product rule and the chain rule as follows:

$$\mathbf{a}\cdot\nabla\left(\frac{\mathbf{x}}{|\mathbf{x}|}\right) = \frac{\mathbf{a}\cdot\nabla\mathbf{x}}{|\mathbf{x}|} - \frac{(\mathbf{a}\cdot\nabla\,|\mathbf{x}|)}{|\mathbf{x}|^2}\,\mathbf{x} = \frac{\mathbf{a}}{|\mathbf{x}|} - \frac{\mathbf{a}\cdot\hat{\mathbf{x}}\mathbf{x}}{|\mathbf{x}|^2}\,.$$

Hence,

$$\mathbf{a}\cdot\nabla\hat{\mathbf{x}} = \frac{\mathbf{a} - \mathbf{a}\cdot\hat{\mathbf{x}}\hat{\mathbf{x}}}{|\mathbf{x}|} = \frac{\hat{\mathbf{x}}\hat{\mathbf{x}}\wedge\mathbf{a}}{|\mathbf{x}|}\,. \tag{8.9}$$

Other important derivatives are evaluated in the exercises.

Taylor's Formula

To approximate arbitrary functions of position by simpler functions, we need Taylor's formula expressed in terms of the directional derivative. To this end, write

$$G(\tau) = F(\mathbf{x} + \mathbf{a}\tau).$$

Then

$$\frac{\mathrm{d}G(0)}{\mathrm{d}\tau} = \frac{\mathrm{d}F}{\mathrm{d}\tau}(\mathbf{x} + \mathbf{a}\tau)\bigg|_{\tau=0} = \mathbf{a}\cdot\nabla F(\mathbf{x}),$$

$$\frac{\mathrm{d}^2G(0)}{\mathrm{d}\tau^2} = (\mathbf{a}\cdot\nabla)\,(\mathbf{a}\cdot\nabla F(\mathbf{x})) \equiv (\mathbf{a}\cdot\nabla)^2F(\mathbf{x}),$$

$$\frac{\mathrm{d}^kG(0)}{\mathrm{d}\tau^k} = (\mathbf{a}\cdot\nabla)^kF(\mathbf{x}).$$

From the Taylor expansion of $G(\tau)$ taken about $\tau = 0$ and evaluated at $\tau = 1$, we get

$$G(1) = G(0) + \frac{\mathrm{d}G(0)}{\mathrm{d}\tau} + \frac{1}{2}\frac{\mathrm{d}^2G(0)}{\mathrm{d}\tau^2} + \ldots = \sum_{k=0}^{\infty}\frac{1}{k!}\frac{\mathrm{d}G^k(0)}{\mathrm{d}\tau^k}\,.$$

Expressed in terms of F, this gives the desired Taylor expansion

$$F(\mathbf{x}+\mathbf{a}) = F(\mathbf{x}) + \mathbf{a}\cdot\nabla F(\mathbf{x}) + \frac{(\mathbf{a}\cdot\nabla)^2}{2!} F(\mathbf{x}) + \ldots$$

$$= \sum_{k=0}^{\infty} \frac{(\mathbf{a}\cdot\nabla)^k}{k!} F(\mathbf{x}) \equiv e^{\mathbf{a}\cdot\nabla} F(\mathbf{x}) . \tag{8.10}$$

Note that this leads to a natural definition of the exponential function $e^{\mathbf{a}\cdot\nabla}$ of the differential operator $\mathbf{a}\cdot\nabla$.

The Differential

The Taylor expansion (8.10) reveals a property of the function $\mathbf{a}\cdot\nabla F(\mathbf{x})$ of such importance that it deserves a special notation and a name. We will use the term *differential* as an alternative to the term *directional derivative*, and we introduce the notation

$$F'(\mathbf{a}, \mathbf{x}) \equiv \mathbf{a}\cdot\nabla F(\mathbf{x}) . \tag{8.11}$$

This notation is intended to emphasize the fact that $F' = F'(\mathbf{a}, \mathbf{x})$ is a function of two variables obtained from $F = F(\mathbf{x})$ by differentiation. When the dependence on $\mathbf{a}$ with $\mathbf{x}$ held fixed is of interest, it is convenient to write $F' = F'(\mathbf{a})$, which still reminds us that this function was obtained from a function F. It must be noted that the differential is a *linear function* of its first variable, which is to say that it has the properties

$$F'(\mathbf{a}+\mathbf{b}) = F'(\mathbf{a}) + F'(\mathbf{b}) , \tag{8.12a}$$

$$F'(\lambda\mathbf{a}) = \lambda F'(\mathbf{a}) \tag{8.12b}$$

for scalar λ. Here we have merely written Equations (8.2a, b) in a different notation.

Now suppose that we are interested in the behavior of a function $F = F(\mathbf{x})$ in the neighborhood of a point $\mathbf{x}_0 + \mathbf{r}$, a Taylor expansion about $\mathbf{x}_0$ gives

$$F(\mathbf{x}) = F(\mathbf{x}_0 + \mathbf{r}) = F(\mathbf{x}_0) + \mathbf{r}\cdot\nabla F(\mathbf{x}_0) + \tfrac{1}{2}(\mathbf{r}\cdot\nabla)^2 F(\mathbf{x}_0)$$

$$= F(\mathbf{x}_0) + |\mathbf{r}|\, \hat{\mathbf{r}}\cdot\nabla F(\mathbf{x}_0) + \frac{|\mathbf{r}|^2}{2!} (\hat{\mathbf{r}}\cdot\nabla)^2 F(\mathbf{x}_0) +$$

$$+ \ldots + \frac{|\mathbf{r}|^k}{k!} (\hat{\mathbf{r}}\cdot\nabla)^k F(\mathbf{x}_0) + \ldots .$$

For $|\mathbf{r}| = |\mathbf{x} - \mathbf{x}_0|$ sufficiently small, the first two terms approximate $F(\mathbf{x})$ to any desired accuracy, and we can write

$$F(\mathbf{x}) - F(\mathbf{x}_0) \approx F'(\mathbf{x} - \mathbf{x}_0) = F'(\mathbf{x}) - F'(\mathbf{x}_0) . \tag{8.13}$$

Thus, we see that the differential provides a *linear approximation* to any differentiable function. Since linear functions are simple enough to be analyzed completely, this establishes the great importance of the differential.

Although we use the terms "differential" and "directional derivative" interchangably, they have different historical roots and emphasize different aspects of the same function. If $\mathbf{a}$ is a fixed vector, the term "directional derivative" is most appropriate to emphasize that $\mathbf{a}\cdot\nabla F$ is the derivative in a particular direction, the direction of $\mathbf{a}$. On the other hand, the term "differential" serves to emphasize that $F'(\mathbf{a})$ is a linear function of $\mathbf{a}$. Unfortunately, the term "differential" is commonly taken to connote a "small quantity", especially in the older literature. It must be realized that the differential $F'(\mathbf{a})$ is defined for all values of $\mathbf{a}$, not just small ones. However, as we have seen in obtaining (8.13), the differential $F'(\mathbf{x}-\mathbf{x}_0)$ may be a good approximation to $F(\mathbf{x})-F(\mathbf{x}_0)$ only if $|\mathbf{x}-\mathbf{x}_0|$ is sufficiently small.

Variation on a Curve

In mechanics we will often be interested in how some function of position $F = F(\mathbf{x})$ varies along the path of a particle $\mathbf{x} = \mathbf{x}(\mathrm{t})$. Strictly speaking $\mathbf{x} = \mathbf{x}(t)$ is a parametric equation for the particle path and not the path itself, which is a set of points, but we suppress such distinctions when they are not at issue. To describe the variation of F along the path we must differentiate the composite function $F = F(\mathbf{x}(t))$. The derivative can be reduced to derivatives of $F(\mathbf{x})$ and $\mathbf{x}(t)$ by proving that

$$\frac{\mathrm{d}F}{\mathrm{d}t}(\mathbf{x}(t)) = \frac{\mathrm{d}F}{\mathrm{d}\tau}(\mathbf{x}(t+\tau))\bigg|_{\tau=0} = \frac{\mathrm{d}F}{\mathrm{d}\tau}(\mathbf{x}(t)+\tau\dot{\mathbf{x}}(t))\bigg|_{\tau=0}.$$

It will be left to the interested reader to fill in the missing mathematical details of the proof. The last term is seen to be the directional derivative, so we can write

$$\frac{\mathrm{d}F}{\mathrm{d}t}(\mathbf{x}(t)) = \dot{\mathbf{x}}(t)\cdot\nabla F(\mathbf{x})\bigg|_{\mathbf{x}=\mathbf{x}(t)}. \tag{8.14}$$

This is the *chain rule* for the composite function $F = F(\mathbf{x}(t))$.

Partial and Total Derivatives

More generally, let $F = F(\mathbf{x}, t)$ be a function of position $\mathbf{x}$ as well as time t. For fixed $\mathbf{x}$, the time derivative is denoted and defined by

$$\partial_t F(\mathbf{x}, t) \equiv \lim_{\Delta t\to 0}\frac{F(\mathbf{x}, t+\Delta t) - F(\mathbf{x}, t)}{\Delta t}. \tag{8.15}$$

This describes how F varies with time at each point $\mathbf{x}$. The derivative $\partial_t F$ is called the "*partial derivative* of F with respect to time." If $\partial_t F = 0$, then F is said to be static or "constant in time" or "*not* an explicit function of time", and we write $F(\mathbf{x}, t) = F(\mathbf{x})$. On the other hand, if $\mathbf{a}\cdot\nabla F = 0$ for all directions $\mathbf{a}$ and all points $\mathbf{x}$, then F is said to be *uniform* or "uniform in space", and we can write $F(\mathbf{x}, t) = F(t)$. Obviously if $F = F(\mathbf{x}, t)$ is both static and uniform, then $\mathrm{d}F/\mathrm{d}t = 0$ on any path $\mathbf{x} = \mathbf{x}(t)$. In this case, F is said to be *constant*.

To describe how F varies along a particle path $\mathbf{x} = \mathbf{x}(t)$, we must differentiate $F = F(\mathbf{x}(t), t)$. Using (8.14) and (8.15), we get

$$\frac{\mathrm{d}F}{\mathrm{d}t}(\mathbf{x}(t), t) = \partial_t F(\mathbf{x}(t), t) + \dot{\mathbf{x}}(t)\cdot\nabla F(\mathbf{x}(t), t). \tag{8.16a}$$

Notice that we have two terms here, because we have used a generalization of the product rule (7.6b) which allows us to separately differentiate the two distinct functional dependencies on time. Our notation enables us to suppress the variables in (8.16a) without confusion and write

$$\frac{\mathrm{d}F}{\mathrm{d}t} = \partial_t F + \dot{\mathbf{x}}\cdot\nabla F = (\partial_t + \dot{\mathbf{x}}\cdot\nabla)F. \tag{8.16b}$$

The derivative $\mathrm{d}F/\mathrm{d}t$ is given many different names in the literature; the term *total derivative* is one of the most common, but the term *convective derivative* is most appropriate, because it suggests change in a function with respect to flow along a path $\mathbf{x} = \mathbf{x}(t)$.

Line Integrals

Having seen how scalar derivatives are related to directional derivatives, we now consider how integrals with respect to a scalar parameter are related to integrals in space.

Let C be a smooth curve in $\mathcal{E}_3$ from point $\mathbf{a}$ to point $\mathbf{b}$, and let $F = F(\mathbf{x})$ be a multivector-valued function defined at each point $\mathbf{x}$ on C. The *line integral* of F on C is defined by

$$\int_C F(\mathbf{x})\,\mathrm{d}\mathbf{x} \equiv \lim_{\substack{\Delta x_i \to 0 \\ n\to\infty}} \sum_{i=1}^{n} F(\mathbf{x}_i)\Delta\mathbf{x}_i. \tag{8.17}$$

The limit can be understood with reference to Figure 8.1. Points $\mathbf{x}_0 = \mathbf{a}, \mathbf{x}_1, \mathbf{x}_2, \ldots, \mathbf{x}_n = \mathbf{b}$ are selected on the curve C; they determine chords $\Delta\mathbf{x}_i \equiv \mathbf{x}_i - \mathbf{x}_{i-1}$ and the sum on the right side of (8.17). The larger the number of points selected and the smaller the $|\Delta\mathbf{x}_i|$, the more closely the sum in (8.17) approximates the integral, if the k-vector parts $\langle F(\mathbf{x})\rangle_k$ do not vary too rapidly along the curve. Although it involves vectors instead of scalars, the limit process defining the line integral (8.17) is formally the same as the one defining the scalar integral in elementary calculus. Indeed, if the curve C is represented parametrically by the equation $\mathbf{x} = \mathbf{x}(t)$ with $\mathbf{x}(\alpha) = \mathbf{a}$ and $\mathbf{x}(\beta) = \mathbf{b}$, then it is not difficult to prove that

$$\int_C F(\mathbf{x})\, \mathrm{d}\mathbf{x} = \int_\alpha^\beta F(\mathbf{x}(t)) \frac{\mathrm{d}\mathbf{x}}{\mathrm{d}t}(\mathrm{t})\, \mathrm{d}t. \tag{8.18}$$

We could have used (8.18) as a definition of the line integral in terms of the scalar integral already studied. However, definition (8.17) has the advantage over (8.18) of being completely independent of any parametric representation of the curve, giving it a certain conceptual and computational simplicity. For example, when $F(\mathbf{x}) = 1$, we can easily evaluate the sum in (8.17), and noting that it is independent of n (see Figure 8.1), we get the result

$$\int_C \mathrm{d}\mathbf{x} = \mathbf{b} - \mathbf{a}. \tag{8.19}$$

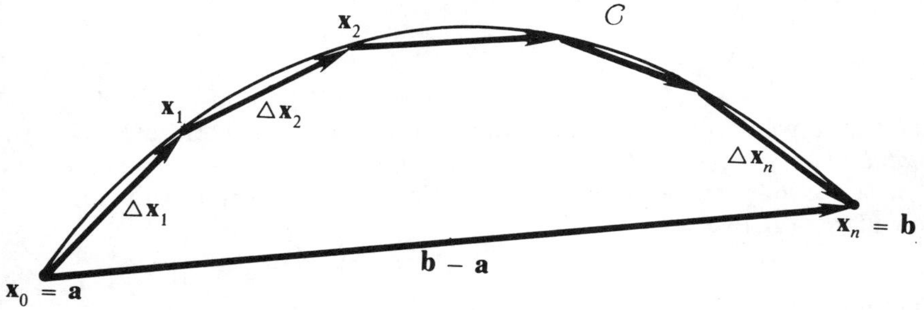

Fig. 8.1. Approximation of a curve C by line segments.

For any closed curve the end points are identical, and the result can be written

$$\oint \mathrm{d}\mathbf{x} = 0. \tag{8.20}$$

This should be interpreted a sum of vectors adding to zero.

Because of the relation (8.18), the general properties of the line integral are so easily determined from those of the scalar integral that it is hardly necessary to write them down. However, some comments on notation and some words of caution are in order. When the relevant variables and the domain of integration are clear from the context, they can be suppressed in the notation for integrals. For instance, (8.18) can be written in the abbreviated form

$$\int F\, \mathrm{d}\mathbf{x} = \int F \frac{\mathrm{d}\mathbf{x}}{\mathrm{d}t} \mathrm{d}t.$$

To indicate the endpoints of the line integral we may write

$$\int_\mathbf{a}^\mathbf{b} F \cdot \mathrm{d}\mathbf{x};$$

this notation is especially apt when the function F is such that the value of the integral is independent of the path (curve) from **a** to **b**. This is obviously the case for the integral (8.19), which accordingly may be written

$$\int_{\mathbf{a}}^{\mathbf{b}} d\mathbf{x} = \mathbf{b} - \mathbf{a}.$$

It must be remembered, however, that in general the value of the line integral depends on the entire curve C and not on the endpoints alone. The caution about ordering factors deserves repeating as well; it must be realized that the integral $\int d\mathbf{x}\ F$ is not necessarily equivalent to $\int F d\mathbf{x}$ unless the product $F\,d\mathbf{x}$ is commutative, as when F is scalar-valued.

Line Integrals of Vector Fields

For a more specific example of a line integral, let $\mathbf{f} = \mathbf{f}(\mathbf{x})$ be vector-valued function on the curve C. Since $d\mathbf{x}$ is also vector-valued we have

$$\int \mathbf{f}\, d\mathbf{x} = \int \mathbf{f}\cdot d\mathbf{x} + \int \mathbf{f}\wedge d\mathbf{x}. \tag{8.21a}$$

This integral has a scalar part

$$\int \mathbf{f}\cdot d\mathbf{x} = \tfrac{1}{2}\int \mathbf{f}\, d\mathbf{x} + \tfrac{1}{2}\int d\mathbf{x}\,\mathbf{f}, \tag{8.21b}$$

and a bivector part

$$\int \mathbf{f}\wedge d\mathbf{x} = \int \tfrac{1}{2}(\mathbf{f}\, d\mathbf{x} - d\mathbf{x}\,\mathbf{f}). \tag{8.21c}$$

Both parts are called line integrals and, because they are of different grade, they can be considered separately. But there are times when they are best considered together as in (8.21a).

Next to constant **f**, the simplest vector-valued function is $f(\mathbf{x}) = \mathbf{x}$, with the line integral

$$\int \mathbf{x}\, d\mathbf{x} = \int \mathbf{x}\cdot d\mathbf{x} + \int \mathbf{x}\wedge d\mathbf{x}. \tag{8.22}$$

Both the scalar and bivector parts of this integral are of independent interest, so let us consider them separately. First, the scalar part. If we represent the curve parametrically by $\mathbf{x} = \mathbf{x}(t)$, then according to (7.8),

$$\mathbf{x}\cdot\frac{d\mathbf{x}}{dt} = \frac{1}{2}\frac{d\mathbf{x}^2}{dt},$$

so

$$\int \mathbf{x}\cdot d\mathbf{x} = \int \mathbf{x}\cdot\frac{d\mathbf{x}}{dt}\,dt = \tfrac{1}{2}\int \frac{d\mathbf{x}^2}{dt}\,dt.$$

But we are at liberty to choose the scalar $\mathbf{x}^2$ as our parameter, because there is no contribution to the integral from portions of the curve where $\mathbf{x}^2$ is constant. Hence we get the specific result

$$\int_{\mathbf{a}}^{\mathbf{b}} \mathbf{x}\cdot d\mathbf{x} = \tfrac{1}{2}\int_{\mathbf{a}^2}^{\mathbf{b}^2} d(\mathbf{x}^2) = \tfrac{1}{2}(\mathbf{b}^2 - \mathbf{a}^2)\,. \tag{8.23}$$

Since its value depends only on the endpoints, this integral is path-independent.

The Area Integral

Now consider the bivector part of (8.22) written in the form

$$\mathbf{A} \equiv \tfrac{1}{2}\int_{\mathbf{a}}^{\mathbf{b}} \mathbf{x}\wedge d\mathbf{x}. \tag{8.24}$$

The significance of this integral can be understood by approximating it by a sum in accordance with the definition (8.17).

$$\begin{aligned}\mathbf{A} &\approx \tfrac{1}{2}\sum_{k=1}^{n} \mathbf{x}_k\wedge\Delta\mathbf{x}_k \\ &= \tfrac{1}{2}\,\mathbf{x}_0\wedge\mathbf{x}_1 + \tfrac{1}{2}\,\mathbf{x}_1\wedge\mathbf{x}_2 + \ldots + \tfrac{1}{2}\,\mathbf{x}_{n-1}\wedge\mathbf{x}_n\,.\end{aligned} \tag{8.25}$$

As illustrated in Figure 8.2, each term in this sum is the directed area of a triangle with one vertex at the origin. The first term $\frac{1}{2}\mathbf{x}_0\wedge\mathbf{x}_1$ approximates the directed area "swept out" by the line segment represented by vector variable $\mathbf{x}$ as its tip "moves" continuously along the curve from $\mathbf{x}_0$ to $\mathbf{x}_1$ while its tail is

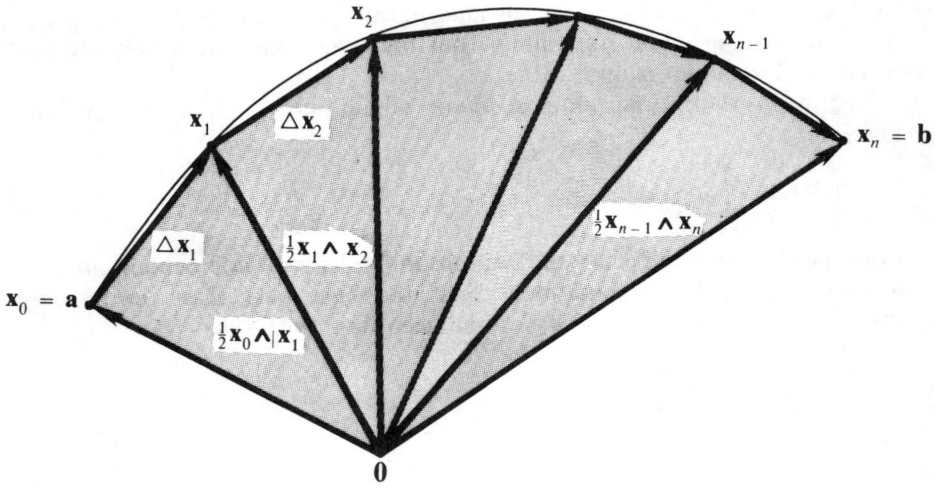

Fig. 8.2. Polygonal area approximation.

anchored at the origin. Therefore, the sum in (8.25) approximates the directed area swept out as the variable **x** moves from **a** to **b**. Accordingly, we arrive at the exact interpretation of the integral (8.24) as the total *directed area* "swept out" by the vector variable **x** as it moves continuously along the curve from **a** to **b**. This interpretation makes it obvious that the value of the integral is *not* path independent, because the area swept out depends on the path from **a** to **b**.

If the curve is represented by the parametric equation $\mathbf{x} = \mathbf{x}(t)$ with $\mathbf{x}(0) = \mathbf{a}$, then the area swept out can also be expressed as a parametric function $\mathbf{A} = \mathbf{A}(t)$ by writing (8.24) in the form

$$\mathbf{A}(t) = \tfrac{1}{2}\int_{\mathbf{x}(0)}^{\mathbf{x}(t)} \mathbf{x}\wedge \mathrm{d}\mathbf{x} = \tfrac{1}{2}\int_{0}^{t} \mathbf{x}\wedge\dot{\mathbf{x}}\, \mathrm{d}t. \tag{8.26}$$

Differentiating with respect to the upper limit of the integral, we get

$$\dot{\mathbf{A}} = \tfrac{1}{2}\,\mathbf{x}\wedge\dot{\mathbf{x}}\,, \tag{8.27}$$

expressing the rate at which area is swept out. This rate depends on the choice of parameter t, although the total area swept out depends only on the curve.

Consider a closed curve C in a plane enclosing the origin, as shown in Figure 8.3a. The line integral

$$\mathbf{A} = \tfrac{1}{2}\oint \mathbf{x}\wedge \mathrm{d}\mathbf{x} \tag{8.28}$$

along C gives us the directed area "enclosed" by the curve. Its magnitude $|\,\mathbf{A}\,|$ is the conventional scalar measure of area enclosed by the curve. This should be evident from the fact that the vector **x** "sweeps through" each of the points enclosed by the curve exactly once, or again by considering approximation of the integral by the areas of triangles as expressed by (8.25), which applies here with $\mathbf{x}_0 = \mathbf{x}_n$. As in Section 2-3, we represent the unit of directed area for the plane by a bivector **i**. Then $\mathbf{A} = A\mathbf{i}$, where $A = |\,\mathbf{A}\,|$ if the curve C has a counterclockwise orientation (as it does in Figure 8.3a), or $A = -|\,\mathbf{A}\,|$ if C has a clockwise orientation. For the situation depicted by Figure 8.3a, we have

$$\tfrac{1}{2}\,\mathbf{x}_k\wedge\Delta\mathbf{x}_k = \tfrac{1}{2}\,|\,\mathbf{x}_k\wedge\Delta\mathbf{x}_k\,|\,\mathbf{i} \tag{8.29a}$$

for the kth "element of area", whence from (8.28),

$$|\,\mathbf{A}\,| = \tfrac{1}{2}\oint |\,\mathbf{x}\wedge \mathrm{d}\mathbf{x}\,|\,. \tag{8.29b}$$

It must be emphasized that (8.29b) follows from (8.27) *only* when all coplanar elements of area have the same orientation, as in (8.29a). This condition is not met if the curve C is self-intersectering or does not enclose the origin.

The area integral (8.27) is independent of the origin, in spite of the fact that the values of the vector variable **x** in the integrand depend on the origin. To

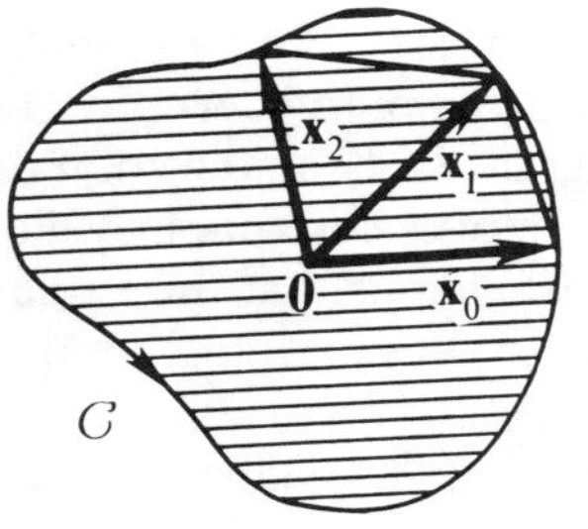

Fig. 8.3a.

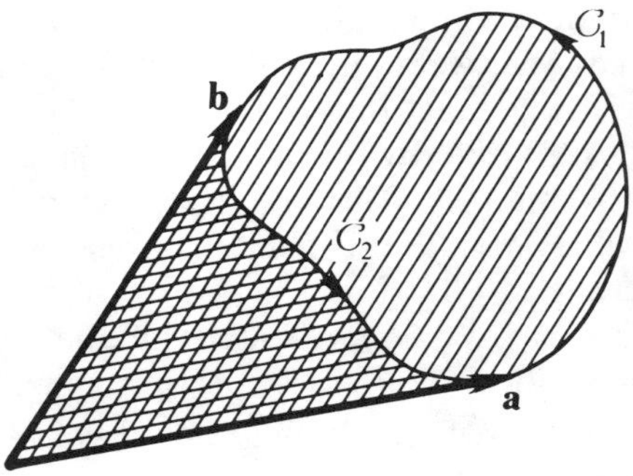

Fig. 8.3b.

Area swept out by radius vector along a closed curve. Crosshatched region is swept out twice in opposite directions, so its area is zero.

understand how this can be, displace the origin from inside the curve C shown in Figure 8.3a to a place outside the curve as shown in Figure 8.3b. Choosing points **a** and **b** on C, as shown in Figure 8.3b, we separate C into two pieces C_1 and C_2, so the area integral can be written

$$\mathbf{A} = \tfrac{1}{2}\oint_C \mathbf{x}\wedge d\mathbf{x} = \tfrac{1}{2}\int_{C_1} \mathbf{x}\wedge d\mathbf{x} + \tfrac{1}{2}\int_{C_2} \mathbf{x}\wedge d\mathbf{x}. \tag{8.30}$$

Referring to Figure 8.3b, we see that the coordinate vector sweeps over the region inside C once as it goes between **a** and **b** along C, but it sweeps over the region to the left of C_2 twice, once as it traverses C_2 and again as it traverses C_1; since the two sweeps over the latter region are in opposite directions the directed area they sweep out have the same magnitude but opposite sign, so their contributions to the integral (8.30) cancel, and we are left with the directed area enclosed by C, as claimed.

For a general proof that *the closed area integral* (8.27) *is independent of the origin*, we displace the origin by an "amount" **c** by making the change of variables $\mathbf{x} \to \mathbf{x}' = \mathbf{x} - \mathbf{c}$. Then,

$$\oint \mathbf{x}'\wedge d\mathbf{x}' = \oint (\mathbf{x} - \mathbf{c})\wedge d\mathbf{x} = \oint \mathbf{x}\wedge d\mathbf{x} - \mathbf{c}\wedge \oint d\mathbf{x}.$$

But the last term vanishes because $\oint d\mathbf{x} = 0$, so the independence of origin is proved. Note that the vector **c** is entirely arbitrary, so our restriction of the origin to the plane of the curve in Figure 8.3b is quite irrelevant to the value of the area integral, though it helped us see how parts of the integral cancel when the origin is not enclosed by the curve; such cancellation occurs even when the origin is outside the plane, as our proof of origin independence implies.

Our discussion shows that the integral (8.28) generalizes the ancient concept of the area enclosed by a simple closed curve to a concept of directed

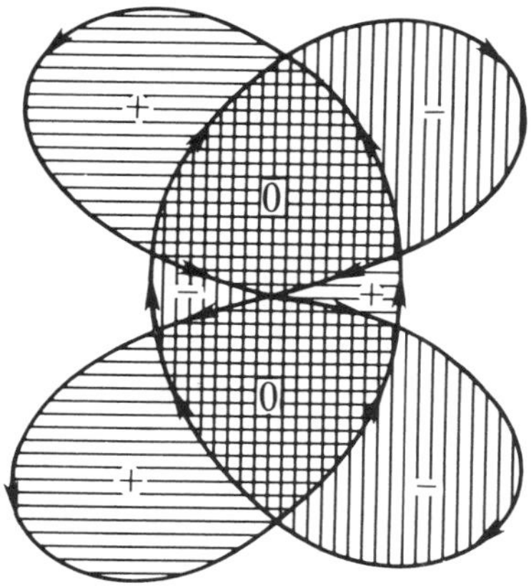

Fig. 8.4. Directed area of a self-intersecting closed plane curve. Vertical and horizontal lines denote areas with opposite orientation, so cross-hatched region has zero area.

area determined by an arbitrary closed curve. Thus, the integral (8.28) defines an "enclosed" area even for self-interesecting plane curves such as the one shown in Figure 8.4; the sign of the area integral for subregions is indicated in the figure, with zero for subregions which are "swept out" twice with cancelling signs. The integral (8.28) also applies to closed curves in space which do not lie in a plane, but we will not need to consider its significance further in this book.

The General Line Integral

Our definition (8.17) of the line integral is not the most general one possible, though it will suffice for most purposes of this book. A word about the general case may be helpful. Consider a multivector-valued function $L(\mathbf{a}, \mathbf{x})$ of two vector variables which is linear in the first variable. *The line integral* of $L(\mathbf{a}, \mathbf{x})$ along some curve C is defined as in (8.17) by

$$\int_C L(\mathrm{d}\mathbf{x}, \mathbf{x}) \equiv \lim_{\substack{\Delta\mathbf{x}_i \to 0 \\ n \to \infty}} \sum_{i=1}^{n} L(\Delta\mathbf{x}_i, \mathbf{x}_i). \tag{8.31}$$

The linearity of the first variable in $L(\mathbf{a}, \mathbf{x})$ is necessary for the limit in (8.31) to be independent of the subdivision of the curve. Equation (8.18) is now generalized to

$$\int_C L(\mathrm{d}\mathbf{x}, \mathbf{x}) = \int_\alpha^\beta L(\dot{\mathbf{x}}(t), \mathbf{x}(t))\, \mathrm{d}t. \tag{8.32}$$

We consider but one special case of such an integral. Suppose that, at every point in some region $\mathscr{R}$ containing C, the function $L(\dot{\mathbf{x}}, \mathbf{x})$ is the differential of some function $F = F(\mathbf{x})$. Then we write

$$L(\dot{\mathbf{x}}, \mathbf{x}) = \dot{\mathbf{x}}\cdot\nabla F(\mathbf{x}), \tag{8.33}$$

and using (8.14) in (8.32), we get

$$\int_C \mathrm{d}\mathbf{x}\cdot\nabla F(\mathbf{x}) = \int_\alpha^\beta \frac{\mathrm{d}F}{\mathrm{d}t}\mathrm{d}t = F(\mathbf{b}) - F(\mathbf{a}). \tag{8.34}$$

Since the value of the integral is completely determined by the value of F at the endpoints, the integral is independent of the path in $\mathscr{R}$. Thus we see that for path-independence of an integral it is sufficient that the integrand be a differential of some function. The terms "perfect differential" or "exact differential" are also used to indicate path-independence in the literature.

The Gradient

Path-independent integrals arise most commonly in connection with an important kind of vector field. If a vector field $\mathbf{f} = \mathbf{f}(\mathbf{x})$ has the property that $\mathbf{a}\cdot\mathbf{f} = \mathbf{a}\cdot\nabla\phi$ is the differential of some scalar field $\phi = \phi(\mathbf{x})$, then we write

$$\mathbf{f} = \nabla\phi \tag{8.35}$$

and say that $\mathbf{f}$ is the *gradient* of ϕ. We say that ϕ is a *potential* of $\mathbf{f}$. The gradient has a simple geometric interpretation which follows from the fact that $\mathbf{a}\cdot\nabla\phi$ is a directional derivative. The directional derivative tells us the rate at which the value of ϕ changes in the direction $\mathbf{a}$. If $\mathbf{a}$ is a unit vector with direction of our choosing, then $\mathbf{a}\cdot\nabla\phi$ has its maximum value when $\mathbf{a}$ and $\nabla\phi$ are in the same direction, that is, when $\mathbf{a}\cdot\nabla\phi = |\nabla\phi|$. Thus, the gradient $\nabla\phi = \nabla\phi(\mathbf{x})$ tells us both the direction and magnitude of maximum change in the value of $\phi = \phi(x)$ at any point $\mathbf{x}$ where it is defined. Furthermore, the change of ϕ in any given direction $\mathbf{a}$ is obtained by taking the inner product of $\mathbf{a}$ with $\nabla\phi$.

As shown in Figure 8.5, the equation $\phi(\mathbf{x}) = k$ defines a one-parameter family of surfaces, called *equipotential surfaces*, one surface for each constant value of k. At a given point $\mathbf{x}$, the gradient $\nabla\phi$ is *normal* (perpendicular) to the surface through that point, and, when not zero, it is directed towards surfaces with larger values of k. Figure 8.5 shows only a 2-dimensional cros-section. As is conventional, the change in k is the same for each pair of neighboring surfaces, so the separation provides a measure of the change in ϕ, the closer the surfaces, the larger the gradient. The change in ϕ between *any* two points $\mathbf{a}$ and $\mathbf{b}$ is given by

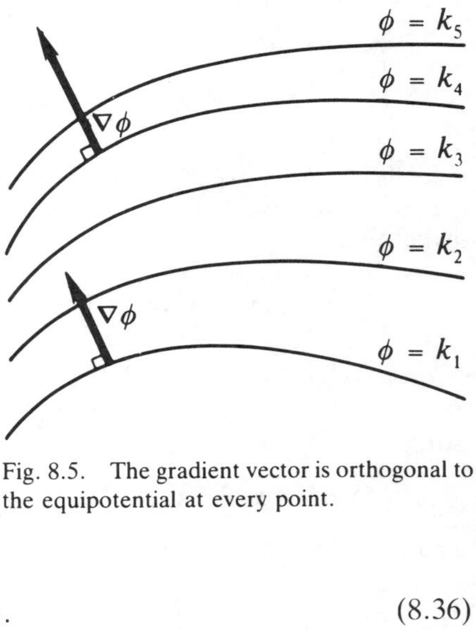

Fig. 8.5. The gradient vector is orthogonal to the equipotential at every point.

$$\int_{\mathbf{a}}^{\mathbf{b}} d\mathbf{x}\cdot\mathbf{f} = \int_{\mathbf{a}}^{\mathbf{b}} d\mathbf{x}\cdot\nabla\phi = \phi(\mathbf{b}) - \phi(\mathbf{a}). \tag{8.36}$$

As we have noted, the value of such an integral is independent of the path.

For a given scalar function $\phi = \phi(\mathbf{x})$, the gradient is easily found from the directional derivative by interpreting $\mathbf{a}\cdot\nabla\phi$ as the inner product of $\nabla\phi$ with an arbitrary vector **a**. Thus, for $\phi(\mathbf{x}) = \mathbf{x}\cdot\mathbf{g}$ where **g** is a constant vector, we get $\mathbf{a}\cdot\nabla(\mathbf{x}\cdot\mathbf{g}) = \mathbf{a}\cdot\mathbf{g}$ from Equation (8.5), hence

$$\nabla\mathbf{x}\cdot\mathbf{g} = \mathbf{g}\,. \tag{8.37}$$

Similarly, from (8.7) and (8.8) we get

$$\nabla\mathbf{x}^2 = 2\mathbf{x}\,, \tag{8.38}$$

$$\nabla\,|\,\mathbf{x}\,| = \hat{\mathbf{x}}\,. \tag{8.39}$$

These formulas enable us to evaluate the gradient of certain functions without referring to the directional derivative at all. Thus, if $F = F(|\,\mathbf{x}\,|)$ is a function of the magnitude of **x** but not its direction, then, by using (8.39) in connection with the chain rule (8.4), we get

$$\nabla F = \hat{\mathbf{x}}\,\frac{\partial F}{\partial\,|\,\mathbf{x}\,|}\;. \tag{8.40}$$

In NFII, we shall see that *the gradient operator* ∇ *can be regarded as the derivative with respect to the vector* **x**. Then we shall see that the directional derivative $\mathbf{a}\cdot\nabla$ can indeed be regarded as the inner product of a vector **a** with a vector operator ∇, just as our notation suggests.

2-8. Exercises

(8.1) Evaluate the derivatives

$$\mathbf{a}\cdot\nabla(\mathbf{x}\times\mathbf{b})\,,$$
$$\mathbf{a}\cdot\nabla(\mathbf{x}\cdot\langle A\rangle_r)\,,$$
$$\mathbf{a}\cdot\nabla[\mathbf{x}\cdot(\mathbf{x}\wedge\mathbf{b})]\,.$$

where **b** and A are independent of **x**.

(8.2) Let $\mathbf{r} = \mathbf{r}(\mathbf{x}) = \mathbf{x} - \mathbf{x}'$ and $r = |\,\mathbf{r}\,| = |\,\mathbf{x} - \mathbf{x}'\,|$, where $\mathbf{x}'$ is a vector independent of **x**. Verify the following derivatives:

(a) $\mathbf{a}\cdot\nabla r = \mathbf{a}\cdot\hat{\mathbf{r}}$

(b) $\mathbf{a}\cdot\nabla\hat{\mathbf{r}} = \dfrac{\hat{\mathbf{r}}\hat{\mathbf{r}}\wedge\mathbf{a}}{r}$

(c) $\mathbf{a}\cdot\nabla(\hat{\mathbf{r}}\cdot\mathbf{a}) = \dfrac{|\,\hat{\mathbf{r}}\wedge\mathbf{a}\,|^2}{r}$

(d) $\mathbf{a}\cdot\nabla(\hat{\mathbf{r}}\wedge\mathbf{a}) = \dfrac{\hat{\mathbf{r}}\cdot\mathbf{a}\mathbf{a}\wedge\hat{\mathbf{r}}}{r}$

(e) $\mathbf{a}\cdot\nabla\,|\,\hat{\mathbf{r}}\wedge\mathbf{a}\,| = -\,\dfrac{\hat{\mathbf{r}}\cdot\mathbf{a}\,|\,\hat{\mathbf{r}}\wedge\mathbf{a}\,|}{r}$

(f) $\mathbf{a}\cdot\nabla\frac{1}{\mathbf{r}} = -\frac{1}{\mathbf{r}}\,\mathbf{a}\,\frac{1}{\mathbf{r}}$

(g) $\mathbf{a}\cdot\nabla\frac{1}{r^2} = -2\,\frac{\mathbf{a}\cdot\hat{\mathbf{r}}}{r^3}$

(h) $\frac{1}{2}(\mathbf{a}\cdot\nabla)^2\,\frac{1}{r^2} = \frac{3(\mathbf{a}\cdot\hat{\mathbf{r}})^2 - |\,\hat{\mathbf{r}}\wedge\mathbf{a}\,|^2}{r^4}$

(i) $\frac{1}{6}(\mathbf{a}\cdot\nabla)^3\,\frac{1}{r^2} = -\,\frac{4(\mathbf{a}\cdot\hat{\mathbf{r}})^3 + 4\,|\,\hat{\mathbf{r}}\wedge\mathbf{a}\,|^2\,\mathbf{a}\cdot\hat{\mathbf{r}}}{r^5}$

(j) $\mathbf{a}\cdot\nabla\log r = \frac{\mathbf{a}\cdot\mathbf{r}}{r^2}$

(k) $\mathbf{a}\cdot\nabla\mathbf{r}^{2k} = 2k\mathbf{a}\cdot\mathbf{r}\mathbf{r}^{2(k-1)}$

(l) $\mathbf{a}\cdot\nabla\mathbf{r}^{2k+1} = \mathbf{r}^{2k}\,(\mathbf{a} + 2k\mathbf{a}\cdot\hat{\mathbf{r}}\hat{\mathbf{r}})$.

In the last two cases, k is any nonzero integer and $\mathbf{r} \neq 0$ if $k < 0$.

(8.3) Show that the Taylor expansion of $(\mathbf{x} - \mathbf{a})^{-1}$ about $\mathbf{x}$ is term by term equivalent to the series

$$\frac{1}{\mathbf{x}-\mathbf{a}} = \frac{1}{\mathbf{x}} + \frac{1}{\mathbf{x}}\mathbf{a}\frac{1}{\mathbf{x}} + \frac{1}{\mathbf{x}}\mathbf{a}\frac{1}{\mathbf{x}}\mathbf{a}\frac{1}{\mathbf{x}} + \ldots .$$

The series is convergent if $|\,\mathbf{a}\,| < |\,\mathbf{x}\,|$.

(8.4) The *Legendre Polynomials* $P_n(\mathbf{xa})$ can be defined as the coefficients in the power series expansion

$$\frac{1}{|\,\mathbf{x}-\mathbf{a}\,|} = \sum_{n=0}^{\infty}\frac{P_n(\hat{\mathbf{x}}\mathbf{a})}{|\,\mathbf{x}\,|^{n+1}} = \sum_{n=0}^{\infty}\frac{P_n(\mathbf{xa})}{|\,\mathbf{x}\,|^{2n+1}}$$

$$= \frac{P_0(\mathbf{xa})}{|\,\mathbf{x}\,|} + \frac{P_1(\mathbf{xa})}{|\,\mathbf{x}\,|^3} + \frac{P_2(\mathbf{xa})}{|\,\mathbf{x}\,|^5} + \frac{P_3(\mathbf{xa})}{|\,\mathbf{x}\,|^7} + \ldots$$

The series converges for $|\,\mathbf{a}\,| < |\,\mathbf{x}\,|$.
Use a Taylor expansion to evaluate the polynomials of lowest order:

$$P_0(\mathbf{xa}) = 1,$$
$$P_1(\mathbf{xa}) = \mathbf{x}\cdot\mathbf{a},$$
$$P_2(\mathbf{xa}) = \tfrac{1}{2}[3(\mathbf{x}\cdot\mathbf{a})^2 - \mathbf{a}^2\,\mathbf{x}^2] = (\mathbf{x}\cdot\mathbf{a})^2 + \tfrac{1}{2}(\mathbf{x}\wedge\mathbf{a})^2,$$
$$P_3(\mathbf{xa}) = \tfrac{1}{2}[5(\mathbf{x}\cdot\mathbf{a})^3 - 3\mathbf{a}^2\,\mathbf{x}^2\,\mathbf{x}\cdot\mathbf{a}] = (\mathbf{x}\cdot\mathbf{a})^3 + \tfrac{3}{2}\mathbf{x}\cdot\mathbf{a}(\mathbf{x}\wedge\mathbf{a})^2.$$

The $P_n(\mathbf{x}\cdot\mathbf{a})$ are polynomials of vectors. Show that they are *homogeneous* functions of degree n, that is,

$$P_n(\mathbf{xa}) = |\,\mathbf{x}\,|^n\,P_n(\hat{\mathbf{x}}\mathbf{a}) = |\,\mathbf{x}\,|^n\,|\,\mathbf{a}\,|^n\,P_n(\hat{\mathbf{x}}\cdot\hat{\mathbf{a}}) .$$

(8.5) Verify the value given for the following line integral:

$$\int_{\mathbf{a}}^{\mathbf{b}} \mathbf{x}^{-1}\, \mathrm{d}\mathbf{x} = \log(\mathbf{a}^{-1}\mathbf{b}),$$

taken along any continuous curve in the $(\mathbf{a}\wedge\mathbf{b})$-plane which does not pass through the origin. Write separate integrals for the k-vector parts. Note that the integral is multivalued, and specify conditions on the curve which give the principal value of the logarithm. (Recall the discussion of the logarithmic function in Section 2-5.)

Introduce the spinor-valued function $z = z(\mathbf{x}) = \mathbf{a}^{-1}\mathbf{x}$ and a parametric equation $\mathbf{x} = \mathbf{x}(t)$ for the curve and show that the integral can be written in the equivalent forms

$$\int_{\mathbf{a}}^{\mathbf{b}} \mathbf{x}^{-1}\, \mathrm{d}\mathbf{x} = \int_0^t \frac{\dot{z}}{z}\, \mathrm{d}t = \int_1^{\mathbf{a}^{-1}\mathbf{b}} \frac{\mathrm{d}z}{z}\,.$$

The last form of the integral is discussed extensively in textbooks on "Functions of a Complex Variable".

Chapter 3

Mechanics of a Single Particle

In this chapter we learn how to use geometric algebra to describe and analyze the motion of a single particle. From a physical point of view, we will be concerned with constructing the simplest models for a physical system and discussing their applications and limitations. From a mathematical point of view we will be concerned with solving the simplest second order vector differential equations and analyzing the geometrical properties of the solutions.

Most of the results of Chapter 2 will be used in this chapter in one way or another. Of course, the main mathematical tool will be the geometric algebra of physical space. Since the reader is presumed to have some familiarity with mechanics already, a number of basic terms and concepts will be used with no more than the briefest introduction. A critical and systematic analysis of the foundations of mechanics will be undertaken in Chapter 9. General methods for solving differential equations will be developed as they are needed.

Although only the simplest models and differential equations are considered in this chapter, the results should not be regarded as trivial, for the simple models provide the starting point for the analyzing and solving complex problems. Consequently, the time required for developing an elegant formulation and thorough analysis of simple models is time well spent.

3-1. Newton's Program

Isaac Newton (1642–1727) is rightly regarded as the founder of the science called mechanics. Of course, he was neither the first nor the last to make important contributions to the subject. He deserves the title of "founder", because he integrated the insights of his predecessors into a comprehensive theory. Furthermore, he inaugurated a program to refine and extend that theory by systematically investigating and classifying the properties of all physical objects. Newtonian mechanics is, therefore, more than a particular scientific theory; it is a well-defined program of research into the structure of the physical world.

This section reviews the major features of Newtonian mechanics as it stands today. Naturally, a modern formulation of mechanics differs somewhat from Newton's, but this is not the place to trace the intricacies of its evolution. Also, for the time being we take the fundamental concepts of space and time for granted, just as Newton did in his *Principia*. It will not be profitable to wrestle with subtleties in the foundations of mechanics until some proficiency with the mathematical formalism has been developed, so we delay the attempt until Chapter 9.

The grand goal of Newton's program is to describe and explain all properties of all physical objects. The approach of the program is determined by two general assumptions: first, that every physical object can be represented as a composite of particles; second, that the behavior of a particle is governed by interactions with other particles. The properties of a physical object, then, are determined by the properties of its parts. For example, structural properties of an object, such as rigidity or plasticity, are determined by the interactions among its parts. The program of mechanics is to explain the diverse properties of objects in our experience in terms of a few kinds of interactions among a few kinds of particles.

The great power of Newtonian mechanics is achieved by formulating the generalities of the last paragraph in specific mathematical terms. It depends on a clear formulation of the key concepts: *particle* and *interaction*. A particle is understood to be an object with a definite orbit in space and time. The *orbit* is represented by a function $\mathbf{x} = \mathbf{x}(t)$ which specifies the particle's *position* $\mathbf{x}$ at each time t. To express the continuous existence of the particle in some interval of time, the function $\mathbf{x}(t)$ must be a continuous function of the variable t in that interval. When specified for all times in an interval, the function $\mathbf{x}(t)$ describes a *motion* of the particle.

The *central* hypothesis of Newtonian mechanics is that variations in the motion of a particle are completely determined by its interactions with other particles. More specifically, the motion is determined by an equation of the general form

$$\mathbf{f} = m\ddot{\mathbf{x}} \tag{1.1}$$

where $\ddot{\mathbf{x}}$ is the acceleration of the particle, the scalar m is a constant called the mass of the particle, and the force $\mathbf{f}$ expresses the influence of other particles. This hypothesis is commonly referred to as "Newton's second law of motion", though it was Euler who finally cast it in the form we use today.

Newton's Law (1.1) becomes a definite differential equation determining the motion of a particle only when the force $\mathbf{f}$ is expressed as a specific function of $\mathbf{x}(t)$ and its derivatives. With this much understood, the thrust of Newton's program can be summarized by the dictum: *focus on the forces*. This should be interpreted as an admonition to study the motions of physical objects and find forces of interaction sufficient to determine those motions. The aim is to classify the kinds of forces and so develop a classification of particles according to the kinds of interactions in which they participate. The classification is not complete today, but it has been carried a long way.

Newton's program has been so successful largely because it has proved possible to account for the motions of physical objects with forces of simple mathematical form. The "forces of nature" appear to have two properties of such universality that they could be regarded as laws, though they were not identified as such by Newton. We shall refer to them as the principles of additivity and analyticity.

According to the principle of *additivity* (or *superposition*) *of forces*, the force $\mathbf{f}$ on a given particle can be expressed as the vector sum of forces $\mathbf{f}_k$ independently exerted by each particle with which it interacts, that is,

$$\mathbf{f} = \mathbf{f}_1 + \mathbf{f}_2 + \ldots = \sum_k \mathbf{f}_k . \tag{1.2}$$

This principle enables us to isolate and study different kinds of forces independently as well as reduce complex forces to a superposition of simple forces. Newton's program could hardly have progressed without it.

According to the *principle of analyticity* (or *continuity*), the force of one particle on another is an exclusive, analytic function of the positions and velocities of both particles. The adjective "exclusive" means that no other variables are involved. The adjective "analytic" means that the function is smoothly varying in the sense that derivatives (with respect to time) of all orders have finite values. The principle of analyticity implies that the force $\mathbf{f}$ on a particle with position $\mathbf{x} = \mathbf{x}(t)$ and velocity $\dot{\mathbf{x}} = \dot{\mathbf{x}}(t)$ is always an analytic function

$$\mathbf{f} = \mathbf{f}(\dot{\mathbf{x}}, \mathbf{x}, t), \tag{1.3}$$

where the explicit time dependence arises from motions of the particles determining the force. Notice that (1.3) assumes that the force is not a function of $\ddot{\mathbf{x}}$ and higher order derivatives. An exception to this rule is the so-called "radiative reaction force" which requires special treatment that we cannot go into here. Mathematical idealizations or approximations that violate the analyticity principle are often useful, as long as their ranges of validity are understood.

A specific functional form for the force on a particle is commonly called a *force law*. For example, a force of the form $\mathbf{f} = -k\mathbf{x}$ is called "Hooke's Law". It should be understood, however, that there is more to a force law than a mathematical formula; it is essential to know the law's *domain of validity*, that is, the circumstances in which it applies and the fidelity with which it represents the phenomena. Much of physics is concerned with determining the domains of validity for specific force laws, so it shall be our concern as well throughout this book. Therefore, we can only sketch a classification of forces here. The major distinction to be made is between *fundamental* and *approximate* force laws.

Physicists have discovered four kinds of fundamental forces, the gravitational, the electromagnetic, the strong and the weak forces. They are called "fundamental", because every known force can be understood as a superposition and approximation of these forces. The strong and the weak forces were discovered only fairly recently, because their effects on ordinary human experience are quite indirect. The strong forces *bind* the atomic nuclei; so they determine the naturally occurring elements, but they do not otherwise come into play in everyday experience. The weak forces govern radioactive decay. The strong and weak forces are not as well understood as the electromagnetic and gravitational forces, so they are major objects of basic research today. Furthermore, they are mathematically formulated in terms of "quantum mechanics" which goes beyond the "classical mechanics" developed here. For these reasons, they will not be discussed further in this book.

The gravitational force of a particle at a point $\mathbf{x}' = \mathbf{x}'(t)$ on a particle at $\mathbf{x} = \mathbf{x}(t)$ is given by

$$\mathbf{f}(\mathbf{x}, t) = -mm'G\frac{\mathbf{x} - \mathbf{x}'(t)}{|\mathbf{x} - \mathbf{x}'(t)|^3}, \tag{1.4}$$

where m and m' are masses of the particles and G is an empirically known constant. This is *Newton's gravitational force law*. Its domain of validity is immense. It applies to all particles with masses, and only minute deviations from it can be detected in the most sensitive astronomical experiments. For practical reasons, we often work with approximations to Newton's law, but the law's great validity enables us to estimate the accuracy of our approximations with great confidence whenever necessary.

The electromagnetic force on a particle with charge q has the form

$$\mathbf{f}(\dot{\mathbf{x}}, \mathbf{x}, t) = q\left(\mathbf{E} + \frac{\dot{\mathbf{x}}}{c} \times \mathbf{B}\right), \tag{1.5}$$

where $\mathbf{E} = \mathbf{E}(\mathbf{x}, t)$ is an electric field, $\mathbf{B} = \mathbf{B}(\mathbf{x}, t)$ is a magnetic field, and c is a constant with value equal to the speed of light. This is commonly called the *Lorentz force law* in honor of the man who first used it extensively to analyze the electromagnetic properties of material media. The charge q in (1.5) is a scalar constant characteristic of the particle; it can be positive, negative or zero; consequently, electromagnetic interactions determine a three-fold classification of particles into positively, negatively or neutrally charged groups. The electric and magnetic fields in (1.5) can in principle be expressed as functions of the positions and velocities of the particles that produce them, but this is totally impractical in most applications, because the number of particles is very large, and often the information would be irrelevant because only a simple approximation to the functional dependence on $\mathbf{x}$ and t is required.

The known consequences and applications of electromagnetic interactions are vastly richer and more numerous than those of the other interactions. Research during the last hundred years or so has established the astounding fact that, aside from simple gravitational attraction, all the manifold properties of familiar physical objects can be explained as consequences of electromagnetic interactions. This includes explanations of solid, liquid and gaseous phases of matter, thermal and electrical conduction and resistance, the varieties of chemical binding and reaction, even the colors of objects are explained as electromagnetic interactions of particles with light. The full explanation requires "quantum mechanics", but the "classical mechanics" of concern in this book is indispensible, and its resources are far from exhausted. In all this the Lorentz force law (1.5) plays a crucial role, so it must be regarded as one of the most important mathematical expressions in physics.

The abstract mathematical framework of Newtonian mechanics does not specify any force laws. Newton began the search for "force laws of nature" with the stunning proposal of his gravitational force law, which has served as a paradigm for force laws ever since. The fundamental force laws were discovered by a combination of empirical study and mathematical analysis. The importance of mathematical analysis in this endeavor should not be underestimated, even in Newton's initial discovery. Textbooks proceed rapidly to Newton's law, but Newton spent years preparing himself mathematically. His preparatory studies in analytic geometry led him to a complete classification of third order algebraic plane curves, which is far beyond what students encounter today. The extent of Newton's mathematical preparation is evident in *The Mathematic Papers of Isaac Newton*, recently published under the careful editorship of D. T. Whiteside. Every serious student of physics should become acquainted with these splendid volumes.

In this chapter we will be engaged in specific and general studies of a variety of force laws. As will be seen, mathematical analysis leads to a classification of force laws according to their mathematical properties. This is of great importance for many reasons: (1) It helps us identify common properties of many force laws and systematically zero in on the specific law appropriate in a given situation. (2) To understand why one force law is fundamental rather than another, we must examine a range of possibilities and learn to distinguish the crucial from the unimportant properties of fundamental laws. (3) If we hope to refine or improve present laws, we must know what reasonable possibilities are available. (4) If we hope to develop a unified theory of the fundamental laws, we must know how they differ and what they have in common. (5) Finally, mathematical analysis is essential for practical approximations and applications of the fundamental laws.

We have mentioned the important distinction between fundamental and approximate laws. In the macroscopic domain of familiar physical objects, we deal mostly with approximate laws, because a reduction to fundamental laws is impractical. We call them "approximate laws" for the obvious reason that they approximate fundamental laws. They are also called "empirical" or "phenomenological" laws, because their relation to empirical evidence is fairly direct, though there is usually reason to believe that the relation is incomplete or approximate in some respect.

It is useful to distinguish between *long range* and *short range* forces. Short range approximate forces are also called *contact forces*, because they are exerted by the surface of one body or medium on the surface of another body in contact with it. The forces exerted by molecules on the two surfaces have a macroscopically short range. The resultant force supposed to act on either body is necessarily approximate, because so many molecules are involved. As examples of contact forces, we list friction, viscosity and bouyant forces. Long range approximate forces are also called *body forces*, because they are exerted on particles throughout a macroscopic body. The gravitational force exerted by the Earth is the most familiar force of this kind.

Once the force law has been determined, the problem of determining the motion of a particle is a strictly mathematical one. According to (1.1) and (1.3), the equation of motion necessarily has the form

$$m\ddot{\mathbf{x}} = \mathbf{f}(\dot{\mathbf{x}}, \mathbf{x}, t). \tag{1.6}$$

This is called a *second order differential equation*, because it contains no derivatives of the dependent variable $\mathbf{x}$ with respect to the independent variable t of order greater than the second. Books on the theory of differential equations prove that if $\mathbf{f}$ is an analytic function, then (1.6) has a *unique general solution* depending only on two *arbitrary vector constants* ("vector", because the dependent variable is a vector.) Designating these constants by $\mathbf{a}$ and $\mathbf{b}$, the general solution of (1.6) can be written as a function of the form

$$\mathbf{x} = \mathbf{x}(t, \mathbf{a}, \mathbf{b}). \tag{1.7}$$

If desired, the constants can be determined from initial conditions, that is, the position $\mathbf{x}_0$ and the velocity $\mathbf{v}_0$ of the particle at time $t = 0$; thus from (1.7) we get the equations

$$\begin{aligned} \mathbf{x}(0, \mathbf{a}, \mathbf{b}) &= \mathbf{x}_0, \\ \dot{\mathbf{x}}(0, \mathbf{a}, \mathbf{b}) &= \mathbf{v}_0. \end{aligned} \tag{1.8}$$

Being two simultaneous vector equations in two vector variables, these equations determine $\mathbf{a}$, $\mathbf{b}$ from $\mathbf{x}_0$ and $\mathbf{v}_0$ or vice-versa.

We shall have many occasions to use this important general theorem, in particular, when solving the equations of motion. There are many ways to solve differential equations, but, whichever way we use, we will know that we have found *all* possible solutions if we have found one solution depending on two arbitrary independent vector constants. By assigning these constants appropriate values we determine a *unique particular* solution, for example, the one with given initial values $\mathbf{x}_0$, $\mathbf{v}_0$. It follows from the general theorem that if the position and velocity of a particle subject to a known force are specified at any time, then the position and velocity at any subsequent time are uniquely determined. For this reason, the position and velocity are commonly called *state variables* and are said to designate the *state* of a particle.

3-2. Constant Force

In this section we study the motion of a particle subject to a constant gravitational force $\mathbf{f} = m\mathbf{g}$. Of course, our results describe the motion of a particle with charge q in a constant electric field $\mathbf{E}$ just as well; it is only necessary to write $\mathbf{g} = (q/m)\mathbf{E}$ for the electrical force per unit mass.

According to the fundamental law of mechanics, a particle subject to a constant force undergoes a constant acceleration. For a force per unit mass $\mathbf{g}$, the particle trajectory $\mathbf{x} = \mathbf{x}(t)$ is determined by the differential equation

$$\ddot{\mathbf{x}} = \dot{\mathbf{v}} = \mathbf{g} \tag{2.1}$$

subject to the initial conditions

$$\dot{\mathbf{x}}(0) \equiv \mathbf{v}(0) \equiv \mathbf{v}_0, \tag{2.2a}$$

$$\mathbf{x}(0) \equiv \mathbf{x}_0. \tag{2.2b}$$

Using (2–7.19) and (2–7.17), Equation (2.1) can be integrated directly to get the velocity $\mathbf{v} = \mathbf{v}(t)$ at any time t; thus,

$$\dot{\mathbf{x}} = \mathbf{v} = \mathbf{g}t + \mathbf{v}_0. \tag{2.3}$$

A second integration gives

$$\mathbf{r} \equiv \mathbf{x} - \mathbf{x}_0 = \tfrac{1}{2}\mathbf{g}t^2 + \mathbf{v}_0 t. \tag{2.4}$$

This is a parametric equation for the displacement of the particle as a function of time. The trajectory is a segment of a parabola, as shown in Figure 2.1. The solution (2.4) presents the parabolic motion of the particle as the superposition of two linear motions. The term $\mathbf{v}_0 t$ can be interpreted as the displacement of a point at rest in a reference system moving with velocity $\mathbf{v}_0$, while the term $\frac{1}{2}\mathbf{g}t^2$ is interpreted as the displacement of the particle *initially at rest* in the moving system. Accordingly, the constants in (2.4) determine a "natural" coordinate system for locating the particle at any time; the origin is determined by $\mathbf{x}_0$, while coordinate directions and scales are given by the vectors $\mathbf{v}_0$ and $\mathbf{g}$. This is a skew coordinate system, but the particle's natural position coordinates t and $\frac{1}{2}\mathbf{g}t^2$ are obviously quadratically related, just as they are in the familiar representation by rectangular coordinates. Mathematically speaking the quantity $\frac{1}{2}\mathbf{g}t^2$ is a particular solution of the equation of motion (2.1), while $\mathbf{v}_0 t + \mathbf{x}_0$ is the general solution of the homogeneous differential equation $\ddot{\mathbf{x}} = 0$.

The description of motion can be simplified by representing it in velocity space instead of position space. A curve traced by the velocity $\mathbf{v} = \mathbf{v}(t)$ is

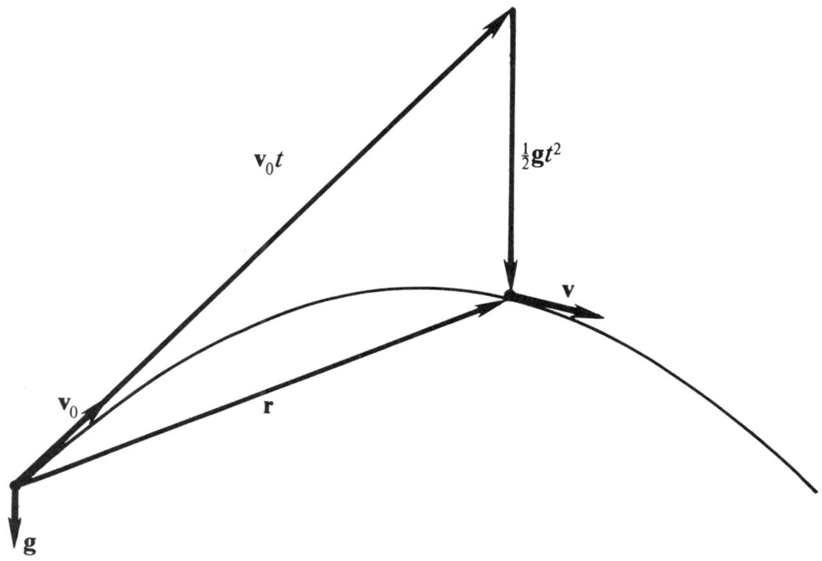

Fig. 2.1 The trajectory in *position space*.

called a *hodograph*. According to (2.3), the hodograph of a particle subject to a constant force is a straight line. In velocity space the location of the particle can be represented by the average velocity $\bar{\mathbf{v}} = \bar{\mathbf{v}}(t)$. The parametric equation for $\bar{\mathbf{v}}$ is obtained by writing (2.4) in the form

$$\bar{\mathbf{v}} = \frac{\mathbf{r}}{t} = \tfrac{1}{2}\mathbf{g}t + \mathbf{v}_0 . \tag{2.5}$$

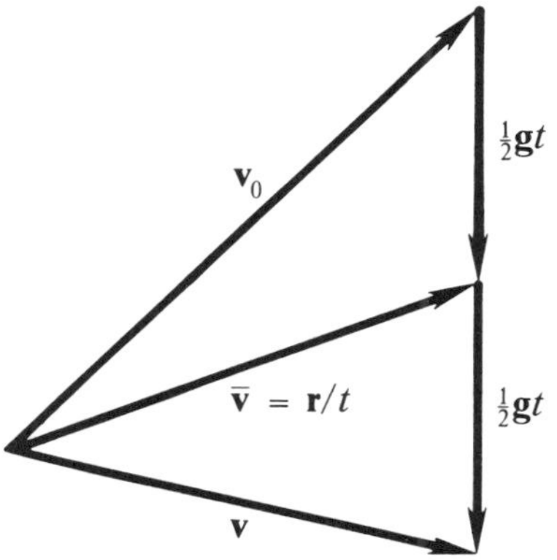

Fig. 2.2. The trajectory in *velocity space*.

As illustrated in Figure 2.2, the trajectory $\bar{\mathbf{v}}(t)$ is simply a vertical straight line beginning at $\mathbf{v}_0$. Figure 2.2 also shows the simple relation of the average velocity (2.5) to the actual velocity (2.3), a relation which is disguised if the equation for displacement (2.4) is used directly. Figure 2.2 contains all the information about the projectile motion, so all questions about the motion can be answered by solving the triangles in the figure, graphically or algebraically, for the relevant variables. Let us see how to do this efficiently.

Projectile Range

Although the displacement $r = |\,\mathbf{r}\,|$ is represented only indirectly in velocity space, it is nonetheless easy to compute. Consider, for example, the problem of determining the range r of a target sighted in a direction $\hat{\mathbf{r}}$, which has been hit by a projectile launched with velocity $\mathbf{v}_0$. This problem can easily be solved by the graphical method illustrated in Figure 2.2, assuming, of course, that the flight of the projectile is adequately represented as that of a particle with constant acceleration $\mathbf{g}$. The method simply exploits properties of Figure 2.2. Having "laid out" $\mathbf{v}_0$ on graph paper, as indicated in Figure 2.3, one extends a line from the base of $\mathbf{v}_0$ in the direction $\hat{\mathbf{r}}$ to its intersection with a vertical line extending from the tip of $\mathbf{v}_0$. The lengths of the two sides of the triangle thus constructed are then measured to get the values of $\frac{1}{2}gt$ and r/t, from which one can compute r and t as well, if desired. Figure 2.2 also shows how the construction can be extended to determine the final velocity $\mathbf{v}$ of the projectile. Angles with the horizontal are indicated in Figure 2.2, since they are commonly used to specify relative direction in practical problems.

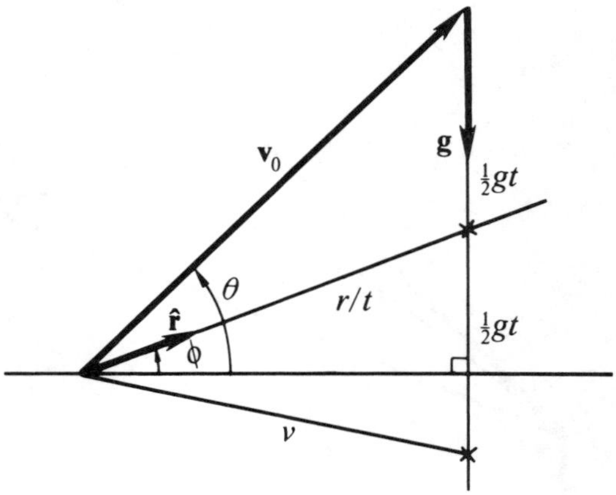

Fig. 2.3. Graphical determination of the displacement r, time of flight t and the final velocity $\mathbf{v}$.

The same problem can be solved by algebraic means. Outer multiplication of (2.5) by $\mathbf{r}$ produces

$$\tfrac{1}{2}t\mathbf{g}\wedge\mathbf{r} = \mathbf{r}\wedge\mathbf{v}_0 .$$

Hence

$$t = \frac{2\hat{\mathbf{r}}\wedge\mathbf{v}_0}{\mathbf{g}\wedge\hat{\mathbf{r}}} = \frac{2v_0}{g}\frac{|\hat{\mathbf{r}}\wedge\hat{\mathbf{v}}_0|}{|\hat{\mathbf{g}}\wedge\hat{\mathbf{r}}|} . \tag{2.6}$$

This completely determines t from the target direction $\hat{\mathbf{r}}$ and the initial velocity $\mathbf{v}_0$. We can get one other relation from (2.5) by "wedging" it with $\mathbf{g}t$, namely

$$\mathbf{g}\wedge\mathbf{r} = t\mathbf{g}\wedge\mathbf{v}_0 . \tag{2.7}$$

We can solve this for r and use (2.6) to eliminate t, thus

$$r = \frac{\mathbf{g}\wedge\mathbf{v}_0}{\mathbf{g}\wedge\hat{\mathbf{r}}}t = \frac{2\mathbf{g}\wedge\mathbf{v}_0\,\hat{\mathbf{r}}\wedge\mathbf{v}_0}{(\mathbf{g}\wedge\hat{\mathbf{r}})^2} = \frac{2(\mathbf{g}\wedge\mathbf{v}_0)\cdot(\mathbf{v}_0\wedge\hat{\mathbf{r}})}{|\,\mathbf{g}\wedge\hat{\mathbf{r}}\,|^2} . \tag{2.8}$$

Thus, the range r has been expressed as an explicit function of the given vectors $\mathbf{v}_0$, $\hat{\mathbf{r}}$ and $\mathbf{g}$.

Maximum Range

The algebraic expression (2.8) for the range supplies more information than the corresponding graphical construction. For example, for a fixed "muzzle velocity" $\mathbf{v}_0$ and target direction $\hat{\mathbf{r}}$, Equation (2.8) gives the range r as a function of the firing direction $\hat{\mathbf{v}}_0$. The variation of r with changes in $\hat{\mathbf{v}}_0$ is unclear from the form of (2.8), because an increase of $\hat{\mathbf{g}}\wedge\mathbf{v}_0$ will be accompanied by a decrease of $\mathbf{v}_0\wedge\hat{\mathbf{r}}$, so their product might either increase or decrease. The functional dependence is made more obvious by using the identity

$$2(\hat{\mathbf{g}}\wedge\mathbf{v}_0)\cdot(\mathbf{v}_0\wedge\hat{\mathbf{r}}) = \mathrm{v}_0^2\,[\hat{\mathbf{g}}\cdot\hat{\mathbf{r}} - \hat{\mathbf{g}}\cdot(\hat{\mathbf{v}}_0\,\hat{\mathbf{r}}\hat{\mathbf{v}}_0)], \tag{2.9}$$

established in Exercise (2–4.8d). The first term on the right side of (2.9) is constant, while the second term is a function of the vector $\hat{\mathbf{v}}_0\hat{\mathbf{r}}\hat{\mathbf{v}}_0$ (the reader is invited to construct a diagram showing the relation of this vector to $\hat{\mathbf{r}}$ and $\hat{\mathbf{v}}_0$). The direction of $\hat{\mathbf{v}}_0$ which gives maximum range is obtained by maximizing the value of (2.9). It is readily verified that $\hat{\mathbf{v}}_0\hat{\mathbf{r}}\hat{\mathbf{v}}_0$ is a unit vector, so (2.9) has its maximum value when

$$-\hat{\mathbf{g}}\cdot(\hat{\mathbf{v}}_0\hat{\mathbf{r}}\hat{\mathbf{v}}_0) = 1.$$

From this we can conclude that

$$-\hat{\mathbf{g}} = \hat{\mathbf{v}}_0\hat{\mathbf{r}}\hat{\mathbf{v}}_0,$$

or equivalently,

$$-\hat{\mathbf{g}}\hat{\mathbf{v}}_0 = \hat{\mathbf{v}}_0\hat{\mathbf{r}}. \tag{2.10}$$

This equation tells us that the angle between $-\hat{\mathbf{g}}$ and $\hat{\mathbf{v}}_0$ must be equal to the angle between $\hat{\mathbf{r}}$ and $\hat{\mathbf{v}}_0$. Thus, the vector $\hat{\mathbf{v}}_0$ bisects the angle between $\hat{\mathbf{r}}$ and $-\hat{\mathbf{g}}$ (Figure 2.4), so we can express $\hat{\mathbf{v}}_0$ as a function of $\hat{\mathbf{r}}$ and $\hat{\mathbf{g}}$ by

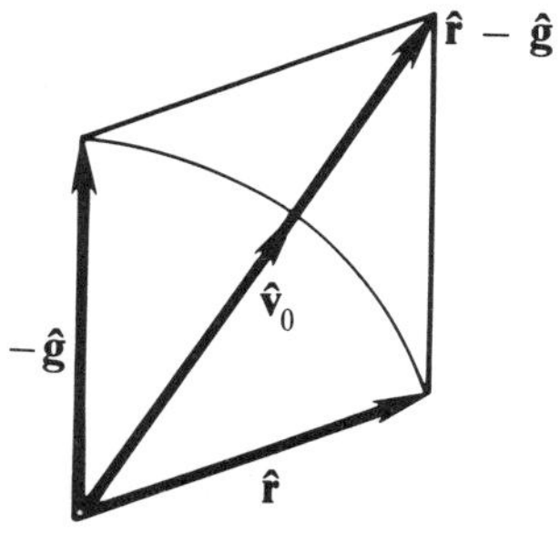

Fig. 2.4. For maximum range, shoot in a direction bisecting the vertical $-\hat{\mathbf{g}}$ and the line of sight $\hat{\mathbf{r}}$.

$$\hat{\mathbf{v}}_0 = \frac{\hat{\mathbf{r}} - \hat{\mathbf{g}}}{|\,\hat{\mathbf{r}} - \hat{\mathbf{g}}\,|}\,. \tag{2.11}$$

Substituting (2.11) in (2.8), one immediately finds the following expression for the maximum range:

$$r_{\max} = \frac{2v_0^2}{g}\,\frac{1}{|\,\hat{\mathbf{r}} - \hat{\mathbf{g}}\,|^2} = \frac{v_0^2}{g}\,\frac{1}{1 - \hat{\mathbf{g}}\cdot\hat{\mathbf{r}}} \tag{2.12}$$

We saw in Section 2-6 that this is an equation for a paraboloid of revolution; here, it expresses $r_{\max}$ as a function of the target direction $\hat{\mathbf{r}}$. As illustrated in Figure 2.5, the

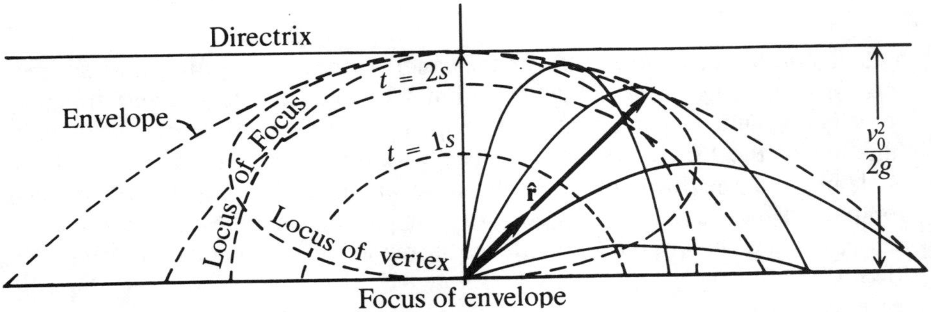

Fig. 2.5. The envelope of all trajectories with the same initial speed is a paraboloid of revolution [Figure redrawn from W. D. MacMillan, *Theoretical Mechanics*, reprinted by Dover, N. Y. (1958)].

paraboloid (2.12) is the envelope of all trajectories emanating from the origin with the same initial speed v_0. Thus, only points under the paraboloid can be reached by a projectile.

The relation between t and r in (2.7) can be expressed differently by eliminating $\mathbf{g}\wedge\mathbf{r}$ between (2.6) and (2.7) to get

$$\tfrac{1}{2}t^2 = \frac{\mathbf{r}\wedge\mathbf{v}_0}{\mathbf{g}\wedge\mathbf{v}_0} \; . \tag{2.13}$$

For maximum range, one sees from either (2.10) or (2.11) that $\hat{\mathbf{r}}\wedge\mathbf{v}_0 = \hat{\mathbf{g}}\wedge\mathbf{v}_0$, in which case (2.12) and (2.13) give

$$t^2 = \frac{2r_{\max}}{g} = \frac{2v_0^2}{g^2}\,\frac{1}{1-\hat{\mathbf{g}}\cdot\hat{\mathbf{r}}} \; . \tag{2.14}$$

So far all results have been obtained by analyzing the upper triangle in Figure 2.2. The lower triangle contains additional information. The information in both triangles can be represented in a symmetrical form by extending Figure 2.2 to a parallelogram, as shown in Figure 2.6a. The parallelogram can be characterized algebraically by the equations for its diagonals;

$$\mathbf{v} - \mathbf{v}_0 = \mathbf{g}t, \tag{2.15}$$

$$\mathbf{v} + \mathbf{v}_0 = \frac{2\mathbf{r}}{t} = 2\bar{\mathbf{v}}. \tag{2.16}$$

These are, of course, equivalent to the basic equations (2.3) and (2.4).

Multiplying equations (2.15) and (2.16) to eliminate t, one obtains

$$2\mathbf{gr} = (\mathbf{v} - \mathbf{v}_0)(\mathbf{v} + \mathbf{v}_0) = v^2 - v_0^2 + 2\mathbf{v}\wedge\mathbf{v}_0 . \tag{2.17}$$

Therefore,

$$2\mathbf{g}\cdot\mathbf{r} = v^2 - v_0^2, \tag{2.18}$$

$$\mathbf{g}\wedge\mathbf{r} = \mathbf{v}\wedge\mathbf{v}_0 . \tag{2.19}$$

Equation (2.18) will be recognized as expressing conservation of energy. It determines the final speed $v = |\,\mathbf{v}\,|$ of the particle in terms of initial data. On the other hand, Equation (2.19) determines the relative direction of initial and final velocities.

A glance at Figure 2.5 suggests that, there are two distinct trajectories with initial speed v_0 passing through any given point $\mathbf{r}$ within the maximum range. The trajectories can be ascertained from equations of the last paragraph. Let $\mathbf{v}_0$ and $\mathbf{v}_0'$ be their initial velocities. By assumption the initial speeds are the same;

$$|\,\mathbf{v}_0\,| = |\,\mathbf{v}_0'\,| = v_0 . \tag{2.20a}$$

By (2.18), the final speeds are the same;

$$|\,\mathbf{v}\,| = |\,\mathbf{v}'\,| = v. \tag{2.20b}$$

And by (2.19), the area of the parallelogram determined by $\mathbf{v}'$ and $\mathbf{v}_0'$ is equal to the area of the parallelogram determined by $\mathbf{v}$ and $\mathbf{v}_0$;

$$\mathbf{v}\wedge\mathbf{v}_0 = \mathbf{v}'\wedge\mathbf{v}_0'. \tag{2.20c}$$

Hence the two parallelograms are similar. They are illustrated in Figures 2.6a and 2.6b. Since corresponding diagonals have equal length, it follows at once that

$$tt' = \frac{2r}{g} . \tag{2.21}$$

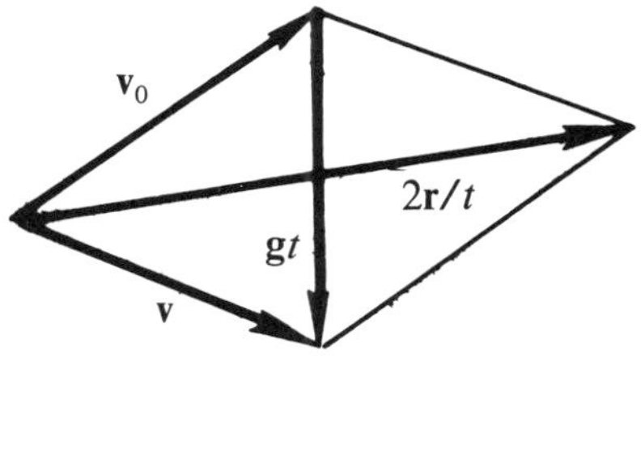

Fig. 2.6a.

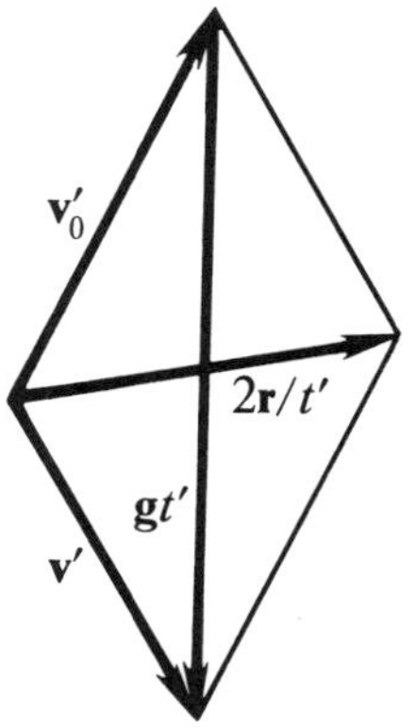

Fig. 2.6b.

This relation can be used to determine one of the two trajectories from the other. The problem is to find the direction $\hat{\mathbf{v}}_0'$ from the directions $\hat{\mathbf{r}}$ and $\hat{\mathbf{v}}_0$. This can be done by using (2.7) to get the relation

$$t'\mathbf{g}\wedge\mathbf{v}_0' = \mathbf{g}\wedge\mathbf{r} = t\mathbf{g}\wedge\mathbf{v}_0 .$$

Eliminating $|\mathbf{v}_0'|$ and t′ with (2.20a) and (2.21), we get

$$\mathbf{g}\wedge\hat{\mathbf{v}}_0' = \frac{t}{t'}\mathbf{g}\wedge\hat{\mathbf{v}}_0 = \frac{gt^2}{2r}\ \mathbf{g}\wedge\hat{\mathbf{v}}_0 .$$

Finally, by substitution from (2.4) we get the simple relation

$$\hat{\mathbf{g}}\wedge\hat{\mathbf{v}}_0' = \hat{\mathbf{r}}\wedge\hat{\mathbf{v}}_0 . \tag{2.22}$$

This says that the angle $\hat{\mathbf{v}}_0'$ makes with the vertical is equal to the angle between $\hat{\mathbf{v}}_0$ and $\hat{\mathbf{r}}$. Equation (2.22) generalizes the relation (2.10) which we found for trajectories of maximum range.

For the case of maximum range, Equation (2.21) must agree with (2.14); hence $t = t'$, and the two parallelograms in Figure 2.6 reduce to a single rectangle. Then $\mathbf{v}\cdot\mathbf{v}_0 = 0$, and (2.19) can be solved for the final velocity:

$$\mathbf{v} = (\mathbf{g}\wedge\mathbf{r})\mathbf{v}_0^{-1} = \frac{(\mathbf{g}\wedge\mathbf{r})\cdot\mathbf{v}_0}{v_0^2} . \tag{2.23}$$

All the significant properties of a trajectory with maximum range have now been determined.

Rectangular and Polar Coordinates

For some purposes it may be convenient to express the above results in terms of rectangular or polar coordinates. The positive vertical direction is represented by the unit vector $-\hat{\mathbf{g}} = -\mathbf{g}/g$. The *vertical coordinate* y of the particle is then defined by

$$y = -\hat{\mathbf{g}}\cdot\mathbf{r}, \tag{2.24}$$

while its *horizontal* coordinate x is defined by

$$\hat{\mathbf{g}}\wedge\mathbf{r} = \mathbf{i}\,|\hat{\mathbf{g}}\wedge\mathbf{r}| = \mathbf{i}x, \tag{2.25a}$$

or

$$x = |\hat{\mathbf{g}}\wedge\mathbf{r}| \geqslant 0. \tag{2.25b}$$

These relations are more useful when combined into the single set of equations

$$\hat{\mathbf{g}}\mathbf{r} = -y + \mathbf{i}x = \mathbf{i}(x+\mathbf{i}y) = \mathbf{i}re^{\mathbf{i}\phi} = re^{\mathbf{i}(\phi + \pi/2)}. \tag{2.26}$$

Similarly, for the initial velocity $\mathbf{v}_0$ one writes

$$\hat{\mathbf{g}}\mathbf{v}_0 = -v_{0y} + \mathbf{i}v_{0x} = \mathbf{i}(v_{0x} + \mathbf{i}v_{0y}) = \mathbf{i}v_0e^{\mathbf{i}\theta} . \tag{2.27}$$

As illustrated in Figure 2.7 the angles ϕ and θ measure inclinations of $\hat{\mathbf{r}}$ and $\hat{\mathbf{v}}_0$ from the horizontal direction specified by the vector $\hat{\mathbf{g}}\mathbf{i}$.

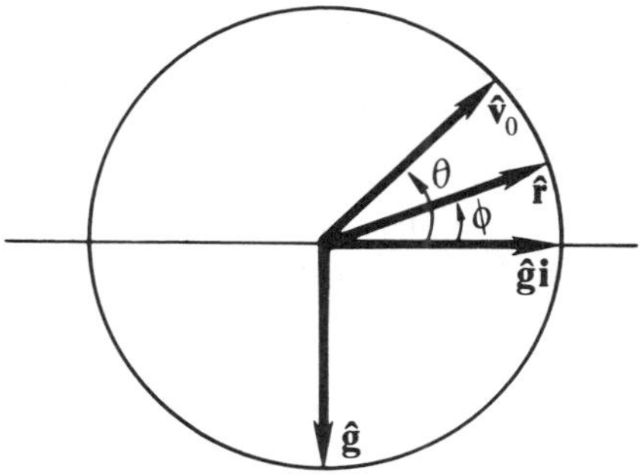

Fig. 2.7.

Equations (2.26) and (2.27) compactly describe how rectangular and polar coordinates are related to the vectors $\mathbf{g}$, $\mathbf{v}_0$ and $\mathbf{r}$. From them one can read off, for example, the relations

$$\hat{\mathbf{g}}\wedge\mathbf{r} = \mathbf{i}r\cos\phi,$$

$$\hat{\mathbf{g}}\wedge\mathbf{v}_0 = \mathbf{i}v_0\cos\theta.$$

Moreover,

$$\mathbf{r}\mathbf{v}_0 = (\mathbf{r}\hat{\mathbf{g}})(\hat{\mathbf{g}}\mathbf{v}_0) = rv_0 e^{\mathbf{i}(\phi-\theta)}. \tag{2.28}$$

This makes it easy to put the range equation (2.8) in the more conventional form

$$r = \left(\frac{2v_0^2}{g}\right)\frac{\cos\theta\sin(\phi-\theta)}{\cos^2\phi}\,. \tag{2.29}$$

Similarly, (2.7) can be put in the form

$$\cos\theta = \frac{r\cos\phi}{v_0 t}\,. \tag{2.30}$$

This formula can be used to find the firing angle θ when the location of the target is given. The time of flight t can be evaluated by using the result of Exercise (2.1) below. In a similar way, other multivectors equations in this section can be put in conventional trigonometric form. But it should be evident by now that the multivector equations are usually easier to manipulate than their trigonometric counterparts.

3-2. Exercises

(2.1) Derive the following expression for time of flight as a function of target location:

$$t = \frac{\sqrt{2}}{g}\{v_0^2 + \mathbf{g}\cdot\mathbf{r} \pm [(v_0^2 + \mathbf{g}\cdot\mathbf{r})^2 - r^2g^2]^{1/2}\}^{1/2}.$$

(2.2) From Equation (2.4) one can get a quadratic equation for t,

$$t^2 + 2\mathbf{v}_0\cdot\mathbf{g}^{-1}\,t - 2\mathbf{r}\cdot\mathbf{g}^{-1} = 0.$$

Discuss the significance of the roots and how they are related to the result of Exercise 1.

(2.3) The vertex of a parabolic trajectory is defined by the equation $\mathbf{v}\cdot\mathbf{g} = 0$. Show that the time of flight to the vertex is given by $t = -\mathbf{v}_0\cdot\mathbf{g}^{-1}$. Use this to determine the location of the vertex.

(2.4) Use equations (2.18) and (2.19) to determine the maximum horizontal range x for a projectile with initial speed v_0 fired at targets on a plateau with (vertical) elevation y above the firing pad.

(2.5) Find the minimum initial speed v_0 needed for a projectile to reach a target with horizontal range x and elevation y. Determine also the firing angle θ, the time of flight t and the final velocity v of the projectile. Specifically, show that

$$v = [g(r-y)]^{1/2}, \qquad v_0 = [g(r+y)]^{1/2}, \qquad \tan\theta = \left[\frac{r+y}{r-y}\right]^{1/2}.$$

(2.6) From Equations (2.15) and (2.16) obtain

$$\mathbf{v}\wedge\mathbf{g} = \mathbf{v}_0\wedge\mathbf{g},$$

$$\mathbf{v}\wedge\mathbf{r} = \mathbf{r}\wedge\mathbf{v}_0.$$

Solve these equations to get

$$\mathbf{v} = \left(\frac{\mathbf{v}_0\wedge\mathbf{r}}{\mathbf{r}\wedge\mathbf{g}}\right)\mathbf{g} + \left(\frac{\mathbf{v}_0\wedge\mathbf{g}}{\mathbf{r}\wedge\mathbf{g}}\right)\mathbf{r}.$$

(2.7) Determine the area swept out in time t by the displacement vector of a particle with constant acceleration $\mathbf{g}$ and initial velocity $\mathbf{v}_0$.

3-3. Constant Force with Linear Drag

We have seen that the trajectory of a particle subject only to a constant force $m\mathbf{g}$ is a parabola. We now consider deviations from parabolic motion due to a linear resistive force, that is, a resistive force directly proportional to the velocity. Expressing the resistive force in the form $-m\gamma\mathbf{v}$, where γ is a positive constant, the equation of motion can be written

$$\dot{\mathbf{v}} = \mathbf{g} - \gamma\mathbf{v}. \tag{3.1}$$

This differential equation is most easily solved by noticing that $e^{\gamma t}$ is an *integrating factor**. Thus,

$$e^{\gamma t}(\dot{\mathbf{v}} + \gamma\mathbf{v}) = \frac{\mathrm{d}}{\mathrm{d}t}(\mathbf{v}e^{\gamma t}) = \mathbf{g}e^{\gamma t}.$$

This integrates directly to

$$\mathbf{v}(t)e^{\gamma t} - \mathbf{v}_0 = \int_0^t \mathbf{g}e^{\gamma t}\,\mathrm{d}t = \mathbf{g}\,\frac{(e^{\gamma t}-1)}{\gamma}.$$

Solving for $\mathbf{v} = \mathbf{v}(t)$, we have

*For a general method of determining integrating factors without guessing, see Exercise (3.3).

$$\mathbf{v} = \mathbf{g}\frac{(1-e^{-\gamma t})}{\gamma} + \mathbf{v}_0 e^{-\gamma t}. \tag{3.2}$$

The constant γ^{-1} is called a *relaxation time*; it provides a measure of the time it takes for the retarding force to make the particle "forget" its initial conditions. If $t \gg \gamma^{-1}$, then $e^{-\gamma t} \ll 1$, so no matter what the value of $\mathbf{v}_0$, the first term on the right side of (3.2) eventually dominates all others; then we have

$$\mathbf{v} \approx \mathbf{v}(\infty) = \gamma^{-1}\,\mathbf{g}. \tag{3.3}$$

This value of the velocity is called the *terminal velocity*. As the particle approaches the terminal velocity its acceleration becomes negligible. Indeed, (3.1) gives the terminal velocity directly when the term $\dot{\mathbf{v}}$ is regarded as negligible.

The displacement $\mathbf{r} = \mathbf{x} - \mathbf{x}_0$ of the particle from its initial position is found by integrating (3.2) directly. The result is

$$\mathbf{r} = \mathbf{g}\,\frac{(e^{-\gamma t} + \gamma t - 1)}{\gamma^2} + \mathbf{v}_0\frac{(1 - e^{-\gamma t})}{\gamma}. \tag{3.4}$$

From the two Equations (3.4) and (3.2), properties of the trajectory can be determined by algebraic means. With the initial conditions $\mathbf{x}_0$, $\mathbf{v}_0$ specified, the general shape of the trajectory can be determined by locating the vertical maximum and asymptote. The location of a vertical maximum is determined by the condition $\mathbf{g}\cdot\mathbf{v} = 0$. (Exercise (3.1)). To locate the asymptote, we deduce from (3.4) that the horizontal displacement of the particle is

$$x(t) \equiv |\,\hat{\mathbf{g}}\wedge\mathbf{r}\,| = \gamma^{-1}\,|\,\hat{\mathbf{g}}\wedge\mathbf{v}_0\,|\,(1 - e^{-\gamma t}),$$

with a maximum

$$x(\infty) = \gamma^{-1}\,|\,\hat{\mathbf{g}}\wedge\mathbf{v}_0\,|. \tag{3.5}$$

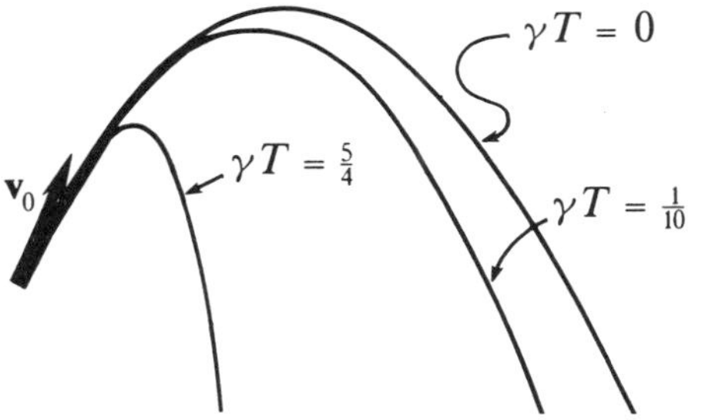

Fig. 3.1. Comparison of trajectories with differing resistance. The parameter T here is the time of flight to vertical maximum for the parabolic trajectory.

Figure 3.1 compares trajectories of particles with the same initial velocities subject to resistive forces of different strengths. Asymptotes are not shown because they do not fit on the figure. (See Exercise (3.2) for analysis.)

Time of Flight

The time of flight to a target specified by $\mathbf{v}_0$ and $\hat{\mathbf{r}}$ can be determined by the same general method used in the parabolic case. The range $|\,\mathbf{r}\,|$ is eliminated from (3.4) to produce an equation for t. To this end, take the outer product of (3.4) with $\gamma^2\hat{\mathbf{r}}$ to get

$$\hat{\mathbf{r}}\wedge\mathbf{g}(e^{-\gamma t} + \gamma t - 1) + \gamma\,\hat{\mathbf{r}}\wedge\mathbf{v}_0\,(1 - e^{-\gamma t}) = 0.$$

Divide this by $\hat{\mathbf{r}}\wedge\mathbf{g}$ and introduce the notation

$$\tfrac{1}{2}T \equiv \frac{\hat{\mathbf{r}}\wedge\mathbf{v}_0}{\mathbf{g}\wedge\hat{\mathbf{r}}} \ , \tag{3.6}$$

to get

$$\gamma t = (1 + \tfrac{1}{2}\gamma T)(1 - e^{-\gamma t}). \tag{3.7}$$

Comparison with (2.6) shows that the scalar T defined by (3.6) is precisely the time of flight in the parabolic case. It should be noted that T is completely determined by the target direction $\hat{\mathbf{r}}$ and the initial velocity $\mathbf{v}_0$.

Equation (3.7) is a transcendental equation for t in terms of T. For $t < \gamma^{-1}$, we can get an approximate solution of the equation by expanding the exponential; thus

$$\gamma t \approx (1 + \tfrac{1}{2}\gamma T)(\gamma t - \tfrac{1}{2}\gamma^2 t^2 + \tfrac{1}{6}\gamma^3 t^3).$$

Dividing by $\tfrac{1}{2}\gamma^2 t\,(1 + \tfrac{1}{2}\gamma T)$ and rearranging terms, we get

$$t = \frac{T}{1 + \tfrac{1}{2}\gamma T} + \tfrac{1}{3}\gamma t^2. \tag{3.8}$$

As it should, this equation reduces to $t = T$ when $\gamma = 0$. The first order approximation of this result is obtained by replacing t by T on the right of (3.8) and using the binomial expansion of the first term to first order; thus,

$$t = T(1 - \tfrac{1}{2}\gamma T) + \tfrac{1}{3}\gamma T^2$$

or,

$$t = T(1 - \tfrac{1}{6}\gamma T). \tag{3.9}$$

The time of flight computed here is less than in the parabolic case, because the range is less, though the target direction is the same.

Range

We can estimate the range by expressing it as a function of t, just as we did in the parabolic case. Taking the outer product of (3.4) with g and solving for $r = |\,\mathbf{r}\,|$, we get

$$r = \frac{\hat{\mathbf{g}}\wedge\mathbf{v}_0}{\hat{\mathbf{g}}\wedge\hat{\mathbf{r}}}\left(\frac{1 - e^{-\gamma t}}{\gamma}\right).$$

Using (3.9) to estimate the time-dependence to first order in γ, we find

$$\begin{aligned}\frac{1 - e^{-\gamma t}}{\gamma} &\approx t - \tfrac{1}{2}\gamma t^2 \approx T(1 - \tfrac{1}{6}\gamma T) - \tfrac{1}{2}\gamma T^2 \\ &= T(1 - \tfrac{2}{3}\gamma T).\end{aligned}$$

So

$$r = \frac{\hat{\mathbf{g}}\wedge\mathbf{v}_0}{\hat{\mathbf{g}}\wedge\hat{\mathbf{r}}}\, T(1 - \tfrac{2}{3}\gamma T)\,.$$

Comparison with (2.8) shows that

$$R \equiv \frac{\hat{\mathbf{g}}\wedge\mathbf{v}_0}{\hat{\mathbf{g}}\wedge\hat{\mathbf{r}}}\, T = \frac{2\mathbf{v}_0\wedge\mathbf{g}\hat{\mathbf{r}}\wedge\mathbf{v}_0}{|\,\mathbf{g}\wedge\hat{\mathbf{r}}\,|^2} \tag{3.10a}$$

is identical to our previous expression for range in the parabolic case. Consequently, it is convenient to write our range formula in the form

$$r = R(1 - \tfrac{2}{3}\gamma T) = R\left(1 - \frac{4\gamma}{3}\,\frac{\hat{\mathbf{r}}\wedge\mathbf{v}_0}{\mathbf{g}\wedge\hat{\mathbf{r}}}\right), \tag{3.10b}$$

showing the first-order correction to the range in the parabolic case.

Ohm's Law

The equations we have been discussing are useful in the analysis of microscopic as well as macroscopic motions. For example, consider an electron (with mass m and charge e) moving in a conductor under the influence of a constant electric field $\mathbf{E}$. The electron's motion will be retarded by collisions with atoms in the conductor. We may attempt to represent the retardation by a resistive force proportional to the velocity. If the resistance is independent of the direction in which the electron moves, we say that the conductor is an isotropic medium, and we can write the resistive force in the form $-\mu\mathbf{v}$, where μ is a constant. We are thus led to consider the equation of motion

$$m\dot{\mathbf{v}} = e\mathbf{E} - \mu\mathbf{v}\,. \tag{3.11}$$

For times large compared with the relaxation time

$$\tau \equiv \frac{m}{\mu}\,, \tag{3.12}$$

the electron will reach the terminal velocity

$$\mathbf{v} = \frac{e\mathbf{E}}{\mu}\,, \tag{3.13}$$

and there will be a steady electric current in the conductor. The electric current density $\mathbf{J}$ is given by

$$\mathbf{J} = Ne\mathbf{v}\,, \tag{3.14}$$

where N is the density of electrons. Substituting (3.13) into (3.14), we get *Ohm's Law*

$$\mathbf{J} = \sigma\mathbf{E}\,, \tag{3.15}$$

where the conductor's *d-c* conductivity σ is given by

$$\sigma = \frac{Ne^2}{\mu}\,. \tag{3.16}$$

Ohm's law holds remarkably well for many conductors over a wide range of currents. The conductivity σ and the electron density N can be measured, so μ can be calculated from (3.16). Then the relaxation time τ can be calculated from (3.12) and compared with measured values. The results are in general agreement with the extremely short relaxation times found for metals. Thus, our selection of Equation (3.11) is vindicated to some degree, and we have come to understand Ohm's law as something more than a mere empirical relation. But we can hardly claim to have a satisfactory explanation or derivation of Ohm's law, because our understanding of (3.11) and its domain of validity is too rudimentary at this point. For one thing, the velocity $\mathbf{v}$ can certainly not be seriously regarded as the velocity of an individual electron. It must be interpreted as some kind of average electron velocity. The trajectory of an individual electron must be very irregular as it collides repeatedly with the much more massive atoms in the conductor. Our equations describe only average motion in the microscopic domain. Derivation and explanation of these equations requires statistical mechanics and equations governing the submicroscopic motion of electrons, specifically, the basic equations of quantum mechanics. This much is certain: Ohm's law is not a fundamental law of physics, it is a macroscopic approximation to complex processes taking place at the atomic level.

3-3. Exercises

(3.1) Show that if $\mathbf{g}\cdot\mathbf{v}_0 < 0$, then a particle subject to Equation (3.1) reaches a maximum height in time

$$t_m = \gamma^{-1}\log(1 - \mathbf{v}_0\cdot\mathbf{v}_\infty^{-1}),$$

where $\mathbf{v}_\infty \equiv \gamma^{-1}\mathbf{g}$ is the terminal velocity. Show that the displacement to maximum $\mathbf{r}_m$ is given by

$$\gamma\mathbf{r}_m = \mathbf{v}_\infty\log(1 - \mathbf{v}_0\cdot\mathbf{v}_\infty^{-1}) - \frac{\mathbf{v}_0\cdot\mathbf{v}_\infty^{-1}(\mathbf{v}_0 - \mathbf{v}_\infty)}{1 - \mathbf{v}_0\cdot\mathbf{v}_\infty^{-1}}\,.$$

Show also that

$$\frac{x_m}{x_\infty} = \frac{-\mathbf{v}_0\cdot\mathbf{v}_\infty^{-1}}{1 - \mathbf{v}_0\cdot\mathbf{v}_\infty^{-1}} = \frac{-\mathbf{v}_0\cdot\mathbf{g}^{-1}}{\gamma^{-1} - \mathbf{v}_0\cdot\mathbf{g}^{-1}}\,,$$

where x_m is the horizontal coordinate of the maximum, and x_∞ is the distance from the initial position to the vertical asymptote. Note that this relation implies that the greater $|\mathbf{v}_0|$ or γ, the more blunted the trajectory.

(3.2) A natural unit of time for a parabolic trajectory is the time of flight $T = -\mathbf{v}_0 \cdot \mathbf{g}^{-1}$ to its vertical maximum. The ratio of T to the relaxation time γ^{-1} for resisted motion is a dimensionless parameter which completely determines the relative shapes of trajectories for resisted and non-resisted motion. A convenient parameter determining the size of the trajectories is the horizontal distance X to the vertical maximum in the parabolic case. To make the comparison quantitative, derive the exact relations

$$x_\infty = \frac{X}{\gamma T},$$
$$x_m = \frac{X}{1 + \gamma T},$$

where x_∞ and x_m are as defined in the preceding exercise. Let x_c denote the coordinate of the point where the trajectory crosses the horizontal. For $t_c < \gamma^{-1}$, derive the approximate relation

$$x_c \approx 2X(1 - \tfrac{2}{3}\gamma T).$$

For the accurate curves in Figure 3.1, check the given values for γT, locate the asymptotes and compare horizontal crossing points with computed values of x_c.

(3.3) An equation with general form

$$\dot{\mathbf{v}} + \gamma \mathbf{v} = \mathbf{g},$$

where $\mathbf{g} = \mathbf{g}(t)$ and scalar $\gamma = \gamma(t)$ are specified functions of t, is said to be a linear first order differential equation with scalar coefficient. A function $\lambda = \lambda(t)$ is an integrating factor of this equation if, when multiplied by λ, the equation can be put in the form

$$\frac{\mathrm{d}}{\mathrm{d}t}(\lambda \mathbf{v}) = \lambda \mathbf{g}.$$

Show that the integrating factor is determined by

$$\lambda = \lambda_0 e^{\int_0^t \gamma(s)\,\mathrm{d}s},$$

and the solution is given by

$$\mathbf{v} = \lambda^{-1}(t) \int_0^t \lambda(s)\mathbf{g}(s)\,\mathrm{d}s + \lambda^{-1}(t)\lambda_0 \mathbf{v}_0.$$

3-4. Constant Force with Quadratic Drag

As a rule, resistance of the atmosphere to motion of a projectile is more accurately described by a quadratic function of the velocity than by the linear function considered in the last section. In this case, the resistive force has the form $-m\alpha v\mathbf{v}$, where $v = |\mathbf{v}|$ and α is a positive constant. For a particle subject to a constant force and quadratic drag, the equation of motion can be written.

$$\dot{\mathbf{v}} = \mathbf{g} - \alpha v\mathbf{v}. \tag{4.1}$$

To solve this equation, we must resort to approximation methods. By such methods we can solve any differential equation to the degree of accuracy required by a given problem. However, one approximation method may be easier to apply than another, or it may yield results in a more useful form.

Before getting involved in details of a calculation, we should find out what we can about general features of the motion. Our analysis of motion with linear drag provides a valuable qualitative guide to the quadratic case. Indeed, if the coefficient αv in (4.1) is replaced by some estimate γ of its average value on the trajectory, the exact solution with linear drag provides a good quantitative approximation to the motion in limited time intervals. About the motion over unlimited time intervals, we can draw the following general conclusions:

(1) The particle eventually "forgets" its initial velocity and reaches a terminal velocity

$$\mathbf{v}_\infty = \hat{\mathbf{g}}\left(\frac{g}{\alpha}\right)^{1/2}. \tag{4.2}$$

This is obtained from (4.1) by neglecting $\dot{\mathbf{v}}$.

(2) As in Figure 3.1, trajectories with different initial velocities have different shapes. The greater the initial speed, the more blunted the trajectory and the less symmetrical its shape.

(3) There will be a maximum horizontal displacement which, by (3.5), cannot exceed $(\alpha v_\infty)^{-1}\,|\,\hat{\mathbf{g}}\wedge\mathbf{v}_0\,|$

(4) The maximum horizontal range for given initial speed v_0 occurs for a firing angle $\theta_{max} < 45°$, and θ_{max} decreases as v_0 increases.

Horizontal and Vertical Components

The constant vector $\mathbf{g}$ in (4.1) determines a preferred direction which will naturally be reflected in the solution. For this reason, it is of some interest to decompose the motion into horizontal and vertical components. We write

$$\hat{\mathbf{g}}\mathbf{v} = \hat{\mathbf{g}}\cdot\mathbf{v} + \hat{\mathbf{g}}\wedge\mathbf{v} = v_\parallel + \mathbf{i}v_\perp, \tag{4.3}$$

where $v_\parallel \equiv \hat{\mathbf{g}}\cdot\mathbf{v}$ is the vertical component of velocity and $v_\perp$ is the horizontal component. The orientation of the unit bivector **i** in (4.3) is fixed by choosing $v_\perp$ positive at some time. The horizontal and vertical coordinates of position are similarly defined by

$$\hat{\mathbf{g}}\mathbf{x} = x_\parallel + \mathbf{i}x_\perp . \tag{4.4}$$

Equation (4.1) can be decomposed into components by separating scalar and bivector parts after multiplying by $\hat{\mathbf{g}}$. The scalar part gives the equation

$$\hat{\mathbf{g}}\cdot\dot{\mathbf{v}} = \frac{\mathrm{d}}{\mathrm{d}t}(\hat{\mathbf{g}}\cdot\mathbf{v}) = g - \alpha v \hat{\mathbf{g}}\cdot\mathbf{v},$$

or

$$\dot{v}_\parallel = g - \alpha v v_\parallel . \tag{4.5a}$$

The bivector part yields

$$\hat{\mathbf{g}}\wedge\dot{\mathbf{v}} = \frac{\mathrm{d}}{\mathrm{d}t}(\hat{\mathbf{g}}\wedge\mathbf{v}) = -\alpha v \hat{\mathbf{g}}\wedge\mathbf{v}$$

$$= \mathbf{i}\dot{v}_\perp + v_\perp \frac{\mathrm{d}\mathbf{i}}{\mathrm{d}t} = -\alpha v v_\perp \mathbf{i}.$$

Since $\mathbf{i}^2 = -1$ implies that $\langle \mathbf{i}\, \mathrm{d}\mathbf{i}/\mathrm{d}t\rangle_0 = 0$, we conclude that

$$\dot{v}_\perp = -\alpha v v_\perp , \tag{4.5b}$$

and

$$\frac{\mathrm{d}\mathbf{i}}{\mathrm{d}t} = 0. \tag{4.5c}$$

In view of (4.3) and (4.4), we see that (4.5c) means that the entire trajectory lies in a vertical plane.

The differential equations (4.5a, b) are not as simple as they look, because they are coupled by the condition $v^2 = v_\parallel^2 + v_\perp^2$. However, along a fairly horizontal trajectory, the condition $v_\perp \gg v_\parallel$ is satisfied, and the equations can be approximated by

$$\dot{v}_\perp = -\alpha v_\perp^2 , \tag{4.6a}$$

$$\dot{v}_\parallel = g - \alpha v_\perp v_\parallel . \tag{4.6b}$$

These equations can be solved exactly (Exercise 4.2).

Perturbation Theory

We turn now to a different method for getting approximate solutions to the equation of motion. To estimate the deviation from a parabolic trajectory due to drag, we write the displacement vector in the form

$$\mathbf{r}(t) = \tfrac{1}{2}\mathbf{g}t^2 + \mathbf{v}_0 t + \mathbf{s}(t), \tag{4.7a}$$

and we impose the initial condition

$$\mathbf{s}(0) = 0, \tag{4.7b}$$

so $\mathbf{s} = \mathbf{s}(t)$ describes the deviation as a function of time. Differentiating, we have

$$\mathbf{v} = \dot{\mathbf{r}} = \mathbf{g}t + \mathbf{v}_0 + \dot{\mathbf{s}}, \tag{4.8a}$$

subject to the initial condition

$$\dot{\mathbf{s}}(0) = 0. \tag{4.8b}$$

Substituting (4.8a) into (4.1), we get an exact equation for the deviation

$$\ddot{\mathbf{s}} = -\alpha \,|\, \mathbf{g}t + \mathbf{v}_0 + \dot{\mathbf{s}} \,|\, (\mathbf{g}t + \mathbf{v}_0 + \dot{\mathbf{s}}). \tag{4.9}$$

Over a time interval in which $|\, \mathbf{g}t + \mathbf{v}_0 \,| \gg |\, \dot{\mathbf{s}} \,|$, Equation (4.9) is well approximated by

$$\ddot{\mathbf{s}} = -\alpha \,|\, \mathbf{g}t + \mathbf{v}_0 \,|\, (\mathbf{g}t + \mathbf{v}_0). \tag{4.10}$$

This equation can be integrated directly and exactly, but the result is unwieldy, so we will be satisfied with solutions which meet the condition

$$|\, \mathbf{g}t + \mathbf{v}_0 \,| = v_0(1 + 2\mathbf{v}_0^{-1}\cdot\mathbf{g}t + v_0^{-2}g^2t^2)^{1/2}$$

$$\approx v_0 \,|\, 1 + \mathbf{v}_0^{-1}\cdot\mathbf{g}t \,|. \tag{4.11}$$

In fact, this relation is exact if $\mathbf{g}\wedge\mathbf{v}_0 = 0$, and it is an excellent approximation as long as $gt < v_0$. Assuming (4.11), we integrate (4.10) to get

$$\dot{\mathbf{s}} = -\alpha v_0 \{\mathbf{g} \,|\, \tfrac{1}{2} t^2 + \tfrac{1}{3} \mathbf{g}\cdot\mathbf{v}_0^{-1} t^3 \,|\, + \mathbf{v}_0 \,|\, \mathrm{t} + \tfrac{1}{2} \mathbf{g}\cdot\mathbf{v}_0^{-1} t^2 \,|\, \}. \tag{4.12}$$

Integrating once more, we get

$$\mathbf{s} = -\alpha v_0 \left\{ \mathbf{g} \left| \frac{t^3}{6} + \mathbf{v}_0^{-1}\cdot\mathbf{g}\, \frac{t^4}{12} \right| + \mathbf{v}_0 \left| \frac{t^2}{2} + \mathbf{v}_0^{-1}\cdot\mathbf{g}\, \frac{t^3}{6} \right| \right\}. \tag{4.13}$$

Substituting this result in (4.7a), we get

$$\mathbf{r} = \tfrac{1}{2}\,\mathbf{g}t^2 \left(1 - \alpha \left| \frac{v_0 t}{3} + \hat{\mathbf{v}}_0\cdot\mathbf{g}\, \frac{t^2}{6} \right|\right)$$

$$+ \mathbf{v}_0 t \left(1 - \alpha \left| \frac{v_0 t}{2} + \hat{\mathbf{v}}_0\cdot\mathbf{g}\, \frac{t^2}{6} \right|\right). \tag{4.14}$$

A graph of the trajectory can be constructed from a parabola by evaluating s at a few points, as indicated in Figure 4.1. It should be noted that it is the second term in solid brackets on the right side of (4.11) that distinguishes quadratic drag from linear drag, for, if it is neglected, our approximation amounts to assuming a linear drag force $-\alpha v_0 \mathbf{v} \approx -\alpha v_0\, (\mathbf{g}t + \mathbf{v}_0)$.

Our calculation of (4.14) illustrates a general method of approximation called *perturbation theory*. The idea is to estimate the deviation (i.e. *perturbation*) from a known (i.e. *unperturbed*) trajectory caused by a (usually small) *perturbing force*. We have estimated the perturbation $\mathbf{s}(t)$ caused by quadratic drag. Our result (4.14) is a *first order* perturbative approximation to the exact trajectory. We can get a (more accurate) *second order* approximation by regarding (4.14) as the unperturbed trajectory and calculating first order deviations from it. A more efficient way to get the same result is to substitute the expression (4.12) for $\dot{\mathbf{s}}$ into the exact Equation (4.9) to get an explicit expression for the time dependence of $\ddot{\mathbf{s}}$ which can be integrated directly.

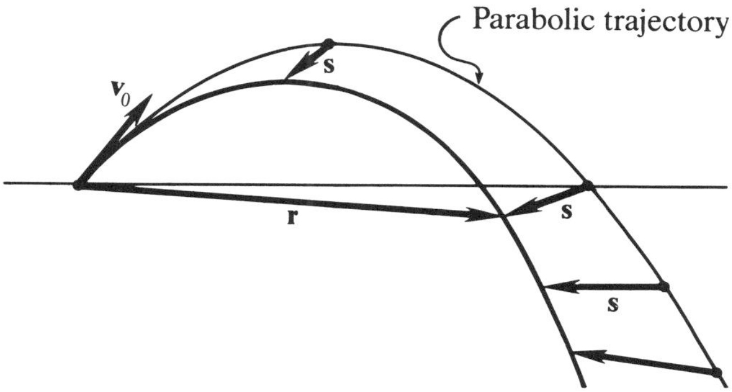

Fig. 4.1. Drag deviation from a parabolic trajectory. The deviation vector $\mathbf{s}$ relates "simultaneous" positions on the two trajectories.

Polygonal Approximation

Another general method commonly used in numerical computation of trajectories is the method of polygonal approximation. The idea is that any curve can be approximated with arbitrary accuracy by a sequence of joined line segments (sides of a polygon if the curve is in a plane). We proceed as follows: Choose a small interval of time τ and consider a succession of times $t_k = k\tau$, where $k = 0, 1, 2, \ldots$. The velocities at successive times are determined by the equation

$$\mathbf{v}_{k+1} = \tau\mathbf{a}_k + \mathbf{v}_k, \tag{4.15}$$

where $\mathbf{v}_k = \mathbf{v}(t_k)$ and $\mathbf{a}_k = \mathbf{a}(t_k) = \dot{\mathbf{v}}(t_k)$. This equation is obtained by regarding the acceleration as constant in the small time interval τ. The acceleration is determined by the equation of motion (4.1), which gives

$$\mathbf{a}_k = \mathbf{g} - \alpha v_k \mathbf{v}_k. \tag{4.16}$$

Because this equation happens to be independent of $\mathbf{x}_k = \mathbf{x}(t_k)$, we can use it at once to find each $\mathbf{v}_k$ from the initial velocity $\mathbf{v}_0$ by iterating (4.15). To find points on the trajectory, we use the *mean value theorem* from differential calculus, which implies that the chord of a small segment on a smooth curve is parallel to the tangent at the segment's midpoint, or, in our case,

$$\mathbf{v}_{k+1} \approx \frac{\mathbf{x}_{k+2} - \mathbf{x}_k}{2\tau}.$$

We use this in the form

$$\mathbf{x}_{k+2} = 2\tau\mathbf{v}_{k+1} + \mathbf{x}_k, \tag{4.17}$$

which, by iteration, enables us to find $\mathbf{x}_k$ for even k from $\mathbf{x}_0$ and the $\mathbf{v}_k$. Drawing a smooth curve through these points, we get the desired approximate trajectory, as illustrated in Figure 4.2.

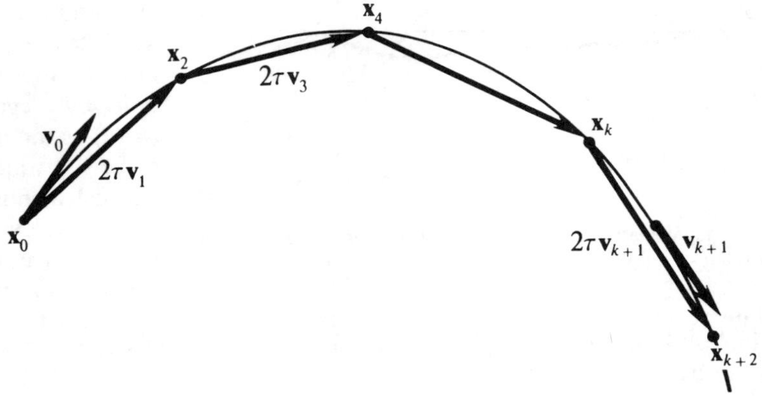

Fig. 4.2. Polygonal approximation.

Clearly, the perturbation method is superior to the polygonal method for the present problem. It requires only one iteration to achieve a useful result whereas, the polygonal method requires many iterations, because it must reproduce the curvature of the parabola as well as the effect of the perturbation. Also, the results of the perturbation method are easier to use and interpret, because they are in analytical instead of graphical or tabular form, and they provide explicit relations between pertùrbed and unperturbed solutions.

3-4. Exercises

(4.1) Integrate the equation $\dot{\mathbf{v}} = -\alpha v\mathbf{v}$ to get the exact solution

$$\mathbf{v} = \frac{\mathbf{v}_0}{1 + v_0\,\alpha t}.$$

How long will it take for the drag to reduce the velocity to half the initial velocity? How far will the particle travel in this time interval?

(4.2) Solve Equation (4.6a, b). (It is helpful to notice that $\lambda = 1 + v_0\alpha t$ is an integrating factor.) Show that the horizontal displacement $r_\perp$ and the vertical displacement $r_\parallel$ are given by

$$r_\perp = \frac{1}{\alpha}\log(1 + v_0\alpha t),$$

$$r_\parallel = \frac{1}{4}gt^2 + \frac{gt}{2v_0\alpha} + \left(\frac{v_{0\parallel}}{v_0\alpha} - \frac{g}{2v_0^2\alpha^2}\right)\log(1 + v_0\alpha t).$$

Compare this result with Equation (4.14), and point out any advantages of one result over the other.

(4.3) Check Figure 4.1 by evaluating the vertical "shortfall" $\mathbf{s}\cdot\hat{\mathbf{g}}$ at distinguished points on the figure. Give a qualitative argument for believing that there is a time at which two particles would cross the same horizontal plane if .they could be launched along the two trajectories simultaneously. Why can't you determine this time from the approximate expression (4.13) for $\mathbf{s}$?

(4.4) Show that if $\mathbf{v}_0\cdot\hat{\mathbf{g}} < 0$, then air resistance will have the effect of hastening the return of a projectile to the horizontal plane from which it is launched by a time interval

$$\Delta t = \frac{\hat{\mathbf{g}}\cdot\mathbf{s}}{-\hat{\mathbf{g}}\cdot\mathbf{v}_0} = \tfrac{2}{3}\alpha v_0(\mathbf{v}_0\cdot\mathbf{g}^{-1})^2.$$

(Hint: First determine the time of flight without air resistance).

(4.5) Use the perturbation method to derive a general estimate for the time of flight to a specified target.

(4.6) Use the perturbation method to find the maximum height of a projectile trajectory. Compare with the exact result from Exercise 4.7.

(4.7) Equation (4.1) can be integrated by separation of variables when $\mathbf{g}\wedge\mathbf{v}_0 = 0$. First show that for an initial downward velocity the equation can be put in the form

$$\frac{\mathrm{d}v}{\mathrm{d}t} = g\left(1 - \frac{v^2}{v_\infty^2}\right).$$

Integrate this to get

$$v = v_\infty \tanh\left(\frac{gt}{v_\infty} - c\right) = v_\infty\,\frac{v_0 - v_\infty\tanh\left(\dfrac{gt}{v_\infty}\right)}{v_\infty - v_0\tanh\left(\dfrac{gt}{v_\infty}\right)},$$

where $c = \tanh^{-1}\left(\dfrac{v_0}{v_\infty}\right)$.

Integrate again to get the displacement

$$r = \frac{v_\infty^2}{g} \log \cosh\left(\frac{gt}{v_\infty} + c \right).$$

To express the displacement as a function of velocity, substitute

$$\frac{dv}{dt} = \frac{dv}{dr}\frac{dr}{dt} = v\frac{dv}{dr}$$

into the equation of motion and integrate to get

$$r = \frac{v_\infty^2}{2g} \log \left(\frac{v_0^2 - v_\infty^2}{v^2 - v_\infty^2} \right)$$

Repeat the calculation for an initial upward velocity. Determine the maximum height of the trajectory.

3-5. Fluid Resistance

In the two preceding sections, we integrated the equations of motion for a particle subject to resistive forces linear and quadratic in the velocity. The solutions are of little practical value unless we know the physical circumstances in which they apply and have some estimate of the numerical factors involved. This section is devoted to such "physical considerations", but we deal with only one among many kinds of resistive forces.

An object moving through a fluid such as water or air is subject to a force exerted by the fluid. This force can be resolved into a *resistive force*, sometimes called *drag*, directly opposing the motion and a component called *lift*, orthogonal to the velocity of the object. The lift vanishes for objects which are sufficiently small or symmetrical, so there are many problems in which the drag component alone is significant.

The analysis of drag, not to mention lift, is a complex problem in fluid dynamics which is even today under intensive study. Our aim in this section is only to summarize some general results pertinent to particle mechanics. In particular, we are interested in specific expressions for the drag along with a rough idea of their physical basis and range of applicability.

Since the drag $\mathbf{D}$ always opposes motion through a fluid, it can be written in the general form

$$\mathbf{D} = -D\hat{\mathbf{V}}, \tag{5.1}$$

where $\mathbf{V}$ is the ambient velocity. The ambient velocity of an object is its velocity relative to the undisturbed fluid. If $\mathbf{v}$ is the velocity of the object and $\mathbf{u}$ is the velocity of the fluid relative to some fixed object or frame, then the ambient velocity is

$$\mathbf{V} = \mathbf{v} - \mathbf{u}. \tag{5.2}$$

For example, **v** might be the velocity of a projectile relative to the earth while **u** is the velocity of the wind.

The magnitude $D = |\mathbf{D}|$ of the drag is commonly written in the standard form

$$D = \tfrac{1}{2}\, C_D \varrho A V^2, \tag{5.3}$$

where ϱ is the mass density of the fluid, A is the cross-sectional area of the object across the line of motion, $V = |\mathbf{V}|$ is the ambient speed, and the *drag coefficient* C_D is a dimensionless quantity measuring the relative strength of the drag. The value of C_D depends on the size, shape and speed of the object in relation to properties of the fluid. It is advantageous to express the ambient speed as a dimensionless variable $\mathcal{R}$ called the *Reynolds number*, because the functional dependence of C_D on $\mathcal{R}$ is the same for all fluids, notably, for the two most common fluids on earth, water and air.

For a sphere with diameter 2a the Reynolds number is defined by the expression

$$\mathcal{R} = \frac{2a\varrho V}{\eta}, \tag{5.4}$$

where η is the viscosity and, as before, ϱ is the mass density of the fluid. The viscosity, like the mass density, is determined empirically; it would take us too far afield to explain how. For small $\mathcal{R}$, the drag coefficient can easily be determined from hydrodynamic theory, with the result

$$C_D = \frac{24}{\mathcal{R}}. \tag{5.5}$$

When this is substituted in (5.3) and (5.4) is used, the drag assumes the form

$$\mathbf{D} = -6\pi\eta a\mathbf{V}. \tag{5.6}$$

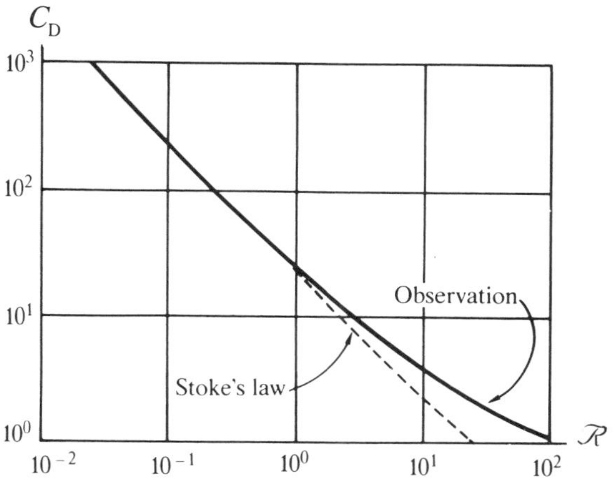

Fig. 5.1. Drag coefficient for a sphere at small Reynolds number [Redrawn from Batchelor (1967)].

This famous result is known as *Stokes' Law* in honor of the man who first derived it. Figure 5.1 shows that Stokes' Law agrees well with experiment for $\mathcal{R} \leqslant 1$. Another condition for its validity is that the sphere's radius a be large compared with the mean free path of molecules in the fluid, which for air under standard conditions is of order 10^{-5} cm. The mean free path is the average distance a molecule travels between collisions.

Note that Stokes' Law (5.6) is the linear drag law assumed in Section 3-3, not the quadratic drag assumed in Section 3-4. From the data in Table 5.1 we can estimate the range of size and speed for which Stokes' Law applies. For a sphere moving through air, the condition $\mathcal{R} \leqslant 1$ implies

$$2aV \leqslant \left(\frac{\eta}{\varrho}\right)_{\text{air}} = 0.15 \text{ cm}^2 \text{ sec}^{-1}. \tag{5.7}$$

For motion through water this number is reduced by a factor of 15. Thus, Stokes' Law applies only to quite small objects at low velocities, such as one encounters in the sedimentation of silt in steams or pollutants in the atmosphere. This gives us some idea of the domain in which the linear resistive force studied in Section 3-3 can be expected to lead to quantitatively accurate results.

TABLE 5.1. Density and viscosity of water and air at a temperature of 20 °C and a pressure of 1 atmosphere.

For water,

$$\varrho = 1.00 \text{ g cm}^{-3} = 10^3 \text{ kg m}^{-3},$$

$$\eta = 1.00 \times 10^{-2} \text{ dyn sec cm}^{-2}.$$

For air,

$$\varrho = 1.20 \times 10^{-3} \text{ g cm}^{-3} = 1.20 \text{ kg m}^{-3},$$

$$\eta = 1.81 \times 10^{-4} \text{ dyn sec cm}^{-2}.$$

Specific viscosity:

$$\left(\frac{\eta}{\varrho}\right)_{\text{air}} = 0.15 \text{ cm}^2 \text{ sec}^{-1} = 15 \left(\frac{\eta}{\varrho}\right)_{\text{water}}$$

To see how velocities compare in magnitude to Reynolds numbers for projectiles, consider a sphere of radius a = 1 cm moving through air. From Table 5.1, we find

$$\text{for } V = 1 \text{ m sec}^{-1},\ \mathcal{R} = 1.3 \times 10^3; \tag{5.8}$$

$$\text{for } V = 330 \text{ m sec}^{-1},\ \mathcal{R} = 4.4 \times 10^5.$$

According to Figure 5.2, for velocities in this range the drag coefficient is nearly constant with the value

$$C_D \approx \tfrac{1}{2}\,. \tag{5.9}$$

The upper limit of 330 m sec^{-1} is the speed of sound in air. It follows that for spherical projectiles with speeds less than sound, the drag is fairly well represented by

$$D = (0.94 \text{ kg m}^{-3})\, a^2V^2 . \tag{5.10}$$

Figure 5.2 shows that there is a pronounced decrease in the value of the drag coefficient in the vicinity of the speed of sound. A well-struck golf ball takes advantage of this. The dimples on a golf ball also reduce drag by disrupting the boundary layer of air that tends to form on the balls surface. Besides this, the trajectory of a golf ball will be affected by its spin which is usually large enough to produce a significant lift.

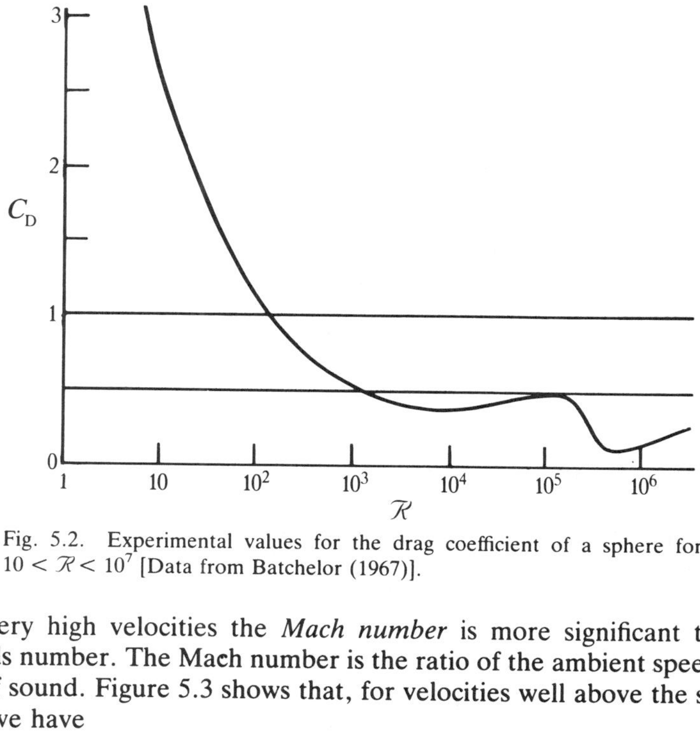

Fig. 5.2. Experimental values for the drag coefficient of a sphere for $10 < \mathcal{R} < 10^7$ [Data from Batchelor (1967)].

For very high velocities the *Mach number* is more significant than the Reynolds number. The Mach number is the ratio of the ambient speed to the speed of sound. Figure 5.3 shows that, for velocities well above the speed of sound, we have

$$C_D \approx 1 , \tag{5.11}$$

so in this regime the drag (5.3) is again a quadratic function of the ambient velocity.

A crude qualitative understanding of fluid drag can be achieved by regarding it as a composite of two effects, *viscous drag* and *pressure drag*. Pure viscous drag is described by Stokes' Law (5.6). It is due to fluid friction as the fluid flows smoothly over the surface of the object, so, as (5.6) shows, viscous drag is proportional to the circumference of the sphere.

As the ambient velocity of the sphere increases, the pressure differential between front and back increases as well, giving rise to pressure drag. Pure pressure drag is characterized by (5.3) when C_D is constant. The form of the

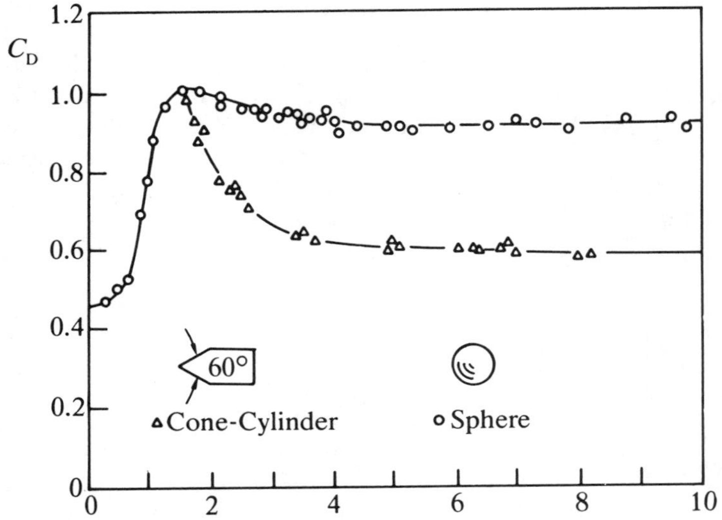

Fig. 5.3. Ballistic range measurements of drag on spheres and cone-cylinders as a function of Mach number (Charters and Thomas (1945), Hodges (1957), Stevens (1950)). [from R. N. Cox and L. F. Crabtree, *Elements of Hypersonic Aerodynamics*, Academic Press, N. Y. (1965), p. 42, with permission].

equation can be interpreted by noting that the rate of collision with molecules in the fluid is proportional to ϱAV, while the average momentum transfer per collision introduces another factor of V. Deviations from pressure drag arise from viscous effects in the fluid that piles up in front of the moving object.

The data introduced above apply only to objects which are large compared to the mean free path of molecules in the fluid. In Chapter 8 we will be concerned with the effect of atmospheric drag on artificial satellites. The dimensions of such a satellite are small compared with the mean free path of molecules in the outer atmosphere, (which at 300 km above the surface of the earth is about 10 km.) In this case fluid mechanics does not apply, and the collisions of individual molecules with the satellite can be regarded as independent of one another. The result is pure pressure drag with

$$C_D \approx 2. \tag{5.12}$$

For nonspherical artificial satellites the average of the lift force over changing orientations will tend to be zero, and an average drag can be employed, but the result will be subject to statistical uncertainties that render a refined dynamical analysis futile. The average drag will differ from that given by (5.3) only in the interpretation of A as the average cross-sectional area. The cross-sectional area of any convex body averaged over all orientations can be shown to be equal to one-fourth its surface area. A "convex" body is one whose surface intersects a straight line no more than twice.

Section 3-5.

(5.1) Find the terminal velocity of a man $m = 73$ kg falling through the atmosphere for two extreme cases:
(a) He spread-eagles, producing a cross-sectional area of 0.6 m^2, in the direction of motion.
(b) He tucks to produce a cross-sectional area of only 0.3 m^2.
What terminal velocity do you get for the first case using Stokes' Law?

(5.2) What parachute diameter is required to reduce the terminal velocity of a parachutist ($m = 73$ kg) to 5 m sec^{-1}

(5.3) What is the terminal velocity of a raindrop with (typical) radius of 10^{-3} m? What result would you get from Stokes' Law?

(5.4) A mortar has a maximum range of 2000 m at sea level. What would be the maximum range in the absence of air resistance?

(5.5) Two iron balls of weights 1 kg and 100 kg are dropped simultaneously and fall through a distance 100 m at sea level. Which ball hits the ground first and how far behind does the other lag? What is the difference between arrival times of the two balls?

3-6. Constant Magnetic Field

The classical equation of motion for a particle with charge q in a magnetic field $\mathbf{B}$ is

$$m\dot{\mathbf{v}} = \frac{q}{c}\, \mathbf{v} \times \mathbf{B}. \tag{6.1}$$

Let us lump the constants together by writing

$$\boldsymbol{\omega} \equiv - \frac{q}{mc}\mathbf{B}, \tag{6.2}$$

so Equation (6.1) takes the form

$$\dot{\mathbf{v}} = \boldsymbol{\omega} \times \mathbf{v}. \tag{6.3}$$

It is convenient to introduce the bivector $\boldsymbol{\Omega}$ dual to $\boldsymbol{\omega}$ by writing

$$\boldsymbol{\Omega} = i\boldsymbol{\omega}. \tag{6.4}$$

Then, since

$$\boldsymbol{\omega} \times \mathbf{v} = - i\boldsymbol{\omega} \wedge \mathbf{v} = - (i\boldsymbol{\omega})\cdot\mathbf{v} = \mathbf{v}\cdot\boldsymbol{\Omega}$$

Equation (6.3) can be written in the alternative form

$$\dot{\mathbf{v}} = \mathbf{v}\cdot\boldsymbol{\Omega}. \tag{6.5}$$

Without solving this equation, we can conclude from it that $|\mathbf{v}|$ is a constant of the motion by an argument already made in Section 2-7. Since the magnitude of $\mathbf{v}$ is constant, the solution of (6.5) is a rotating velocity vector. The solution will show that $\boldsymbol{\Omega}$ can be regarded as the angular velocity of this rotation.

Equation (6.5) can be solved by generalizing the method of integrating factors used in Section 3-3. This is made evident by writing (6.5) in the form

$$\dot{\mathbf{v}} + \tfrac{1}{2}\boldsymbol{\Omega}\mathbf{v} + \mathbf{v}(-\tfrac{1}{2}\boldsymbol{\Omega}) = 0. \tag{6.6}$$

Suppose we can find a function $R = R(t)$ with the property that

$$\dot{R} = R\tfrac{1}{2}\boldsymbol{\Omega}. \tag{6.7a}$$

Since $\boldsymbol{\Omega}^\dagger = -\boldsymbol{\Omega}$ the reverse of (6.7a) is the equation

$$\dot{R}^\dagger = -\tfrac{1}{2}\boldsymbol{\Omega}R^\dagger. \tag{6.7b}$$

Now multiplying (6.6) on the left by R and on the right by $R^\dagger$ and using (6.7a, b) we put it in the form

$$\frac{\mathrm{d}}{\mathrm{d}t}(R\mathbf{v}R^\dagger) = 0. \tag{6.8}$$

Thus we see that R and $R^\dagger$ are integrating factors for (6.6). Two integrating factors instead of one are needed, because $\boldsymbol{\Omega}$ does not commute with $\mathbf{v}$. We say that R is a *left integrating factor* while $R^\dagger$ is a *right integrating factor* for the equation.

Our method of integrating factors has replaced the problem of solving the equation for $\mathbf{v}$ by the problem of solving the differential equation (6.7a) for the integrating factor. This is a significant simplification when $\boldsymbol{\Omega}$ is constant, for then (6.7a) will be recognized as the derivative of the exponential function

$$R = e^{(1/2)\boldsymbol{\Omega}t}. \tag{6.9}$$

It follows that

$$R^\dagger = e^{(1/2)\boldsymbol{\Omega}^\dagger t} = e^{-(1/2)\boldsymbol{\Omega}t}.$$

Moreover,

$$R(0) = R^\dagger(0) = 1, \tag{6.10}$$

and

$$R^\dagger R = RR^\dagger = 1. \tag{6.11}$$

With the initial conditions (6.10) for the integrating factors, Equation (6.8) integrates immediately to

$$R\mathbf{v}R^\dagger - \mathbf{v}_0 = 0,$$

and, using (6.11) to solve for $\mathbf{v}$, we get

$$\mathbf{v} = R^\dagger \mathbf{v}_0 R = e^{-(1/2)\boldsymbol{\Omega} t}\, \mathbf{v}_0\, e^{(1/2)\boldsymbol{\Omega} t}. \tag{6.12}$$

By squaring both sides of this equation, it is readily checked that the solution has the property $\mathbf{v}^2 = \mathbf{v}_0^2$, so the magnitude of $\mathbf{v}$ is constant, as we had anticipated.

We can combine the two exponential factors in (6.12) into a single term by decomposing $\mathbf{v}_0$ into components that commute or anticommute with $\boldsymbol{\Omega}$. Let $\mathbf{v}_{0\|}$ be the component of $\mathbf{v}_0$ parallel to the magnetic field while $\mathbf{v}_{0\perp}$ is the component perpendicular to the magnetic field, as indicated in Figure 6.1.

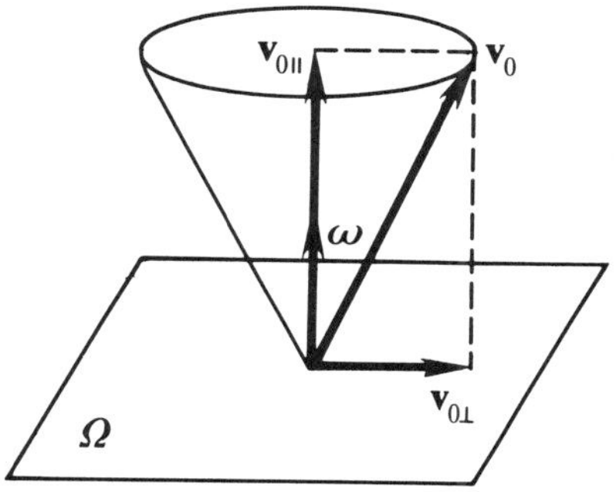

Fig. 6.1.

Algebraically, the decomposition of $\mathbf{v}$ at any time t is expressed by the equations

$$\mathbf{v} = \mathbf{v}_\| + \mathbf{v}_\perp,$$

where

$$\boldsymbol{\omega}\mathbf{v}_\| = \boldsymbol{\omega}\cdot\mathbf{v} = \mathbf{v}_\|\,\boldsymbol{\omega}, \tag{6.13a}$$

$$\boldsymbol{\omega}\mathbf{v}_\perp = \boldsymbol{\omega}\wedge\mathbf{v} = -\mathbf{v}_\perp\boldsymbol{\omega}, \tag{6.13b}$$

or, alternatively

$$\boldsymbol{\Omega}\mathbf{v}_\| = \boldsymbol{\Omega}\wedge\mathbf{v} = \mathbf{v}_\|\,\boldsymbol{\Omega}, \tag{6.14a}$$

$$\boldsymbol{\Omega}\mathbf{v}_\perp = \boldsymbol{\Omega}\cdot\mathbf{v} = -\mathbf{v}_\perp\boldsymbol{\Omega}. \tag{6.14b}$$

Using (6.14a, b) and the series definition of the exponential function, it is readily established that

$$R^\dagger\mathbf{v}_{0\|} = e^{-(1/2)\boldsymbol{\Omega} t}\,\mathbf{v}_{0\|} = \mathbf{v}_{0\|}\,e^{-(1/2)\boldsymbol{\Omega} t} = \mathbf{v}_{0\|}\,R^\dagger \tag{6.15a}$$

$$R^\dagger\mathbf{v}_{0\perp} = e^{-(1/2)\boldsymbol{\Omega} t}\,\mathbf{v}_{0\perp} = \mathbf{v}_{0\perp}\,e^{(1/2)\boldsymbol{\Omega} t} = \mathbf{v}_{0\perp}\,R. \tag{6.15b}$$

Now, with (6.15a, b) and (6.11), we can put (6.12) in the form

$$\mathbf{v} = \mathbf{v}_{0\perp}R^2 + \mathbf{v}_{0\|} = \mathbf{v}_{0\perp}\,e^{\boldsymbol{\Omega} t} + \mathbf{v}_{0\|}. \tag{6.16}$$

Here we see that $\mathbf{v}_\| = \mathbf{v}_{0\|}$ is fixed while $\mathbf{v}_\perp = \mathbf{v}_{0\perp}\,e^{\boldsymbol{\Omega} t}$ rotates through an angle $\boldsymbol{\Omega} t$ in time t; so the resultant velocity vector $\mathbf{v}$ sweeps out a portion of a cone, as indicated in Figure 6.1.

We find the particle trajectory by substituting $\mathbf{v} = \dot{\mathbf{x}}$ into (6.16) and integrating directly, with the result

$$\mathbf{x} - \mathbf{x}_0 = \mathbf{v}_{0\perp}\,\boldsymbol{\Omega}^{-1}(e^{\boldsymbol{\Omega} t} - 1) + \mathbf{v}_{0\|}t.$$

The form of the solution can be simplified by an appropriate choice of origin. Introducing the variable

$$\mathbf{r} \equiv \mathbf{x} - \mathbf{x}_0 + \mathbf{v}_0\cdot\boldsymbol{\Omega}^{-1} = \mathbf{x} - \mathbf{x}_0 + \mathbf{v}_0 \times \boldsymbol{\omega}^{-1} \tag{6.17}$$

the solution can be cast in the equivalent forms

$$\mathbf{r} = \mathbf{v}_0\cdot\boldsymbol{\Omega}^{-1}\,\mathbf{e}^{\boldsymbol{\Omega} t} + \mathbf{v}_{0\|}\,t = \mathbf{v}_0 \times \boldsymbol{\omega}^{-1}\,e^{i\boldsymbol{\omega} t} + \mathbf{v}_0\cdot\boldsymbol{\omega}^{-1}\boldsymbol{\omega} t. \tag{6.18}$$

This is a parametric equation for a helix with *radius*

$$\mathbf{a} \equiv \mathbf{v}_0 \times \boldsymbol{\omega}^{-1}, \tag{6.19a}$$

and *pitch*

$$b \equiv \mathbf{v}_0 \cdot \boldsymbol{\omega}^{-1}. \tag{6.19b}$$

Adopting the angle of rotation

$$\theta = |\boldsymbol{\omega}|\, t \tag{6.19c}$$

as parameter, and writing $\boldsymbol{\theta} = \theta\hat{\boldsymbol{\omega}}$, Equation (6.18) can be put in the "standard form" for a helix

$$\mathbf{r}(\boldsymbol{\theta}) = \mathbf{a}e^{i\boldsymbol{\theta}} + b\boldsymbol{\theta} \tag{6.20}$$

with it understood that $\mathbf{a}\cdot\boldsymbol{\theta} = 0$. The helix is said to be *right-handed* if $b > 0$ and *left-handed* if $b < 0$. (See Figure 6.2a, b).

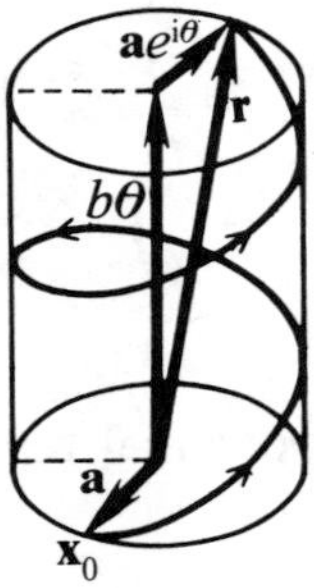

Fig. 6.2a. Righthanded Helix.

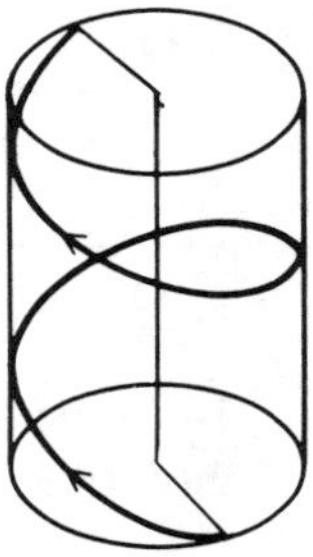

Fig. 62b. Lefthanded Helix.

The trajectory (6.18) reduces to a circle when $\mathbf{v}_{0\parallel} = 0$. The radius vector $\mathbf{r}$ rotates with an angular speed $|\boldsymbol{\omega}| = |q\mathbf{B}|/mc$ called the *cyclotron frequency*. According to (6.2), ω has the same direction as the magnetic field $\mathbf{B}$ when the charge q is negative and the opposite direction when the charge is positive. As shown in Figure 6.3, the circular motion of a negative charge is right-handed relative to $\mathbf{B}$ while that of a positive charge is left-handed.

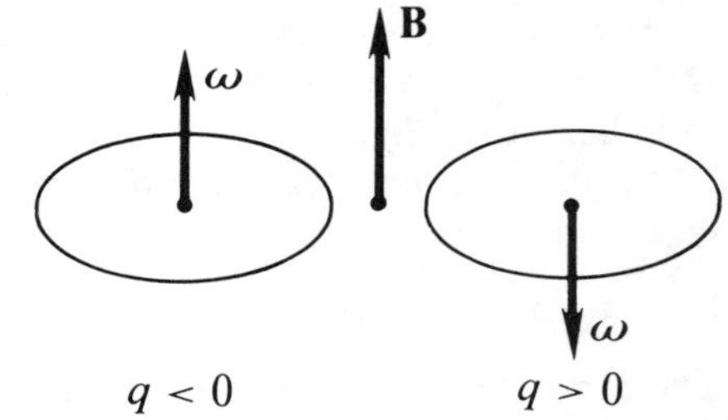

Fig. 6.3.

The Effect of Linear Drag

Now let us see how the motion just considered is modified by a linear resistive force. We seek solutions of the equation

$$\dot{\mathbf{v}} + \boldsymbol{\Omega}\cdot\mathbf{v} + \gamma\mathbf{v} = 0, \tag{6.21}$$

where γ is a positive constant. With all quantities defined as before, by introducing appropriate integrating factors, we put (6.21) in the form

$$\frac{\mathrm{d}}{\mathrm{d}t}(e^{\gamma t}R\mathbf{v}R^{\dagger}) = 0.$$

Integrating and solving for $\mathbf{v} = \mathbf{v}(t)$, we get

$$\mathbf{v} = e^{-\gamma t}R^{\dagger}\mathbf{v}_0 R \quad = e^{-\gamma t}(\mathbf{v}_{0\perp}\, e^{\boldsymbol{\Omega} t} + \mathbf{v}_{0\|}). \tag{6.22}$$

Writing $\mathbf{v} = \dot{\mathbf{x}}$ and integrating once more, we find that the equation for the trajectory has the same form as (6.22), specifically,

$$\mathbf{r} = e^{-\gamma t}(\mathbf{a}e^{\boldsymbol{\Omega} t} + \mathbf{b}), \tag{6.23}$$

where

$$\mathbf{a} = \mathbf{v}_{0\perp}\,(\boldsymbol{\Omega} - \gamma)^{-1},$$

$$\mathbf{b} = -\,\mathbf{v}_{0\|}\,\gamma^{-1},$$

$$\mathbf{r} = \mathbf{x} - \mathbf{x}_0 + \mathbf{a} + \mathbf{b}.$$

Equation (6.23) describes a particle spiraling to rest at $\mathbf{r} = 0$. The displacement in time $t \gg \gamma^{-1}$ is given by

$$\mathbf{x}(\infty) - \mathbf{x}_0 = -\mathbf{a} - \mathbf{b} = \frac{\mathbf{v}_{0\perp}(\boldsymbol{\Omega}+\gamma)}{\boldsymbol{\Omega}^2 - \gamma^2} + \frac{\mathbf{v}_{0\|}}{\gamma}.$$

The trajectory lies in a plane if $\mathbf{v}_{0\|} = 0$. In this case, (6.22) and (6.23) imply the following simple relation between velocity and radius vector:

$$\mathbf{v}\mathbf{r}^{-1} = \boldsymbol{\Omega} - \gamma.$$

The angle ϕ between $\mathbf{v}$ and $\mathbf{r}$ is therefore given by

$$\tan\phi = \frac{|\,\mathbf{v}\wedge\mathbf{r}\,|}{\mathbf{v}\cdot\mathbf{r}} = \frac{|\,\boldsymbol{\Omega}\,|}{-\gamma}.$$

Since this angle is the same at all points of the trajectory, the curve is commonly called an equiangular spiral.

The spiral tracks of electrons in a bubble chamber shown in Figure 6.4 are not accurately described as equiangular, because the retarding force is not linear in the velocity and velocities of interest are high enough for relativistic effects to be significant.

3-7. Uniform Electric and Magnetic Fields

In this section we develop a general method for solving the equation of motion of a charged particle in uniform electric and magnetic fields. The equation of motion is

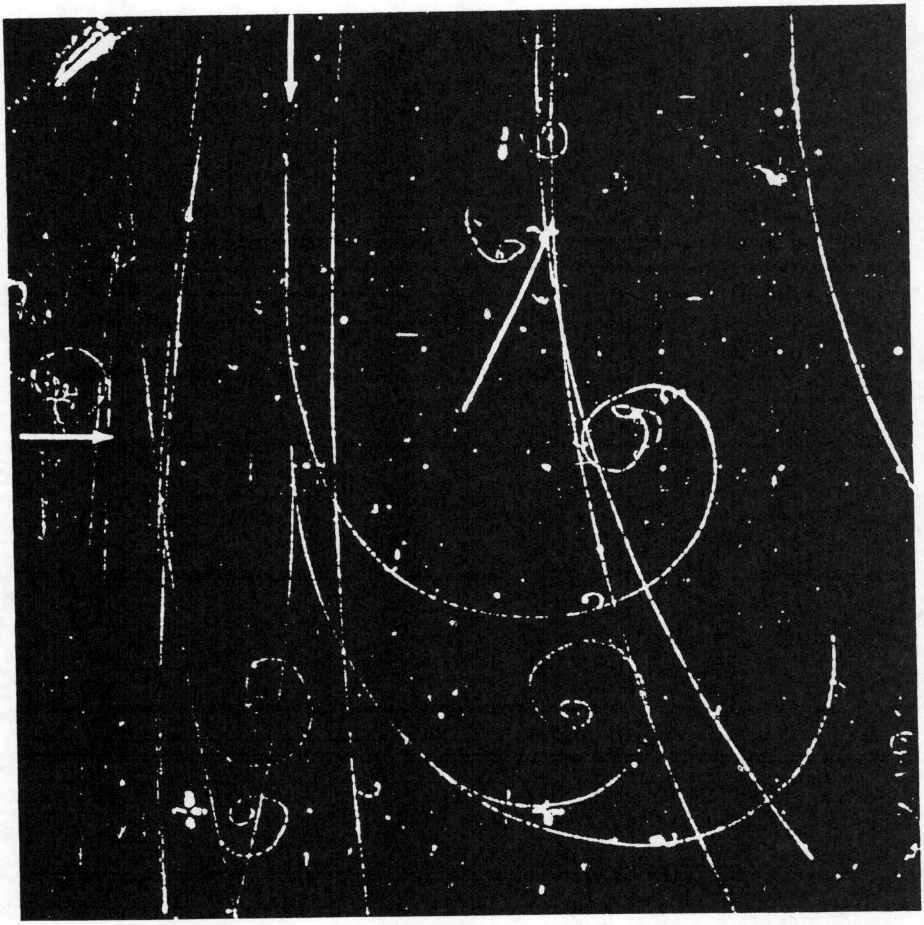

Fig. 6.4. An electron loses kinetic energy by emitting a photon (light quantum), as shown by the sudden increase in the curvature of its trajectory (track) in a propane bubble chamber. The emitted photon creates an electron-position pair. The curvature of an electron trajectory in a magnetic field perpendicular to the photograph increases as the electron loses energy by collisions. Since electrons and positrons have the same mass but opposite charges, their trajectories in a magnetic field curve in opposite directions. The smaller curvature of the positron was created with more kinetic energy than the electron.

$$m\dot{\mathbf{v}} = q\left(\mathbf{E} + \frac{\mathbf{v}}{c} \times \mathbf{B}\right). \tag{7.1}$$

As before, we suppress the constants by writing

$$\mathbf{g} = \frac{q}{m}\mathbf{E}, \tag{7.2a}$$

$$\boldsymbol{\Omega} = i\boldsymbol{\omega} = -\frac{iq\mathbf{B}}{mc}, \tag{7.2b}$$

so the equation of motion takes the form

$$\dot{\mathbf{v}} = \mathbf{g} + \mathbf{v}\cdot\boldsymbol{\Omega} = \mathbf{g} + \boldsymbol{\omega}\times\mathbf{v}. \tag{7.3}$$

For uniform fields, $\mathbf{g} = \mathbf{g}(t)$ and $\boldsymbol{\Omega} = \boldsymbol{\Omega}(t)$ are functions of time but not of position.

We aim to solve Equation (7.3) by the *method of integrating factors*. As in the preceding section, we introduce an integrating factor $R = R(t)$ defined by the equation

$$\dot{R} = \tfrac{1}{2}R\boldsymbol{\Omega} \tag{7.4a}$$

and the initial condition

$$R(0) = 1. \tag{7.4b}$$

This enables us to put Equation (7.3) in the form

$$\frac{\mathrm{d}}{\mathrm{d}t}(R\mathbf{v}R^\dagger) = R\mathbf{g}R^\dagger.$$

Integrating and solving for $\mathbf{v}$, we get the general solution

$$\mathbf{v} = R^\dagger\{\mathbf{v}_0 + \int_0^t R\mathbf{g}R^\dagger\,\mathrm{d}t\}\,R, \tag{7.5}$$

or, with the arguments made explicit

$$\mathbf{v}(t) = R^\dagger(t)\,\{\mathbf{v}_0 + \int_0^t R(s)\mathbf{g}(s)R^\dagger(s)\,\mathrm{d}s\}R(t)\}.$$

After this has been used to evaluate $\mathbf{v}$ as an explicit function of t, the trajectory is found directly by integrating $\dot{\mathbf{x}} = \mathbf{v}(t)$.

Now let us examine some specific solutions. For constant $\boldsymbol{\Omega}$ we know that (7.4) has the solution

$$R = e^{(1/2)\boldsymbol{\Omega}t} = e^{(1/2)i\boldsymbol{\omega}t}. \tag{7.6}$$

In this case, we can simplify the integral in (7.5) with a procedure we have used before. We decompose $\mathbf{g}$ into components parallel and perpendicular to $\boldsymbol{\omega}$, which have the algebraic properties

$$\mathbf{g}_\parallel\boldsymbol{\Omega} = \mathbf{g}\wedge\boldsymbol{\Omega} = \boldsymbol{\Omega}\mathbf{g}_\parallel, \tag{7.7a}$$

$$\mathbf{g}_\perp\boldsymbol{\Omega} = \mathbf{g}\cdot\boldsymbol{\Omega} = -\boldsymbol{\Omega}\mathbf{g}_\perp; \tag{7.7b}$$

whence,

$$R\mathbf{g}_\parallel = \mathbf{g}_\parallel R, \tag{7.7c}$$

$$R\mathbf{g}_\perp = \mathbf{g}_\perp R^\dagger. \tag{7.7d}$$

So the integral can be put in the form

$$\int_0^t RgR^\dagger \, dt = \int_0^t \mathbf{g}_\perp e^{-\boldsymbol{\Omega} t} \, dt + \int_0^t \mathbf{g}_\| \, dt, \tag{7.8}$$

ready to be integrated when the functional form of $\mathbf{g}$ is specified.

When $\mathbf{g}$ is constant, the integral (7.8) has the specific value

$$\int_0^t R\mathbf{g}R^\dagger \, dt = \mathbf{g}_\perp \boldsymbol{\Omega}^{-1} (1 - e^{-\boldsymbol{\Omega} t}) + \mathbf{g}_\| \, t.$$

Inserting this into (7.5) and using (7.7), we get the velocity in the form

$$\mathbf{v} = \mathbf{a}e^{\boldsymbol{\Omega} t} + \mathbf{b}t + \mathbf{c}, \tag{7.9}$$

where the coefficients are given by

$$\mathbf{a} = \mathbf{v}_{0\perp} + \mathbf{g}\cdot\boldsymbol{\Omega}^{-1} = \mathbf{v}_{0\perp} + \mathbf{g} \times \boldsymbol{\omega}^{-1} = \mathbf{v}_{0\perp} - c\mathbf{E} \times \mathbf{B}^{-1}, \tag{7.10a}$$

$$\mathbf{b} = \mathbf{g}_\| = \mathbf{g}\cdot\boldsymbol{\omega}^{-1}\boldsymbol{\omega} = \frac{q}{m}\mathbf{E}_\|, \tag{7.10b}$$

$$\mathbf{c} = \mathbf{v}_{0\|} - \mathbf{g}\cdot\boldsymbol{\Omega}^{-1} = \mathbf{v}_{0\|} - \mathbf{g} \times \boldsymbol{\omega}^{-1} = \mathbf{v}_{0\|} + c\mathbf{E} \times \mathbf{B}^{-1}. \tag{7.10c}$$

Since $\mathbf{a}$ lies in the $\boldsymbol{\Omega}$ – plane and $\mathbf{b}$ is orthogonal to that plane, Equation (7.9) is a parametric equation for a helix with radius $|\,\mathbf{a}\,|$.

By integrating (7.9) we find the trajectory

$$\mathbf{r} = \mathbf{a}\boldsymbol{\Omega}^{-1} e^{\boldsymbol{\Omega} t} + \tfrac{1}{2}\mathbf{b}t^2 + \mathbf{c}t, \tag{7.11}$$

where an appropriate choice of origin has been made by writing

$$\mathbf{r} \equiv \mathbf{x} - \mathbf{x}_0 + \mathbf{a}\boldsymbol{\Omega}^{-1}.$$

Equation (7.11) is best interpreted by regarding it as a composite of two motions. First, a parabolic motion of the *guiding center* described by the equation

$$\mathbf{r}_1 = \tfrac{1}{2}\mathbf{b}t^2 + \mathbf{c}t. \tag{7.12a}$$

Second a uniform circular motion about the guiding center described by

$$\mathbf{r}_2 = \mathbf{a}\boldsymbol{\Omega}^{-1} e^{\boldsymbol{\Omega} t} \tag{7.12b}$$

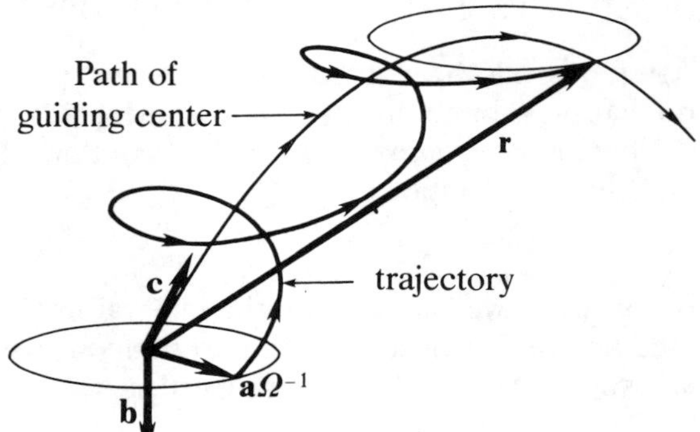

Fig. 7.1. Trajectory of a charged particle in uniform electric and magnetic fields.

The composite motion $\mathbf{r} = \mathbf{r}_1 + \mathbf{r}_2$ can be visualized as the motion of a point on a spinning disk traversing a parabola with its axis aligned along the vertical, as illustrated in Figure 7.1. Corresponding directions of electric and magnetic fields are shown in Figure 7.2.

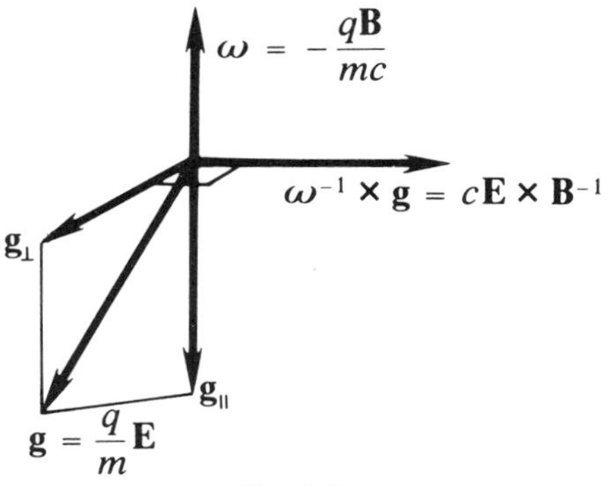

Fig. 7.2.

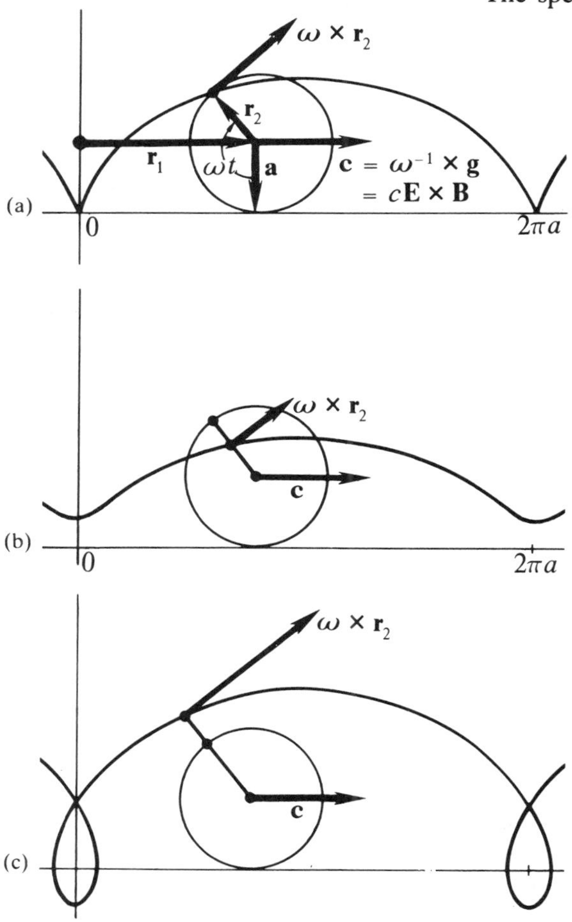

Fig. 7.3. Trajectories of a charged particle in orthogonal electric and magnetic fields.

Motion about the guiding center averages to zero over a time period of $2\pi|\mathbf{\Omega}|^{-1} = 2\pi mc|q\mathbf{B}|^{-1}$, so motion of the guiding center can be regarded as an average motion of the particle. Accordingly, the velocity of the guiding center is called the *drift velocity* of the particle.

Motion in Orthogonal Electric and Magnetic Fields

The special case of motion in orthogonal electric and magnetic fields is of particular interest. From (7.10b) we see that if $\mathbf{E}\cdot\mathbf{B} = 0$, then $\mathbf{b} = 0$, so according to (7.11) or (7.12), the parabolic path of the guiding center reduces to a straight line. If, in addition, the initial velocity is orthogonal to the magnetic field, then, by (7.10c) and (7.12a), the drift velocity is

$$\dot{\mathbf{r}}_1 = \mathbf{c} = \boldsymbol{\omega}^{-1} \times \mathbf{g} = c\mathbf{E} \times \mathbf{B}^{-1}. \quad (7.13)$$

Surprisingly, the drift velocity in this case is orthogonal to the electric as well as the magnetic field. The particle trajectory can be visualized as the path of a point on a disk spinning with angular speed $\omega = -q|\mathbf{B}|/mc$ as it moves in the plane of the disk with constant speed $|\mathbf{c}| = c|\mathbf{E} \times \mathbf{B}^{-1}| = c|\mathbf{E}|/|\mathbf{B}|$. The Greeks long ago gave the name trochoid (trochas = wheel) to curves of this kind. The trochoids fall into three classes characterized by the conditions $|\boldsymbol{\omega} \times \mathbf{a}| = |\mathbf{c}|$, $|\boldsymbol{\omega} \times \mathbf{a}| < |\mathbf{c}|$, $|\boldsymbol{\omega} \times \mathbf{a}| > |\mathbf{c}|$, with respective curves illustrated by Figures 7.3a, b, c. As the figures suggest, the three curves can also be interpreted as the paths of particles rigidly attached to a wheel which is rolling without slipping, the three conditions being that the particle is attached on the rim, inside the rim or outside the rim, respectively.

According to (7.11) or (7.12), the particle motion coincides with the guiding center motion when $\mathbf{a} = 0$, which, according to (7.10a), occurs when

$$\mathbf{v}_0 = \boldsymbol{\omega}^{-1} \times \mathbf{g} = c\mathbf{E} \times \mathbf{B}^{-1}. \tag{7.14}$$

The trajectory is a straight line if $\mathbf{E}\cdot\mathbf{B} = 0$. This suggests a practical way to construct a velocity filter for charged particles. Only a particle which satisfies the condition (7.14) will continue undeflected along its initial line of motion. The **E** and **B** fields can be adjusted to select any velocity in a wide range. The selection is independent of the sign of the charge or the mass of the particle.

The Hall Effect

The above results can be used to analyze the effect of an external magnetic field on an electric current in a conductor. Suppose we have a conductor with a current I immersed in a uniform magnetic field **B**. As Figure 7.4 shows, whether the current is due to a flow of positive charges or negative charges, the charge carriers in both cases will be deflected to the right by the magnetic force $\mathbf{F}_\pm = q_\pm c^{-1}\, \mathbf{v}_\pm \times \mathbf{B}$. Consequently, carriers will accumulate on the right wall of the conductor until they produce an electric field $\mathbf{E}_H$ sufficient to cancel the magnetic force exactly. The condition (7.14) for a velocity filter has then been met and charges flow undeflected in the conductor with speed

$$v = cE_H B^{-1}.$$

The electric field E_H is manifested by a readily measured potential difference

$$\phi_H = aE_H$$

between the right and left sides of the conductor. The appearance of such a transverse potential difference induced in a conductor by a magnetic field is known as the *Hall Effect*.

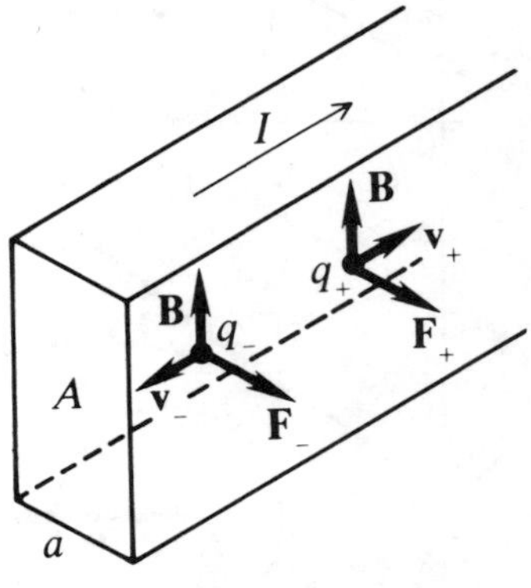

Fig. 7.4. The Hall effect.

The sign of the Hall potential ϕ_H indicates the sign of the charge on the carriers. The magnitude of ϕ_H permits an estimate of the density N of charge carriers. The current density is given by

$$J = \frac{I}{A} = Nqv,$$

where A is the cross-sectional area of the conductor and q is the unit of charge. Hence,

$$N = \frac{J}{qv} = \frac{JB}{qcE_H} = \frac{JBa}{qc\phi_H} \ .$$

The quantities on the right are known or readily measured. The Hall effect has been a valuable aid in the study of semiconductors, such as silicon and germanium, in which N is small compared with its value in metals, for which the Hall potential is correspondingly large and easily measured.

We can estimate the magnitude of the Hall potential by taking resistance into account. Suppose, as expected from Ohm's law, that the resistive force exerted by the medium on the charge carriers is linear. The equation of motion (7.3) is then generalized to

$$\dot{\mathbf{v}} = \mathbf{g} + \mathbf{v}\cdot\boldsymbol{\Omega} - \gamma\mathbf{v}. \tag{7.15}$$

The condition $\mathbf{v}\cdot\mathbf{B} = 0$ implies that $\mathbf{v}\cdot\boldsymbol{\Omega} = \mathbf{v}\boldsymbol{\Omega}$, so for terminal velocity Equation (7.15) gives

$$\mathbf{v}(\gamma - \boldsymbol{\Omega}) = \mathbf{g},$$

or

$$\mathbf{v} = \mathbf{g}(\gamma - \boldsymbol{\Omega})^{-1} = \mathbf{g}\left(\frac{\gamma + \boldsymbol{\Omega}}{\gamma^2 + |\boldsymbol{\Omega}|^2}\right) . \tag{7.16}$$

From this we can read off immediately that the direction of $\mathbf{v}$ differs from that of $\mathbf{g}$ by the angle $\tan^{-1}(\gamma^{-1}|\boldsymbol{\Omega}|)$. In the case of the Hall effect, the electric field $\mathbf{E}$ has a component $\mathbf{E}_0$ collinear with $\mathbf{v}$ and component $\mathbf{E}_H$ orthogonal to it, so

$$\mathbf{g} = \frac{q}{m}\mathbf{E} = \frac{q}{m}(\mathbf{E}_0 + \mathbf{E}_H).$$

The condition that components orthogonal to $\mathbf{v}$ cancel in (7.16) entails

$$\mathbf{E}_H\gamma + \mathbf{E}_0\boldsymbol{\Omega} = 0,$$

whence

$$\mathbf{E}_H = -\gamma^{-1}\mathbf{E}_0\boldsymbol{\Omega} = \frac{q}{mc\gamma}\mathbf{B}\times\mathbf{E}_0 . \tag{7.17}$$

By substituting this back into (7.16), it can be verified that $\mathbf{v} = q\mathbf{E}_0/m\gamma$ in accordance with Ohm's law. Equation (7.17) can be used for experimental determination of the resistive coefficient γ, and for some metals (eg. Bismuth) the resistance is found to depend strongly on the magnetic field and its orientation of crystal axes in the conductor, revealing another valuable application of the Hall effect.

Cyclotron Resonance

As another application of our general method, let us study the motion of a charged particle in a constant magnetic field subject to a periodic electric field $\mathbf{E} = \mathbf{E}_0 \cos \omega' t$ as well. To find the velocity of the particle, we substitute

$$\mathbf{g} = m^{-1} q\mathbf{E} = \mathbf{g}_0 \cos \omega' t \tag{7.18}$$

into (7.8) and evaluate the integrals

$$\int_0^t \mathrm{d}t \cos \omega' t = \frac{\sin \omega' t}{\omega'} \ ,$$

$$\int_0^t \mathrm{d}t\, e^{-\mathbf{i}\omega t} \cos \omega' t = \int_0^t \mathrm{d}t\, \tfrac{1}{2} \left(e^{\mathbf{i}(\omega' - \omega)t} + e^{-\mathbf{i}(\omega' + \omega)t}\right)$$

$$= \frac{1}{2\mathbf{i}} \left(\frac{e^{\mathbf{i}(\omega' - \omega)t}}{\omega' - \omega} - \frac{e^{-\mathbf{i}(\omega' + \omega)t}}{\omega' + \omega} \right)$$

$$= e^{-\mathbf{i}\omega t} \left(\frac{\omega' \sin \omega' t - \mathbf{i}\omega \cos \omega' t}{\omega'^2 - \omega^2} \right) ,$$

where we have written the "cyclotron frequency" of the magnetic field in the form $\boldsymbol{\Omega} = i\boldsymbol{\omega} = \mathbf{i}\omega$ for the convenience of operating with the unit bivector $\mathbf{i}$. With the integral evaluated, Equation (7.5) gives us

$$\mathbf{v} = e^{-(1/2)i\boldsymbol{\omega} t} \left\{ \mathbf{v}_0 + \mathbf{g}_{0\perp}\, e^{-i\boldsymbol{\omega} t} \left(\frac{\omega' \sin \omega' t - i\boldsymbol{\omega} \cos \omega' t}{\omega'^2 - \omega^2} \right) \right.$$

$$\left. + \, \mathbf{g}_{0\|} \frac{\sin \omega' t}{\omega'} \right\} e^{(1/2)i\boldsymbol{\omega} t},$$

which simplifies to

$$\mathbf{v} = \mathbf{g}_{0\perp} \left(\frac{\omega' \sin \omega' t - i\boldsymbol{\omega} \cos \omega' t}{\omega'^2 - \omega^2} \right) + \mathbf{g}_{0\|} \frac{\sin \omega' t}{\omega'} + \mathbf{v}_{0\perp}\, e^{i\boldsymbol{\omega} t} + \mathbf{v}_{0\|} . \tag{7.19}$$

This can be integrated to get the trajectory, but the feature of greatest interest is apparent right here, namely the fact that the first term on the right side of (7.19) is infinite when $\omega' = \omega$. This implies the existence of a *resonance* when the "driving frequency" ω' is in the neighborhood of the cyclotron frequency ω. As we shall see later, dissipative effects must be taken into account near resonance, with the consequence that the infinity in (7.19) is averted, but the velocity $\mathbf{v}$ has a maximum magnitude at resonance. At the same time the terms depending on $\mathbf{v}_0$ in (7.19) are "damped out". The chief significance of all this is that at resonance the particle extracts energy from the driving electric field at a maximum rate, energy which is lost in collisions or by reradiation.

These considerations help explain attenuation of radio waves in the ionosphere. The driving field (7.18) can be attributed to a plane polarized electromagnetic wave with frequency $\omega'/2\pi$. For free electrons in the ionosphere the earth's magnetic field strength of 5×10^{-1} gauss yields a cyclotron frequency of magnitude $\omega = -qB/mc = 8.5 \times 10^6$. So we expect resonant absorption of electromagnetic waves with frequency near $\omega/2\pi = 1400$ kHz. The ionosphere does indeed exhibit a marked absorption of radio waves in that frequency range.

The same considerations are important for describing the propagation of electromagnetic waves in the vicinity of stars and through plasmas generated in the laboratory.

We have not considered problems with time-varying magnetic fields, but it should be pointed out that, for uniform magnetic fields with fixed direction but time-dependent magnitude, the solution of (7.4) has the general form

$$R = e^{\int_0^t \boldsymbol{\Omega}\, dt} = e^{i\int_0^t \boldsymbol{\omega}\, dt}, \tag{7.20}$$

which can be evaluated by straightforward integration. Solutions of (7.4) when the direction of $\boldsymbol{\Omega}$ is time-dependent will be obtained later.

3-7. Exercises

(7.1) Solve Equation (7.3) for constant fields by the method of undetermined coefficients. Since the equation is linear, we expect the solution to be expressible as a sum of solutions for the special cases $\mathbf{g} = 0$ and $\boldsymbol{\Omega} = 0$ which we found earlier. Therefore we expect a solution of the form

$$\mathbf{v} = \mathbf{a}e^{\boldsymbol{\Omega}t} + \mathbf{b}t + \mathbf{c}.$$

Verify this and evaluate the coefficients by substitution in (7.3) and imposition of initial conditions.

(7.2) A charged particle in constant, orthogonal electric and magnetic fields is at rest at the origin at time $t = 0$. Determine the time and place of its next "rest stop".

(7.3) Show that the parametric equation (7.11) for the trajectory in constant fields can be put in the form

$$\begin{aligned}\mathbf{r} = {} & \mathbf{v}_0 t + \tfrac{1}{2}\mathbf{g}t^2 + \\ & + \boldsymbol{\omega}\times\left[\left(\frac{\omega t - \sin\omega t}{\omega^3}\right)\mathbf{g} + \left(\frac{1-\cos\omega t}{\omega^2}\right)\mathbf{v}_0\right] + \\ & + \boldsymbol{\omega}\times\left\{\boldsymbol{\omega}\times\left[\left(\frac{1-\frac{1}{2}\omega^2t^2-\cos\omega t}{\omega^4}\right)\mathbf{g} + \left(\frac{\sin\omega t - \omega t}{\omega^3}\right)\mathbf{v}_0\right]\right\},\end{aligned}$$

showing the deviation from the parabolic trajectory obtained when $\boldsymbol{\omega} = 0$. Evaluate the deviation to fourth order in $\omega = |\boldsymbol{\omega}|$.

(7.4) Use the method of integrating factors to solve the equation of motion

$$\dot{\mathbf{v}} = \mathbf{g} + \mathbf{v}\cdot\boldsymbol{\Omega} - \gamma\mathbf{v},$$

for a charged particle in constant electric and magnetic fields subject to a linear resistive force. Verify that the trajectory is given by

$$\mathbf{r} = (\mathbf{g}_\perp Z^{-1} + \mathbf{v}_{0\perp})Z^{-1}\, e^{Zt} - \mathbf{g}_\perp\, Z^{-1}t + {}$$

$$+\, \mathbf{g}_\| \frac{(e^{-\gamma t} - 1 + \gamma t)}{\gamma^2} + \mathbf{v}_{0\|}\,\frac{(1 - e^{-\gamma t})}{\gamma}\ ,$$

where $Z = \boldsymbol{\Omega} - \gamma$, the parallel and perpendicular components of the vectors are defined as in Equation (7.7), and $\mathbf{r}$ is related to the position $\mathbf{x}$ by

$$\mathbf{r} = \mathbf{x} - \mathbf{x}_0 + (\mathbf{g}_\perp Z^{-1} + \mathbf{v}_{0\perp})Z^{-1}.$$

Draw a rough sketch and describe the solution, taking the last three terms to describe the guiding center.

3-8. Linear Binding Force

The binding of an electron to an atom or of an atom to a molecule, these are examples of the ubiquitous and general phenomenon of binding. To understand this phenomenon, we need an equally general mathematical theory of binding forces and the motion of bound particles.

A binding force is understood to be a force which tends to confine a particle to some finite region of space. The force function of a bound particle must be a function of position, and we can develop a theory of binding without considering velocity-dependent forces. Let us determine some general properties of binding forces. A point $\mathbf{x}_0$ is said to be an equilibrium point of a force $\mathbf{f} = \mathbf{f}(\mathbf{x})$ if

$$\mathbf{f}(\mathbf{x}_0) = 0. \tag{8.1}$$

The equilibrium point $\mathbf{x}_0$ is said to be isolated if there is some neighborhood of $\mathbf{x}_0$ which does not contain any other equilibrium points. Let us focus attention on such a point and such a neighborhood.

Near an equilibrium point $\mathbf{x}_0$, a simple approximation to any force function is obtained by using the Taylor expansion (as explained in Section 2-8)

$$\mathbf{f}(\mathbf{x}) = \mathbf{f}(\mathbf{x}_0 + \mathbf{r}) = \mathbf{f}(\mathbf{x}_0) + \mathbf{r}\cdot\nabla\,\mathbf{f}(\mathbf{x}_0) + \tfrac{1}{2}\,(\mathbf{r}\cdot\nabla)^2\,\mathbf{f}(\mathbf{x}_0) + \ldots . \tag{8.2}$$

The equation of motion $m\ddot{\mathbf{x}} = \mathbf{f}$ for a bound particle is then given in terms of the displacement from equilibrium $\mathbf{r} = \mathbf{x} - \mathbf{x}_0$ by

$$m\ddot{\mathbf{x}} = m\ddot{\mathbf{r}} = \mathbf{L}(\mathbf{r}) + \mathbf{Q}(\mathbf{r}) + \ldots, \tag{8.3}$$

where $\mathbf{L}(\mathbf{r}) \equiv \mathbf{r}\cdot\nabla\mathbf{f}(\mathbf{x}_0)$ and $\mathbf{Q}(\mathbf{r}) \equiv \frac{1}{2}(\mathbf{r}\cdot\nabla)^2\,\mathbf{f}(\mathbf{x}_0)$. The force function $\mathbf{L}(\mathbf{r})$ is *linear*, that is, it has the property

$$\mathbf{L}(\alpha_1\mathbf{r}_1 + \alpha_2\mathbf{r}_2) = \alpha_1\,\mathbf{L}(\mathbf{r}_1) + \alpha_2\,\mathbf{L}(\mathbf{r}_2), \tag{8.4}$$

where α_1 and α_2 are constant scalars. The force function $\mathbf{Q}(\mathbf{r})$ is a quadratic vector function of $\mathbf{r}$. Whatever the exact form of $\mathbf{f}(\mathbf{x})$, if $\mathbf{L}(\mathbf{r}) \neq 0$, there is a (sufficiently small) neighborhood of $\mathbf{x}_0$, defined roughly by the condition $|\,\mathbf{L}(\mathbf{r})\,| \gg |\,\mathbf{Q}(\mathbf{r})\,|$, in which the equation

$$m\ddot{\mathbf{r}} = \mathbf{L}(\mathbf{r}) \tag{8.5}$$

provides an accurate description of particle motion. The particle will remain in a neighborhood of the equilibrium point if $\mathbf{r}\cdot\ddot{\mathbf{r}} < 0$, which is to say that the particle's acceleration must be directed toward the interior of the neighborhood. This can be expressed as a binding property of the force by using (8.5); thus,

$$\mathbf{r}\cdot\mathbf{L}(\mathbf{r}) \leqslant 0, \tag{8.6}$$

with equality only if $\mathbf{r} = 0$. This relation is also called a *stability condition*, because it is a necessary condition for the particle to remain bound.

A particle subject to a linear *binding force* $\mathbf{L}(\mathbf{r})$ is called a *harmonic oscillator*, because its motion is similar to vibrations in musical instruments. Indeed, if a plucked violin string is represented as a system of particles, the harmonic oscillator provides a good description of each particle. The quadratic force $\mathbf{Q}(\mathbf{r})$ and higher order terms in the Taylor expansion of the force are called *anharmonic perturbations*, and when they are included in its description, the particle is said to be an *anharmonic oscillator*. The most significant difference between harmonic and anharmonic oscillations is that the former obey a linear superposition principle; specifically, if $\mathbf{r}_1 = \mathbf{r}_1(t)$ and $\mathbf{r}_2 = \mathbf{r}_2(t)$ are solutions of the equation of motion (8.5), then, as a consequence of (8.4), so is the "linear superposition" $\mathbf{r}_3 = \alpha_1\mathbf{r}_1 + \alpha_2\mathbf{r}_2$. This superposition principle makes the analysis of harmonic motions easy, and the lack of any such general principle makes the analysis of anharmonic motions difficult. Anharmonic motions are most easily analyzed by perturbation theory as small deviations from harmonic motion.

The Isotropic Oscillator

A particle subject to an isotropic binding force is called an *isotropic oscillator*. In this case $\mathbf{L}(\mathbf{r})$ has the simple form

$$\mathbf{L}(\mathbf{r}) = -k\mathbf{r}, \tag{8.7}$$

where k is a constant positive scalar called the *force constant*. Equation (8.7) will be recognized as *Hooke's Law*, but the term "Hooke's Law" is commonly used to refer to any linear binding force. Obviously, Hooke's Law is so ubiquitous in physics, because any binding force can be approximated by a linear force under some conditions. By the same token, it is evident that Hooke's Law is not a fundamental force law, but only a useful approximation.

Let us now examine the motion of an isotropic oscillator. Its equation of motion can be put in the form

$$\ddot{\mathbf{r}} + \frac{k}{m}\mathbf{r} = 0. \tag{8.8}$$

Our experience with exponential functions suggests that this equation might have a solution of the form

$$\mathbf{r} = \mathbf{a}e^{\lambda t}$$

Inserting this trial solution into (8.8) and carrying out the differentiation, we find that the equation of motion is satisfied if and only if

$$\lambda^2 + \frac{k}{m} = 0.$$

This algebraic equation is called the *characteristic equation* of (8.8); it has the roots

$$\lambda = \pm\, \mathbf{i}\omega_0,$$

where $\mathbf{i}^2 = -1$ and

$$\omega_0 \equiv \sqrt{\frac{k}{m}}\,. \tag{8.9}$$

To each of these roots there corresponds a distinct solution of (8.8), namely

$$\mathbf{r}_+ = \mathbf{a}_+ e^{\mathbf{i}\omega_0 t} = \mathbf{a}_+(\cos\omega_0 t + \mathbf{i}\sin\omega_0 t), \tag{8.10a}$$

and

$$\mathbf{r}_- = \mathbf{a}_- e^{-\mathbf{i}\omega_0 t} = \mathbf{a}_-(\cos\omega_0 t - \mathbf{i}\sin\omega_0 t), \tag{8.10b}$$

The reader will recognize that we are freely using properties of the exponential function determined in Section 2-5. Notice that these equations imply that $\mathbf{i}$ is a bivector satisfying

$$\mathbf{a}_\pm \wedge \mathbf{i} = 0, \tag{8.11}$$

because $\mathbf{r}_\pm$ and $\mathbf{a}_\pm$ must all be vectors. This tells us the significance of the "imaginary roots" of the characteristic equation. The "imaginary" $\mathbf{i}$ is the unit bivector for a plane in which the solution vectors $\mathbf{r}_\pm$ lie at all times.

According to the super principle, we can add solutions $\mathbf{r}_+$ and $\mathbf{r}_-$ to get a solution.

$$\mathbf{r} = \mathbf{a}_+ e^{\mathbf{i}\omega_0 t} + \mathbf{a}_- e^{-\mathbf{i}\omega_0 t} . \tag{8.12}$$

Alternatively, this solution can be written in the form

$$\mathbf{r} = \mathbf{a}_0 \cos \omega_0 t + \mathbf{b}_0 \sin \omega_0 t , \tag{8.13}$$

where

$$\mathbf{a}_0 = \mathbf{a}_+ + \mathbf{a}_- , \tag{8.14}$$
$$\mathbf{b}_0 = (\mathbf{a}_+ - \mathbf{a}_-)\mathbf{i}.$$

The two constant vectors $\mathbf{a}_0$ and $\mathbf{b}_0$ can have any values, so we know that (8.12) is the general solution of (8.8); thus, the orbit of any harmonic oscillator can be described by an equation of the form (8.13) or (8.12). Note that if either $\mathbf{a}_0 = 0$ or $\mathbf{b}_0 = 0$, then $\mathbf{a}_+ \wedge \mathbf{a}_- = 0$ and the bivector $\mathbf{i}$ in (8.12) is not uniquely determined by the condition (8.11); however, any unit bivector $\mathbf{i}$ satisfying (8.10) will do.

Equation (8.13) has the advantage of being directly related to initial conditions, for the vector coefficients are related to the initial position and velocity by

$$\begin{aligned} \mathbf{r}_0 &\equiv \mathbf{r}(0) = \mathbf{a}_0 . \\ \mathbf{v}_0 &= \dot{\mathbf{r}}(0) = \omega_0 \mathbf{b}_0 . \end{aligned} \tag{8.15}$$

Equation (8.13) represents the motion as a superposition of independent simple harmonic motions along the lines determined by $\mathbf{r}_0$ and $\mathbf{v}_0$. The resultant motion is elliptical, and of course, it reduces to one dimensional simple harmonic motion if $\mathbf{r}_0 = 0$ or $\mathbf{v}_0 = 0$ or, more generally, if $\mathbf{r}_0 \wedge \mathbf{v}_0 = \omega_0 \mathbf{a}_0 \wedge \mathbf{b}_0 = 0$. This may be more obvious if (8.13) is recast in the standard form (see Exercise 8.1):

$$\mathbf{r} = \mathbf{a} \cos \phi + \mathbf{b} \sin \phi, \tag{8.16a}$$

where $\mathbf{a}^2 \geqslant \mathbf{b}^2$, $\mathbf{a}\cdot\mathbf{b} = 0$, and

$$\phi = \phi(\mathrm{t}) = \omega_0 \mathrm{t} + \phi_0 . \tag{8.16b}$$

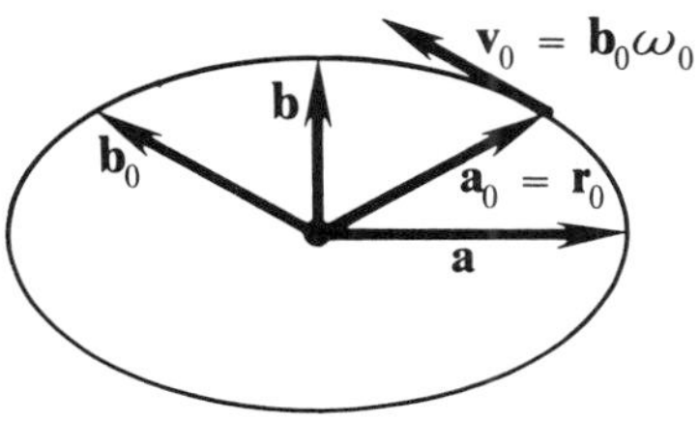

Fig. 8.1. Orbit of an isotropic harmonic oscillator.

The scalar constant ϕ_0 in (8.16b) can be eliminated, if desired, by writing $\phi_0 = \omega_0 t_0$ and shift the zero of time by t_0. It will be recognized that (8.16a) is a parametric equation for an ellipse with major axis $\mathbf{a}$ and minor axis $\mathbf{b}$ (Figure 8.1).

The elliptical motion of an isotropic oscillator is periodic in time. A particle motion is said to be *periodic* if its *state variables* $\mathbf{r}$ and $\dot{\mathbf{r}}$ have exactly the same values at any two times separated by a definite time interval T called the *period* of the motion. For the elliptical motion (8.16), the period T is related to the *natural frequency* ω_0 by

$$\omega_0 T = 2\pi. \tag{8.17}$$

The motion during a single period or cycle is called an *oscillation* or, if it is one dimensional, a *vibration*. The constant ϕ_0 in (8.16b) is referred to as the *phase* of an oscillation beginning at $t = 0$. The maximum displacement from equilibrium during an oscillation is called the *amplitude* of the oscillator. For the elliptical motion (8.16), the amplitude is $a = |\mathbf{a}|$.

Equation (8.12) represents the elliptical motion as a superposition of two *uniform circular motions* (8.10a, b) with opposite senses. This is illustrated in Figure 8.2 for $\phi = \omega_0 t$. As the figure suggests, this relation provides a practical means for constructing an ellipse from two circles.

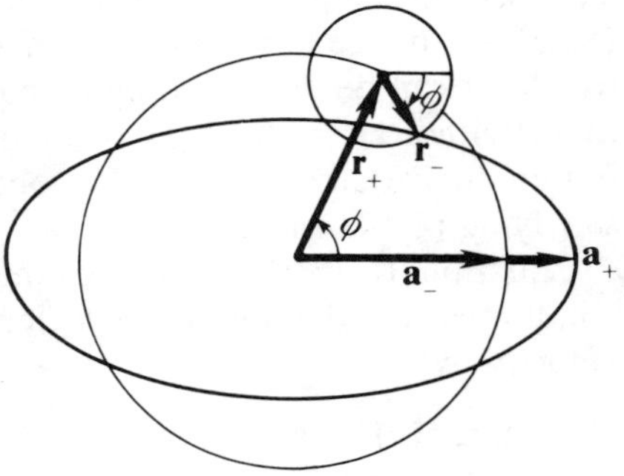

Fig. 8.2. Elliptical motion as a superposition of coplanar circular motions.

Elliptical motion can also be represented as a superposition of two circular motions with the same amplitude, frequency and phase. As shown in Figure 8.3, the circular motions are in two distinct planes intersecting along the major axis of the resultant ellipse. This relation is described by the equation

$$\mathbf{r} = \tfrac{1}{2}\,\mathbf{a}(e^{\mathbf{i}_+\phi} + e^{\mathbf{i}_-\phi}), \tag{8.18}$$

where $\mathbf{i}_+$ and $\mathbf{i}_-$ are unit bivectors for the two planes. It can be shown (Exercise 8.3) that Equations (8.18) and (8.16a) are equivalent if

$$\mathbf{b} = \tfrac{1}{2}\,\mathbf{a}(\mathbf{i}_+ + \mathbf{i}_-) = \tfrac{1}{2}\,\mathbf{a}\cdot(\mathbf{i}_+ + \mathbf{i}_-). \tag{8.19}$$

We will encounter other significant forms for the equation of an ellipse later on.

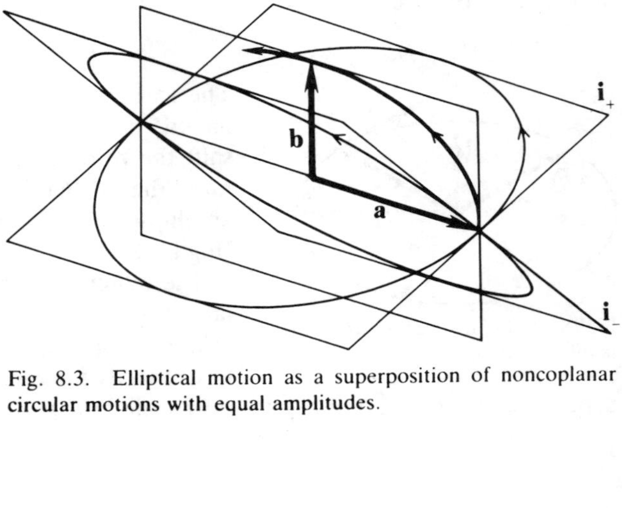

Fig. 8.3. Elliptical motion as a superposition of noncoplanar circular motions with equal amplitudes.

The Anisotropic Oscillator

Let us turn now to a brief consideration of the anisotropic oscillator. An anisotropic linear binding force

$\mathbf{L}(\mathbf{r})$ is characterized by the existence of three orthogonal *principal vectors* $\mathbf{a}_1$, $\mathbf{a}_2$, $\mathbf{a}_3$ with the properties

$$\begin{aligned}\mathbf{L}(\mathbf{a}_1) &= -k_1\mathbf{a}_1,\\ \mathbf{L}(\mathbf{a}_2) &= -k_2\mathbf{a}_2,\\ \mathbf{L}(\mathbf{a}_3) &= -k_3\mathbf{a}_3,\end{aligned} \tag{8.20}$$

where k_1, k_2 and k_3 are positive force constants describing the strength of the binding force along the three principle directions. Linear functions with the property (8.20) will be studied systematically in Chapter 5. Our problem, now, is to solve the equation of motion (8.5) subject to (8.20). The superposition principle enables us to decompose the general motion into independent one dimensional motions along the three principle directions; for if $\mathbf{r}_i$ is the component of displacement proportional to $\mathbf{a}_i$, then (8.5), (8.20) and (8.4) imply

$$m\ddot{\mathbf{r}} = m\ddot{\mathbf{r}}_1 + m\ddot{\mathbf{r}}_2 + m\ddot{\mathbf{r}}_3 = -k_1\mathbf{r}_1 - k_2\mathbf{r}_2 - k_3\mathbf{r}_3;$$

but the $\mathbf{r}_i$ are orthogonal, so each component must independently satisfy the equation

$$m\ddot{\mathbf{r}}_i = -k_i\mathbf{r}_i .$$

The solutions to this equation must have the same general form as those for the isotropic oscillator restricted to one dimensional motion. Hence, the general solution for the anisotropic oscillator is given by

$$\mathbf{r} = \mathbf{a}_1\cos(\omega_1 t + \phi_1) + \mathbf{a}_2\cos(\omega_2 t + \phi_2) + \mathbf{a}_3\cos(\omega_3 t + \phi_3), \tag{8.21a}$$

with the three natural frequencies given by

$$\omega_i = \left(\frac{k_i}{m}\right)^{1/2} . \tag{8.21b}$$

The motion described by (8.21a) will not be periodic and along a closed curve unless the ratios ω_1/ω_2 and ω_2/ω_3 are rational numbers. In general, the orbit will not lie in a plane, but it will lie within an ellipsoid centered at the equilibrium point with principal axes $\mathbf{a}_1$, $\mathbf{a}_2$, $\mathbf{a}_3$. If $\mathbf{a}_3 = 0$, then the orbit will lie in the $\mathbf{a}_1\mathbf{a}_2$–plane; and it is commonly known as a *Lissajous figure*.

An atom in a crystalline solid is typically subject to an anisotropic binding force determined by the structure of the solid. The modeling of such an atom as an anisotropic oscillator is one of the basic theoretical techniques of solid state physics. The amplitude of such an oscillator is of the order of atomic dimensions 10^{-8} cm, while the vibrational frequencies in solids range from 10^{12} to 10^{14} Hz (1 Hertz = 1 cycle sec^{-1}). It has been said that quantum mechanics is needed to describe interactions at the atomic level. Quantum mechanics is, indeed, required to calculate force constants and vibrational frequencies. However, given force constants determined either experimentally or theoretically, much can be inferred from the model of an atom as a classical oscillator. The oscillator model is so useful because, as we have seen, it does not commit us to definite assumptions about the "true nature" of the binding force.

Energy Conservation

When an oscillator is used as an atomic or molecular model, the actual orbit of the oscillator is only of peripheral interest. Considering the small amplitudes and high frequencies of atomic vibrations, there is evidently no hope of directly observing the orbits. Only general features of the motion are susceptible to measurement, namely, the frequencies and amplitudes of oscillation. Of course, the amplitude cannot be measured directly, but measurements of a closely related quantity, the energy, are possible. To determine the energy of an isotropic oscillator, we multiply (8.8) by $\dot{\mathbf{r}}$ and observe that

$$\dot{\mathbf{r}}\cdot(m\ddot{\mathbf{r}}+k\mathbf{r}) = \frac{\mathrm{d}}{\mathrm{d}t}\left(\tfrac{1}{2}m\dot{\mathbf{r}}^2 + \tfrac{1}{2}k\mathbf{r}^2\right) = 0.$$

Hence, the quantity

$$E = \tfrac{1}{2}m\dot{\mathbf{r}}^2 + \tfrac{1}{2}k\mathbf{r}^2 \tag{8.22}$$

is a *constant of the motion*, that is, a function of the state variables $\mathbf{r}$ and $\dot{\mathbf{r}}$ which is independent of time. The quantity E is called the (total) *energy* of the oscillator, and Equation (8.22) expresses it as a sum of kinetic and potential energies respectively. The energy is related to the amplitude by substituting the solution (8.16) into (8.22), with the result

$$E = \tfrac{1}{2}k(\mathbf{a}^2 + \mathbf{b}^2). \tag{8.23}$$

Of course, energy is an important state variable even for a macroscopic oscillator, but for an atomic oscillator it is indispensible.

The Damped Isotropic Oscillator

A physical system which can be modeled as an oscillator is never isolated from other interactions besides the binding force. Invariably there are interactions which resist the motion of the oscillator. Such interactions can be accounted for, at least qualitatively, by introducing a linear resistive force $-m\gamma\dot{\mathbf{r}}$, so that the equation for the motion of an isotropic oscillator becomes

$$\ddot{\mathbf{r}} + \gamma\dot{\mathbf{r}} + \omega_0^2\mathbf{r} = 0, \tag{8.24}$$

where $\omega_0^2 = k/m$ as before. The resistive force is also called a *damping* force, because it reduces the amplitude of oscillation, or a *dissipative* force, because it dissipates the energy of the oscillator.

To solve Equation (8.24), we substitute into it the trial solution $\mathbf{r} = \mathbf{a}e^{\lambda t}$

which worked before. After carrying out the differentiations, we find that the solution works if λ is a root of the *characteristic equation*

$$\lambda^2 + \gamma\lambda + \omega_0^2 = 0.$$

Thus,

$$\lambda = -\tfrac{1}{2}\gamma \pm (\tfrac{1}{4}\gamma^2 - \omega_0^2)^{1/2}. \tag{8.25}$$

It is readily verified that a linear superposition of solutions to (8.24) is again a solution. Hence, we get the general solution by adding solutions corresponding to the two roots of (8.25), namely,

$$\mathbf{r} = e^{-(1/2)\gamma t}(\mathbf{a}_+ e^{(\gamma^2/4-\omega_0^2)^{1/2}t} + \mathbf{a}_- e^{-(\gamma^2/4-\omega_0^2)^{1/2}t}). \tag{8.26}$$

Actually, we have three types of solutions corresponding to positive, negative and zero values of the quantity $\frac{1}{4}\gamma^2 - \omega_0^2$. Let us consider each type separately.

(a) *Light damping* is defined by the condition $\frac{1}{2}\gamma < \omega_0$. In this case we write

$$(\tfrac{1}{4}\gamma^2 - \omega_0^2)^{1/2} = \mathbf{i}(\omega_0^2 - \tfrac{1}{4}\gamma^2)^{1/2} \equiv \mathbf{i}\Omega, \tag{8.27a}$$

for we know that the unit imaginary must be a bivector $\mathbf{i}$ specifying the plane of motion. The solution, therefore, has the form

$$\begin{aligned}\mathbf{r} &= e^{-(1/2)\gamma t}(\mathbf{a}_+ e^{\mathbf{i}\Omega t} + \mathbf{a}_- e^{-\mathbf{i}\Omega t}) \\ &= e^{-(1/2)\gamma t}(\mathbf{a}\cos\Omega t + \mathbf{b}\sin\Omega t).\end{aligned} \tag{8.28b}$$

This can be interpreted as an ellipse with decaying amplitude. The exponential factor shows that in time $t = 2\gamma^{-1}$ the amplitude will be damped by the significant factor e^{-1}. If $\frac{1}{2}\gamma \ll \omega_0 \approx \Omega$, then $2\gamma^{-1} \gg \Omega^{-1}$, so the amplitude of the ellipse will be nearly constant during single period, and many periods will pass before its amplitude has been damped significantly.

We have noted that, in general, the energy of an oscillator is a more significant state variable than the amplitude. Using the solution (8.28b) in the form $\mathbf{r} = e^{-(1/2)\gamma t}\mathbf{s}$, we find that the energy E defined by (8.22) can be written in the form

$$E = \tfrac{1}{2}m(\tfrac{1}{4}\gamma^2\mathbf{s}^2 - \gamma\mathbf{s}\cdot\dot{\mathbf{s}} + \dot{\mathbf{s}}^2)\, e^{-\gamma t}.$$

The factor in parenthesis is bounded and oscillatory, with a constant value when averaged over a period of the oscillator. Therefore, the average decrease in energy with time is determined by the exponential factor $e^{-\gamma t} = e^{-t/\tau}$, where

$$\tau \equiv \frac{1}{\gamma} \tag{8.29}$$

is referred to as the *lifetime* of the oscillator's initial state of motion.

(b) *Heavy damping* is defined by the condition $\frac{1}{2}\gamma > \omega_0$. In this case,

$$\alpha \equiv (\tfrac{1}{4}\gamma^2 - \omega_0^2)^{1/2} \tag{8.30a}$$

is a positive scalar and the solution (8.26) assumes the form

$$\mathbf{r} = \mathbf{a}_+ e^{-(\gamma/2 + \alpha)t} + \mathbf{a}_- e^{-(\gamma/2 - \alpha)t} . \tag{8.30b}$$

The first term decays more rapidly than the second one, and the orbit does not encircle the equilibrium point as it does in the case of light damping.

(c) *Critical damping* is defined by the condition $\omega_0 = \frac{1}{2}\gamma$. For this case, the characteristic equation corresponding to our trial solution has only one distinct root, so we get the solution $\mathbf{r} = \mathbf{a}e^{-\gamma t/2}$, which cannot be the most general solution, because it contains only one of the two required constant vectors. However, we can find the general solution by allowing the coefficient to be a function of time. (This is called the *method of variable coefficients*.) After substituting $\mathbf{r}(t) = \mathbf{a}(t)e^{-\gamma t/2}$ into the equation of motion (8.24) with $\omega_0 = \frac{1}{2}\gamma$, we find that our trial solution works if $\ddot{\mathbf{a}} = 0$, so $\mathbf{a} = \mathbf{a}_0 + \mathbf{b}t$, where $\mathbf{a}_0$ and $\mathbf{b}$ are constants.

Thus we arrive at the general solution

$$\mathbf{r} = e^{-\gamma t/2}(\mathbf{a}_0 + \mathbf{b}t). \tag{8.31}$$

The condition for critical damping is unlikely to be met in naturally occurring systems, but it is built into certain detection devices such as the galvanometer. A detection device may consist of a damped oscillator which is displaced from its equilibrium position by an impulsive force (signal). One wants it to have sufficient damping to return to the equilibrium position without oscillation so it will be ready to respond to another signal as soon as possible. On the other hand, for the sake of sensitivity, one wants it to respond significantly to weak impulses, which requires that the damping be as light as possible. The maximal compromise between these two conflicting criteria is the condition for critical damping.

3-8. Exercises

(8.1) For an oscillator with orbit

$$\mathbf{r}(t) = \mathbf{a}\cos(\omega_0 t + \phi_0) + \mathbf{b}\sin(\omega_0 t + \phi_0),$$

determine the major axis a and the minor axis b from the initial conditions $\mathbf{r}_0 = \mathbf{r}(0)$, $\mathbf{v}_0 = \dot{\mathbf{r}}(0)$.
Specifically, show that, for $0 < \phi_0 < \frac{\pi}{2}$,

$$\mathbf{a} = \mathbf{r}_0 \cos\phi_0 - \frac{\mathbf{v}_0}{\omega_0}\sin\phi_0 .$$

$$\mathbf{b} = \mathbf{r}_0 \sin\phi_0 + \frac{\mathbf{v}_0}{\omega_0}\cos\phi_0 ,$$

while ϕ_0 is determined by

$$\tan 2\phi_0 = \frac{2\omega_0 \mathbf{r}_0 \cdot \mathbf{v}_0}{\mathbf{v}_0^2 - \omega_0^2 \mathbf{r}_0^2} .$$

(8.2) Ancient properties of the ellipse: On the ellipse $\mathbf{r}(\phi) = \mathbf{a}\cos\phi + \mathbf{b}\sin\phi$, the point $\mathbf{s}(\phi) \equiv \mathbf{r}(\phi + \pi/2)$ is said to be conjugate to the point $\mathbf{r}(\phi)$. Prove that the following holds for any pair of conjugate points.

(a) The tangent to the ellipse at $\mathbf{r}$ is parallel to the conjugate radius $\mathbf{s}$ (Figure 8.1).

(b) The first theorem of Appolonius:

$$\mathbf{r}^2 + \mathbf{s}^2 = \mathbf{a}^2 + \mathbf{b}^2 .$$

(c) The second theorem of Appolonius:

$$\mathbf{r}\wedge\mathbf{s} = \mathbf{a}\wedge\mathbf{b} ,$$

that is, the parallelograms determined by conjugate radii all have the same area.

Interpret these theorems of Appolonius as conservation laws for an oscillator (see Section 3.10).

(8.3) Establish the equivalence of Equations (8.18) and (8.16a). Show the planes of circular motion are determined from a given ellipse by the equations

$$\mathbf{i}_{\pm} = \mathbf{a}^{-1}\mathbf{b} \pm i\varepsilon\hat{\mathbf{b}}' ,$$

where the *eccentricity* ε is determined by the equation $\varepsilon^2 a^2 = a^2 - b^2$. Show that the dihedral angle α between the planes is determined by

$$\cos\alpha = 1 - 2\varepsilon^2 .$$

(8.4) Show that the effect of a constant force on a harmonic oscillator is equivalent to the displacement of equilibrium point of the oscillator. Use this to find a parametric equation for the orbit of an isotropic oscillator in a constant gravitational field.

(8.5) Find a parametric equation for the orbit of a charged isotropic oscillator in a uniform magnetic field, as characterized by the equation

$$m\ddot{\mathbf{r}} = -m\omega_0^2 \mathbf{r} + \frac{q}{c}\dot{\mathbf{r}} \times \mathbf{B} .$$

(Suggestion: Write $\mathbf{r} = \mathbf{x} + z\hat{\mathbf{B}}$ where $\mathbf{x}\cdot\mathbf{B} = 0$ and separate differential equations for $\mathbf{x}$ and z.)

(8.6) Show that the general solution to the equation

$$m\ddot{\mathbf{r}} + k\mathbf{r} = 0 ,$$

for both positive and negative values of k, can be put in the form

$$\mathbf{r}(t) = \mathbf{c} \cosh z(t),$$

where $z = \mu + \mathbf{i}\phi$ and $\mathbf{i}$ is a unit bivector. Determine the time dependence of scalars μ and ϕ for the two cases.

Note that $\mathbf{r}(\mu, \phi) = \mathbf{c} \cosh (\mu + \mathbf{i}\phi)$ is a parametric equation for an ellipse if μ is held constant or a hyperbola if ϕ is held constant. Express the major axis $\mathbf{a}$ and the minor axis $\mathbf{b}$ of the ellipse in terms of $\mathbf{c}$ and μ. For given $\mathbf{a}$ and ϕ determine the directance from the origin to each asymptote of the hyperbola.

Every point $\mathbf{r} = \mathbf{r}(\mu, \phi)$ in the $\mathbf{i}$-plane is designated by unique values of μ and ϕ in the ranges $-\infty < \mu < \infty, 0 \leqslant \phi < 2\pi$. The parameters μ, ϕ are called elliptical coordinates for the plane.

(8.7) Evaluate constants $\mathbf{a}_+$ and $\mathbf{a}_-$ in Equation (8.30b) in terms of initial conditions $\mathbf{r}_0$, $\mathbf{v}_0$, and sketch a representative trajectory.

3-9. Forced Oscillations

In this section we study the response of a bound particle to a periodic force. We concentrate on the very important case of a bound charge driven by an electromagnetic plane wave. But our results are quite characteristic of driven oscillatory systems in general. The properties of electromagnetic waves which we use in this section will be established in NFII.

For a charged isotropic oscillator in an "external" electromagnetic field, we have the equation of motion

$$m\ddot{\mathbf{r}} + m\omega_0^2 \mathbf{r} = q(\mathbf{E} + c^{-1}\dot{\mathbf{r}} \times \mathbf{B}). \tag{9.1}$$

If the external field is an electromagnetic plane wave, then $\mathbf{E}^2 = \mathbf{B}^2$, so

$$\frac{|\, c^{-1}\dot{\mathbf{r}} \times \mathbf{B} \,|}{|\, \mathbf{E} \,|} < \frac{|\, \dot{\mathbf{r}} \,|}{c}.$$

Hence, for velocities small compared to the speed of light, the magnetic force is negligible compared to the electric force. At any point $\mathbf{r}$ in space, the electric field $\mathbf{E}$ of a *circularly polarized* plane wave is a rotating vector of the form

$$\mathbf{E}(\mathbf{r}, t) = \mathbf{E}_0 e^{\mathbf{i}(\omega t - \mathbf{k}\cdot\mathbf{r})}. \tag{9.2}$$

Here, $\mathbf{E}_0$ is a constant vector and $|\, \mathbf{E}_0 \,|$ is the amplitude of the wave; the (circular) frequency of the wave is $\omega = c\,|\, \mathbf{k} \,|$; the *wavelength* is $\lambda = 2\pi/|\, \mathbf{k} \,|$; the plane of the rotating vector $\mathbf{E}$ is perpendicular to the direction $\hat{\mathbf{k}}$ of the propagating wave, and it is specified by the unit bivector $\mathbf{i} = i\hat{\mathbf{k}}$. For $\omega > 0$, Equation (9.2) describes a *left circularly* polarized plane wave; (for $\omega < 0$, it describes a *right circularly* polarized wave). We are most interested in applying (9.2) to a region of atomic or molecular dimensions; such a region is small compared to the wavelength of visible light, in which case $|\, \mathbf{k}\cdot\mathbf{r} \,|$

$\leqslant 2\pi r/\lambda \approx 0$. Furthermore, we shall see that the motion of a driven oscillator tends to lie in the plane $\mathbf{k}\cdot\mathbf{r} = 0$. For these reasons, it is an acceptible approximation, besides being a considerable mathematical simplification, to neglect any effect of the factor $\mathbf{k}\cdot\mathbf{r}$ in (9.2) on oscillator motion. Thus, our equation of motion (9.1) assumes the specific form

$$\ddot{\mathbf{r}} + \omega_0^2\mathbf{r} = \boldsymbol{\varepsilon}e^{\mathbf{i}\omega t}, \tag{9.3}$$

where $\boldsymbol{\varepsilon} \equiv q/m\,\mathbf{E}_0$.

To solve (9.3), we try a solution of the form $\mathbf{r} = \mathbf{A}e^{\lambda t}$ and, after carrying out the differentiation, find that

$$\mathbf{A}e^{\lambda t}(\lambda^2 + \omega_0^2) = \boldsymbol{\varepsilon}e^{\mathbf{i}\omega t}.$$

This equation will hold for all values of t if and only if $\lambda = \mathbf{i}\omega$. The equation also determines $\mathbf{A}$ uniquely, whence

$$\mathbf{r} = \frac{\boldsymbol{\varepsilon}e^{\mathbf{i}\omega t}}{\omega_0^2 - \omega^2},$$

is a *particular solution* of (9.3). We get the *general solution* with two arbitrary constants by adding to the particular solution the solution of the homogeneous equation $\ddot{\mathbf{r}} + \omega_0^2\mathbf{r} = 0$, thus,

$$\mathbf{r} = \mathbf{a}\cos\omega_0 t + \mathbf{b}\sin\omega_0 t + \frac{\boldsymbol{\varepsilon}e^{\mathbf{i}\omega t}}{\omega_0^2 - \omega^2}.$$

The last term in this equation describes the displacement of the oscillator due to the driving force exerted by the electromagnetic wave. It is a rotating vector in the $\mathbf{E}\wedge\mathbf{B}$ plane. Its amplitude is infinite when the driving frequency ω of the wave *matches* the natural frequency ω_0 of the oscillator, a condition called *resonance*. As usual, an "unphysical" infinity such as this points to a deficiency in our model of the interacting systems. At resonance it becomes essential to take into account the omnipresent resistive forces which otherwise might be negligible.

Forced Oscillator with Linear Resistance

We can improve our model of an electromagnetic wave interacting with a bound charge by adding a linear resistive force. Thus, we consider the equation of motion

$$\ddot{\mathbf{r}} + \gamma\dot{\mathbf{r}} + \omega_0^2\mathbf{r} = \boldsymbol{\varepsilon}e^{\mathbf{i}\omega t}. \tag{9.4}$$

This equation can be solved in the same way as (9.3). The solution can be written as a sum $\mathbf{r} = \mathbf{r}_1 + \mathbf{r}_2$ of "transient" and "forced" displacements. The transient displacement $\mathbf{r}_1$ is a solution of the homogeneous equation

$$\ddot{\mathbf{r}}_1 + \gamma\dot{\mathbf{r}}_1 + \omega_0^2\mathbf{r}_1 = 0,$$

determined by the initial conditions of the oscillator. We have seen that the amplitude of this solution decreases with a relaxation time $2\gamma^{-1}$, so eventually it will be negligible compared to the *steady state* displacement supported by the continually applied driving force.

The forced displacement $\mathbf{r}_2$ is the particular solution of (9.4) determined entirely by the driving force. Ignoring the transient solution, we write $\mathbf{r} = \mathbf{r}_2$ and we insert the trial solution $\mathbf{r} = \mathbf{a}e^{\mathbf{i}\omega t}$ into (9.4), with the result

$$\mathbf{r}(-\omega^2 + \gamma\mathbf{i}\omega + \omega_0^2) = \boldsymbol{\varepsilon}e^{\mathbf{i}\omega t}.$$

Solving for $\mathbf{r}$, we put the solution in the form

$$\mathbf{r}(t) = \mathbf{A}e^{\mathbf{i}(\omega t - \delta)} \tag{9.5a}$$

with vector amplitude

$$\mathbf{A} = \mathbf{A}(\omega) = \frac{\boldsymbol{\varepsilon}}{[(\omega_0^2 - \omega^2)^2 + \gamma^2\omega^2]^{1/2}} \tag{9.5b}$$

and *phase angle*

$$\delta = \delta(\omega) = \tan^{-1}\left(\frac{\gamma\omega}{\omega_0^2 - \omega^2}\right), \tag{9.5c}$$

where $0 \leqslant \delta \leqslant \pi$ accounts for the full range of the parameter ω.

The solution (9.5) shows that $\mathbf{r}$ is a rotating vector lagging behind the driving force $q\mathbf{E}$ by the phase angle $\delta = \delta(\omega)$ (Figure 9.1). Let us examine the limiting cases in the range of driving frequencies.

(a) *Low frequencies* $\omega \ll \omega_0$. (A scale distinguishing large from small is determined by the ratio ω_0/γ). Equation (9.5c) gives $\tan\delta \approx \delta \approx 0$, and the solution reduces to

$$\mathbf{r}(t) \approx \frac{\boldsymbol{\varepsilon}}{\omega_0^2}\, e^{\mathbf{i}\omega t}.$$

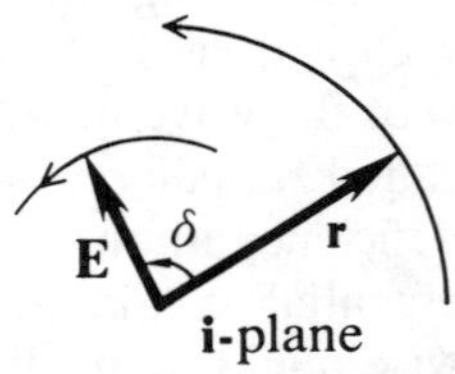

Fig. 9.1. Phase angle between the driving field **E** and the position response **r**.

This shows that resistance is not important in slowly driven motion. A gradually varying force gives the oscillator time to respond and follow exactly in phase ($\delta = 0$).

(b) *High frequencies* $\omega \gg \omega_0$. Then $\tan\delta \approx -\gamma/\omega \approx 0$, $\delta = \pi$ and

$$\mathbf{r}(t) \approx \frac{\boldsymbol{\varepsilon}}{\omega^2} e^{\mathbf{i}\omega t}.$$

Thus, the response to a rapidly varying force lags behind the force by the maximum phase angle $\delta = \pi$, and it is weaker in amplitude than the response to a slowly varying force.

(c) *Resonance* $\omega = \omega_0$. Then $\tan\delta = \infty$, $\delta = \pi/2$ and

$$\mathbf{r}(t) = -\frac{\boldsymbol{\varepsilon} \mathrm{i} e^{\mathrm{i}\omega t}}{\gamma\omega_0}.$$

Thus, the response at resonance is orthogonal to the driving force, and it is stronger than the low frequency response if and only if $\gamma < \omega_0$.

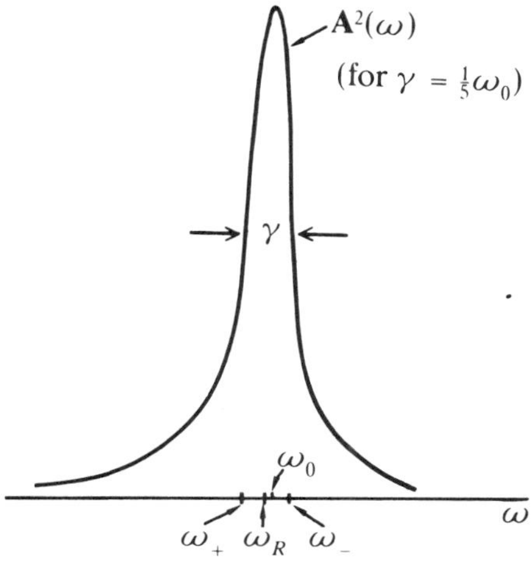

Fig. 9.2a. Squared amplitude near resonance.

The squared amplitude $\mathbf{A}^2 = |\mathbf{A}(\omega)|^2$ and the phase angle $\delta(\omega)$ are graphed in Figure 9.2. As the graph shows, maximum amplitude is not attained at the resonant frequency ω_0, but rather at a resonant amplitude frequency ω_R defined by the condition

$$\left.\frac{d\mathbf{A}^2}{d\omega}\right|_{\omega = \omega_R} = 0.$$

Carrying out the differentiation on (9.5b), we find that

$$\omega_R = (\omega_0^2 - \tfrac{1}{2}\gamma^2)^{1/2}. \tag{9.6}$$

To get a measure of the width of the resonance, we locate the points $\omega_\pm$ at which $\mathbf{A}^2$ has half its maximum value.

After some algebra, we find that

$$\omega_\pm^2 = \omega_R^2 \pm \gamma(\omega_R^2 + \tfrac{1}{4}\gamma^2)^{1/2}. \tag{9.7}$$

The *resonance width* $\Delta\omega$ is defined by

$$\Delta\omega = |\omega_+ - \omega_-|. \tag{9.8}$$

For *light damping* ($\gamma \ll \omega_0$), Equation (9.6) gives $\omega_R \approx \omega_0$, and (9.7) yields

$$\omega_\pm - \omega_R \approx \frac{\pm\gamma\omega_R}{\omega_\pm + \omega_R} \approx \pm\tfrac{1}{2}\gamma.$$

Therefore, the resonance width is given by the simple expression

$$\Delta\omega \approx \gamma = \frac{1}{\tau}, \tag{9.9}$$

where, as was established in the last section, τ is the *lifetime* of the oscillator state if the driving force is suddenly removed.

The inverse relation (9.9) between resonance width and lifetime is a very general and important property of unstable bound states of motion. Equation (9.5b) shows that the narrower the width, the higher the resonance peak (which would be infinite if $\gamma = 0$). Thus *long-lived bound states are characterized by narrow resonances with relatively high peaks.*

Energy storage and dissipation

The potential energy of the oscillator has the value $\frac{1}{2}m\omega_0\mathbf{r}^2 = \frac{1}{2}m\omega_0\mathbf{A}^2$, so the graph of A^2 in Figure 9.2 is equivalent to a graph of potential energy. The graph, therefore, describes the relative amount of *energy stored* in the oscillator for the various driving frequencies, and ω_R is the frequency for which the stored energy is a maximum. However, as we have noted before, the stored energy of an atomic system is not directly observable. Rather, it is the energy absorption as a function of frequency that can be directly measured. We must determine this function before our results can be fully interpreted.

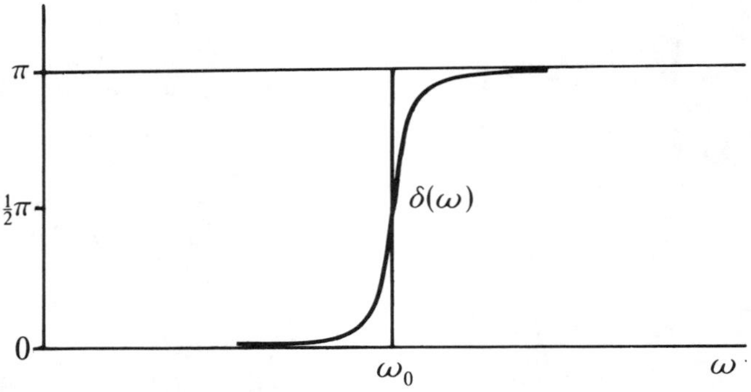

Fig. 9.2b. Phase angle near resonance.

An isotropic oscillator subject to an arbitrary external force **f** obeys the equation of motion

$$m\ddot{\mathbf{r}} + m\omega_0^2\mathbf{r} = \mathbf{f}.$$

Multiplying this by $\dot{\mathbf{r}}$, we get the equation

$$\frac{\mathrm{d}E}{\mathrm{d}t} = \mathbf{f}\cdot\dot{\mathbf{r}} \tag{9.10}$$

for the rate of energy change (Power) induced by external forces. For the case we have been considering,

$$\mathbf{f} = -m\gamma\dot{\mathbf{r}} + q\mathbf{E},$$

so

$$\frac{\mathrm{d}E}{\mathrm{d}t} = -m\gamma\dot{\mathbf{r}}^2 + q\mathbf{E}\cdot\dot{\mathbf{r}}. \tag{9.11}$$

The first term on right describes the power loss due to dissipative forces. The second term describes the power delivered by the electromagnetic field. For a steady state of motion the energy is constant so the power lost must equal power supplied, and from (9.11) we conclude that the *power supply P* is determined by

$$P = m\gamma\dot{\mathbf{r}}^2 = 2\gamma(\tfrac{1}{2}m\dot{\mathbf{r}}^2). \tag{9.12}$$

Obviously, this relation applies to any driving force producing a steady state of motion.

Our analysis shows that a damped oscillator in a steady state of motion dissipates energy continuously, but the motion persists because an equal amount of energy is continuously supplied by the driving force. According to (9.12), the rate at which energy is absorbed and dissipated is proportional to the kinetic energy of the oscillator. For the motion described in (9.5), the power absorbed as a function of driving frequency is specifically given by

$$P(\omega) = \frac{2m\gamma\boldsymbol{\varepsilon}^2\omega^2}{(\omega_0^2 - \omega^2)^2 + \gamma^2\omega^2} . \tag{9.13}$$

This function has a maximum at the resonance frequency ω_0, as the reader may verify. Thus, the maximum of the kinetic energy is at the frequency ω_0 for which the *energy dissipated* is a maximum, whereas, as we have seen, the maximum of the potential energy is at the frequency ω_R for which the *energy stored* is a maximum. The general shape of the graph for $P(\omega)$ is similar to that for the potential energy, so it would be repetitious to discuss it.

Resonance occurs in every oscillatory system, so it is a phenomenon of great practical importance. If the damping is weak, a small periodic force can set up large oscillations. Consequently, it is undesirable to build a boat with a natural pitching frequency which might be close to a likely frequency of waves. The same basic principle was overlooked in the design of the Tacoma Narrows Bridge, which was destroyed when it resonated with periodic wind gusts. Undesirable resonances in machinery of all kinds can be avoided by damping devices such as shock absorbers in cars.

The Faraday Effect

Undoubtedly the most common and important example of resonance occurs in the interaction of light with matter. The mathematical theory of optical properties of matter was pioneered by H. A. Lorentz. He supposed that the atoms in a given material can be regarded as charged harmonic oscillators when considering their interactions with light. He was then able to derive mathematical expressions for the index of refraction of gases, dielectrics, and metals as well as explain a number of other optical properties. Our model of a damped oscillator is used in his theory of dielectrics. From the model it is clear that incident light with frequencies close to the natural frequencies of the atoms will be strongly absorbed, while a dielectric will be transparent to light with frequencies outside the range of its natural frequencies. A modern introduction to the *Lorentz electron theory* is given by Feynman (1963, Vol. II, Chap. 32).

As an application of the Lorentz electron theory, let us see how it explains the *Faraday effect*. In 1845 Faraday showed that the polarization plane of light passing through a glass rod will be rotated by an external magnetic field directed along the line of propagation. This was the first experimental demonstration of a connection between magnetism and light. Today the Faraday effect has important technological applications. And in astronomy, it explains the polarization of light passing through the strong magnetic fields of pulsars.

To understand the Faraday effect, we consider the effect of a constant magnetic field **B** on an oscillator driven by a linearly polarized plane wave. The equation of motion is

$$m\ddot{\mathbf{r}} + m\omega_0^2\mathbf{r} = \frac{q}{c}\dot{\mathbf{r}} \times \mathbf{B} + q\mathbf{E}_0 \cos \omega t\,, \tag{9.14}$$

where $\mathbf{B}\cdot\mathbf{E}_0 = 0$ for a plane wave propagating along the line of the magnetic field. We know that it is unnecessary to include damping terms in order to locate resonances, and we know that steady state solutions will be orthogonal to the direction of wave propagation. Consequently, we can assume that $\mathbf{r}\cdot\mathbf{B} = 0$ and rewrite (9.14) in the form

$$\ddot{\mathbf{r}} - 2\dot{\mathbf{r}}\mathbf{i}\omega_L + \omega_0^2\mathbf{r} = \boldsymbol{\varepsilon} \cos \omega t\,, \tag{9.15}$$

where $\boldsymbol{\varepsilon}$ is defined as before, the bivector $\mathbf{i}$ is defined by $\mathbf{i} = i\hat{\mathbf{B}}$, and the so-called Larmor frequency ω_L *is defined by*

$$\omega_L \equiv \frac{q\,|\,\mathbf{B}\,|}{2mc}\ . \tag{9.16}$$

Note that $\omega_L = -\,|\,\omega_L\,|$ for an electron with charge q. The linearly polarized plane wave can be expressed as a superposition of left and right circularly polarized waves:

$$\boldsymbol{\varepsilon} \cos \omega t = \tfrac{1}{2}\boldsymbol{\varepsilon}(e^{\mathbf{i}\omega t} + e^{-\mathbf{i}\omega t}), \tag{9.17}$$

where ω is positive. Consequently, the steady state solution must have the form

$$\mathbf{r} = \mathbf{a}_+ e^{\mathbf{i}\omega t} + \mathbf{a}_- e^{-\mathbf{i}\omega t}. \tag{9.18}$$

To determine the amplitudes $\mathbf{a}_\pm$, we substitute (9.18) into (9.15) and equate coefficients with the same time dependence, with the result

$$\mathbf{a}_\pm = \frac{\tfrac{1}{2}\boldsymbol{\varepsilon}}{\omega_0^2 - \omega^2 \mp \omega_L\omega}\ . \tag{9.19}$$

This shows that there will be resonance when

$$\omega = (\omega_0^2 + \omega_L^2)^{1/2} \mp \omega_L\,. \tag{9.20}$$

For atomic systems in feasible laboratory fields it can be shown that $\omega_0 \gg \omega_L$. Consequently, for propagation in the direction of **B**, the left circularly polarized component of the wave has a resonance at $\omega = \omega_0 + |\,\omega_L|$, while the right circularly polarized component has a resonance at $\omega = \omega_0 - |\,\omega_L\,|$. The locations of the resonances are interchanged if **B** is opposite to the direction of propagation. In summary, the effect of a magnetic field is to shift the resonance frequencies of circularly polarized waves by opposite amounts $\pm\, \omega_L$.

According to the Lorentz theory, the index of refraction for a material depends on the locations of its resonances. For more details see Feynman (1963). In the present case, the left and right circularly polarized waves resonate at different frequencies, so they have different indices of refraction and different phase velocities in the medium. The net effect is a rotation of the polarization plane of the propagating wave.

3-9. Exercises

(9.1) Derive Equations (9.6) and (9.7). Show that the maximum kinetic energy and power dissipation is attained at a driving frequency equal to the natural frequency of an oscillator.

(9.2) Solve the damped isotropic oscillator with a sinusoidal driving force $\boldsymbol{\varepsilon} \sin \omega t$. Determine the phase angle and the average potential energy (where the average is taken over a period of the driving force). Compare with the solution (9.5a, b, c).

(9.3) Determine the velocity $\mathbf{v} = \mathbf{v}(t)$ of a charged particle in a constant magnetic field $\mathbf{B}$ and a plane electromagnetic wave with circular frequency ω. Use your result to explain the fact that electromagnetic waves with linear frequencies in the neighborhood of the "cyclotron frequency" $\omega_c/2\pi = 1400$ kHz are sharply attenuated when passing through the ionosphere. (The charge to mass ratio of an electron, $|\, q/m \,| = 1.76 \times 10^{11}$ Coul kg^{-1}. The strength of the Earth's magnetic field in the ionosphere is given by $B/c = 5 \times 10^{-5}$ webers m^{-2}, where c is the speed of light.)

3-10. Conservative Forces and Constraints

So far we have studied only specific force laws of the simplest mathematical form. To survey the broad range of force laws with physical significance, we must procede systematically, classifying forces according to general principles. The general approach which has proved to be most powerful is to distinguish forces by identifying *conservation laws* or *constants of motion* which they admit or disallow. In this section we examine conditions under which *energy conservation* holds and some of its implications for single particle motion.

We have already analyzed energy conservation and dissipation for an isotropic oscillator. For a more general analysis of energy conservation, we multiply the equation of motion $m\dot{\mathbf{v}} = \mathbf{f}$ by $\mathbf{v} = \dot{\mathbf{x}}$ to get

$$\frac{\mathrm{d}}{\mathrm{d}t}(\tfrac{1}{2}m\mathbf{v}^2) = \mathbf{v}\cdot\mathbf{f}. \tag{10.1}$$

This is an equation for the change of the so-called *kinetic energy* $\frac{1}{2}m\mathbf{v}^2$ due to the action of the force $\mathbf{f}$. If $\mathbf{v}\cdot\mathbf{f} = 0$, the kinetic energy is a constant of motion. The magnetic force $(q/c)\,\mathbf{v} \times \mathbf{B}$ has this property, irrespective of how the magnetic field $\mathbf{B} = \mathbf{B}(\mathbf{x}, t)$ depends on position and time. Since it never alters the energy of a particle, the *magnetic* force is said to be a *conservative force.*

A more general concept of conservative force can be developed by considering a force $\mathbf{f}$ with the property

$$\mathbf{v}\cdot\mathbf{f} = -\dot{\mathbf{x}}\cdot\nabla V, \tag{10.2}$$

where $V = V(\mathbf{x}, t)$. According to the identity (2–8.16),

$$\dot{\mathbf{x}}\cdot\nabla V = \frac{\mathrm{d}V}{\mathrm{d}t} - \partial_t V,$$

so, substitution of (10.2) into (10.1) yields

$$\frac{\mathrm{d}}{\mathrm{d}t}(\tfrac{1}{2}m\mathbf{v}^2 + V) = \partial_t V. \tag{10.3}$$

Hence, the quantity

$$E \equiv \tfrac{1}{2}m\mathbf{v}^2 + V, \tag{10.4}$$

is conserved if and only if $\partial_t V = 0$. We refer the quantity V as the *potential energy* and to E as the (total) *energy* of the particle.

The force $\mathbf{f}$ is said to be conservative if the associated energy given by (10.4) is *conserved*. There is no commonly accepted term to refer to the more general case when $\partial_t V \neq 0$, for the good reason that (10.3) then gives little useful information. However, it should be noted that the explicit time dependence of the potential $V(x, t)$ arises from the motion of "its source," namely, the particles "producing" the force $\mathbf{f} = -\nabla V(\mathbf{x}, t)$. We shall see later on that the explicit time dependence often disappears when the potential is expressed as a function of relative positions of interacting particles so the conservative case is more general that it might appear at first. Now, from (10.2) we can conclude that a *conservative* force $\mathbf{f}$ has the general form

$$\mathbf{f} = -\nabla V(\mathbf{x}) + \mathbf{N} \quad \text{where} \quad \mathbf{N}\cdot\mathbf{v} = 0, \tag{10.5}$$

but $\mathbf{N} = \mathbf{N}(\mathbf{v}, \mathbf{x}, t)$ can otherwise have any functional dependence on $\mathbf{v}$, $\mathbf{x}$ and t. Let us refer to $\mathbf{N}$ as the *normal* component of the conservative force, because the condition $\mathbf{v}\cdot\mathbf{N} = 0$ implies that it is always normal (or perpendicular) to the particle path. The normal force changes the direction of particle motion without affecting the speed (or kinetic energy). We have seen that the magnetic force has this property. So do forces of constraint, as we shall see below.

It is customary to define a conservative force as one which can be put in the form $\mathbf{f} = -\nabla V(\mathbf{x})$. The more general definition (10.5) has been adopted here to emphasize the most general conditions under which energy conservation obtains. Of course, by the superposition principle both $-\nabla V$ and $\mathbf{N}$ can be regarded as distinct forces, and in specific applications they have independent sources, so it is perfectly reasonable to consider them separately.

Work

It is instructive to put the energy conservation law in *integral form*, as distinct from its differential form (10.3). Integrating from an initial state $\mathbf{x}_0 = \mathbf{x}(0)$, $\mathbf{v}_0 = \mathbf{v}(0)$ to a final state $\mathbf{x} = \mathbf{x}(t)$, $\mathbf{v} = \mathbf{v}(t)$, we put (10.1) in the form

$$\tfrac{1}{2}m\mathbf{v}^2 - \tfrac{1}{2}m\mathbf{v}_0^2 = \int_0^t dt\, \mathbf{v}\cdot\mathbf{f} = \int_{\mathbf{x}_0}^{\mathbf{x}} d\mathbf{x}\cdot\mathbf{f}. \tag{10.6}$$

The integral here is referred to as the *work done by* force $\mathbf{f}$ on the particle in the time interval t. *Work* can be regarded as a transfer of energy from one physical system to another – in the present case, from the system producing the force $\mathbf{f}$ to the particle. The work is positive if the particle gains energy and negative if the particle loses energy.

Now, substituting (10.2) into (10.6) and using (2–8.34), we find that, for a *conservative force*,

$$\tfrac{1}{2}m\mathbf{v}^2 - \tfrac{1}{2}m\mathbf{v}_0^2 = -\int_{\mathbf{x}_0}^{\mathbf{x}} d\mathbf{x}\cdot\nabla V = V(\mathbf{x}_0) - V(\mathbf{x}). \tag{10.7a}$$

Thus, though kinetic and potential energies may differ, the total energy

$$\tfrac{1}{2}m\mathbf{v}^2 + V(\mathbf{x}) = \tfrac{1}{2}m\mathbf{v}_0^2 + V(\mathbf{x}_0) \tag{10.7b}$$

has the same value for the initial and final states, as well as for all intermediate states. From (10.7a) it is evident that energy conservation depends only on the potential difference $V(\mathbf{x}_0) - V(\mathbf{x})$ and not on the absolute value of the potential function $V(\mathbf{x})$. Therefore, we are free to assign any convenient value to the potential at one point, say $\mathbf{x}_0$, and the value at any other point $\mathbf{x}$ will then be determined by an integral as in Equation (10.7a).

We have seen in Section 2-8 that an integral like the one in (10.7a) is independent of the path between initial and final states. We can conclude, therefore, that the work done by a conservative force is path-independent. The notion of path independence involves the concept of a force field, which is a more general concept of force than we started out with, so a few words about force fields are in order.

Conservative Fields of Force

The concept of a force field arises naturally from an examination of the possible mathematical forms for the force function $\mathbf{f} = \mathbf{f}(\mathbf{v}, \mathbf{x}, t)$ in the equation of motion $\mathbf{f} = m\dot{\mathbf{v}}$. A velocity independent force function $\mathbf{f}(\mathbf{x}, t)$ is a time dependent *vector field*, and it can be characterized mathematically without reference to any particle on which the force acts. We imagine, then, that at each point $\mathbf{x}$ there is a time varying vector $\mathbf{f}(\mathbf{x}, t)$, which is the force that would be exerted on a particle if there were one at that point. This conception of a *force field*, obtained by separating the concept of force from the concept of particle, has proved to be one of the most profound and fruitful ideas in physics. Later on we shall discuss implications of attributing an independent physical existence to force fields. For the time being, however, the concept of force field can be regarded merely as a convenient mathematical abstraction. It should be evident, now, that a conservative force $\mathbf{f}(\mathbf{x}) = -\nabla V(\mathbf{x})$ is actually a *conservative force field*, because its properties, such as path-independence, relate values of the function at more than one point.

The path-independence of the energy conservation law is a major reason for its importance. Thus, from (10.7a) we can deduce the change in speed of a particle passing from $\mathbf{x}$ to $\mathbf{x}_0$ without bothering to solve the equations of motion to determine its path. On the other hand, since kinetic energy is necessarily positive, the energy conservation law in the form (10.4) implies $E - V(\mathbf{x}) \geqslant 0$, from which we can conclude that a particle with energy E will be confined to a region bounded by the surface $E - V(\mathbf{x}) = 0$ whatever its trajectory. This shows that the path-independence is limited when the energy is assigned a specific value. We will make good use of this fact in the next chapter.

Though the energy is path-independent, the conservative force $-\nabla V$ allows only a limited selection of paths connecting given points. However, any path consistent with energy conservation can be achieved in principle simply by specifying an appropriate normal force $\mathbf{N}$. Let us see how the problem of finding such a force can be formulated mathematically.

Surface Constraints

Suppose the particle is constrained to move on a surface determined by the scalar equation

$$\phi(\mathbf{x}, t) = 0\,. \tag{10.8}$$

In a physical application, the equation of constraint (10.8) might describe the surface of a solid body. The explicit time dependence of the equation then allows for the possibility that the solid body may be moving. The body will exert a force on a particle in contact with its surface. If the surface is frictionless, the contact force $\mathbf{N}$ will be exerted along the direction of the surface normal $\nabla\phi \neq 0$, so we can write

$$\mathbf{N} = \lambda\nabla\phi, \tag{10.9}$$

where the proportionality factor $\lambda = \lambda(\mathbf{v}, \mathbf{x}, t)$ is a scalar function which can be determined only by using the equation of motion.

If $\mathbf{x} = \mathbf{x}(t)$ is the particle path on the surface, then differentiation of (10.8) gives

$$\dot{\phi} = \dot{\mathbf{x}}\cdot\nabla\phi + \partial_t\phi = 0,$$

or by virtue of (10.9),

$$\mathbf{v}\cdot\mathbf{N} = -\lambda\,\partial_t\phi. \tag{10.10}$$

The quantity $\mathbf{v}\cdot\mathbf{N}$ is the rate at which the constraining force $\mathbf{N}$ does work on the particle, and, according to (10.10) it vanishes if and only if $\partial_t\phi = 0$. Therefore, the constraining force is conservative if and only if the surface of constraint remains at rest. For this reason, it is appropriate to say that an equation of the form $\phi(\mathbf{x}) = 0$ determines a *conservative constraint*. The more general time dependent equation (10.8) is said to determine a *holonomic constraint*. A conservative constraint is therefore a time-independent holonomic constraint.

Before completing our general discussion of constrainted motion, let us put some flesh on these abstractions by considering some examples.

EXAMPLE 1: *Particle in a Constant Gravitation Field*

As we shall verify below, the potential energy for a particle in constant gravitational (force) field can be written

$$V(\mathbf{x}) = -m\mathbf{g}\cdot\mathbf{x} = mgh, \tag{10.11}$$

where

$$h = -\hat{\mathbf{g}}\cdot\mathbf{x}, \tag{10.12}$$

is the height of the particle above some arbitrarily chosen "ground level". Note that Equation (10.12) can be interpreted in two ways: either as an equation for the height h as a function of position $\mathbf{x}$, or as an equation for a 1-parameter family of horizontal planes, which are the "*equipotential surfaces*" of the gravitational field. The gravitational force is obtained by differentiating (10.11) with the help of (2–8.37); thus,

$$-\nabla V = m\nabla\mathbf{g}\cdot\mathbf{x} = m\mathbf{g}.$$

The energy conservation law (10.7a) now takes the specific form

$$\tfrac{1}{2}m(\mathbf{v}^2 - \mathbf{v}_0^2) = m\mathbf{g}\cdot(\mathbf{x} - \mathbf{x}_0). \tag{10.13}$$

This is equivalent to (2.18), which we found before only after determining the general solution to the equation of motion.

In Section 3-2 we saw that a particle in a constant gravitational field follows a parabolic trajectory, and if it is launched with a specific initial speed there are at most two such trajectories connecting a pair of given points. Let us consider some alternative trajectories that result from adding holonomic constraints.

EXAMPLE 2: *Particle on a Stationary Plane Surface*
A block placed on a fixed frictionless plane will be subject to the equation of constraint

$$\phi(\mathbf{x}) = (\mathbf{x}-\mathbf{y})\cdot\mathbf{n} = 0\,, \tag{10.14}$$

which is the equation for a plane with unit normal $\nabla\phi = \mathbf{n}$ passing through a given point $\mathbf{y}$. If the block is regarded as a particle, its equation of motion is, in accordance with (10.9),

$$m\dot{\mathbf{v}} = m\mathbf{g} + \lambda\mathbf{n}\,. \tag{10.15}$$

Before we can solve this equation, we must determine the magnitude λ of the force exerted by the plane. Since the plane is at rest, the normal $\mathbf{n}$ is constant, and, according to (10.10), $\mathbf{v}\cdot\mathbf{n} = 0$. So, multiplying (10.15) by $\mathbf{n}$, we find that λ is given by

$$0 = m\mathbf{g}\cdot\mathbf{n} + \lambda\,.$$

Using this to eliminate λ from (10.15), we get

$$m\dot{\mathbf{v}} = m(\mathbf{g}-\mathbf{g}\cdot\mathbf{n}\mathbf{n}) \equiv m\mathbf{g}_{\parallel}\,. \tag{10.16}$$

Thus, the net force is merely the component of the gravitational force in the plane. The general solution of (10.16) is a parabola in the plane, provided of course, the initial velocity $\mathbf{v}_0$ satisfies the condition of constraint $\mathbf{v}_0\cdot\mathbf{n} = 0$. Since the constraint is conservative, the energy conservation law (10.13) still applies.

EXAMPLE 3: *Particle on a Moving Plane Surface*
The equation of constraint (10.14) can be generalized to describe a rigidly moving plane simply by allowing $\mathbf{y} = \mathbf{y}(t)$ to be a given function describing the motion. It can be further generalized to describe a plane rotating about an axis through the point $\mathbf{y}$ by allowing the unit normal to be a function of time $\mathbf{n} = \mathbf{n}(t)$. Let us consider implications of the first generalization. The equation of constraint is

$$\phi(\mathbf{x}, t) = (\mathbf{x} - \mathbf{y}(t))\cdot\mathbf{n} = 0.$$

Differentiation implies

$$\dot{\phi} = (\mathbf{v} - \dot{\mathbf{y}})\cdot\mathbf{n} = 0$$

and

$$\ddot{\phi} = (\dot{\mathbf{v}} - \ddot{\mathbf{y}})\cdot\mathbf{n} = 0\,.$$

Using this to solve the equation of motion (10.15) for λ, we get

$$\lambda = m(\ddot{\mathbf{y}} - \mathbf{g})\cdot\mathbf{n}$$

The equation of motion can therefore be written

$$\dot{\mathbf{v}} = \mathbf{g} + (\ddot{\mathbf{y}} - \mathbf{g})\cdot\mathbf{n}\mathbf{n} \tag{10.17}$$

This is the generalization of (10.16) for a moving plane. Integration of (10.17) is trivial since $\mathbf{y} = \mathbf{y}(t)$ is supposed to be a given function. Note that the orbit is again a parabola if the acceleration $\ddot{\mathbf{y}}$ is constant. However, (10.17) will fail to apply if the plane moves in such a way as to "break contact" with the particle.

EXAMPLE 4: *Particle on a Stationary Spherical Surface*
Now consider a particle constrained to move on the surface of a sphere of radius a. If the sphere's center is chosen as the origin, the equation of constraint can be written in the form

$$\mathbf{x}^2 - a^2 = 0,$$

or, in the form

$$\phi(\mathbf{x}) = |\,\mathbf{x}\,| - a = 0. \tag{10.18a}$$

The second form is a little more convenient because by (2–8.39) its gradient is equal to the unit exterior normal to the sphere

$$\hat{\mathbf{x}} = \nabla\,|\,\mathbf{x}\,| = \nabla\phi. \tag{10.18b}$$

The equation of motion is

$$m\dot{\mathbf{v}} = m\mathbf{g} + \lambda\hat{\mathbf{x}}. \tag{10.19}$$

Whence,

$$\lambda = m(\dot{\mathbf{v}} - \mathbf{g})\cdot\hat{\mathbf{x}}.$$

According to (10.10), Equation (10.18b) implies that $\mathbf{v}\cdot\mathbf{x} = 0$. Differentiating $\mathbf{v}\cdot\mathbf{x} = 0$, we get $\dot{\mathbf{v}}\cdot\mathbf{x} + \mathbf{v}^2 = 0$ or

$$\dot{\mathbf{v}}\cdot\hat{\mathbf{x}} = -\frac{v^2}{a}. \tag{10.20}$$

This enables us to express λ in the form

$$\lambda = -m\left(\frac{v^2}{a} + \mathbf{g}\cdot\hat{\mathbf{x}}\right). \tag{10.21}$$

This can be reduced to a function of position alone by using energy conservation. The result is obtained in terms of initial conditions by using (10.13) to eliminate v^2 from (10.21); thus,

$$\lambda = -\frac{m}{a}(3\mathbf{g}\cdot\mathbf{x} - 2\mathbf{g}\cdot\mathbf{x}_0 + v_0^2)\,. \tag{10.22}$$

Either (10.21) or (10.22) can be substituted into (10.19) to get a well-defined equation of motion; however, the result looks complicated, and we shall find more convenient ways to express it later on. In the meantime, we can draw some significant conclusions directly from (10.22).

Equation (10.22) allows both positive and negative values for λ. However, if the particle is a small object on the surface of solid ball, only positive values of λ are significant, because the force exerted by the ball must be outward. At a point where $\lambda = 0$ the constraining force vanishes, so the particle is no longer in contact with the surface. Therefore, from (10.22) we can conclude that the particle will break contact with the surface at a height h above the "equitorial plane" of the ball given by

$$h \equiv -\,\hat{\mathbf{g}}\cdot\mathbf{x} = \tfrac{1}{3}\left(\frac{v_0^2}{g} - 2\hat{\mathbf{g}}\cdot\mathbf{x}_0\right).$$

In particular, if the particle starts from rest at the top of the ball, then

$$h = \tfrac{2}{3}(-\hat{\mathbf{g}}\cdot\mathbf{x}_0) = \tfrac{2}{3}a\,.$$

If the particle is constrained by the inside instead of the outside of the spherical surface, then the constraining force must point inward so $\lambda \leqslant 0$. A common example of this kind of constraint is a pendulum consisting of a bob supported by a massless *flexible string*. On the other hand, for a pendulum consisting of a bob supported by a massless rigid rod, the constraining force may be either inward or outward, so λ can be either positive or negative. Constraints of this kind, which do not allow the particle to leave the surface of constraint, are called *bilateral*. Constraints which confine a particle to one side of a surface are called *unilateral*. The same general equations can be applied to both kinds of constraint, but, as we have seen, they must be interpreted differently in each case.

Our analysis of spherical constraints, in particular the derivation and application of (10.22), is readily generalized to handle constraints exerted by any smooth surface. However, a complete generalization involves the differential geometry of surfaces, which is beyond the scope of this text.

Lagrange's Equations for Constrained Motion

We are still faced with the problem of developing a systematic method for solving the equations of motion subject to holonomic constraints. This problem was solved by Lagrange. His method employs constraints expressed in parametric form rather than the nonparametric form (10.8).

The parametric equation for a surface has the form $\mathbf{x} = \mathbf{x}(q_1, q_2, t)$, where q_1 and q_2 are independent scalar parameters (or *coordinates*), and the explicit

t dependence allows for the possibility that the surface is moving. For fixed q_2 and variable q_1, the parametric equation describes a "coordinate curve" on the surface with a tangent vector $\mathbf{e}_1$ at each point defined by

$$\mathbf{e}_1 \equiv \partial_{q_1}\mathbf{x}, \tag{10.23}$$

where ∂_{q_1} denotes the derivative with respect to q_1. Similarly, the variable q_2 determines a tangent vector $\mathbf{e}_2$ at each point on the surface, as shown in Figure 10.1. Both cases are for covered by the "free index notation"

$$\mathbf{e}_\alpha \equiv \partial_{q_\alpha}\mathbf{x} \tag{10.23}$$

where $\alpha = 1, 2$.

The equation of motion has the form

$$m\ddot{\mathbf{x}} = \mathbf{F} + \mathbf{N}, \tag{10.24}$$

where $\mathbf{N}$ is the force of constraint and $\mathbf{F}$ is some given external force. Since $\mathbf{N}$ is normal to the surface of constraint, we have

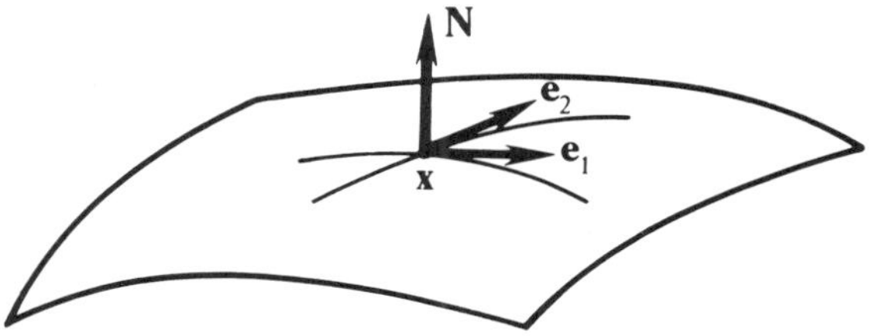

Fig. 10.1 Coordinate curves and tangent vectors on a surface of constraint.

$$\mathbf{N}\cdot\mathbf{e}_\alpha = 0. \tag{10.25}$$

Therefore, multiplication of (10.24) by the $\mathbf{e}_\alpha$ gives us independent components of the equation of motion on the surface.

$$m\ddot{\mathbf{x}}\cdot\mathbf{e}_\alpha = \mathbf{F}\cdot\mathbf{e}_\alpha. \tag{10.26}$$

These equations can be expressed as a set of differential equations for the coordinates q_α alone. This is merely an exercise using the chain rule of differentiation.

The trajectory of the particle on the surface of constraint is described by the parametric equation

$$\mathbf{x} = \mathbf{x}(q_1(t), q_2(t), t).$$

Therefore, its velocity is given by the parametric equation

$$\dot{\mathbf{x}} = \sum_\alpha \dot{q}_\alpha \partial_{q_\alpha}\mathbf{x} + \partial_t\mathbf{x}, \tag{10.27}$$

where the sum is over all values of α. Hence

$$\partial_{\dot{q}_\alpha}\dot{\mathbf{x}} = \partial_{q_\alpha}\mathbf{x} = \mathbf{e}_\alpha,$$

and

$$\dot{\mathbf{e}}_\alpha = \Big(\sum_\beta \dot{q}_\beta \partial_{q_\beta} + \partial_t\Big)\partial_{q_\alpha}\mathbf{x} = \partial_{q_\alpha}\dot{\mathbf{x}},$$

because $\partial_{q_\alpha}\dot{q}_\beta = 0$ for all values of α and β. We can use these facts to rewrite the left side of (10.26); thus,

$$m\ddot{\mathbf{x}}\cdot\mathbf{e}_\alpha = m\left\{\frac{\mathrm{d}}{\mathrm{d}t}(\dot{\mathbf{x}}\cdot\mathbf{e}_\alpha) - \dot{\mathbf{x}}\cdot\dot{\mathbf{e}}_\alpha\right\}.$$

But,

$$\dot{\mathbf{x}}\cdot\mathbf{e}_\alpha = \dot{\mathbf{x}}\cdot\partial_{\dot{q}_\alpha}\dot{\mathbf{x}} = \tfrac{1}{2}\partial_{\dot{q}_\alpha}(\dot{\mathbf{x}}^2),$$

and

$$\dot{\mathbf{x}}\cdot\dot{\mathbf{e}}_\alpha = \dot{\mathbf{x}}\cdot(\partial_{q_\alpha}\dot{\mathbf{x}}) = \tfrac{1}{2}\partial_{q_\alpha}(\dot{\mathbf{x}}^2).$$

Hence

$$m\ddot{\mathbf{x}}\cdot\mathbf{e}_\alpha = \frac{\mathrm{d}}{\mathrm{d}t}(\partial_{\dot{q}_\alpha}K) - \partial_{q_\alpha}K,$$

where $K \equiv \frac{1}{2}m\dot{\mathbf{x}}^2$ is the kinetic energy of the particle.

Now introducing the notation

$$Q_\alpha \equiv \mathbf{F}\cdot\mathbf{e}_\alpha = \mathbf{F}\cdot(\partial_{\dot{q}_\alpha}\mathbf{x}) \tag{10.28}$$

for the component of force in the $\mathbf{e}_\alpha$ direction, we can write (10.26) in the form

$$\frac{\mathrm{d}}{\mathrm{d}t}(\partial_{\dot{q}_\alpha}K) - \partial_{q_\alpha}K = Q_\alpha. \tag{10.29}$$

This is called *Lagrange's equation*. There is one such equation for each of the coordinates q_α. To use Lagrange's equation, it is necessary to express K and Q_α as functions of q_α and $\dot{q}_\alpha$ by using the parametric equations of constraint $\mathbf{x} = \mathbf{x}(q, t)$, where, for brevity, the symbol q is used to denote the whole set of coordinates q_α. Thus, using (10.27), we find

$$K = \tfrac{1}{2}m\dot{\mathbf{x}}^2 = \tfrac{1}{2}\sum_\alpha\sum_\beta a_{\alpha\beta}\dot{q}_\alpha\dot{q}_\beta + \sum_\alpha b_\alpha\dot{q}_\alpha + c, \tag{10.30}$$

where the coefficients are functions of the coordinates given by

$$\begin{aligned} a_{\alpha\beta} &= a_{\alpha\beta}(q, t) \equiv m\mathbf{e}_\alpha\cdot\mathbf{e}_\beta, \\ b_\alpha &= b_\alpha(q, t) \equiv m\mathbf{e}_\alpha\cdot\partial_t\mathbf{x}, \\ c &= c(q, t) \equiv \tfrac{1}{2}m(\partial_t\mathbf{x})^2. \end{aligned} \tag{10.31}$$

If the external force is conservative so that $\mathbf{F} = -\nabla V(\mathbf{x})$, then the equation of constraint gives us $V = V(\mathbf{x}(q, t)) = V(q, t)$, and

$$Q_\alpha = -(\partial_{q_\alpha}\mathbf{x})\cdot\nabla V = -\partial_{q_\alpha}V.$$

In this case, Lagrange's equation (10.29) can be written in the form

$$\frac{\mathrm{d}}{\mathrm{d}t}(\partial_{\dot{q}_\alpha}L) - \partial_{q_\alpha}L = 0, \tag{10.32}$$

where the so-called Lagrangian $L = L(\dot{q}, q, t)$ is defined by

$$L = K - V \tag{10.33}$$

Once Lagrange's equations have been solved for the functions $q_\alpha = q_\alpha(t)$, the result can be substituted in the equation of constraint to get an explicit equation for the orbit $\mathbf{x}(t) = \mathbf{x}(q(t), t)$. Lagrange's method has the advantage of totally eliminating the need to consider forces of constraint, which are often of no interest in themselves. However, once the orbit has been found by Lagrange's method, the force of constraint can be computed directly from $\mathbf{N} = m\ddot{\mathbf{x}} - \mathbf{F}$. Just the same, our previous method using a nonparametric constraint is often a more efficient way to find $\mathbf{N}$.

It will be noted that our derivation of Lagrange's equation is actually independent of the number of coordinates in the equation of constraint. It is only necessary to sum over the appropriate number of coordinates in (10.27) and (10.30). If the particle is constrained to a curve rather than a surface, then the parametric equation involves only one coordinate and only one Lagrange equation is needed to determine the motion. On the other hand, if there are no constraints, then three coordinates are needed and Lagrange's method is simply a way of writing the equivalent of Newton's equation in terms of coordinates.

EXAMPLE 5: *The Plane Pendulum*
Now let us consider an application of Lagrange's method. The best way of writing a parametric equation for a sphere will not be evident until we have discussed rotations in chapter 5. So let us limit our considerations here to the special case of a particle constrained to a vertical circle. This is the so-called *simple pendulum*. The equation of constraint can be written in the parametric form

$$\mathbf{x} = \mathbf{x}(\phi) = \mathbf{a}e^{\mathbf{i}\phi},$$

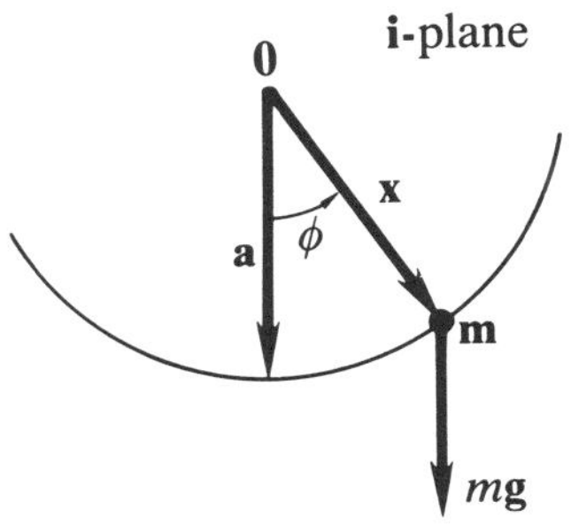

Fig. 10.2. Vector diagram for the simple pendulum.

where, as shown in Figure 10.2, the constant $\mathbf{a}$ is the vertical radius vector, and $\mathbf{i}$ is the unit bivector for the plane of the circle. The external force is the conservative gravitational force, so we can get Lagrange's equation from a Lagrangian. Our first task is to express the Lagrangian as an explicit function of ϕ and $\dot{\phi}$. The equation of constraint gives

$$\dot{\mathbf{x}} = \mathbf{x}\mathbf{i}\dot{\phi} = -\mathbf{i}\mathbf{x}\dot{\phi}.$$

Hence,

$$K = \tfrac{1}{2}m(\mathbf{x}\mathbf{i}\dot{\phi})^2 = \tfrac{1}{2}ma^2\dot{\phi}^2,$$

where $a = |\,\mathbf{x}\,| = |\,\mathbf{a}\,|$. Also,

$$V = -\langle m\mathbf{g}\mathbf{x}\rangle = -m\langle \mathbf{g}\mathbf{a}e^{\mathbf{i}\phi}\rangle = -mga\langle e^{\mathbf{i}\phi}\rangle = -mga\cos\phi .$$

Hence

$$L = \tfrac{1}{2}ma^2\dot{\phi}^2 + mga\cos\phi .$$

Now

$$\partial_{\dot{\phi}}L = ma^2\dot{\phi},$$

$$\partial_{\phi}L = -mga\sin\phi .$$

So Lagrange's equation

$$\frac{\mathrm{d}}{\mathrm{d}t}(\partial_{\dot{\phi}}L) - \partial_{\phi}L = 0 .$$

takes the explicit form

$$ma^2\ddot{\phi} + mga\sin\phi = 0 .$$

The general solution of this equation involves elliptic functions. However, for small oscillations the approximation $\sin\phi \approx \phi$ is often satisfactory, in which case Lagrange's equation reduces to

$$\ddot{\phi} + \frac{g}{a}\phi = 0 .$$

This is the familiar equation for harmonic motion, so we can write down the solution at once. If $\phi = 0$ at time $t = 0$, the solution is

$$\phi = \sin\omega_0 t$$

where $\omega_0 = (g/a)^{1/2}$. Therefore, the parametric equation for the orbit is

$$\mathbf{x}(t) = \mathbf{a}e^{\mathbf{i}\sin\omega_0 t} .$$

We shall study the pendulum from other points of view later on.

Frictional Forces

Let us conclude this section with a suggestion on how to account for friction. An object sliding on the surface of a solid body is subject to a frictional force $\mathbf{F}$ described by the empirical formula

$$\mathbf{F} = -\hat{\mathbf{v}}\mu N, \tag{10.34}$$

where $\mathbf{v}$ is the velocity relative to the surface, μ is a constant called "the coefficient of sliding friction", and $N = |\mathbf{N}|$ is the force of constraint normal to the surface. The method of Lagrange by itself cannot handle such a frictional force, because N is not a known function. However, we can evaluate N from Newton's Law, just as we have done before. Indeed, for a particle on a spherical surface, we can use (10.21) to get the frictional force in the form

$$\mathbf{F} = \hat{\mathbf{v}}\frac{\mu m}{a}(v^2 + \mathbf{g}\cdot\mathbf{x}). \tag{10.35}$$

This can be used in the Lagrange equations to determine the orbit on the sphere. However, approximation methods must be used to solve the resulting equations for even the simplest problems.

3-10. Exercises

(10.1) The bob on a pendulum with flexible support of length a moves with speed v_0 at the bottom of a vertical circle. What is the minimum value of v_0 needed for the bob to reach the top of the circle? Why can't this result be obtained by energy conservation assuming speed $v = 0$ at the top?

(10.2) For a particle subject to a force **f**;

(a) Show that if **a** is a constant vector, then $\mathbf{f}\cdot\mathbf{a} = 0$ implies that $\mathbf{v}\cdot\mathbf{a}$ is a constant of motion. What does this imply about the trajectory?

(b) Find a constant of motion when $\mathbf{a}\wedge\mathbf{f} = 0$.

(c) Show that if $\mathbf{v}_0\wedge\mathbf{f} = 0$, then the particle moves in a straight line. Though $\hat{\mathbf{v}}$ is a constant of motion here, this would not be called a conservation law, because it holds only for particular values of the initial velocity $\mathbf{v}_0$.

(10.3) A bead is constrained to move on a frictionless right circular helical wire described by the parametric equation

$$\mathbf{x}(\theta) = \mathbf{a}e^{\mathbf{i}\theta} + \mathbf{b}\theta.$$

The wire is placed upright in a constant gravitational field (Figure 10.3). Evaluate and solve Lagrange's equation for θ. Determine how the height of the bead varies with time, and compare it with the motion of a particle on an inclined plane.

(10.4) A bead moves on a frictionless hoop rotating in a horizontal plane with a constant angular speed ω about a fixed point on the hoop. Show that the bead oscillates about a diameter of the hoop like a simple pendulum of length g/ω^2. Begin by establishing the equation of constraint

$$\mathbf{x}(\phi, t) = \mathbf{a}e^{\mathbf{i}\omega t}(1 + e^{\mathbf{i}\phi}),$$

as suggested by Figure 10.4.

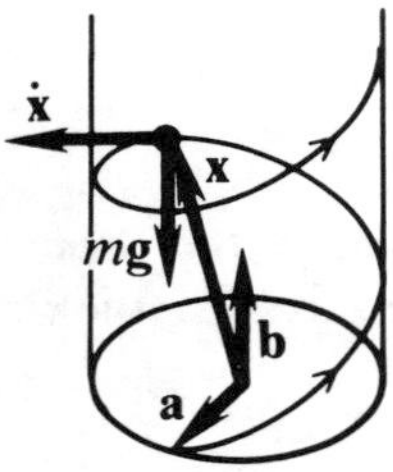

Fig. 10.3. Bead on a helical wire.

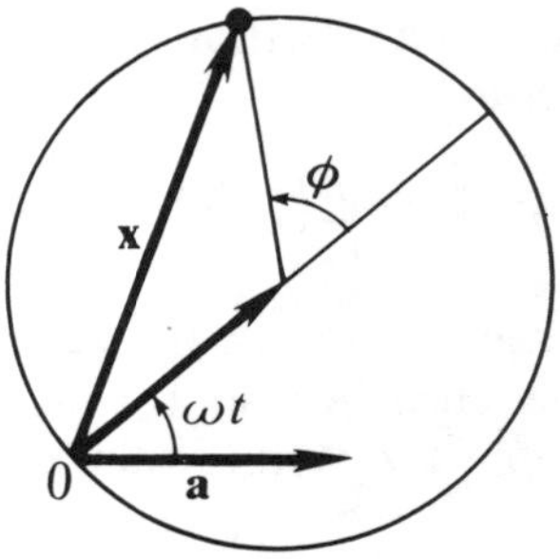

Fig. 10.4. Bead on a rotating hoop.

(10.5) The equation of constraint for a particle in a plane can be written

$$\mathbf{x} = \mathbf{x}(r, \phi) = r\boldsymbol{\sigma}_1 e^{\mathbf{i}\phi}.$$

The variables r and ϕ are *polar coordinates*. Express the kinetic energy in terms of these variables, and determine the form of the Lagrange equations if the particle is subject to an arbitrary external force **F**. Derive the same equations directly from Newton's law.

(10.6) For a conservative force, expand the potential $V(\mathbf{x})$ in a Taylor series about an equilibrium point $\mathbf{x}_0$, and show that, in the first approximation, it is equivalent to the anisotropic oscillator potential

$$V(\mathbf{r}) \approx -\tfrac{1}{2}\mathbf{r}\cdot\mathbf{L}(\mathbf{r}),$$

where $\mathbf{r} = \mathbf{x} - \mathbf{x}_0$, and

$$\mathbf{L}(\mathbf{r}) = -\nabla V(\mathbf{r})$$

is the linear binding force characterized by Equation (8.20). The general shape of an equipotential $V(\mathbf{r}) =$ constant is an ellipsoid. Use Equation (8.20) to show the energy of the oscillator has the constant value

$$E = \tfrac{1}{2}(k_1\mathbf{a}_1^2 + k_2\mathbf{a}_2^2 + k_3\mathbf{a}_3^2).$$

Chapter 4

Central Forces and Two-Particle Systems

The simple laws of force studied in Chapter 3 are said to be *phenomenological* laws, which is to say that they are only *ad hoc* or approximate descriptions of *real* forces in nature. As a rule, they describe resultants of forces exerted by a very large number of particles. A fundamental force law describes the force exerted by a single particle. The simplest candidates for such a law are central forces with the particle at the center of force. This is reason enough for the systematic study of central forces in this chapter. And it should be no surprise that the results are of great practical value.

The investigation of fundamental forces is actually a two-particle problem, for, as Newton's third law avers, a particle cannot act without being acted upon. Fortunately, the two-particle central force problem can be reduced to a mathematically equivalent one-particle problem, greatly simplifying the solution. However, a complete description of central force motion must include an account of the "two-body effects" involved in this reduction.

4-1. Angular Momentum

We have seen that motion of a particle in any conservative force field is characterized by a conserved quantity called energy. Now we shall show that general motion in a central force field is characterized by another conserved quantity called angular momentum. Then we shall derive the basic properties of angular momentum which will be helpful in a detailed analysis of central force motion.

A force field $\mathbf{f} = \mathbf{f}(\mathbf{x})$ is said to be *central* if it is everywhere directed along a line through a fixed point $\mathbf{x}'$ called the *center of force*. This property can be expressed by the equation

$$(\mathbf{x}-\mathbf{x}')\wedge\mathbf{f} = \mathbf{r}\wedge\mathbf{f} = 0, \tag{1.1}$$

where $\mathbf{r} = \mathbf{x} - \mathbf{x}'$ is introduced as the convenient position variable with the center of force as origin. The *angular momentum* $\mathbf{L}$ *about the center of force* for a particle with mass m is defined by

$$\mathbf{L} \equiv m\mathbf{r}\wedge\dot{\mathbf{r}} = m(\mathbf{x}-\mathbf{x}')\wedge\dot{\mathbf{x}}. \tag{1.2}$$

It is customary to define the angular momentum as a vector quantity

$$\mathbf{l} \equiv m\mathbf{r}\times\dot{\mathbf{r}}. \tag{1.3}$$

However, the bivector $\mathbf{L}$ is more fundamental than the vector $\mathbf{l}$ and will be somewhat more convenient in our study of central forces. In any case, it is easy to switch from one quantity to the other, because they are related by duality; specifically,

$$\mathbf{L} = \mathbf{il}. \tag{1.4}$$

We shall use the term *angular momentum* for either $\mathbf{L}$ or $\mathbf{l}$ and add the term "bivector" or "vector" if it is necessary to specify one or the other.

Now, from the equation of motion we have

$$\mathbf{r}\wedge\mathbf{f} = \mathbf{r}\wedge(m\ddot{\mathbf{r}}) = \frac{\mathrm{d}}{\mathrm{d}t}(m\mathbf{r}\wedge\dot{\mathbf{r}}),$$

because $\dot{\mathbf{r}}\wedge\dot{\mathbf{r}} = 0$. Hence,

$$\mathbf{r}\wedge\mathbf{f} = 0 \quad \text{if and only if} \quad \dot{\mathbf{L}} = 0, \tag{1.5}$$

that is, the *angular momentum is conserved if and only if the force is central*. It should be noted that this conclusion holds even if the force is velocity dependent, though central forces of this type are not common enough to merit special attention here.

Angular momentum has a simple geometrical interpretation. According to Equation (2–8.26), the directed area $\mathbf{A} = \mathbf{A}(t)$ swept out by the *radius vector* $\mathbf{r}$ in time t is given by

$$\mathbf{A}(t) = \tfrac{1}{2}\int_{\mathbf{r}(0)}^{\mathbf{r}(t)} \mathbf{r}\wedge\mathrm{d}\mathbf{r} = \tfrac{1}{2}\int_0^t \mathbf{r}\wedge\dot{\mathbf{r}}\,\mathrm{d}t. \tag{1.6}$$

Therefore, the rate at which area is swept out is determined by the angular momentum according to

$$\dot{\mathbf{A}} = \tfrac{1}{2}\mathbf{r}\wedge\dot{\mathbf{r}} = \frac{\mathbf{L}}{2m}\,. \tag{1.7}$$

For constant angular momentum this can be integrated immediately, with the result

$$\mathbf{A}(t) = \frac{1}{2m}\,\mathbf{L}t. \tag{1.8}$$

Thus, we conclude that *the radius vector of a particle in a central force field sweeps out area at a constant rate*. This is a generalization of Kepler's Second Law of planetary motion.

The orbit of a particle in a central field lies in a plane through the center of force with direction given by the angular momentum; for, from (1.2) we deduce that every point **r** on the orbit satisfies

$$\mathbf{r}\wedge\mathbf{L} = 0, \tag{1.9}$$

which, as we have seen in Section 2-6, is a necessary and sufficient condition for a point **r** to lie in the **L**-plane.

If the orbit in a central field is closed, then the particle will return to its starting point in a definite time T called the *period* of the motion. From (1.6) and (1.8), we conclude that the period is given by

$$\mathbf{A}(T) = \tfrac{1}{2}\oint \mathbf{r}\wedge d\mathbf{r} = \frac{1}{2m}\,\mathbf{L}T. \tag{1.10}$$

As we shall see, this formula leads to Kepler's third law of planetary motion.

A major reason for the importance of angular momentum is the fact that it determines the rate at which the radius vector changes direction. To show this, we differentiate $\mathbf{r} = r\hat{\mathbf{r}}$ to get

$$\dot{\mathbf{r}} = \dot{r}\,\hat{\mathbf{r}} + r\dot{\hat{\mathbf{r}}}. \tag{1.11}$$

Whence,

$$\mathbf{r}\wedge\dot{\mathbf{r}} = r\mathbf{r}\wedge\dot{\hat{\mathbf{r}}} = r^2\hat{\mathbf{r}}\dot{\hat{\mathbf{r}}}. \tag{1.12}$$

or

$$\dot{\hat{\mathbf{r}}} = \frac{\hat{\mathbf{r}}\mathbf{L}}{mr^2} = \frac{\mathbf{r}\mathbf{L}}{m\,|\,\mathbf{r}\,|^3} = -\frac{\mathbf{L}\mathbf{r}}{mr^3}\,. \tag{1.13}$$

When **L** is constant, this gives $\dot{\hat{\mathbf{r}}}$ as an explicit function of **r**. Substituting (1.13) into (1.11), we get the velocity in the form

$$\dot{\mathbf{r}} = \hat{\mathbf{r}}\left(\dot{r} + \frac{\mathbf{L}}{mr}\right) = \left(\dot{r} - \frac{\mathbf{L}}{mr}\right)\hat{\mathbf{r}}. \tag{1.14}$$

For any central force motion, this can be used to determine the velocity as a function of direction $\dot{\mathbf{r}} = \mathbf{v}(\hat{\mathbf{r}})$ whenever the orbit is expressed as an equation of the form $r = r(\hat{\mathbf{r}})$ specifying the radial distance as a function of direction.

For planar motion, we can express the radial direction $\hat{\mathbf{r}}$ in the parametric form

$$\hat{\mathbf{r}} = \hat{\boldsymbol{\varepsilon}}e^{\mathbf{i}\theta}, \tag{1.15}$$

where $\hat{\boldsymbol{\varepsilon}}$ is a fixed unit vector, $\mathbf{i} \equiv \hat{\mathbf{L}}$ is the unit bivector for the orbital plane, and $\theta = \theta(t)$ is the scalar measure for the angle of rotation. Differentiating (1.15) and equating to (1.13), we get

$$\dot{\hat{\mathbf{r}}} = \hat{\mathbf{r}}\mathbf{i}\dot{\theta} = \hat{\mathbf{r}}\,\frac{\mathbf{L}}{mr^2}$$

So,

$$\mathbf{i}\dot{\theta} = \frac{\mathbf{L}}{mr^2} . \tag{1.16}$$

Whence

$$L = |\,\mathbf{L}\,| = mr^2\dot{\theta} . \tag{1.17}$$

The derivation of this result assumes only that $\hat{\mathbf{L}}$ is constant. If $|\,\mathbf{L}\,|$ is a constant also, then (1.17) implies that $\dot{\theta} \geqslant 0$ always, so $\theta = \theta(t)$ increases monotonically with time if $|\,\mathbf{L}\,| \neq 0$. Therefore, the orbit of a particle in a central field never changes the direction of its circulation about the center of force. We could also have reached this conclusion directly from (1.13).

4-1. Exercises

(1.1) Prove that $\dot{\mathbf{L}} = 0$ implies that the magnitude $|\,\mathbf{L}\,|$ and the direction $\hat{\mathbf{L}}$ of the angular momentum $\mathbf{L}$ are separately constants of motion.

4-2. Dynamics from Kinematics

The science of motion can be subdivided into kinematics and dynamics. *Kinematics* is concerned with the *description of motion* without considering conditions or interactions required to bring particular motions about. Dynamics is concerned with the *explanation of motion* by specifying forces or other laws of interaction to describe the influence of one physical system on another.

If the dynamics is known, the kinematics of particle motion can be determined by solving the equation of motion. However, the converse problem of determining dynamics from kinematics is far more difficult, and it is rarely solved without considerable prior knowledge about force laws likely to be operative. Historically one of the first and still the most significant solution to such a problem was *Newton's deduction of the law of gravitation from Kepler's laws*. Let us see how the problem can be formulated and solved in modern language.

Kepler's Laws of Planetary Motion can be formulated as follows:

(1) The planets move in ellipses with the sun at one focus.
(2) The radius vector sweeps out equal areas in equal times.
(3) The square of the period of revolution is proportional to the cube of the semi-major axis.

The first and second laws are illustrated in Figure 2.1; the elliptical orbit is divided into six segments of equal area, showing how a planet's speed decreases with increasing distance from the sun.

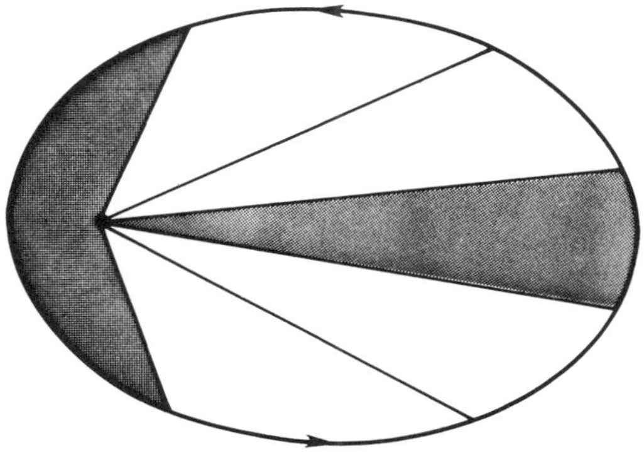

Fig. 2.1. Kepler's first and second laws.

After the discussion in the last section, we recognize Kepler's second law immediately as a statement of angular momentum conservation, and we conclude that the planets move in a central field with the sun at the center of force. For constant angular momentum, we can compute the acceleration of a particle by using (1.13) to differentiate the expression (1.14) for its velocity; thus,

$$\ddot{\mathbf{r}} = \dot{\hat{\mathbf{r}}}\left(\dot{r} + \frac{\mathbf{L}}{mr}\right) + \hat{\mathbf{r}}\left(\ddot{r} - \frac{\dot{r}\mathbf{L}}{mr^2}\right)$$
$$= \hat{\mathbf{r}}\left\{\frac{\mathbf{L}}{mr^2}\left(\dot{r} + \frac{\mathbf{L}}{mr}\right) + \ddot{r} - \frac{\dot{r}\mathbf{L}}{mr^2}\right\}$$

from which we obtain

$$m\ddot{\mathbf{r}} = \left(m\ddot{r} - \frac{L^2}{mr^3}\right)\hat{\mathbf{r}}\ . \tag{2.1}$$

The right side of this equation has the form of a central force as required, and we can determine the magnitude of the force by evaluating the coefficient.

Recalling the equation for an ellipse discussed in Section 2.6, we can express Kepler's first law as an equation of the form

$$r = \frac{\ell}{1 + \boldsymbol{\varepsilon}\cdot\hat{\mathbf{r}}} \tag{2.2}$$

where ℓ is a positive constant and $\boldsymbol{\varepsilon}$ is a fixed vector in the orbital plane. As an aid to differentiating (2.2), we multiply (1.13) by $\boldsymbol{\varepsilon}$ and note that, since $\boldsymbol{\varepsilon}\wedge\mathbf{L} = 0$, the scalar part of the result can be written

$$\boldsymbol{\varepsilon}\cdot\dot{\hat{\mathbf{r}}} = \frac{(\boldsymbol{\varepsilon}\wedge\hat{\mathbf{r}})\cdot\mathbf{L}}{mr^2} = \frac{\boldsymbol{\varepsilon}\wedge\hat{\mathbf{r}}\mathbf{L}}{mr^2}\ , \tag{2.3a}$$

while the bivector part has the form

$$\boldsymbol{\varepsilon}\wedge\dot{\hat{\mathbf{r}}} = \frac{\boldsymbol{\varepsilon}\cdot\hat{\mathbf{r}}\mathbf{L}}{mr^2}\ . \tag{2.3b}$$

Now, by differentiating (2.2) we obtain

$$\dot{r} = -\frac{\ell\boldsymbol{\varepsilon}\cdot\dot{\hat{\mathbf{r}}}}{(1 + \boldsymbol{\varepsilon}\cdot\hat{\mathbf{r}})^2} = -\frac{\boldsymbol{\varepsilon}\wedge\hat{\mathbf{r}}\mathbf{L}}{\ell m}\ .$$

Differentiating again, we obtain

$$\ddot{r} = -\frac{\boldsymbol{\varepsilon}\cdot\hat{\mathbf{r}}\mathbf{L}^2}{\ell m^2 r^2} = \left(1 - \frac{\ell}{r}\right)\frac{\mathbf{L}^2}{\ell m^2 r^2} \quad ,$$

which, since $\mathbf{L}^2 = -L^2$, gives

$$m\ddot{r} - \frac{L^2}{mr^3} = -\frac{k}{r^2}, \tag{2.4}$$

where

$$k = \frac{L^2}{\ell m} > 0. \tag{2.5}$$

Comparison of (2.4) with (2.1) leads to the attractive central force law

$$\mathbf{f} = -\frac{k\hat{\mathbf{r}}}{r^2} \; . \tag{2.6}$$

Thus, we have arrived at Newton's inverse square law for gravitational force. The problem remains to evaluate the constant k in terms of measurable quantities. Kepler's third law can be used for this purpose.

According to (1.10) the period T of a planet's motion depends on the area enclosed by its orbit. The area integral in (1.10) is most easily evaluated by taking advantage of the fact, proved in Section 2-8, that it is independent of the choice of origin. As we have seen before, with the origin at the center instead of at a focus, an ellipse can be described by the parametric equation

$$\mathbf{x} = \mathbf{a}\cos\phi + \mathbf{b}\sin\phi, \tag{2.7}$$

where $a = |\,\mathbf{a}\,|$ is the *semi-major axis* referred to in Kepler's Third Laws. Using this to carry out the integration, we get

$$\mathbf{A} = \tfrac{1}{2}\oint \mathbf{r}\wedge \mathrm{d}\mathbf{r} = \tfrac{1}{2}\oint \mathbf{x}\wedge \mathrm{d}\mathbf{x} = \tfrac{1}{2}\int_0^{2\pi} \mathbf{x}\wedge\frac{\mathrm{d}\mathbf{x}}{\mathrm{d}\phi}\,\mathrm{d}\phi = \pi\mathbf{ab}. \tag{2.8}$$

According to (1.10), therefore, the period is given by

$$T = \frac{2\pi m\mathbf{ab}}{\mathbf{L}} = \frac{2\pi mab}{L} \; . \tag{2.9}$$

Squaring this, using (2.5) and the fact that $b^2 = a\ell$ (Exercise (2.2)), we obtain

$$\frac{T^2}{a^3} = \frac{4\pi^2 \ell m^2}{L^2} = 4\pi^2\,\frac{m}{k} \; . \tag{2.10}$$

Kepler's third law says that this ratio has the same numerical value for all planets. Therefore, *the constant k must be proportional to the mass m*. Also, note the surprising fact that (2.10) implies that the period T does not depend on the eccentricity.

Universality of Newton's Law

We have learned all that Kepler's laws can tell us about dynamics. Having recognized as much, Newton set about investigating the possibility that the inverse square law (2.6) is a universal law of attraction between all massive particles. He hypothesized that each planet exerts a force on the Sun equal and opposite to the force exerted by the Sun on the planet. From Kepler's third law, then, he could conclude that the constant k is proportional to the mass M of the Sun as well as the mass m of the planet, that is,

$$k = GmM, \tag{2.11}$$

where G is a universal constant describing the strength of gravitational attraction between all bodies. Substitution of (2.11) into (2.10) gives

$$M = \frac{4\pi^2 a^3}{GT^2} \; . \tag{2.12}$$

The constant G can be determined by measuring the gravitational force between objects on Earth, so (2.12) gives the mass of the Sun from astronomical measurements of a and T.

Kepler presented his three laws as independent empirical propositions about regularities he had observed in planetary motions. He did not possess the conceptual tools needed to recognize that the laws are related to one another, or indeed, to recognize that they are more significant than many other propositions he proposed to describe planetary motion. Though we have seen how to infer Newton's universal law of gravitation from Kepler's laws, and conversely, we can derive all three of Kepler's laws from Newton's law, it would be a mistake to think that Newton's law is merely a summary of information in Kepler's laws. Actually, Kepler's laws are only approximately true, and they can be derived only by neglecting the forces exerted by the planets on one another and the Sun. But we shall see that the appropriate corrections can be derived from Newton's law of gravitation, which is so close to being an exact law of nature that only the most minute deviations from it have been detected. These deviations have been explained only in this century by Einstein's theory of gravitation.

Epicycles of Ptolemy

Long before Kepler proposed elliptical orbits about the Sun, Ptolemy described the orbits of the planets as epitrochoids centered near the Earth. It is of some interest, therefore, to deduce the force required to produce such a motion. An *epitrochoid* is a curve described by the parametric equation

$$\mathbf{r}(t) = \mathbf{r}_1 + \mathbf{r}_2 = \mathbf{a}_1 e^{i\omega_1 t} + \mathbf{a}_2 e^{i\omega_2 t}, \tag{2.13}$$

where

$$\boldsymbol{\omega}_1 \wedge \boldsymbol{\omega}_2 = 0. \tag{2.14}$$

and $\boldsymbol{\omega}_1 \cdot \boldsymbol{\omega}_2 > 0$. If, instead, $\boldsymbol{\omega}_1 \cdot \boldsymbol{\omega}_2 < 0$, the curve is called a *hypotrochoid* or retrograde epitrochoid. Equation (2.13) is a superposition of two vectors rotating with constant angular velocities in the same plane.

In astronomical literature, the circle generated by one of these vectors is called the *epicycle* while the circle generated by the other is called the *deferent*, and the orbit is traced out by a particle moving uniformly on the epicycle while the center of the epicycle moves uniformly along the deferent. A constant vector could be added to (2.13) to express the fact that Ptolemy displaced the center of the planetary orbits slightly from the earth to account for observed variations in speed along the orbits.

Now to deduce the force, we differentiate (2.13) twice; thus,

$$\dot{\mathbf{r}} = \boldsymbol{\omega}_1 \times \mathbf{r}_1 + \boldsymbol{\omega}_2 \times \mathbf{r}_2,$$
$$\ddot{\mathbf{r}} = \boldsymbol{\omega}_1 \times (\boldsymbol{\omega}_1 \times \mathbf{r}_1) + \boldsymbol{\omega}_2 \times (\boldsymbol{\omega}_2 \times \mathbf{r}_2). \tag{2.15}$$

We must eliminate $\mathbf{r}_1$ and $\mathbf{r}_2$ from this last expression to get a force law as a function of $\mathbf{r}$ and $\dot{\mathbf{r}}$. To do this, note that

$$(\boldsymbol{\omega}_1 + \boldsymbol{\omega}_2) \times \dot{\mathbf{r}} = \boldsymbol{\omega}_1 \times (\boldsymbol{\omega}_1 \times \mathbf{r}_1) + \boldsymbol{\omega}_2 \times (\boldsymbol{\omega}_2 \times \mathbf{r}_2) - \boldsymbol{\omega}_1 \cdot \boldsymbol{\omega}_2 \mathbf{r} + \boldsymbol{\omega}_1 \boldsymbol{\omega}_2 \cdot \mathbf{r}_1 + \\ + \boldsymbol{\omega}_2 \boldsymbol{\omega}_1 \cdot \mathbf{r}_2 .$$

But the last two terms vanish when we apply the condition that $\mathbf{r}_1$ and $\mathbf{r}_2$ be orthogonal to a common axis of rotation along $\boldsymbol{\omega}_1$ and $\boldsymbol{\omega}_2$. Consequently we write (2.15) in the form

$$\ddot{\mathbf{r}} = \boldsymbol{\omega} \times \dot{\mathbf{r}} + k\mathbf{r}, \tag{2.16}$$

where the vector

$$\boldsymbol{\omega} = \boldsymbol{\omega}_1 + \boldsymbol{\omega}_2 \tag{2.17a}$$

and the scalar

$$k = \boldsymbol{\omega}_1 \cdot \boldsymbol{\omega}_2 \tag{2.17b}$$

can be specified independently.

The force law expressed by (2.16) does indeed arise in physical applications, though not from gravitational forces. For $k < 0$, (2.16) will be recognized as the equation for an isotropic oscillator in a magnetic field, encountered before in Exercise (3–8.5). Equation (2.16) with $k > 0$ describes the motion of electrons in the magnetron, a device for generating microwave radiation.

4-2. Exercises

(2.1) Carry out the integration in Equation (2.8) to determine the area of an ellipse

(2.2) To relate the constants in the two different Equations (2.2) and (2.7) for an ellipse (See Figure 2–6.14a),
(a) evaluate r at the points $\mathbf{x} = \pm\mathbf{a}$ to show that

$$\ell = a(1 - \varepsilon^2),$$

(b) and that $a\varepsilon = |\,\mathbf{a}\,|\,|\,\boldsymbol{\varepsilon}\,|$ is the distance from the center of the ellipse to the foci;
(c) evaluate r at $\mathbf{x} = \mathbf{b}$ to show that $b^2 = (1 - \varepsilon^2)a^2 = a\ell$.

(2.3) Show that for a circular orbit $\mathbf{v} = r\dot{\hat{\mathbf{r}}}$, and that for circular motion under a force $\mathbf{f}$,

$$\mathbf{f}\hat{\mathbf{r}} = \mathbf{f}\cdot\hat{\mathbf{r}} = -\frac{mv^2}{r}.$$

Show further, that if Kepler's third law is satisfied, then $\mathbf{f}$ must be a central attractive force varying inversely with the square of the distance. An argument like this led Robert Hooke and others to suspect an inverse square gravitational force before Newton; however, they were unable to generalize the argument to elliptical motion and account for Kepler's first two laws.

(2.4) To establish the universality of his law of gravitation, Newton had to relate the laws of falling objects on the surface of the Earth to the laws of planetary motion. He was able to accomplish this after establishing the theorem that the gravitational force exerted by a spherically symmetric planet is equivalent to the force that would be exerted if all its mass were concentrated at its center. It follows, then, that the readily measured gravitational acceleration at the surface of the Earth is related to the mass M and radius R of the Earth by

$$\frac{GM}{R^2} = g = 9.80\,\frac{\text{m}}{\text{sec}^2}.$$

Even without taking into account the oblateness and rotation of the earth, this value for g is accurate to better than one percent. Newton also knew that the radius of the Moon's orbit is about 60 times the radius of the Earth, as the Greek's had established by a geometrical analysis of the lunar eclipse. And he possessed a fairly good value for the radius of the Earth:

$$R = 6.40 \times 10^6\ \text{m}.$$

Use these facts to calculate the period of the Moon's orbit and compare the result with the observed value of 27.32 days.

(2.5) Find the period and velocity of an object in a circular orbit just skimming the surface of the Earth.

(2.6) Find the height above the earth of a "synchronous orbit", circling the Earth in 24 hours. Such orbits are useful for "communications satellites." How many such satellites would be needed so that every point on the equator is in view of at least one of them?

(2.7) Estimate the Sun-Earth mass ratio from the length of the year and the lunar month (27.3 days), and the mean radii of the Earth's orbit (1.49×10^8 km) and the moon's orbit (3.80×10^5 km).

(2.8) Solve Equations (2.17a, b) subject to (2.14) for $\boldsymbol{\omega}_1$ and $\boldsymbol{\omega}_2$ in terms of $\boldsymbol{\omega}$ and k. Show thereby that every epitrochoid and every hypertrochoid can be regarded as an ellipse precessing (i.e. rotating) with angular velocity $\frac{1}{2}\boldsymbol{\omega}$, and determine those values of ω and k for which such motions are impossible.

(2.9) What central force will admit circular orbits passing through the center of force? What is the value of $\mathbf{L}$ at $\mathbf{r} = 0$?
Hint: Show that $r = 2\mathbf{a}\cdot\hat{\mathbf{r}}$ is an appropriate equation for such an orbit.

(2.10) Show that the central force under which a particle describes the cardioid

$$r = a(1 + \hat{\mathbf{a}}\cdot\hat{\mathbf{r}})$$

is

$$\mathbf{f} = -\frac{3aL^2}{mr^4}\hat{\mathbf{r}}.$$

(2.11) Show that the central force under which a particle describes the lemniscate

$$r^2 = \langle(\mathbf{a}\hat{\mathbf{r}})^2\rangle_0$$

is

$$f = -\frac{3a^4L^2}{mr^7} .$$

4-3. The Kepler Problem

The problem of describing the motion of a particle subject to a central force varying inversely with the square of the distance from the center of force is commonly referred to as the *Kepler Problem*. It is the beginning for investigations in atomic theory as well as celestial mechanics, so it deserves to be studied in great detail. Basically, the problem is to solve the equation of motion

$$m\ddot{\mathbf{r}} = m\dot{\mathbf{v}} = -\frac{k}{r^2}\hat{\mathbf{r}}. \tag{3.1}$$

The "*coupling constant*" k depends on the kind of force and describes the strength of interaction. As we saw in the last section, in celestial mechanics the force in (3.1) is *Newton's law of gravitation* and $k = GmM$. In atomic theory, Equation (3.1) is used to describe the motion of a particle with charge q in the electric field of a particle with charge q'; then $k = -qq'$, and the force is known as *Coulomb's Law*. The Newtonian force is always attractive $(k > 0)$, whereas the Coulomb force may be either attractive or repulsive $(k < 0)$.

There are a number of ways to solve Equation (3.1), but the most powerful and insightful method is to determine its constants of motion. We have already seen that angular momentum is conserved by any central force, so we can immediately write down the constant of motion

$$\mathbf{L} = m\mathbf{r} \wedge \mathbf{v} = mr^2\hat{\mathbf{r}}\dot{\hat{\mathbf{r}}}. \tag{3.2}$$

When $\mathbf{L} \neq 0$, we can use this to eliminate $\hat{\mathbf{r}}$ from (3.1) as follows:

$$\mathbf{L}\dot{\mathbf{v}} = -\frac{k\mathbf{L}}{mr^2}\hat{\mathbf{r}} = k\dot{\hat{\mathbf{r}}}.$$

Since $\mathbf{L}$ is constant, this can be written

$$\frac{\mathrm{d}}{\mathrm{d}t}(\mathbf{L}\mathbf{v} - k\hat{\mathbf{r}}) = 0.$$

Therefore, we can write

$$\mathbf{L}\mathbf{v} = k(\hat{\mathbf{r}} + \boldsymbol{\varepsilon}), \tag{3.3}$$

where $\boldsymbol{\varepsilon}$ is a dimensionless constant vector. It should be evident that this new vector constant of motion $\boldsymbol{\varepsilon}$ is peculiar to the inverse square law, distinguishing it from all other central forces. This constant of motion is called the *Laplace vector* by astronomers, since Laplace was the first of many to discover it. It is sometimes referred to as the "Runge-Lenz vector" in the physics literature. We shall prefer the descriptive name *eccentricity vector* suggested by Hamilton.

Since $\mathbf{L} = i\mathbf{l}$, Equation (3.3) can be expressed in terms of the angular momentum vector $\mathbf{l}$, with the result

$$\mathbf{v} \times \mathbf{l} = k(\hat{\mathbf{r}} + \boldsymbol{\varepsilon}),$$

along with the condition $\mathbf{l}\cdot\mathbf{v} = 0$. However, Equation (3.3) is much easier to manipulate, because the geometric product $\mathbf{L}\mathbf{v}$ is associative while the cross product $\mathbf{l} \times \mathbf{v}$ is not.

Energy and Eccentricity

Besides $\mathbf{L}$ and $\boldsymbol{\varepsilon}$, Equation (3.1) conserves energy, for

$$\nabla \frac{k}{r} = \left(\partial_r \frac{k}{r} \right) \nabla r = -\frac{k}{r^2}\, \hat{\mathbf{r}},$$

so the force is conservative with potential $-kr^{-1}$. It follows, then, from our general considerations in Section 3-10, that the energy

$$E = \tfrac{1}{2} m v^2 - \frac{k}{r}\,, \tag{3.4}$$

is a constant of the motion. However, if $\mathbf{L} \neq 0$, this is not a new constant, because E is determined by $\mathbf{L}$ and $\boldsymbol{\varepsilon}$. Thus, from (3.3) we have

$$k^2\boldsymbol{\varepsilon}^2 = (\mathbf{L}\mathbf{v} - k\hat{\mathbf{r}})^2 = L^2 v^2 - k(\mathbf{L}\mathbf{v}\hat{\mathbf{r}} + \hat{\mathbf{r}}\mathbf{L}\mathbf{v}) + k^2 .$$

But, by (3.2)

$$(\mathbf{L}\mathbf{v}\hat{\mathbf{r}} + \hat{\mathbf{r}}\mathbf{L}\mathbf{v}) = 2\mathbf{L}\mathbf{v}\wedge\hat{\mathbf{r}} = \frac{2\mathbf{L}\mathbf{L}^\dagger}{mr} = \frac{2L^2}{mr}\,.$$

Hence,

$$k^2(\varepsilon^2 - 1) = L^2 \left(v^2 - \frac{2k}{mr} \right).$$

The last factor in this equation must be constant, because the other factors are constant. Indeed, using (3.4) to express this factor in terms of energy, we get the relation

$$\varepsilon^2 - 1 = \frac{2L^2 E}{mk^2}\,. \tag{3.5}$$

When $\mathbf{L} = 0$, this relation tells us nothing about energy, and our derivation of the energy equation (3.4) from (3.2) and (3.3) fails. But we know from our previous derivation that energy conservation holds nevertheless. Indeed, for $\mathbf{L} = 0$, Equation (3.2) implies that the orbit lies on a straight line through the origin, and the energy equation (3.4) must be used instead of (3.3) to describe motion along that line.

The Orbit

For $\mathrm{L} \neq 0$, the algebraic equations (3.2) and (3.3) for the constants $\mathbf{L}$ and $\boldsymbol{\varepsilon}$ determine the orbit and all its geometrical properties without further integrations. This is to be expected, because we know that the general solution of (3.1) is determined by two independent vector constants. Now, to find an equation for the orbit, we use (3.2) to eliminate $\mathbf{v}$ from (3.3). Thus,

$$k(\boldsymbol{\varepsilon} + \hat{\mathbf{r}})\mathbf{r} = \mathbf{L}\mathbf{v}\mathbf{r}\,.$$

The scalar part of this equation is

$$k(\boldsymbol{\varepsilon}\cdot\mathbf{r} + r) = \mathbf{L}\mathbf{v}\wedge\mathbf{r} = \frac{\mathbf{L}\mathbf{L}^\dagger}{m} = \frac{L^2}{m}$$

For L ≠ 0, this yields,

$$r = \frac{\pm \ell}{1 + \boldsymbol{\varepsilon}\cdot\hat{\mathbf{r}}}\ , \tag{3.6a}$$

where,

$$\ell = \frac{L^2}{m\,|\,k\,|}\ , \tag{3.6b}$$

so the sign in (3.6a) distinguishes between attractive and repulsive forces. As we have noted in Section 2-6, Equation (3.6a) describes a conic with *eccentricity* $\varepsilon = |\,\boldsymbol{\varepsilon}\,|$ and an axis of symmetry with direction $\hat{\boldsymbol{\varepsilon}}$. The relation (3.5) enables us to classify the various orbits according to values either of the geometrical parameter ε or the physical parameter E, as shown in Table 3.1.

TABLE 3.1. Classification of Orbits with $L \neq 0$.

Geometrical name	Eccentricity	Energy	Hodograph center
Hyperbola	$\varepsilon > 1$	$E > 0$	$u > \frac{\|k\|}{L}$
Parabola	$\varepsilon = 1$	$E = 0$	$u = \frac{\|k\|}{L}$
Ellipse	$0 \leq \varepsilon < 1$	$E < 0$	$u < \frac{\|k\|}{L}$
Circle	$\varepsilon = 0$	$E = -\frac{mk^2}{2L^2}$	$u = 0$

The Hodograph

Since the orbit equation (3.6) is a consequence of the equation for eccentricity conservation (3.3), we should be able to describe the motion directly with (3.3) itself. Indeed, we can interpret (3.3) as a parametric equation $\mathbf{v} = \mathbf{v}(\hat{\mathbf{r}})$ for the velocity as a function of direction by writing it in the form

$$\mathbf{v} = \frac{k}{\mathbf{L}}\,(\boldsymbol{\varepsilon} + \hat{\mathbf{r}}). \tag{3.7}$$

This equation describes a circle of radius k/L centered at the point

$$\mathbf{u} \equiv k\mathbf{L}^{-1}\,\boldsymbol{\varepsilon}. \tag{3.8}$$

In standard non-parametric form, the equation for this circle is

$$(\mathbf{v} - \mathbf{u})^2 = \frac{k^2}{L^2}\ . \tag{3.9}$$

Since the center of the circle is determined by the eccentricity vector in (3.8), the distance $u = |\mathbf{u}|$ of the center from the origin can be used to classify the orbits, as shown in Table 3.1. Thus, the orbit is an ellipse if the origin is inside the circle or a hyperbola if the origin is outside the circle. For an elliptical orbit the hodograph described by (3.7) is a complete circle, as shown in Figure 3.1. The hodograph and the orbit are drawn with common directions in the figure, so for any velocity $\mathbf{v}$ on the hodograph, the position $\mathbf{r}$ on the orbit can be determined, or vice-versa. Of course, the relations between $\mathbf{v}$, $\hat{\mathbf{r}}$ and $\boldsymbol{\varepsilon}$ shown on the figure are expressed algebraically by (3.7).

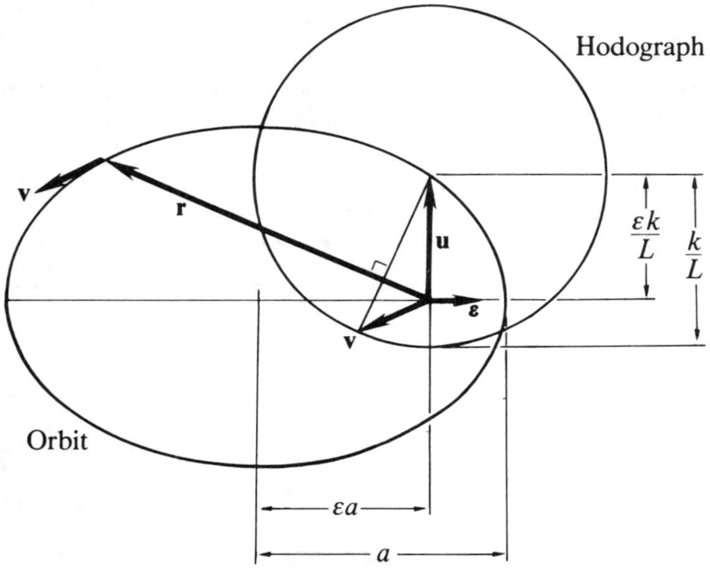

Fig. 3.1. Elliptical orbit and hodograph.

The hodograph for hyperbolic motion is shown in Figure 3.3a, and should be compared with the corresponding orbit in Figure 3.3b. The figure shows that, in this case, the hodograph is only a portion of a circle. This fact cannot be expressed by (3.9), but (3.7) implies that $\mathbf{v}$ is restricted to a circular arc if $\hat{\mathbf{r}}$ is so restricted. Details of the hyperbolic motion will be worked out below.

The Initial Value Problem

We have seen how the orbit and its hodograph are determined by the equations for angular momentum and eccentricity conservation. The same equations can be used to calculate the constants $\mathbf{L}$ and $\boldsymbol{\varepsilon}$ for objects of known mass when the velocity $\mathbf{v}$ is known at one point $\hat{\mathbf{r}}$. This solves the *initial value problem*, a basic problem in celestial mechanics. The main part of the problem is to determine the eccentricity and orientation of the orbit, that is, to determine the eccentricity vector $\boldsymbol{\varepsilon}$. From (3.3) we immediately get

$$\boldsymbol{\varepsilon} = \frac{\mathbf{L}\mathbf{v}}{k} - \hat{\mathbf{r}} = \frac{m}{k}(\mathbf{r}\wedge\mathbf{v})\cdot\mathbf{v} - \hat{\mathbf{r}}. \tag{3.10}$$

This determines $\boldsymbol{\varepsilon}$ from $\mathbf{v}$ and $\hat{\mathbf{r}}$, but comparison with observation will be facilitated if this result is expressed in terms of angles. The angle α between the velocity and the radial direction can be introduced by

$$\hat{\mathbf{r}}\hat{\mathbf{v}} = e^{\mathbf{i}\alpha}. \tag{3.11}$$

Since the bivector $\mathbf{i}$ specifies a specific orientation for the orbital plane, this equation will describe orbits with both orientations if the angle has the range $0 \leqslant \alpha < 2\pi$. Now we can write the first term on the right side of (3.10) in the form

$$\frac{\mathbf{L}\mathbf{v}}{k} = \frac{mv^2r}{k}\,\hat{\mathbf{r}}\wedge\hat{\mathbf{v}}\hat{\mathbf{v}} = \mathbf{i}\hat{\mathbf{v}}\,\lambda\sin\alpha,$$

where

$$\lambda = \frac{mv^2r}{k} = \frac{2K}{-V}, \tag{3.12}$$

with K the kinetic energy and V the potential energy.

Consequently, Equation (3.10) can be written in the form

$$\boldsymbol{\varepsilon} = \mathbf{i}\hat{\mathbf{v}}\,\lambda\sin\alpha - \hat{\mathbf{r}}. \tag{3.13}$$

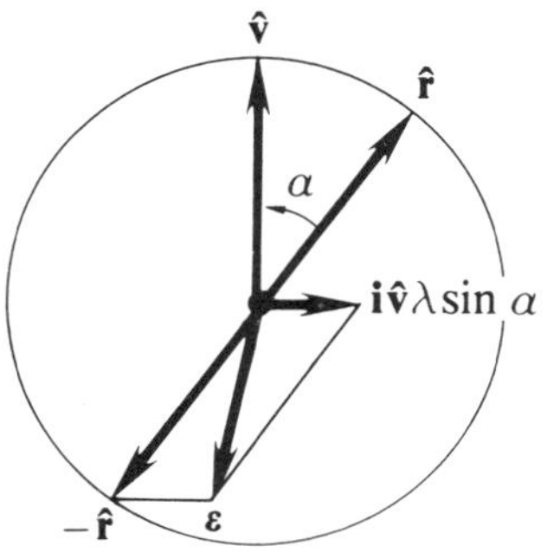

Fig. 3.2. Determination of the eccentricity vector.

The determination of $\boldsymbol{\varepsilon}$ by this equation is illustrated in Figure 3.2* According to (3.12), the parameter λ is determined by the speed and radial distance, or, if you will, the ratio K/V of kinetic to potential energies. If desired, the right side of (3.13) can be expressed in terms of orthogonal components by using (3.11) to eliminate $\hat{\mathbf{v}}$, with the result

$$\boldsymbol{\varepsilon} = \hat{\mathbf{r}}(\lambda\sin^2\alpha - 1) + \mathbf{i}\hat{\mathbf{r}}\,\lambda\sin\alpha\cos\alpha. \tag{3.14}$$

This can be used to compute the angle θ between the radial direction and the direction to the pericenter of the orbit. We write

$$\boldsymbol{\varepsilon}\hat{\mathbf{r}} = \varepsilon e^{\mathbf{i}\theta}. \tag{3.15}$$

then, from (3.14) we obtain

$$\tan\theta = -\frac{\mathbf{i}\boldsymbol{\varepsilon}\wedge\hat{\mathbf{r}}}{\boldsymbol{\varepsilon}\cdot\hat{\mathbf{r}}} = \frac{\lambda\sin\alpha\cos\alpha}{1-\lambda\sin^2\alpha} \tag{3.16}$$

*Graphical methods for constructing orbits based on Equation (3.13) are discussed by W. G. Harter, *Am. J. Phys.* **44**, 348 (1976).

Scattering

Our solution of the initial value problem works for both bounded and unbounded orbits. However, for unbounded motion we are often interested instead in the *scattering problem*, which can be formulated as follows: Given the angular momentum $\mathbf{L}$ and the initial velocity $\mathbf{v}_0$ of a particle approaching the center of force from a great distance, find the final velocity $\mathbf{v}_f$ of the particle receding from the center of force at a great distance. A particle that has thus traversed an unbounded orbit is said to be *scattered*.

The scattering problem for hyperbolic motion can be solved by applying the conservation laws in the *asymptotic region* (defined by the condition that the distance from the center of force be very large). From energy and angular momentum conservation, we have

$$E = \tfrac{1}{2}m\mathbf{v}^2 - \frac{k}{r} \xrightarrow[r \to \infty]{} \tfrac{1}{2}mv_0^2,$$

and

$$\hat{\mathbf{r}} \wedge \mathbf{v} = \frac{\mathbf{L}}{mr} \xrightarrow[r \to \infty]{} 0.$$

The first condition implies that initial and final speeds are equal, that is,

$$v_0 = |\,\mathbf{v}_0\,| = |\,\mathbf{v}_f\,| = \left(\frac{2E}{m}\right)^{1/2} \tag{3.17}$$

The second condition implies that

$$\hat{\mathbf{r}}_0 = -\hat{\mathbf{v}}_0 \quad \text{and} \quad \hat{\mathbf{r}}_f = \hat{\mathbf{v}}_f. \tag{3.18}$$

The Equation (3.3) for eccentricity conservation can be cast in the equivalent form

$$\mathbf{L}\mathbf{v}_2 - k\hat{\mathbf{r}}_2 = \mathbf{L}\mathbf{v}_1 - k\hat{\mathbf{r}}_1, \tag{3.19}$$

where $\mathbf{r}_1$ and $\mathbf{r}_2$ are any two points on the orbit. Using the asymptotic relations (3.17) and (3.18) in this equation, we deduce

$$(\mathbf{L}v_0 - k)\mathbf{v}_f = (\mathbf{L}v_0 + k)\mathbf{v}_0.$$

Hence,

$$\mathbf{v}_f = \left(\frac{\mathbf{L}v_0 + k}{\mathbf{L}v_0 - k}\right)\mathbf{v}_0. \tag{3.20}$$

This solves the scattering problem, because it gives $\mathbf{v}_f$ in terms of the initial velocity and angular momentum. However, for comparison with experiment, it is desirable to express this relation in terms of different parameters.

The angle Θ between the initial and final velocity is called the *scattering angle*; it is defined by the equation

$$\mathbf{v}_f = \mathbf{v}_0 e^{\mathbf{i}\Theta}\,, \tag{3.21}$$

where $0 \leqslant \Theta < \pi$. The bivector $\mathbf{i}$ here is related to the direction of angular momentum by

$$\hat{\mathbf{L}} = \pm\mathbf{i}\,, \tag{3.22a}$$

where the positive sign refers to the attractive case ($k > 0$), and the negative sign refers to the repulsive case ($k < 0$). We have assumed that the angular momentum has opposite orientations in the two cases so we can use (3.21) for both cases, as shown in Figure 3.3a.

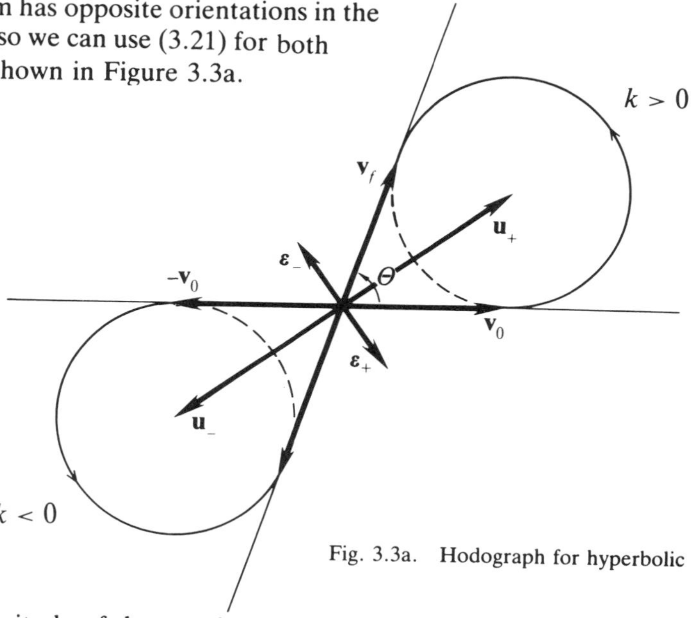

Fig. 3.3a. Hodograph for hyperbolic motion.

The magnitude of the angular momentum can be written in the form

$$L = bmv_0 = \frac{2bE}{v_0}\,, \tag{3.22b}$$

where the so-called *impact parameter b* is the distance of the center of force from the asymptotes of the hyperbola, as shown in Figure 3.3b.

Now, substituting (3.21) and (3.22) into (3.20), we get an expression for the scattering angle as a function of energy and impact parameter. Thus,

$$e^{\mathbf{i}\Theta} = \frac{\mathbf{L}v_0 - k}{\mathbf{L}v_0 + k} = \frac{2Eb\mathbf{i} - |\,k\,|}{2Eb\mathbf{i} + |\,k\,|}\,. \tag{3.23}$$

This equation can be solved for the impact parameter, with the result

$$b = \left|\frac{k}{2E}\right| \cot\tfrac{1}{2}\Theta\,. \tag{3.24}$$

We will use this relation in Section 4-8.

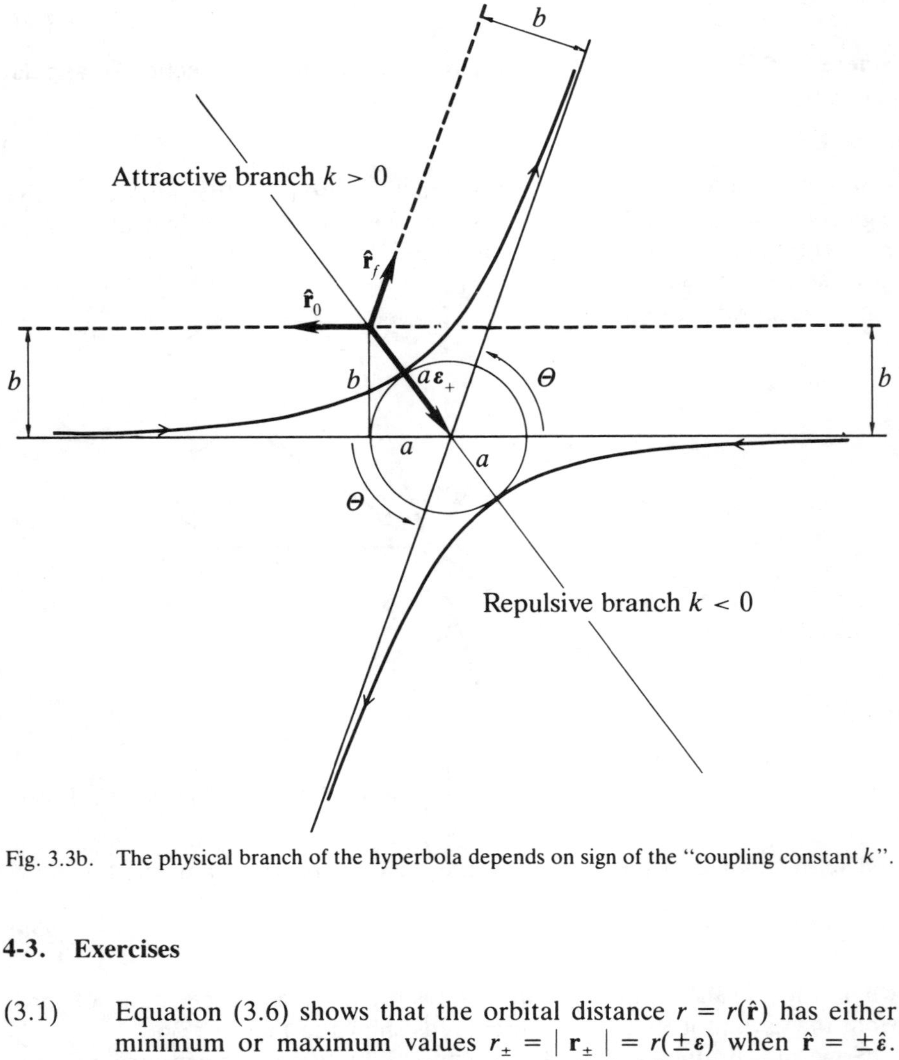

Fig. 3.3b. The physical branch of the hyperbola depends on sign of the "coupling constant k".

4-3. Exercises

(3.1) Equation (3.6) shows that the orbital distance $r = r(\hat{\mathbf{r}})$ has either minimum or maximum values $r_\pm = |\mathbf{r}_\pm| = r(\pm\boldsymbol{\varepsilon})$ when $\hat{\mathbf{r}} = \pm\hat{\boldsymbol{\varepsilon}}$. The *semi-major axis a* is defined as half the distance between the points $\mathbf{r}_+$ and $\mathbf{r}_-$. The *semi-latus rectum* ℓ is defined as the orbital distance when $\mathbf{r}\cdot\boldsymbol{\varepsilon} = 0$ in the attractive case or $\mathbf{r}\cdot\boldsymbol{\varepsilon} = -2$ in the repulsive case. Verify the following relations:

For an ellipse

$$r_\pm = \frac{\ell}{1 \pm \varepsilon} = a(1 \mp \varepsilon)\,,$$

$$a = \tfrac{1}{2}(r_+ + r_-)\,,$$

$$\frac{1}{\ell} = \frac{1}{2}\left(\frac{1}{r_+} + \frac{1}{r_-}\right),$$

$$\varepsilon = \frac{r_- - r_+}{r_- + r_+},$$

$$\frac{v_+}{v_-} = \frac{1+\varepsilon}{1-\varepsilon} \quad \text{where } v_\pm \equiv |\, \mathbf{v}(\pm\hat{\boldsymbol{\varepsilon}})\,| .$$

These relations are of interest in astronomy, where the point of closest approach $\mathbf{r}_+$ is called the *pericenter*, or
the *perihelion* (for an orbit about the Sun),
the *perigee* (for an orbit about the Earth),
the *periastron* (for an orbit about a star).
The point of greatest distance $\mathbf{r}_-$ is called the *apocenter*, or *aphelion* (Sun), *apogee* (Earth), *apastron* (star).
For a hyperbola,

$$2a = r_- - r_+$$

is the distance between the two branches of a hyperbola. For hyperbolic orbits, formulas involving both r_+ and r_- are of no physical interest, because motions along the two branches are not related.

For both elliptical and hyperbolic orbits, the geometrical and physical parameters are related by

$$a = \left|\frac{k}{2E}\right|,$$

$$\ell = \left|\frac{L^2}{mk}\right| = a\,|\,1-\varepsilon^2\,|,$$

(3.2) The *turning points* of an orbit are defined by the condition $\mathbf{v}\cdot\mathbf{r} = 0$. Show that, for a turning point, Equation (3.3) gives the relation

$$\mathbf{r} = \frac{k}{2E}(\boldsymbol{\varepsilon} - \hat{\mathbf{r}}).$$

Verify that, for both elliptic and hyperbolic orbits, this relation gives the points $\mathbf{r}_+$ and $\mathbf{r}_-$ specified in Exercise (3.1)

(3.3) Show that the eccentricity vectors for the attractive and repulsive hyperbolic branches are given in terms of asymptotic initial conditions by

$$\boldsymbol{\varepsilon}_\pm = \left(\frac{2Eb\mathbf{i}}{|\,k\cdot|} + 1\right)\hat{\mathbf{v}}_0$$

whence $\boldsymbol{\varepsilon}_- = -\boldsymbol{\varepsilon}_+$ because initial velocities are opposite on the two branches.

(3.4) Show that orbital distance can be expressed as a function of velocity by

$$r = r(\mathbf{v}) = \frac{-k}{2E - m\mathbf{u}\cdot\mathbf{v}} .$$

For elliptical motion, compare with the orbit Equation (3.6) at points where $\mathbf{v}\cdot\boldsymbol{\varepsilon} = 0$ and $\mathbf{v}\wedge\boldsymbol{\varepsilon} = 0$.

(3.5) *Escape velocity*. Estimate the minimum initial velocity required for an object to escape from the surface of the Earth.

(3.6) *Orbital Transfer*. It is desired to transfer a spaceship from an orbit about Earth to an orbit about Mars. Estimate the minimum launch velocity required to make passage on an elliptical orbit in the sun's gravitational field. Neglect the gravitational fields of the planets, and assume that their orbits are circular. The orbit should be designed to take advantage of the motion of both planets. Estimate the time of passage, and so determine the relative position of the planets at launching.
Data:

$$r_{\text{Earth}} = 1 \text{ AU} \equiv 150 \times 10^6 \text{ km} = 93 \times 10^6 \text{ miles}$$

$$r_{\text{Mars}} = 1.5 \text{ AU}$$

$$GM_{\text{Sun}} = 1.3 \times 10^{20} \text{ m}^3 \text{ s}^{-2} .$$

(3.7) Halley's comet moves in an orbit with an eccentricity of 0.97 and a period of about 76 years. Determine the distance of its perihelion and aphelion from the Sun in units of the Earth's radius.

(3.8) *Impulsive change in orbit*. An impulsive force, such as the firing of a rocket, will produce a change $\Delta\mathbf{v}$ in the velocity of a satellite without a significant change in position during a short time interval. Show that, to first order, the resulting change in the eccentricity vector of the satellite's orbit is given by

$$\Delta\boldsymbol{\varepsilon} = \frac{(\Delta\mathbf{L})}{k}\cdot\mathbf{v} + \frac{\mathbf{L}}{k}\cdot\Delta\mathbf{v},$$

where $\Delta\mathbf{L} = m\mathbf{r}\wedge\Delta\mathbf{v}$. Use this to determine qualitatively the effect of a radial impulse on a circular orbit (See Figure 3.4). Draw figures to show the effect of a tangential impulse. What is the effect of an impulse perpendicular to the orbital plane?

(3.9) A satellite circles the Earth in an orbit of radius equal to twice the radius of the Earth. The direction of motion is changed impulsively

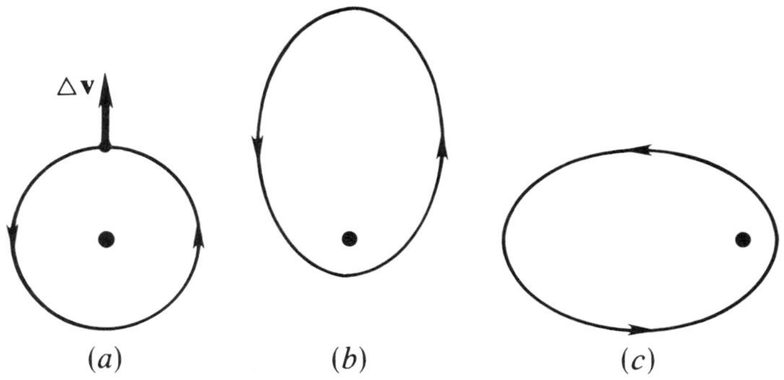

Fig. 3.4. Will the radial impulse on orbit (a) produce (b) or (c)?

through an angle δ towards the Earth. Determine δ so the orbit just skims the Earth's surface.

(3.10) *Ballistic Trajectory*. Neglecting air drag and the like, the trajectory of a ballistic missile is a segment of an ellipse beginning and terminating on the surface of the Earth, as in Figure 3.5. Show that the missile's range βR_E is determined by the formula

$$\tan \tfrac{1}{2}\beta = \frac{\sin\alpha\cos\alpha}{gR_E/v_0^2 - \sin^2\alpha}$$

where α is the firing angle measured from the vertical, as in the figure. Determine the maximum height of the missile above the Earth on a given orbit. Determine the firing angle α_0 that gives maximum range for $0 < \beta < \pi$ and given initial speed v_0.

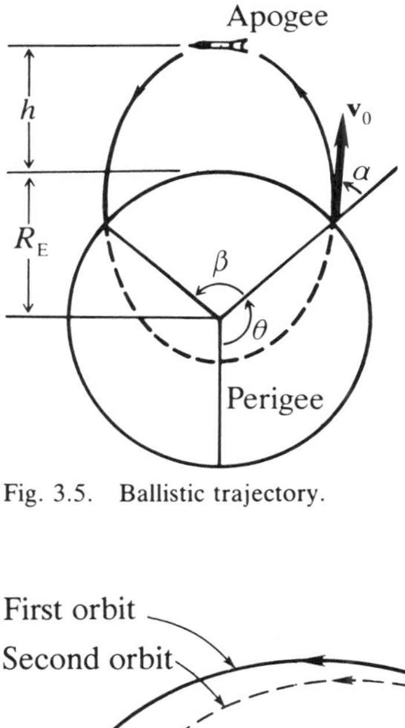

Fig. 3.5. Ballistic trajectory.

(3.11) Atmospheric drag tends to reduce the orbit of a satellite to a circle, as shown in (Figure 3.6). For a rough estimate of this effect, suppose that the net effect of the atmosphere is a small impulse at perigee which reduces the satellite

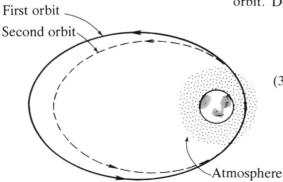

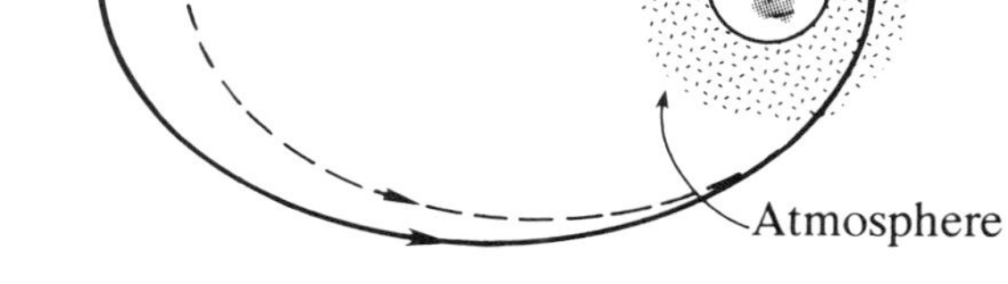

Fig. 3.6. Atmospheric drag on a satellite circularizes its orbit.

velocity by a factor α. Show that the resulting change in eccentricity is

$$\Delta \boldsymbol{\varepsilon} = -2\alpha(\varepsilon + 1)\hat{\boldsymbol{\varepsilon}}.$$

For $\varepsilon = 0.9$ and $\alpha = 0.01$, estimate the number of orbits required to reduce the orbit to a circle. Show that the speed at apogee actually increases with each orbit.

4-4. The Orbit in Time

Although we have learned how to determine elliptic and hyperbolic orbits from arbitrary initial conditions, the Kepler Problem will not be completely solved until we can describe how a particle moves along the orbit in time. In astronomy it is not enough to determine the size and shape of a planetary orbit; it is necessary to be able to locate the planet on the orbit at any specified time. Kepler himself solved this problem in an ingenious way, and we cannot do better than simplify his argument a little using the modern algebraic apparatus at our disposal.

Kepler's Equation

In our study of the harmonic oscillator we saw that elliptical orbits can be described by the parametric equation

$$\mathbf{x} = \mathbf{a}\cos\phi + \mathbf{b}\sin\phi. \tag{4.1}$$

This is related to the radius vector $\mathbf{r}$ from a focus of the ellipse by

$$\mathbf{r} = \mathbf{x} - a\boldsymbol{\varepsilon} = \mathbf{x} - \varepsilon\mathbf{a}. \tag{4.2}$$

Hence,

$$\mathbf{r} = \mathbf{a}(\cos\phi - \varepsilon) + \mathbf{b}\sin\phi \tag{4.3}$$

is a parametric equation $\mathbf{r} = \mathbf{r}(\phi)$ for the orbit. Kepler introduced the angle ϕ into his description of an ellipse by a geometrical construction involving an auxilliary circle, as shown in Figure 4.1.

If, now, we can determine the parameter ϕ as a function of time $\phi = \phi(t)$, then (4.3) gives us the desired function $\mathbf{r} = \mathbf{r}(t)$ at once. From (4.2), we obtain

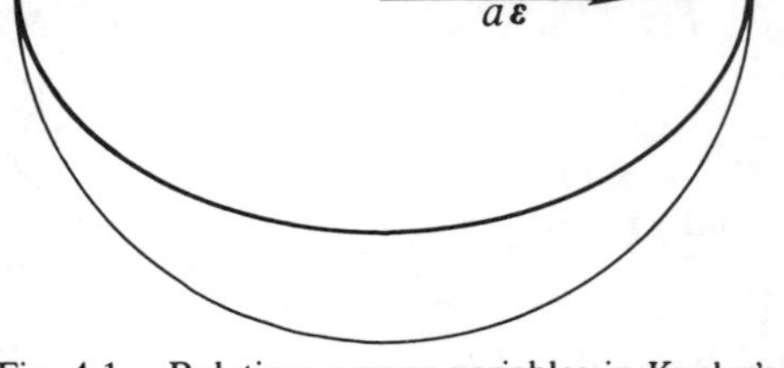

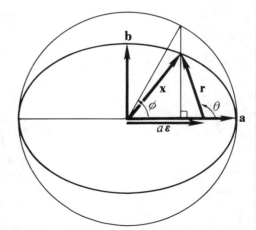

Fig. 4.1. Relations among variables in Kepler's problem.

$$\mathbf{L} = m\mathbf{r}\wedge\dot{\mathbf{r}} = m\mathbf{x}\wedge\dot{\mathbf{x}} - m\varepsilon\mathbf{a}\wedge\dot{\mathbf{x}}.$$

This can be integrated with respect to time using (4.1) to get a relation between t and ϕ. Associating the zero of time with the position at pericenter, we obtain

$$\frac{\mathbf{L}t}{m} = \int_{\mathbf{a}}^{\mathbf{x}} \mathbf{x}\wedge d\mathbf{x} - \varepsilon\mathbf{a}\wedge(\mathbf{x} - \mathbf{a})$$

$$= \mathbf{ab}\phi - \varepsilon\mathbf{ab}\sin\phi.$$

But in Section 4-2 we saw that

$$\frac{\mathbf{L}}{m} = \frac{2\pi\mathbf{ab}}{T},$$

where T is the orbital period. Hence,

$$\frac{2\pi t}{T} = \phi - \varepsilon \sin\phi. \tag{4.4}$$

This is known as *Kepler's equation* for planetary motion.

Solutions of Kepler's Equation

Kepler's equation gives time t as a function of ϕ, so it must be solved for ϕ as a function of t. Unfortunately, the solution cannot be expressed in terms of standard functions, so we must solve the equation by some approximation method. Newton devised a mechanical method which is easy to visualize. He noticed that Kepler's equation can be solved by projection from a trochoid, the curve traced out by a point on a rolling wheel (See Figure 3–7.3b). As shown in Figure 4.2, one simply marks a point a distance ε below the center of a wheel of unit radius. Then the solution is generated by rolling the wheel until that point has moved a horizontal distance $M = 2\pi t/T$; the value of ϕ at the time t is then the measured distance that the wheel has moved.

Although Newton's mechanical method enables us to visualize the solution to Kepler's equation, a numerical solution to any desired accuracy is easily obtained with a computer. There are many ways to do this, but the simplest when ε is small is to treat the last term in (4.4) as a perturbation. Thus, the *zeroth order* approximation to (4.4) is

$$\phi = M \equiv \frac{2\pi t}{T}.$$

Substituting this into the "perturbing term" in (4.4), we get the *first order* approximation

$$\phi = M + \varepsilon \sin M.$$

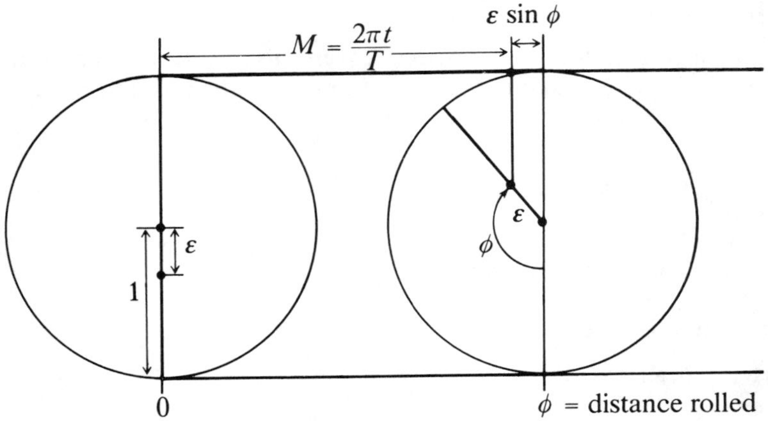

Fig. 4.2. Mechanical solution of Kepler's equation.

The *second order* approximation is, then,

$$\begin{aligned}\phi &= M + \varepsilon \sin(M + \varepsilon \sin M) \\ &= M + \varepsilon \sin M \cos(\varepsilon \sin M) + \varepsilon \cos M \sin(\varepsilon \sin M),\end{aligned}$$

or, since ε is small,

$$\phi = M + \varepsilon \sin M + \frac{\varepsilon^2}{2} \sin 2M. \tag{4.5}$$

This approximate solution would have been quite sufficient for Kepler's observations of planetary motion.

The angle θ defined by

$$\hat{\mathbf{r}} = \hat{\boldsymbol{\varepsilon}} e^{\mathbf{i}\theta} \tag{4.6}$$

has a more direct observational significance than the angle ϕ, so we can make best use of our solution to Kepler's equation by expressing θ in terms of $\phi = \phi(t)$ and so get $\theta = \theta(t)$. To do this, we first square (4.3) and, using the relation $b^2 = a^2(1 - \varepsilon^2)$ derived in Exercise (2.2), we find that

$$r = a(1 - \varepsilon \cos \phi). \tag{4.7}$$

Using this after substituting (4.6) into (4.3) we find

$$e^{\mathbf{i}\theta} = \frac{\cos \phi - \varepsilon + \mathbf{i}(1 - \varepsilon^2)^{1/2} \sin \phi}{1 - \varepsilon \cos \phi}. \tag{4.8}$$

A more convenient relation between θ and ϕ can be found as follows: The scalar part of (4.8) is

$$\cos \theta = \frac{\cos \phi - \varepsilon}{1 - \varepsilon \cos \phi},$$

whence

$$\frac{1 - \cos\theta}{1 + \cos\theta} = \left(\frac{1 + \varepsilon}{1 - \varepsilon}\right) \frac{1 - \cos\phi}{1 + \cos\phi} ,$$

or, using the half angle formula for the tangent,

$$\tan \tfrac{1}{2}\theta = \left(\frac{1 + \varepsilon}{1 - \varepsilon}\right)^{1/2} \tan \tfrac{1}{2}\phi. \tag{4.9}$$

This given $\theta(t)$ uniquely in terms of $\phi(t)$, because both half angles always lie in the same quadrant.

4-4. Exercises

(4.1) Show that Kepler's equation can be put in the form

$$t\left(\frac{k}{ma^3}\right)^{1/2} = \phi - \varepsilon \sin \phi.$$

Derive the corresponding equation for hyperbolic motion (with $k > 0$):

$$t\left(\frac{k}{ma^3}\right)^{1/2} = \phi - \varepsilon \sinh \phi.$$

Use the parametric equation for a hyperbola

$$\mathbf{x} = \mathbf{a} \cosh \phi + \mathbf{b} \sinh \phi.$$

4-5. Conservative Central Forces

The method used in Section 4-3 to analyze the motion of a particle subject to an inverse square law of force will not work for arbitrary central forces owing to the absence of a simple constant of motion like the eccentricity vector. We turn, therefore, to a more general method which exploits the constants of motion that are available.

We have already determined in Section 4-1 that the angular momentum

$$\mathbf{L} = m\mathbf{r}\wedge\dot{\mathbf{r}} = mr^2\hat{\mathbf{r}}\dot{\hat{\mathbf{r}}} \tag{5.1}$$

is a constant of motion in a central field of force. And we know from Section 3-10 that for a conservative force the energy

$$E = \tfrac{1}{2}m\dot{\mathbf{r}}^2 + V \tag{5.2}$$

is a constant of motion. When the potential V is specified, the orbit can be found from the Equations (5.1) and (5.2) without referring to the equation of motion

$$m\ddot{\mathbf{r}} = -\nabla V = -\hat{\mathbf{r}}\partial_r V(r) \tag{5.3}$$

from whence the constants of motion are derived.

The potential of a central conservative force is called a *central potential*. The potential in (5.3) has been written in the special form $V(r) = V(|\,\mathbf{r}\,|)$ instead of the general form $V(\mathbf{r}) = V(r, \hat{\mathbf{r}})$ to indicate that it must be independent of the directional variable $\hat{\mathbf{r}}$. This follows from the requirement that variations in V expressed by ∇V be in the radial direction $\hat{\mathbf{r}} = \nabla r$, whereas $\hat{\mathbf{r}}$ can vary only in directions orthogonal to itself. Thus a *central potential is necessarily spherically symmetric*.

To determine characteristics of the motion common to all central potentials, we endeavor to carry the solution of the equation of motion as far as possible without assuming a specific functional form for the potential V. Equation (5.1) suggests that we should separate the radial variable $r = |\,\mathbf{r}\,|$ from the directional variable $\hat{\mathbf{r}}$. As we have already observed in Section 4-1, from (5.1) it follows that

$$\dot{\mathbf{r}} = \hat{\mathbf{r}}\left(\dot{r} + \frac{\mathbf{L}}{mr}\right) = \left(\dot{r} - \frac{\mathbf{L}}{mr}\right)\hat{\mathbf{r}},$$

whence

$$\dot{\mathbf{r}}^2 = \left(\dot{r} - \frac{\mathbf{L}}{mr}\right)\left(\dot{r} + \frac{\mathbf{L}}{mr}\right) = \dot{r}^2 + \frac{L^2}{m^2r^2}, \tag{5.4}$$

where $L^2 = |\,\mathbf{L}\,|^2 = -\mathbf{L}^2$. Substituting (5.4) into the energy function (5.2), we obtain the *radial energy equation*

$$\tfrac{1}{2}m\dot{r}^2 + \frac{L^2}{2mr^2} + V(r) = E. \tag{5.5}$$

This is identical to the energy equation for motion of a particle along a line in an *effective potential*

$$U(r) = \frac{L^2}{2mr^2} + V(r) \tag{5.6}$$

with the restriction $r > 0$. Thus, the 3-dimensional central force problem has been reduced to an equivalent 1-dimensional problem.

Putting (5.5) in the form

$$\dot{r}^2 = \frac{2}{m}(E - U(r)), \tag{5.7}$$

we see that the variables r and t are separable after taking the square root, so we can integrate immediately to get

$$t - t_0 = \int_{r_0}^{r} \frac{\mathrm{d}r}{[2(E-U)/m]^{1/2}} \tag{5.8}$$

This integral is not well defined until we have specified the range of values for r. The range can be determined by noting that since $\dot{r}^2 \geqslant 0$, Equation (5.7) implies that the allowed values of r are restricted by the inequality

$$U(r) = \frac{L^2}{2mr^2} + V(r) \leqslant E. \tag{5.9}$$

When $\dot{r} = 0$, the inequality reduces to an equation $U(r) = E$ whose roots are maximum and minimum values of r specifying *turning points* (or *apses*) of the motion. A minimum allowed value $r_{\min}$ always exists, since r must be positive. If there is no maximum allowed value, the motion is *unbounded*. If there is a maximum value $r_{\max}$ the motion is *bounded*, and at the distance $r_{\max}$, the particle will change from retreat to approach towards the center of force. Accordingly, the integral (5.8) can be taken over increasing values of r only to the point $r_{\max}$, from which point it must be taken over decreasing values of r. Of course, the integral can be taken repeatedly over the range from $r_{\min}$ to $r_{\max}$ corresponding to repeated oscillations of the particle. The *period* of a single complete oscillation in the value of r is therefore given by

$$T = 2\int_{r_{\min}}^{r_{\max}} \frac{\mathrm{d}r}{[2(E-U)/m]^{1/2}} \,. \tag{5.10}$$

This defines a period for all bounded central force motion. By considering diagrams for the orbits it can be seen that this definition of period must agree with the usual one for a particle in a Coulomb potential, but it gives only half the usual one for a harmonic oscillator, because it is a period of the radial motion rather than a period of angular motion.

To determine the particle's orbit, we must use (5.1) to relate changes in the direction $\hat{\mathbf{r}}$ to changes in the radial distance r. If we parametrize $\hat{\mathbf{r}}$ as a function of angle θ by writing

$$\hat{\mathbf{r}} = \hat{\boldsymbol{\varepsilon}} e^{i\theta},$$

then, as we have seen in Section 4-1, Equation (5.1) reduces to

$$L = mr^2\dot{\theta}. \tag{5.11}$$

We can use this to make the change of variables

$$\dot{r} = \frac{\mathrm{d}r}{\mathrm{d}\theta}\,\dot{\theta} = \frac{L}{mr^2}\,\frac{\mathrm{d}r}{\mathrm{d}\theta}$$

which, on substitution into (5.7) gives

$$\left(\frac{\mathrm{d}r}{\mathrm{d}\theta}\right)^2 = 2m(E-U)\left(\frac{r^2}{L}\right)^2.$$

Separating variables and integrating, we get an equation for the orbit in the form

$$\theta - \theta_0 = \int_{r_0}^{r} \frac{L\,dr}{r^2[2m(E-U)]^{1/2}}$$

$$= \int_{r_0}^{r} \frac{dr}{r^2\left[\dfrac{2mE}{L^2} - \dfrac{2mV}{L^2} - \dfrac{1}{r^2}\right]^{1/2}} . \qquad (5.12)$$

The integral can be evaluated in terms of known functions only for special potentials, such as those of the form

$$V = \frac{-k}{r^n} , \qquad (5.13)$$

where k is a constant and n is a nonzero integer. For $n = -2, 1, 2$ the integral can be evaluated in terms of inverse trigonometric functions, corresponding to the linear, the inverse square and the inverse cube force laws. For $n = -6, -4, -1, 3, 4, 6$ the integral can be evaluated in terms of elliptic functions (Appendix B). For other values of n the integral cannot be expressed in terms of tabulated functions.

We can ascertain the general features of the orbit without actually evaluating the integral (5.12). When integrated over a period of bounded motion, (5.12) gives

$$2\pi + \Delta\theta = \int_{r_{\text{min}}}^{r_{\text{max}}} \frac{2L\,dr}{r^2[2m(E-U(r))]^{1/2}} , \qquad (5.14)$$

where $\Delta\theta$ is the deviation from an angular period of 2π. As shown in Figure 5.1, $\Delta\theta$ can be regarded as the angular displacement of the apse in one period. For Kepler's elliptical orbits $\Delta\theta = 0$, so the other central force orbits can be regarded as ellipses precessing (not necessarily at a constant rate) through a total angle $\Delta\theta$ in one period.

According to (5.14), the angular precession is determined by integrating over any half period of the motion beginning at an apse. Indeed, from the symmetry of the integral we can infer that the orbit must be symmetric with respect to reflection about each apsidal line. So if we have a segment of the orbit from one apse

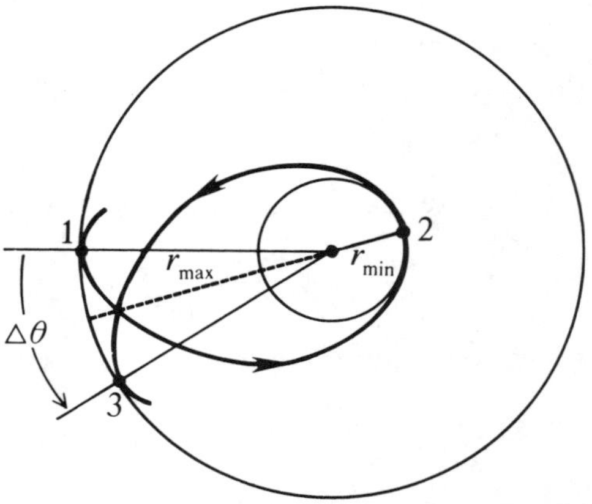

Fig. 5.1. Orbital symmetry of bounded motion in a central potential.

to the next, say from apse 1 to apse 2 in Figure 5.1, then we can get the next segment from apse 2 to apse 3 by reflecting the first segment through apsidal line 2. The segment after that is obtained by reflection through apsidal line 3, and so on. In this way, the entire orbit can be generated from a single segment.

An orbit that eventually repeats itself exactly is said to be closed. In the present case, the condition for a closed orbit is

$$\Delta\theta = 2\pi \frac{m}{n} , \tag{5.15}$$

where n and m are integers. This is the condition that the functions $r = r(t)$ and $\theta = \theta(t)$ have commensurable periods. Examples of closed orbits are shown in Figure 5.2.

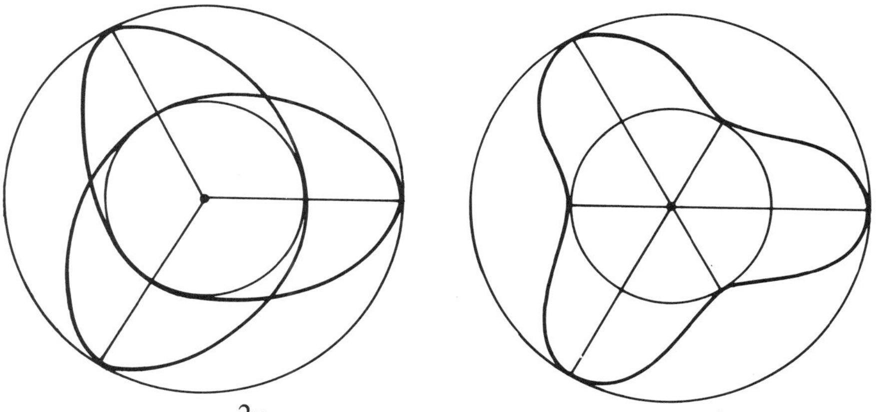

$\Delta\theta = -\frac{2\pi}{3}$ $\Delta\theta = -\frac{4\pi}{3}$

Fig. 5.2. Central force motions with three symmetry axes.

Energy Diagrams

As we have noted before, in atomic physics individual orbits cannot be observed, so it is necessary to characterize the state of motion in terms of conserved quantities. In particular, it is important to know how the general states of bounded and unbounded motion allowed by a particular potential depend on energy and angular momentum. This would also provide a useful classification of allowed orbits. The desired information is contained in the radial energy equation (5.5), which describes the state of motion explicitly in terms of L and E. The information can be put in an easily surveyable form by constructing *energy diagrams*. We learn best how to do this by considering a specific example.

Let us classify the allowed motions in the potential

$$V(r) = \frac{-ke^{-r/a}}{r}, \tag{5.16}$$

where k and a are positive constants. This is called the *screened Coulomb potential* in atomic physics, where the exponential factor $e^{-r/a}$ describes a partial screening (or cancellation) of the nuclear Coulomb potential $-k/r$ by a cloud of electrons surrounding the nucleus. A potential of the same form (5.16) arises also in nuclear physics, where it is called the *Yukawa potential* in honor of the Japanese Nobel Laureate who was the first to use it to describe nuclear interactions.

To interpret the radial energy equation

$$E = \tfrac{1}{2}m\dot{r}^2 + U(r) \tag{5.17}$$

graphically, we need a graph of the effective potential

$$U(r) = -\frac{ke^{-r/a}}{r} + \frac{L^2}{2mr^2} \tag{5.18}$$

The term $L^2/2mr^2$ is called the *centrifugal potential* because of its relation to the centrifugal force discussed in Section 5-6. To graph the function, $U(r)$, we first examine its asymptotic values. For small $r \ll a$, we have

$$\frac{e^{-r/a}}{r} \approx \frac{1}{r} \ll \frac{1}{r^2} \quad \text{so} \quad U \approx \frac{L^2}{2mr^2} > 0,$$

or for large $r \gg$ a, we have

$$\frac{e^{-r/a}}{r} = \frac{1}{re^{r/a}} \ll \frac{1}{r^2} \quad \text{so} \quad U \approx \frac{L^2}{2mr^2} > 0.$$

Thus, when $L \neq 0$ the centrifugal potential $L^2/2mr^2$ dominates the Yukawa potential in the asymptotic regions (where r is very large or very small).

Next we determine the inflection points of the effective potential. To simplify the notation let us write

$$s = r/a \tag{5.19a}$$

and

$$\alpha = \frac{L^2}{2mka} \tag{5.19b}$$

so that

$$U(r) = U(as) = \frac{k}{a}\left(\frac{-e^{-s}}{s} + \frac{\alpha}{s^2}\right).$$

At an extremum we have

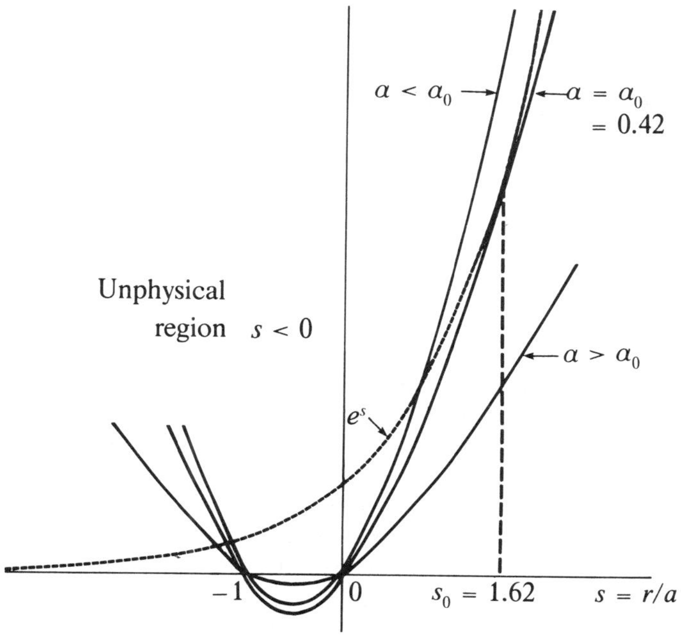

Fig. 5.3. Graphical solutions of a transcendental equation.

$$\partial_r U \propto \left(\frac{1}{s} + \frac{1}{s^2}\right) e^{-s} - \frac{2\alpha}{s^3} = 0,$$

or

$$\frac{s(s+1)}{2\alpha} = e^s. \tag{5.20}$$

This is a transcendental equation for s as a function of α. Its solutions are points of intersection of the exponential curve $y = e^s$ with the parabola $y = s(s+1)/2\alpha$. As graphs of the functions in Figure 5.3 show, there are three distinct possibilities. (a) For a particular constant α_0, the parabola intersects the exponential curve at exactly one point $s_0 > 0$, a point of tangency. In this case, then, $U(r)$ has a single inflection point. (b) For $\alpha > \alpha_0$ the curves do not intersect for positive s, in which case $U(r)$ has no inflection points. (c) For $\alpha < \alpha_0$, the curves intersect and they must intersect exactly twice for $s > 0$, because e^s increases with s faster than any power of s. Therefore, $U(s)$ has two inflection points in this case.

The constants s_0 and α_0 are determined by the requirement that the exponential curve be tangent to the parabola, so by differentiating (5.20) we obtain

$$\frac{2s_0 + 1}{2\alpha_0} = e^{s_0}.$$

Equating this to (5.20) we obtain

$$s_0^2 - s_0 - 1 = 0,$$

which has the single positive solution

$$s_0 = r_0/a = \tfrac{1}{2}(1 + \sqrt{5}) \approx 1.62. \tag{5.21a}$$

Furthermore,

$$\alpha_0 = (s_0 + \tfrac{1}{2})e^{-s_0} \approx 0.42. \tag{5.21b}$$

Note that the values of these constants are independent of the values for the physical constants in the problem. It is amusing to note also that the number $\frac{1}{2}(1 + \sqrt{5})$ is the famous *golden ratio*, to which the ancient Greeks attributed a mystical significance. The golden ratio has many remarkable mathematical properties (among which might be numbered its appearance in this problem), and it appears repeatedly in art and science in a variety of peculiar ways. The Greeks, for example, believed that a perfect rectangle is one whose sides are in proportion to the golden ratio, and this ratio runs throughout their art and architecture, including the Parthenon.

Graphs of the effective potential for the three possible cases are shown in Figure 5.4. The case with two inflection points is most interesting, so let us examine it in detail. A typical graph is shown in Figure 5.5. Allowed orbits are represented in the figure as lines of constant energy in regions where they pass above the effective potential. Characteristics of the various allowed states of motion can be read off the figure. Thus

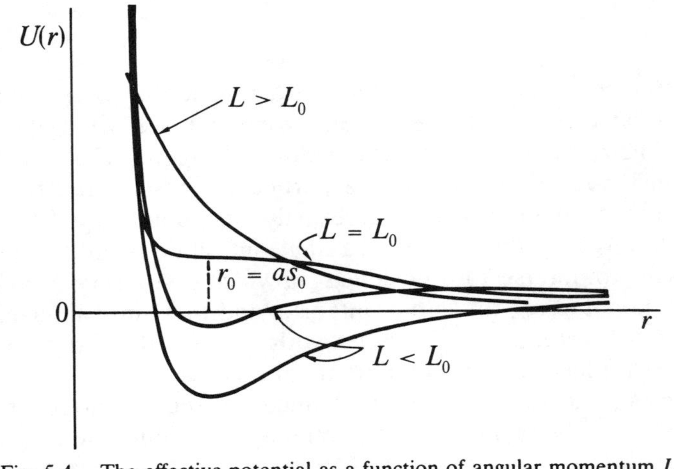

Fig. 5.4. The effective potential as a function of angular momentum L.

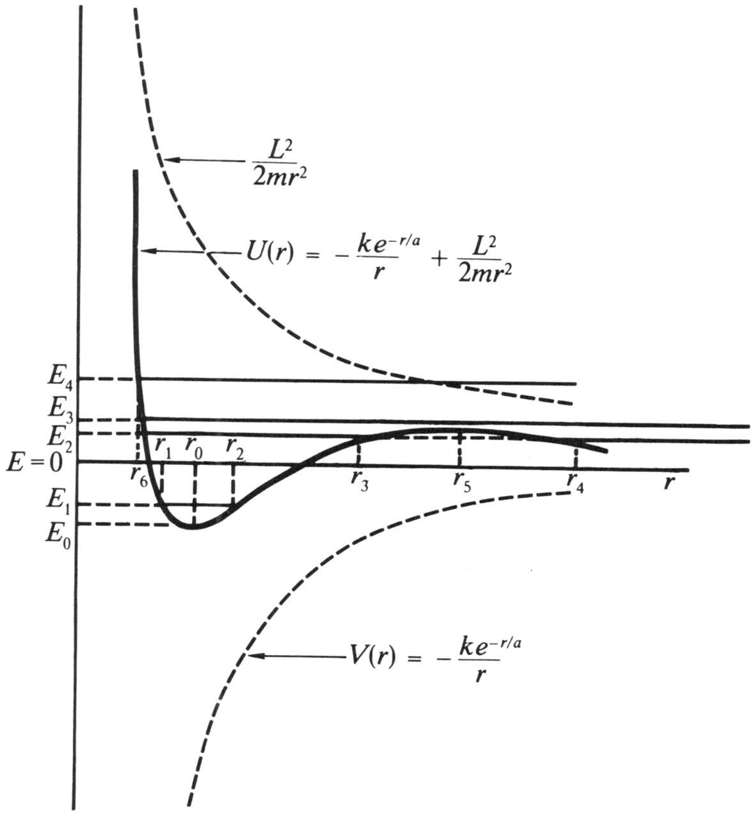

Fig. 5.5. Energy diagram for motion in a screened Coulomb potential.

(a) A particle with energy E_1 will oscillate between the turning points r_1 and r_2 in Figure 5.5. According to the radial energy equation (5.14), the radial kinetic energy at any point is given by $E_1 - U(r)$, with a maximum value at r_0.
(b) A particle with energy E_0 is allowed only at the radius r_0 where it has zero radial kinetic energy, so this state corresponds to a circular orbit. If the particle is given a small impetus raising its radial kinetic energy and its total energy from E_0 to E_1, say, then the radius of its orbit will oscillate without deviating far from r_0. For this reason we say that the circular orbit is stable.
(c) A particle with energy E_2 will be *bound* if its initial radius is less than r_3 and *free* (in an unbounded orbit) if its initial radius is greater than r_4. Note that two possibilities like this can occur only for positive energies. They cannot occur in a Coulomb potential for which, as we have shown in Section 4-3, all bound states have negative energies. Bound states with positive energy have a special significance in quantum theory. In quantum

theory the energy of a particle is subject to fluctuations of short duration, so a particle with energy E_2 has a finite probability of temporarily increasing its energy to more than E_3, enabling it to "jump" the *potential barrier* from r_3 to r_4 and so pass from a bound to a free state with no net change in its energy. Thus, in quantum theory, bound states with positive energy have a finite lifetime.

(d) No particle with energy greater than E_3 in Figure 5.5 has bound states of motion. If a particle with energy E_4, for instance, has an initial negative radial speed, it will approach the origin until it "collides with the potential wall" at r_6, after which it will retreat to infinity.

To summarize, Figure 5.5 shows an effective potential with a potential well of depth E_0. The bound states are composed of particles "trapped in the well" with energy less than E_3. All other states are unbound.

Returning now to Figure 5.4 and recalling (5.19b), we see that there is a critical value of the angular momentum

$$L_0 = (2mka\alpha_0)^{1/2} \tag{5.22}$$

For $L < L_0$, the effective potential has a dip, allowing bound states. For $L > L_0$, the effective potential has no dip so there can be no bound states. For $L = L_0$ there is a single bound state, a circular orbit of radius $r_0 = 1.62a$; however, this orbit is unstable, because a small increase in energy will free the particle completely.

Stability of Circular Orbits

Every attractive central potential admits circular orbits for certain values of the energy and angular momentum. We have seen, however, that stability of a circular orbit against small disturbances depends on the curvature of the effective potential. This insight enables us to ascertain the stability of circular orbits in any given central potential with ease. The question of stability is more than academic, for the only circular orbits we can hope to observe in nature are stable ones, and only motions along stable orbits can be controlled in the laboratory.

Let us investigate the stability of circular orbits in the important class of attractive potentials with the form $V = -k/r^n$. The effective potential is

$$U(r) = -\frac{k}{r^n} + \frac{L^2}{2mr^2}\,. \tag{5.23}$$

Now, for a circular orbit the radius r must be constant so $\dot{r} = 0$, and it must stay that way so $\ddot{r} = 0$. Differentiating the radial energy equation (5.17), we find

$$m\dot{r}\ddot{r} = -\dot{r}\,\partial_r U,$$

or

$$m\ddot{r} = -\partial_r U. \tag{5.24}$$

Hence, *the radii of the possible circular orbits are the extrema of the effective potential*, determined by the condition $\partial_r U = 0$. In Figure 5.5, for example, the inflection point r_5 is the radius of a circular orbit with energy E_3, although the orbit is obviously unstable. Applying the condition for circular orbits to (5.23) we obtain

$$\partial_r U = \frac{nk}{r^{n+1}} - \frac{L^2}{mr^3} = 0.$$

Hence,

$$r^{n-2} = \frac{nkm}{L^2}\ . \tag{5.25}$$

The orbit will be stable if this inflection point is a minimum and unstable if it is a maximum. Differentiating the effective potential once more, we obtain

$$\partial_r^2 U = -\frac{n(n+1)k}{r^{n+2}} + \frac{3L^2}{mr^4}\ ,$$

and using (5.25) we find that

$$\partial_r^2 U = \frac{L^2}{mr^4}(2-n) \tag{5.26}$$

at the inflection point. From (5.26) we can conclude that *the attractive potential $-k/r^n$ admits stable circular orbits if $n < 2$ but not if $n > 2$*. The case $n = 2$ requires further examination (See Exercise 5.3).

4-5. Exercises

Central force problems are also found in Section 4-2.

(5.1) Show that a particle with nonzero angular momentum in a central potential can fall to the origin only if

$$V(r) \xrightarrow[r \to 0]{} \frac{-k}{r^n}\ , \quad \text{where} \quad n \geqslant 2.$$

(5.2) For the cases $n = -2, 1, 2$, integrate (5.12) with the potential (5.13) and invert the result to get an equation for the orbit in the form $r = r(\theta)$. Compare with previously obtained orbits for these cases. Determine the precession angle $\Delta\theta$ for each case.

(5.3) Use energy diagrams to classify orbits in the attractive central potential $-k/r^2$. Can circular orbits be stable in this potential?

(5.4) Determine the necessary conditions for stable circular orbits in the potential

$$V(r) = \frac{-k}{r} + \frac{k'}{r^2} \,.$$

4-6. Two-Particle Systems

Consider a system of two particles with masses m_1, m_2 and positions $\mathbf{x}_1$, $\mathbf{x}_2$ respectively. The equations of motion for the particles can be written in the form

$$m_1\ddot{\mathbf{x}}_1 = \mathbf{f}_{12} + \mathbf{F}_1 \,, \tag{6.1a}$$

$$m_2\ddot{\mathbf{x}}_2 = \mathbf{f}_{21} + \mathbf{F}_2 \,, \tag{6.1b}$$

where $\mathbf{f}_{12}$ is the force exerted on particle 1 by particle 2, and $\mathbf{F}_1$ is the force exerted on particle 1 by agents external to the system, with a similar description of the forces on particle 2. To write (6.1a, b) we have appealed to the superposition principle to separate the *internal forces* exerted by the particles on one another from the *external forces* $\mathbf{F}_1$ and $\mathbf{F}_2$.

Our aim now is to describe the system by distinguishing the *external motion* of the system as a whole from the *internal motions* of its parts. This aim is greatly facilitated by Newton's third law, which holds that the mutual forces of two particles on one another are equal and opposite, that is,

$$\mathbf{f}_{12} = -\mathbf{f}_{21} \,. \tag{6.2}$$

We shall see later that Newton's third law is not universally true in this form. Nevertheless, it is usually an excellent approximation, and we can easily tell if it is violated when the form of the force $\mathbf{f}_{12}$ is specified, so we sacrifice little by adopting it.

Notice now that, because of (6.2), the internal forces cancel when we add the two Equations (6.1a, b). The result can be written in the form

$$M\ddot{\mathbf{X}} = \mathbf{F}_1 + \mathbf{F}_2 \equiv \mathbf{F}, \tag{6.3}$$

where

$$M \equiv m_1 + m_2 \tag{6.4}$$

is said to be the *mass* of the system, and

$$\mathbf{X} \equiv \frac{m_1\mathbf{x}_1 + m_2\mathbf{x}_2}{m_1 + m_2} \tag{6.5}$$

is called its *center of mass*.

Equation (6.3) can be regarded as an equation of motion for the system as a whole. It is like the equation of motion for a single particle with mass M and position $\mathbf{X}$, except that the total external force $\mathbf{F} = \mathbf{F}_1 + \mathbf{F}_2$ generally depends on the structure of the system (i.e. the positions and velocities of its parts). Among the few structurally independent external forces we have the following: If the system is immersed in a uniform force field, then the total external force is obviously constant. Thus, for a constant gravitational field, we have

$$\mathbf{F}_{\text{grav}} = m_1\mathbf{g} + m_2\mathbf{g} = M\mathbf{g}. \tag{6.6}$$

Or, if the system is subject to a linear resistive force proportional to the mass, then the external force is

$$\mathbf{F}_{\text{res}} = -\gamma(m_1\dot{\mathbf{x}}_1 + m_2\dot{\mathbf{x}}_2) = -\gamma M\dot{\mathbf{X}}. \tag{6.7}$$

Finally, for particles with a constant charge to mass ratio $\alpha = q_1/m_1 = q_2/m_2$ in a uniform magnetic field, the external force is

$$\begin{aligned}\mathbf{F}_{\text{mag}} &= \frac{q_1}{c}\dot{\mathbf{x}}_1 \times \mathbf{B} + \frac{q_2}{c}\dot{\mathbf{x}}_2 \times \mathbf{B} = \frac{\alpha}{c}(m_1\dot{\mathbf{x}}_1 + m_2\dot{\mathbf{x}}_2) \times \mathbf{B} \\ &= \frac{\alpha}{c}\, M\dot{\mathbf{X}} \times \mathbf{B} = \frac{Q}{c}\,\dot{\mathbf{X}} \times \mathbf{B},\end{aligned} \tag{6.8}$$

where $Q = q_1 + q_2$ is the total charge of the system. In general, the total force on a system is independent of structure only when it can be expressed as a function of the form $\mathbf{F} = \mathbf{F}(\dot{\mathbf{X}}, \mathbf{X}, t)$, as in the examples just mentioned.

The internal structure can be described in terms of the particle positions $\mathbf{r}_1$ and $\mathbf{r}_2$ with respect to the center of mass; they are given by

$$\mathbf{r}_1 = \mathbf{x}_1 - \mathbf{X} = \frac{m_2}{M}\mathbf{r}, \tag{6.9a}$$

$$\mathbf{r}_2 = \mathbf{x}_2 - \mathbf{X} = -\frac{m_1}{M}\mathbf{r}, \tag{6.9b}$$

where

$$\mathbf{r} = \mathbf{x}_1 - \mathbf{x}_2 = \mathbf{r}_1 - \mathbf{r}_2, \tag{6.10}$$

is the relative position of the particles with respect to one another (see Figure 6.1). Differentiating (6.9a, b), we have

$$\dot{\mathbf{r}}_1 = \dot{\mathbf{x}}_1 - \dot{\mathbf{X}} = \frac{m_2}{M}\dot{\mathbf{r}}, \tag{6.11a}$$

$$\dot{\mathbf{r}}_2 = \dot{\mathbf{x}}_2 - \dot{\mathbf{X}} = -\frac{m_1}{M}\dot{\mathbf{r}}, \tag{6.11b}$$

and

$$\ddot{\mathbf{r}}_1 = \ddot{\mathbf{x}}_1 - \ddot{\mathbf{X}} = \frac{m_2}{M}\ddot{\mathbf{r}},$$

$$\ddot{\mathbf{r}}_2 = \ddot{\mathbf{x}}_2 - \ddot{\mathbf{X}} = -\frac{m_1}{M}\ddot{\mathbf{r}}.$$

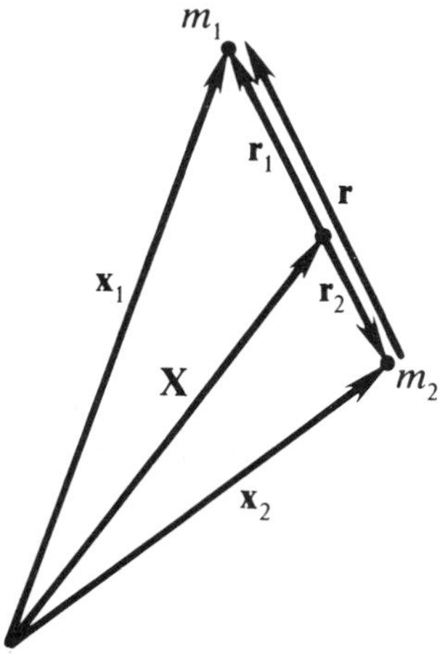

Fig. 6.1. Variables for a two-particle system.

Using these equations to eliminate $\mathbf{x}_1$ and $\mathbf{x}_2$ after we subtract (6.1b) from (6.1a), we get

$$\frac{2m_1m_2}{M}\ddot{\mathbf{r}} + (m_1 - m_2)\ddot{\mathbf{X}} = 2\mathbf{f}_{12} + \mathbf{F}_1 - \mathbf{F}_2 .$$

and using (6.3) to eliminate $\ddot{\mathbf{X}}$, we obtain

$$\mu\ddot{\mathbf{r}} = \mathbf{f}_{12} + \frac{m_2\mathbf{F}_1 - m_1\mathbf{F}_2}{M} , \tag{6.12}$$

where the quantity

$$\mu = \frac{m_1m_2}{M} = \frac{m_1m_2}{m_1 + m_2} \tag{6.13}$$

is called the *reduced mass*.

Equation (6.12) describes the internal motion of the two particle system. The effect of external forces on the internal motion is determined by the term $(m_2\mathbf{F}_1 - m_1\mathbf{F}_2)/M$. Note that this term vanishes for the external gravitational force (6.6) and it is a function of $\dot{\mathbf{r}}$ only for the forces (6.7) and (6.8). For such external forces, therefore, if the internal force $\mathbf{f}_{12}$ depends only on the relative motions of the particles, then (6.12) is an equation of the form

$$\mu\ddot{\mathbf{r}} = \mathbf{f}(\dot{\mathbf{r}}, \mathbf{r}). \tag{6.14}$$

But this is the equation of motion for a single particle of mass μ subject to the force $\mathbf{f}(\dot{\mathbf{r}}, \mathbf{r})$. In this way, the problem of solving the coupled equations of motion for a two-particle system can often be reduced to an equivalent pair of one-particle problems with equations of motion (6.3) and (6.14).

External forces often have a relatively small effect on the internal motion, even when the force function cannot be put in the functional form of (6.14). In such cases, the external forces can be handled by perturbation theory, as will be demonstrated in detail in Chapter 8. Before taking external forces into account, we should analyze the effect of internal forces acting alone.

Isolated Systems

A system of particles subject to a negligible external force is said to be isolated. From (6.3) it follows that the center of mass of an isolated two-particle system moves with constant velocity $\dot{\mathbf{X}}$. The only problem then is to solve the equations of motion for the internal motion. Evidently for an isolated pair of particles, the relative position $\mathbf{r} = \mathbf{x}_1 - \mathbf{x}_2$ and the relative velocity $\dot{\mathbf{r}}$ are the only relevant kinematical variables if the particles are structureless objects themselves. So we expect that an internal equation of motion of the general form (6.14) will be valid quite generally. Certainly the force law will be of the form $\mathbf{f}(\dot{\mathbf{r}}, \mathbf{r})$ if each particle is the center of force for the force it exerts on the other particle. Our previous analysis of central force motion for a single particle can therefore be applied immediately to the more realistic case of two interacting particles. The only mathematical difference in the two cases is that, as (6.14) shows, the reduced mass μ must be used in the equation of motion instead of the mass of a single particle. If the mass of one particle, say m_1, is negligible compared to m_2, then according to (6.13), $\mu \approx m_1$ and (6.14) reduces to a single particle equation.

Two-Body Effects on the Kepler Problem

Let us see now how our previous analysis of motion under a gravitational or Coulomb force must be modified to take into account the "two-body effects" arising from the finite mass of both particles. As before, the force law is

$$\mathbf{f} = -\frac{k(\mathbf{x}_1 - \mathbf{x}_2)}{|\mathbf{x}_1 - \mathbf{x}_2|^3} = -\frac{k\mathbf{r}}{r^3} = -\frac{k\hat{\mathbf{r}}}{r^2}, \tag{6.15}$$

so (6.14) takes the specific form

$$\mu\ddot{\mathbf{r}} = -\frac{k\hat{\mathbf{r}}}{r^2}. \tag{6.16}$$

Obviously, Kepler's first two laws still follow from (6.16), but his third law must be modified, because it depends on value of the mass μ.

From (2.10) and (2.11), we see that Kepler's third law should be modified to read

$$\frac{T^2}{a^3} = 4\pi^2\frac{\mu}{k} = \frac{4\pi^2}{G(m_1 + m_2)}. \tag{6.17}$$

For planetary motion, the mass of a planet is too small compared to the mass of the Sun to produce an observable deviation from Kepler's third law, except in the case of Jupiter, where $m_{\text{Jupiter}} = (0.001)m_{\text{Sun}}$. For binary stars, Equation (6.17) is used to deduce the masses from observations of periods.

Newton was the first to derive (6.17) and he used it to estimate the Moon-Earth mass ratio. In terms of quantities readily measured in Newton's day, Equation (6.17) gives the mass ratio

$$\frac{m_1}{m_2} = \frac{4\pi^2 a^3}{gR_E^2 T^2} - 1, \tag{6.18}$$

where R_E is the Earth's radius and g is the gravitational acceleration at the surface of the Earth. Even with the best data available today Equation (6.18) gives a Moon–Earth mass ratio with no better than 30% accuracy. The main reason for this is neglect of the effect of the Sun. The force of the Sun on the Moon is in fact more than twice as great as the force of the Earth on the Moon. Even so, we have seen that if the Sun's gravitational field were uniform over the dimensions of the Earth–Moon system, it would have no effect on the internal motion of the system, and (6.17) would be valid. The variations of the Sun's force on the moon are responsible for deviations from (6.17). They have a maximum value of about one hundredth the force of the Earth on the Moon. Newton realized all this, so he embarked on a long program of analyzing the dynamics of the Earth–Moon system in ever increasing detail, a program that has continued into this century. However, we shall see below that a fairly good estimate of the Moon–Earth mass ratio can be achieved without calculating the effect of the Sun.

Returning now to the interpretation of solutions to Equation (6.16), we know that for a bound system the vector $\mathbf{r}$, representing the relative separation of the particles, traverses an ellipse. However, this is not a complete description of the system's internal motion, for both particles move relative to the center of mass. The complete internal motion is easily ascertained from (6.9a, b), which gives the internal particle positions $\mathbf{r}_1$ and $\mathbf{r}_2$ directly from $\mathbf{r}$. From these equations we can conclude immediately that both particles move on ellipses with focus at the center of mass and eccentricity vectors $\boldsymbol{\varepsilon}_1 = -\boldsymbol{\varepsilon}_2 = \boldsymbol{\varepsilon}$, where $\boldsymbol{\varepsilon}$ is eccentricity vector for the r-orbit; the orbits differ in scale, and both particles move so as to remain in opposition relative to the center of mass, more specifically, from (6.9a, b) we have

$$m_1\mathbf{r}_1 = -m_2\mathbf{r}_2 = \mu\mathbf{r}, \tag{6.19}$$

for all particle positions on the orbits. These relations are shown in Figure 6.2.

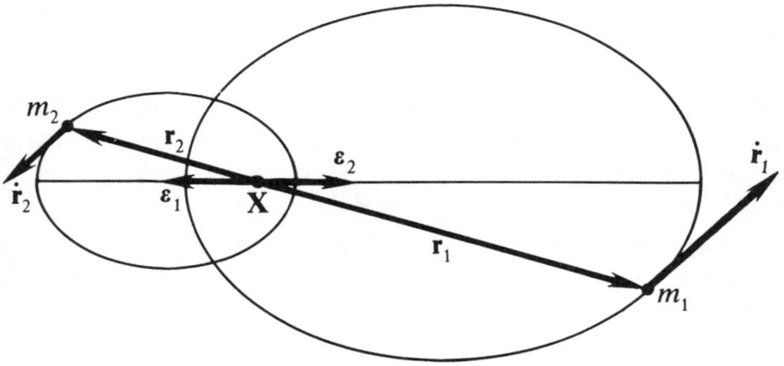

Fig. 6.2. Two-particle Kepler motion for $m_2 = 2m_1$.

Internal motion relative to the center of mass is observable only by reference to some external object. For example, internal motion of the Earth-Moon gives rise to a small oscillation in the apparent direction of the sun, as indicated in Figure 6.3. This gives us another method of determining the Earth–Moon mass ratio. Observations give a value of 6.5″ for the angle α in Figure 6.3, and $R = 1.5 \times 10^8$ km for the distance to the Sun. From this we

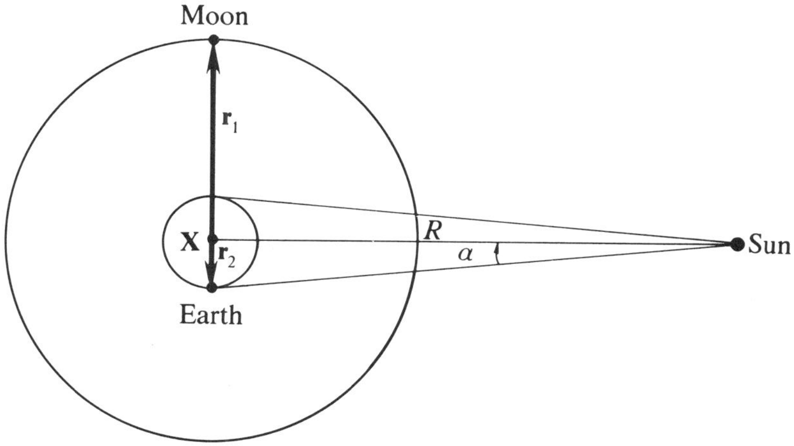

Fig. 6.3. Internal motion of the Earth–Moon system (not to scale).

obtain $r_2 = \alpha R = 4.7 \times 10^3$ km for the distance of the Earth's center from the center of mass **X**. It will be noted that **X** lies well within the Earth's surface. Now, using the value $r = 3.8 \times 10^5$ km for the average Earth–Moon distance, from (6.19) we obtain the Earth–Moon mass ratio

$$\frac{m_2}{m_1} \approx 81, \tag{6.20}$$

close to the accepted value of 81.25 obtained by more refined methods. The result obtained by this method is evidently so much better than the result obtained from (6.18), because it involves only a lateral cross-section of the Earth–Moon orbits along which the Sun's field is nearly uniform, whereas the period T in (6.18) depends on variations of the Sun's gravitational field over the entire orbit. The discrepancy between these two results indicates that the Sun has a measurable effect on the Moon's period and challenges us to account for it. A method for solving this problem will be developed in chapter 8.

4-6. Exercises

(6.1) For internal motion governed by Equation (6.16) show that the energies and angular momenta of the two particles are related by

$$E_2 = \frac{m_1}{m_2} E_1, \quad \mathbf{L}_2 = \frac{m_1}{m_2} \mathbf{L}_1,$$

while the total internal energy $E = E_1 + E_2$ and angular momentum $\mathbf{L} = \mathbf{L}_1 + \mathbf{L}_2$ can be attributed to an equivalent single particle of mass μ.

(6.2) Show that two particle elliptical orbits intersect, as in Figure 6.2, when

$$\varepsilon > \left| \frac{m_1 - m_2}{m_1 + m_2} \right| .$$

4-7. Elastic Collisions

In Section 4-2 we saw how the gravitational force law could be ascertained from experimental information about planetary orbits. For *atomic particles*, however, it is quite impossible to observe the orbits directly, so we must resort to more indirect methods to ascertain the atomic force laws. Experimental information about atomic (and nuclear) forces is gained by *scattering experiments* in which binary (two particle) collisions are arranged. Measurements on the particles can be made only in the *asymptotic regions* before and after collision, where the interparticle interaction is negligible. The problem is to determine the forces which will produce the observed relations between the initial and final states. To approach this problem systematically, we first determine the consequences of the most general conservation laws for unbounded motion of two-particle systems. Then (in Section 4-8) we investigate the consequence of specific laws, like the Coulomb force, which are believed to describe atomic forces.

Conserved Quantities

We are concerned here with unbounded motion of an isolated two-particle system. The general equations we need were developed in the last section. For an isolated system, Equation (6.3) tells us that the center of mass velocity $\dot{\mathbf{X}}$ is constant, so according to (6.3),

$$M\dot{\mathbf{X}} = m_1\dot{\mathbf{x}}_1 + m_2\dot{\mathbf{x}}_2 \tag{7.1}$$

is a constant of the motion for a 2-particle system. The product of a particle mass with its velocity is called its *momentum*. Thus, the vector $m_1\dot{\mathbf{x}}_1$ is the momentum of particle 1. Let $\mathbf{p}_1$ and $\mathbf{p}'_1$ denote *initial* and *final* momenta of particle 1, that is, the asymptotic value of $m_1\dot{\mathbf{x}}_1$ before and after collision. With a similar notation for the momentum of particle 2, we obtain from (7.1) the law of *momentum conservation*

$$\mathbf{p}_1 + \mathbf{p}_2 = M\dot{\mathbf{X}} = \mathbf{p}'_1 + \mathbf{p}'_2 . \tag{7.2}$$

The additivity of momenta in this conservation law shows that momentum is an important physical concept.

Momentum conservation is the most general principle governing collisions, because it is independent of the forces involved. Next in generality we have *energy conservation*, which holds if the interaction force is conservative. To determine its consequences, we introduce the internal or *center of mass* (CM) momentum $\mu\dot{\mathbf{r}}$, which, according to (6.11a, b) and (6.13), is related to the external variables by

$$\mu\dot{\mathbf{r}} = m_1(\dot{\mathbf{x}}_1 - \dot{\mathbf{X}}) = -m_2(\dot{\mathbf{x}}_2 - \dot{\mathbf{X}}). \tag{7.3}$$

If $\mathbf{p}$ and $\mathbf{p}'$ are the initial and final values of the CM momentum $\mu\dot{\mathbf{r}}$, then

$$\mathbf{p} = \mathbf{p}_1 - m_1\dot{\mathbf{X}} = -(\mathbf{p}_2 - m_2\dot{\mathbf{X}}), \tag{7.4a}$$

$$\mathbf{p}' = \mathbf{p}_1' - m_1\dot{\mathbf{X}} = -(\mathbf{p}_2' - m_2\dot{\mathbf{X}}). \tag{7.4b}$$

Now, if the internal force is conservative, then the internal energy is a constant of the motion and equal to the value of the internal kinetic energy $\frac{1}{2}\mu\dot{\mathbf{r}}^2$ in the asymptotic region. Thus, *energy conservation* is expressed by

$$\frac{1}{2\mu}\,\mathbf{p}^2 = \frac{1}{2\mu}\,\mathbf{p}'^2$$

or simply

$$\mathbf{p}^2 = \mathbf{p}'^2 \tag{7.5}$$

This can be related to external variables by using the relation

$$\tfrac{1}{2}m_1\dot{\mathbf{x}}_1^2 + \tfrac{1}{2}m_2\dot{\mathbf{x}}_2^2 = \tfrac{1}{2}M\dot{\mathbf{X}}^2 + \tfrac{1}{2}\mu\dot{\mathbf{r}}^2, \tag{7.6}$$

which follows from (7.3) and holds whether energy is conserved or not. Evaluating (7.6) in the asymptotic regions and using (7.5), we obtain energy conservation in the form

$$\frac{\mathbf{p}_1^2}{2m_1} + \frac{\mathbf{p}_2^2}{2m_2} = \frac{\mathbf{p}_1'^2}{2m_1} + \frac{\mathbf{p}_2'^2}{2m_2}\,. \tag{7.7}$$

Of course, it is easier to apply (7.5) than (7.7).

A collision is said to be *elastic* if it conserves energy and the masses of the particles involved. The above equations apply to any binary elastic collision. It follows from (7.5) that an elastic collision has the effect of simply rotating the initial CM momentum $\mathbf{p}$ through some angle Θ into its final value $\mathbf{p}'$, as described by the equation

$$\mathbf{p}' = \mathbf{p}e^{\mathbf{i}\Theta}, \tag{7.8}$$

where the unit bivector $\mathbf{i}$ specifies the scattering plane. The angle Θ is called the *CM scattering angle*. The internal (CM) velocities of the colliding particles are $\dot{\mathbf{r}}_1 = \dot{\mathbf{x}}_1 - \dot{\mathbf{X}}$ and $\dot{\mathbf{r}}_2 = \dot{\mathbf{x}}_2 - \dot{\mathbf{X}}$, which are always oppositely directed according (7.3), so the relation between initial and final states can be represented as in Figure 7.1.

If the interparticle force is central as well as conservative, then angular momentum $\mathbf{L}$ is conserved. This, in turn, implies that the direction $\hat{\mathbf{L}}$ and magnitude $|\mathbf{L}|$ are separately conserved. Conservation of $\hat{\mathbf{L}}$ implies that the orbits of both particles as well as the center of mass lie in a single plane, so $\hat{\mathbf{L}}$ can be identified with the $\mathbf{i}$ in (7.8). This condition is usually taken for granted in scattering experiments. Conservation of $|\mathbf{L}|$ does not supply helpful relations between initial and final states, because it pertains to details of the orbits which are not observed in scattering. However, the initial value of $|\mathbf{L}|$ must be given to determine the scattering angle Θ from a given interparticle force law.

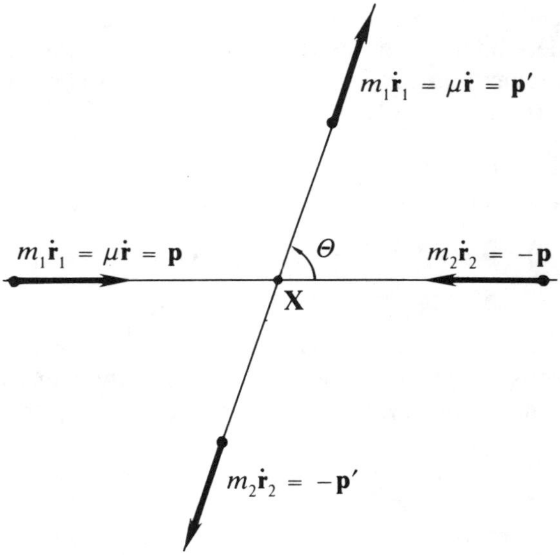

Fig. 7.1. Center of mass variables.

Momentum and Energy Transfer

Note that all the above equations for elastic collisions hold for any value given to the center of mass velocity $\dot{\mathbf{X}}$. From (7.4a, b) we seen that the momentum transfer $\Delta\mathbf{p}$ defined by

$$\Delta\mathbf{p} \equiv \mathbf{p}' - \mathbf{p} = \mathbf{p}_1' - \mathbf{p}_1 = -(\mathbf{p}_2' - \mathbf{p}_2)\,, \tag{7.9}$$

is independent of $\dot{\mathbf{X}}$. However, the energy transfer ΔE defined by

$$\Delta E \equiv \frac{1}{2m_2}(\mathbf{p}_2'^2 - \mathbf{p}_2^2) = \frac{1}{2m_1}(\mathbf{p}_1^2 - \mathbf{p}_1'^2), \tag{7.10}$$

depends on the value of $\dot{\mathbf{X}}$, for, with the help of (7.4a, b), we can express it in the form

$$\Delta E = -\dot{\mathbf{X}}\cdot\Delta\mathbf{p}. \tag{7.11}$$

According to (7.10), ΔE is positive if particle 1 loses energy in the collision and negative if particle 1 gains energy.

Equation (7.11) has important applications in astromechanics as well as atomic physics. A spacecraft travelling from Earth to the outer planets Uranus, Neptune and Pluto can be given a large boost in velocity by scattering it off Jupiter. According to (7.11), the boost can be maximized by maximizing $\dot{\mathbf{X}}\cdot\Delta\mathbf{p} = \dot{\mathbf{X}}\cdot(\mathbf{p}' - \mathbf{p})$. For the Jupiter-spacecraft system, we can

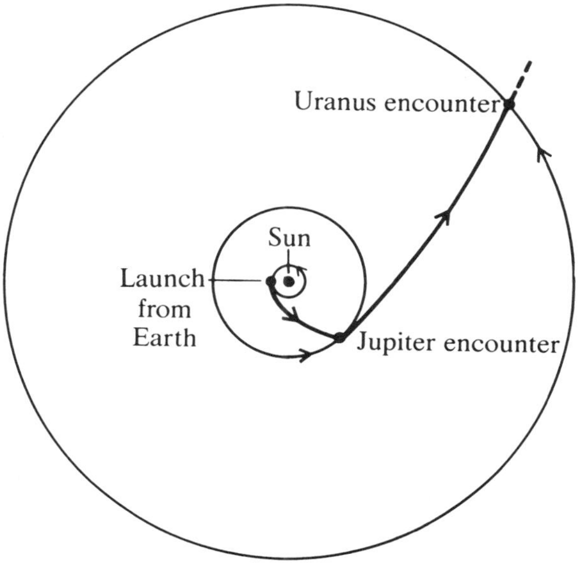

Fig. 7.2. Trajectory from Earth to Uranus with "gravity-assist" by Jupiter.

identify $\dot{\mathbf{X}}$ with the velocity of Jupiter relative to the Sun. The initial spacecraft momentum $\mathbf{p} = \mathbf{p}_1 - m_1\dot{\mathbf{X}}$ is determined by the launch of the spacecraft from Earth. With $\mathbf{p}$ fixed, therefore, the maximum boost is achieved by adjusting the impact parameter for the collision so that $\mathbf{p}'$ is parallel to $\dot{\mathbf{X}}$. This can be arranged by appropriate timing of the launch and manuevering of the spacecraft. A "gravity-assist" trajectory from Earth to Uranus is shown in Figure 7.2. The transit time from Earth to Uranus is about 5 years on the assisted orbit as compared with 16 years on an unassisted orbit with the same initial conditions. (Exercise (7.6)).

LAB Scattering Variables for Elastic Collisions

In a typical scattering experiment with atomic particles, one particle with initial "LAB momentum" $\mathbf{p}_1$ is fired at a target particle at rest in the laboratory with initial momentum $\mathbf{p}_2 = 0$. Scattering variables which can be measured fairly directly in the laboratory are indicated in Figure 7.3 The angle θ is called the *LAB scattering angle*, and ϕ is called the *recoil angle*. With the "LAB condition" $\mathbf{p}_2 = 0$, Equation (7.4a) gives the relation between LAB and CM momenta before collision

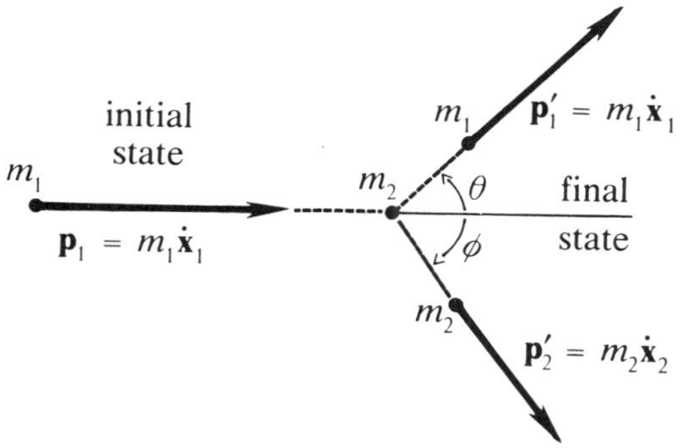

Fig. 7.3. LAB variables.

$$\mathbf{p}_1 = \frac{M}{m_2}\mathbf{p} = M\dot{\mathbf{X}}. \quad (7.12a)$$

Using this to eliminate $\dot{\mathbf{X}}$ from (7.4b), we can solve for the final LAB momenta in terms of CM momenta, with the result

$$\mathbf{p}_1' = \frac{m_1}{m_2}\mathbf{p} + \mathbf{p}', \tag{7.12b}$$

$$\mathbf{p}_2' = \mathbf{p} - \mathbf{p}' = -\Delta\mathbf{p}. \tag{7.12c}$$

These three equations (7.12a, b, c) describe all the relations between LAB and CM variables. These relations along with the conservation laws (7.2) and (7.5) are represented in Figure 7.4.

The laboratory energy transfer ΔE and the angle of deflection θ are of direct interest in scattering, so let us express them in terms of CM variables and see what that tells us. The total energy E_0 of the 2-particle system is, by (7.7), equal to the initial kinetic energy of the projectile, which, by (7.12a), can be expressed in terms of the CM momentum; thus,

$$E_0 = \frac{\mathbf{p}_1^2}{2m_1} = \frac{M^2\mathbf{p}^2}{2m_1 m_2^2}. \tag{7.13}$$

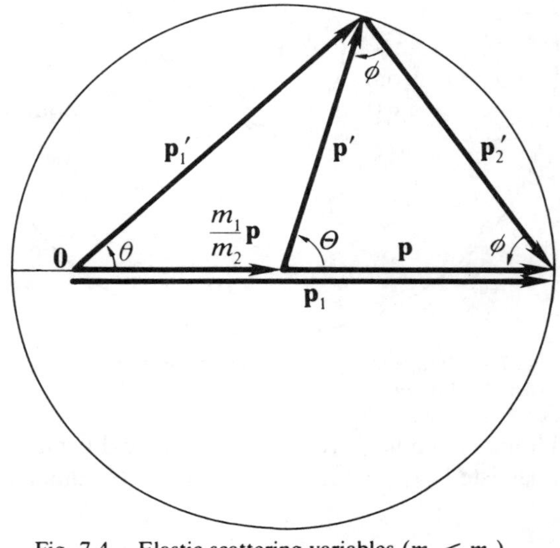

Fig. 7.4. Elastic scattering variables ($m_1 < m_2$).

In the collision this energy will be redistributed among the particles. Since the target particle is at rest initially, according to (7.10) the *energy transfer* ΔE is equal to its kinetic energy after collision; by (7.12c), then,

$$\Delta E = \frac{\mathbf{p}_2'^2}{2m_2} = \frac{(\mathbf{p}' - \mathbf{p})^2}{2m_2}. \tag{7.14}$$

The fractional energy transfer is therefore

$$\frac{\Delta E}{E_0} = \frac{\mu(\mathbf{p}' - \mathbf{p})^2}{M\mathbf{p}^2} = \frac{4m_1 m_2}{(m_1 + m_2)^2}\sin^2\tfrac{1}{2}\Theta, \tag{7.15}$$

where we have used (7.5) and (7.8) to get

$$(\mathbf{p}' - \mathbf{p})^2 = 2p^2(1 - \cos\Theta) = 4p^2\sin^2\tfrac{1}{2}\Theta. \tag{7.16}$$

Note that (7.15) would be unaffected by a change in the center of mass velocity for the two-particle system, so it must be applicable to moving targets as well as the stationary targets we are considering.

Some important conclusions can be drawn from (7.15). The energy transfer has its maximum value when $\Theta = \pi$, so

$$\left(\frac{\Delta E}{E_0}\right)_{\max} = \frac{4m_1m_2}{(m_1+m_2)^2} \leqslant 1. \tag{7.17}$$

This tells us that all of the energy can be transferred to the target only if $m_1 = m_2$. For this reason, since the mass of hydrogen is nearly the same as the mass of the neutron, hydrogen-rich materials are more effective for slowing down neutrons than heavy materials like lead. Also, when electrons pass through a material they lose most of their energy to other electrons rather than atomic nuclei. To see how little energy electrons lose to nuclei, we need only know that the proton-electron mass ratio is 1836, so $(\Delta E/E_0)_{\max} \approx 4/1836 = 0.2\%$. A nucleus hardly budges when an electron bounces off it, just as bowling ball will hardly be budged by collision with a ping-pong ball.

To relate the LAB scattering angle θ to the CM scattering angle Θ, we note that θ can be defined algebraically by

$$\hat{\mathbf{p}}_1' = \hat{\mathbf{p}}_1 e^{i\theta}. \tag{7.18}$$

According to (7.12a), $\hat{\mathbf{p}}_1 = \hat{\mathbf{p}}$, so if we multiply (7.12b) by $\hat{\mathbf{p}}$ and introduce the scattering angles by (7.8) and (7.18), we obtain,

$$p_1' e^{i\theta} = p\left(\frac{m_1}{m_2} + e^{i\Theta}\right). \tag{7.19}$$

We can eliminate p and p_1' from this relation by observing that it implies

$$\left(\frac{p_1'}{p}\right)^2 = \left(\frac{m_1}{m_2} + e^{i\Theta}\right)\left(\frac{m_1}{m_2} + e^{-i\Theta}\right) = 1 + \left(\frac{m_1}{m_2}\right)^2 + \frac{2m_1}{m_2}\cos\Theta,$$

which, on substitution back into (7.19), gives

$$e^{i\theta} = \frac{m_1/m_2 + e^{i\Theta}}{[1 + (m_1/m_2)^2 + 2(m_1/m_2)\cos\Theta]^{1/2}}. \tag{7.20}$$

A somewhat simpler relation between scattering angles can be obtained by taking the ratio of bivector to scalar parts of (7.19) or (7.20) to get

$$\tan\theta = \frac{\sin\Theta}{\dfrac{m_1}{m_2} + \cos\Theta}. \tag{7.21}$$

To interpret this formula for $m_1 < m_2$, refer to Figure 7.4. For fixed initial CM momentum $\mathbf{p}$, the final momentum $\mathbf{p}'$ must lie on a sphere of radius $|\mathbf{p}|$. It is clear from Figure 7.4 that there is a unique value of θ for every value of Θ in the range $0 \leqslant \Theta \leqslant \pi$, which covers all possibilities. Indeed, for the limiting case of a stationary target ($m_1 \ll m_2$), Equation (7.21) reduces to $\tan\Theta \approx \tan\theta$, whence $\theta = \Theta$. However, in the equal mass case $m_1 = m_2$, the origin **0** in Figure 7.4 lies on the circle, and (7.21) reduces to $\tan\theta = \tan\frac{1}{2}\Theta$, whence $\theta = \frac{1}{2}\Theta$.

For a light target ($m_1 > m_2$), the origin **0** lies outside the circle, as shown in

Figure 7.5. In this case, there are two values say Θ_a and Θ_b, of the CM scattering angle for each value of the LAB scattering angle. The two values Θ_a and Θ_b can be distinguished in the lab by measuring the kinetic energy of the scattered particle.

The LAB scattering angle has a maximum value Θ_{max} given by

$$\sin \theta_{max} = \frac{m_2}{m_1}, \quad (7.22)$$

which can be read directly off Figure 7.5. From this we can deduce, for example, that a proton cannot be scattered by more than 0.03° by an electron. Therefore, any significant deflection of protons or heavier atomic nuclei passing through matter is due to collisions with nuclei rather than electrons.

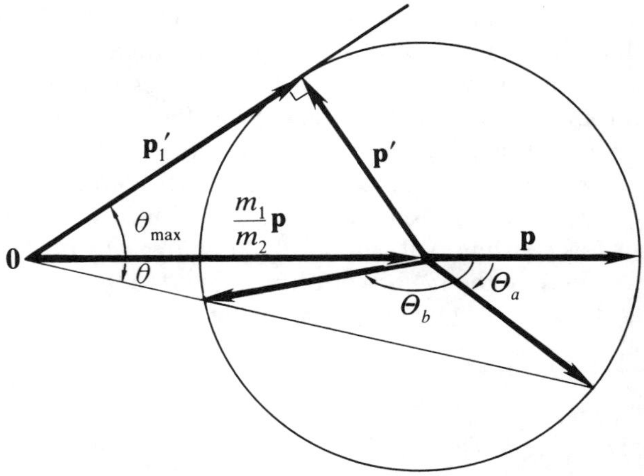

Fig. 7.5. Range of scattering angles for $m_1 > m_2$.

4-7. Exercises

(7.1) Establish Equations (7.6) and (7.11)

(7.2) Prove that for elastic scattering of equal mass particles the sum of the scattering angle and recoil angle is always 90°.

(7.3) Alpha particles (i.e. Helium nuclei ^4_2He) are scattered elastically from protons at rest. Show that for a scattered particle the maximum angle of deflection is 14.5°, and the maximum fractional energy loss is 64%.

(7.4) A particle of mass m_1 collides elastically with a particle of mass m_2 at rest. Determine the mass ratio m_1/m_2 from the scattering angle θ and the recoil angle ϕ.

(7.5) An unstable particle of mass $m = m_1 + m_2$ decays into particles with masses m_1 and m_2, releasing energy Q to products.

(a) Determine the CM kinetic energies of the two particles produced.

(b) If the unstable particle has an initial kinetic energy K, determine the maximum and minimum kinetic energies of the products.

(7.6) Evaluate the advantage of *gravitational assist* by Jupiter for a mission from Earth to Uranus (Figure 7.2) as follows:

(a) Calculate the time of passage Δt_0 from Earth to Uranus on an orbit of minimum energy. Determine the speed v of the spacecraft as it crosses Jupiter's orbit, and the angle of intersection α between the two orbits. (Data on the planetary orbits are given in Appendix C).

(b) Suppose that the launch is arranged so the spacecraft encounters Jupiter on the orbit specified by (a). Calculate the maximum speed Δv that the satellite can gain from scattering off Jupiter, and the corresponding scattering angle θ in the rest system of Jupiter.

(c) Determine the distance d of closest approach to the surface of Jupiter for maximum speed gain. (The radius of Jupiter is $R_J = 71\ 400$ km).

(d) Determine the eccentricity ε of the spacecraft's orbit in the heliocentric system after escape from Jupiter's influence. Calculate the transit time Δt_2 from Jupiter to Uranus on this orbit by evaluating the integral

$$\Delta t_2 = \int_{t_1}^{t_2} \mathrm{d}t = \frac{m}{L}\int_0^{\theta_2} r^2\,\mathrm{d}\theta .$$

The upper limit θ_2 can be determined from the orbit Equation (3.6a). Similarly, calculate the transit time Δt_1 from Earth to Jupiter to get the total transit time for the mission to Uranus. Of course, these estimates are only approximate, since the influences of Jupiter and the Sun were evaluated separately.

4-8. Scattering Cross Sections

In a typical scattering experiment an incident beam of monoenergetic particles is directed at a small sample containing the target particles, and the scattered particles are collected in detectors, as shown in Figure 8.1. Even solid material is mostly "empty space" at the atomic level, so it is not difficult to prepare a sample thin enough so that multiple collisions with incident particles are negligibly rare compared to single collisions. Consequently, we can restrict our attention to the scattering of the beam by a single target in the sample.

All the particles in the incident beam have the same energy E_0 and direction of motion. The beam has a uniform cross section with an *intensity* N_0 defined as the number of particles per unit area per unit time incident on the sample. Let N be the number of incident particles scattered by a single target particle in a unit time. The *total cross section* σ is defined by

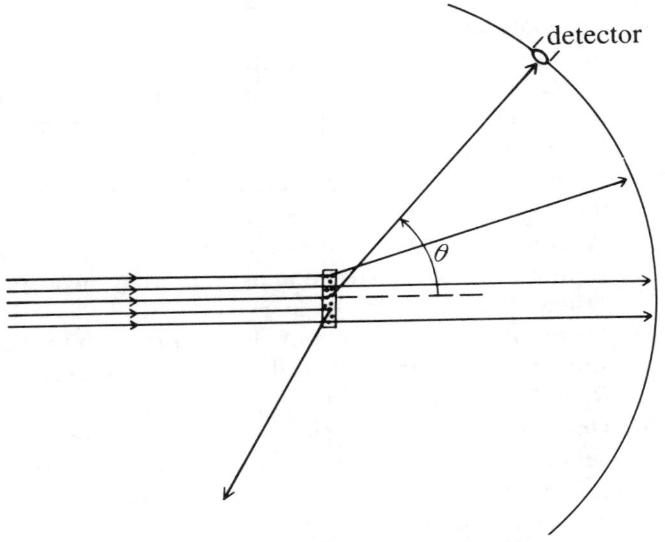

Fig. 8.1. Arrangement for a scattering experiment.

$$\sigma = \frac{N}{N_0} . \tag{8.1}$$

It has the dimensions of area. It can be interpreted as the area of an imaginary disk in the asymptotic region transverse to the beam and centered on a line through the center of the target, so that only the incident particles which pass through the disk are scattered. Now, consider an annulus on this disk of radius b and width $\mathrm{d}b$. All incident particles intercepting this annulus have the same impact parameter b, so, for a central force, they will all be scattered by the same angle θ, as shown in Figure 8.2. Particles intercepting a segment of the annulus with area $\mathrm{d}\sigma = b\ \mathrm{d}b\ \mathrm{d}\phi$ will be scattered into a segment on a unit sphere centered at the target with the $\mathrm{d}\omega = \sin\theta\ \mathrm{d}\theta\ \mathrm{d}\phi$ called the *solid angle*. The quantity

$$\frac{\mathrm{d}\sigma}{\mathrm{d}\omega} = \frac{b}{\sin\theta}\left|\frac{\mathrm{d}b}{\mathrm{d}\theta}\right| = \frac{1}{2}\,\frac{\mathrm{d}b^2}{\mathrm{d}(\cos\theta)} \tag{8.2}$$

is called the (differential) *scattering cross section*. Since all scattered particles must pass through the sphere somewhere, the total cross section will be obtained by integrating the cross section over the unit sphere, that is,

$$\sigma = \oint \left(\frac{\mathrm{d}\sigma}{\mathrm{d}\omega}\right)\mathrm{d}\omega = 2\pi \int_0^{b_{\max}} b\ \mathrm{d}b . \tag{8.3}$$

The rate at which particles are "scattered into $\mathrm{d}\omega$" is

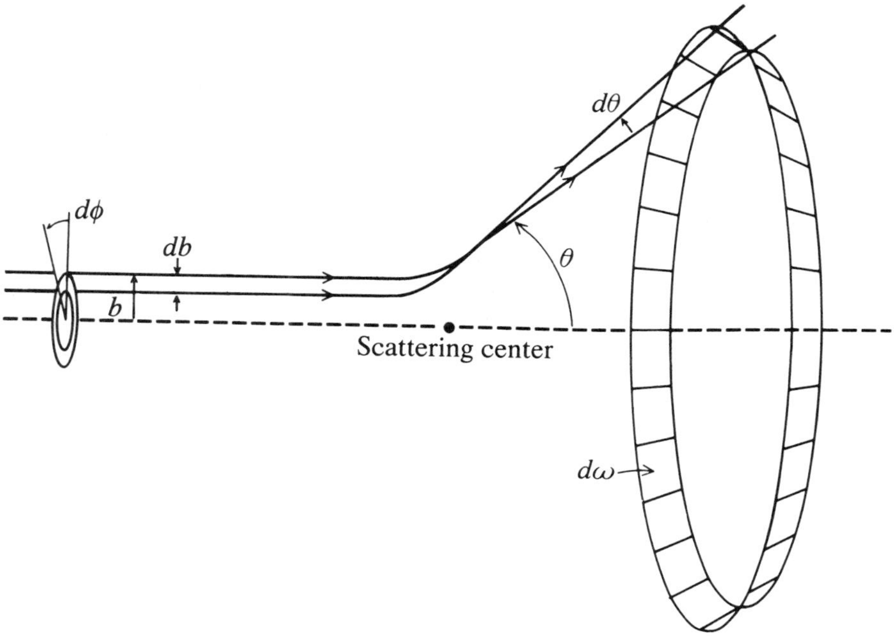

Fig. 8.2. Central force scattering for a given impact parameter or scattering angle.

$$\frac{\mathrm{d}N}{\mathrm{d}\omega} = N_0 \frac{\mathrm{d}\sigma}{\mathrm{d}\omega} , \tag{8.4}$$

which gives (8.1) when substituted into (8.3). Thus, the differential cross section can be measured quite directly simply by counting particles scattered through each angle, as indicated in Figure 8.1.

Given the force law which determines the scattering, we can deduce the deflection function $b = b(E_0, \theta)$ for the impact parameter as a function of initial energy and scattering angle. Then we compute $\mathrm{d}b/\mathrm{d}\theta$ and obtain $\mathrm{d}\sigma/\mathrm{d}\omega$ from (8.2). (Note that $\mathrm{d}b/\mathrm{d}\theta$ is negative, because, as Figure 8.2 shows, an increase in scattering angle corresponds to a decrease in impact parameter; for this reason the absolute value $|\,\mathrm{d}b/\mathrm{d}\theta\,|$ has been used in Equation (8.2).) It is much easier to predict $\mathrm{d}\sigma/\mathrm{d}\omega$ from a given force law than it is to determine a force law from the observed values of $\mathrm{d}\sigma/\mathrm{d}\omega$. So, when the force law is unknown, the usual approach is to guess at the form of the force law, compare the predicted $\mathrm{d}\sigma/\mathrm{d}\omega$ with experimental data, and then look for simple modifications of the force law which will account for the discrepancies. Of course some discrepancies are not due to the force law at all, but to other effects such as multiple scattering. The theory of atomic scattering has reached a high level of sophistication, and physicists are able to distinguish a wide variety of subtle effects. Although the classical theory of interactions which we have been studying must be modified to account for quantum effects at the atomic level, we must first know the consequences of the classical theory before we can understand the modifications of quantum theory. Moreover, the consequences of classical and quantum theories are practically equivalent in many situations. Therefore, it is well worthwhile to continue applying classical concepts to the analysis of atomic phenomena.

We have reduced the problem of analyzing a scattering experiment to the determination of the deflection function $b = b(E_0, \theta)$ for the impact parameter as a function of angle. Let us carry out the calculation of the deflection function and scattering cross section for some important force laws.

Hard Sphere Scattering

Let us first calculate the deflection function and cross section for the simplest kind of scattering, the scattering of particles by a stationary hard-sphere. Later we shall see that the general problem of hard-sphere scattering can be solved by reducing it to this one. As shown in Figure 8.3, an incident particle is scattered impulsively on contact with the surface of the sphere. Let $\mathbf{p}$ and $\mathbf{p}'$ respectively be the initial and final momentum of the scattered particle. Energy conservation implies that

$$|\mathbf{p}| = |\mathbf{p}'|. \tag{8.5}$$

Now, the force of interaction is central, though discontinuous at the surface of the sphere. Consequently, angular momentum is conserved in the collision, and

$$\mathbf{R} \wedge \mathbf{p} = \mathbf{R} \wedge \mathbf{p}', \tag{8.6}$$

where $\mathbf{R}$ is the radius vector to the point where the particle makes contact with the sphere. The left side of (8.6) is the angular momentum immediately before collision, while the right side is the angular momentum immediately after. From (8.5) and (8.6) we conclude that $\sin\alpha = \sin\alpha'$ and

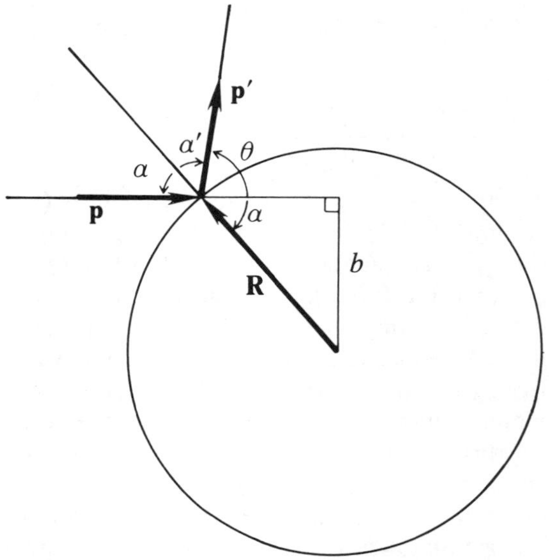

Fig. 8.3. Scattering by a stationary hard sphere.

$$\alpha = \alpha', \tag{8.7}$$

that is, the angle of incidence α equals the angle of reflection α'.

As Figure 8.3 shows, the scattering angle is given by $\theta = \pi - \alpha - \alpha' = \pi - 2\alpha$, and the deflection function is given by

$$b = R \sin \alpha = R \sin\left(\frac{\pi}{2} - \frac{\theta}{2}\right) = R \cos \frac{\theta}{2} . \tag{8.8}$$

Therefore, $\mathrm{d}b/\mathrm{d}\theta = -\frac{1}{2} R \sin \frac{1}{2}\theta$, and (8.2) gives

$$\frac{\mathrm{d}\sigma}{\mathrm{d}\omega} = \tfrac{1}{4} R^2 . \tag{8.9}$$

Thus, the differential cross section is *isotropic*, which is to say that particles are scattered at the same rate in all directions.

Substituting (8.9) into (8.3), we get

$$\sigma = \tfrac{1}{4} R^2 \oint \mathrm{d}\omega = \pi R^2 . \tag{8.10}$$

Thus, the total cross-section is exactly equal to the cross-sectional area of the sphere, as we expect from geometrical considerations.

Coulomb Scattering

We derived the deflection function for scattering by a Coulomb force in Section 4-3. According to (3.24) the Coulomb deflection function is

$$b = a \cot \tfrac{1}{2}\theta , \tag{8.11}$$

where

$$a = \frac{|q_1 q_2|}{2E_0} \tag{8.12}$$

for interacting particles with charges q_1 and q_2. Now, the derivative of (8.11) is

$$\frac{\mathrm{d}b}{\mathrm{d}\theta} = -\tfrac{1}{2} a \left(\frac{1}{\sin^2 \frac{1}{2}\theta} \right) .$$

So, according to (8.2),

$$\frac{\mathrm{d}\sigma}{\mathrm{d}\omega} = \frac{a^2 \cot \frac{1}{2}\theta}{2 \sin \theta \sin^2 \frac{1}{2}\theta} .$$

But, $\sin \theta = 2 \sin \frac{1}{2}\theta \cos \frac{1}{2}\theta$. Hence,

$$\frac{\mathrm{d}\sigma}{\mathrm{d}\omega} = \frac{a^2}{4} \frac{1}{\sin^4 \frac{1}{2}\theta} . \tag{8.13}$$

This formula is the justly famous *Rutherford Scattering Cross-section*. Ernest Rutherford derived it in 1911 and showed that it accurately described the angular distribution of α particles (${}^4_2$He nuclei) scattered from heavy nuclei in experiments by Geiger and Marsden. The $\sin^{-4} \frac{1}{2}\theta$ dependence was verified over a range of angles on which $d\sigma/d\omega$ varied by a factor of 250 000. The energy dependent factor a was varied by a factor of 10. In particular, the value $d\sigma/d\omega = a^2/4$ agreed well with the experimental counts for backscattering (when $\theta \approx \pi$). Backscattering can occur only for a head on collision, and the parameter a is the distance of closest approach. The experiments attained sufficient energy to give values of $a \approx 10^{-12}$ cm for which the Rutherford formula (8.13) held good. From this Rutherford was able to conclude that the positive charge in an atom is concentrated in a nucleus of radius no more than 10^{-12} cm, about one ten thousandth of the known diameter of an atom.

The Rutherford cross-section (8.13) is infinite at $\theta = 0$. As (8.11) shows, the Coulomb force gives some scattering no matter how large the impact parameter. In atomic scattering, however, the concentrated charge of a nucleus is screened by the cloud of atomic electrons surrounding it. Consequently, the atom will appear neutral and the scattering of α particles will be negligible for impact parameters greater than the radius of an atom (about 10^{-8} cm.). Coulomb's law provides a good description of the atomic force only for α particles that penetrate the electron cloud. We have seen that the electrons themselves cannot significantly scatter an α particle, because they are so much lighter.

Lab and CM Cross Sections

So far we have evaluated scattering cross sections only under the assumption that the target is stationary. Target recoil is most easily taken into account by evaluating the cross section in terms of the center of mass (CM) variables and then transforming the result to LAB variables. We have seen the 2-particle scattering problem is reduced to an equivalent 1-particle problem by using CM variables.

The relation between the LAB scattering angle θ and the CM scattering angle Θ was determined in Section 4–7. In particular, the scalar part of Equation (7.20) gives the relation

$$\cos\theta = \frac{m_1/m_2 + \cos\Theta}{[1 + (m_1/m_2)^2 + 2(m_1/m_2)\cos\Theta]^{1/2}} . \tag{8.14}$$

LAB scattering through an angle θ into $d\omega = \sin\theta\, d\theta\, d\phi$ corresponds to CM scattering through an angle Θ into $d\Omega = \sin\Theta\, d\Theta\, d\phi$. According to (8.2), therefore, the lab scattering cross section $d\sigma/d\omega$ is related to the CM scattering cross section $d\sigma/d\Omega$ by

$$\frac{d\sigma}{d\Omega} = \frac{d\sigma}{d\omega}\frac{d\omega}{d\Omega} = \frac{d\sigma}{d\omega}\frac{d(\cos\theta)}{d(\cos\Theta)} . \tag{8.15}$$

From (8.14) we obtain, after some algebra,

$$\frac{d\omega}{d\Omega} = \frac{d(\cos\theta)}{d(\cos\Theta)} = \frac{1 + (m_1/m_2)\cos\Theta}{[1 + (m_1/m_2)^2 + 2(m_1/m_2)\cos\Theta]^{3/2}}$$

$$= \frac{(1 + (m_1/m_2)^2 \sin^2\theta)^{1/2}}{[(m_1/m_2)\cos\theta + (1 - (m_1/m_2)^2\sin^2\theta)^{1/2}]^2} \ . \quad (8.16)$$

For $m_1 = m_2$ this reduces to

$$\frac{d\omega}{d\Omega} = \frac{1}{4\cos\frac{1}{2}\Theta} = \frac{1}{4\cos\theta} \ . \quad (8.17)$$

Recall from Section 8–7 that for $m_1 > m_2$ there are two distinct CM deflections for each lab angle θ, but the experimenter can distinguish between them by an energy analysis of the scattered particles. Of course, for a heavy target ($m_1 \ll m_2$), the expression (8.16) reduces to $d\omega/d\Omega \approx 1$, so the lab and CM cross-sections are nearly equal.

The expression (8.16) for the factor $d\omega/d\Omega$ in (8.15) shows that even when the CM cross section is simple, the angular dependence of the lab cross section can be quite complex. Consider for example, the CM scattering of smooth hard spheres illustrated in Figure 8.4. The adjectives "hard" and

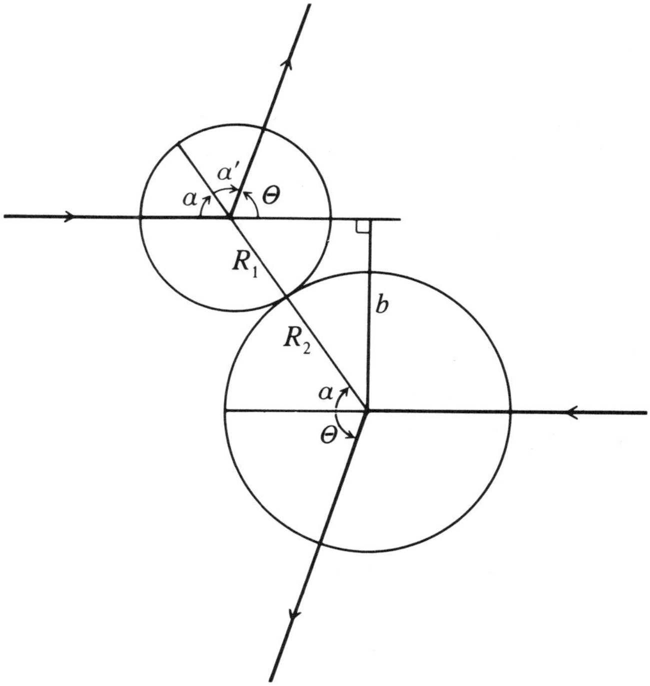

Fig. 8.4. CM collision of hard spheres.

"smooth" express the assumption that the collision does not excite any significant internal vibration or rotation of the spheres. It should be evident from Figure 8.4 that the CM scattering of spheres with radii R_1 and R_2 is equivalent to the scattering of a particle from a stationary sphere of radius $R = R_1 + R_2$. According to (8.9), therefore, the CM cross section has the constant value $\mathrm{d}\sigma/\mathrm{d}\Omega = \frac{1}{4}R^2$, and, by (8.15) and (8.17), the LAB cross section for the equal mass case is

$$\frac{\mathrm{d}\sigma}{\mathrm{d}\omega} = R^2 \cos\theta, \tag{8.18}$$

where $0 \leqslant \theta \leqslant \frac{1}{2}\pi$, because $\theta = \frac{1}{2}\Theta$ when $m_1 = m_2$. Thus, all the scattering is in the forward direction.

4-8. Exercises

(8.1) For a particle with mass m_1 scattered by a hard-sphere with mass m_2, show that the angle of incidence α is related to the angle of reflection α' by

$$\tan\alpha' = \left(\frac{m_1 + m_2}{m_1 - m_2}\right)\tan\alpha.$$

(8.2) In proton-proton scattering, the incident particles scattered cannot be distinguished from recoiling targets. Show, therefore, that for classical Coulomb scattering the angular distribution of protons detected in the LAB should be given by

$$\frac{\mathrm{d}\sigma}{\mathrm{d}\omega} = \left(\frac{e^2}{2E_0}\right)^2 (\sin^{-4}\theta + \cos^{-4}\theta)\cos\theta.$$

(8.3) The CM energy distribution of scattered particles with final LAB energy E is given by

$$\frac{\mathrm{d}\sigma}{\mathrm{d}E} = \frac{\mathrm{d}\sigma}{\mathrm{d}(\cos\Theta)}\ \frac{\mathrm{d}(\cos\Theta)}{\mathrm{d}E}\ .$$

Show that for hard-sphere scattering

$$\frac{\mathrm{d}\sigma}{\mathrm{d}E} = \frac{(m_1 + m_2)^2}{4m_1m_2}\ \frac{\sigma}{E_0}\ ,$$

and for Coulomb scattering

$$\frac{\mathrm{d}\sigma}{\mathrm{d}E} = \frac{m_1\pi}{m_2E_0}\left(\frac{q_1\, q_2}{E_0 - E}\right)^2$$

if $m_1 \ll m_2$.

(8.4) The screening of nuclear charge by atomic electrons can be taken into account in a rough way by using the *Cutoff Coulomb force* defined by

$$\mathbf{f} = q_1 q_2 \frac{\hat{\mathbf{r}}}{r^2} \quad \text{for} \quad r \leqslant R,$$

$$\mathbf{f} = 0 \qquad\qquad \text{for} \quad r > R.$$

For a stationary target, use the eccentricity conservation law of Section 4-3 to derive the following expression for momentum transfer to the scattered particle

$$\mathbf{p}' - \mathbf{p} = \frac{p\,|q_1 q_2|}{2Eb\mathbf{i}} \cdot \left[\frac{(R^2 - b^2)^{1/2}}{R} (\hat{\mathbf{p}} + \hat{\mathbf{p}}') - \mathbf{i}\,\frac{b}{R}\,(\hat{\mathbf{p}} - \hat{\mathbf{p}}') \right],$$

where b is the impact parameter and $E = p^2/2m$ is the energy. Express this result in terms of the scattering angle θ and derive the deflection function

$$b^2 = \frac{R^2}{1 + (1 + a^{-1}R)^2 \tan^2 \frac{1}{2}\theta}$$

where $a = |\, q_1 q_2 \,|/2E$. Derive the differential scattering cross section

$$\frac{d\sigma}{d\omega} = \frac{(a^{-1} + R^{-1})^2}{4[R^{-2} + a^{-1}(a^{-1} + 2R^{-1}) \sin^2 \frac{1}{2}\theta]^2}\,.$$

Note that this reduces to the Rutherford cross section for $R \gg$ a and to the hard-sphere cross section for $R \ll a$. What is the total cross section?

Chapter 5

Operators and Transformations

This chapter develops a system of mathematical concepts of great utility in all branches of physics. Linear operators and transformations are represented in terms of geometric algebra to facilitate computation. The group theory of rotations, reflections and translations is discussed in detail. The most important result is a compact spinor representation of finite rotations, which is shown to be a powerful computational device. This representation is used to develop the kinematics of rigid motions, which, in turn, is applied to the description of reference frames and motion with respect to moving frames.

5-1. Linear Functions and Matrices

Linear functions arise so frequently in physics that it is worthwhile to study their mathematical properties systematically.

A function $F = F(X)$ is said to be *linear* if

$$F(\alpha X + \beta Y) = \alpha F(X) + \beta F(Y), \tag{1.1}$$

where α and β are scalars. The condition (1.1) is equivalent to the two independent conditions

$$F(X + Y) = F(X) + F(Y), \tag{1.2a}$$

$$F(\alpha X) = \alpha F(X). \tag{1.2b}$$

We have been using a variety of linear functions all along, of course. For example, the function $s(\mathbf{x}) = \mathbf{a}\cdot\mathbf{x}$ is linear function of a vector variable. The linearity of this function comes from the distributive property of the inner product; thus,

$$\begin{aligned} s(\alpha\mathbf{x} + \beta\mathbf{y}) &= \mathbf{a}\cdot(\alpha\mathbf{x} + \beta\mathbf{y}) = \mathbf{a}\cdot(\alpha\mathbf{x}) + \mathbf{a}\cdot(\beta\mathbf{y}) \\ &= \alpha(\mathbf{a}\cdot\mathbf{x}) + \beta(\mathbf{a}\cdot\mathbf{y}) = \alpha s(\mathbf{x}) + \beta s(\mathbf{y}). \end{aligned}$$

Similarly, a linear bivector-valued function of a vector variable is defined by $b(\mathbf{x}) = \mathbf{a}\wedge\mathbf{x}$. And a spinor-valued function of a vector variable is defined by $S(\mathbf{x}) = s(\mathbf{x}) + b(\mathbf{x}) = \mathbf{a}\mathbf{x}$. In general, the sum of linear functions with the same domain is also a linear function.

We shall be concerned primarily with vector-valued linear functions of a vector variable, more specifically, with linear functions which transform (or map) vectors in Euclidean 3-space $\mathcal{E}_3$ into vectors in $\mathcal{E}_3$. Such functions are commonly called linear transformations, linear operators or tensors. Strictly speaking, the tensors we deal with here are tensors of rank 2, but we can ignore that, since we will not need a more general concept of tensor. We have, of course, encountered specific tensors before; for example, the projection

$$\mathcal{P}_{\mathbf{a}}(\mathbf{x}) = \mathbf{a}^{-1}\mathbf{a}\cdot\mathbf{x} = \tfrac{1}{2}(\mathbf{x} + \mathbf{a}^{-1}\mathbf{x}\mathbf{a}). \tag{1.3}$$

and its generalizations introduced in Section 2-4.

Although the terms "linear transformation" and "tensor" refer to mathematical functions of the same kind, they are not completely synonymous, because they have different connotations in applications. The term "tensor" is used when describing certain properties of physical systems. For example, the *inertia tensor* is a property of a rigid body to be discussed in Chapter 6. It is never called the "inertia linear transformation". On the other hand, the term "transformation" generally suggests some *change* of state in a physical system or an *equivalence* of one system with another. The term "linear operator" is fairly free of such connotations, so it may be preferred when the emphasis is on mathematical structure.

To handle linear transformations efficiently, we need a suitable notation and formulation of general properties. For a linear transformation it is a common practice to write $f(\mathbf{x}) = f\mathbf{x}$, allowing the parenthesis to be dropped in writing f as a function of $\mathbf{x}$. Accordingly, the composite function $g(f(\mathbf{x}))$ of linear functions f and g can be written in any one of the forms

$$g(f(\mathbf{x})) = g(f\mathbf{x}) = gf(\mathbf{x}) = gf\mathbf{x}. \tag{1.4}$$

The composite gf of linear operators is often called the *product* of g and f. There is some danger of confusing this kind of product with the geometric product AB of multivectors A and B, because we will have occasion to use both kinds of product in the same equation. However, to help keep the distinction between multivectors and linear operators clear, we shall *usually* use *script type to denote linear operators*. An important exception to this convention is the most elementary kind of linear transformation

$$\alpha(\mathbf{x}) = \alpha\mathbf{x}, \tag{1.5}$$

obtained by multiplying vectors by a scalar α. Here the same symbol α is used to denote both a scalar and the associated linear operator.

Both the product gf and the sum $f + g$ of linear operators are themselves linear operators. The product is associative, that is, for linear operators f, g and h we have the rule of composition

$$h(gf) = (hg)f. \tag{1.6}$$

From (1.2a) it follows that the operator product is also distributive with respect to addition;

$$h(g + f) = hg + hf. \tag{1.7}$$

The product of linear operators is not generally commutative. However, from (1.2b) it follows that

$$f\alpha = \alpha f, \tag{1.8}$$

that is, all linear operators commute with the operation of scalar multiplication.

Note that in Equations (1.6, 7, 8) the linear operators can be regarded as combining with other operators rather than operating directly on vectors. The general rules for adding and multiplying operators are the same as rules of elementary scalar algebra, except for the restrictions on the commutative law. For this reason, the study and application of linear operators is called *linear algebra* or operator algebra.

The reader cannot have failed to notice that the abstract algebraic rules governing linear algebra are identical to rules governing geometric algebra. This identity is no accident. Every specific linear operator can be constructed from multivectors using the geometric sum and product alone. Equation (1.3), for example, gives the construction or, if you will, the definition of projection operators in terms of geometric algebra. We will find similar constructions for all the important linear operators. It will be apparent then that the associativity (1.6) and distributivity (1.7) of linear operators can be regarded as consequences of the associativity and distributivity of the geometric product. Thus, linear algebra can be regarded as an important application of geometric algebra rather than an independent mathematical system.

Let us now turn to the development of some general concepts useful for characterizing and classifying linear operators.

Adjoint Operators

To every linear operator f on $\mathcal{E}_3$ there corresponds another linear operator $\bar{f}$ on $\mathcal{E}_3$ uniquely defined by the condition that

$$\mathbf{y}\cdot f(\mathbf{x}) = \bar{f}(\mathbf{y})\cdot\mathbf{x} \tag{1.9}$$

for all vectors $\mathbf{x}$ and $\mathbf{y}$ in $\mathcal{E}_3$. To emphasize that f operates before the inner product, we may write (1.9) in the form

$$\mathbf{y}\cdot(f\mathbf{x}) = (\bar{f}\mathbf{y})\cdot\mathbf{x}. \tag{1.10}$$

The operator $\overline{f}$ is called the *adjoint* or *transpose* of f. Its significance will become clear after we have seen how it can be used.

Outermorphisms

Recall from Section 2–3 that a vector space $\mathcal{E}_3$ generates a geometric algebra $\mathcal{G}_3$ with $\mathcal{E}_3 = \langle \mathcal{G}_3 \rangle_1$. We shall now show how every linear transformation f on $\mathcal{E}_3$ *induces* a natural linear transformation $\underline{f}$ on $\mathcal{G}_3$. We define the induced transformation $\underline{f}(\mathbf{x}\wedge\mathbf{y})$ of a bivector $\mathbf{x}\wedge\mathbf{y}$ by

$$\underline{f}(\mathbf{x}\wedge\mathbf{y}) = f(\mathbf{x})\wedge f(\mathbf{y}) = (f\mathbf{x})\wedge(f\mathbf{y}). \tag{1.11}$$

Thus, $\underline{f}$ transforms bivectors into bivectors. The fact that it is a linear transformation of bivectors follows from the linearity of the outer product. In particular, the distributive rule for the outer product gives

$$\underline{f}(\mathbf{x}\wedge\mathbf{y} + \mathbf{x}\wedge\mathbf{z}) = \underline{f}(\mathbf{x}\wedge\mathbf{y}) + \underline{f}(\mathbf{x}\wedge\mathbf{z}) \tag{1.12}$$

Naturally, the induced transformation of a trivector into a trivector is defined by

$$\underline{f}(\mathbf{x}\wedge\mathbf{y}\wedge\mathbf{z}) = (f\mathbf{x})\wedge(f\mathbf{y})\wedge(f\mathbf{z}) \tag{1.13}$$

Since every trivector is proportional to the dextral unit pseudoscalar i, we can write

$$\underline{f}(\mathbf{x}\wedge\mathbf{y}\wedge\mathbf{z}) = (\det f)\mathbf{x}\wedge\mathbf{y}\wedge\mathbf{z}, \tag{1.14}$$

where $\det f$, called the *determinant* of f, is a scalar depending on f. Since $\mathbf{x}\wedge\mathbf{y}\wedge\mathbf{z}$ is the oriented volume of a parallelepiped with "edges" $\mathbf{x}$, $\mathbf{y}$, $\mathbf{z}$, we can interpret (1.14) as an induced change in scale of the volume by the factor $\det f$. If $\det f$ is negative, then the orientation as well as the magnitude of the volume is changed.

Supposing that $\mathbf{x}\wedge\mathbf{y}\wedge\mathbf{z} = i\mathbf{x}\cdot(\mathbf{y}\times\mathbf{z})$ is not zero, we can solve to get several equivalent expressions for the determinant:

$$\begin{aligned}\det f &= i^{-1}\underline{f}(i) = \frac{\underline{f}(\mathbf{x}\wedge\mathbf{y}\wedge\mathbf{z})}{\mathbf{x}\wedge\mathbf{y}\wedge\mathbf{z}} = \frac{(f\mathbf{x})\cdot[(f\mathbf{y})\times(f\mathbf{z})]}{\mathbf{x}\cdot(\mathbf{y}\times\mathbf{z})} \\ &= (\mathbf{z}^{-1}\wedge\mathbf{y}^{-1}\wedge\mathbf{x}^{-1})\cdot\underline{f}(\mathbf{x}\wedge\mathbf{y}\wedge\mathbf{z}).\end{aligned} \tag{1.15}$$

The first equality can be regarded as a definition of the determinant. This is consistent with the more general definition of a determinant given in Exercise (1.8).

The induced transformation of a trivector is simpler than that of a bivector, because it involves a change of scale only. However, since $\mathbf{x}\wedge\mathbf{y}$ is a directed area, the linear transformation $\underline{f}(\mathbf{x}\wedge\mathbf{y})$ can be interpreted as a change of scale together with a change in direction of the directed area.

We can extend the induced transformation of the entire geometric algebra by adopting the notation $\underline{f}\mathbf{x} = f\mathbf{x}$ for vectors and defining the induced transformation of a scalar α by

$$\underline{f}(\alpha) = \alpha . \tag{1.16}$$

Then $\underline{f}$ is defined uniquely on all multivectors $X, Y, \ldots$ in $\mathcal{G}_3$. To sum up, the operator $\underline{f}$ has the following properties: It is linear,

$$\underline{f}(\alpha X + \beta Y) = \alpha\underline{f}(X) + \beta\underline{f}(Y); \tag{1.17a}$$

It is grade-preserving,

$$\underline{f}(\langle X\rangle_k) = \langle\underline{f}(X)\rangle_k; \tag{1.17b}$$

and it preserves (i.e. commutes with) outer products,

$$\underline{f}(X\wedge Y) = \underline{f}(X)\wedge\underline{f}(Y). \tag{1.17c}$$

It does not preserve the inner product, that is, $\underline{f}(X\cdot Y)$ is not generally equal to $\underline{f}(X)\cdot\underline{f}(Y)$.

The transformation $\underline{f}$ induced by f is called an outermorphism of $\mathcal{G}_3$. The root "morphism" is widely used in mathematics with reference to functions which preserve some sort of mathematical structure. Thus, the name "outermorphism" expresses the fact that $\underline{f}$ preserves the outer product.

Since the adjoint $\bar{f}$ of f is also a linear transformation, it too induces an outermorphism, which we designate by the same symbol $\bar{f}$ and define by writing

$$\bar{f}(\mathbf{x}_1\wedge\ldots\wedge\mathbf{x}_k) = (\bar{f}\mathbf{x}_1)\wedge\ldots\wedge(\bar{f}\mathbf{x}_k). \tag{1.18}$$

Obviously, the general properties of the outermorphism $\bar{f}$ are the same as those written down for $\underline{f}$ in (1.17a, b, c).

Nonsingular Linear Operators

A linear operator f on $\mathcal{E}_3$ is said to be *nonsingular* if and only if $\det f \neq 0$ or, equivalently, $\underline{f}(i) \neq 0$.

If f is a nonsingular linear operator, there exists a linear operator f^{-1}, called the *inverse* of f, such that

$$f^{-1}f = 1, \tag{1.19}$$

where 1 is the *identity operator* defined, in accordance with (1.5), by

$$1(\mathbf{x}) = \mathbf{x} \tag{1.20}$$

Thus, for any vector $\mathbf{x}$ in $\mathcal{E}_3$, we have

$$f^{-1}f(\mathbf{x}) = \mathbf{x}. \tag{1.21}$$

The inverse operator can be computed from $\bar{f}$ by using the equation

$$f^{-1}(\mathbf{y}) = \bar{f}(\mathbf{y}i)\,/\,\bar{f}(i) = \frac{\bar{f}(\mathbf{y}i)i^{-1}}{\det f}, \tag{1.22}$$

which is obviously valid only if f is nonsingular. Note the role of the adjoint outermorphism and the double dual in (1.22). The right side of (1.22) shows that $f^{-1}(\mathbf{y})$ is obtained from the induced transformation of the bivector $\mathbf{y}i$ dual to $\mathbf{y}$.

To prove (1.22), we employ the factorization $i = \boldsymbol{\sigma}_1\boldsymbol{\sigma}_2\boldsymbol{\sigma}_3$ of the pseudoscalar and proceed as follows:

$$\begin{aligned}\mathbf{x}\overline{f}(i) &= \mathbf{x}\cdot[(\overline{f}\boldsymbol{\sigma}_1)\wedge(\overline{f}\boldsymbol{\sigma}_2)\wedge(\overline{f}\boldsymbol{\sigma}_3)]\\ &= \mathbf{x}\cdot(\overline{f}\boldsymbol{\sigma}_1)\ \overline{f}(\boldsymbol{\sigma}_2\wedge\boldsymbol{\sigma}_3) - \mathbf{x}\cdot(\overline{f}\boldsymbol{\sigma}_2)\ \overline{f}(\boldsymbol{\sigma}_1\wedge\boldsymbol{\sigma}_3) + \mathbf{x}\cdot(\overline{f}\boldsymbol{\sigma}_3)\ \overline{f}(\boldsymbol{\sigma}_1\wedge\boldsymbol{\sigma}_2)\\ &= (f\mathbf{x})\cdot\boldsymbol{\sigma}_1\ \overline{f}(i\boldsymbol{\sigma}_1) + (f\mathbf{x})\cdot\boldsymbol{\sigma}_2\ \overline{f}(i\boldsymbol{\sigma}_2) + (f\mathbf{x})\cdot\boldsymbol{\sigma}_3\ \overline{f}(i\boldsymbol{\sigma}_3)\\ &= \overline{f}(if\mathbf{x}).\end{aligned}$$

Dividing this by $\overline{f}(i)$ and using the defining identity (1.21), we get (1.22) as required. The student should carefully consider the justification for each step in this proof.

Matrix Representations of Linear Operators

For some kinds of computation it is convenient to employ a standard basis $\boldsymbol{\sigma}_1$, $\boldsymbol{\sigma}_2$, $\boldsymbol{\sigma}_3$ for $\mathscr{E}_3$ defined by the orthonormality condition

$$\boldsymbol{\sigma}_i\cdot\boldsymbol{\sigma}_j = \delta_{ij} \tag{1.23}$$

($i, j = 1, 2, 3$), and the relation

$$\boldsymbol{\sigma}_1\wedge\boldsymbol{\sigma}_2\wedge\boldsymbol{\sigma}_3 = \boldsymbol{\sigma}_1\boldsymbol{\sigma}_2\boldsymbol{\sigma}_3 = i \tag{1.24}$$

of the base vectors to the dextral unit pseudoscalar.

Any vector $\mathbf{x}$ in $\mathscr{E}_3$ can be expanded in a standard basis; thus,

$$\mathbf{x} = \sum_k \boldsymbol{\sigma}_k x_k \equiv \sum_{k=1}^{3} \boldsymbol{\sigma}_k x_k\ . \tag{1.25a}$$

The scalar components x_k in the expansion (1.25a) are given by

$$x_k = \boldsymbol{\sigma}_k\cdot\mathbf{x}, \tag{1.25b}$$

for $k = 1, 2, 3$.

A linear operator f transforms each vector $\boldsymbol{\sigma}_k$ in the standard basis into a vector $\mathbf{f}_k$ which can be expanded in the standard basis, as expressed by the equation

$$\mathbf{f}_k = f\boldsymbol{\sigma}_k = \sum_j \boldsymbol{\sigma}_j f_{jk} \tag{1.26a}$$

Each of the scalar coefficients f_{jk} is called a *matrix element* of the operator f, and the set of all such matrix elements denoted by $[f] = [f_{jk}]$ is called *the matrix of* f in the standard basis. The matrix is called a 3×3 matrix to indicate the range of the indices $j, k = 1, 2, 3$. The matrix elements of f are given by

$$f_{jk} = \boldsymbol{\sigma}_j \cdot (f\boldsymbol{\sigma}_k) = \boldsymbol{\sigma}_j \cdot \mathbf{f}_k . \tag{1.26b}$$

The complete matrix can be written as an array of the matrix elements in the following way:

$$[f] = [f_{jk}] = \begin{bmatrix} f_{11} & f_{12} & f_{13} \\ f_{21} & f_{22} & f_{23} \\ f_{31} & f_{32} & f_{33} \end{bmatrix} .$$

A linear operator is completely determined by its matrix in a given basis, for the matrix determines the transformation of the basis by (1.26a), which, in turn, determines the transformation of any given vector.

$$f\mathbf{x} = \sum_k (f\boldsymbol{\sigma}_k)x_k = \sum_j \sum_k \boldsymbol{\sigma}_j f_{jk} x_k . \tag{1.27}$$

Consequently, the equation

$$f\mathbf{x} = \mathbf{y} \tag{1.28a}$$

is equivalent to the *matrix equation*

$$\sum_k f_{jk} x_k = y_j, \tag{1.28b}$$

which is actually a set of 3 simultaneous equations obtained by dotting (1.28a) with each of the vectors $\boldsymbol{\sigma}_j$ and using (1.27). This can be expressed by writing the matrix equation (1.28b) as an array of the form

$$\begin{bmatrix} f_{11} \, f_{12} \, f_{13} \\ f_{21} \, f_{22} \, f_{23} \\ f_{31} \, f_{32} \, f_{33} \end{bmatrix} \begin{bmatrix} x_1 \\ x_2 \\ x_3 \end{bmatrix} = \begin{bmatrix} f_{11}\, x_1 + f_{12}\, x_2 + f_{13}\, x_3 \\ f_{21}\, x_1 + f_{22}\, x_2 + f_{23}\, x_3 \\ f_{31}\, x_1 + f_{32}\, x_2 + f_{33}\, x_3 \end{bmatrix} = \begin{bmatrix} y_1 \\ y_2 \\ y_3 \end{bmatrix} .$$

The set of 3×3 matrices can be made into a matrix algebra which is equivalent to the linear algebra of operators on $\mathcal{E}_3$. Thus, the operator sum $f + g$ corresponds to the *matrix sum*

$$f_{jk} + g_{jk} = \boldsymbol{\sigma}_j \cdot (f\boldsymbol{\sigma}_k + g\boldsymbol{\sigma}_k) . \tag{1.29}$$

The operator product gf corresponds to the matrix product

$$\sum_j g_{ij} f_{jk} = \boldsymbol{\sigma}_i \cdot (gf\boldsymbol{\sigma}_k), \tag{1.30a}$$

since

$$gf\boldsymbol{\sigma}_k = \sum_j (g\boldsymbol{\sigma}_j) f_{jk} = \sum_i \boldsymbol{\sigma}_i \Big(\sum_j g_{ij} f_{jk} \Big). \tag{1.30b}$$

Thus, the product of matrices is equal to the matrix of the operator product:

$$[g][f] = [gf] . \tag{1.30c}$$

According to (1.21) and (1.23), the *identity matrix* corresponding to the identity operator is determined by

$$\boldsymbol{\sigma}_i \cdot 1(\boldsymbol{\sigma}_k) = \boldsymbol{\sigma}_i \cdot \boldsymbol{\sigma}_k = \delta_{ik} . \tag{1.31}$$

Consequently, the operator equation $1f = f$ corresponds to the matrix equation

$$\sum_j \delta_{ij} f_{jk} = f_{ik}, \qquad \text{or} \qquad [1][f] = [f] \tag{1.32}$$

And the equation $f^{-1}f = 1$ corresponds to

$$\sum_j f^{-1}{}_{ij} f_{jk} = \delta_{ik}, \qquad \text{or} \qquad [f^{-1}][f] = [1]. \tag{1.33}$$

Any other operator equation can be converted into a matrix equation in a similar way, and vice-versa.

For any matrix $[f]$ we define the determinant as equal to the determinant of the corresponding operator f. Thus, using (1.26b) and (1.15) we can write

$$\begin{aligned} \det f &= \det\,[f] = \det\,[\boldsymbol{\sigma}_j \cdot \mathbf{f}_k] \\ &= (\boldsymbol{\sigma}_3 \wedge \boldsymbol{\sigma}_2 \wedge \boldsymbol{\sigma}_1) \cdot (\mathbf{f}_1 \wedge \mathbf{f}_2 \wedge \mathbf{f}_3) \end{aligned} \tag{1.34}$$

The value of the determinant is a scalar which can be computed from the matrix elements by the Laplace expansion (Exercise (1.9)).

Matrix algebra is widely used in mathematics and physics to carry out calculations with linear operators. Since the matrix elements are scalars, matrix algebra has the advantage of reducing all such calculations to addition and multiplication of real numbers. It has the disadvantage, however, of requiring that a basis be introduced which may be quite irrelevant to the problem at hand, and this often obscures the geometrical meaning of the transformations involved.

Geometric algebra is a more general and efficient computational tool than matrix algebra. In the next two sections we shall see how the most important linear transformations can be expressed in terms of geometric algebra so that computations can be carried out without introducing an arbitrary basis. This is not to say that we shall dispense with matrix algebra. Rather we shall regard it as subsidiary to geometric algebra. In some problems a basis is natural or information is given in matrix form, so matrices should be used. In other problems, which we shall formulate and solve without matrices, the results will be put in matrix form for comparison with standard treatments using matrix algebra. Matrix algebra itself is simplified and clarified when used in conjunction with the operations of geometric algebra, because geometric algebra enables us to operate directly with vectors without decomposing them into components.

5-1. Exercises

(1.1) Prove that $\underline{f}(\mathbf{x}\wedge\mathbf{y}) = 0$ for $\mathbf{x}\wedge\mathbf{y} \neq 0$ if and only if $f(\mathbf{z}) = 0$ for some nonzero vector $\mathbf{z}$ in the $\mathbf{x}\wedge\mathbf{y}$ - plane.

(1.2) Prove that $\underline{f}(\alpha\mathbf{X}) = \alpha\underline{f}(\mathbf{X})$ when $\mathbf{X}$ is a k-blade.

(1.3) Prove that $(\mathbf{u}\wedge\mathbf{v})\cdot\underline{f}(\mathbf{x}\wedge\mathbf{y}) = \overline{f}(\mathbf{u}\wedge\mathbf{v})\cdot(\mathbf{x}\wedge\mathbf{y})$. Generalize the proof to show that $\det f = \det \overline{f}$.

(1.4) Prove that the following propositions about a linear transformation f on $\mathcal{E}_3$ are equivalent:
   (a) f is nonsingular.
   (b) $f(\mathbf{x}) = 0$ if and only if $\mathbf{x} = 0$.
   (c) To every vector $\mathbf{y}$ there corresponds a unique vector $\mathbf{x}$ such that $\mathbf{y} = f(\mathbf{x})$.

(1.5) Prove the following identities:
   (a) $\det(gf) = (\det g)(\det f)$.
   (b) $\det(f^{-1}) = (\det f)^{-1}$.

(1.6) To find the inverse of a linear transformation, Equation (1.22) can always be used, but a more direct approach is often better. Find the inverse of

$$f\mathbf{x} = \alpha\mathbf{x} + \mathbf{a}\mathbf{b}\cdot\mathbf{x}$$

by solving the algebraic equation $\mathbf{y} = f\mathbf{x}$ for $\mathbf{x}$ as a function of $\mathbf{y}$.

(1.7) Find the inverse of the linear transformation

$$g\mathbf{x} = \alpha\mathbf{x} + \mathbf{x}\cdot\mathbf{B} = \alpha\mathbf{x} + \mathbf{b}\times\mathbf{x}$$

where $\mathbf{B} = i\mathbf{b}$ is, of course, a bivector.

(1.8) The entire treatment of linear operators and matrices in this section is easily generalized to vector spaces of any finite dimension. Details are given in the book *Geometric Calculus* (1984), but let us look at some of the basic ideas.

A set of linearly independent vectors $\mathbf{a}_1, \mathbf{a}_2, \ldots, \mathbf{a}_n$ is a *frame* (or basis) for n-dimensional vector space. By generalizing the argument in Exercise (2-1.2), it can be proved that $\mathbf{a}_1\wedge\mathbf{a}_2\wedge\ldots\wedge\mathbf{a}_n \neq 0$ is a necessary and sufficient condition for the vectors to be linearly independent.

Any matrix of scalars α_{ij}, with $i, j = 1, \ldots, n$, can be expressed in the form $\alpha_{ij} = \mathbf{a}_i\cdot\mathbf{b}_j$, where the $\mathbf{a}_i$ and $\mathbf{b}_j$ are vectors. The *determinant* of the matrix is defined by

$$\det \alpha_{ij} = \det \mathbf{a}_i\cdot\mathbf{b}_j = (\mathbf{a}_n\wedge\ldots\wedge\mathbf{a}_1)\cdot(\mathbf{b}_1\wedge\ldots\wedge\mathbf{b}_n).$$

The determinant is commonly represented as an array; thus,

$$\det \alpha_{ij} = \begin{vmatrix} \alpha_{11}\ \alpha_{12}\ \ldots\ \alpha_{1n} \\ \alpha_{21} & \cdot \\ \cdot & \cdot \\ \cdot & \cdot \\ \alpha_{ni} & \alpha_{nn} \end{vmatrix} = \begin{vmatrix} \mathbf{a}_1\cdot\mathbf{b}_1\ \mathbf{a}_1\cdot\mathbf{b}_2\ \ldots\ \mathbf{a}_1\cdot\mathbf{b}_n \\ \mathbf{a}_2\cdot\mathbf{b}_1 & \cdot \\ \cdot & \cdot \\ \cdot & \cdot \\ \mathbf{a}_n\cdot\mathbf{b}_1 & \mathbf{a}_n\cdot\mathbf{b}_n \end{vmatrix}$$

The number of rows and columns in a determinant is called its rank.

All the properties of determinants are consequences of general properties of inner and outer products established in Section 2.1. Establish the following properties: A determinant

(a) changes sign if any two rows are interchanged;
(b) is unchanged by an interchange of rows and columns;
(c) vanishes if two rows are equal;
(d) vanishes if the rows are linearly dependent.

A determinant of *rank n* can be reduced to determinants of lower rank by using the *Laplace expansion*:

$$(\mathbf{a}_n\wedge \ldots \wedge\mathbf{a}_1)\cdot(\mathbf{b}_1\wedge \ldots \wedge\mathbf{b}_n)$$
$$= \sum_{k=1}^{n} (-1)^{k+1}\mathbf{a}_1\cdot\mathbf{b}_k(\mathbf{a}_n\wedge \ldots \wedge\mathbf{a}_2)\cdot(\mathbf{b}_1\wedge \ldots \check{\mathbf{b}}_k \ldots \wedge\mathbf{b}_n).$$

Derive the result.

(1.9) Use the Laplace expansion to evaluate the determinant of a linear operator in terms of its matrix elements. Specifically, from Equation (1.34), derive the result

$$\det f_{jk} = f_{11}(f_{22}f_{33} - f_{32}f_{23}) - f_{12}(f_{21}f_{33} - f_{31}f_{23}) + f_{13}(f_{21}f_{32} - f_{22}f_{31}).$$

(1.10) The equation

$$\alpha_1\mathbf{a}_1 + \alpha_2\mathbf{a}_2 + \ldots + \alpha_n\mathbf{a}_n = \mathbf{c}$$

can be solved for the scalars α_k in terms of the vectors $\mathbf{a}_k$ and $\mathbf{c}$ if the $\mathbf{a}_k$ are linearly independent, that is, if $A_n \equiv \mathbf{a}_1\wedge \ldots \wedge\mathbf{a}_n \neq 0$. Derive the solution

$$\alpha_k = \frac{\mathbf{a}_1\wedge \ldots (\mathbf{c})_k \ldots \wedge\mathbf{a}_n}{\mathbf{a}_1\wedge \ldots \wedge\mathbf{a}_n}$$

where $(\mathbf{c})_k$ indicates that $\mathbf{c}$ has been substituted for $\mathbf{a}_k$ in the product $\mathbf{a}_1\wedge \ldots \wedge\mathbf{a}_n$. Suppose that $B_n \equiv \mathbf{b}_1\wedge \ldots \wedge\mathbf{b}_n \neq 0$ is proportional to A_n. Derive *Cramer's rule*, expressing the α_k as a ratio of determinants:

$$\alpha_k = \frac{(\mathbf{b}_n\wedge \ldots \wedge\mathbf{b}_1)\cdot(\mathbf{a}_1\wedge \ldots (\mathbf{c})_k \ldots \wedge\mathbf{a}_n)}{(\mathbf{b}_n\wedge \ldots \wedge\mathbf{b}_1)\cdot(\mathbf{a}_1\wedge \ldots \wedge\mathbf{a}_n)} .$$

(1.11) *Frames and Reciprocal Frames*

A frame $\{\mathbf{e}_k,\ k = 1, 2, 3\}$ of vectors in $\mathcal{G}_3$ (i) determines a *pseudoscalar* $\mathbf{e}_1\wedge\mathbf{e}_2\wedge\mathbf{e}_3$ which is necessarily a non-vanishing scalar multiple of the righthanded unit pseudoscalar i; thus, $\mathbf{e}_1\wedge\mathbf{e}_2\wedge\mathbf{e}_3 = ei$. The *determinant* of the frame $e = -i\mathbf{e}_1\wedge\mathbf{e}_2\wedge\mathbf{e}_3$ is positive (negative) if the frame is right (left) handed. The *reciprocal frame* $\{\mathbf{e}^k, k = 1, 2, 3\}$ is determined by the set of equations

$$\mathbf{e}^k\cdot\mathbf{e}_j = \delta_j^k,$$

for $j, k = 1, 2, 3$, where $\delta_j^k = 1$ if $k = j$ *and* $\delta_j^k = 0$ if $k \neq j$. Show that the unique solution of these equations are given by

$$\mathbf{e}^1 = \frac{\mathbf{e}_2\wedge\mathbf{e}_3}{\mathbf{e}_1\wedge\mathbf{e}_2\wedge\mathbf{e}_3} = \frac{\mathbf{e}_2\times\mathbf{e}_3}{e},$$

$$\mathbf{e}^2 = \frac{\mathbf{e}_3\wedge\mathbf{e}_1}{\mathbf{e}_1\wedge\mathbf{e}_2\wedge\mathbf{e}_3} = \frac{\mathbf{e}_3\times\mathbf{e}_1}{e},$$

$$\mathbf{e}^3 = \frac{\mathbf{e}_1\wedge\mathbf{e}_2}{\mathbf{e}_1\wedge\mathbf{e}_2\wedge\mathbf{e}_3} = \frac{\mathbf{e}_1\times\mathbf{e}_2}{e}.$$

Note that the orthonormal frame $\{\boldsymbol{\sigma}_k\}$ is reciprocal to itself.

Any vector $\mathbf{a}$ in $\mathcal{G}_3$ (i) can be expressed as the linear combination

$$\mathbf{a} = a^1\mathbf{e}_1 + a^2\mathbf{e}_2 + a^3\mathbf{e}_3 = a^k\mathbf{e}_k,$$

where the summation convention has been used to abbreviate the sum on the right. The scalar coefficients a^k are commonly called *contravariant components* of the vector $\mathbf{a}$ (with respect to the frame $\{\mathbf{e}_k\}$). The reciprocal frame simplifies the problem of determining these coefficients from $\mathbf{a}$ and $\{\mathbf{e}_k\}$. Show that

$$a^k = \mathbf{e}^k\cdot\mathbf{a},$$

and that this solution is merely an applications of Cramer's rule (Exercise (1.11)). Similarly show that the covariant components a_k of $\mathbf{a}$, which are defined by the equation $\mathbf{a} = a_k\mathbf{e}^k$, are determined by the equations $a_k = \mathbf{e}_k\cdot\mathbf{a}$.

(1.12) Let $\mathcal{E}_n$ be an n-dimensional vector space with an orthonormal basis $\boldsymbol{\sigma}_1, \boldsymbol{\sigma}_2, \ldots \boldsymbol{\sigma}_n$ and pseudoscalar $i = \boldsymbol{\sigma}_1\boldsymbol{\sigma}_2 \ldots \boldsymbol{\sigma}_n$. For a linear operator f on $\mathcal{E}_n$, the matrix elements of the adjoint operator $\overline{f}$ are given by

$$\overline{f}_{ij} = \boldsymbol{\sigma}_i\cdot(\overline{f}\boldsymbol{\sigma}_j) = \boldsymbol{\sigma}_j\cdot(f\boldsymbol{\sigma}_i) = f_{ji}.$$

Thus, the matrix element $\overline{f}_{ij}$ is obtained simply by transposing the indices on f_{ij}. The transformation of the basis by $\overline{f}$ is therefore given by

$$\overline{\mathbf{f}}_k = \overline{f}\boldsymbol{\sigma}_k = \sum_j \boldsymbol{\sigma}_j\,\overline{f}_{jk} = \sum_j f_{kj}\sigma_j.$$

Show that the matrix elements of the inverse operator f^{-1} are given by the ratio of determinants

$$f_{jk}^{-1} = \frac{\bar{\mathbf{f}}_1\wedge \ldots (\boldsymbol{\sigma}_j)_k \ldots \wedge \bar{\mathbf{f}}_n}{\mathbf{f}_1\wedge \ldots \wedge \mathbf{f}_n}$$

$$= \frac{(\boldsymbol{\sigma}_n\wedge \ldots \wedge \boldsymbol{\sigma}_1)\cdot(\bar{\mathbf{f}}_1\wedge \ldots (\boldsymbol{\sigma}_j)_k \ldots \wedge \bar{\mathbf{f}}_n}{\det f_{ij}} .$$

where $(\boldsymbol{\sigma}_j)_k$ indicates that $\mathbf{f}_k$ has been replaced by $\boldsymbol{\sigma}_j$.

5-2. Symmetric and Skewsymmetric Operators

In this section we study the properties of symmetric and skewsymmetric linear operators. Although the mathematical results have many physical applications, they will be needed in this book only to determine the forms of inertia tensors for rigid bodies. So the student can skip this section until that information is required.

A linear operator $\mathcal{S}$ is said to be *symmetric* (or *self-adjoint*) if $\bar{\mathcal{S}} = \mathcal{S}$, that is, if it is equivalent to its adjoint. Similarly, a linear operator $\mathcal{A}$ is said to be *skewsymmetric* (or *antisymmetric*) if $\bar{\mathcal{A}} = -\mathcal{A}$. Indeed, any linear operator f can be uniquely expressed as the sum of a symmetric operator f_+ and a skewsymmetric part f_-. One simply forms the operator identity

$$f = \tfrac{1}{2}(f + \bar{f}\,) + \tfrac{1}{2}(f - \bar{f}\,).$$

Hence,

$$f = f_+ + f_-, \tag{2.1a}$$

where

$$f_\pm = \tfrac{1}{2}(f \pm \bar{f}\,). \tag{2.1b}$$

Skewsymmetric Operators

We consider skewsymmetric operators first, because they are so easy to characterize completely. Indeed, any skewsymmetric transformation $\mathcal{A}$ can be put in the *canonical* (or standard) *form*

$$\mathcal{A}\mathbf{x} = \mathbf{x}\cdot\mathbf{A}\,, \tag{2.2}$$

where $\mathbf{A}$ is a unique bivector. All the properties of $\mathcal{A}$ are therefore determined by the algebraic properties of the bivector $\mathbf{A}$; the skewsymmetry, for example, follows from

$$\mathbf{y}\cdot(\mathcal{A}\mathbf{x}) = \mathbf{y}\cdot(\mathbf{x}\cdot\mathbf{A}) = (\mathbf{y}\wedge\mathbf{x})\cdot\mathbf{A} = -(\mathbf{x}\wedge\mathbf{y})\cdot\mathbf{A} = -\mathbf{x}\cdot(\mathbf{y}\cdot\mathbf{A}) = -\mathbf{x}\cdot(\mathcal{A}\mathbf{y})\,.$$

We can prove (2.2) by using the fact that $\mathcal{A}$ is completely determined by the transformation $\mathbf{a}_k = \mathcal{A}\boldsymbol{\sigma}_k = \Sigma_j\,\boldsymbol{\sigma}_j A_{jk}$ of a standard basis. In terms of the standard basis, the bivector $\mathbf{A}$ is given by

$$\mathbf{A} = \sum_k \tfrac{1}{2}\boldsymbol{\sigma}_k \wedge \mathbf{a}_k = \tfrac{1}{2}\sum_k \boldsymbol{\sigma}_k \wedge (\mathscr{A}\boldsymbol{\sigma}_k) = \sum_{k,j} \tfrac{1}{2}\boldsymbol{\sigma}_k \wedge \boldsymbol{\sigma}_j A_{jk} \,. \tag{2.3}$$

This is proved by

$$\boldsymbol{\sigma}_j \cdot \mathbf{A} = \tfrac{1}{2}\sum_k \boldsymbol{\sigma}_j \cdot (\boldsymbol{\sigma}_k \wedge \mathbf{a}_k) = \tfrac{1}{2}\sum_k (\boldsymbol{\sigma}_j \cdot \boldsymbol{\sigma}_k\, \mathbf{a}_k - \boldsymbol{\sigma}_k \boldsymbol{\sigma}_j \cdot \mathbf{a}_k)$$

$$= \tfrac{1}{2}\sum_k (\delta_{jk}\mathscr{A}\boldsymbol{\sigma}_k - \boldsymbol{\sigma}_k A_{jk}) = \tfrac{1}{2}\,(\mathscr{A}\boldsymbol{\sigma}_j - \bar{\mathscr{A}}\boldsymbol{\sigma}_j) = \mathscr{A}\boldsymbol{\sigma}_j \,.$$

This establishes (2.2) for a standard basis,·whence, by linearity, the result is generally true.

By way of example, note that the magnetic force on a charged particle is a skewsymmetric linear function $\mathscr{B} = \mathscr{B}\mathbf{v}$ of the particle velocity. Thus,

$$\mathscr{B}\mathbf{v} = \frac{q}{c}\,\mathbf{v} \times \mathbf{B} = -\frac{qi}{c}\mathbf{v}\wedge\mathbf{B} = \mathbf{v}\cdot\left(-\frac{qi\mathbf{B}}{c}\right).$$

Another important skewsymmetric operator will be seen to arise from differentiating a rotation.

Eigenvectors and Eigenvalues

If a nonzero vector $\mathbf{e}$ is transformed into a scalar multiple of itself by a linear operator f, we have the equation

$$f\mathbf{e} = \lambda\mathbf{e}\,, \tag{2.4}$$

where λ is a scalar. We say that $\mathbf{e}$ is an *eigenvector* of f corresponding to the *eigenvalue* λ. Obviously, any nonzero scalar multiple of $\mathbf{e}$ is also an eigenvector of f. The problem of finding the eigenvalues and/or the eigenvectors for a given operator is called the *eigenvalue* or *eigenvector* problem.

The simplicity of the "eigenvalue equation" (2.4) shows that very basic properties of a linear transformation are described by its eigenvectors and eigenvalues. Therefore, it is often important to determine these properties if they are not evident from the form in which the transformation is given. For example, if we are given the matrix f_{jk} of an operator f, then we have the vectors

$$\mathbf{f}_k = f\boldsymbol{\sigma}_k = \sum_{j=1}^{3} \boldsymbol{\sigma}_j f_{jk}\,.$$

To develop a general method for solving the eigenvalue problem from this information, note that (2.4) can be written in the form

$$(f - \lambda)\mathbf{e} = 0\,, \tag{2.5}$$

showing that the operator $(f - \lambda)$ is singular. But every singular operator has a vanishing determinant, hence

$$\det(f-\lambda) = \frac{(\mathbf{f}_1-\lambda\boldsymbol{\sigma}_1)\wedge(\mathbf{f}_2-\lambda\boldsymbol{\sigma}_2)\wedge(\mathbf{f}_3-\lambda\boldsymbol{\sigma}_3)}{\boldsymbol{\sigma}_1\boldsymbol{\sigma}_2\boldsymbol{\sigma}_3} = 0. \tag{2.6}$$

This is commonly called the *secular equation* for f. The left side of (2.6) is a third degree polynomial in λ, with coefficients composed of the f_{ij}. The reader is invited to expand the numerator and show that (2.6) can be put in the form

$$\lambda^3 - \alpha_1\lambda^2 + \alpha_2\lambda - \alpha_3 = 0, \tag{2.7a}$$

where the scalar coefficients are given by

$$\alpha_1 = \sum_k \boldsymbol{\sigma}_k\cdot\mathbf{f}_k = f_{11} + f_{22} + f_{33}, \tag{2.7b}$$

$$\alpha_2 = (\boldsymbol{\sigma}_3\wedge\boldsymbol{\sigma}_2)\cdot(\mathbf{f}_2\wedge\mathbf{f}_3) + (\boldsymbol{\sigma}_3\wedge\boldsymbol{\sigma}_1)\cdot(\mathbf{f}_1\wedge\mathbf{f}_3) + (\boldsymbol{\sigma}_2\wedge\boldsymbol{\sigma}_1)\cdot(\mathbf{f}_1\wedge\mathbf{f}_2) \tag{2.7c}$$

$$\alpha_3 = \det f = (\boldsymbol{\sigma}_3\wedge\boldsymbol{\sigma}_2\wedge\boldsymbol{\sigma}_1)\cdot(\mathbf{f}_1\wedge\mathbf{f}_2\wedge\mathbf{f}_3). \tag{2.7d}$$

Since the secular equation is an algebraic equation of the third degree, the fundamental theorem of algebra tells us that it has at most three distinct roots, some of which may be complex numbers. The real roots are the desired eigenvalues. Complex roots are also regarded as eigenvalues in conventional treatments of linear algebra, but geometric algebra makes this unnecessary, as explained below and at the end of Section 5–3.

After the eigenvalues have been determined, the corresponding eigenvectors can be found from (2.5). To do this, it is convenient to write (2.5) in the form

$$\mathbf{g}_1 e_1 + \mathbf{g}_2 e_2 + \mathbf{g}_3 e_3 = 0 \tag{2.8a}$$

where the vectors

$$\mathbf{g}_k = \mathbf{f}_k - \lambda\boldsymbol{\sigma}_k \tag{2.8b}$$

are known for each eigenvalue λ and the scalar components $e_k = \mathbf{e}\cdot\boldsymbol{\sigma}_k$ of the eigenvector are to be determined. Equation, (2.8a) can be solved for ratios of the e_k (Cramer's rule). Thus, we can "wedge" (2.8a) with $\mathbf{g}_3$ to get

$$e_1\mathbf{g}_1\wedge\mathbf{g}_3 + e_2\mathbf{g}_2\wedge\mathbf{g}_3 = 0.$$

Whence

$$\frac{e_2}{e_1} = \frac{\mathbf{g}_3\wedge\mathbf{g}_1}{\mathbf{g}_2\wedge\mathbf{g}_3}. \tag{2.9a}$$

Similarly,

$$\frac{e_3}{e_1} = \frac{\mathbf{g}_1\wedge\mathbf{g}_2}{\mathbf{g}_2\wedge\mathbf{g}_3}. \tag{2.9b}$$

Since the length and sense (or orientation) of the eigenvector $\mathbf{e}$ is not determined by the eigenvector equation (2.5), we are free to fix them by assigning any convenient value to the component e_1; then e_2 and e_3 are uniquely determined by (2.9a, b).

If λ is a single root of the secular equation, then 2 of the 3 vectors $\mathbf{g}_k = \mathbf{f}_k - \lambda\boldsymbol{\sigma}_k$ are necessarily linearly independent, and only one component of the eigenvector $\mathbf{e}$ can be specified arbitrarily, as we have seen. However, if λ is a double root of the secular equation, the $\mathbf{g}_k$ are not linearly independent and two components of $\mathbf{e}$ can be specified arbitrarily. In this case, Equations (2.9a, b) cannot be used to obtain e_2 and e_3. However, we are free to set $e_1 = e_2 = 1$, so going back to (2.8a) we get

$$\mathbf{g}_1 + \mathbf{g}_2 + e_3\mathbf{g}_3 = 0. \tag{2.10a}$$

from which e_3 can be obtained trivially. Alternatively, we can set $e_1 = 1$ and $e_3 = 0$, so (2.8a) reduces to

$$\mathbf{g}_1 + e_2\mathbf{g}_2 = 0. \tag{2.10b}$$

The eigenvector we get from (2.10b) is linearly independent of the eigenvector we get from (2.10a). Any other eigenvector obtained by a different choice of components will be a linear combination of these two eigenvectors. Thus, the eigenvectors corrresponding to a double root of the secular equation form a plane, so the eigenvector problem is solved when two independent vectors in that plane have been found.

A secular equation with a multiple root is said to be *degenerate*; more specifically, it is said to be k-fold degenerate if the root has multiplicity k. To an eigenvalue with multiplicity k there corresponds exactly k linearly independent eigenvectors, which can be found in the manner described for $k = 2$.

Example

Let us see how the general method works on a specific example. Let us solve the eigenvalue problem for the linear transformation specified by the matrix

$$[f] = \begin{bmatrix} 4 & -1 & -1 \\ -1 & 4 & -1 \\ -1 & -1 & 4 \end{bmatrix}.$$

Operating on a standard basis, this matrix gives

$$\begin{aligned} f\boldsymbol{\sigma}_1 &= 4\boldsymbol{\sigma}_1 - \boldsymbol{\sigma}_2 - \boldsymbol{\sigma}_3 = \mathbf{f}_1, \\ f\boldsymbol{\sigma}_2 &= -\boldsymbol{\sigma}_1 + 4\boldsymbol{\sigma}_2 - \boldsymbol{\sigma}_3 = \mathbf{f}_2, \\ f\boldsymbol{\sigma}_3 &= -\boldsymbol{\sigma}_1 - \boldsymbol{\sigma}_2 + 4\boldsymbol{\sigma}_3 = \mathbf{f}_3. \end{aligned} \tag{2.11}$$

From these vectors we calculate

$$\mathbf{f}_1 \wedge \mathbf{f}_2 = (4\boldsymbol{\sigma}_1 - \boldsymbol{\sigma}_2 - \boldsymbol{\sigma}_3) \wedge (-\boldsymbol{\sigma}_1 + 4\boldsymbol{\sigma}_2 - \boldsymbol{\sigma}_3),$$

which after expansion and collection of like terms takes the form

$$\mathbf{f}_1\wedge\mathbf{f}_2 = 15\boldsymbol{\sigma}_1\wedge\boldsymbol{\sigma}_2 + 5\boldsymbol{\sigma}_2\wedge\boldsymbol{\sigma}_3 + 5\boldsymbol{\sigma}_3\wedge\boldsymbol{\sigma}_1 .$$

Similarly, we find

$$\mathbf{f}_2\wedge\mathbf{f}_3 = 5\boldsymbol{\sigma}_1\wedge\boldsymbol{\sigma}_2 + 15\boldsymbol{\sigma}_2\wedge\boldsymbol{\sigma}_3 + 5\boldsymbol{\sigma}_3\wedge\boldsymbol{\sigma}_1 ,$$

$$\mathbf{f}_3\wedge\mathbf{f}_1 = 5\boldsymbol{\sigma}_1\wedge\boldsymbol{\sigma}_2 + 5\boldsymbol{\sigma}_2\wedge\boldsymbol{\sigma}_3 + 15\boldsymbol{\sigma}_3\wedge\boldsymbol{\sigma}_1 ,$$

as well as

$$\mathbf{f}_1\wedge\mathbf{f}_2\wedge\mathbf{f}_3 = 50\boldsymbol{\sigma}_1\wedge\boldsymbol{\sigma}_2\wedge\boldsymbol{\sigma}_3 .$$

We use these multivectors in (2.7b, c, d) to evaluate the coefficients in the secular equation; thus,

$$\alpha_1 = 4 + 4 + 4 = 12,$$

$$\alpha_2 = 15 + 15 + 15 = 45,$$

$$\alpha_3 = 50.$$

Hence the secular equation (2.7a) takes the specific form

$$\lambda^3 - 12\lambda^2 + 45\lambda - 50 = 0 .$$

This polynomial has the factored form

$$(\lambda - 2)\,(\lambda - 5)^2 = 0 .$$

Hence the eigenvalues are 2 and 5 with double degeneracy.

To determine the eigenvector corresponding to the eigenvalue $\lambda = 2$, we use (2.11) in (2.8b) to get

$$\mathbf{g}_1 = \mathbf{f}_1 - 2\boldsymbol{\sigma}_1 = 2\boldsymbol{\sigma}_1 - \boldsymbol{\sigma}_2 - \boldsymbol{\sigma}_3$$

$$\mathbf{g}_2 = \mathbf{f}_2 - 2\boldsymbol{\sigma}_2 = - \boldsymbol{\sigma}_1 + 2\boldsymbol{\sigma}_2 - \boldsymbol{\sigma}_3$$

$$\mathbf{g}_3 = \mathbf{f}_3 - 2\boldsymbol{\sigma}_3 = - \boldsymbol{\sigma}_1 - \boldsymbol{\sigma}_2 + 2\boldsymbol{\sigma}_3 .$$

From this we obtain

$$\mathbf{g}_1\wedge\mathbf{g}_2 = 3(\boldsymbol{\sigma}_1\wedge\boldsymbol{\sigma}_2 + \boldsymbol{\sigma}_2\wedge\boldsymbol{\sigma}_3 + \boldsymbol{\sigma}_3\wedge\boldsymbol{\sigma}_1)$$

$$= \mathbf{g}_2\wedge\mathbf{g}_3 = \mathbf{g}_3\wedge\mathbf{g}_1 .$$

Using this in (2.9a, b) with $e_1 = 1$, we get $e_2 = e_3 = 1$. Hence,

$$\mathbf{e}_1 \equiv \boldsymbol{\sigma}_1 + \boldsymbol{\sigma}_2 + \boldsymbol{\sigma}_3 \tag{2.12}$$

is the desired eigenvector.

To find an eigenvector corresponding to $\lambda = 5$, we evaluate $\mathbf{g}_k = \mathbf{f}_k - 5\boldsymbol{\sigma}_k$ and find that

$$\mathbf{g}_1 = \mathbf{g}_2 = \mathbf{g}_3 = - (\boldsymbol{\sigma}_1 + \boldsymbol{\sigma}_2 + \boldsymbol{\sigma}_3) .$$

Using this in (2.10a), we find $e_3 = -2$ when $e_1 = e_2 = 1$; hence,

$$\mathbf{e}_2 \equiv \boldsymbol{\sigma}_1 + \boldsymbol{\sigma}_2 - 2\boldsymbol{\sigma}_3 \tag{2.13a}$$

is an eigenvector. On the other hand, from (2.10b), we find that $e_2 = -1$ when $e_1 = 1$ and $e_3 = 0$; hence,

$$\mathbf{e}_3 \equiv \boldsymbol{\sigma}_1 - \boldsymbol{\sigma}_2 . \tag{2.13b}$$

is another eigenvector. Therefore, every vector in the plane determined by the bivector.

$$\mathbf{e}_2 \wedge \mathbf{e}_3 = 2(\boldsymbol{\sigma}_2 - \boldsymbol{\sigma}_3) \wedge (\boldsymbol{\sigma}_2 - \boldsymbol{\sigma}_1)$$

is an eigenvector with eigenvalue $\lambda = 5$.

The method we have developed for finding eigenvectors and eigenvalues is sufficiently general to apply to any problem. However, the generality of the method can be a drawback, because it may require more work than necessary for special problems. For example, it often happens that an eigenvector is known at the beginning. In this case it would be foolish to use the secular equation to find the eigenvalue. Rather the eigenvalue should be obtained directly from

$$\lambda = \frac{f\mathbf{e}}{\mathbf{e}} = \mathbf{e}^{-1} \cdot (f\mathbf{e}) . \tag{2.14}$$

Often it is easy to identify an eigenvector from symmetries in the given information. Thus, perusing (2.11), we see that if we add the three equations we get

$$f(\boldsymbol{\sigma}_1 + \boldsymbol{\sigma}_2 + \boldsymbol{\sigma}_3) = 2(\boldsymbol{\sigma}_1 + \boldsymbol{\sigma}_2 + \boldsymbol{\sigma}_3) .$$

This tells us immediately that 2 is the eigenvalue corresponding to the eigenvector $\mathbf{e} = \boldsymbol{\sigma}_1 + \boldsymbol{\sigma}_2 + \boldsymbol{\sigma}_3$, in agreement with what we found by the general method after much labor. It may be a little more difficult to identify the eigenvectors (2.13a) and (2.13b) by examining (2.11). But remember, any other vectors in the $\mathbf{e}_2 \wedge \mathbf{e}_3$-plane will serve as well. Actually, as will be proved below, all we need to do is to find a vector orthogonal to $\mathbf{e}_1$. Thus, we can write $\mathbf{e}_2 = \boldsymbol{\sigma}_1 + \boldsymbol{\sigma}_2 + e_3\boldsymbol{\sigma}_3$ and choose e_3 so that

$$\mathbf{e}_1 \cdot \mathbf{e}_2 = (\boldsymbol{\sigma}_1 + \boldsymbol{\sigma}_2 + \boldsymbol{\sigma}_3) \cdot (\boldsymbol{\sigma}_1 + \boldsymbol{\sigma}_2 + e_3\boldsymbol{\sigma}_3) = 2 + e_3 = 0 .$$

Clearly $e_3 = -2$, so $\mathbf{e}_2 = \boldsymbol{\sigma}_1 + \boldsymbol{\sigma}_2 - 2\boldsymbol{\sigma}_3$, in agreement with (2.13a). From (2.11) then, we find $f\mathbf{e}_2 = 5\mathbf{e}_2$, so the eigenvalue is 5. The vector

$$\mathbf{e}_1 \times \mathbf{e}_2 = -3(\boldsymbol{\sigma}_1 - \boldsymbol{\sigma}_2)$$

is orthogonal to both the eigenvectors $\mathbf{e}_1$ and $\mathbf{e}_2$ and is, in fact, proportional to the eigenvector (2.13b).

In the example just considered, all the roots of the secular equation are real. To understand the significance of complex roots, consider the skewsymmetric transformation

$$f\mathbf{x} = \mathbf{i}\cdot\mathbf{x} = -\,\mathbf{x}\cdot\mathbf{i}\,,$$

where $\mathbf{i} = \boldsymbol{\sigma}_1\boldsymbol{\sigma}_2$ is a unit bivector. Operating on a standard basis we get

$$f\boldsymbol{\sigma}_1 = \mathbf{i}\boldsymbol{\sigma}_1 = -\boldsymbol{\sigma}_2\,, \tag{2.15a}$$

$$f\boldsymbol{\sigma}_2 = \mathbf{i}\boldsymbol{\sigma}_2 = \boldsymbol{\sigma}_1\,, \tag{2.15b}$$

$$f\boldsymbol{\sigma}_3 = 0\,. \tag{2.15c}$$

It is readily shown that the secular equation for this transformation is

$$\lambda(\lambda^2 + 1) = 0\,.$$

The root $\lambda = 0$ corresponds to the eigenvector $\boldsymbol{\sigma}_3$ in (2.15c). The point of interest, however, is that the roots of $\lambda^2 + 1 = 0$ are "imaginary", and it is natural to identify them as bivectors $\lambda = \pm\,\mathbf{i}$, since they must be related to (2.15a, b) and have the form of the eigenvalue equation (2.4), with $\boldsymbol{\sigma}_1$ and $\boldsymbol{\sigma}_2$ as eigenvectors and the bivector $\mathbf{i}$ as eigenvalue. The effect of the "imaginary eigenvalue" $\mathbf{i}$ is to rotate the eigenvectors $\boldsymbol{\sigma}_1$ and $\boldsymbol{\sigma}_2$ by 90°, so we conclude that, in general, complex roots in the secular determinant indicate that rotations are involved.

Although "complex roots" of the secular equation can be interpreted in the manner just described, we shall continue to regard only real roots as eigenvalues, because an analysis of eigenvalues is not the best way to approach problems in which eigenvalues are complex. Already we have developed a general method for finding the canonical form of a skewsymmetric transformation which is clearly superior to the "method of eigenvalues". In Section 5-3 we shall come to a similar conclusion about the best method for handling rotations. By then it should be evident that the "method of eigenvalues" is best reserved for symmetric operators, to which we now turn.

Symmetric Operators

The terms *principal vectors* and *principal values* are sometimes used for the eigenvectors and eigenvalues of a symmetric operator. The scalar multiples of a principal vector compose a line called *principal axis* of the operator. A principal axis is thus a set of equivalent principal vectors.

The chief structural property of symmetric operators is described by the following fundamental theorem: *Every symmetric operator on $\mathcal{E}_3$ has three orthogonal principal axes*. This implies, of course, that all three roots of the secular equation for a symmetric operator must be real. Let us accept this much without proof and see what it implies about the principal vectors. If $\mathbf{e}_1$ and $\mathbf{e}_2$ are principal vectors of a symmetric operator $\mathcal{S}$, then we have

$$\begin{aligned} \mathcal{S}\mathbf{e}_1 &= \lambda_1\mathbf{e}_1\,, \\ \mathcal{S}\mathbf{e}_2 &= \lambda_2\mathbf{e}_2\,. \end{aligned} \tag{2.16}$$

Dotting these equations by $\mathbf{e}_2$ and $\mathbf{e}_1$ respectively, we obtain

$$\lambda_1\mathbf{e}_2\cdot\mathbf{e}_1 = \mathbf{e}_2\cdot(\mathcal{S}\mathbf{e}_1) = \mathbf{e}_1\cdot(\mathcal{S}\mathbf{e}_2) = \lambda_2\mathbf{e}_1\cdot\mathbf{e}_2\,. \tag{2.17}$$

This implies that $\mathbf{e}_1 \cdot \mathbf{e}_2 = 0$ if $\lambda_1 \neq \lambda_2$. Thus the *principal vectors of a symmetric operator which correspond to distinct principal values are necessarily orthogonal.* If $\lambda_1 = \lambda_2$, then (2.17) tells us nothing, but (2.16) tells us that any linear combination of $\mathbf{e}_1$ and $\mathbf{e}_2$ is also a principal vector of $\mathcal{S}$, so we are free to choose any combination that gives us a pair of orthogonal principal vectors. The third principal axis is given immediately by $\mathbf{e}_1 \times \mathbf{e}_2$. Furthermore, *the principal axes are unique if all the principal values are distinct.*

In mathematical terms, the theorem that a symmetric operator $\mathcal{S}$ has 3 orthogonal principal axes is expressed by the equations

$$\mathcal{S}\mathbf{e}_k = \lambda_k \mathbf{e}_k \quad \text{for} \quad k = 1, 2, 3 \tag{2.17}$$

and

$$\mathbf{e}_j \cdot \mathbf{e}_k = 0 \quad \text{if} \quad j \neq k. \tag{2.18}$$

The operator $\mathcal{S}$ is uniquely determined by the "*spectrum*" of its eigenvalues and eigenvectors. Indeed, the operator $\mathcal{S}$ can be written in the *canonical form*

$$\mathcal{S}\mathbf{x} = \sum_{k=1}^{3} \lambda_k \mathbf{e}_k^{-1} \mathbf{e}_k \cdot \mathbf{x}, \tag{2.19}$$

or, more abstractly,

$$\mathcal{S} = \sum_k \lambda_k \mathcal{P}_k \tag{2.20a}$$

where

$$\mathcal{P}_k \mathbf{x} \equiv \mathbf{e}_k^{-1} \mathbf{e}_k \cdot \mathbf{x} \tag{2.20b}$$

is the *projection* of $\mathbf{x}$ onto the kth principal axis. The canonical form (2.19) or (2.20a) is sometimes called the *spectral decomposition* or *spectral form* of a symmetric operator, by analogy with the decomposition of light into a spectrum of colors. Note that the eigenvalue equations (2.17) follow trivially from the spectral form (2.19), so (2.19) can be regarded as the result of solving (2.17) for the operator $\mathcal{S}$ in terms of the λ_k and the $\mathbf{e}_k$.

From the spectral form (2.20a) for a nonsingular symmetric operator $\mathcal{S}$, the inverse operator is given immediately by

$$\mathcal{S}^{-1} = \sum_k \frac{1}{\lambda_k} \mathcal{P}_k. \tag{2.21}$$

To verify this by showing that $\mathcal{S}^{-1}\mathcal{S} = 1$, one needs the following basic properties of projection operators:

(a) *orthogonality*

$$\mathcal{P}_j \mathcal{P}_k = 0 \quad \text{if} \quad j \neq k, \tag{2.22a}$$

(b) *idempotence*

$$\mathcal{P}_k^2 = \mathcal{P}_k, \tag{2.22b}$$

(c) *completeness*

$$\mathcal{P}_1 + \mathcal{P}_2 + \mathcal{P}_3 = 1\,. \qquad (2.22c)$$

If the principal values λ_k are positive, then $\mathcal{S}$ has a unique *square root*

$$\mathcal{S}^{1/2} = \sum_k \lambda_k^{\ 1/2}\,\mathcal{P}_k \qquad (2.23)$$

This is a square root operator in the sense that $(\mathcal{S}^{1/2})^2 = \mathcal{S}^{1/2}\mathcal{S}^{1/2} = \mathcal{S}$, as is readily verified. An operator f is said to be positive if $\mathbf{x}\cdot(f\mathbf{x}) > 0$ for every nonzero vector $\mathbf{x}$. This implies that any eigenvalues of f are positive. So with this nomenclature, we can assert that *every positive symmetric operator has a unique square root* which is also a positive symmetric operator.

A positive symmetric operator $\mathcal{S}$ can be given a geometric interpretation by considering its effect on vectors in a principal plane (i.e. a plane determined by two principal vectors). As Figure 2.1 shows, $\mathcal{S}$ transforms (i.e. *stretches*) the points on a square into points on a parallelogram. Similarly, $\mathcal{S}$ transforms circles into ellipses. In particular, $\mathcal{S}$ stretches the unit circle into an ellipse for which the lengths of the semi-axes are the principle values of $\mathcal{S}$. A positive symmetric operator $\mathcal{S}$ on $\mathcal{E}_3$ transforms the unit sphere into an ellipsoid, as specified by

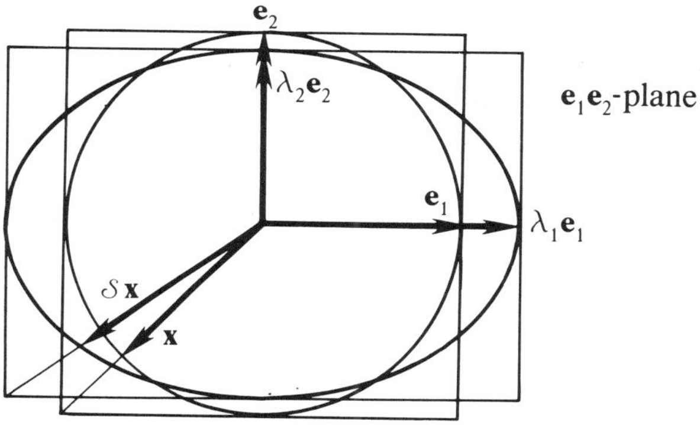

Fig. 2.1. Symmetric transformation with principle values $\lambda_1 = 3/2$, $\lambda_2 = 2/3$.

$$\mathbf{x} = \mathcal{S}\mathbf{u}\,, \qquad (2.24)$$

where $\mathbf{u}$ is any unit vector. This is a parametric equation for the ellipsoid with parameter vector $\mathbf{u}$. We obtain a nonparametric equation for the ellipsoid by elliminating $\mathbf{u}$ as follows:

$$(\mathcal{S}^{-1}\,\mathbf{x})^2 = \mathbf{u}^2 = 1\,.$$

Since $\mathcal{S}^{-1}$ is a symmetric operator, this equation can be put in the form

$$\mathbf{x}\cdot(\mathcal{S}^{-2}\,\mathbf{x}) = 1\,, \qquad (2.25)$$

where $\mathcal{S}^{-2} = \mathcal{S}^{-1}\mathcal{S}^{-1}$. Using the spectral decomposition of $\mathcal{S}^{-1}$ (see Equation (2.21)), we can write (2.25) in the form

$$\frac{x_1^2}{\lambda_1^2} + \frac{x_2^2}{\lambda_2^2} + \frac{x_3^2}{\lambda_3^2} = 1\,, \qquad (2.26)$$

where $x_k = \mathbf{x}\cdot\hat{\mathbf{e}}_k$. Equation (2.26) will be recognized as the standard "coordinate form" for an ellipsoid with "semi-axes" λ_1, λ_2, λ_3 (Figure 2.2).

We have now found a canonical form for arbitrary symmetric operators and supplied it with a geometrical interpretation. In some problems the eigenvectors and eigenvalues are given in the intial information so an appropriate operator can be constructed directly from its spectral form. We shall encounter variants of the canonical form which are more convenient in certain applications, but all variants must, of course, be constructed from the eigenvectors and eigenvalues.

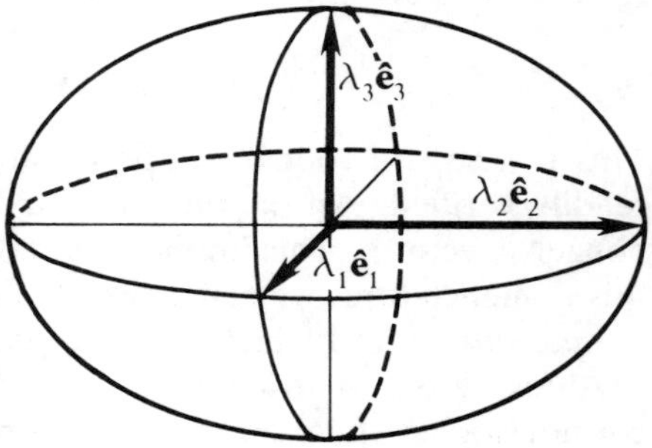

Fig. 2.2. An ellipsoid with semi-axes λ_1, λ_2, λ_3.

The Eigenvector Problem in 2 Dimensions

We have seen that the "secular method" for solving the eigenvector problem can be quite laborious. For operators acting on a 2-dimensional vector space there is an easier method, which we now derive.

For a positive symmetric operator $\mathcal{S}$ on a plane $\mathcal{E}_2$, the eigenvector equations can be written

$$\mathcal{S}\mathbf{e}_\pm = \lambda_\pm\,\mathbf{e}_\pm \tag{2.27}$$

where $\mathbf{e}_+$ and $\mathbf{e}_-$ are the principal vectors corresponding to the principal values λ_+ and λ_- respectively.

We assume that $\mathcal{S}$ is known, so its action $\mathcal{S}\mathbf{u}$ on any specified unit vector $\mathbf{u}$ in the plane can be determined. Now write $\mathbf{e} = \hat{\mathbf{e}}_+$ and decompose $\mathbf{u}$ into a component $\mathbf{u}_\parallel$ collinear with $\mathbf{e}$ and a component $\mathbf{u}_\perp$ orthogonal with $\mathbf{e}$. Then we can write

$$\begin{aligned}\mathcal{S}\mathbf{u} &= \mathcal{S}(\mathbf{u}_\parallel + \mathbf{u}_\perp) = \lambda_+\mathbf{u}_\parallel + \lambda_-\mathbf{u}_\perp \\ &= \lambda_+\mathbf{e}\mathbf{e}\cdot\mathbf{u} + \lambda_-\,\mathbf{e}\mathbf{e}\wedge\mathbf{u} = \tfrac{1}{2}\lambda_+(\mathbf{u} + \mathbf{e}\mathbf{u}\mathbf{e}) + \tfrac{1}{2}\lambda_-(\mathbf{u} - \mathbf{e}\mathbf{u}\mathbf{e})\,.\end{aligned}$$

Therefore,

$$\mathcal{S}\mathbf{u} = \tfrac{1}{2}(\lambda_+ + \lambda_-)\mathbf{u} + \tfrac{1}{2}(\lambda_+ - \lambda_-)\mathbf{e}\mathbf{u}\mathbf{e}\,. \tag{2.28}$$

The angle ϕ between the unit vectors $\mathbf{e}$ and $\mathbf{u}$ is given by the equation

$$\mathbf{u}\mathbf{e} = e^{\mathbf{i}\phi} \quad\text{or}\quad \mathbf{e} = \mathbf{u}e^{\mathbf{i}\phi}, \tag{2.29}$$

where $\mathbf{i}$ is the unit bivector for the plane. Therefore, Equation (2.28) involves three unknowns λ_+, λ_- and ϕ, so we need another equation before we can solve for them. This is most easily obtained by operating on the vector $\mathbf{ui} = -\mathbf{iu}$ which is orthogonal to $\mathbf{u}$. Thus, from (2.28) we obtain

$$\mathbf{i}\mathcal{S}(\mathbf{ui}) = \tfrac{1}{2}(\lambda_+ + \lambda_-)\mathbf{u} - \tfrac{1}{2}(\lambda_+ - \lambda_-)\mathbf{eue} \tag{2.30}$$

Combining (2.28) and (2.30), we get

$$\mathbf{u}_+ \equiv \mathcal{S}\mathbf{u} + \mathbf{i}\mathcal{S}(\mathbf{ui}) = (\lambda_+ + \lambda_-)\mathbf{u} \tag{2.31a}$$

$$\mathbf{u}_- \equiv \mathcal{S}\mathbf{u} - \mathbf{i}\mathcal{S}(\mathbf{ui}) = (\lambda_+ - \lambda_-)\mathbf{eue} \tag{2.31b}$$

Without loss of generality we may assume $\lambda_+ \geqslant \lambda_-$, so (2.31a, b) shows that the principal values are determined by the magnitudes $|\mathbf{u}_\pm| = \lambda_+ \pm \lambda_-$ of the known vectors $\mathbf{u}_+$ and $\mathbf{u}_-$. In addition, we obtain the unit vector equation $\hat{\mathbf{u}}_- = \mathbf{eue}$ from (2.31b). When reexpressed in the form $\mathbf{e}\hat{\mathbf{u}}_- = \mathbf{ue} = e^{\mathbf{i}\phi}$, this tells us that the direction $\mathbf{e}$ is half way between the directions $\hat{\mathbf{u}}_-$ and $\mathbf{u} = \hat{\mathbf{u}}_+$. Therefore

$$\mathbf{e}_+ = \alpha(\hat{\mathbf{u}}_+ + \hat{\mathbf{u}}_-) \tag{2.32}$$

is an eigenvector of $\mathcal{S}$ for any nonzero scalar α. If $\mathbf{u}_+\wedge\mathbf{u}_- \neq 0$, then $\mathbf{e}_- = \alpha(\hat{\mathbf{u}}_+ - \hat{\mathbf{u}}_-)$ is the other eigenvector we want since $\mathbf{e}_+\cdot\mathbf{e}_- = 0$.

Our results are summarized by *Mohr's algorithm*: To solve the eigenvector problem for a positive symmetric operator $\mathcal{S}$ on a plane with direction $\mathbf{i}$, choose any convenient unit vector $\mathbf{u}$ in the plane and compute the two vectors

$$\mathbf{u}_\pm \equiv \mathcal{S}\mathbf{u} \pm \mathbf{i}\mathcal{S}(\mathbf{ui}). \tag{2.33a}$$

Then, for $\mathbf{u}_+\wedge\mathbf{u}_- \neq 0$, the vectors

$$\mathbf{e}_\pm = \alpha(\hat{\mathbf{u}}_+ \pm \hat{\mathbf{u}}_-) \tag{2.33b}$$

are principal vectors of $\mathcal{S}$ with corresponding principal values

$$\lambda_\pm = \tfrac{1}{2}(|\mathbf{u}_+| \pm |\mathbf{u}_-|). \tag{2.33c}$$

(See Figure 2.3) If $\mathbf{u}$ happens to be collinear with one of the principal vectors, then $\mathbf{u}_+\wedge\mathbf{u}_- = 0$ and (2.33b) yields that vector only. Of course, the other vector is orthogonal to it.

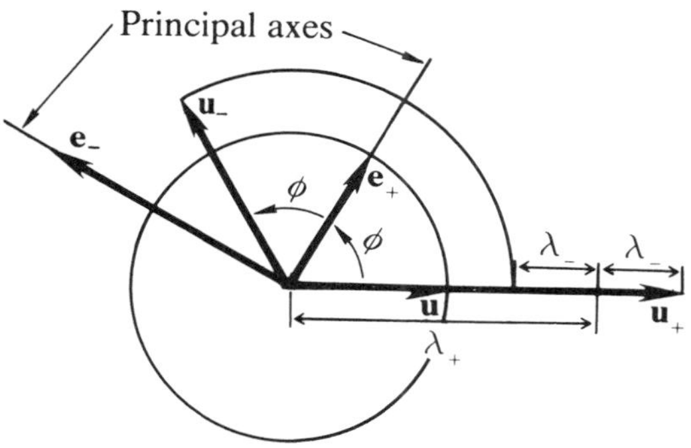

Fig. 2.3. Parameters in Mohr's Algorithm.

The principal vectors can be found in an alternative manner. Multiplying (2.31a) with (2.31b) and using (2.29), we obtain

$$\mathbf{u}_+\mathbf{u}_- = (\lambda_+^2 - \lambda_-^2)\, e^{\mathbf{i}2\phi}. \tag{2.34}$$

Whence the angle ϕ is determined by

$$\mathbf{i}\tan 2\phi = \frac{\mathbf{u}_+\wedge\mathbf{u}_-}{\mathbf{u}_+\cdot\mathbf{u}_-}. \tag{2.35}$$

Then $\hat{\mathbf{e}}_+ = \mathbf{e}$ is determined by (2.29).

The name for Mohr's algorithm is taken from Mohr's circle (Figure 2.4), which is used in engineering textbooks to solve the eigenvalue problem by graphical means. A parametric equation $Z = Z(\phi)$ for Mohr's circle can be obtained directly from (2.28) and (2.29); thus

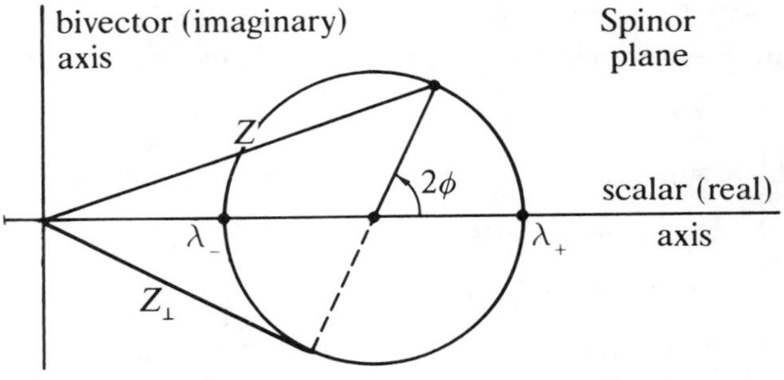

Fig. 2.4. Mohr's Circle.

$$Z(\mathbf{u}) = \mathbf{u}\mathcal{S}\mathbf{u} = \tfrac{1}{2}(\lambda_+ + \lambda_-) + \tfrac{1}{2}(\lambda_+ - \lambda_-)\,(\mathbf{ue})^2 = Z(\phi)$$

$$= \tfrac{1}{2}(\lambda_+ + \lambda_-) + \tfrac{1}{2}(\lambda_+ - \lambda_-)e^{\mathbf{i}2\phi} \tag{2.36}$$

To solve for the unknowns, Z must be known for two values of ϕ. The choice corresponding to (2.30) is

$$Z_\perp(\phi) \equiv Z\left(\phi + \frac{\pi}{2}\right) = \tfrac{1}{2}(\lambda_+ + \lambda_-) - \tfrac{1}{2}(\lambda_+ - \lambda_-)e^{\mathbf{i}2\phi}\,. \tag{2.37}$$

The solution of these two equations is of course equivalent to Mohr's algorithm. But the formulation of Mohr's algorithm by (2.33a, b, c) has the advantage of involving vectors only.

Example

To demonstrate the effectiveness of Mohr's algorithm, let us solve the eigenvalue problem for the tensor

$$\mathcal{S}\mathbf{u} = \mathbf{a}\mathbf{a}\wedge\mathbf{u} + \mathbf{b}\mathbf{b}\wedge\mathbf{u}\,. \tag{2.38}$$

As will be seen in Chapter 7, this is a general form for the inertia tensor of a plane lamina. Now

$$\mathcal{S}\mathbf{a} = \mathbf{b}\mathbf{b}\wedge\mathbf{a} = b^2\mathbf{a} - \mathbf{b}\mathbf{a}\cdot\mathbf{b}$$

and, for $\mathbf{a}\wedge\mathbf{b} = |\,\mathbf{a}\wedge\mathbf{b}\,|\,\mathbf{i}$,

$$\mathbf{i}\mathcal{S}(\mathbf{ai}) = \mathbf{i}[\mathbf{aa}\cdot\mathbf{ai} + \mathbf{bb}\cdot\mathbf{ai}] = \mathbf{a}^3 + \mathbf{bb}\cdot\mathbf{a}$$

Hence, for $\mathbf{u} = \hat{\mathbf{a}}$ in (2.33a),

$$\mathbf{a}_+ = \mathcal{S}\hat{\mathbf{a}} + \mathbf{i}\mathcal{S}(\hat{\mathbf{a}}\mathbf{i}) = (a^2 + b^2)\hat{\mathbf{a}}\,,$$

$$\mathbf{a}_- = \mathcal{S}\hat{\mathbf{a}} - \mathbf{i}\mathcal{S}(\hat{\mathbf{a}}\mathbf{i}) = (b^2 - a^2)\hat{\mathbf{a}} - 2\hat{\mathbf{a}}\cdot\mathbf{bb}\,,$$

and

$$|\mathbf{a}_-| = [(b^2 - a^2)^2 + 4(\mathbf{a}\cdot\mathbf{b})^2]^{1/2}.$$

Therefore, by (2.33c) the principal values are

$$\lambda_\pm = a^2 + b^2 \pm [(b^2 - a^2)^2 + 4(\mathbf{a}\cdot\mathbf{b})^2]^{1/2} \tag{2.39}$$

By (2.33b) the corresponding principal vectors are

$$\mathbf{e}_\pm = \mathbf{a} \pm \frac{(b^2 - a^2)\mathbf{a} - 2\mathbf{a}\cdot\mathbf{b}\mathbf{b}}{[(b^2 - a^2)^2 + 4(\mathbf{a}\cdot\mathbf{b})^2]^{1/2}}. \tag{2.40}$$

It will be noted in this example how the free choice of **u** in Mohr's algorithm enabled us to simplify computations by taking the special structure of $\mathcal{S}$ into account.

Unfortunately, there is no known generalization of Mohr's algorithm to solve the eigenvector problem in 3-dimensions. However, whenever one eigenvector is already known, Mohr's algorithm can be applied to the plane orthogonal to it. For example, any tensor constructed from two vectors **a** and **b** necessarily has **a** × **b** as an eigenvector. Thus, for (2.38) we find

$$\mathcal{S}(\mathbf{a}\times\mathbf{b}) = (a^2 + b^2)\,\mathbf{a}\times\mathbf{b}.$$

5-2. Exercises

(2.1) Find the adjoint as well as the symmetric and skewsymmetric parts of the linear transformation

$$f\mathbf{x} = \alpha\mathbf{x} + \mathbf{a}\mathbf{b}\cdot\mathbf{x} + \mathbf{x}\cdot\mathbf{A}.$$

(2.2) Derive Equations (2.7a, b, c, d) from Equation (2.6).

(2.3) Find the eigenvectors and eigenvalues for operators with the following matrices

(a) $\begin{bmatrix} 1 & 0 & 5 \\ 0 & -2 & 0 \\ 5 & 0 & 1 \end{bmatrix}$ (b) $\begin{bmatrix} 7 & \sqrt{6} & -\sqrt{3} \\ \sqrt{6} & 2 & -5\sqrt{2} \\ -\sqrt{3} & -5\sqrt{2} & -3 \end{bmatrix}$.

(2.4) We write $\mathcal{S}^n$ for the n-fold product of an operator $\mathcal{S}$ with itself. Prove that if $\mathcal{S}$ is symmetric with eigenvalues λ_k, then $\mathcal{S}^n$ is symmetric with eigenvalues $(\lambda_k)^n$ and the same eigenvectors as $\mathcal{S}$.

(2.5) A linear operator $\mathcal{S}$ is given by

$$\mathcal{S}\boldsymbol{\sigma}_1 = 7\boldsymbol{\sigma}_1 + 2\boldsymbol{\sigma}_2$$

$$\mathcal{S}\boldsymbol{\sigma}_2 = 2\boldsymbol{\sigma}_1 + 6\boldsymbol{\sigma}_2 - 2\boldsymbol{\sigma}_3$$

$$\mathcal{S}\boldsymbol{\sigma}_3 = -2\boldsymbol{\sigma}_2 + 5\boldsymbol{\sigma}_3$$

Determine its eigenvalues and eigenvectors.

(2.6) Describe the eigenvalue spectrum of a symmetric operator $\mathcal{S}$ so that the equation

$$\mathbf{x}\cdot(\mathcal{S}\mathbf{x}) = 1$$

is equivalent to the standard coordinate forms for each of the following *quadratic surfaces*:

(a) *Ellipsoid*:

$$\frac{x_1^2}{a^2} + \frac{x_2^2}{b^2} + \frac{x_3^2}{c^2} = 1 .$$

(b) *Hyperboloid of one sheet*:

$$\frac{x_1^2}{a^2} + \frac{x_2^2}{b^2} - \frac{x_3^2}{c^2} = 1 .$$

(c) *Hyperboloid of two sheets*:

$$\frac{x_1^2}{a^2} - \frac{x_2^2}{b^2} - \frac{x_3^2}{c^2} = 1 .$$

(2.7) Describe the solution set $\{\mathbf{x}\}$ of the equation

$$[f(\mathbf{x} - \mathbf{a})]^2 = 1 ,$$

where f is any linear operator.

(2.8) If $\mathbf{a}$, $\mathbf{b}$, $\mathbf{c}$ are mutually perpendicular and $\mathcal{S}$ is a symmetric tensor, prove that the three vectors $\mathbf{a} \times \mathcal{S}\mathbf{a}$, $\mathbf{b} \times \mathcal{S}\mathbf{b}$, $\mathbf{c} \times \mathcal{S}\mathbf{c}$ are coplanar.

(2.9) Prove the basic properties of projection operators formulated by Equations (2.22a, b, c), and verify that the inverse of a nonsingular symmetric operator is given by (2.21).

(2.10) Find the eigenvalues and eigenvectors of the tensors

(a) $\mathcal{S}\mathbf{u} = \mathbf{a}\mathbf{a}\cdot\mathbf{u} + \mathbf{b}\mathbf{b}\cdot\mathbf{u}$.

(b) $\mathcal{T}\mathbf{u} = \mathbf{a}\mathbf{b}\cdot\mathbf{u} + \mathbf{b}\mathbf{a}\cdot\mathbf{u}$.

(2.11) For an operator f specified by the symmetric matrix

$$[f] = \begin{bmatrix} f_{11} & f_{12} \\ f_{12} & f_{22} \end{bmatrix}$$

with respect to an orthonormal basis $\boldsymbol{\sigma}_1$, $\boldsymbol{\sigma}_2$, show that

$$\lambda_\pm = \tfrac{1}{2}|\, f_{11} + f_{22} \,| \pm \tfrac{1}{2}|\, (f_{11} - f_{22})^2 + 4f_{12}^2 \,|^{1/2}$$

are eigenvalues, and the angle ϕ from $\boldsymbol{\sigma}_1$ to the eigenvector $\mathbf{e}_+$ is given by

$$\tan 2\phi = \frac{2f_{12}}{f_{22} - f_{11}} .$$

(2.12) Solve the eigenvector problem for the tensor

$$\mathcal{S}\mathbf{u} = \mathbf{a}\mathbf{a}\wedge\mathbf{u} + \mathbf{b}\mathbf{b}\wedge\mathbf{u} + \mathbf{c}\mathbf{c}\wedge\mathbf{u}$$

where $\mathbf{a} + \mathbf{b} + \mathbf{c} = 0$.

5-3. The Arithmetic of Reflections and Rotations

A linear transformation can be classified according to specific relations among vectors which it leaves unchanged. Such relations are said to be *invariants* of the transformation. Transformations for which the inner product is an invariant are called *orthogonal transformations*. Thus, an orthogonal transformation f on $\mathcal{E}_3$ has the property

$$(f\mathbf{x})\cdot(f\mathbf{y}) = \mathbf{x}\cdot\mathbf{y} \tag{3.1}$$

for all vectors $\mathbf{x}$ and $\mathbf{y}$ in $\mathcal{E}_3$. On the other hand, the property (1.9) of the adjoint implies that

$$(f\mathbf{x})\cdot(f\mathbf{y}) = \mathbf{x}\cdot(\bar{f}f\mathbf{y}),$$

This is equivalent to (3.1) if and only if

$$\bar{f} = f^{-1}. \tag{3.2}$$

Therefore, an orthogonal operator is a nonsingular operator for which the inverse is equal to the adjoint.

From (3.1) it follows that

$$(f\mathbf{x})^2 = \mathbf{x}^2 = |\,\mathbf{x}\,|^2. \tag{3.3}$$

Thus, the magnitude of every vector in $\mathcal{E}_3$ is invariant under an orthogonal transformation. The orthogonality of vectors in a standard basis is another invariant. Specifically for $\mathbf{f}_k = f\boldsymbol{\sigma}_k$, (3.1) implies

$$\mathbf{f}_j\cdot\mathbf{f}_k = \boldsymbol{\sigma}_j\cdot\boldsymbol{\sigma}_k = \delta_{jk}. \tag{3.4}$$

This can be used to prove that the magnitude of the unit pseudoscalar is invariant under orthogonal transformations. Since

$$\underline{f}(i) = \underline{f}(\boldsymbol{\sigma}_1\boldsymbol{\sigma}_2\boldsymbol{\sigma}_3) = \mathbf{f}_1\mathbf{f}_2\mathbf{f}_3,$$

we have

$$|\,\underline{f}(i)\,|^2 = (\mathbf{f}_3\mathbf{f}_2\mathbf{f}_1)\,(\mathbf{f}_1\mathbf{f}_2\mathbf{f}_3) = 1.$$

But $\underline{f}(i) = (\det \underline{f})i$. Hence,

$$|\,\underline{f}(i)\,|^2 = (\det f)^2 = 1,$$

and

$$\det f = \pm 1. \tag{3.5}$$

This condition distinguishes two kinds of orthogonal transformations. An orthogonal transformation f is said to be *proper* if $\det f = 1$, and *improper* if $\det f = -1$. Proper orthogonal transformations are usually called *rotations*.

Our problem now is to find the canonical form for orthogonal transformations, that is, the simplest expression for an arbitrary orthogonal transformation in terms of geometric algebra. The general solution can be constructed from the simplest examples.

Reflections

The simplest nonsingular linear transformation that can be built out of a single nonzero vector $\mathbf{u}$ is

$$\mathcal{U}\mathbf{x} = -\mathbf{u}^{-1}\mathbf{x}\mathbf{u} = -\mathbf{u}\mathbf{x}\mathbf{u}^{-1} = -\hat{\mathbf{u}}\mathbf{x}\hat{\mathbf{u}}. \tag{3.6}$$

The magnitude of $\mathbf{u}$ does not actually play a role in any of these equivalent expressions, so we might as well suppose that $\mathbf{u}$ is a unit vector and write

$$\mathcal{U}\mathbf{x} = -\mathbf{u}\mathbf{x}\mathbf{u}. \tag{3.7}$$

The effect of this transformation is made clear by decomposing $\mathbf{x}$ into a component $\mathbf{x}_\parallel$ collinear with $\mathbf{u}$ plus a component $\mathbf{x}_\perp$ orthogonal to $\mathbf{u}$:

$$\mathbf{x} = \mathbf{x}_\parallel + \mathbf{x}_\perp,$$

where

$$\mathbf{x}_\parallel = \mathbf{x}\cdot\mathbf{u}\mathbf{u}$$

and

$$\mathbf{x}_\perp = \mathbf{x}\wedge\mathbf{u}\mathbf{u}.$$

Now, $\mathbf{u}$ commutes with $\mathbf{x}_\parallel$ and anticommutes with $\mathbf{x}_\perp$. Hence,

$$\mathbf{u}\mathbf{x}\mathbf{u} = \mathbf{u}(\mathbf{x}_\parallel + \mathbf{x}_\perp)\mathbf{u} \doteq \mathbf{x}_\parallel - \mathbf{x}_\perp,$$

and (3.7) yields

$$\mathbf{x}' = \mathcal{U}\mathbf{x} = -\mathbf{u}\mathbf{x}\mathbf{u} = \mathbf{x}_\perp - \mathbf{x}_\parallel. \tag{3.8}$$

Thus $\mathcal{U}$ transforms each vector $\mathbf{x}$ into a vector $\mathbf{x}'$ by reversing the sign of the component of $\mathbf{x}$ along $\mathbf{u}$. In $\mathcal{E}_3$ the vector $\mathbf{x}'$ is the "mirror image" of $\mathbf{x}$ in the plane (through the origin) with normal $\mathbf{u}$, as shown in Figure 3.1. Accordingly, the transformation (3.8) is called the *reflection* along $\mathbf{u}$.

Equation (3.8) obviously describes a linear transformation; in fact, a reflection is an improper orthogonal transformation as we shall now show. Consider the product of two transformed vectors:

$$(\mathcal{U}\mathbf{x})(\mathcal{U}\mathbf{y}) = (-\mathbf{u}\mathbf{x}\mathbf{u})(-\mathbf{u}\mathbf{y}\mathbf{u}) = \mathbf{u}\mathbf{x}\mathbf{y}\mathbf{u}.$$

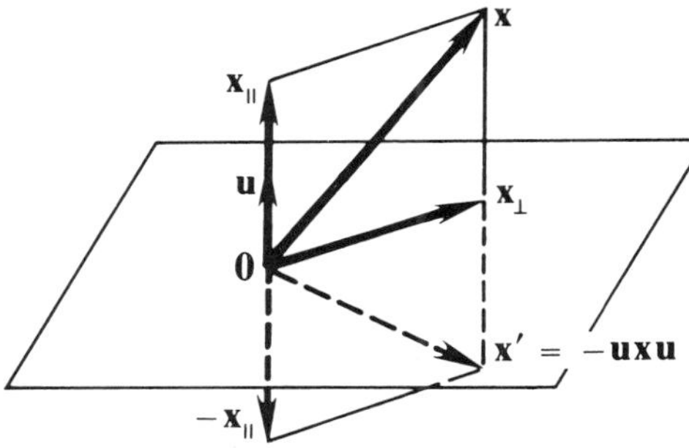

Fig. 3.1. Reflection of **x** along **u**.

The scalar part of this equation gives us

$$(\mathcal{U}\mathbf{x})\cdot(\mathcal{U}\mathbf{y}) = \mathbf{u}\mathbf{x}\cdot\mathbf{y}\mathbf{u} = \mathbf{x}\cdot\mathbf{y},$$

which proves that $\mathcal{U}$ is orthogonal. The bivector part of the equation gives us the outermorphism

$$\underline{\mathcal{U}}(\mathbf{x}\wedge\mathbf{y}) = (\mathcal{U}\mathbf{x})\wedge(\mathcal{U}\mathbf{y})$$
$$= \mathbf{u}(\mathbf{x}\wedge\mathbf{y})\mathbf{u}. \quad (3.9)$$

Now, for transformations on $\mathcal{E}_3$, the determinant is obtained from the outermorphism of trivectors, which are the pseudoscalars. Thus, from the product of three transformed vectors

$$(\mathcal{U}\mathbf{x})\,(\mathcal{U}\mathbf{y})\,(\mathcal{U}\mathbf{z}) = (-\mathbf{uxu})\,(-\mathbf{uyu})\,(-\mathbf{uzu}) = -\,\mathbf{u}(\mathbf{xyz})\mathbf{u},$$

we take the trivector part to get

$$\underline{\mathcal{U}}(\mathbf{x}\wedge\mathbf{y}\wedge\mathbf{z}) = -\,\mathbf{u}(\mathbf{x}\wedge\mathbf{y}\wedge\mathbf{z})\mathbf{u} = -\,\mathbf{x}\wedge\mathbf{y}\wedge\mathbf{z}. \quad (3.10)$$

The last step in (3.10) follows from the fact that all pseudoscalars commute with the vectors in $\mathcal{E}_3$. From (3.10) it follows that

$$\det\mathcal{U} = -1. \quad (3.11)$$

Hence, the transformation is improper as claimed.

Example

Reflections occur frequently in physics. For example, for a particle rebounding elastically from a fixed plane with normal **u**, the final momentum **p**′ is related to the initial momentum **p** by

$$\mathbf{p}' = -\,\mathbf{upu}. \quad (3.12)$$

This, of course, has the same form as (3.8), but the Figure 3.2 which we associate with it is a little different than the Figure 3.1 associated with (3.8). Now, Equation (3.12) implies that $|\,\mathbf{p}'\,| = |\,\mathbf{p}\,|$, as required by kinetic energy conservation. Consequently, we can put (3.12) in the form

$$\mathbf{u}\hat{\mathbf{p}}' = -\,\hat{\mathbf{p}}\mathbf{u}. \quad (3.13)$$

We know from Section 2-4 that the product of unit vectors can be expressed as the exponential of the angle between them. So (3.13) can be written

$$e^{\mathbf{i}\theta'} = e^{\mathbf{i}\theta}, \quad (3.14)$$

where **i** is the unit bivector for the plane of reflection and the angles θ and θ' are as indicated in Figure 3.2.

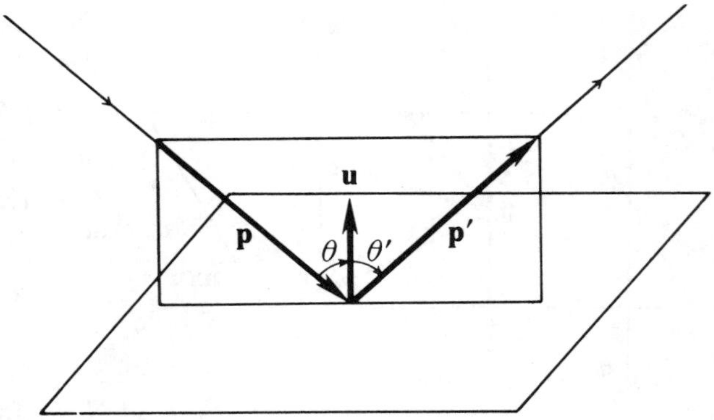

Fig. 3.2. Elastic reflection from a plane.

From (3.14) it follows that the *angle of reflection θ' equals the angle of incidence θ*. This is the elementary statement of the "law of reflection". But a full description of a reflection must specify the plane of reflection as well as the angles. All this is expressed by (3.12), which can be regarded as a complete statement of the *law of reflection*. Equation (3.12) provides an approximate description of the rebound of a ball from a wall and quite an accurate description for the change in direction of light reflected from a plane surface. It provides an especially efficient means of computing the net effect of several successive reflections, as will be obvious after we have examined the composition of reflections.

Rotations

Now let us consider the product of the reflection (3.7) with another reflection

$$\mathcal{V}\mathbf{x} = -\mathbf{vxv},$$

where $\mathbf{v}$ is a unit vector. We have

$$\mathcal{V}\mathcal{U}\mathbf{x} = \mathbf{vuxuv}.$$

This determines a new linear transformation $\mathcal{R} = \mathcal{V}\mathcal{U}$ of the form

$$\mathcal{R}\mathbf{x} = R^\dagger\mathbf{x}R, \tag{3.15}$$

where R can be written in the form

$$R = \mathbf{uv} = \mathbf{u}\cdot\mathbf{v} + \mathbf{u}\wedge\mathbf{v} = e^{(1/2)\mathbf{A}}, \tag{3.16}$$

$$R^\dagger = \mathbf{vu} = \mathbf{u}\cdot\mathbf{v} - \mathbf{u}\wedge\mathbf{v} = e^{-(1/2)\mathbf{A}} \tag{3.17}$$

The reason for writing $\frac{1}{2}\mathbf{A}$ for the bivector angle between vectors $\mathbf{u}$ and $\mathbf{v}$ will be made clear below. According to Section 2-3, R is a spinor, or quaternion. Since

$$R^\dagger R = 1, \tag{3.18}$$

it has unit magnitude $|R| = 1$, and it is said to be a *unitary* or *unimodular* spinor. Note from (3.16) that the bivector of R can be written

$$\langle R\rangle_2 = \mathbf{u}\wedge\mathbf{v} = \langle e^{(1/2)\mathbf{A}}\rangle_2 = \sinh \tfrac{1}{2}\mathbf{A} = \hat{\mathbf{A}} \sin \tfrac{1}{2} | \mathbf{A} |,$$

so it has the direction

$$\hat{\mathbf{A}} = \frac{\mathbf{u}\wedge\mathbf{v}}{|\mathbf{u}\wedge\mathbf{v}|}. \tag{3.19}$$

We are using here important properties of exponential and trigonometric functions established in Section 2-5. We shall now show that this bivector specifies a plane of rotation.

To determine the effect of $\mathcal{R}$ on a vector $\mathbf{x}$, we decompose $\mathbf{x}$ into a component $\mathbf{x}_\parallel$ in the $\mathbf{A}$-plane and a component $\mathbf{x}_\perp$ orthogonal to the $\mathbf{A}$-plane

$$\mathbf{x} = \mathbf{x}_\parallel + \mathbf{x}_\perp$$

where, for $\mathbf{A} \neq 0$,

$$\mathbf{x}_\parallel = \mathbf{x}\cdot\mathbf{A}\mathbf{A}^{-1}$$

and

$$\mathbf{x}_\perp = \mathbf{x}\wedge\mathbf{A}\mathbf{A}^{-1}.$$

It follows that $\mathbf{A}$ anticommutes with $\mathbf{x}_\parallel$ and commutes with $\mathbf{x}_\perp$, that is,

$$\mathbf{x}_\parallel\mathbf{A} = \mathbf{x}\cdot\mathbf{A} = -\mathbf{A}\cdot\mathbf{x} = -\mathbf{A}\mathbf{x}_\parallel$$

and

$$\mathbf{x}_\perp\mathbf{A} = \mathbf{x}\wedge\mathbf{A} = \mathbf{A}\wedge\mathbf{x} = \mathbf{A}\mathbf{x}_\perp .$$

Using these relations, we find from (3.16) and (3.17) that

$$R^\dagger\mathbf{x}_\parallel = \mathbf{x}_\parallel R$$

and

$$R^\dagger\mathbf{x}_\perp = \mathbf{x}_\perp R^\dagger .$$

Therefore,

$$R^\dagger\mathbf{x}\mathrm{R} = R^\dagger(\mathbf{x}_\perp + \mathbf{x}_\parallel)R = \mathbf{x}_\perp + \mathbf{x}_\parallel R^2 = \mathbf{x}_\perp + \mathbf{x}_\parallel e^{\mathbf{A}}.$$

Hence, from (3.15) we obtain

$$\mathbf{x}' = \mathcal{R}\mathbf{x} = e^{-(1/2)\mathbf{A}}\mathbf{x}e^{(1/2)\mathbf{A}} = \mathbf{x}_\perp + \mathbf{x}_\parallel e^{\mathbf{A}} \tag{3.20}$$

We have already seen in Section 2-3 that an expression of the form $\mathbf{x}_\parallel e^{\mathbf{A}}$ describes a rotation of $\mathbf{x}_\parallel$ in the $\mathbf{A}$-plane through an angle of magnitude $|\mathbf{A}|$. Therefore, the transformation (3.20) can be depicted as in Figure 3.3.

It is easy to prove that $\det \mathcal{R} = 1$ in the same way that we proved $\det \mathcal{U} = -1$. Therefore $\mathcal{R}$ is a rotation. More specifically, we say that (3.20) describes ***a rotation by (or through) an angle* A**. In $\mathcal{G}_3$ we can write

$$\mathbf{A} = i\mathbf{a}, \tag{3.21}$$

expressing **A** as the dual of a vector **a**. Then the direction of **a** specifies the axis of rotation, as shown in Figure 3.3, and the magnitude $|\mathbf{a}| = |\mathbf{A}|$ gives the scalar angle of rotation. It is important to note that the vector **a** has been defined so that the following *right hand rule for rotation* applies: If the thumb of the right hand points along the direction of **a**, then the rotation in the plane "follows" the fingers, as indicated in Figure 3.3.

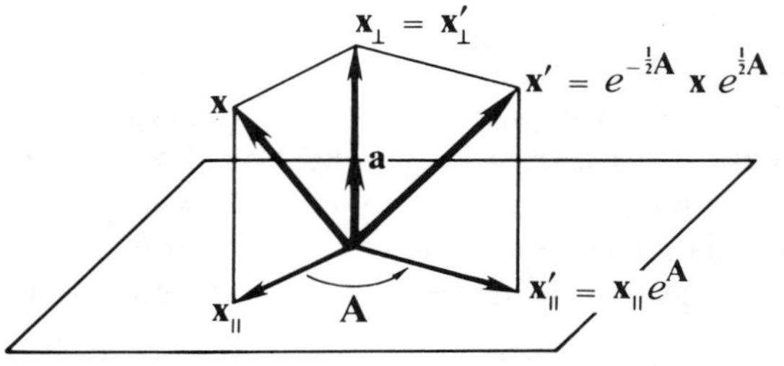

Fig. 3.3. Rotation by an angle $\mathbf{A} = i\mathbf{a}$.

The equation $\mathcal{R}\mathbf{x} = R^\dagger \mathbf{x} R$ where R is a unitary spinor is, in fact, the desired *canonical form* for any rotation on $\mathcal{E}_3$. We can always express R in the form (3.16), as is proved below. For a rotation $\mathcal{R}\mathbf{x} = R^\dagger \mathbf{x} R$, it is usually most convenient to express the spinor R in one of the two *parametric forms*,

$$R = \alpha + i\boldsymbol{\beta} \tag{3.22a}$$

$$= e^{(1/2)i\mathbf{a}}, \tag{3.22b}$$

rather than in the form (3.16). These forms exploit the fact that in $\mathcal{G}_3$ every bivector can be expressed as the dual of a vector. Since

$$e^{(1/2)i\mathbf{a}} = \cos \tfrac{1}{2}\mathbf{a} + i \sin \tfrac{1}{2}\mathbf{a},$$

the parameters α, $\boldsymbol{\beta}$ and **a** in (3.22a, b) are therefore related by

$$\alpha = \cos \tfrac{1}{2}\mathbf{a} = \cos \tfrac{1}{2}|\mathbf{a}|, \tag{3.23a}$$

$$\boldsymbol{\beta} = \sin \tfrac{1}{2}\mathbf{a} = \hat{\mathbf{a}} \sin \tfrac{1}{2}|\mathbf{a}|. \tag{3.23b}$$

The parameter α is, of course, not independent of $\boldsymbol{\beta}$, because

$$R^\dagger R = (\alpha - i\boldsymbol{\beta})(\alpha + i\boldsymbol{\beta}) = \alpha^2 + \boldsymbol{\beta}^2 = 1.$$

This will be recognized as a familiar trigonometric identity if expressed in terms of the angle by (3.23a, b). Since the form (3.22b) expresses the spinor R and therefore the rotation R as a function of the angle and axis of rotation represented by a vector **a**, it is appropriate to refer to it as the *angular form* or the *angular parametrization* of the rotation. The four parameters α, $\beta_k = \boldsymbol{\sigma}_k \cdot \boldsymbol{\beta}$ for $k = 1, 2, 3$ are called *Euler parameters* in the literature, so let us refer to α and $\boldsymbol{\beta}$ respectively as the *Euler scalar* and *Euler vector* of the rotation. Other parametrizations which are useful for various special purposes are given in the exercises.

In the canonical form for a rotation $\mathbf{x}' = R^\dagger \mathbf{x} R$, it is obvious that the same rotation will result if we change the spinor R to its negative $-R$. To understand the significance of this ambiguity, use (3.22b) to write

$$-R = e^{-i\hat{\mathbf{a}}\pi}\, e^{(1/2)i\mathbf{a}} = e^{(1/2)i(-\hat{\mathbf{a}})(2\pi - a)} = e^{(1/2)i\mathbf{a}'} \tag{3.24}$$

We may interpret R as specifying a rotation in the righthanded sense about the 'axis' $\hat{\mathbf{a}}$ through an angle $a = |\,\mathbf{a}\,|$ in the range $0 \leqslant a \leqslant 2\pi$. So (3.24) shows that $-R$ specifies the rotation with opposite sense through the complimentary angle $a' = 2\pi - a$. Thus *R and $-R$ represent equivalent rotations with opposite senses* as shown in Figure 3.4. The representation of a rotation as a linear transformation in the form $R^{\dagger}\mathbf{x}R$ or as an orthogonal matrix does not distinguish between these two possibilities. Therefore, spinors provide a more general representation of rotations than orthogonal matrices. Specifically, each unimodular spinor represents a unique *oriented rotation*, whereas each orthogonal matrix represents an *unoriented rotation*. Note that the vector $\boldsymbol{\beta}$ in (3.22a) specifies the *oriented axis* of the rotation directly. Also, from (3.23a) we see that $\alpha = \langle R\rangle_0$ is always positive for rotations through angles less than π and always negative for their complementary rotations. So the "shortest" of two complementary rotations is represented by the spinor with positive scalar part.

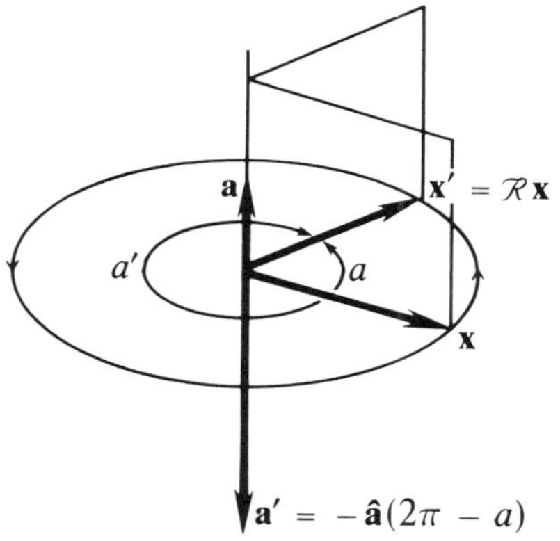

Fig. 3.4. Equivalent Rotations.

Composition of Rotations

The product (or composite) of a rotation

$$\mathcal{R}_1\mathbf{x} = R_1^{\dagger}\mathbf{x}R_1$$

with a rotation

$$\mathcal{R}_2\mathbf{x} = R_2^{\dagger}\mathbf{x}R_2$$

is a linear transformation

$$\mathcal{R}_3\mathbf{x} = R_3^{\dagger}\mathbf{x}R_3 \tag{3.25a}$$

where

$$\mathcal{R}_3 = \mathcal{R}_2\mathcal{R}_1 \tag{3.25b}$$

and

$$R_3 = R_1R_2\,. \tag{3.25c}$$

In Section 2-3 we proved that the product of two spinors produces a spinor. Furthermore,

$$R_3^{\dagger}R_3 = (R_1R_2)^{\dagger}R_1R_2 = R_2^{\dagger}R_1^{\dagger}R_1R_2 = 1\,,$$

since we have assumed $R_1^{\dagger}R_1 = R_2^{\dagger}R_2 = 1$. Therefore R_3 is a unitary spinor and $\mathcal{R}_3$ is a rotation. This will provide us with a proof that the product of two rotations is always a rotation, once we have established that every rotation can be written in the canonical form we have used.

Equations (3.25a, b, c) show that the problem of determining the rotation $\mathcal{R}_3$ which is equivalent to the product of rotations $\mathcal{R}_2\mathcal{R}_1$, can be reduced to a straightforward computation of the geometric product of spinors. If the spinor equation (3.25c) is expressed in terms of rotation angles, it becomes

$$e^{(1/2)i\mathbf{a}_3} = e^{(1/2)i\mathbf{a}_1}\, e^{(1/2)i\mathbf{a}_2}. \tag{3.26}$$

This equation determines the rotation angle $\mathbf{a}_3$ in terms of angles $\mathbf{a}_1$ and $\mathbf{a}_2$. The same equation is basic to spherical trigonometry (Appendix A). Indeed, the problem of determining the product of two rotations is mathematically equivalent to the problem of solving a spherical triangle.

For computational purposes, it is more convenient to express (3.25c) in the *Eulerian form*

$$\alpha_3 + i\boldsymbol{\beta}_3 = (\alpha_1 + i\boldsymbol{\beta}_1)\,(\alpha_2 + i\boldsymbol{\beta}_2) \tag{3.27}$$

rather than the angular form (3.26). Expanding the product in (3.27) and equating scalar and bivector parts separately, we obtain the following expressions for the Euler parameters of $R_3 = \alpha_3 + i\boldsymbol{\beta}_3$:

$$\alpha_3 = \alpha_1\alpha_2 - \boldsymbol{\beta}_1\cdot\boldsymbol{\beta}_2, \tag{3.28a}$$

$$\boldsymbol{\beta}_3 = \alpha_1\boldsymbol{\beta}_2 + \alpha_2\boldsymbol{\beta}_1 - \boldsymbol{\beta}_1 \times \boldsymbol{\beta}_2. \tag{3.28b}$$

These equations can be expressed as relations among rotation angles by using (3.23a, b), but the results are so complicated that it is clearly much easier to avoid angles and work directly with Euler parameters whenever possible. Note that (3.28b) gives us the rotation axis $\hat{\boldsymbol{\beta}}_3 = \hat{\mathbf{a}}_3$ without requiring that we use angles.

Example

To illustrate the composition of rotations, let us compute the product of rotations by 90° about orthogonal axes, as described by the spinors

$$R_1 = e^{(1/2)i\boldsymbol{\sigma}_1\pi/2} = \cos\frac{\pi}{4} + i\boldsymbol{\sigma}_1 \sin\frac{\pi}{4} = \frac{1}{\sqrt{2}}(1 + i\boldsymbol{\sigma}_1)\,, \tag{3.29a}$$

$$R_2 = e^{(1/2)i\boldsymbol{\sigma}_2\pi/2} = \frac{1}{\sqrt{2}}(1 + i\boldsymbol{\sigma}_2)\;. \tag{3.29b}$$

We can compute the product directly without using (3.28a, b). Thus, using $\boldsymbol{\sigma}_1\boldsymbol{\sigma}_2 = i\boldsymbol{\sigma}_3$,

$$R_3 = R_1R_2 = \frac{(1+i\boldsymbol{\sigma}_1)}{\sqrt{2}}\,\frac{(1+i\boldsymbol{\sigma}_2)}{\sqrt{2}} = \tfrac{1}{2}[1+i(\boldsymbol{\sigma}_1 + \boldsymbol{\sigma}_2 - \boldsymbol{\sigma}_3)]$$

$$= \tfrac{1}{2} + i\hat{\mathbf{a}}_3 \tfrac{\sqrt{3}}{2} = e^{(1/2)i\hat{\mathbf{a}}_3 2\pi/3}. \tag{3.30}$$

Thus, the composite rotation is by $2\pi/3 = 120°$ about the "diagonal axis" $\hat{\mathbf{a}}_3 = (\boldsymbol{\sigma}_1 + \boldsymbol{\sigma}_2 - \boldsymbol{\sigma}_3)/\sqrt{3}$. The reader can check this result by performing the rotations on some solid object, as shown in Figure 3.5.

It is worth emphasizing that the product of two rotations is generally not commutative, that is, $\mathcal{R}_2\mathcal{R}_1 \neq \mathcal{R}_1\mathcal{R}_2$. This fact is perfectly expressed by the noncommutativity of the corresponding spinors, $R_1R_2 \neq R_2R_1$. The result of performing the rotations specified by (3.29a, b) in both orders are illustrated in Figure 3.5.

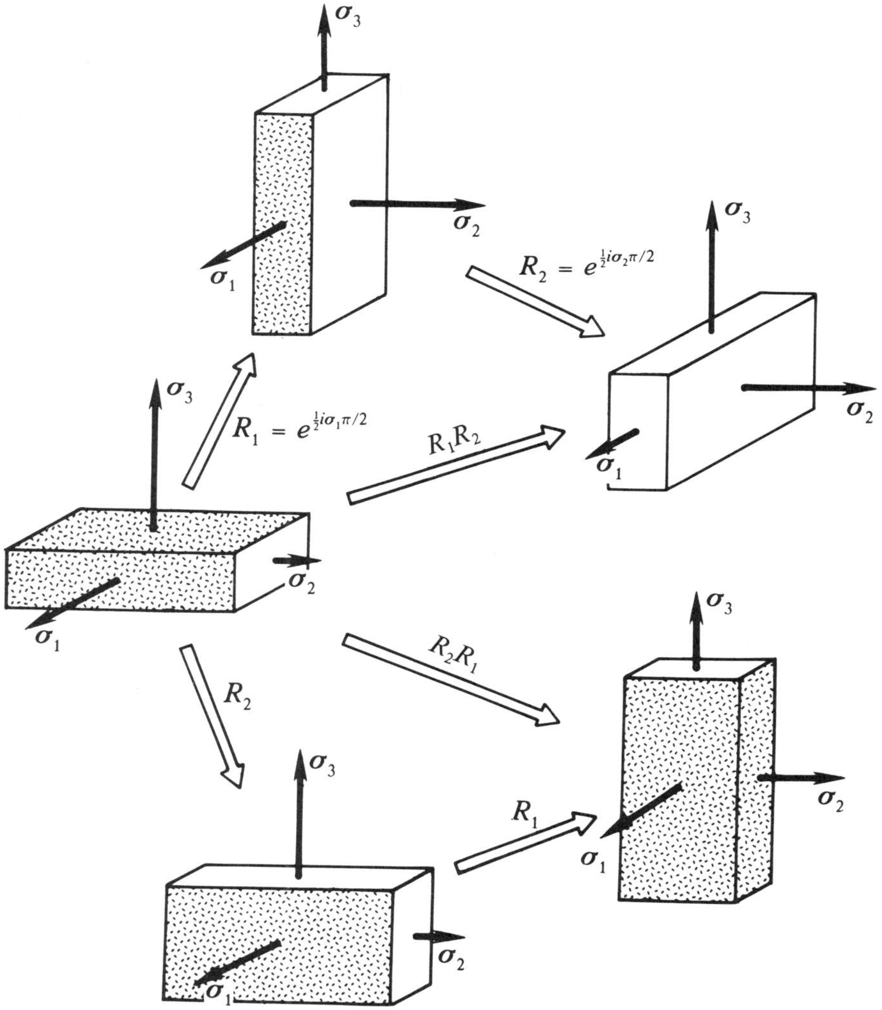

Fig. 3.5. Composition of Rotations.

The theory of rotations presented here was developed in the mid-nineteenth century by the mathematical physicist William Rowen Hamilton, who called it the *quaternion theory of rotations*.

Our formulation differs from Hamilton's in utilizing the entire geometric algebra $\mathcal{G}_3$. Hamilton employed only the quaternions, which we identify with the even subalgebra of $\mathcal{G}_3$. Geometric algebra integrates quaternions with the rest of vector and matrix algebra and thus makes Hamilton's powerful theory of rotations available without translation to a different mathematical language. Hamilton's theory has been little used in this century just because its relation to conventional vector algebra was obscure.

In this century Hamilton's theory has been independently rediscovered in an equivalent matrix form called the *spinor theory of rotations*. The spinor theory is widely used by physicists in advanced quantum mechanics. We have preferred the term "spinor" to "quaternion" to call attention to the fact that our concept of spinor is equivalent to the concept of spinor in quantum mechanics. Proof of this equivalence will be given in a subsequent book, NFII.

Matrix Elements of a Rotation

A rotation $\mathcal{R}$ transforms a standard frame $\{\boldsymbol{\sigma}_k\}$ into a new set of orthonormal vectors

$$\mathbf{e}_k = \mathcal{R}\boldsymbol{\sigma}_k = R^\dagger \boldsymbol{\sigma}_k R = \sum_j \boldsymbol{\sigma}_j e_{jk}. \tag{3.31}$$

The matrix elements of the rotation are therefore given by

$$e_{jk} = \boldsymbol{\sigma}_j \cdot \mathbf{e}_k = \boldsymbol{\sigma}_j \cdot (\mathcal{R}\boldsymbol{\sigma}_k) = \langle \boldsymbol{\sigma}_j R^\dagger \boldsymbol{\sigma}_k R\rangle_0. \tag{3.32}$$

This enables us to compute the matrix elements directly from the spinor R and express them in terms of any parameters used to parametrize R. For example, to express the matrix elements in terms of Euler parameters, we substitute (3.22a) into (3.32) to get

$$e_{jk} = \langle \alpha^2\, \boldsymbol{\sigma}_j \boldsymbol{\sigma}_k + \boldsymbol{\sigma}_j \boldsymbol{\beta} \boldsymbol{\sigma}_k \boldsymbol{\beta} + \alpha i \boldsymbol{\sigma}_j (\boldsymbol{\sigma}_k \boldsymbol{\beta} - \boldsymbol{\beta} \boldsymbol{\sigma}_k)\rangle_0 .$$

Evaluating the scalar parts in terms of inner products, we obtain the desired result

$$e_{jk} = (2\alpha^2 - 1)\delta_{jk} + 2\beta_j \beta_k - 2\alpha\,(\sum_i \beta_i \varepsilon_{ijk}) \tag{3.33}$$

where $\varepsilon_{ijk} \equiv -\langle i\boldsymbol{\sigma}_i \boldsymbol{\sigma}_j \boldsymbol{\sigma}_k\rangle_0 = \boldsymbol{\sigma}_i \cdot (\boldsymbol{\sigma}_j \times \boldsymbol{\sigma}_k)$.

Equation (3.32) enables us to translate from spinor to matrix representations for rotations. To translate the other way from given matrix elements a spinor, we need to solve (3.31) for R in terms of the $\mathbf{e}_k$ and $\boldsymbol{\sigma}_k$. This can be done most easily by constructing the quaternion

$$T = \sum_k \boldsymbol{\sigma}_k \mathbf{e}_k = \sum_j \sum_k e_{jk} \boldsymbol{\sigma}_j \boldsymbol{\sigma}_k , \tag{3.34}$$

which is uniquely determined by the $\boldsymbol{\sigma}_k$ and $\mathbf{e}_k$. According to (3.31), then,

$$T = \sum_k \boldsymbol{\sigma}_k R^\dagger \boldsymbol{\sigma}_k R .$$

So our problem is to solve this equation for R as a function of T. To that end, notice that

$$\sum_k \boldsymbol{\sigma}_k \boldsymbol{\sigma}_k = 3$$

and

$$\sum_k \boldsymbol{\sigma}_k \boldsymbol{\beta} \boldsymbol{\sigma}_k = \sum_k \boldsymbol{\sigma}_k (2\boldsymbol{\beta}\cdot\boldsymbol{\sigma}_k - \boldsymbol{\sigma}_k \boldsymbol{\beta}) = 2\boldsymbol{\beta} - 3\boldsymbol{\beta} = -\boldsymbol{\beta}$$

Therefore, for $R = \alpha + i\boldsymbol{\beta}$, we have

$$T = \sum_k \boldsymbol{\sigma}_k (\alpha - i\boldsymbol{\beta}) \boldsymbol{\sigma}_k R = (3\alpha + i\boldsymbol{\beta})R = (4\alpha - R^\dagger)R = 4\alpha R - 1 .$$

Thus,

$$4\alpha R = 1 + T . \tag{3.35}$$

We can solve this equation for $\alpha = \langle R \rangle_0$ by taking the scalar part or by computing the norm of both sides. Doing both, we obtain

$$16\alpha^2 = 4\langle 1 + T \rangle_0 = | 1 + T |^2 . \tag{3.36}$$

So, if $\alpha \neq 0$, we can solve (3.35) for R in the form

$$\pm R = \frac{1 + T}{2\langle 1 + T \rangle_0^{1/2}} = \frac{1 + T}{| 1 + T |} . \tag{3.37}$$

This enables us to compute R from the matrix elements e_{jk}, or any other specification of the $\mathbf{e}_k$ and the $\boldsymbol{\sigma}_k$, with the help of the definition of T in (3.34). Unfortunately (3.37) is undefined for any rotation through an angle of 180°, in which case $\alpha = \langle R \rangle_0 = 0$ and $T = -1$. To handle this case, we need an alternative parametrization of R in terms of the matrix elements.

Let us see how much we can find out about R given the transformation of a single vector, say $\boldsymbol{\sigma}_3$ to $\mathbf{e}_3$. From (3.31) we obtain

$$R\mathbf{e}_3 = \boldsymbol{\sigma}_3 R ,$$

so

$$(\alpha + i\boldsymbol{\beta})\mathbf{e}_3 = \boldsymbol{\sigma}_3(\alpha + i\boldsymbol{\beta}) .$$

Since we already have a general expression for α in (3.36), we seek to solve this equation for $\boldsymbol{\beta}$. Using the trick $\boldsymbol{\sigma}_3\boldsymbol{\beta} = -\boldsymbol{\beta}\boldsymbol{\sigma}_3 + 2\boldsymbol{\beta}\cdot\boldsymbol{\sigma}_3$ to reorder the geometric product, we obtain

$$i\boldsymbol{\beta}(\boldsymbol{\sigma}_3 + \mathbf{e}_3) = 2i\beta_3 + \alpha(\boldsymbol{\sigma}_3 - \mathbf{e}_3).$$

We solve this for $\boldsymbol{\beta}$ by using

$$(\boldsymbol{\sigma}_3 + \mathbf{e}_3)^{-1} = \frac{\boldsymbol{\sigma}_3 + \mathbf{e}_3}{|\,\boldsymbol{\sigma}_3 + \mathbf{e}_3\,|^2} = \frac{\boldsymbol{\sigma}_3 + \mathbf{e}_3}{2(1 + e_{33})}$$

and

$$(\boldsymbol{\sigma}_3 - \mathbf{e}_3)\,(\boldsymbol{\sigma}_3 + \boldsymbol{\sigma}_3)^{-1} = \frac{i\boldsymbol{\sigma}_3 \times \mathbf{e}_3}{1 + e_{33}}$$

Thus, we obtain

$$\boldsymbol{\beta} = (1 + e_{33})^{-1}\,[\beta_3(\boldsymbol{\sigma}_3 + \mathbf{e}_3) + \alpha\boldsymbol{\sigma}_3 \times \mathbf{e}_3]. \tag{3.38}$$

This gives us $\boldsymbol{\beta}$ from $\boldsymbol{\sigma}_3$ and $\mathbf{e}_3$ provided we know β_3 and α. From (3.34) and (3.36) we have

$$4\alpha^2 = 1 + e_{11} + e_{22} + e_{33}. \tag{3.39}$$

Using this in (3.33) we obtain

$$4\beta_3^2 = 1 + e_{33} - e_{11} - e_{22}. \tag{3.40}$$

Unfortunately, (3.39) and (3.40) do not determine the correct sign of α relative to β_3. However, this can be taken care of by using (3.33) again to get

$$\alpha = \frac{e_{12} - e_{21}}{4\beta_3}. \tag{3.41}$$

Equation (3.38) supplemented by (3.40) and (3.41) provides a practical means of computing $R = \alpha + i\boldsymbol{\beta}$ from the matrix elements though it is not so neat as (3.37).

Equation (3.38) is singular only when $e_{33} = -1$, that is, when $\boldsymbol{\sigma}_3$ is rotated by 180°. Of course, we get two similar equations by changing the subscripts from 3 to 2 or 1 in (3.38). At least one of these three equations will be nonsingular for any rotation. For numerical purposes, the optimal choice among the three possibilities corresponds to the most positive among e_{11}, e_{22}, and e_{33}. This amounts to selecting from $\boldsymbol{\sigma}_1$, $\boldsymbol{\sigma}_2$, and $\boldsymbol{\sigma}_3$ the vector which is closest to the rotation axis.

By deriving explicit formulas for calculating the spinor of a rotation from any given rotation matrix, we have proved, as a byproduct, the earlier assertion that *every rotation R can be written in the canonical form* $\mathcal{R}\mathbf{x} = R^\dagger \mathbf{x} R$. For we know that every rotation is determined by its matrix or the transformation of a standard basis.

With geometric algebra at our disposal, it should be obvious that the spinor representation of rotations is superior to the matrix representation for both theoretical and computational purposes. Rotations are characterized so much more simply and directly by spinors than by matrices! Even in problems where a rotation matrix is given as part of the initial data, the most efficient way to use it is usually to convert it to an equivalent spinor using (3.37) or (3.38). This, for example, is the most efficient general method for finding the rotation axis and angle from the matrix elements; for, according to (3.22), the angle and axis can be read off directly from the spinor. In many problems an appropriate spinor can be written down directly or calculated from the given data without introducing matrices. It is best to avoid the matrix representation of a rotation whenever possible, but we have developed the apparatus to handle it if necessary.

Euler Angles

We have considered several different parametrizations for spinors and rotations. Another parametrization which has been widely used by physicists and astronomers is specified by the equation

$$R = e^{(1/2)i\boldsymbol{\sigma}_3\psi}e^{(1/2)i\boldsymbol{\sigma}_1\theta}e^{(1/2)i\boldsymbol{\sigma}_3\phi}. \tag{3.42}$$

Introducing the notation

$$R_\phi = e^{(1/2)i\boldsymbol{\sigma}_3\phi}, \quad R_\psi = e^{(1/2)i\boldsymbol{\sigma}_3\psi}, \tag{3.43a}$$

$$Q_\theta = e^{(1/2)i\boldsymbol{\sigma}_1\theta}, \tag{3.43b}$$

we can write the parametrized spinor R in the form

$$R = R_\psi Q_\theta R_\phi. \tag{3.43c}$$

The scalar parameters ψ, θ, ϕ introduced in this way are called *Euler angles*. The advantage of using Euler angles is that every rotation is reduced to a product of rotations about fixed axes of a *standard basis*. It is especially easy, then to calculate the matrix elements of a rotation in terms of Euler angles. The rotation of a standard basis is given by

$$\mathbf{e}_k = \mathcal{R}\boldsymbol{\sigma}_k = R^\dagger\boldsymbol{\sigma}_k R = R_\phi^\dagger Q_\theta^\dagger R_\psi^\dagger \boldsymbol{\sigma}_k R_\psi Q_\theta R_\phi. \tag{3.44}$$

Consider the rotation of $\boldsymbol{\sigma}_3$, for example. From (3.43a) we have

$$R_\psi^\dagger \boldsymbol{\sigma}_3 R_\psi = \boldsymbol{\sigma}_3,$$

and, since $i\boldsymbol{\sigma}_3\boldsymbol{\sigma}_1 = -\boldsymbol{\sigma}_2$,

$$Q_\theta^\dagger \boldsymbol{\sigma}_3 Q_\theta = \boldsymbol{\sigma}_3 Q_\theta^2 = \boldsymbol{\sigma}_3 e^{i\boldsymbol{\sigma}_1\theta} = \boldsymbol{\sigma}_3 \cos\theta - \boldsymbol{\sigma}_2 \sin\theta.$$

Therefore,

$$\begin{aligned}\mathbf{e}_3 &= R_\phi^\dagger(\boldsymbol{\sigma}_3\cos\theta - \boldsymbol{\sigma}_2\sin\theta)R_\phi \\ &= \boldsymbol{\sigma}_3\cos\theta - \boldsymbol{\sigma}_2 e^{i\boldsymbol{\sigma}_3\phi}\sin\theta \\ &= \boldsymbol{\sigma}_3\cos\theta + (\boldsymbol{\sigma}_1\sin\phi - \boldsymbol{\sigma}_2\cos\phi)\sin\theta.\end{aligned} \tag{3.45}$$

From this the matrix elements $e_{j3} = \boldsymbol{\sigma}_j\cdot\mathbf{e}_3$ can be read off directly (Exercise

3.8). Figure 3.6 shows the effect on a standard basis of the successive rotations by Euler angles as specified by (3.42).

From the final diagram in Figure 3.6, we can see a different way to describe rotations with Euler angles. We can read off the following spinor for the rotation directly from the diagram:

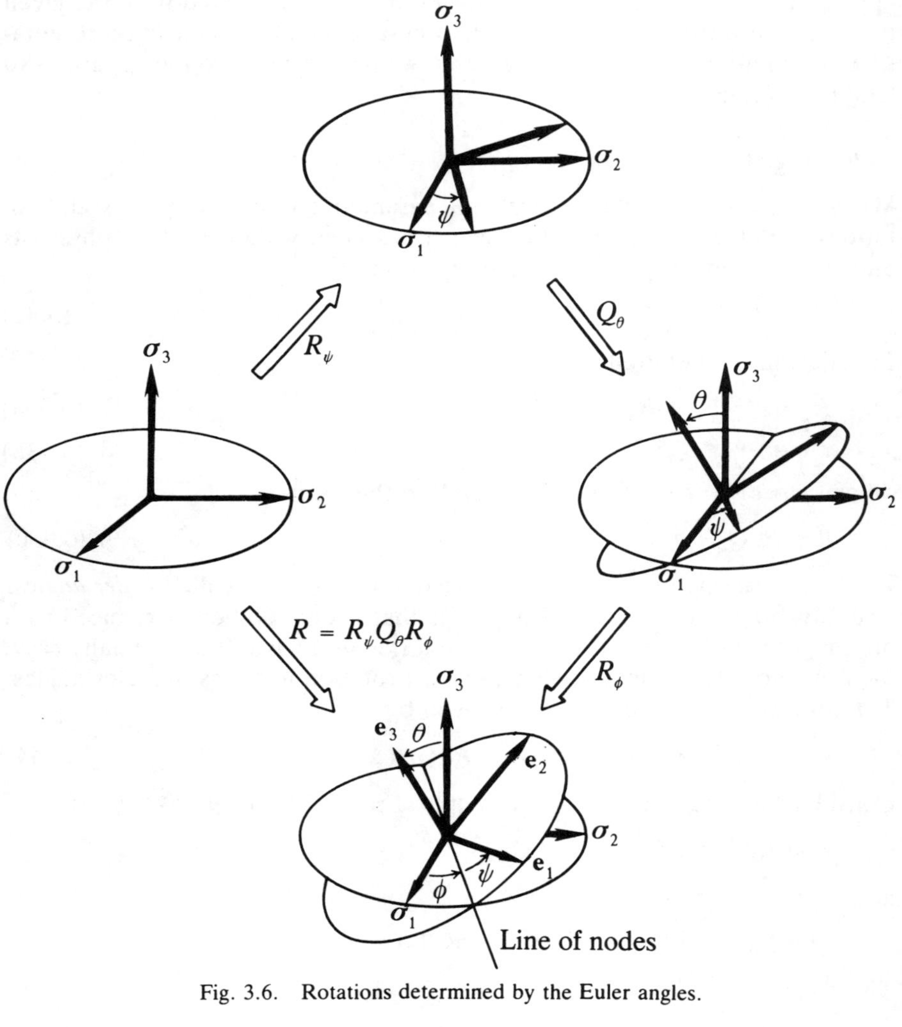

Fig. 3.6. Rotations determined by the Euler angles.

$$R = e^{(1/2)i\boldsymbol{\sigma}_3\phi}e^{(1/2)i\mathbf{n}\theta}e^{(1/2)i\mathbf{e}_3\psi}. \qquad (3.46)$$

This expression tells us that the net rotation can be achieved by a rotation of angle ϕ about the $\boldsymbol{\sigma}_3$–axis, followed by a rotation of angle θ about the so-called *line of nodes*, which has direction

$$\mathbf{n} = R_\phi^\dagger \boldsymbol{\sigma}_1 R_\phi = \boldsymbol{\sigma}_1 R_\phi^2 = \boldsymbol{\sigma}_1 e^{i\boldsymbol{\sigma}_3\phi} = \frac{\boldsymbol{\sigma}_3 \times \mathbf{e}_3}{|\boldsymbol{\sigma}_3 \times \mathbf{e}_3|}, \tag{3.47}$$

followed finally by a rotation of angle ψ about $\mathbf{e}_3$.

Note that the order of Euler angles in (3.46) is opposite that in (3.42). Nevertheless, both expressions describe the same rotation; they show that the parametrization by Euler angles can be given two different geometrical interpretations. As we shall see in Chapter 8, the form (3.46) is preferred by astronomers, because $\boldsymbol{\sigma}_3$ and $\mathbf{e}_3$ can be associated with easily measured directions. On the other hand, (3.42) has the advantage of fixed rotation axes even for Euler angles changing with time.

To prove algebraically that (3.46) is indeed equivalent to (3.42), note that

$$e^{(1/2)i\mathbf{n}\theta} = R_\phi^\dagger Q_\theta R_\phi$$

and

$$e^{(1/2)i\mathbf{e}_3\psi} = R^\dagger R_\psi R = R_\phi^\dagger Q_\theta^\dagger R_\psi Q_\theta R_\phi .$$

Substituting this into (3.46) and using the unitarity condition $R^\dagger R = 1$ for the various spinors, we get (3.42) as desired. The student should carry this step out to see how it works.

Canonical Forms for Linear Operators

We have found canonical forms for rotations and simple reflections in this section and for symmetric operators in the preceding section. These results determine canonical forms for all nonsingular linear operators, we can now easily show.

To complete our characterization of orthogonal operators, we need canonical forms for both proper and improper operators. The canonical form for a proper operator (i.e. rotation) is given by (3.15). The canonical form for an arbitrary improper orthogonal operator $\mathcal{I}$ is determined by the following theorem: Let $\mathcal{U}$ be a simple reflection along *any* direction $\mathbf{u}$, as expressed by the canonical form (3.6). Then, there is a unique rotation $\mathcal{R}$ such that

$$\mathcal{I} = \mathcal{R}\mathcal{U}. \tag{3.48}$$

A canonical form for $\mathcal{I}$ is therefore determined by the canonical forms for $\mathcal{U}$ and $\mathcal{R}$. The proof of (3.48) is easy. One simply uses the fact that $\mathcal{U}^2 = 1$ to write $\mathcal{I} = (\mathcal{I}\mathcal{U})\mathcal{U}$ and so define $\mathcal{R}$ by $\mathcal{R} = \mathcal{I}\mathcal{U}$. The fact that $\mathcal{R}$ is a rotation is proved by det $\mathcal{R}$ = (det $\mathcal{I}$) (det $\mathcal{U}$) = (–1) (–1) = 1.

Canonical forms for an arbitrary nonsingular operators are determined by the *Polar Decomposition Theorem*: Every nonsingular tensor f has a *unique* decomposition of the form

$$f = \mathcal{R}\mathcal{S} = \mathcal{I}\mathcal{R} \tag{3.49}$$

where $\mathcal{R}$ is a rotation and $\mathcal{S}$ and $\mathcal{I}$ are positive symmetric tensors given by

$$\mathcal{S} = (\bar{f}f)^{1/2} \quad \text{and} \quad \mathcal{T} = (f\bar{f})^{1/2}. \tag{3.50}$$

A canonical form for f is therefore determined by the canonical forms for $\mathcal{R}$ and $\mathcal{S}$.

To prove (3.49), first note that

$$\mathbf{y}\cdot(\bar{f}f\mathbf{x}) = (f\mathbf{y})\cdot(f\mathbf{x}) = \mathbf{x}\cdot(\bar{f}f\mathbf{y})$$

and

$$\mathbf{x}\cdot(\bar{f}f\mathbf{x}) = (f\mathbf{x})^2 > 0 \quad \text{if} \quad \mathbf{x} \neq 0.$$

Therefore, the operator $\bar{f}f$ is symmetric, so its square root $\mathcal{S}$ specified by (3.50) is well defined and unique. Since $\mathcal{S}$ is nonsingular, we can solve (3.49) for the rotation

$$\mathcal{R} = f\mathcal{S}^{-1} = f(\bar{f}f)^{-1/2} \tag{3.51}$$

It is easily verified that $\mathcal{R}$ is indeed a rotation, and the properties of the operator $\mathcal{T}$ can be determined in much the same way as the properties of $\mathcal{S}$.

The eigenvalues and eigenvectors of $\mathcal{S}$ describe basic structural features of f. They are sometimes called *principal vectors* and *principal values* of f to distinguish them from eigenvectors and eigenvalues of f. There is, of course, no distinction if f itself is symmetric. The principal values of $\mathcal{S}$ are always real numbers and, in general, they are not related in a simple way to the eigenvalues of f, some of which may be "complex numbers". The polar decomposition theorem (3.49) tells us that the complex eigenvalues arise from rotations with the imaginary numbers corresponding to bivectors for planes of rotations, as we noted earlier in the special case of (2.15a, b). But we have seen that geometric algebra enables us to characterize rotations completely and effectively without reference to secular equations and complex eigenvalues. Thus, the polar decomposition enables us to characterize any linear transformation completely without introducing complex eigenvalues and eigenvectors.

The polar decomposition (3.49) provides us with a simple geometrical interpretation for any linear operator f. Consider the action of f on the points $\mathbf{x}$ of a 3-dimensional body of geometrical figure. According to (3.49), then, the body is first stretched and/or reflected along the principle directions of f. Then the distorted body is rotated through some angle specified by $\mathcal{R}$. In contrast to the clear geometrical interpretation of principle directions and principle values, in conventional treatments of linear algebra complex eigenvectors and eigenvalues do not have an evident interpretation.

We shall not have the occasion to apply the polar decomposition theorem to physics problems in this book. The theorem has been discussed only to provide the student with a general perspective on the structure of linear operators.

5-3. Exercises

(3.1) Show that the transformation $\mathcal{R}\mathbf{x} = \mathbf{u}^{-1}\mathbf{x}\mathbf{u}$, determined by a nonzero vector $\mathbf{u}$ is a rotation. Find the axis, the angle and the spinor for this rotation.

(3.2) Find the inverse of a reflection.

(3.3) Prove that the product of three successive elementary reflections in orthogonal planes is an *inversion*, the linear transformation that reverses the direction of every vector.
Explain how this fact made it easy to place mirrors on the moon which reflect laser signals from Earth back to their source. What precision measurements can be made with such signals?

(3.4) A unitary spinor R can be given the following parametrizations

$$R = e^{(1/2)i\mathbf{a}} = \alpha + i\boldsymbol{\beta} = \alpha(1 + i\boldsymbol{\gamma}) = \frac{1 + i\mathbf{b}}{1 - i\mathbf{b}},$$

where the parameters $\mathbf{a}$, $\boldsymbol{\beta}$, γ and $\mathbf{b}$ are all vectors. Establish the following relations among the parameters:

$$\alpha = \cos\tfrac{1}{2}\mathbf{a} = \frac{1}{\sqrt{1 + \gamma^2}} = \frac{1 - \mathbf{b}^2}{1 + \mathbf{b}^2}$$

$$\boldsymbol{\gamma} = \tan\tfrac{1}{2}\mathbf{a} = \hat{\mathbf{a}}\tan\tfrac{1}{2}a = \frac{2\mathbf{b}}{1 - \mathbf{b}^2}$$

$$\mathbf{b} = \tan\tfrac{1}{4}\mathbf{a}.$$

In the following problems

$$\mathbf{x}' = \mathcal{R}\mathbf{x} = R^{\dagger}\mathbf{x}R,$$

where R has the parametrizations just described.

(3.5) Derive "Rodrigues formula"

$$\mathbf{x}' - \mathbf{x} = \boldsymbol{\gamma} \times (\mathbf{x}' + \mathbf{x})$$

(3.6) Establish the following "vector forms" for a rotation:

$$\mathbf{x}' = \mathbf{x} + 2\alpha\boldsymbol{\beta} \times \mathbf{x} + 2\boldsymbol{\beta} \times (\boldsymbol{\beta} \times \mathbf{x})$$

$$= \mathbf{x} + \hat{\mathbf{a}} \times \mathbf{x}\sin a + \hat{\mathbf{a}} \times (\hat{\mathbf{a}} \times \mathbf{x})(1 - \cos a).$$

(3.7) Derive the following expression for the matrix elements of a rotation by an arbitrary vector angle $\mathbf{a}$:

$$e_{jk} = \delta_{jk}\cos a - \varepsilon_{jkm}\hat{a}_m \sin a + \hat{a}_j\hat{a}_k (1 - \cos a),$$

where $\varepsilon_{jkm} = i^{\dagger}\boldsymbol{\sigma}_j\wedge\boldsymbol{\sigma}_k\wedge\boldsymbol{\sigma}_m$ and the $\hat{a}_k = \hat{\mathbf{a}}\cdot\boldsymbol{\sigma}_k$ are "direction cosines" of the rotation axis.

Evaluate matrix elements for

(a) $\hat{\mathbf{a}} = \boldsymbol{\sigma}_3$,

(b) $\mathbf{a} = \frac{2\pi}{3\sqrt{3}}(\boldsymbol{\sigma}_1 + \boldsymbol{\sigma}_2 - \boldsymbol{\sigma}_3)$.

(3.8) Evaluate the matrix elements of a rotation in terms of Euler angles to get the matrix

$$\begin{bmatrix} \cos\psi\cos\phi - \cos\theta\sin\phi\sin\psi & -\sin\psi\cos\phi - \cos\theta\sin\phi\cos\psi & \sin\theta\sin\phi \\ \cos\psi\sin\phi + \cos\theta\cos\phi\sin\psi & \sin\psi\sin\phi + \cos\theta\cos\phi\cos\psi & -\sin\theta\cos\phi \\ \sin\theta\sin\psi & \sin\theta\cos\psi & \cos\theta \end{bmatrix}.$$

(3.9) Show that any spinor R can be written in the form

$$R = \pm(\mathbf{uv})^{1/2} = \frac{1 + \mathbf{uv}}{[2(1+\mathbf{u}\cdot\mathbf{v})]^{1/2}}$$

where **u** and **v** are unit vectors. Derive therefrom the trigonometric formulas

$$\cos\tfrac{1}{2}a = \frac{1 + \cos a}{[2(1+\cos a)]^{1/2}}$$

$$\sin\tfrac{1}{2}a = \frac{\sin a}{[2(1+\cos a)]^{1/2}}.$$

(3.10) Given that a rotation $\mathcal{R}(\mathbf{x}) = R^{\dagger}\mathbf{x}R$ has the properties

$$\mathcal{R}(\mathbf{a}\times\mathbf{b}) = \mathbf{a}\times\mathbf{b},$$

$$\mathcal{R}(\mathbf{a}) = \mathbf{b},$$

show that

$$\pm R = \frac{1 + \mathbf{a}^{-1}\mathbf{b}}{[2(1 + \mathbf{a}^{-1}\cdot\mathbf{b})]^{1/2}}.$$

(3.11) For the composition of rotations described by the spinor equation

$$R_1R_2 = R_3$$

where

$$R_k = e^{(1/2)i\mathbf{a}_k} = \alpha_k(1 + i\boldsymbol{\gamma}_k),$$

derive the "law of tangents"

$$\tan\tfrac{1}{2}\mathbf{a}_3 = \boldsymbol{\gamma}_3 = \frac{\boldsymbol{\gamma}_1 + \boldsymbol{\gamma}_2 + \boldsymbol{\gamma}_2\times\boldsymbol{\gamma}_1}{1 - \boldsymbol{\gamma}_1\cdot\boldsymbol{\gamma}_2}.$$

(3.12) The sum of the diagonal matrix elements f_{kk} of a linear transformation f is called the *trace* of f and denoted by Tr f. Show that the trace of a rotation $\mathcal{R}$ is given by

$$\text{Tr}\ \mathcal{R} = \sum_k \boldsymbol{\sigma}_k \cdot (\mathcal{R}\boldsymbol{\sigma}_k) = 1 + 2\cos \mathbf{a},$$

where **a** is the vector angle of rotation.

(3.13) (a) Prove that every reflection is a symmetric transformation.
(b) Under what condition will the product of two reflections be a symmetric transformation?
(c) Prove that every symmetric transformation can be expressed as the product of a symmetric orthogonal transformation and a positive symmetric transformation.

(3.14) Explain how results developed in the text can be used to prove Hamilton's theorem: Every rotation can be expressed as a product of two elementary reflections.

(3.15) Find the polar decompositon of the skewsymmetric transformation

$$f\mathbf{x} = \mathbf{x} \cdot \mathbf{A}.$$

(3.16) The linear transformation

$$f\mathbf{x} = \mathbf{x} + 2\alpha \boldsymbol{\sigma}_1 \boldsymbol{\sigma}_2 \cdot \mathbf{x}$$

is called a *shear*. Draw a diagram (similar to Figure 2.1) showing the effect of f on a rectangle in the $\boldsymbol{\sigma}_1\boldsymbol{\sigma}_2$ - plane. Find the eigenvectors, eigenvalues, principle vectors and principle values of f in this plane. Determine also the angle of rotation in the polar decomposition of f.

5-4. Transformation Groups

So far in this chapter we have concentrated our attention on properties of individual linear transformations. However, in physical applications transformations often arise in families. For example, the change of a physical system from one state to another may be described by a transformation, so the set of all changes in physical state is a family of transformations. If the changes are reversible, then this family has the general structure of a mathematical group. Transformation groups are so common and significant in physics that they deserve to be studied systematically in their own right.

As there is a great variety of different groups, it will be conceptually efficient to begin with the abstract definition of a mathematical group describing the common properties of all groups. Then we shall examine the structure of specific groups of particular importance in physics.

An abstract group is a set of elements $\mathcal{F}, \mathcal{G}, \mathcal{H}, \ldots$ interrelated by a binary function called the *group product* with the following properties:

(1) *Closure*: To every ordered pair of group elements $(\mathcal{G}, \mathcal{H})$ the group product associates a unique group element denoted by $\mathcal{G}\mathcal{H}$.
(2) *Associativity*: For any three group elements $(\mathcal{F}\mathcal{G})\mathcal{H} = \mathcal{F}(\mathcal{G}\mathcal{H})$.
(3) *Identity*: There is a unique identity element $\mathcal{E}$ in the group with the property $\mathcal{E}\mathcal{G} = \mathcal{G}$ for every element $\mathcal{G}$.
(4) *Inverse*: To every element $\mathcal{G}$ there corresponds a unique element $\mathcal{G}^{-1}$ such that

$$\mathcal{G}^{-1}\mathcal{G} = \mathcal{E}.$$

The Rotation Group $\mathcal{O}^+(3)$

From the properties of rotations discussed in Section 5-3, one can easily show that the set of all rotations on Euclidian 3-space $\mathcal{E}_3$ forms a group for which the group product is the composition of two rotations. This group is called the rotation group on $\mathcal{E}_3$ and denoted by $\mathcal{O}^+(3)$. The rotation group deserves detailed study, first, because it is the most common and generally useful group in physics, and, second, because it exhibits most of the interesting properties of groups in general. After one has become familiar with specific properties of the rotation group, other groups can be efficiently analyzed by comparing them with the rotation group.

As we have noted before, by selecting a standard basis $\{\boldsymbol{\sigma}_k\}$ we can associate with each rotation $\mathcal{R}$ a unique matrix with matrix elements

$$r_{jk} = \boldsymbol{\sigma}_j \cdot (\mathcal{R}\boldsymbol{\sigma}_k). \tag{4.1}$$

These matrices form a group for which the matrix product is the group product. This group is called a *matrix representation* of the rotation group. These two groups are isomorphic. Two groups are said to be *isomorphic* if their elements and group products are in one-to-one correspondence. The *isomorphism* between the rotation group and its matrix representation determined by Equation (4.1) is shown in Table 4.1.

Besides the matrix representation there is another important representation of the rotation group. As shown in Section 5-3, the equation

$$\mathcal{R}\boldsymbol{\sigma}_k = R^\dagger \boldsymbol{\sigma}_k R = \sum_j \boldsymbol{\sigma}_j r_{jk} \tag{4.2}$$

determines a correspondence between each rotation $\mathcal{R}$ on $\mathcal{E}_3$ with matrix $[r_{jk}]$ and a pair of unitary spinors $\pm R$. The unitary spinors form a group which we dub the *dirotation group* and denote by $2\mathcal{O}^+(3)$, though this nomenclature is not standard. For this group the geometric product is the group product. The dirotation group is commonly referred to as the *spin-$\frac{1}{2}$ representation* of the rotation group. The two-to-one correspondence of elements in $2\mathcal{O}^+(3)$ with elements in $\mathcal{O}^+(3)$ is called a *homomorphism*. Table 4.1 shows the correspon-

TABLE 4.1. Homomorphism of the rotation group $O^+(3)$ with its matrix representation and the group of unitary spinors $2O^+(3)$

	Dirotation group		Rotation group		Matrix representation
		2 to 1		1 to 1	
Elements	$\{\pm R, \pm S, \ldots\}$	$\longleftrightarrow$	$\{\mathcal{R}, \mathcal{S}, \ldots\}$	$\longleftrightarrow$	$\{r_{jk}, s_{jk}, \ldots\}$
Closure	$SR = T$	$\longleftrightarrow$	$\mathcal{R}\mathcal{S} = \mathcal{T}$	$\longleftrightarrow$	$\sum_j r_{ij}s_{jk} = t_{ik}$
Associativity	$S(RQ) = (SR)Q$	$\longleftrightarrow$	$(\mathcal{Q}\mathcal{R})\mathcal{S} = \mathcal{Q}(\mathcal{R}\mathcal{S})$	$\longleftrightarrow$	$\sum_k (\sum_j q_{ij}r_{jk})s_{kl} = \sum_j q_{ij} (\sum_k r_{jk}s_{kl})$
Identity	$1R = R$	$\longleftrightarrow$	$1\mathcal{R} = \mathcal{R}$	$\longleftrightarrow$	$\sum_j \delta_{ij}r_{jk} = r_{ik}$
Inverse	$R^\dagger R = \bar{1}$	$\longleftrightarrow$	$\mathcal{R}^{-1}\mathcal{R} = 1$	$\longleftrightarrow$	$\sum_j r_{ij}^{-1}r_{jk} = \delta_{ik}$

dence between elements and operations of these groups. In general, a homomorphism is a many-to-one correspondence between groups, and the special case of a one-to-one correspondence is called an *isomorphism*.

Homomorphic groups can be regarded as different ways of representing the same system of mathematical relations. Mathematical groups derive physical significance from correspondences with actual (or imagined) groups of operations on physical systems. The displacement of a solid body with one point fixed is a *physical rotation*, and the set of all such displacements is the *physical rotation group*. This group can be represented mathematically by any one of the three groups in Table 4.1, a group of linear transformations (the *mathematical rotation group* $O^+(3)$), a group of matrices, or the group of unitary spinors $2O^+(3)$. From our experience in Section 5-3, we know that for computational purposes the spinor group is the most convenient representation of the physical rotation group, so we shall make great use of it.

Table 4.1 gives three different mathematical representations of the group product for the rotation group. There are many others. For example, equation (3.28b) gives a representation of the group product in the form

$$\phi(\boldsymbol{\beta}_1, \boldsymbol{\beta}_2) = (1-\boldsymbol{\beta}_2^2)^{1/2}\boldsymbol{\beta}_1 + (1-\boldsymbol{\beta}_1^2)^{1/2}\boldsymbol{\beta}_2 - \boldsymbol{\beta}_1 \times \boldsymbol{\beta}_2, \tag{4.3}$$

where $\boldsymbol{\beta}_1$ and $\boldsymbol{\beta}_2$ are vectors with magnitude less than one which determine the axis and angle of rotation. The group product has been written in the functional from $\phi(\boldsymbol{\beta}_1, \boldsymbol{\beta}_2)$ so as not to confuse it with the geometric product

$\boldsymbol{\beta}_1\boldsymbol{\beta}_2$ of vectors. With this notation the properties of the group product take the form:
Closure:

$$\phi(\boldsymbol{\beta}_1, \boldsymbol{\beta}_2) = \boldsymbol{\beta}_3, \tag{4.4a}$$

Associativity:

$$\phi(\phi(\boldsymbol{\beta}_1, \boldsymbol{\beta}_2), \boldsymbol{\beta}_3) = \phi(\boldsymbol{\beta}_1, \phi(\boldsymbol{\beta}_2, \boldsymbol{\beta}_3)), \tag{4.4b}$$

Identity:

$$\phi(\mathbf{0}, \boldsymbol{\beta}) = \boldsymbol{\beta}, \tag{4.4c}$$

Inverse:

$$\phi(-\boldsymbol{\beta}, \boldsymbol{\beta}) = \mathbf{0}. \tag{4.4d}$$

The set $\{\boldsymbol{\beta}\}$ of all vectors in the unit ball $|\boldsymbol{\beta}| \leqslant 1$ is a group under the product (4.3). According to (4.4c), the identity element in this group is the zero vector, and, according to (4.4d), the negative of a vector in the group is its "group inverse". The reader can verify by substitution that the product $\phi(\boldsymbol{\beta}_1, \boldsymbol{\beta}_2)$ defined by (4.3) has the group properties (4.4a, b, c, d). However, from the derivation of (4.3) in Section 5-3 we know without calculation that the group properties must be satisfied, because the equation

$$R = (1 - \beta^2)^{1/2} + i\boldsymbol{\beta} \tag{4.5}$$

determines homomorphisms of $2\mathcal{O}^+(3)$ and $\mathcal{O}^+(3)$ with the group of vectors. Each vector $\boldsymbol{\beta} = \sin\frac{1}{2}\mathbf{a}$ in the unit ball determines a clockwise rotation through an angle $|\mathbf{a}|$ about an axis with direction $\hat{\boldsymbol{\beta}} = \hat{\mathbf{a}}$. Note that $\boldsymbol{\beta}$ and $-\boldsymbol{\beta}$ determine equivalent rotations when $|\boldsymbol{\beta}| = 1$.

Properties of the rotation group can be established by establishing corresponding properties for any of the groups homomorphic to it. As a rule it will be most convenient to work with the spinor group and its various parametrizations such as (4.5) or the parametrization by angle

$$R = e^{(1/2)i\mathbf{a}}. \tag{4.6}$$

According to (4.6), every spinor R is a continuous function of the "angle vector" $\mathbf{a}$. Since $\mathbf{a}$ can vary continuously to the value $\mathbf{0}$, *every spinor is continuously connected to the identity element*

$$e^{(1/2)i\mathbf{0}} = 1.$$

Because of this property, $2\mathcal{O}^+(3)$ is said to be a *continuous group*. It follows that the homomorphic rotation group $\mathcal{O}^+(3)$ is also a continuous group. The continuity property makes it possible to differentiate the elements of a continuous group. In Section 5-6 we shall see how to differentiate a rotation by reducing it to the derivative of the corresponding spinor.

If we keep the direction $\hat{\mathbf{a}} = \mathbf{n}$ fixed and allow the magnitude $a = |\mathbf{a}|$ of the angle vector in (4.6) to range over the values $0 \leqslant a \leqslant 2\pi$, then we get a group of spinors with the parametric form

$$R = e^{(1/2)i\mathbf{n}a} \tag{4.7}$$

showing that they have a common axis $\mathbf{n}$ in $\mathcal{E}_3$. This is the spinor group $2\mathcal{O}^+(2)$, the *dirotation group* in the Euclidean plane $\mathcal{E}_2$. According to (4.7) all group elements are determined by the values of a single scalar parameter a, so $2\mathcal{O}^+(2)$ and $\mathcal{O}^+(2)$ are said to be 1-*parameter* groups. $2\mathcal{O}^+(3)$ and $\mathcal{O}^+(3)$ are 3-*parameter groups* because every element can be specified by the values of three scalar parameters, for example, the three components of the vector $\mathbf{a}$ in (4.6) relative to a standard basis, or the values of the three Euler angles in Equation (3.34).

$2\mathcal{O}^+(2)$ is a 1-parameter subgroup of $2\mathcal{O}^+(3)$ while $\mathcal{O}^+(2)$ is a subgroup of $\mathcal{O}^+(3)$. A *subgroup* of a group is a subset of group elements which is closed under the group product, so it is itself a group. Obviously, $\mathcal{O}^+(3)$ contains infinitely many 1-parameter subgroups of the type $\mathcal{O}^+(2)$, one such subgroup for each distinct axis of rotation.

The Orthogonal Group $\mathcal{O}(3)$

In Section 5-3 we saw that there are two kinds of orthogonal transformations, proper and improper. We have seen that the proper transformations (rotations) form a group $\mathcal{O}^+(3)$. The improper transformations do not form a group, because the product of two improper transformations is a proper one. However, this shows that the set of orthogonal transformations on $\mathcal{E}_3$ is closed under composition, so it is a group. This group is called the *orthogonal group* $\mathcal{O}(3)$. The rotation group $\mathcal{O}^+(3)$ is obviously a subgroup of $\mathcal{O}(3)$.

According to Section 5-3, every rotation has the canonical form

$$\mathcal{R}\mathbf{x} = R^\dagger \mathbf{x} R \tag{4.8}$$

where R is a spinor, that is, an even multivector. On the other hand, every improper orthogonal transformation has the form

$$\mathcal{R}\mathbf{x} = -R^\dagger \mathbf{x} R \tag{4.9}$$

where R is an odd multivector. In both cases

$$R^\dagger R = 1\,. \tag{4.10}$$

Equations (4.8) and (4.9) can be combined into a single equation

$$\mathcal{R}\mathbf{x} = \tilde{R}\mathbf{x}R\,, \tag{4.11}$$

where

$$\tilde{R} = R^\dagger \quad \text{if } R \text{ is even}, \tag{4.12a}$$

$$\tilde{R} = -R^\dagger \quad \text{if } R \text{ is odd}. \tag{4.12b}$$

Thus, (4.11) subject to (4.10) is the canonical form for any element of the orthogonal group.

In order to refer to the multivector R in (4.11) as a *spinor*, we must enlarge our concept of spinor. We can then distinguish two distinct kinds of spinor, a *proper* (or even) spinor satisfying (4.12a) or an *improper* (or odd) spinor satisfying (4.12b). These spinors form the *diorthogonal* group $2\mathcal{O}(3)$ which, by (4.11), is two-to-one homomorphic to the orthogonal group $\mathcal{O}(3)$. Obviously, the even spinors form the subgroup $2\mathcal{O}^+(3)$ homomorphic to the rotation group $\mathcal{O}^+(3)$. Notice also that odd spinors cannot be continuously connected to the identity element 1, because 1 is even; hence, $2\mathcal{O}(3)$ is not a continuous group. It does, however, contain two connected subsets, namely, the even and the odd spinors. Likewise, $\mathcal{O}(3)$ is not a continuous group, though it consists of two continuous subsets, the proper and the improper orthogonal transformations. Only one of these two subsets is a subgroup.

The complete orthogonal group $\mathcal{O}(3)$ does not play so important a role in dynamics as the subgroup of rotations $\mathcal{O}^+(3)$, because any physical transformation of a rigid body must be continuously connected to the identity transformation. However, improper orthogonal transformations are needed for a full description of the symmetries of a physical system, as we shall see in Section. 5-5.

The Translation Groups

A translation $\mathcal{T}_{\mathbf{a}}$ on $\mathcal{E}_3$ is a transformation of each point $\mathbf{x}$ in $\mathcal{E}_3$ to another point

$$\mathcal{T}_{\mathbf{a}}\mathbf{x} = \mathbf{x} + \mathbf{a}. \tag{4.13}$$

This equation is mathematically defined for all points in $\mathcal{E}_3$, but in physical applications we shall usually be concerned only with applying it to points which designate positions of physical particles. Thus, we can regard (4.13) as describing a shift $\mathbf{a}$ in the position of each particle in a physical object, as shown in Figure 4.1. We have already made similar interpretations of equations describing rotations without saying so explicitly. It should be clear from the context when such an interpretation is made in the future.

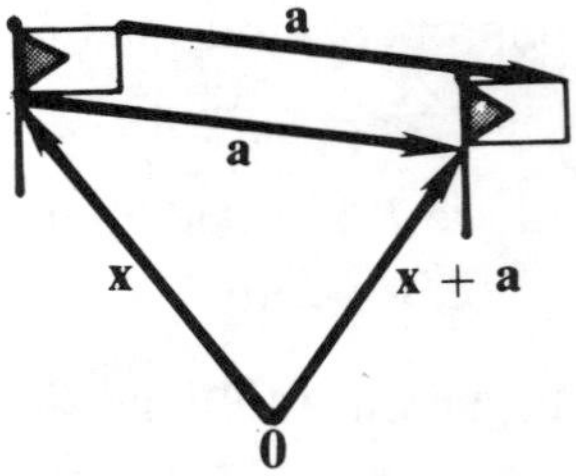

Fig. 4.1. Translation by **a** of a physical object.

Although translations are point transformations and they transform straight lines into straight lines, in contrast to rotations, they are not linear transformations. However, the translations do form a group, so it is convenient to use the operator notation we have adopted for groups as well as linear transformations, dropping parentheses to write $\mathcal{T}_{\mathbf{a}}\,\mathbf{x}$ in (4.13) instead of $\mathcal{T}_{\mathbf{a}}(\mathbf{x})$. The main reason for this convention is the simplicity it gives to the associative rule

$$\mathcal{T}_{\mathbf{a}}(\mathcal{T}_{\mathbf{b}}\mathcal{T}_{\mathbf{c}}) = (\mathcal{T}_{\mathbf{a}}\mathcal{T}_{\mathbf{b}})\,\mathcal{T}_{\mathbf{c}},$$

This rule and the other group properties follow directly from the definition of a translation by (4.13). In addition, from (4.13) it follows that

$$\mathcal{T}_{\mathbf{a}}\mathcal{T}_{\mathbf{b}} = \mathcal{T}_{\mathbf{a}+\mathbf{b}} = \mathcal{T}_{\mathbf{b}}\mathcal{T}_{\mathbf{a}}. \tag{4.14}$$

Thus, all translations commute with one another. For this reason, the *translation group* is said to be a *commutative group*. The commutivity of translations is clearly a consequence of the commutativity of vector addition. Indeed, it is readily verified that the function $\phi(\mathbf{a}, \mathbf{b}) = \mathbf{a} + \mathbf{b}$ has the properties (4.4a, b, c, d) of a group product. Hence, the vectors of $\mathcal{E}_3$ form a *group under addition* which is isomorphic to the translation group on $\mathcal{E}_3$.

The Euclidean Group

An isometry of Euclidean space $\mathcal{E}_3$ is a point transformation of $\mathcal{E}_3$ onto $\mathcal{E}_3$ which leaves the distance between every pair of points invariant. Thus, if f is an isometry taking each point $\mathbf{x}$ into a point

$$\mathbf{x}' = f(\mathbf{x}), \tag{4.15}$$

then for every pair of points $\mathbf{x}$ and $\mathbf{y}$ we have

$$(\mathbf{x}' - \mathbf{y}')^2 = (\mathbf{x} - \mathbf{y})^2. \tag{4.16}$$

The condition (4.16) tells us that there is a transformation $\mathcal{R}$ of each vector $\mathbf{x} - \mathbf{y}$ to a vector

$$\mathbf{x}' - \mathbf{y}' = f(\mathbf{x}) - f(\mathbf{y}) = \mathcal{R}(\mathbf{x} - \mathbf{y}) \tag{4.17}$$

with the same length as $\mathbf{x} - \mathbf{y}$. What kind of a function is $\mathcal{R}$?

Since (4.17) must apply to every pair of vectors, we have

$$\mathbf{y}' - \mathbf{z}' = \mathcal{R}(\mathbf{y} - \mathbf{z})$$

which, when added to (4.17), gives

$$\mathbf{x}' - \mathbf{z}' = \mathcal{R}(\mathbf{x} - \mathbf{y}) + \mathcal{R}(\mathbf{y} - \mathbf{z}) = \mathcal{R}(\mathbf{x} - \mathbf{z}).$$

Setting $\mathbf{y} = 0$ in this expression, we find that $\mathcal{R}$ has the distributive property

$$\mathcal{R}(\mathbf{x} - \mathbf{z}) = \mathcal{R}(\mathbf{x}) + \mathcal{R}(-\mathbf{z}). \tag{4.18}$$

If we can prove also that

$$\mathcal{R}(\alpha\mathbf{x}) = \alpha\mathcal{R}(\mathbf{x}) \tag{4.19}$$

for any scalar α, then we will know that $\mathcal{R}$ is necessarily a linear transformation. Moreover, $\mathcal{R}$ must be an orthogonal transformation, because it leaves the length of vectors unchanged.

Equation (4.19) can be proved in the following way. Setting $-\mathbf{z} = \mathbf{x}$ in

(4.18), we get $\mathcal{R}(2\mathbf{x}) = 2\mathcal{R}(\mathbf{x})$. Then setting $-\mathbf{z} = 2\mathbf{x}$ in (4.18), we get $\mathcal{R}(3\mathrm{x}) = 3\mathcal{R}(\mathbf{x})$. Continuing in this way we establish

$$\mathcal{R}(m\mathbf{x}) = m\mathcal{R}(\mathbf{x})$$

where m is any integer. If $\mathbf{x} = n\mathbf{y}$ where n is a nonzero integer, then

$$\mathcal{R}\left(\frac{m}{n}\,\mathbf{x}\right) = \mathcal{R}(m\mathbf{y}) = m\mathcal{R}(\mathbf{y}) = \frac{m}{n}\mathcal{R}(\mathbf{x}),$$

which proves that (4.19) holds when α is a rational number. Since any real number can be approximated to arbitrary accuracy by a rational number, it follows from the continuity of $\mathcal{R}$ that (4.19) must hold for any real number. Since $[\mathcal{R}(\mathbf{x})]^2 = \mathbf{x}^2$ is a continuous function of $\mathbf{x}$, the function $\mathcal{R}(\mathbf{x})$ must be continuous. This completes the proof that $\mathcal{R}$ is a linear transformation.

Now, setting $\mathbf{y} = \mathbf{0}$ in (4.17) and writing $f(\mathbf{0}) = \mathbf{a}$, we have

$$f(\mathbf{x}) = \mathcal{R}\mathbf{x} + \mathbf{a} = \mathcal{T}_{\mathbf{a}}\mathcal{R}\mathbf{x}. \tag{4.20}$$

Thus, we have proved that every isometry of Euclidean space is the product of an orthogonal transformation and a translation. Using (4.11), we can write (4.20) in the form

$$\{R \mid \mathbf{a}\}\mathbf{x} = \mathcal{T}_{\mathbf{a}}\mathcal{R}\mathbf{x} = \tilde{R}\mathbf{x}R + \mathbf{a}. \tag{4.21}$$

The new notation $\{R \mid \mathbf{a}\}$ has been introduced to indicate that each isometry is uniquely determined by a spinor R and a vector $\mathbf{a}$. In this notation

$$\mathcal{R} = \{R \mid \mathbf{0}\} \tag{4.22}$$

denotes an orthogonal transformation, and

$$\mathcal{T}_{\mathbf{a}} = \{1 \mid \mathbf{a}\} \tag{4.23}$$

denotes a translation. Note that

$$\{R \mid \mathbf{a}\} = \{-R \mid \mathbf{a}\}, \tag{4.24}$$

because of the 2-to-1 homomorphism between spinors and orthogonal transformations. The right side of (4.21) reduces an isometry to multiplication and addition in geometric algebra.

The isometries of Euclidean space form a group. The student can verify the following group properties. The *group product* is given by

$$\{S \mid \mathbf{b}\}\{R \mid \mathbf{a}\} = \{RS \mid \tilde{S}\mathbf{a}S + \mathbf{b}\}. \tag{4.25}$$

The *identity* element is

$$\{1 \mid \mathbf{0}\} = \{-1 \mid \mathbf{0}\}. \tag{4.26}$$

The *inverse* of an isometry is given by

$$\{R \mid \mathbf{a}\}^{-1} = \{\tilde{R} \mid -R\mathbf{a}\tilde{R}\}. \tag{4.27}$$

The most significant result here is that computations of composite isometries can be carried out explicitly with (4.25), without referring to (4.21).

Isometries describing displacements of a rigid body are especially important. A rigid body is a system of particles with fixed distances from one another, so every displacement of a rigid body must be an isometry. But a finite rigid body displacement must unfold continuously, so it must be continuously connected to the identity. From our discussion of the orthogonal group, it should be evident that only isometries composed of a rotation and a translation have this property. An isometry of this kind is called a *rigid displacement*. Of course a body need not be rigid to undergo a rigid displacement (see Figure 4.2).

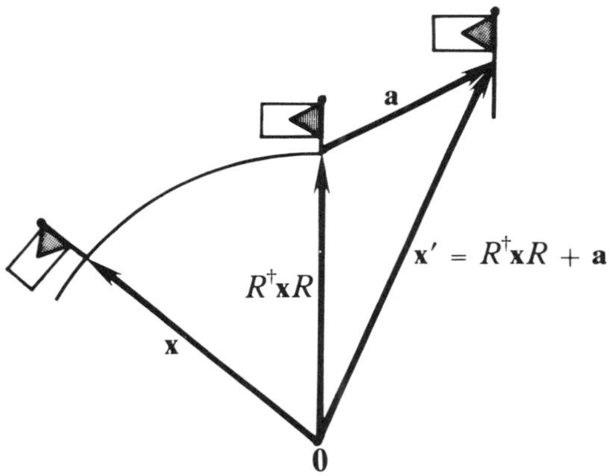

Fig. 4.2. A rigid displacement is the composite of a rotation and a translation. (The translation vector **a** need not be in the plane of rotation).

The set of all rigid displacements is a continuous group called the *Euclidean Group*. This group underlies the geometrical concept of congruence. Two figures are said to be *congruent* if one can be superimposed on the other by a rigid displacement. Thus the Euclidean Group describes all possible relations of congruency. These relations underlie all physical measurements. A ruler is a rigid body, and any measurement of length involves rigid displacements to compare a ruler with the object being measured.

Insight into the structure of the Euclidean group can be developed by examining specific properties of the rigid displacements. We have proved that any rigid displacement can be put in the canonical form

$$\{R|\mathbf{a}\}\mathbf{x} = \mathcal{T}_{\mathbf{a}}\mathcal{R}\mathbf{x} = R^\dagger \mathbf{x} R + \mathbf{a}\,. \tag{4.28}$$

Note that the rotation here is about an axis through the origin, so the origin is a distinguished point in this representation of a rigid displacement. But the choice of origin was completely arbitrary in our derivation of (4.28), so different choices of origin give different decompositions of a rigid displacement into a rotation and a translation. Let us see how they are related.

Let $\mathcal{R}_{\mathbf{b}}$ denote a rotation about a point **b**. This rotation can be expressed in the notation of (4.28) by using $\mathcal{T}_{-\mathbf{b}} = \mathcal{T}_{\mathbf{b}}^{-1}$ to shift the point **b** to the origin, performing the rotation $\mathcal{R}$ about the origin and finally using $\mathcal{T}_{\mathbf{b}}$ to shift the origin back to the point **b**. With the help of (4.21) we obtain

$$\mathcal{R}_{\mathbf{b}} = \mathcal{T}_{\mathbf{b}}\mathcal{R}\mathcal{T}_{\mathbf{b}}^{-1} = \{R|\mathbf{b} - R^\dagger \mathbf{b} R\}\,. \tag{4.29}$$

The *rotation axis* of $\mathcal{R}_{\mathbf{b}}$ is the set of all points invariant under $\mathcal{R}_{\mathbf{b}}$, that is, the points $\mathbf{x}$ satisfying the equation

$$\mathcal{R}_{\mathbf{b}}\mathbf{x} = \mathbf{x}, \tag{4.30a}$$

or equivalently,

$$R^{\dagger}(\mathbf{x}-\mathbf{b})R + \mathbf{b} = \mathbf{x}. \tag{4.30b}$$

If $\mathcal{R}_{\mathbf{b}}$ is not the identity transformation, Equation (4.30a or b) determines a straight line passing through the point $\mathbf{b}$. The rotations $\mathcal{R}_{\mathbf{b}}$ and $\mathcal{R} = \mathcal{R}_{\mathbf{0}}$ rotate points through equal angles about parallel axes passing through the points $\mathbf{b}$ and $\mathbf{0}$ respectively. The vector $\mathbf{b}$ can be decomposed into a component $\mathbf{b}_{\parallel}$ parallel to the rotation axis and a component $\mathbf{b}_{\perp}$ perpendicular to it, as described by the equation

$$R^{\dagger}\mathbf{b}R = R^{\dagger}(\mathbf{b}_{\parallel} + \mathbf{b}_{\perp})R = \mathbf{b}_{\parallel} + \mathbf{b}_{\perp}R^2. \tag{4.31}$$

Substitution of (4.31) into (4.29) gives

$$\mathcal{R}_{\mathbf{b}} = \{R|\mathbf{b}_{\perp}(1-R^2)\}. \tag{4.32}$$

If $\mathbf{b}_{\perp} = 0$, then $\mathbf{b} = \mathbf{b}_{\parallel}$ and (4.32) reduces to

$$\mathcal{R}_{\mathbf{b}} = \{R|\mathbf{0}\} = \mathcal{R}.$$

Thus, rotations differing only by a shift of origin along the axis of rotation are equivalent.

The vector $\mathbf{b}_{\perp}(1 - R^2)$ is perpendicular to the axis of rotation determined by R. We can conclude from (4.30b), therefore, that a *rigid displacement* $\{R|\mathbf{a}\}$ *is a rotation if and only if* $\mathbf{a}R = R^{\dagger}\mathbf{a}$, that is, if and only if the translation vector $\mathbf{a} = \mathbf{a}_{\perp}$ is perpendicular to the axis of rotation. Moreover, a fixed point $\mathbf{b}_{\perp}$ of a rotation $\{R \mid \mathbf{a}\}$ is determined by the equation

$$\mathbf{a}_{\perp} = \mathbf{b}_{\perp}(1 - R^2).$$

Whence,

$$\mathbf{b}_{\perp} = \mathbf{a}_{\perp}(1-R^2)^{-1}. \tag{4.33}$$

For rotation by an angle ϕ about an axis with direction $\mathbf{n}$, the spinor has the form

$$R = e^{(1/2)i\mathbf{n}\phi},$$

When this is substituted into (4.33) a little calculation gives

$$\mathbf{b}_{\perp} = \tfrac{1}{2}\mathbf{a}_{\perp}\left(1 + i\mathbf{n}\cot\frac{\phi}{2}\right) = \tfrac{1}{2}\left(\mathbf{a}_{\perp} + \mathbf{n}\times\mathbf{a}\cot\frac{\phi}{2}\right). \tag{4.34}$$

Since $\mathbf{a}_{\perp}$ is perpendicular to the axis of rotation, the transformation $\{R \mid \mathbf{a}_{\perp}\}$ leaves every plane perpendicular to the rotation axis invariant, and it consists of a rotation-translation in each plane. Therefore, we have proved that every rotation-translation $\{R \mid \mathbf{a}_{\perp}\}$ in a plane is equivalent to rotation centered at the point $\mathbf{b}_\perp$ specified by (4.34), as shown in Figure 4.3. Our proof fails in the case $R^2 = 1$, for then (4.33) is not defined. In this case we have a pure translation $\{1 \mid \mathbf{a}\}$. Hence, we have proved that *every rigid displacement in a plane is either a rotation or a translation*. This, of course, is just a byproduct of our general results on rotations in three dimensions.

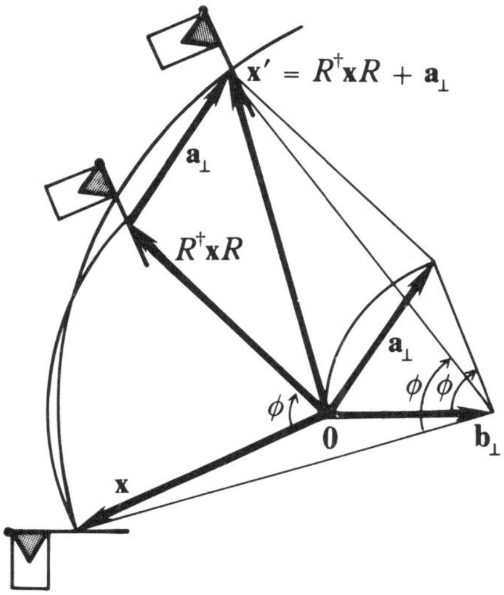

Fig. 4.3. Equivalence of a rotation-translation in a plane to a pure rotation.

Having determined how rotations about different points are related, we are equipped to choose a center of rotation yielding the simplest possible decomposition of a rigid displacement into a rotation and a translation. Given a general rigid displacement $\{R \mid \mathbf{a}\}$, we decompose $\mathbf{a}$ into components $\mathbf{a}_\parallel$ and $\mathbf{a}_\perp$ parallel and perpendicular to the "axis" of the spinor R, so that

$$\begin{aligned}\{R|\mathbf{a}\} &= \{R|\mathbf{a}_\perp + \mathbf{a}_\parallel\}\\ &= \{1|\mathbf{a}_\parallel\}\{R|\mathbf{a}_\perp\}. \qquad (4.35)\end{aligned}$$

By comparison with (4.32), the last factor in (4.35) can be identified as a rotation

$$\mathcal{R}_\mathbf{b} = \{R|\mathbf{a}_\perp\}, \tag{4.36}$$

where the center of rotation $\mathbf{b}_\perp$ is given by (4.33) or (4.34). Therefore, Equation (4.35) can be written

$$\{R|\mathbf{a}\} = \mathcal{T}_{\mathbf{a}_\parallel}\mathcal{R}_\mathbf{b}, \tag{4.37}$$

where $\mathcal{T}_{\mathbf{a}_\parallel} = \{1 \mid \mathbf{a}_\parallel\}$ is a translation parallel to the rotation axis of $\mathcal{R}_\mathbf{b}$. This result proves the *theorem of Chasles* (1830): *Any rigid displacement can be expressed as a screw displacement*. A *screw displacement* consists of the product of a rotation with a translation parallel to or, if you will, along the axis of rotation (the *screw axis*). We have done more than prove Chasles' theorem; we have shown how to find the screw axis for a given rigid displacement. Our result shows that the screw axis is a unique line in $\mathcal{E}_3$, except when the rotation is the identity transformation so the screw displacement reduces to a pure translation.

In spite of the uniqueness and simplicity of the representation of a rigid displacement as a screw displacement, no one has shown that it has any great practical advantages, so it is seldom used. The representation $\{R \mid \mathbf{a}\}$ is usually more useful, because the center of rotation (the origin) can be specified at will to simplify the problem at hand.

5.4. Exercises

(4.1) Prove that the translations satisfy each of the four group properties.

(4.2) Derive Equation (4.25) and prove that the isometries of Euclidean space form a group.

(4.3) Prove that any rigid displacement with a fixed point is a rotation.

(4.4) Prove that rotations with parallel axes do not generally commute unless the axes coincide.

(4.5) Derive Equation (4.34) from Equation (4.33).

(4.6) A rigid displacement $\{R \mid \mathbf{a}\}$ can be expressed as the product of a translation $\mathcal{T}_\mathbf{c}$ and a rotation $\mathcal{R}_\mathbf{b}$ centered at a specified point $\mathbf{b}$, i.e.

$$\{R|\mathbf{a}\} = \mathcal{T}_\mathbf{c}\mathcal{R}_\mathbf{b}.$$

Determine the translation vector $\mathbf{c}$.

(4.7) A subgroup $\{\mathcal{T}\}$ of a group $\{\mathcal{G}\}$ is said to be an *invariant subgroup* if $\mathcal{G}^{-1}\mathcal{T}\mathcal{G}$ is in $\{\mathcal{T}\}$ for each $\mathcal{T}$ in $\{\mathcal{T}\}$ and every $\mathcal{G}$ in $\{\mathcal{G}\}$.
Prove that the translations comprise an invariant subgroup of the Euclidean isometry group.

(4.8) Let $\mathcal{S}$ denote the reflection along a (non-zero) vector $\mathbf{a}$. if $\mathcal{T}_\mathbf{a}$ is the translation by $\mathbf{a}$, then $\mathcal{S}_\mathbf{a} = \mathcal{T}_\mathbf{a}\mathcal{S}\mathcal{T}_\mathbf{a}^{-1}$ is the reflection $\mathcal{S}$ shifted to the point $\mathbf{a}$. Show that

$$\mathcal{S}\mathcal{S}_{(-\mathbf{a})} = \mathcal{T}_{2\mathbf{a}}$$

Thus, a translation by $\mathbf{a}$ can be expressed as a product of reflections in parallel planes separated by a directance $\frac{1}{2}\mathbf{a}$.

5-5. Rigid Motions and Frames of References

Having determined a general mathematical form for rigid displacements in Section. 5-4, we are prepared to develop a mathematical description of rigid motions.

Let $\mathbf{x} = \mathbf{x}(t)$ designate the position of a particle in a rigid body at time t. According to (4.21), a rigid displacement $\mathcal{D}_t$ of the body from an initial position at time $t = 0$ to a position at time t is described by the equation

$$\begin{aligned}\mathbf{x}(t) &= \mathcal{D}_t\mathbf{x}(0) = \mathcal{R}_t\mathbf{x}(0) + \mathbf{a}(t)\\ &= R^\dagger(t)\mathbf{x}(0)R(t) + \mathbf{a}(t).\end{aligned} \tag{5.1}$$

This gives the displacement of each particle in the body as $\mathbf{x}(0)$ ranges over the initial positions of particles in the body. The displacement operator $\mathcal{D}_t$ is the same for all particles. Regarded as a function of time, $\mathcal{D}_t$ is a 1-parameter family of displacement operators, one operator for each time. A *rigid motion* is 1-parameter family of rigid displacements, described by a time-dependent displacement operator $\mathcal{D}_t$. Since the path $\mathbf{x} = \mathbf{x}(t)$ of a material particle is a differentiable function of time the displacement operator $\mathcal{D} = \mathcal{D}_t$ must also be a differentiable function of time. Our next task is to compute the derivative of a displacement operator.

Kinematics of Rigid Motions

It will usually be convenient to suppress the time variable and write (5.1) in the form

$$\mathbf{x} = \mathcal{D}\mathbf{x}_0 = \mathcal{R}\mathbf{x}_0 + \mathbf{a} = R^\dagger \mathbf{x}_0 R + \mathbf{a}. \tag{5.2}$$

If $\mathbf{x}$ and $\mathbf{y}$ designate the positions of two particles in the body, then

$$\mathbf{x} - \mathbf{y} = \mathcal{R}\mathbf{x}_0 - \mathcal{R}\mathbf{y}_0 = \mathcal{R}(\mathbf{x}_0 - \mathbf{y}_0) = R^\dagger (\mathbf{x}_0 - \mathbf{y}_0) R. \tag{5.3}$$

So time-dependence of the relative position $\mathbf{r} = \mathbf{x} - \mathbf{y}$ of two particles in the body is

$$\mathbf{r} = \mathcal{R}\mathbf{r}_0 = R^\dagger \mathbf{r}_0 R. \tag{5.4}$$

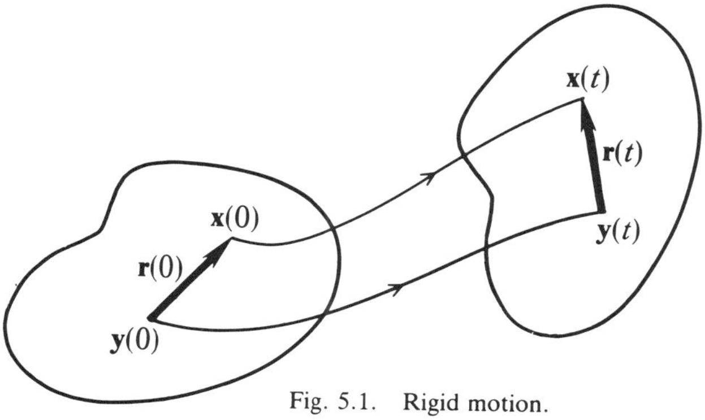

Fig. 5.1. Rigid motion.

Since the rotation operator $\mathcal{R}$ is independent of the choice of particles, it follows that *the motion of a rigid body relative to any of its particles is a rotation*. (Euler's Theorem). This is illustrated in Figure 5.1.

Equation (5.4) expresses the rotation operator $\mathcal{R}$ in terms of a spinor R. This enables us to compute the derivative of a rotation from the derivative of a spinor. To carry this out we need to prove that *the derivative $\dot{R}$ of a unitary spinor R has the form*

$$\dot{R} = \tfrac{1}{2} R\boldsymbol{\Omega} = \tfrac{1}{2} R i\boldsymbol{\omega} \tag{5.5}$$

where $\boldsymbol{\Omega} = i\boldsymbol{\omega}$ is the bivector dual of the vector $\boldsymbol{\omega}$. The bivector property of $\boldsymbol{\Omega}$ implies that

$$\boldsymbol{\Omega}^\dagger = -\boldsymbol{\Omega} = -i\boldsymbol{\omega}. \tag{5.6}$$

By appealing to the definition of reversion and the derivative, one can easily prove that

$$\dot{R}^\dagger = \frac{\mathrm{d}}{\mathrm{d}t}(R^\dagger) = \left(\frac{\mathrm{d}R}{\mathrm{d}t}\right)^\dagger, \tag{5.7}$$

which says the operations of reversion and differentiation commute. Consequently, we can get the derivative of $R^\dagger$ from (5.5) by reversion, with the result

$$\dot{R}^\dagger = -\tfrac{1}{2}\boldsymbol{\Omega} R^\dagger = -\tfrac{1}{2} i\boldsymbol{\omega} R^\dagger . \tag{5.8}$$

The unitarity of R is expressed by the equation

$$R^\dagger R = 1 . \tag{5.9}$$

Using this property, we can solve (5.5) for

$$\boldsymbol{\Omega} = 2R^\dagger \dot{R} . \tag{5.10}$$

This can be taken as the definition of $\boldsymbol{\Omega}$, so all we need to prove is that $\boldsymbol{\Omega}$ is necessarily a bivector. Differentiation of (5.9) gives

$$\dot{R}^\dagger R + R^\dagger \dot{R} = 0 .$$

Because of (5.7), then,

$$\boldsymbol{\Omega} = 2R^\dagger \dot{R} = -2\dot{R}^\dagger R = -(2R^\dagger \dot{R})^\dagger = -\boldsymbol{\Omega}^\dagger ,$$

which proves (5.6). The property $\boldsymbol{\Omega}^\dagger = -\boldsymbol{\Omega}$ implies that $\boldsymbol{\Omega}$ cannot have scalar or vector parts, leaving the possibility that $\boldsymbol{\Omega}$ has nonvanishing bivector and pseudoscalar parts. On the other hand, the requirement that the spinor R must be an even multivector implies that $\dot{R}$ and $R^\dagger$ are even, so, by (5.7), $\boldsymbol{\Omega}$ must be even, and it cannot have a nonvanishing pseudoscalar part. This completes our proof that $\boldsymbol{\Omega}$ is a bivector.

Now we can evaluate the derivative of $\mathcal{R}$ Differentiating (5.4) and using (5.5) and (5.8), we have

$$\begin{aligned}\dot{\mathbf{r}} = \dot{\mathcal{R}}\mathbf{r}_0 &= R^\dagger \mathbf{r}_0 \dot{R} + \dot{R}^\dagger \mathbf{r}_0 R = \tfrac{1}{2} R^\dagger \mathbf{r}_0 R \boldsymbol{\Omega} - \tfrac{1}{2}\boldsymbol{\Omega} R^\dagger \mathbf{r}_0 R \\ &= \tfrac{1}{2}(\mathbf{r}\boldsymbol{\Omega} - \boldsymbol{\Omega}\mathbf{r}) = \tfrac{1}{2} i(\mathbf{r}\boldsymbol{\omega} - \boldsymbol{\omega}\mathbf{r}) .\end{aligned}$$

Hence,

$$\dot{\mathbf{r}} = \mathbf{r}\cdot\boldsymbol{\Omega} = \boldsymbol{\omega} \times \mathbf{r} , \tag{5.11}$$

or, in terms of the rotation operator $\mathcal{R}$ and its derivative $\dot{\mathcal{R}}$,

$$\dot{\mathcal{R}}\mathbf{r}_0 = (\mathcal{R}\mathbf{r}_0)\cdot\boldsymbol{\Omega} = \boldsymbol{\omega} \times (\mathcal{R}\mathbf{r}_0) . \tag{5.12}$$

Equation (5.12) shows that $\dot{\mathcal{R}}$ is a linear operator, for it is the composite of $\mathcal{R}$ and the skewsymmetric linear function $\mathbf{r}\cdot\boldsymbol{\Omega} = \boldsymbol{\omega} \times \mathbf{r}$.

We will refer to the time dependent vector $\boldsymbol{\omega} = \boldsymbol{\omega}(t)$, or the equivalent bivector $\boldsymbol{\Omega} = \boldsymbol{\Omega}(t)$, as the *rotational velocity* of the time-dependent spinor $R = R(t)$ and the family of rotations it determines. The alternative term "angular velocity" is common, but it is misleading, because it suggests that ω is the derivative of the rotation angle, and this is true only if the direction of the rotation axis is time-independent (Exercise (5.3)). So let us use the term angular velocity only when the rotation axis is fixed. When the rotational velocity is a known function of time, the spinor equation (5.5) is a well-defined differential equation, which can be integrated to find the time-dependence of the spinor describing the rotational motion. We have, in fact, encountered this equation already in Section 3-6, and we know that when $\dot{\boldsymbol{\omega}} = 0$ it has the elementary solution

$$R = e^{(1/2)i\boldsymbol{\omega}t}$$

Equation (5.5) is a kinematical equation describing the rotational motion of a rigid body. In Chapter 7 we will study the dynamical equation describing the influence of forces on a rigid body and use (5.5) to determine the resulting rotational motion.

With the Equation (5.12) for the derivative of a rotation at our disposal, we can ascertain the functional form of the derivative of a general rigid motion (5.2). Differentiating (5.2), we have

$$\dot{\mathbf{x}} = \dot{\mathcal{D}}\mathbf{x}_0 = \dot{\mathcal{R}}\mathbf{x}_0 + \dot{\mathbf{a}} = \boldsymbol{\omega} \times (\mathcal{R}\mathbf{x}_0) + \dot{\mathbf{a}}\,.$$

Therefore, in terms of $\boldsymbol{\omega}$ or $\boldsymbol{\Omega}$, we have

$$\dot{\mathbf{x}} = \boldsymbol{\omega} \times (\mathbf{x}-\mathbf{a}) + \dot{\mathbf{a}} = (\mathbf{x}-\mathbf{a})\cdot\boldsymbol{\Omega} + \dot{\mathbf{a}}\,. \tag{5.13}$$

This is an equation of the form

$$\dot{\mathbf{x}} = \mathbf{v}(\mathbf{x}, t)\,,$$

where $\mathbf{v}(\mathbf{x}, t)$ is a time-dependent vector field giving the velocity at time t of a particle in the rigid body located at any point $\mathbf{x}$. The vector $\mathbf{a} = \mathbf{a}(t)$ designates the center of rotation for the rigid motion, so it is natural to refer to $\dot{\mathbf{a}}$ as the *translational velocity*. If we are given the translational velocity $\dot{\mathbf{a}}$ and the rotational velocity $\boldsymbol{\omega}$ as functions of time, then the rigid motion can be determined by direct integration.

Reference Frames

We have been using the concept of position without defining it fully. For applications it is essential that we make its meaning more explicit.

The position of a particle is a relation of the particle to some rigid body called a *reference body* or *reference frame*. The position of a particle with respect to a given reference frame at a specified time is represented by a position vector $\mathbf{x}$ in a reference system attached to (associated with) the frame. The set of all possible position vectors $\{\mathbf{x}\}$ is called the *position space* of the reference system or reference frame.

Often the reference body presumed in a physical application is not mentioned explicitly. For example, in Figure 5.1 illustrating the displacement of a rigid body, the paper on which the figure is drawn is the tacitly assumed reference body. The rigid body illustrated is displaced relative to the paper. A rigid body cannot be displaced in relation to itself. A displacement is a change in the relation to a reference body — a change of position.

The position space of a reference frame is a 3-dimensional Euclidean space. The points (vectors) of a position space are "rigidly related" to one another like the particles of a rigid body. Furthermore, at any given time the points in one reference system can be put in coincidence with the points in any other reference system by a rigid displacement. This is a consequence of our theorem that the most general transformation leaving the distance between points invariant is a rigid displacement. Thus, the points (position vectors) $\{\mathbf{x}\}$ in one reference system called the *unprimed system* are related to the points $\{\mathbf{x}'\}$ in another reference system called the *primed system* by a transformation of the form

$$\mathbf{x}' = \mathcal{R}\mathbf{x} + \mathbf{a}. \tag{5.14}$$

If $\mathbf{x}$ designates the position of a particle in the unprimed system, then (5.14) determines the position $\mathbf{x}'$ of the particle in the primed system. Thus, the vectors $\mathbf{x}$ and $\mathbf{x}'$ in (5.14) designate the same physical place in relation to two different reference bodies. The vector $\mathbf{a}$ locates the origin of the primed system in the unprimed system.

The reference bodies may be moving relative to one another so in general the transformation (5.14) is time-dependent. The *rotational velocity*

$$\boldsymbol{\omega}' = \mathcal{R}\boldsymbol{\omega} = R^\dagger\boldsymbol{\omega}R \tag{5.15}$$

of the unprimed frame (or reference body) relative to the primed frame (or reference body) is defined by

$$\dot{R} = \tfrac{1}{2}Ri\boldsymbol{\omega}' = \tfrac{1}{2}i\boldsymbol{\omega}R. \tag{5.16}$$

From this the derivative of the rotation $\mathcal{R}\mathbf{x} = R^\dagger\mathbf{x}R$ is found to be

$$\dot{\mathcal{R}}\mathbf{x} = \mathcal{R}(\boldsymbol{\omega}\times\mathbf{x}) = (\mathcal{R}\boldsymbol{\omega})\times(\mathcal{R}\mathbf{x}) = \boldsymbol{\omega}'\times(\mathbf{x}'-\mathbf{a}). \tag{5.17}$$

Note that equation (5.15) is actually superfluous, because it is a consequence of (5.16). It has been written down to emphasize that there are two equivalent ways to represent the angular velocity, either as a vector $\boldsymbol{\omega}$ in the unprimed system or as a vector $\boldsymbol{\omega}'$ in the primed system. Only one of these vectors is needed, and the choice is a matter of convenience. We will use $\boldsymbol{\omega}$ rather than $\boldsymbol{\omega}'$ in order to express the rotational kinematics in the unprimed system.

Now suppose that $\mathbf{x} = \mathbf{x}(t)$ is the trajectory of a particle in the unprimed system. Substitution of $\mathbf{x} = \mathbf{x}(t)$ into (5.14) gives the corresponding trajectory $\mathbf{x}' = \mathbf{x}'(t)$ in the primed system. The relation between velocities in the two frames is then determined by differentiating (5.14); thus,

$$\dot{\mathbf{x}}' = \dot{\mathcal{R}}\mathbf{x} + \mathcal{R}\dot{\mathbf{x}} + \dot{\mathbf{a}},$$

By (5.17), therefore,

$$\dot{\mathbf{x}}' = \mathcal{R}(\dot{\mathbf{x}} + \boldsymbol{\omega} \times \mathbf{x}) + \dot{\mathbf{a}} \tag{5.18}$$

If the bivector $\boldsymbol{\Omega} = i\boldsymbol{\omega}$ preferred, this takes the form

$$\dot{\mathbf{x}}' = \mathcal{R}(\dot{\mathbf{x}} + \mathbf{x}\cdot\boldsymbol{\Omega}) + \dot{\mathbf{a}}. \tag{5.19}$$

If, as "initial conditions" on the rotation operator $\mathcal{R} = \mathcal{R}_t$ and the translation vector $\mathbf{a} = \mathbf{a}(t)$, it is required that

$$\mathbf{x}' = \mathbf{x} \quad \text{at time } t, \tag{5.20}$$

then (5.14) implies that $\mathcal{R} = 1$ at time t, and (5.18) becomes

$$\dot{\mathbf{x}}' = \dot{\mathbf{x}} + \boldsymbol{\omega} \times \mathbf{x} + \dot{\mathbf{a}}. \tag{5.21}$$

This equation is commonly found in physics books without mention of the fact that it can hold only at a single time. It is not a differential equation which can be integrated, nor can it be differentiated to find the relation between accelerations.

To relate accelerations in the two frames we differentiate (5.18); thus

$$\ddot{\mathbf{x}}' = \dot{\mathcal{R}}(\dot{\mathbf{x}} + \boldsymbol{\omega} \times \mathbf{x}) + \mathcal{R}(\ddot{\mathbf{x}} + \dot{\boldsymbol{\omega}} \times \mathbf{x} + \boldsymbol{\omega} \times \dot{\mathbf{x}}) + \ddot{\mathbf{a}}.$$

So, using (5.17), we get the desired result

$$\ddot{\mathbf{x}}' = \mathcal{R}(\ddot{\mathbf{x}} + 2\,\boldsymbol{\omega} \times \dot{\mathbf{x}} + \boldsymbol{\omega} \times (\boldsymbol{\omega} \times \mathbf{x}) + \dot{\boldsymbol{\omega}} \times \mathbf{x}) + \ddot{\mathbf{a}}. \tag{5.22}$$

Equivalently, in terms of $\boldsymbol{\Omega} = i\boldsymbol{\omega}$, we have

$$\ddot{\mathbf{x}}' = \mathcal{R}(\ddot{\mathbf{x}} + 2\dot{\mathbf{x}}\cdot\boldsymbol{\Omega} + (\mathbf{x}\cdot\boldsymbol{\Omega})\cdot\boldsymbol{\Omega} + \mathbf{x}\cdot\dot{\boldsymbol{\Omega}}) + \ddot{\mathbf{a}}. \tag{5.23}$$

This is the general relation between the accelerations of a particle with respect to two reference frames with arbitrary relative motion.

Inertial Systems

Among the possible reference systems, inertial systems are especially significant. An *inertial* system is distinguished by the property that within the system the equation of motion of a *free particle* is

$$\ddot{\mathbf{x}} = 0.$$

The frame to which an inertial system is "attached' is called an *inertial frame*. Thus, with respect to an inertial frame every free particle moves in a straight line with constant speed. The transformation from one inertial system to another is determined by (5.22). We simply require that $\ddot{\mathbf{x}}' = 0$ when $\ddot{\mathbf{x}} = 0$ for any position $\mathbf{x}$ or velocity $\dot{\mathbf{x}}$ of a free particle. This condition can be met in (5.22) only if $\boldsymbol{\omega} = 0$ and $\ddot{\mathbf{a}} = 0$, that is, only if the two frames are moving with a constant relative velocity and without a time-dependent relative rotation.

If the primed system is inertial, then Newton's law of motion for a particle of mass m has the usual form

$$m\ddot{\mathbf{x}}' = \mathbf{f}', \tag{5.24}$$

where $\mathbf{f}'$ is the applied force. Although we have not mentioned it before, Newton's law has the form (5.24) *only in an inertial system*. By substituting (5.22) in (5.24) we get the modified form of Newton's law in an arbitrary system namely

$$m(\ddot{\mathbf{x}} + 2\boldsymbol{\omega} \times \dot{\mathbf{x}} + \boldsymbol{\omega} \times (\boldsymbol{\omega} \times \mathbf{x}) + \dot{\boldsymbol{\omega}} \times \mathbf{x} + \mathcal{R}^{-1}\ddot{\mathbf{a}}) = \mathbf{f}. \tag{5.25}$$

where

$$\mathbf{f} = \mathcal{R}^{-1}\mathbf{f}'. \tag{5.26}$$

The additional terms in (5.25) arise from the motion of the reference body for the unprimed system. The term $\boldsymbol{\omega} \times (\boldsymbol{\omega} \times \mathbf{x})$ is called the *centripetal* (center-seeking) acceleration; as Figure 5.2 shows, it is always directed toward the axis of rotation. The term $2\boldsymbol{\omega} \times \dot{\mathbf{x}}$ is called the *Coriolis acceleration*. The other terms do not have generally accepted names.

Of course, the terms in (5.25) can be rearranged to get an equation of motion formally in the Newtonian form

$$m\ddot{\mathbf{x}} = \mathbf{f}_{\text{eff}}, \tag{5.27a}$$

where the effective force is given by

$$\mathbf{f}_{\text{eff}} = \mathbf{f} - m(\boldsymbol{\omega} \times (\boldsymbol{\omega} \times \mathbf{x}) + 2\boldsymbol{\omega} \times \dot{\mathbf{x}} + \dot{\boldsymbol{\omega}} \times \mathbf{x} + \mathcal{R}^{-1}\ddot{\mathbf{a}}). \tag{5.27b}$$

The "fictitious force" $-m\boldsymbol{\omega} \times (\boldsymbol{\omega} \times \mathbf{x})$ is called the *centrifugal* (center-fleeing) *force*, and $-2m\boldsymbol{\omega} \times \dot{\mathbf{x}}$ is called the *Coriolis force*.

Equation (5.27a) for motion in a noninertial system is as well-defined and solvable as Newton's equation for motion in an inertial system. However, the "fictitious forces" in (5.27b) can in principle be distinguished from the "real force" $\mathbf{f}$ by virtue of their particular functional dependence on m, $\mathbf{x}$ and $\dot{\mathbf{x}}$. In practice, $\boldsymbol{\omega}$ can be measured directly by observing rotation of the frame relative to the "fixed stars". Ideally, fictitious forces can be measured by observing accelerations of free particles with respect to a noninertial frame, though this is seldom practical. Considerations of practicality aside, the point is that a real force (field) is distinguished from a fictitious force by the fact that it depends on

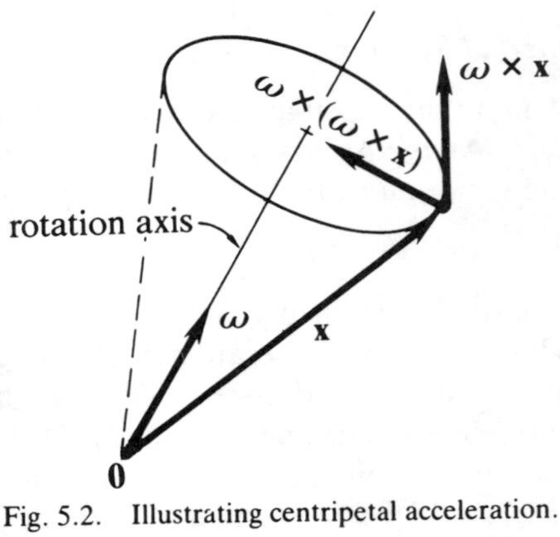

Fig. 5.2. Illustrating centripetal acceleration.

the presence of material bodies to produce it, and it cannot be "transformed away" in a finite region of space by a change of frame. On the basis of this distinction, then, it can be said that Equation (5.27a) does not have the form of Newton's law unless $\boldsymbol{\omega} = 0$ and $\ddot{\mathbf{a}} = 0$ in (5.27b).

Einstein observed that a uniform gravitational force field $\mathbf{f} = m\mathbf{g}$ can be cancelled by transforming to a frame with constant acceleration so that (5.27b) becomes

$$\mathbf{f}_{\text{eff}} = m\mathbf{g} - m\ddot{\mathbf{a}} = 0 . \tag{5.28}$$

This observation played an important heuristic role in the development of Einstein's theory of gravitation. It does indeed show that a uniform gravitational force cannot be distinguished from a fictitious force due to constant acceleration. However, the nature of gravitation is such that there is actually no such thing as completely uniform gravitational field, and the deviations from uniformity are sufficient to uphold the distinction between real and fictitious forces.

Returning to Equation (5.25), we note that if the unprimed frame is inertial, then $\boldsymbol{\omega} = 0$ and $\ddot{\mathbf{a}} = 0$, so it reduces to

$$m\ddot{\mathbf{x}} = \mathbf{f} . \tag{5.29}$$

This proves that a *transformation between inertial systems is the most general transformation preserving the form of Newton's law*. The general form of such a transformation can be deduced from (5.18), which, for vanishing rotational velocity and constant translational velocity $\dot{\mathbf{a}} = \mathbf{v}$, reduces to

$$\dot{\mathbf{x}}' = \mathcal{R}\dot{\mathbf{x}} + \mathbf{v} \tag{5.30}$$

where $\mathcal{R}$ is a time-independent rotation. Integrating (5.30), we obtain

$$\mathbf{x}' = \mathcal{R}\mathbf{x} + \mathbf{a}_0 + \mathbf{v}t . \tag{5.31}$$

Since this applies to particles at rest as well as in motion, it gives us the general relation between points in two inertial systems.

The group of transformations that leave the form of Newton's law invariant is called the *Galilean Group*. This is the group of transformations relating inertial systems. Every element of the group can be put in the form (5.31) which shows that it can be expressed as a composite of a *rotation*

$$\mathbf{x} \to \mathcal{R}\mathbf{x} = R^{\dagger}\mathbf{x}R , \tag{5.32a}$$

a *space translation*

$$\mathbf{x} \to \mathbf{x} + \mathbf{a}_0 \tag{5.32b}$$

and a so-called *Galilean transformation*,

$$\mathbf{x} \to \mathbf{x} + \mathbf{v}t . \tag{5.32c}$$

Note also that the form of (5.30) is unchanged by a *time translation*

$$t \to t + \alpha, \tag{5.32d}$$

so this transformation belongs to the group. Clearly the Galilean Group consists of Euclidean Group of rigid displacements extended to include Galilean transformations and time translations. The Galilean Group is a 10-parameter continuous group with 3 parameters to determine the rotations, 3 parameters for space translations, 3 parameters for Galilean transformations, and one parameter for time translations.

We have interpreted (5.31) as describing the location of a single particle relative to two different reference bodies. It can be interpreted alternatively as a displacement of one or more particles relative to a single reference body. Newton's law is *form invariant* in either case. The important thing is that (5.31) describes a change in the relation of particles relative to some inertial frame.

The form invariance of Newton's law under the Euclidean Group means that it provides a description of particle motions and interactions that is independent of the relative position and orientation of the reference body, which implies further that the reference body does not interact with the particles described by Newton's law. Invariance under translations means that all places in position space are equivalent, that is, *position space is homogeneous*. Invariance under rotations means that all directions in physical space are equivalent, that is, *position space is isotropic*. Similarly, invariance under time translations means that *time is homogeneous*. Thus, the *laws of physics are the same at all times and all places*. This crucial property of physical laws enables us to compare and integrate experimental results from laboratories all over the world without worrying about when and where the experiments were done. The astronomer uses it to infer what is happening on stars many light-years away, and the geologist uses it to ascertain how the Earth's surface was formed. Scientific laws are valuable precisely because they describe features common to all experience.

Note that for inertial frames related by a Galilean transformation

$$\mathbf{x}' = \mathbf{x} + \mathbf{v}t, \tag{5.33}$$

each particle of the unprimed frame is moving with constant velocity $\mathbf{v}$ with respect to the primed frame. Therefore *each particle in the reference body of an inertial system is a free particle*. Thus, the physical requirements of no net force on particles of the reference body distinguishes inertial frames from other reference frames. Of course, this is an idealization that can never be perfectly met in practice.

The derivative of the Galilean transformation (5.31) yields the *velocity addition formula*

$$\dot{\mathbf{x}}' = \dot{\mathbf{x}} + \mathbf{v} \tag{5.34}$$

relating the velocities of a particle with respect to two inertial frames moving with relative velocity $\mathbf{v}$. The important thing about this formula is the fact that, according to (5.26), the force $\mathbf{f}' = \mathbf{f}$ on the particles is the same in both reference systems. This enables us to do such things as analyze the motion of an object in the atmosphere or a river without considering motion relative to the Earth until the analysis is complete. Motion relative to the Earth can then be accounted for trivially with (5.33) or (5.34).

5-5. Exercises

(5.1) For a unitary spinor $R = R(t)$ with the parametrizations

$$R = e^{(1/2)i\mathbf{a}} = \alpha + i\boldsymbol{\beta} = \alpha(1 + i\boldsymbol{\gamma})$$

show that the rotational velocity $i\boldsymbol{\omega} = 2R^\dagger \dot{R}$ has the following parametric expressions:

$$\boldsymbol{\omega} = 2(\alpha\dot{\boldsymbol{\beta}} - \dot{\alpha}\boldsymbol{\beta} + \boldsymbol{\beta} \times \dot{\boldsymbol{\beta}}),$$

where $\alpha^2 + \boldsymbol{\beta}^2 = 1$ and $\alpha\dot{\alpha} = -\boldsymbol{\beta}\cdot\dot{\boldsymbol{\beta}}$;

$$\boldsymbol{\omega} = 2\left(\frac{\dot{\boldsymbol{\gamma}} + \boldsymbol{\gamma} \times \dot{\boldsymbol{\gamma}}}{1 + \boldsymbol{\gamma}^2}\right);$$

$$\boldsymbol{\omega} = \mathbf{n}\dot{a} + \dot{\mathbf{n}} \sin a + \mathbf{n} \times \dot{\mathbf{n}}\,(1 - \cos a),$$

where $\mathbf{n}^2 = 1$, $\mathbf{a} = a\mathbf{n}$ and $\mathbf{n} \times \dot{\mathbf{n}} = i\dot{\mathbf{n}}\mathbf{n}$.

Note that $\boldsymbol{\omega} = \dot{\mathbf{a}}$ if and only if $\dot{\mathbf{n}} = 0$.

(5.2) Four time-dependent unitary spinors R_k satisfying $\dot{R}_k = \frac{1}{2}R_k\boldsymbol{\Omega}_k$ are related by the equation

$$R_4 = R_1 R_2 R_3 .$$

Show that their rotational velocities are related by the equation

$$\boldsymbol{\Omega}_4 = R_3^\dagger(R_2^\dagger\boldsymbol{\Omega}_1 R_2 + \boldsymbol{\Omega}_2)R_3 + \boldsymbol{\Omega}_3 .$$

(This will be useful in Exercises (5.3) and (5.7)).

(5.3) For a unitary spinor $R = R(t)$ with the Eulerian parametrization

$$R = e^{(1/2)i\boldsymbol{\sigma}_3\psi}\, e^{(1/2)i\boldsymbol{\sigma}_1\theta}\, e^{(1/2)i\boldsymbol{\sigma}_3\phi}$$

show the rotational velocity $i\boldsymbol{\omega} = 2R^\dagger\dot{R}$ has the parametric form

$$\boldsymbol{\omega} = \dot{\phi}\boldsymbol{\sigma}_3 + \dot{\theta}\,\widehat{\boldsymbol{\sigma}_3 \times \mathbf{n}} + \dot{\psi}\mathbf{n},$$

where

$$\mathbf{n} = R^\dagger\boldsymbol{\sigma}_3 R = \boldsymbol{\sigma}_3 \cos\theta - \boldsymbol{\sigma}_2 e^{i\boldsymbol{\sigma}_3\phi} \sin\theta$$

and

$$\widehat{\boldsymbol{\sigma}_3 \times \mathbf{n}} = \frac{\boldsymbol{\sigma}_3 \times \mathbf{n}}{|\boldsymbol{\sigma}_3 \times \mathbf{n}|} = \boldsymbol{\sigma}_1\, e^{i\boldsymbol{\sigma}_3\phi} = \boldsymbol{\sigma}_1 \cos\phi + \boldsymbol{\sigma}_2 \sin\phi .$$

(5.4) The derivative of a time-dependent linear operator $\mathcal{R} = \mathcal{R}_t$ is defined by

$$\dot{\mathcal{R}} = \frac{\mathrm{d}\mathcal{R}}{\mathrm{d}t} = \lim_{\Delta\to 0} \frac{\mathcal{R}_{t+\Delta t} - \mathcal{R}_t}{\Delta t} .$$

Show from this that $\dot{\mathcal{R}}$ is a linear operator. Show that the usual rule for differentiating a product holds for the composite of time-dependent linear operators $\mathcal{R}$ and $\mathcal{S}$, that is, show that

$$\frac{\mathrm{d}}{\mathrm{d}t}(\mathcal{R}\mathcal{S}) = \dot{\mathcal{R}}\mathcal{S} + \mathcal{R}\dot{\mathcal{S}} .$$

This rule does not hold for the composite of arbitrary functions. Show, however, that it also holds for the composite of rigid displacements.

(5.5) Fill in the steps in the derivation of Equation (5.17).

(5.6) Derive Equations (5.19) and (5.23) directly using

$$\dot{R} = \tfrac{1}{2}\boldsymbol{\Omega} R .$$

(5.7) A wheel of radius b is rolling upright with constant speed v on a circular track of radius a (Figure 5.3). The motion of a point $\mathbf{x}$ on the wheel can be described by the equation

$$\mathbf{x} = R_2^\dagger(R_1^\dagger \mathbf{r}_0 R_1 + \mathbf{a}_0)R_2 = R^\dagger \mathbf{r}_0 R + \mathbf{a} = \mathbf{r} + \mathbf{a} ,$$

which can be interpreted as follows: a fixed point $\mathbf{r}_0$ on the wheel is rotated about the axis with constant angular velocity $\boldsymbol{\omega}_1$ by a spinor R_1; then the wheel is translated by $\mathbf{a}_0$ along its axis from the center to the edge of the circular track where it is rotated with constant angular velocity $\boldsymbol{\omega}_2$ by a spinor R_2. Consequently,

$$R = R_1 R_2 = e^{(1/2)i\boldsymbol{\omega}_1 t}\, e^{(1/2)i\boldsymbol{\omega}_2 t} .$$

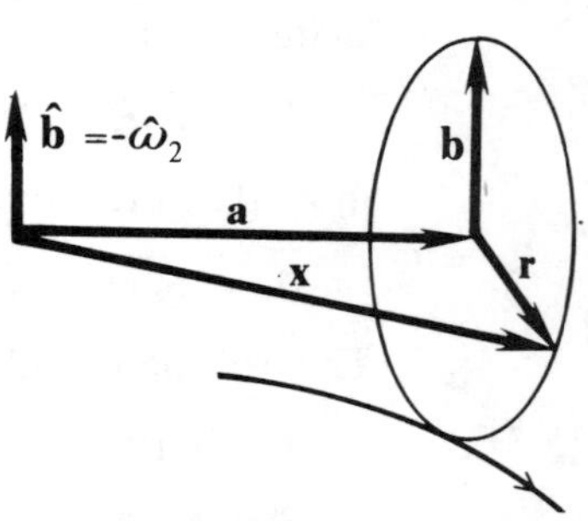

Fig. 5.3. Wheel rolling on a circular track.

(a) Determine $\boldsymbol{\omega}_1$, $\hat{\boldsymbol{\omega}}_2$ and the rotational velocity $\boldsymbol{\omega} = -2iR^\dagger \dot{R}$ of the wheel about its moving center.

(b) Calculate the velocity and acceleration of an arbitrary point $\mathbf{x}$ on the wheel.

(c) Evaluate the velocity and acceleration at the top and bottom of the wheel.

5-6. Motion in Rotating Systems

A choice of reference frame will be dictated by the problem under consideration. The frames which we have most occasion to use are distinguished by their choice of origin. A reference system (and its associated frame) is said to be *heliocentric* if its origin is at the center of the Sun, *geocentric* if its origin is at the center of the Earth, or *topocentric* if its origin is fixed on the surface of the Earth. None of these frames is inertial, since a topocentric frame rotates with the Earth, the Earth revolves about the Sun, and the Sun orbits in a galaxy. Let us evaluate the effect of these relative motions on the observed motion of objects on the Earth.

The relative directions of the distant stars observed on the "celestial sphere" vary so little in "human time intervals" that they can be used as an absolute standard of rotationless motion. With respect to an inertial frame, then, the directions of the distant stars must be fixed in time.

Let $\{\mathbf{x}'\}$ be the position space of an inertial system in which the Earth is initially at rest with its center at the origin. Let $\{\mathbf{x}\}$ be the position space of a geocentric system with the Earth as a body of reference. If we neglect, for the time being, the earth's acceleration due to motion about the Sun, the two frames are related by

$$\mathbf{x}' = \mathcal{R}\mathbf{x} = R^\dagger \mathbf{x} R, \tag{6.1}$$

where the operator $\mathcal{R}$ or the spinor R describes the rotation of the Earth relative to the fixed stars. The consequences of this relation were derived in Section 5-5, so we need only to summarize the relevant relations here. The Earth's rotational velocity $\boldsymbol{\omega}$ (in the rotating Earth system) is defined by the spinor equation

$$\dot{R} = \tfrac{1}{2} i\boldsymbol{\omega} R \tag{6.2}$$

or the corresponding operator equation

$$\dot{\mathcal{R}}\mathbf{x} = \mathcal{R}(\boldsymbol{\omega} \times \mathbf{x}). \tag{6.3}$$

The equation of motion

$$m\ddot{\mathbf{x}}' = \mathbf{f}' \tag{6.4}$$

in the inertial frame corresponds to the equation of motion

$$m\ddot{\mathbf{x}} = \mathbf{f} - m\boldsymbol{\omega} \times (\boldsymbol{\omega} \times \mathbf{x}) - 2m\boldsymbol{\omega} \times \dot{\mathbf{x}} - m\dot{\boldsymbol{\omega}} \times \mathbf{x}, \tag{6.5}$$

in the Earth frame, with

$$\mathbf{f}' = \mathcal{R}\mathbf{f} = R^\dagger \mathbf{f} R. \tag{6.6}$$

Let us examine, now, the effect of the real and fictitious forces in (6.5) on a particle near the Earth's surface. The term $m\dot{\boldsymbol{\omega}} \times \mathbf{x}$ is entirely negligible compared to the other forces, because variation in the Earth's rotation period

$T = 2\pi/\omega$ is of the order of milliseconds per year and variation in the direction of $\boldsymbol{\omega}$ is comparably small. We shall see how to calculate $\boldsymbol{\omega}$ later on when we examine the rotational motion of the Earth itself in more detail.

According to Newton's law of gravitation, in the aproximation of a spherical Earth, the gravitational force on a particle outside the surface of the Earth is

$$\mathbf{f} = -\frac{GmM\mathbf{r}}{r^3} \equiv m\mathbf{G} \tag{6.7}$$

where M is the mass of the Earth. Observe that, by (6.1)

$$\mathcal{R}\mathbf{f} = \mathcal{R}\left(\frac{-GmM\mathbf{r}}{r^3}\right) = -\frac{GmM\mathbf{r}'}{r'^3} = \mathbf{f}',$$

showing that (6.1) is consistent with (6.6).

True and Apparent Weight

The gravitational force $\mathbf{f} = m\mathbf{G}$ exerted by the earth on an object is called the *true weight* of the object. The object's apparent weight $\mathbf{W}$ is

$$\mathbf{W} = m\mathbf{g} = m(\mathbf{G} - \omega \times (\omega \times \mathbf{r}))\,. \tag{6.8}$$

This is the resultant of the gravitational and centrifugal forces (Figure 6.1), which are difficult to separate near the surface of the Earth, because they are slowly varying functions of position. To estimate the contribution of the centrifugal force, we use

$$\mathbf{g} = \mathbf{G} - \omega \times (\omega \times \mathbf{r}) = \mathbf{G} + \omega\cdot(\omega\wedge\mathbf{r}) = \mathbf{G} + \omega^2\mathbf{r} - \omega\mathbf{r}\cdot\omega\,. \tag{6.9}$$

From this we see that $g = |\mathbf{g}|$ has the value

$$g_{\text{pole}} = |\mathbf{G}|$$

at the Earth's pole and the value

$$g_{\text{eq}} = |\mathbf{G}| - \omega^2 r$$

at the Equator. Hence

$$g_{\text{pole}} - g_{\text{eq}} = \omega^2 r \simeq 3.4 \text{ cm s}^{-2}, \tag{6.10}$$

where we have used

$$\omega = 2\pi \text{ radians/day} = 7.29 \times 10^{-5}\,\text{s}^{-1}, \tag{6.11}$$

for the angular speed and

$$r_{\text{E}} = 6370 \text{ km} \tag{6.12}$$

for the mean radius of the Earth. The measured value of $g_{\text{pole}} - g_{\text{eq}}$ is 5.2 cm s^{-2}. The discrepancy between this value and the calculated value is due to the oblateness of the Earth. Indeed, it can be used to estimate the oblateness of the Earth. The value of g at sea level is about 980 cm s^{-2}, so the relative contribution of the centrifugal force varies by only half a percent from pole to Equator.

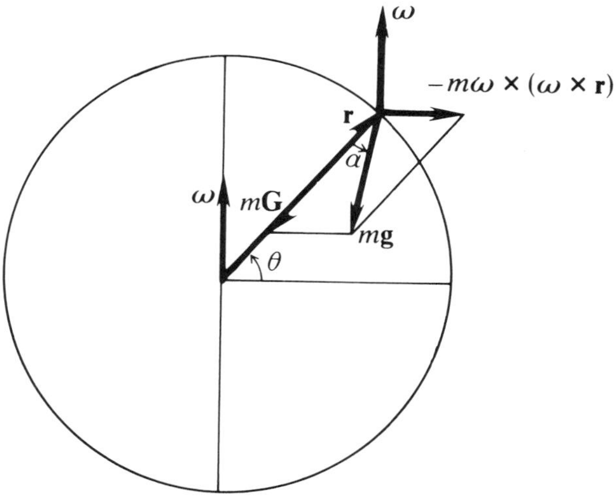

Fig. 6.1. Relation between true and apparent weights (not to scale).

Coriolis Force

The transformation from the geocentric system {**x**} to a topocentric system {**r**}, as shown in Figure 6.2, is a simple translation

$$\mathbf{x} = \mathbf{r} + \mathbf{a}\,, \tag{6.13}$$

where **a** is a fixed point on the Earth's surface. Substituting (6.13) into (6.5), we get, for constant $\boldsymbol{\omega}$, the equation of motion

$$\ddot{\mathbf{r}} = \mathbf{g} - 2\boldsymbol{\omega} \times \dot{\mathbf{r}}\,, \tag{6.14}$$

where the centrifugal force has been incorporated into the "gravitational force" $m\mathbf{g}$ and non-gravitation forces have been omitted for the time being.

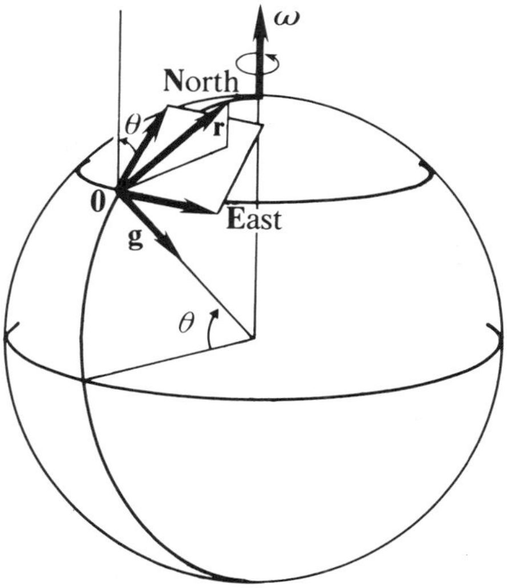

Fig. 6.2. A topocentric frame with latitude θ.

From (6.14) we can calculate the effect of the Coriolis force on projectile motion in the approximation where **g** is constant. Actually, in Section 3-7 we found the exact solution to (6.14) for constant **g** and $\boldsymbol{\omega}$. In the present case, however, for typical velocities we have $|2\dot{\mathbf{r}} \times \boldsymbol{\omega}| \ll g$ because of the relatively small value (6.11) for the angular velocity of the Earth. Consequently, a perturbative solution to (6.14) is more useful than the exact solution. We could, of course, get the appropriate approximation by expanding the exact solution, but it is at least as easy to get it directly from (6.14) in the following way.

Let us write $\mathbf{v} = \dot{\mathbf{r}}$ so (6.14) takes the form

$$\dot{\mathbf{v}} = \mathbf{g} + 2\mathbf{v} \times \boldsymbol{\omega}. \tag{6.15}$$

Regarding the last term in (6.15) as a small perturbing force, the equation can be solved by the method of *successive approximations*. We write the velocity as an expansion of successive orders in $\boldsymbol{\omega}$,

$$\mathbf{v} = \mathbf{v}_1 + \mathbf{v}_2 + \mathbf{v}_3 + \ldots. \tag{6.16}$$

The zeroth order term $\mathbf{v}_1$ is required to satisfy the unperturbed equation $\dot{\mathbf{v}}_1 = \mathbf{g}$, which integrates to

$$\mathbf{v}_1 = \mathbf{g}t + \mathbf{v}_0, \tag{6.17a}$$

where $\mathbf{v}_0$ is the initial velocity. Inserting $\mathbf{v}$ to first order in Equation (6.15) we get

$$\dot{\mathbf{v}} = \dot{\mathbf{v}}_1 + \dot{\mathbf{v}}_2 = \mathbf{g} + 2(\mathbf{v}_1 + \mathbf{v}_2) \times \boldsymbol{\omega}.$$

Neglecting the second order term $2\mathbf{v}_2 \times \boldsymbol{\omega}$, this equation reduces to an equation for $\mathbf{v}_2$ when (6.17a) is used;

$$\dot{\mathbf{v}}_2 = 2\mathbf{v}_1 \times \boldsymbol{\omega} = 2(\mathbf{g}t + \mathbf{v}_0) \times \boldsymbol{\omega}.$$

This integrates to

$$\mathbf{v}_2 = (\mathbf{g}t^2 + 2\mathbf{v}_0 t) \times \boldsymbol{\omega}. \tag{6.17b}$$

In a similar way, we can determine the second order correction $\mathbf{v}_3$ and higher order terms if desired.

Substituting (6.17a, b) into (6.16), we have the velocity to first order in ω

$$\mathbf{v} = \mathbf{v}_0 + \mathbf{g}t + (2\mathbf{v}_0 t + \mathbf{g}t^2) \times \boldsymbol{\omega} + \ldots \tag{6.18}$$

Integrating this, we get a parametric equation for the displacement

$$\mathbf{r} = \tfrac{1}{2}\mathbf{g}t^2 + \mathbf{v}_0 t + \Delta\mathbf{r}, \tag{6.19}$$

where the deviation $\Delta\mathbf{r}$ from a parabolic trajectory is given to first order by

$$\Delta\mathbf{r} = (\mathbf{v}_0 + \tfrac{1}{3}\mathbf{g}t) \times \boldsymbol{\omega}t^2 + \ldots \tag{6.20}$$

To estimate the magnitude of the correction $\Delta\mathbf{r}$, we observe from (6.19) and (6.20) that

$$\frac{|\Delta\mathbf{r}|}{|\mathbf{r}|} \approx \omega t. \tag{6.21}$$

For the correction to be as much as one percent, then, we must have $\omega t \geqslant 0.01$, and from the value (6.11) for ω we find that the time of flight t must be at least two min. As the time of flight in a typical projectile problem is less than 2 min., we need not consider corrections of higher order than the first. Indeed, before higher order corrections are considered, the assumption that $\mathbf{g}$ is constant should be examined.

As it stands, the expression (6.20) for Coriolis deflection $\Delta\mathbf{r}$ is not in the most convenient form, because it is not given as a function of target location $\mathbf{r}$. This can be remedied by using the zeroth order approximation

$$\mathbf{r} \approx \tfrac{1}{2}\mathbf{g}t^2 + \mathbf{v}_0 t \tag{6.22}$$

to eliminate $\mathbf{v}_0$ in (6.20), with the result

$$\Delta\mathbf{r} = -t\boldsymbol{\omega} \times (\mathbf{r} - \tfrac{1}{6}\mathbf{g}t^2)\,. \tag{6.23}$$

This shows the directional dependence of $\Delta\mathbf{r}$ on $\mathbf{r}$. If needed, the dependence of t on $\mathbf{r}$ can be obtained from (6.22); thus

$$t = \frac{\mathbf{r}\wedge\mathbf{g}}{\mathbf{v}_0\wedge\mathbf{g}} \quad \text{and} \quad \tfrac{1}{2}t^2 = \frac{\mathbf{r}\wedge\mathbf{v}_0}{\mathbf{g}\wedge\mathbf{v}_0}\,. \tag{6.24}$$

Notice that

$$\mathbf{r} - \tfrac{1}{6}\mathbf{g}t^2 = r\left(\hat{\mathbf{r}} - \tfrac{1}{3}\left(\frac{\hat{\mathbf{r}}\wedge\hat{\mathbf{v}}_0}{\hat{\mathbf{g}}\wedge\hat{\mathbf{v}}_0}\right)\hat{\mathbf{g}}\right),$$

showing that the two terms in (6.23) are of the same order of magnitude.

From (6.23) we find that the *change* in *range* due to the Coriolis force is

$$\hat{\mathbf{r}}\cdot\Delta\mathbf{r} = \frac{t^3}{6}\hat{\mathbf{r}}\cdot(\boldsymbol{\omega}\times\mathbf{g})\,. \tag{6.25}$$

Similarly, the *vertical deflection* is found to be

$$\hat{\mathbf{g}}\cdot\Delta\mathbf{r} = t\mathbf{r}\cdot(\boldsymbol{\omega}\times\hat{\mathbf{g}})\,. \tag{6.26}$$

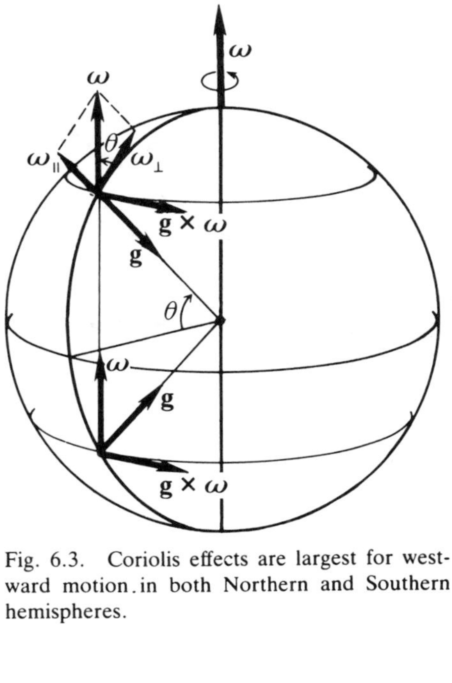

Fig. 6.3. Coriolis effects are largest for westward motion in both Northern and Southern hemispheres.

The vector $\boldsymbol{\omega}\times\mathbf{g}$ is directed West (Figure 6.3), except at the poles, so both (6.25) and (6.26) vanish for trajectories to the North or South. They have maximum values for trajectories to the West. This is due to rotation of the Earth in the opposite direction while the projectile is in flight, as can be seen by examining the trajectory in an inertial frame.

In most circumstances, resistive forces will have a greater effect on the range and vertical deflection than the Coriolis force. The *lateral Coriolis deflection* is more significant, because it will not be masked by resistive forces. So let us examine it. For a target on the horizontal plane $\mathbf{g}\cdot\mathbf{r} = 0$ and $\hat{\mathbf{g}}\times\hat{\mathbf{r}}$ is a unit rightward vector. From

(6.23), therefore, the *rightward deflection* ΔR *is given by*

$$\Delta R \equiv (\hat{\mathbf{g}} \times \hat{\mathbf{r}}) \cdot \Delta \mathbf{r} = -t(\hat{\mathbf{g}} \times \hat{\mathbf{r}}) \cdot \left(\boldsymbol{\omega} \times \left(\mathbf{r} - \frac{t^2}{6} \mathbf{g} \right) \right)$$

$$= t(\mathbf{g} \wedge \mathbf{r}) \cdot \left(\boldsymbol{\omega} \wedge \left(\mathbf{r} - \frac{t^2}{6} \mathbf{g} \right) \right) = t \left(-r\hat{\mathbf{g}} \cdot \boldsymbol{\omega} - g \frac{t^2}{6} \hat{\mathbf{r}} \cdot \boldsymbol{\omega} \right)$$

$$= -rt\boldsymbol{\omega} \cdot \left(\hat{\mathbf{g}} + \frac{t^2 g}{6r} \hat{\mathbf{r}} \right).$$

Using (6.24) we have

$$\Delta R \equiv (\hat{\mathbf{g}} \times \hat{\mathbf{r}}) \cdot \Delta \mathbf{r} = -rt\boldsymbol{\omega} \cdot \left(\hat{\mathbf{g}} + \tfrac{1}{3} \left(\frac{\hat{\mathbf{r}} \wedge \hat{\mathbf{v}}_0}{\hat{\mathbf{g}} \wedge \hat{\mathbf{v}}_0} \right) \hat{\mathbf{r}} \right). \tag{6.27}$$

For nearly horizontal trajectories $\mathbf{r} \wedge \mathbf{v}_0 \approx 0$, so $\Delta R \approx -rt\boldsymbol{\omega} \cdot \mathbf{g}$, which is positive in the Northern hemisphere and negative in the Southern hemisphere (Figure 6.3). As a general rule, therefore, the *Coriolis force tends to deflect particles to the right in the Northern hemisphere and to the left in the Southern hemisphere*. This rule is violated, however, by highly arched trajectories, and from (6.27) one can determine the trajectory without deflection to a given target.

Explicit dependence of the lateral Coriolis deflection on latitude θ, magnetic azimuth ϕ and firing angle ε can be ascertained by reading off the necessary relations from Figure 6.4 to put (6.27) in the form

$$\Delta R = rt\omega \cos\theta \, (\tan\theta - \tfrac{1}{3} \tan\varepsilon \cos\phi). \tag{6.28}$$

The condition for vanishing deflection is therefore

$$\tan \varepsilon_0 = \frac{3 \tan \theta}{\cos \phi} \tag{6.29}$$

and in the Northern hemisphere deflection will be to the left for $\varepsilon > \varepsilon_0$ and to the right for $\varepsilon < \varepsilon_0$.

The Coriolis force plays a significant role in a variety of natural processes, most prominently the weather. It is, for example, responsible for the circular motion of cyclones. To see how this comes about, consider the following equation of motion for a small parcel of air:

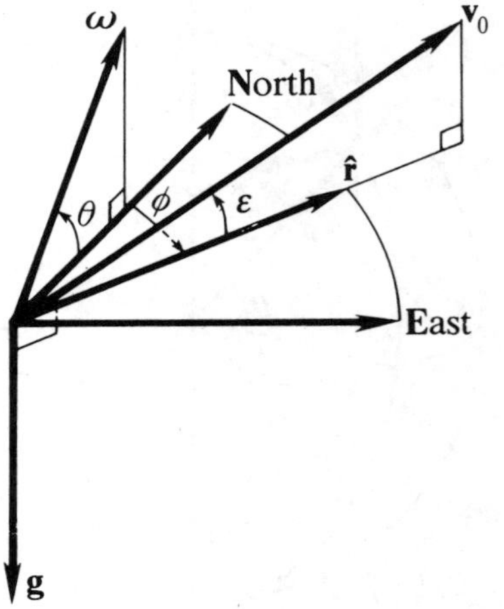

Fig. 6.4. Topocentric directional parameters.

$$\dot{\mathbf{v}} = 2\mathbf{v} \times \boldsymbol{\omega} - \frac{1}{\varrho}\nabla P.$$

Here ϱ is the mass density of the air and $P = P(\mathbf{x})$ is the air pressure, so ∇P describes the local direction and magnitude of the change in air pressure. A cyclone is a system of concentric isobars (lines of constant pressure), as shown in Figure 6.5. As the figure suggests, a parcel of air at rest will be accelerated in the direction $-\nabla P$ of lower pressure. The Coriolis force increases with velocity deflecting the air from the direction $-\nabla P$ until the condition of equilibrium is reached

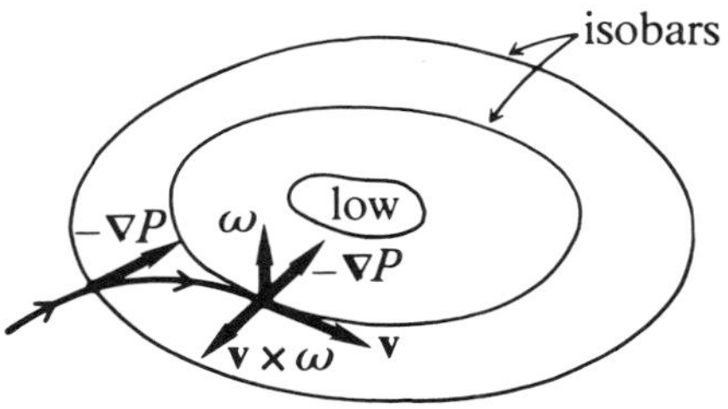

Fig. 6.5. Circulation of air in a cyclone.

$$2\mathbf{v} \times \boldsymbol{\omega} - \frac{1}{\varrho}\nabla P = 0,$$

where the air is circulating with constant speed along the isobars. This motion tends to preserve the pressure gradient. The circulation is counterclockwise in the Northern hemisphere and clockwise in the Southern hemisphere. Naturally, cyclones arise most frequently in regions of the Earth's surface where the Coriolis force is greatest. Of course, our description here is highly idealized, and as a result of effects we have neglected the air flow in a cyclone is not precisely along the isobars.

Cyclones do not occur on the Earth's Equator. However, heating at the Equator causes air to rise, and the air rushing in to replace it is affected by the Coriolis force. Consequently, the "trade winds" come from the North-East just North of the Equator and from the South-East just South of the Equator.

Foucault Pendulum

The Coriolis force produces a small precession, or rotation with time, of a pendulum's plane of oscillation. To exhibit this effect for the first time and thereby demonstrate that the Earth is rotating, Leon Foucault constructed a heavy pendulum of great length in 1851. Accurate measurements of the precession were not made until 1879 by Kamerlingh Onnes in his doctoral thesis.

The equation of motion for the bob of such a pendulum is

$$\ddot{\mathbf{r}} + 2\boldsymbol{\omega} \times \dot{\mathbf{r}} = \mathbf{g} + \frac{\mathbf{T}}{m} = \mathbf{g} - \mathbf{r}\frac{T}{mr}, \tag{6.30}$$

where $r = |\mathbf{r}|$ is the length of the pendulum and $\mathbf{T}$ is the tension in the suspension, as shown in Figure 6.6.

We are interested only in the horizontal component of motion. To separate horizontal and vertical components in the equation of motion, we write

$$\mathbf{r} = \mathbf{x} + z\hat{\mathbf{g}}, \qquad (6.31)$$

where $\mathbf{x}\cdot\mathbf{g} = 0$ (Figure 6.6). Similarly, we decompose $\boldsymbol{\omega}$ into horizontal and vertical components,

$$\boldsymbol{\omega} = \boldsymbol{\omega}_\perp + \boldsymbol{\omega}_\parallel \qquad (6.32)$$

with

$$\boldsymbol{\omega}_\parallel = \hat{\mathbf{g}}\hat{\mathbf{g}}\cdot\boldsymbol{\omega}$$
$$= -\hat{\mathbf{g}}\omega\sin\theta \qquad (6.33)$$

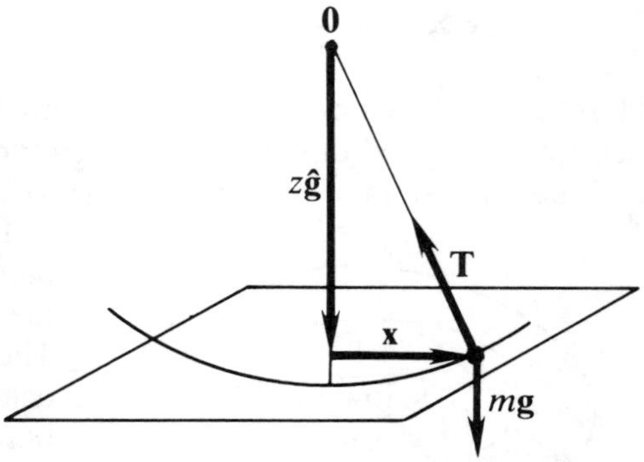

Fig. 6.6. Parameters for a pendulum.

where θ is the latitude, as indicated in Figure 6.3. The components of the Coriolis acceleration are given by

$$\begin{aligned}\omega \times \dot{\mathbf{r}} &= (\omega'_\perp + \boldsymbol{\omega}_\parallel) \times (\dot{\mathbf{x}} + \dot{z}\hat{\mathbf{g}}) \\ &= \boldsymbol{\omega}_\perp \times \dot{\mathbf{x}} + \boldsymbol{\omega}_\parallel \times \dot{\mathbf{x}} + \boldsymbol{\omega} \times \hat{\mathbf{g}}\dot{z}\,.\end{aligned}$$

Thus, when (6.31) and (6.32) are substituted into (6.30), the equation of motion can be separated into the following pair of coupled differential equations for the vertical and horizontal motions:

$$\ddot{z} + 2\dot{\mathbf{x}}\cdot(\hat{\mathbf{g}} \times \omega) = g - z\frac{T}{mr}\,, \qquad (6.34)$$

$$\ddot{\mathbf{x}} + 2\boldsymbol{\omega}_\parallel \times \dot{\mathbf{x}} + 2\omega \times \hat{\mathbf{g}}\dot{z} = -\mathbf{x}\frac{T}{mr}\,. \qquad (6.35)$$

For the small amplitude oscillations of a Focault pendulum, these equations can be decoupled to a good approximation.

We can simplify (6.35) before trying to solve it by using information about the unperturbed periodic motion of a pendulum. The term $2\omega \times \hat{\mathbf{g}}\dot{z}$ *in (6.35)* is periodic as well as small, for $\boldsymbol{\omega}\times\hat{\mathbf{g}}$ has a fixed direction and $\dot{z}$ is positive on the upswing and negative on the downswing. Its average value over half of any period of the pendulum is obviously zero. Therefore we can drop that term from the equation of motion (6.35), because we are interested not in details of the pendulum motion during a single swing but in the cumulative effect of the Coriolis force over many swings. Now consider the "driving force term" $-xT/mr$ on the right side of (6.35). This term is already an explicit function of $\mathbf{x}$, so we get the first order effect of this term on small amplitude oscillations by regarding the coefficient

$$\omega_0^2 \equiv \frac{T}{mr} \qquad (6.36)$$

as constant. Thus (6.35) can be put in the approximate form

$$\ddot{\mathbf{x}} + 2\boldsymbol{\omega}_{\parallel} \times \dot{\mathbf{x}} + \omega_0^2\mathbf{x} = 0\,. \tag{6.37}$$

This is an equation we have encountered and solved before. It is identical to the equation for a charged harmonic oscillator in a uniform magnetic field.

The solution of (6.37) for the initial conditions $\mathbf{x}(0) = \mathbf{a}$ and $\dot{\mathbf{x}}(0) = \mathbf{0}$ is

$$\mathbf{x} = \mathbf{a}e^{-i\boldsymbol{\omega}_{\parallel}t} \cos \frac{2\pi}{T}t\,, \tag{6.38}$$

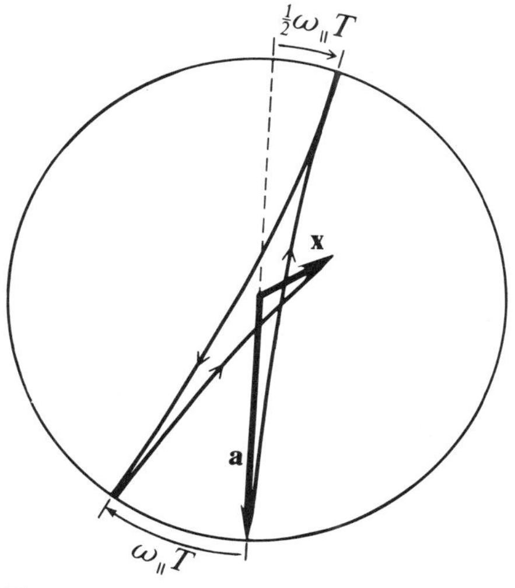

Fig. 6.7. Projection of the path of a pendulum bob on a horizontal plane showing the (exaggerated) Coriolis precession.

where

$$T = \frac{2\pi}{(\omega_{\parallel}^2 + \omega_0^2)^{1/2}} \tag{6.39}$$

is the period of the pendulum. The solution (6.38) describes an oscillator precessing with constant angular velocity

$$-\boldsymbol{\omega}_{\parallel} = \hat{\mathbf{g}}\,\omega \sin\theta \tag{6.40}$$

In the Northern hemisphere $\sin\theta > \theta$, so the precession is clockwise about $\hat{\mathbf{g}}$. Thus, the bob is continually deflected to its right as it swings, with an angular displacement

$$\omega_{\parallel}T = \frac{2\pi\omega}{(\omega^2 + \omega_0^2)^{1/2}}$$

in a single period, and cusps in the orbit at $t = \frac{1}{2}T, T, \frac{3}{2}T, \ldots$, as shown in Figure 6.7.

Rotation and Orbital Motion

As the Earth rotates about its axis, it also revolves about the Sun. Let us see how these motions contribute to the resultant rotational motion of the Earth. Let $\{\mathbf{x}''\}$ be a heliocentric inertial system and let $\{\mathbf{x}\}$, as above, be a geocentric system fixed with respect to the Earth. These two reference systems are related by the equation,

$$\mathbf{x}'' = R_S^{\dagger}(R_E^{\dagger}\mathbf{x}R_E + \mathbf{a}_0)R_S = R^{\dagger}\mathbf{x}R + \mathbf{a}\,, \tag{6.41}$$

where

$$\mathbf{a} = R_S^{\dagger}\mathbf{a}_0R_S \tag{6.42}$$

and

$$R = R_E R_S .\tag{6.43}$$

The various quantities involved require some explanation.

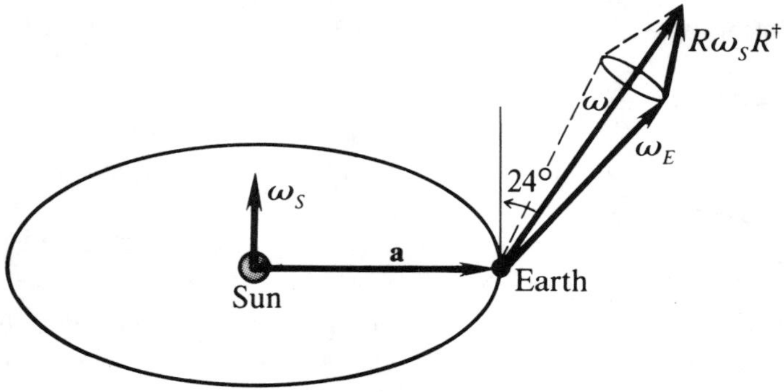

Fig. 6.8 Contribution of orbital motion to the Earth's rotation.

The translation vector **a** designates the location of the Earth's center. The spinor R_S determines the rotation of **a** and so the rotation of the Earth about the Sun, as expressed by (6.42). The vector $\mathbf{a}_0$ is constant if the Earth's orbit is regarded as circular, but its length varies slightly with time for an elliptical orbit. The rotational velocity ω_S of the Earth about the Sun in the heliocentric inertial system $\{\mathbf{x}''\}$ is given by

$$\dot{R}_S = \tfrac{1}{2} R_S i\omega_S \tag{6.44}$$

The period of this rotation is, of course,

$$T = \frac{2\pi}{\omega_S} \approx 365.25 \text{ solar day.}\tag{6.45}$$

Forces of the various planets on the Earth cause small time variations of $\boldsymbol{\omega}_S$, but, for present purposes, $\boldsymbol{\omega}_S$ can be regarded as constant.

The spinor R_E describes the rotation of the Earth about its axis with respect to a frame orbiting the Sun with the Earth. The corresponding rotational velocity in the Earth system $\{\mathbf{x}\}$ is given by

$$\dot{R}_E = \tfrac{1}{2}\, i\omega_E R_E .\tag{6.46}$$

The corresponding period of rotation is

$$T_E = \frac{2\pi}{\omega_E} = 1 \text{ solar day.}\tag{6.47}$$

This period is directly observed on Earth as the time it takes for the Sun to repeat its apparent position relative to the Earth during one revolution of the Earth.

The spinor R describes the rotation of the Earth *in any inertial system* regardless of how the origin has been chosen. The corresponding angular velocity in the Earth frame is given by

$$\dot{R} = \tfrac{1}{2} i\omega R. \tag{6.48}$$

The corresponding period of rotation is

$$T = \frac{2\pi}{\omega} = 1 \text{ sidereal day}. \tag{6.49}$$

This period is directly observed on Earth as the time it takes for the fixed stars to repeat their positions relative to the Earth during one rotation of the Earth.

The relation among the various rotational velocities is determined by differentiating (6.43). Thus, using (6.44) and (6.46) as well as the unitarity property of the spinors, we obtain

$$\dot{R} = \dot{R}_E R_S + R_E \dot{R}_S = \tfrac{1}{2} i\boldsymbol{\omega}_E R_E R_S + R_E R_S \tfrac{1}{2} i\boldsymbol{\omega}_S$$

$$= \tfrac{1}{2} i(\boldsymbol{\omega}_E + R\boldsymbol{\omega}_S R^\dagger)R.$$

Hence, by (6.48),

$$\boldsymbol{\omega} = \boldsymbol{\omega}_E + R\boldsymbol{\omega}_S R^\dagger. \tag{6.50}$$

This is the desired relation among the various angular velocities referred to the Earth's frame. The spinor R appears explicitly in (6.50), because $\boldsymbol{\omega}$ was defined in an inertial system, so it must be transformed to the corresponding angular velocity $R\boldsymbol{\omega}_S R^\dagger$ in the Earth frame. Since $\boldsymbol{\omega}$ is essentially constant, equation (6.48) has the solution

$$R = e^{(1/2)i\boldsymbol{\omega}t}.$$

Therefore, the vector $R\boldsymbol{\omega}_S R^\dagger$ precesses about $\boldsymbol{\omega}$ with a period of one day. Equation (6.50) implies that $\boldsymbol{\omega}_E$ must also precess about $\boldsymbol{\omega}$, as illustrated in Figure 6.8. By measuring to the ecliptic (the apparent path of the Sun) it is found that $\hat{\boldsymbol{\omega}}\cdot\hat{\boldsymbol{\omega}}_S = \cos(23.50) = 0.91$, so from (6.50),

$$\omega \approx \omega_E + 0.91\omega_S \approx \left(1 + \frac{0.91}{365}\right)\omega_E. \tag{6.51}$$

This implies that orbital motion about the Sun is responsible for about $0.91/365 = 0.25\%$ of the Earth's rotational velocity. It is equivalent to saying that the solar day is 3.6 minutes longer than the sidereal day.

Larmor's Theorem

The Coriolis force is important in atomic physics as well as terrestial mechanics. For example, the vibrational and rotational motions of polyatomic molecules are coupled by the Coriolis force. Here we employ it in the proof of a general result of great utility, Larmor's Theorem: *The effect of a weak uniform magnetic field* **B** *on the motion of a charged particle bound by a central force is to cause a precession of the unperturbed orbit with rotational velocity* $\boldsymbol{\omega}_L = -q\mathbf{B}/2mc$, where q is the charge and m is the mass of the particle, the constant c is the speed of light and $\omega_L = |\boldsymbol{\omega}_L|$ is called the *Larmor frequency*.

The proof of Larmor's theorem is based on the formal similarity of the Coriolis force to the magnetic force. We are concerned with a particle subject to the equation of motion

$$m\ddot{\mathbf{r}}' = f(\mathbf{r}')\mathbf{r}' + \frac{q}{c}\dot{\mathbf{r}}' \times \mathbf{B}'. \tag{6.52}$$

Here the system $\{\mathbf{r}'\}$ with origin at the center of force, typically the center of an atomic nucleus, is regarded as an inertial system. In an attempt to simplify the equation of motion by a change of variables, we introduce a rotating system $\{\mathbf{r}\}$ defined by the equations

$$\mathbf{r}' = \mathcal{R}\mathbf{r} = R^\dagger \mathbf{r} R \tag{6.53}$$

and

$$\dot{R} = \tfrac{1}{2}i\omega R = \tfrac{1}{2}R\omega i \tag{6.54a}$$

or

$$\dot{\mathcal{R}}\mathbf{r} = \mathcal{R}(\boldsymbol{\omega} \times \mathbf{r}) = \boldsymbol{\omega}' \times \mathbf{r}'. \tag{6.54b}$$

The motion of the particle in the rotating frame is described by $\mathbf{r} = \mathbf{r}(t)$, while the operator $\mathcal{R}$ describes the motion of the frame itself and $\mathbf{r}' = \mathbf{r}'(t)$ describes the composite of these two motions. In the rotating frame, the equation of motion (6.52) becomes

$$\begin{aligned}&\ddot{\mathbf{r}} + 2\boldsymbol{\omega} \times \dot{\mathbf{r}} + \boldsymbol{\omega} \times (\boldsymbol{\omega} \times \mathbf{r}) + \dot{\boldsymbol{\omega}} \times \mathbf{r}\\ &= \frac{f(r)}{m}\mathbf{r} + \frac{q}{mc}(\dot{\mathbf{r}} + \boldsymbol{\omega} \times \mathbf{r}) \times \mathbf{B},\end{aligned} \tag{6.55}$$

where $\mathbf{B} = \mathcal{R}^{-1}\mathbf{B}'$.

Evidently, the Coriolis force can be made to cancel the magnetic force in (6.55) by selecting the rotating frame so that

$$\boldsymbol{\omega} = \boldsymbol{\omega}_L \equiv -\frac{q}{2mc}\mathbf{B}. \tag{6.56}$$

Whereupon (6.55) becomes

$$\ddot{\mathbf{r}} = \frac{f(r)}{m}\,\hat{\mathbf{r}} + \boldsymbol{\omega} \times (\boldsymbol{\omega} \times \mathbf{r}) - \dot{\boldsymbol{\omega}} \times \mathbf{r}\,. \tag{6.57}$$

If the magnetic field is weak and slowly varying in time, the last two terms in (6.57) are small compared to the binding force and we have, approximately,

$$m\ddot{\mathbf{r}} = f(r)\hat{\mathbf{r}}\,. \tag{6.58}$$

Thus, we have succeeded in transforming away the perturbing force from the equation of motion (6.52). So we have proved Larmor's theorem.

Larmor's theorem has the great advantage of decoupling the effect of an external magnetic field from that of the binding force so they can be studied separately. According to (6.58), the motion in the rotating frame is the same as motion in an inertial frame without a perturbing magnetic force, so we call this the *unperturbed motion*.

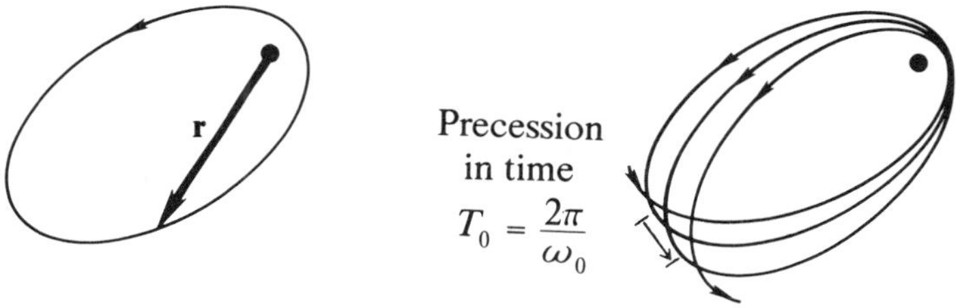

Fig. 6.9(a) Unperturbed elliptical orbit = orbit in rotating frame. (b) Perturbed orbit in inertial frame for the case $\mathbf{B}\cdot\mathbf{v} = 0$.

Let us compare the perturbed and unperturbed motions. The condition that (6.58) be a good approximation of (6.57) can be expressed in the form

$$\left(\frac{q\mathbf{B}}{2mc}\right)^2 = \omega_L^2 \ll \omega_0^2 \equiv -\left\langle \frac{f(r)}{mr} \right\rangle_{\text{ave}} \tag{6.59}$$

Here the average value of $f(r)/mr$ is evaluated over a period of the unperturbed motion. The resulting constant ω_0 can be interpreted as the frequency of the unperturbed motion. Indeed, it will be noted that for circular orbits Equation (6.58) can be put in the form $\ddot{\mathbf{r}} = -\,\omega_0^2\mathbf{r}$, confirming consistency with the interpretation we have previously given to ω_0 for the harmonic oscillator. Thus, the condition (6.59) for the validity of *Larmor's theorem requires that the frequency of the unperturbed motion be much larger than the Larmor frequency*. From our study of central forces in Section 4-5, we know that the unperturbed orbit is elliptical or nearly so. So the perturbed orbit can be visualized as a slowly precessing ellipse, as illustrated in Figure 6.9b.

What are typical values for ω_L and ω_0 in actuality? The rate of orbital precession can easily be estimated from known values of the constants q/mc; thus

$$\omega_L = \left|\frac{q\mathbf{B}}{2mc}\right| = (0.9 \times 10^7 \text{ sec}^{-1} \text{ gauss}^{-1})B. \tag{6.60}$$

The largest magnetic fields attained in the laboratory (with superconducting magnets) are of order 10^6 gauss, in which case

$$\omega_L \approx 10^{13} \text{ sec}^{-1}, \tag{6.61}$$

so the orbit makes a complete precession in about 10^{-12} sec. To estimate ω_0, we need some results from the quantum theory of atoms. According to quantum theory, the orbital angular momentum of an atomic electron is an integral multiple of Planck's constant $\hbar$, so

$$l = |\, m\mathbf{r} \times \dot{\mathbf{r}} \,| \geqslant \hbar = 1.05 \times 10^{-27} \text{ erg sec.}$$

For a circular orbit we have $l = mr\dot{r}$ and $\dot{r} = \omega_0 r$, so with an order of magnitude estimate of the atomic radius, we find

$$\omega_0 \geqslant \frac{\hbar}{mr^2} \approx \frac{10^{-27} \text{ erg sec}}{9 \times 10^{-28}\text{g} \times 10^{-14}\text{cm}^2} \approx 10^{16} \text{ sec}^{-1}. \tag{6.62}$$

We can conclude, then, that only high precision experiments will reveal deviations from Larmor's theorem.

The orbits of electrons in atoms cannot be observed. But angular momentum and energy are constants of motion in a central field, and changes in their values for atoms can be measured. So let us see how these constants of motion are affected by a magnetic perturbation. The velocities of perturbed and unperturbed motions are related by

$$\dot{\mathbf{r}}' = \mathcal{R}(\dot{\mathbf{r}} + \boldsymbol{\omega} \times \mathbf{r}). \tag{6.63}$$

So the angular momenta $\mathbf{l}' = m\mathbf{r}' \times \dot{\mathbf{r}}'$ and $\mathbf{l} = m\mathbf{r} \times \dot{\mathbf{r}}$ are related by

$$\mathbf{l}' = \mathcal{R}(\mathbf{l} + m\mathbf{r} \times (\boldsymbol{\omega} \times \mathbf{r})) = \mathcal{R}\mathbf{l} + m\mathbf{r}' \times (\boldsymbol{\omega}' \times \mathbf{r}'). \tag{6.64}$$

The term $m\mathbf{r}' \times (\boldsymbol{\omega}' \times \mathbf{r}')$ is called the *induced angular momentum*. It is of prime importance when $\mathbf{l} = 0$. However, according to our above estimates of ω_0 and $\omega = \omega_L$, when $\mathbf{l} \neq 0$ the induced angular momentum is usually smaller in magnitude than $\mathbf{l}$ by a factor of 10^3 or more, so

$$\mathbf{l}' \approx \mathcal{R}\mathbf{l} = R^\dagger \mathbf{l} R. \tag{6.65}$$

Differentiating, we have

$$\dot{\mathbf{l}}' = \mathcal{R}(\dot{\mathbf{l}}' + \boldsymbol{\omega} \times \mathbf{l}). \tag{6.66}$$

For central binding forces $\dot{\mathbf{l}}' = 0$, so

$$\dot{\mathbf{l}}' = \mathcal{R}(\boldsymbol{\omega} \times \mathbf{l}) = \boldsymbol{\omega}' \times \mathbf{l}'. \tag{6.67}$$

Notice that this equation of motion for angular momentum is completely independent of any details as to how electrons are bound to atoms.

We have become very familiar with the solution of an equation with the form (6.67) for constant ω'. Thus, from the spinor equation (6.54a) we get

$$R = e^{(1/2)i\boldsymbol{\omega}t}$$

with

$$\boldsymbol{\omega}' = \boldsymbol{\omega} = \boldsymbol{\omega}_L = -\frac{q\mathbf{B}}{2mc}$$

and the initial condition $\mathbf{l}' = \mathbf{l}$ at $t = 0$. In this case, (6.65) becomes the explicit equation

$$\mathbf{l}' = e^{-(1/2)i\boldsymbol{\omega}t}\,\mathbf{l}e^{(1/2)i\boldsymbol{\omega}t}. \tag{6.68}$$

This describes a uniform precession of $\mathbf{l}'$ about $\boldsymbol{\omega}$, as shown in Figure 6.10. In Chapter 7, we will investigate solutions of (6.67) for time varying magnetic fields leading to the phenomenon of magnetic resonance, a phenomenon of great importance for investigating the atomic and molecular structure of matter.

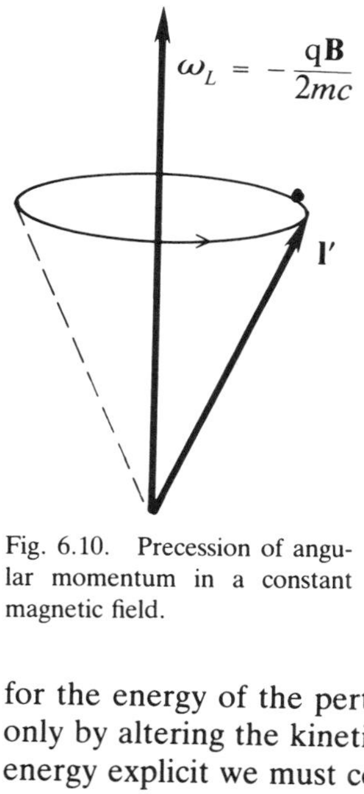

Fig. 6.10. Precession of angular momentum in a constant magnetic field.

Let us turn now to energy considerations. We know that a conservative central force is derivable from a potential, so writing

$$f(r)\hat{\mathbf{r}} = -\Delta V(r),$$

we get the familiar expression

$$E = \tfrac{1}{2}m\dot{\mathbf{r}}^2 + V(r)$$

for the energy of the unperturbed system. We have observed before that the magnetic field does not contribute to the potential energy, so from (6.52) we get

$$E' = \tfrac{1}{2}m\dot{\mathbf{r}}'^2 + V(r')$$

for the energy of the perturbed motion. A magnetic field affects the energy only by altering the kinetic energy, so to make the influence of the magnetic energy explicit we must compare the energies of perturbed and unperturbed motions. From (6.63) we have

$$\dot{\mathbf{r}}'^2 = (\dot{\mathbf{r}} + \boldsymbol{\omega}\times\mathbf{r})^2 = \dot{\mathbf{r}}^2 + 2\boldsymbol{\omega}\cdot(\mathbf{r}\times\dot{\mathbf{r}}) + (\boldsymbol{\omega}\times\mathbf{r})^2.$$

Hence,

$$E' = E + \boldsymbol{\omega}\cdot\mathbf{l} + \tfrac{1}{2}m(\boldsymbol{\omega}\times\mathbf{r})^2. \tag{6.69}$$

The last term in (6.69) is neglected in the Larmor approximation, so we can conclude that a magnetic field induces a shift in the energy of a bound charge particle by the amount

$$\boldsymbol{\omega}\cdot\mathbf{l} = -\frac{q}{2mc}\,\mathbf{B}\cdot\mathbf{l}. \tag{6.70}$$

This energy shift in atoms is easily observed by modern methods, and it is known as the *Zeeman effect*. From (6.70) it follows that the energy $\omega\cdot\mathbf{l}$ is constant for a constant magnetic field. For large magnetic fields the shift in energy due to the last term in (6.69) can also be observed. This is known as the *Quadratic Zeeman* (or Paschen-Bach) *effect*, since it varies quadratically with the magnetic field strength. Like its relative, magnetic resonance (discussed in Section 7-3), the Zeeman effect is of great value for probing the structure of matter.

5-6. Exercises

(6.1) On the surface of the Earth, the *true vertical* is directed along a line through the center of the Earth, and the apparent vertical is directed along a plumb line. Determine how the angle α between true and apparent verticals varies with latitude θ (Figure 6.1). Estimate the maximum value for α and the latitude at which it occurs.

(6.2) At about sea level and a latitude of 45° a 16 pound (7.27 kg) steel ball is dropped from a height of 45 m.

(a) Calculate the displacement of its point of impact to first order in the angular velocity of the earth.

(b) It is argued that while the ball is falling, the Earth will rotate under it to the East, so the displacement will be to the West. Show what is wrong with this argument by describing what happens in an inertial frame.

(c) Estimate the effect of the Coriolis force to order ω^2.

(6.3) A free particle is constrained to move in a horizontal plane of a topocentric frame, but it is confined to a region with circular walls from which it rebounds elastically (Figure 6.11). Show that the direction of the particle motion precesses at exactly twice the rate of a Foucault pendulum. How can this difference in precession rates be accounted for?

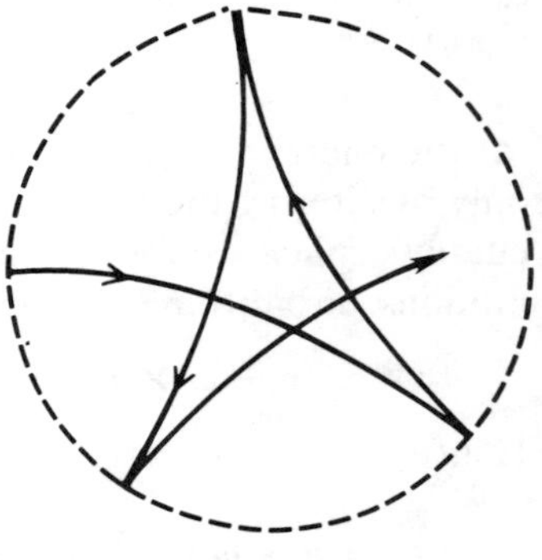

Fig. 6.11. A free particle, reflected by circular walls, precesses at twice the Foucault rate.

(6.4) At a point on the Earth's Equator, determine the relative magnitude of centrifugal forces due to orbital motion about the Sun and rotation of the Earth. (The mean radius of the Earth's orbit is 1.495×10^8 km; see Appendix C for the radius of the Earth.)

(6.5) List all the forces and effects you can think of (at least 10) which are neglected when a projectile launched from the surface of the Earth is described as moving with constant acceleration. Estimate their magnitudes and describe the conditions under which they will be significant.

Chapter 6

Many-Particle Systems

This chapter develops general concepts, theorems and techniques for modeling complex, mechanical systems. The three main theorems on system energy, momentum, and angular momentum are proved in Section 6-1. These theorems provide the starting point for rigid body mechanics in Chapter 7, as well as for other kinds of mechanical system discussed in this chapter. The method of Lagrange formulated in Section 6-2 provides a systematic means for expressing the equations of motion for any mechanical system in terms of any convenient set of variables. The method proves to be of great value in the theory of small oscillations as well as applications to molecular vibrations treated in Section 6-4. The general theory of small oscillations in Section 6-4 raises the level of sophistication required of the student, so some examples of small oscillations are treated in Section 6-3 by more elementary means.

The final section of this chapter discusses the most venerable unsolved problem in celestial mechanics, the Newtonian 3-body problem. It complements the development of celestial mechanics in Chapter 8.

6-1. General Properties of Many-Particle Systems

Classical mechanics provides us with general principles for modeling any material body, be it a solid, liquid or gas, as a system of interacting particles. To analyze the behavior of a system, we must separate it from its environment. This is done by distinguishing external and internal variables. The *external variables* describe the system as whole and its interaction with other (external) systems. The *internal variables* describe the (internal) structure of the system and the interactions among its parts. The analysis of internal and external variables for a two-particle system was carried out in Section 4-6. Here we analyze the general case of an N-particle system. The results include *three major theorems*: (1) the center-of-mass theorem, (2) the angular momentum theorem, and (3) the work-energy theorem. These theorems provide a starting point for the modeling of any complex mechanical system.

The superposition principle allows us to separate external and internal forces; so the equation of motion for ith particle in an N-particle system has the form

$$m_i \ddot{\mathbf{x}}_i = \mathbf{F}_i + \sum_{j=1}^{N} \mathbf{f}_{ij}, \tag{1.1}$$

where the *external force* $\mathbf{F}_i$ is the resultant force exerted by objects external to the system, and the *interparticle force* $\mathbf{f}_{ij}$ is the force exerted on the ith particle by the jth particle. Also, we assume that $\mathbf{f}_{ii} = 0$, that is, that a particle does not exert a force on itself.

According to the *weak form* of *Newton's 3rd law*, mutual forces of any two particles on one another are *equal and opposite* so the interparticle forces are related by

$$\mathbf{f}_{ij} = -\mathbf{f}_{ji}. \tag{1.2}$$

The *strong form of Newton's 3rd law* holds also that all two-particle forces are *central forces*, that is, directed along a straight line connecting the particles. The condition that interparticle forces be central is expressed by

$$(\mathbf{x}_i - \mathbf{x}_j)\wedge\mathbf{f}_{ij} = 0. \tag{1.3}$$

We adopt the strong form of Newton's 3rd law, because it holds for a large class of systems, and it greatly simplifies the analysis. Deviations from the 3rd law arise principally in systems composed of particles which are not accurately described as structureless point particles but have some internal structure which significantly affects their interactions with other particles. However, a deeper analysis may show that the structure of such a particle can be described by assuming that the particle is itself composed of structureless point particles. The general results derived below provide the foundation for such analysis.

Translational Motion

Now, to develop an equation describing the translational motion of the system as a whole, we add the equations of motion (1.1) for each particle; thus,

$$\sum_i m_i \ddot{\mathbf{x}}_i = \sum_i \mathbf{F}_i + \sum_i \sum_j \mathbf{f}_{ij}. \tag{1.4}$$

The weak form of Newton's third law (1.2) implies that the internal forces in the sum cancel; formally,

$$\sum_i \sum_j \mathbf{f}_{ij} = \sum_j \sum_i \mathbf{f}_{ji} = -\sum_i \sum_j \mathbf{f}_{ij} = 0,$$

$$\frac{\mathrm{d}^2}{\mathrm{d}t^2}\Big(\sum_i m_i \mathbf{x}_i\Big) = \sum_i \mathbf{F}_i. \tag{1.5}$$

To put this in the standard form for a particle equation of motion, we define the following set of external variables for the system:

Total mass,

$$M \equiv \sum_i m_i, \tag{1.6}$$

Center of mass (CM),

$$\mathbf{X} \equiv \frac{1}{M}\sum_i m_i\mathbf{x}_i = \frac{\sum_i m_i\mathbf{x}_i}{\sum_i m_i}, \tag{1.7}$$

Total momentum,

$$\mathbf{P} \equiv M\dot{\mathbf{X}} = \sum_i m_i\dot{\mathbf{x}}_i, \tag{1.8}$$

Total external force,

$$\mathbf{F} = \sum_i \mathbf{F}_i. \tag{1.9}$$

In terms of these variables, Equation (1.5) has the form

$$\dot{\mathbf{P}} = M\ddot{\mathbf{X}} = \mathbf{F}. \tag{1.10}$$

This equation, along with its interpretation, is the *Center of Mass (CM) Theorem*.

The system is said to be *isolated* if $\mathbf{F}_i = 0$, that is if the external force on each particle vanishes. For an isolated system, then $\dot{\mathbf{P}} = 0$, so the momentum of the system is a constant of the motion, in other words, *the momentum of an isolated system is conserved.*

According to (1.7), the *CM* $\mathbf{X}$ is a kind of average position of the particles in a system. The *CM* theorem (1.10) tells us that the motion of the CM is determined by the total external force alone, irrespective of the internal forces. This independence of internal force is a consequence of the weak form of the 3rd law (1.2), so empirical verification of the CM theorem supports the 3rd law.

The *CM* theorem describes the average motion of a system as equivalent to that of a single particle of mass M located at the *CM* $\mathbf{X}$. Thus, it separates external and internal motions, allowing us to study them independently. And when we are not interested in internal structure, the *CM* theorem justifies treating the entire system as a single particle.

Internal and external properties of the system can be separated by introducing the *internal variables*

$$\mathbf{r}_i = \mathbf{x}_i - \mathbf{X} \tag{1.11}$$

describing the position of each particle relative to the *CM*. The *internal velocities* are therefore

$$\dot{\mathbf{r}}_i = \dot{\mathbf{x}}_i - \dot{\mathbf{X}}$$

This leads us to the following decomposition of the *total kinetic energy K* for the system:

$$K \equiv \sum_i \tfrac{1}{2} m_i \dot{\mathbf{x}}_i^2 = \sum_i \tfrac{1}{2} m_i (\dot{\mathbf{X}} + \dot{\mathbf{r}}_i)\cdot(\dot{\mathbf{X}} + \dot{\mathbf{r}}_i)$$

$$= \tfrac{1}{2} M\dot{\mathbf{X}}^2 + \sum_i \tfrac{1}{2} m_i \dot{\mathbf{r}}_i^2 + \dot{\mathbf{X}}\cdot \frac{\mathrm{d}}{\mathrm{d}t}\Big(\sum_i m_i \mathbf{r}_i\Big).$$

But

$$\sum_i m_i \mathbf{r}_i = \sum_i m_i(\mathbf{x}_i - \mathbf{X}) = \sum_i m_i \mathbf{x}_i - M\mathbf{X} = \mathbf{0}.$$

Hence,

$$K \equiv \sum_i \tfrac{1}{2} m_i \dot{\mathbf{x}}_i^2 = K_{CM} + K_{\text{int}}, \tag{1.12}$$

where

$$K_{\text{int}} = \sum_i \tfrac{1}{2} m_i \dot{\mathbf{r}}_i^2 \tag{1.13a}$$

is the *internal kinetic energy*, and

$$K_{CM} = \tfrac{1}{2} M\dot{\mathbf{X}}^2 \tag{1.13b}$$

is the *CM kinetic energy*, also known as the *translational kinetic energy* of the system. Thus, the total kinetic energy is the sum of internal and external (*CM*) kinetic energies.

Similarly, the *total angular momentum* **J** for the system submits to the decomposition

$$\mathbf{J} \equiv \sum_i m_i \mathbf{x}_i \times \dot{\mathbf{x}}_i = \sum_i m_i(\mathbf{X} + \mathbf{r}_i) \times (\dot{\mathbf{X}} + \dot{\mathbf{r}}_i)$$

$$= \Big(\sum_i m_i\Big)\mathbf{X} \times \dot{\mathbf{X}} + \Big(\sum_i m_i \mathbf{r}_i\Big) \times \dot{\mathbf{X}} + \mathbf{X} \times \Big(\sum_i m_i \dot{\mathbf{r}}_i\Big) + \sum_i m_i \mathbf{r}_i \times \dot{\mathbf{r}}_i .$$

Hence,

$$\mathbf{J} = \sum_i m_i \mathbf{x}_i \times \dot{\mathbf{x}}_i = M\mathbf{X} \times \dot{\mathbf{X}} + \mathbf{l} \tag{1.14}$$

where

$$\mathbf{l} \equiv \sum_i m_i \mathbf{r}_i \times \dot{\mathbf{r}}_i \tag{1.15}$$

is the *internal angular momentum*, and $\mathbf{X} \times \mathbf{P} = M\mathbf{X} \times \dot{\mathbf{X}}$ is known as the *orbital angular momentum* of the system. Thus the total angular momentum is simply the sum of internal and external (orbital) angular momenta. To conform to standard usage, we have defined the angular momentum with cross products instead of outer products, so it is a vector instead of a bivector. However, as explained in Section 4-1, the bivector form is more fundamental and we shall switch to it later when it is advantageous.

Rotational Motion

To describe rotational motion of the system as a whole, we derive an equation of motion for the total angular momentum. Differentiating (1.14) and using (1.1) and (1.2), we have

$$\begin{aligned}\dot{\mathbf{J}} &= \frac{\mathrm{d}}{\mathrm{d}t}\Big(\sum_i m_i \mathbf{x}_i \times \dot{\mathbf{x}}_i\Big) = \sum_i m_i \mathbf{x}_i \times \ddot{\mathbf{x}}_i \\ &= \sum_i \mathbf{x}_i \times \Big(\mathbf{F}_i + \sum_j \mathbf{f}_{ij}\Big) \\ &= \sum_i \mathbf{x}_i \times \mathbf{F}_i + \tfrac{1}{2}\sum_i \sum_j (\mathbf{x}_i - \mathbf{x}_j) \times \mathbf{f}_{ij}.\end{aligned}$$

For central forces the last term vanishes, and we have the *rotational equation of motion*

$$\dot{\mathbf{J}} = \sum_i \mathbf{x}_i \times \mathbf{F}_i \equiv \boldsymbol{\Gamma}_0, \tag{1.16}$$

where $\boldsymbol{\Gamma}_0$ is known as the *torque* (about the origin). It is readily verified that a displacement of the origin changes the value of $\mathbf{J}$ and $\boldsymbol{\Gamma}_0$ without altering the form of (1.16).

The most notable property of the rotational equation of motion (1.16) is the fact that internal forces do not contribute directly to the torque. Our derivation shows that this is a consequence of Newton's 3rd law in its strong form. Note, however, that the torque in (1.16) depends on the values of the $\mathbf{x}_i$ whose time variations depend on the internal forces. Therefore, the torque is indirectly dependent on the internal forces. Of course, for an isolated system the torque vanishes for arbitrary internal forces. Hence *the angular momentum of an isolated system is conserved.*

It is usually desirable to separate the total angular momentum into its external and internal parts, for the parts satisfy independent equations of motion. Time variation of the external (orbital) angular momentum is determined by the *CM* equation (1.10); thus,

$$\frac{\mathrm{d}}{\mathrm{d}t}(M\mathbf{X} \times \dot{\mathbf{X}}) = M\mathbf{X} \times \ddot{\mathbf{X}} = \mathbf{X} \times \mathbf{F}. \tag{1.17}$$

On the other hand,

$$\dot{\mathbf{J}} \doteq \frac{\mathrm{d}}{\mathrm{d}t}(M\mathbf{X} \times \dot{\mathbf{X}} + \mathbf{l}) = \mathbf{X} \times \mathbf{F} + \dot{\mathbf{l}}.$$

Substituting this into (1.16), we get the *equation of motion for the internal angular momentum*:

$$\dot{\mathbf{l}} = \sum_i (\mathbf{x}_i - \mathbf{X}) \times \mathbf{F}_i = \sum_i \mathbf{r}_i \times \mathbf{F}_i \equiv \boldsymbol{\Gamma}. \tag{1.18}$$

In contrast to (1.16), this equation is independent of the origin, or rather, it differs from (1.16) by a shift of the arbitrary fixed origin to an origin intrinsic to the system, the center of mass.

The angular momentum equation of motion (1.18) is the second major result of general many-particle systems theory. Let us call it the *angular momentum theorem*. To make use of this theorem, however, we need further results relating the angular momentum to kinematical variables of the system.

Often we wish to separate the rotational motion of the system as a whole from the relative motions of its parts. This can be accomplished by introducing a body frame rotating with the system. The position $\mathbf{r}'_i$ of a particle in the body frame is related to the internal position variable $\mathbf{r}_i = \mathbf{x}_i - \mathbf{X}$ by a rotation

$$\mathbf{r}_i = R^\dagger \mathbf{r}'_i R. \tag{1.19}$$

As we saw in Section 5–5, the time dependence of the rotation is determined by the differential equation

$$\dot{R} = \tfrac{1}{2} Ri\boldsymbol{\omega}, \tag{1.20}$$

so that

$$\dot{\mathbf{r}}_i = \boldsymbol{\omega} \times \mathbf{r}_i + R^\dagger \dot{\mathbf{r}}'_i R. \tag{1.21}$$

The vector $\boldsymbol{\omega}$ is called the *rotational velocity* of the system or body, if you will.

Substituting (1.21) into the expression (1.15) for the internal angular momentum, we obtain

$$\mathbf{l} = \sum_i m_i \mathbf{r}_i \times (\boldsymbol{\omega} \times \mathbf{r}_i) + R^\dagger \Big(\sum_i m_i \mathbf{r}'_i \times \dot{\mathbf{r}}'_i\Big) R. \tag{1.22}$$

The first set of terms on the right side of (1.22) defines a linear function,

$$\mathcal{J}\boldsymbol{\omega} \equiv \sum_i m_i \mathbf{r}_i \times (\boldsymbol{\omega} \times \mathbf{r}_i), \tag{1.23}$$

so (1.22) can be put in the form

$$\mathbf{l} = \mathcal{J}\boldsymbol{\omega} + R^\dagger \Big(\sum_i m_i \mathbf{r}'_i \times \dot{\mathbf{r}}'_i\Big) R. \tag{1.24}$$

The linear operator $\mathcal{J}$ is called the *inertia tensor* of the body. Using the identities

$$\mathbf{r} \times (\boldsymbol{\omega} \times \mathbf{r}) = -\mathbf{r}\cdot(\boldsymbol{\omega} \wedge \mathbf{r}) = \mathbf{r}\mathbf{r} \wedge \boldsymbol{\omega} = r^2 \boldsymbol{\omega} - \mathbf{r}\mathbf{r}\cdot\boldsymbol{\omega}$$

we can write the inertia tensor in the alternative forms

$$\mathcal{J}\boldsymbol{\omega} \equiv \sum_i m_i \mathbf{r}_i \mathbf{r}_i \wedge \boldsymbol{\omega} = \sum_i m_i (r_i^2 \boldsymbol{\omega} - \mathbf{r}_i \mathbf{r}_i \cdot \boldsymbol{\omega}). \tag{1.25}$$

For most purposes, (1.25) is more convenient than (1.23).

We have not yet explained how the body frame is to be determined. For a rigid body, it is determined by the condition that all particles of the body be at rest in the body frame. Thus $\dot{\mathbf{r}}'_i = 0$, so the distances between particles

$$r_{ij} = |\mathbf{x}_i - \mathbf{x}_j| = |\mathbf{r}_i - \mathbf{r}_j| = |\mathbf{r}'_i - \mathbf{r}'_j|$$

are constants of the motion. Consequently, from (1.24) it follows that, for a rigid body,

$$\mathbf{l} = \mathcal{J}\boldsymbol{\omega}. \tag{1.26}$$

This reduces the complex "dynamical variable" $\mathbf{l}$ to the simple "kinematical variable" $\boldsymbol{\omega}$, because the inertia tensor $\mathscr{I}$ depends only on the fixed internal structure of the body. Since $\mathscr{I}$ is a linear operator, when (1.26) is substituted into the equation of motion (1.18), one obtains

$$\dot{\mathbf{l}} = \dot{\mathscr{I}}\boldsymbol{\omega} + \mathscr{I}\dot{\boldsymbol{\omega}} = \boldsymbol{\Gamma}, \tag{1.27}$$

The time derivative of the inertia tensor can be computed from the definition (1.25) using $\dot{\mathbf{r}} = \boldsymbol{\omega} \times \mathbf{r}_i$. Thus, for an arbitrary vector argument $\mathbf{a}$ we have

$$\begin{aligned}\dot{\mathscr{I}}\mathbf{a} &= \sum_i m_i(-\dot{\mathbf{r}}_i\mathbf{r}_i\cdot\mathbf{a} - \mathbf{r}_i\dot{\mathbf{r}}_i\cdot\mathbf{a})\\ &= -\sum_i m_i(\omega \times \mathbf{r}_i\mathbf{r}_i\cdot\mathbf{a} + \mathbf{r}_i(\omega \times \mathbf{r}_i)\cdot\mathbf{a}\\ &= \sum_i m_i[\omega \times (\mathbf{r}_i^2\mathbf{a} - \mathbf{r}_i\mathbf{r}_i\cdot\mathbf{a}) - \mathbf{r}_i^2(\omega \times \mathbf{a}) + \mathbf{r}_i\mathbf{r}_i\cdot(\omega \times \mathbf{a})]\\ &= \sum_i m_i[\omega \times (\mathbf{r}_i\mathbf{r}_i\wedge\mathbf{a}) - \mathbf{r}_i\mathbf{r}_i\wedge(\omega \times \mathbf{a})].\end{aligned}$$

Whence, the general result

$$\dot{\mathscr{I}}\mathbf{a} = \boldsymbol{\omega} \times (\mathscr{I}\mathbf{a}) - \mathscr{I}(\boldsymbol{\omega} \times \mathbf{a}). \tag{1.28}$$

Using this in (1.27), we obtain

$$\dot{\mathbf{l}} = \mathscr{I}\dot{\boldsymbol{\omega}} + \boldsymbol{\omega} \times (\mathscr{I}\boldsymbol{\omega}) = \boldsymbol{\Gamma}. \tag{1.29}$$

All the internal motion of a rigid body is rotational, and it is completely determined by the equation of motion (1.29), with appropriate initial conditions, of course. This will be the starting point for the study of rigid body motion in Chapter 7.

For an arbitrary system of particles, such as a gas, the body frame may be difficult to determine. Actually, we have defined the body frame only for a rigid body, and we are free to define the body frame in any convenient way for other systems. For example, for a system consisting of a gas in a rigid box, we could choose the body frame of the box as body frame for the entire system. Then, Equation (1.24) would be interpreted as a separation of the angular momentum into a part $\mathscr{I}\boldsymbol{\omega}$ for "the system as a whole" and a residual angular momentum $\Sigma_i m_i \mathbf{r}'_i \times \dot{\mathbf{r}}'_i$ of the gas in the box. As a rule, however, it is more natural to define the body frame by imposing the condition

$$\sum_i m_i \mathbf{r}'_i \times \dot{\mathbf{r}}'_i = 0 \tag{1.30}$$

on motions in the body frame so $\mathbf{l} = \mathscr{I}\boldsymbol{\omega}$ describes the resultant angular momentum of the entire system. This agrees with our definition for rigid bodies. Of course, for an arbitrary system the derivative of the inertia tensor will not be given by the expression (1.28) for a rigid body; it will include terms describing change in the structure of the system as well as the terms in (1.28) which are due solely to rotation.

Separation of rotational motion from vibrational motion is of great importance in the theory of polyatomic molecules. A molecule can be modeled as a system of atoms (point particles) vibrating about equilibrium points $\mathbf{a}_i$ in the body frame. For small vibrations, then, the condition (1.30) can be approximated by

$$\sum_i m_i \mathbf{a}_i \times \dot{\mathbf{r}}'_i = 0. \tag{1.31}$$

This can be integrated directly, and the integration constant can be chosen so that

$$\sum_i m_i \mathbf{a}_i \times \mathbf{r}'_i = 0. \tag{1.32}$$

This is the appropriate condition on the relative atomic displacements determining the body frame. It must be used along with the center of mass condition

$$\sum_i m_i \mathbf{r}'_i = 0. \tag{1.33}$$

Because of its simplicity, the condition (1.31) or (1.32) is preferable to (1.30), to which it is equivalent only in the first approximation.

To separate the rotational energy from the rest of the internal energy in a system, it is convenient to use (1.21) in the form

$$\dot{\mathbf{r}}_i = R^\dagger(\boldsymbol{\omega}' \times \mathbf{r}'_i + \dot{\mathbf{r}}'_i)R, \tag{1.34}$$

where

$$\boldsymbol{\omega}' = R\boldsymbol{\omega}R^\dagger = \frac{2}{i}\dot{R}R^\dagger \tag{1.35}$$

is the angular velocity in the body frame. Then the internal kinetic energy K_{int} can be written

$$K_{\text{int}} = \tfrac{1}{2}\sum_i m_i \dot{\mathbf{r}}_i^2 = \tfrac{1}{2}\sum_i m_i[(\boldsymbol{\omega}' \times \mathbf{r}'_i)^2 + 2\boldsymbol{\omega}'\cdot(\mathbf{r}'_i \times \dot{\mathbf{r}}'_i) + \dot{\mathbf{r}}'^2_i].$$

Using the identity

$$(\boldsymbol{\omega} \times \mathbf{r})^2 = (\boldsymbol{\omega}\wedge\mathbf{r})\cdot(\mathbf{r}\wedge\boldsymbol{\omega}) = \boldsymbol{\omega}\cdot(\mathbf{r}\mathbf{r}\wedge\boldsymbol{\omega}),$$

we can write

$$\sum_i m_i(\boldsymbol{\omega}' \times \mathbf{r}'_i)^2 = \boldsymbol{\omega}'\cdot(\mathcal{I}'\boldsymbol{\omega}'), \tag{1.36}$$

where

$$\mathcal{I}'\boldsymbol{\omega}' = \sum_i m_i \mathbf{r}'_i \mathbf{r}'_i \wedge \boldsymbol{\omega}'. \tag{1.37}$$

is the inertia tensor in the body frame. Then the internal kinetic energy assumes the form

$$K_{\text{int}} = \tfrac{1}{2}\boldsymbol{\omega}'\cdot(\mathcal{I}'\boldsymbol{\omega}') + \boldsymbol{\omega}'\cdot\Big(\sum_i m_i \mathbf{r}'_i \times \dot{\mathbf{r}}'_i\Big) + \tfrac{1}{2}\sum_i m_i \dot{\mathbf{r}}'^2_i. \tag{1.38}$$

For a polyatomic molecule or, more generally, for a solid body, we can write

$$\mathbf{r}'_i = \mathbf{a}_i + \mathbf{s}_i,$$

where $\mathbf{s}_i$ is the displacement from the equilibrium position $\mathbf{a}_i$. Then, because of the condition (1.31), the kinetic energy (1.38) reduces to

$$K_{\text{int}} = \tfrac{1}{2}\,\boldsymbol{\omega}'\cdot\mathcal{I}'\boldsymbol{\omega}' + \boldsymbol{\omega}'\cdot(\sum_i m_i\mathbf{s}'_i \times \dot{\mathbf{s}}_i) + \tfrac{1}{2}\sum_i m_i\dot{\mathbf{s}}_i^2. \tag{1.39}$$

The first term in (1.39) is the *rotational energy* and the last term is the *vibrational energy*. The middle term is the *Coriolis energy*, coupling the rotational and vibrational motions. In many circumstances it is small, so rotational and vibrational energies can be considered separately. Of course, the internal energy of an ideal rigid body is all rotational, but a more realistic model of a solid body is obtained by including the other terms in (1.39). The vibrational energy of a solid is manifested in thermodynamic as well as elastic properties of the solid.

Internal Energy and Work

To determine how the kinetic energy K evolves with time, we differentiate it and use the equation of motion (1.1); thus,

$$\dot{K} = \frac{\mathrm{d}}{\mathrm{d}t}(\sum_i \tfrac{1}{2}m_i\dot{\mathbf{x}}_i^2) = \sum_i \dot{\mathbf{x}}_i\cdot(m_i\ddot{\mathbf{x}}_i) = \sum_i (\mathbf{F}_i - \sum_j \mathbf{f}_{ij})\cdot\dot{\mathbf{x}}_i.$$

By virtue of the 3rd law (1.2),

$$\begin{aligned}\sum_i\sum_j \mathbf{f}_{ij}\cdot\dot{\mathbf{x}}_i &= \tfrac{1}{2}\sum_i\sum_j (\mathbf{f}_{ij} - \mathbf{f}_{ji})\cdot\dot{\mathbf{x}}_i = \tfrac{1}{2}\sum_i\sum_j \mathbf{f}_{ij}\cdot(\dot{\mathbf{x}}_i - \dot{\mathbf{x}}_j)\\ &= \tfrac{1}{2}\sum_i\sum_j \mathbf{f}_{ij}\cdot\dot{\mathbf{r}}_{ij} = \sum_{i<j}\mathbf{f}_{ij}\cdot\dot{\mathbf{r}}_{ij},\end{aligned}$$

where

$$\mathbf{r}_{ij} \equiv \mathbf{x}_i - \mathbf{x}_j = \mathbf{r}_i - \mathbf{r}_j. \tag{1.40}$$

Hence, changes in the total energy are determined by the equation

$$\dot{K} = \sum_i \mathbf{F}_i\cdot\dot{\mathbf{x}}_i + \sum_{i<j}\mathbf{f}_{ij}\cdot\dot{\mathbf{r}}_{ij}. \tag{1.41}$$

Equation (1.41) can be formally integrated to get

$$\Delta K \equiv K(t_2) - K(t_1) = \sum_i \int_1^2 \mathbf{F}_i\cdot\mathrm{d}\mathbf{x}_i + \sum_{i<j}\int_1^2 \mathbf{f}_{ij}\cdot\mathrm{d}\mathbf{r}_{ij}. \tag{1.42}$$

The limits on the integrals have been abbreviated; for example,

$$\int_1^2 \mathbf{F}_i \cdot \mathrm{d}\mathbf{x}_i = \int_{\mathbf{x}_i(t_1)}^{\mathbf{x}_i(t_2)} \mathbf{F}_i \cdot \mathrm{d}\mathbf{x}_i . \tag{1.43}$$

This integral quantity is called *work*, specifically, *the work done by external forces on the ith particle* during a displacement from the position $\mathbf{x}_i(t_i)$ to the position $\mathbf{x}_i(t_2)$. The right side of (1.42) is the total work on the system, consisting of the sum of the works done on each particle by external and internal forces. Thus, the mathematical equation (1.42) can be expressed in the following words: For any system of particles, in a specified time interval *the change in total kinetic energy is equal to the total work done on the system by external and internal forces*. This is the *general work-energy theorem*. A special case of the theorem is more useful in practice, so we turn to that next.

If the internal forces are conservative, then they are derivable from a potential energy function V, that is.

$$\mathbf{f}_{ij} = -\nabla_{\mathbf{r}_{ij}} V.$$

We have already assumed that the internal forces are central, and we know from Section 4-5 that a conservative central potential can be a function only of the distance between particles. Hence,

$$V = V(r_{12}, r_{13}, r_{23}, \ldots), \tag{1.44}$$

where $r_{ij} = |\mathbf{r}_{ij}| = |\mathbf{r}_i - \mathbf{r}_j| = |\mathbf{x}_i - \mathbf{x}_j|$. In most applications the potential is a sum of 2-particle potentials V_{ij} like so,

$$V = \sum_{i<j} V_{ij}(r_{ij}).$$

However, this stronger hypothesis is unnecessary for present purposes. Now, by the chain rule,

$$\mathbf{f}_{ij} = -(\nabla_{\mathbf{r}_{ij}} r_{ij}) \frac{\partial V}{\partial r_{ij}} ,$$

Hence,

$$\dot{\mathbf{r}}_{ij} \cdot \mathbf{f}_{ij} = -(\dot{\mathbf{r}}_{ij} \cdot \nabla_{\mathbf{r}_{ij}} r_{ij}) \frac{\partial V}{\partial r_{ij}} = -\dot{r}_{ij} \frac{\partial V}{\partial r_{ij}} ,$$

and

$$\sum_{i<j} \dot{\mathbf{r}}_{ij} \cdot \mathbf{f}_{ij} = -\sum_{i<j} \dot{r}_{ij} \frac{\partial V}{\partial r_{ij}} = -\frac{\mathrm{d}V}{\mathrm{d}t} . \tag{1.45}$$

Using this in (1.41) and separating the kinetic energy into external and internal parts, we obtain

$$\dot{K}_{CM} + \dot{E} = \sum_i \mathbf{F}_i \cdot \dot{\mathbf{x}}_i , \tag{1.46}$$

where

$$E = K_{\text{int}} + V \tag{1.47}$$

is the total internal energy of the system. Thus, (1.46) describes the rate at which the internal energy is altered by external forces. In integral form,

$$\Delta K_{CM} + \Delta E = \sum_i \int_1^2 \mathbf{F}_i \cdot \mathrm{d}\mathbf{x}_i, \tag{1.48}$$

it describes the change in energy resulting from work on the system by external forces. Equation (1.48) is the most useful version of the *work-energy theorem*.

We can, however separate the changes in external and internal energies. From the *CM* equation of motion (1.10) we get a separate equation for the external kinetic energy:

$$\dot{K}_{CM} = \frac{\mathrm{d}}{\mathrm{d}t}\left(\tfrac{1}{2}M\dot{\mathbf{X}}^2\right) = M\dot{\mathbf{X}}\cdot\ddot{\mathbf{X}} = \mathbf{F}\cdot\dot{\mathbf{X}}. \tag{1.49}$$

Substituting this into (1.46), we obtain

$$\dot{E} = \sum_i \mathbf{F}_i \cdot \dot{\mathbf{x}}_i - \mathbf{F}\cdot\dot{\mathbf{X}} = \sum_i \mathbf{F}_i \cdot (\dot{\mathbf{x}}_i - \dot{\mathbf{X}}) = \sum_i \mathbf{F}_i \cdot \dot{\mathbf{r}}_i. \tag{1.50}$$

Or, in integral form,

$$\Delta E = \sum_i \int_1^2 \mathbf{F}_i \cdot \mathrm{d}\mathbf{r}_i. \tag{1.51}$$

This looks simpler than (1.48); however, the integral here is not usually as convenient as the total work integral in (1.48).

The theorem that the energy of an isolated system is conserved follows trivially from (1.48). But it has nontrivial consequences. For one thing it helps us formulate the concept of work correctly. From (1.48) alone, one might interpret the work integral on the right as a measure of energy production. However, the correct interpretation is that *work is a measure of energy transfer*. To establish this, we separate the entire universe into two parts, the system of interest and its environment, the rest of the universe. The universe is isolated, so its energy is conserved. The energy of the universe can only be redistributed among the parts of the universe. Therefore, (1.48) must describe an exchange of energy between the system and its environment.

The First Law of Thermodynamics

Thermodynamics is concerned with the transfer and storage of energy among objects. Statistical Mechanics is the branch of physics concerned with deriving the laws and equations of thermodynamics from the properites of atoms and other particles composing macroscopic objects. In a word, statistical mechanics aims to reduce thermodynamics to mechanics. Let us consider, in a qualitative way, the mechanical basis for some fundamental thermodynamic concepts.

The number of atoms in a macroscopic object is immense, about 10^{25} in a golf ball, for example. It is not only impossible to keep track of individual atoms, it is undesirable, because the glut of information would be unwieldy. The best that can be done is to express the internal energy and the work done on a macroscopic object as functions of a few macroscopic variables, such as the volume of the system, while microscopic variables are controlled only partially and indirectly. It is the job of statistical mechanics and thermodynamics to specify precisely how this is done. However, the result is a separation of the work done on a system into two parts, the work $-W$ done by altering macroscopic variables and the remainder of the work Q due to changes in microscopic variables. Thus,

$$\sum_i \int_1^2 \mathbf{F}_i \cdot d\mathbf{x}_i = Q - W.$$

So the work-energy theorem (1.48) is given the form

$$\Delta K_{CM} + \Delta E = Q - W. \tag{1.52}$$

This is the famous *first law of thermodynamics*. The term K_{CM} is often carelessly neglected in the statements of this law. But it is essential in some of the most elementary problems, for example, in determining the rise in temperature of a sliding block as frictional forces reduce its translational kinetic energy.

The negative sign appears in (1.52) because, as is customary in thermodynamics, W denotes the *macroscopic work done by the system* rather than the work done "on" the system. The term Q is commonly referred to as the *heat transferred* to the system, thereby perpetuating in language the old misconception that "heat" is a physical entity of some kind. Rather, heat transfer is a particular mode of energy transfer. It would be better to refer to Q as *microscopic work* to indicate what that mode is.

There are two distinct ways that microscopic work is performed. The first is by "thermal contact"; when two macroscopic objects are in contact, energy is transferred from one to the other by interactions among atoms at the surface of contact. The second is by "*radiation*". Electromagnetic radiation (light) may be emitted when atoms in the system collide or, according to quantum mechanics, by a spontaneous process within a single atom. Emission and absorption of light involves energy transfer between the system and the radiation field, that is, a mode of work.

In thermodynamics, when macroscopic parameters are held fixed, the internal energy of an object is expressed as a function of a single variable, the thermodynamics temperature. The temperature variable can be identified with the internal energy per particle in a perfect gas. A perfect gas is a system of noninteracting identical particles. Hence, all of its internal energy is kinetic, and the *temperature* T of the gas can be defined by

$$\tfrac{3}{2}kT = \frac{1}{N}\sum_{i=1}^{N} \tfrac{1}{2}m\dot{\mathbf{x}}_i^2 , \tag{1.53}$$

where k is known as Boltzmann's constant. The specific value of Boltzmann's constant is of no concern to us here. Boltzmann's constant is merely a conversion factor changing the temperature unit into the energy unit; it is a relic of times before it was realized that temperature is a measure of energy.

The internal energy of a perfect gas provides a standard to which the internal energy E of any other macroscopic object can be compared. This leads to an expression $E = E(T)$ for the object's internal energy as a function of temperature. The function $E = E(T)$ compares the energy of the object to the energy of a perfect gas under "equivalent conditions". To be sure, the perfect gas is an imperfect model of a real gas, just as the rigid body is an imperfect model of a solid body. Nevertheless, the perfect gas provides a theoretical standard for measurements of internal energy, just as the rigid body provides a standard for measurements of length.

Open Systems

So far we have considered systems composed of a definite set of particles. Such systems are said to be *closed*. An *open system* is one which is free to exchange particles with its surroundings. It can be defined as a set of particles within some specified spatial region, usually a region enclosed by the boundaries of some macroscopic object. All macroscopic objects are open systems, but often the rate at which they exchange particles with their surroundings is so small that they can be regarded as closed systems. To handle systems for which particle exchange is significant, our theorems for closed systems must be generalized. The full generalization is best carried out within the domain of continuum mechanics, so only a special case will be considered here.

Suppose that in a short time Δt a body with small mass ΔM and velocity $\mathbf{U}$ coalesces with a larger body of mass M and velocity $\mathbf{V}$, as shown in Figure 1.1. This is an inelastic collision, so energy of the macroscopic motion is not conserved; most of it is converted to internal energy of the body. Let us suppose also that the collision imparts no significant angular momentum to the body. Now, mass is conserved in the collision, and, in the absence of external forces, so is momentum. Hence,

$$M\mathbf{V} + \Delta M\mathbf{U} = (M + \Delta M)(\mathbf{V} + \Delta\mathbf{V}) , \tag{1.54}$$

where $\Delta\mathbf{V}$ is the change in velocity of the larger body, which we regard as an

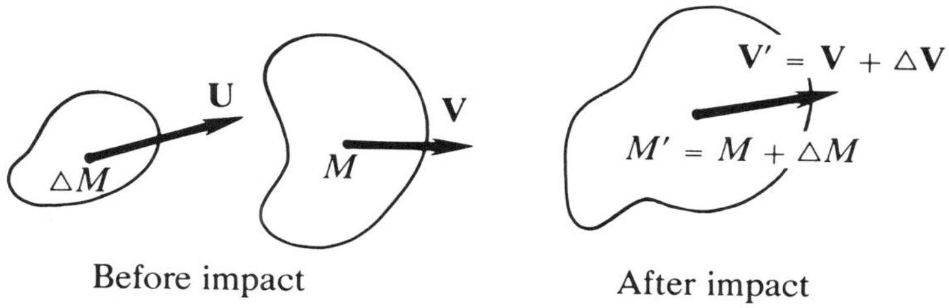

Fig. 1.1. Inelastic collision.

open system. We can pass to the case of a system accreting mass continuously by dividing (1.54) by Δt and passing to the limit $\Delta t = 0$. Thus, we find

$$M\dot{\mathbf{V}} + \dot{M}(\mathbf{V} - \mathbf{U}) = \frac{\mathrm{d}}{\mathrm{d}t}(M\mathbf{V}) - \dot{M}\mathbf{U} = 0 . \tag{1.55}$$

Evidently this equation generalizes to

$$\frac{\mathrm{d}}{\mathrm{d}t}(M\mathbf{V}) = \mathbf{F} + \dot{M}\mathbf{U} \tag{1.56}$$

if the system is subject to an external force $\mathbf{F}$.

Equation (1.56) describes the rate of momentum change in an open system (on the left side) as a result of momentum transfer from its surroundings (on the right side). The momentum transfer is accomplished in two ways, by the action of an external force and by momentum flux. The term $\dot{M}\mathbf{U}$ is called the *momentum flux*, because it describes the rate that momentum is carried into the system by particles crossing the boundary of the system.

Example 1.1

As an example of an open system, consider a raindrop falling through a stationary cloud. It will accrete mass at a rate proportional to its velocity V and cross-sectional area πr^2. For spherical raindrops, the mass M is proportional to r^3, hence, the rate of accretion is described by

$$\dot{M} = cM^{2/3}V, \tag{1.57}$$

where c is a constant. If x is the displacement of the drop in time t, then $V = \dot{x}$ and $\dot{M} = V\,\mathrm{d}M/\mathrm{d}x$, so

$$\frac{\mathrm{d}M}{\mathrm{d}x} = cM^{2/3} .$$

The variables are separable, so we can integrate to get the mass as a function of displacement:

$$M^{1/3} - M_0^{1/3} = \frac{c}{3}x, \tag{1.58}$$

where M_0 is the initial mass.

In accordance with (1.56), if resistive forces are neglected, the equation of motion of a vertically falling raindrop is

$$\frac{\mathrm{d}}{\mathrm{d}t}(MV) = Mg. \tag{1.59}$$

Since M is given as a function of x by (1.58), a change of the independent variable from t to x is indicated. Multiplying (1.59) by M, we obtain

$$M\frac{\mathrm{d}}{\mathrm{d}t}(MV) = \frac{1}{2}\frac{\mathrm{d}}{\mathrm{d}x}(MV)^2 = M^2g.$$

Hence,

$$M^2V^2 - M_0^2V_0^2 = 2g\int_0^x M^2\,\mathrm{d}x. \tag{1.60}$$

This is readily integrated after inserting (1.58), but the result is much cleaner if we neglect M_0 in relation to M, whence

$$x^6V^2 = 2g\int_0^x x^6\,\mathrm{d}x = \frac{2g}{7}x^7,$$

so

$$V^2 = \frac{2g}{7}x. \tag{1.61}$$

Differentiating with respect to time, we get

$$\ddot{x} = \frac{g}{7}, \tag{1.62}$$

from which we can easily find the time dependence of the motion.

Example 1.2

Rocket propulsion affords another example of open system dynamics. In this case mass is expelled from the system instead of accreted, so (1.56) applies with $\dot{M}$ negative instead of positive. For a rocket in a uniform gravitational field, the equation of motion (1.56) can be put in the form

$$M\dot{\mathbf{V}} = \dot{M}(\mathbf{U} - \mathbf{V}) + M\mathbf{g}. \tag{1.63}$$

Here the vector $\mathbf{e} = \mathbf{U} - \mathbf{V}$ is the *exhaust velocity*, the average velocity of exhaust gases relative to the rocket. For constant $\mathbf{e}$, (1.63) integrates to

$$\mathbf{V} - \mathbf{V}_0 = \mathbf{e}\log\frac{M}{M_0} + \mathbf{g}t, \tag{1.64}$$

where $M_0 = M(0)$ is the initial rocket mass. Of course the time dependence of the mass $M = M(t)$ is determined by the *burning rate* $\dot{M}$ programmed in the rocket. Once this is specified, the time dependence of the velocity is determined by (1.64), and the displacement of the rocket can be found by direct integration.

The reader should be cautioned against the mistaken assumption that the velocity $\mathbf{V}$ in our general equation of motion (1.56) is necessarily equal to the center of mass velocity $\dot{\mathbf{X}}$. To understand the difference between $\mathbf{V}$ and $\dot{\mathbf{X}}$, consider the momentum of the system.

$$M\mathbf{V} = \sum_i m_i\dot{\mathbf{x}}_i . \tag{1.65}$$

If internal motion is negligible then the system can be regarded as a rigid body and every particle has the same velocity $\dot{\mathbf{x}}_i = \mathbf{V}$. However, from an open system like a rocket, particles are suddenly expelled, with consequent shifts in the rocket's center of mass. Thus, the difference between $\dot{\mathbf{X}}$ and $\mathbf{V}$ is due to motion of $\dot{\mathbf{X}}$ within the body as a result of mass flux through the boundary. Therefore, the error made by computing displacement under the assumption that $\dot{\mathbf{X}} = \mathbf{V}$ cannot exceed the dimensions of the body. Usually we are more interested in the motion of particles in a body than in the motion of center of mass, so $\mathbf{V}$ may be of more interest than $\dot{\mathbf{X}}$. However, for a system with significant internal motion, such as a spinning body, we cannot identify $\mathbf{V}$ with the velocity of the individual particles, so the relation of $\mathbf{V}$ to $\dot{\mathbf{X}}$ is important.

6-1. Exercises

(1.1) Justify the interpretation of force as the rate of momentum transfer from one system to another and of torque as the rate of angular momentum transfer.

(1.2) For a closed system of particles with internal energy E in a conservative external field of force $\mathbf{F}(\mathbf{x}) = -\nabla V(\mathbf{x})$, show that the total energy

$$\tfrac{1}{2}M\dot{\mathbf{X}}^2 + \sum_i V(\mathbf{x}_i) + E$$

is a constant of the motion, reducing to

$$\tfrac{1}{2}M\dot{\mathbf{X}}^2 + M\mathbf{X}\cdot\mathbf{g} + E$$

in a uniform gravitational field.

(1.3) A uniform flexible chain of mass m and length a is initially at rest on a smooth table with one end just hanging over the side. How long will it take the chain to slide off the table and how much energy has been dissipated in this time? What will its velocity be when it loses contact with the table?

(1.4) A rocket launched from rest is programmed to maintain a constant exhaust velocity **e** and burning rate $\dot{M} = -k$ until its fuel is exhausted. The mass of the fuel is a fraction f of the initial rocket mass M_0. Neglecting air resistance and assuming a constant gravitational acceleration **g**, show that the maximum height h attained is

$$h = \frac{1}{2g}[\hat{\mathbf{g}}\mathbf{e}\log(1-f)]^2 + \hat{\mathbf{g}}\cdot\mathbf{e}\frac{M_0}{k}[f + \log(1-f)].$$

(1.5) For an open system with momentum $M\mathbf{V}$, continuously accreting mass with momentum flux $\dot{M}_+\mathbf{U}_+$ and ejecting mass with momentum flux $-\dot{M}_-\mathbf{U}_-$, show that momentum conservation leads to the equation of motion

$$M\dot{\mathbf{V}} = \mathbf{F} + \dot{M}_+(\mathbf{U}_+ - \mathbf{V}) - \dot{M}_-(\mathbf{U}_- - \mathbf{V}),$$

while mass conservation gives

$$\dot{M} = \dot{M}_+ - \dot{M}_-.$$

(1.6) An open-topped freight car of mass M is initially coasting on smooth rails with speed v_0. Rain is falling vertically.

(a) Determine the speed v of the car after a mass m of rain water has accumulated in the car.

(b) If the water leaks out as fast as it enters, determine the speed of the car after a mass m of rain water has passed through it.

6-2. The Method of Lagrange

The bookkeeping required to describe and analyze the motion of an N-particle system can often be simplified by a judicious choice of variables. Lagrange developed a systematic method for describing a system in terms of an arbitrary set of variables. In Section 3-10 the method was explained in detail for a 1-particle system, and its generalization to an N-particle system is straightforward, so we can treat it concisely.

In the Newtonian approach, an N-particle system is described by specifying the position $\mathbf{x}_i = \mathbf{x}_i(t)$ of each particle as a function of time. To describe the system, instead, in terms of some set of scalar variables $\{ q_\alpha = q_\alpha(t); \alpha = 1, 2, \ldots, n\}$, we must express the positions as functions of the new variables,

$$\mathbf{x}_i = \mathbf{x}_i(q_1, q_2, \ldots, q_n; t), \tag{2.1}$$

where $i = 1, \ldots, N$. The scalar variables q_α are called *generalized coordinates*.

The set of particle positions $\{\mathbf{x}_1, \mathbf{x}_2, \ldots, \mathbf{x}_N\}$ is called a *configuration* of the system. Since each of the N position vectors $\mathbf{x}_i$ is a vector in a 3-dimensional space, it takes $n = 3N$ generalized coordinates q_α to specify all possible configurations of an N-particle system. Therefore, the q_α are coordinates of a point $\mathbf{q} = \mathbf{q}(t)$ in a $3N$-dimensional space. This space is called *configuration space*. The motion of an entire system is thus described by a single trajectory $\mathbf{q} = \mathbf{q}(t)$ in the $3N$-dimensional configuration space instead of a set of N trajectories $\mathbf{x}_i = \mathbf{x}_i(t)$ in the 3-dimensional position space. This helps us apply our intuition and knowledge of the dynamics of a single particle to the dynamics of a many-particle system. It is an important conceptual advantage of Lagrange's method.

It may happen that the position variables $\mathbf{x}_i$ are related by *holonomic constraints* specified by K scalar equations

$$\phi_J(\mathbf{x}_1, \mathbf{x}_2, \ldots, \mathbf{x}_N; t) = 0, \tag{2.2}$$

where $J = 1, 2, \ldots, K$. We saw in Section 3–10 that a holonomic constraint on a single particle confines the particle trajectory to a 2-dimensional surface in position space. Similarly, for an N-particle system each equation of constraint (2.2) determines a $(3N - 1)$-dimensional surface in configuration space and confines the trajectory $\mathbf{q} = \mathbf{q}(t)$ of the system to that surface. The set of K constraints confines the system to a $(3N - K)$-dimensional surface. Consequently, we can use the equations of constraint to eliminate K variables and specify the system by $n = 3N - K$ independent *generalized coordinates* $q_1, \ldots, q_n$. Such independent coordinates are sometimes called *degrees of freedom*, so $n = 3N - K$ is the *number of degrees of freedom* of the system.

Our problem now is to convert the Newtonian equations of motion in position space to an equation of motion for the system in configuration space. Newton's equation for the i-th particle can be put in the form

$$m_i \ddot{\mathbf{x}}_i = -\nabla_{\mathbf{x}_i} V + \mathbf{f}_i + \mathbf{N}_i, \tag{2.3}$$

where $V = V(\mathbf{x}_1, \mathbf{x}_2, \ldots, \mathbf{x}_N, t)$ is the potential for conservative forces, $\mathbf{f}_i = \mathbf{f}_i(\mathbf{x}_1, \ldots, \mathbf{x}_N; \dot{\mathbf{x}}_1, \ldots, \dot{\mathbf{x}}_N, t)$ is the force function for nonconservative forces, and $\mathbf{N}_i$ is the resultant force of constraint.

The J-th equation of constraint (2.2) determines a constraining force $\lambda_J \nabla_{\mathbf{x}_i} \phi_J$, which can be interpreted as the force required to keep the i-th particle on the J-th surface of constraint. The resultant constraining force is therefore

$$\mathbf{N}_i = \sum_J \lambda_J \nabla_{\mathbf{x}_i} \phi_J = \nabla_{\mathbf{x}_i} \Big(\sum_J \lambda_J \phi_J \Big). \tag{2.4}$$

The equations of constraint $\phi_J = \phi_J(\mathbf{x}_1, \ldots, \mathbf{x}_N, t) = 0$ are among the givens of the problem, however the scalars λ_J are among the unknowns and must be obtained from the solution if the constraining force is to be found.

Since the q_α are taken to be variables independent of the constraint equations, we must have $\partial_{q_\alpha}\phi_J = 0$, which by application of the chain rule, gives us K equations

$$\partial_{q_\alpha}\phi_J \equiv \frac{\partial\phi_J}{\partial q_\alpha} = \sum_i (\partial_{q_\alpha}\mathbf{x}_i)\cdot\nabla_{\mathbf{x}_i}\phi_J = 0. \tag{2.5}$$

These equations can be used to eliminate the constraining forces from the equations of motion; for

$$\sum_i \mathbf{N}_i\cdot(\partial_{q_\alpha}\mathbf{x}_i) = \sum_J \lambda_J \sum_i (\partial_{q_\alpha}\mathbf{x}_i)\cdot\nabla_{\mathbf{x}_i}\phi_J = 0,$$

hence, from (2.3) we get

$$\sum_i (m_i\ddot{\mathbf{x}}_i + \nabla_{\mathbf{x}_i}V - \mathbf{f}_i)\cdot(\partial_{q_\alpha}\mathbf{x}_i) = 0, \tag{2.6}$$

for $\alpha = 1, 2, \ldots, n = 3N - K$.

The n equations (2.6) can be re-expressed as equations of motion for the q_α by employing the chain rule for differentiation.

Thus, from (2.1) we obtain

$$\dot{\mathbf{x}}_i = \sum_\alpha \dot{q}_\alpha\partial_{q_\alpha}\mathbf{x}_i + \partial_t\mathbf{x}_i. \tag{2.7}$$

Differentiating this, we establish

$$\partial_{\dot{q}_\alpha}\dot{\mathbf{x}}_i = \partial_{q_\alpha}\mathbf{x}_i,$$

$$\partial_{q_\alpha}\dot{\mathbf{x}}_i = \frac{\mathrm{d}}{\mathrm{d}t}(\partial_{q_\alpha}\mathbf{x}_i).$$

Hence, the first term in (2.6) can be brought into the form

$$\begin{aligned}\sum_i m_i\ddot{\mathbf{x}}_i\cdot(\partial_{q_\alpha}\mathbf{x}_i) &= \sum_i \left\{\frac{\mathrm{d}}{\mathrm{d}t}(m_i\dot{\mathbf{x}}_i\cdot(\partial_{q_\alpha}\mathbf{x}_i)) - m_i\dot{\mathbf{x}}_i\cdot\left(\frac{\mathrm{d}}{\mathrm{d}t}\partial_{q_\alpha}\mathbf{x}_i\right)\right\}\\ &= \sum_i \left\{\frac{\mathrm{d}}{\mathrm{d}t}\partial_{\dot{q}_\alpha}(\tfrac{1}{2}m_i\dot{\mathbf{x}}_i^2) - \partial_{q_\alpha}(\tfrac{1}{2}m_i\dot{\mathbf{x}}_i^2)\right\}\\ &= \frac{\mathrm{d}}{\mathrm{d}t}(\partial_{\dot{q}_\alpha}K) - \partial_{q_\alpha}K,\end{aligned} \tag{2.8}$$

where

$$K = \sum_i \tfrac{1}{2}m_i\dot{\mathbf{x}}_i^2 \tag{2.9}$$

is the total kinetic energy of the system.

By substitution, the potential can be expressed as a function of the generalized coordinates

$$V(\mathbf{x}_1(q_1, \ldots, q_n; t), \ldots, \mathbf{x}_N(q_1, \ldots, q_n; t)) \equiv V(q_1, \ldots, q_n; t). \tag{2.10}$$

Hence, the second term in (2.6) becomes

$$\sum_i (\partial_{q_\alpha}\mathbf{x}_i) \cdot \nabla_{\mathbf{x}_i} V = -\,\partial_{q_\alpha} V. \tag{2.11}$$

This can be combined with the first term (2.8), giving

$$\sum_i (m_i \ddot{\mathbf{x}}_i + \nabla_{\mathbf{x}_i} V) \cdot (\partial_{q_\alpha}\mathbf{x}_i) = \frac{\mathrm{d}}{\mathrm{d}t}(\partial_{\dot{q}_\alpha} L) - \partial_{q_\alpha} L, \tag{2.12}$$

where we have introduced a new function

$$L = K - V = L(\dot{q}_1, \ldots, \dot{q}_n; q_1, \ldots, q_n; t) \tag{2.13}$$

called the *Lagrangian* of the system.

For the third term in (2.6) we introduce the notation

$$F_\alpha = F_\alpha(q_1, \ldots, q_n; q_1, \ldots, q_n; t) \equiv \sum_i \mathbf{f}_i \cdot (\partial_{q_\alpha}\mathbf{x}_i). \tag{2.14}$$

The quantity F_α is called the q_α-component of the *generalized force*. The right side of (2.14) shows that F_α can be interpreted as the component of force on the system in the "direction" of a change in q_α; this "direction" is a direction in configuration space rather than position space.

Using (2.12) and (2.14), we get (2.6) finally in the form

$$\frac{\mathrm{d}}{\mathrm{d}t}(\partial_{\dot{q}_\alpha} L) - \partial_{q_\alpha} L = F_\alpha \tag{2.15}$$

where $\alpha = 1, \ldots, n$. These are called *Lagrange's equations* for the system. These are the desired equations of motion for the system in terms of generalized coordinates.

Before Lagrange's equations can be used, the Lagrangian and the generalized force must be expressed in terms of the generalized coordinates. For the potential this is done by simple substitution, as in (2.10). To do it for the kinetic energy, we must use the chain rule; thus, from

$$K = \tfrac{1}{2} \sum_i \Big(\sum_\alpha \dot{q}_\alpha \partial_{q_\alpha}\mathbf{x}_i + \partial_t \mathbf{x}_i\Big)^2,$$

we get the kinetic energy $K = K(\dot{q}_1, \ldots, \dot{q}_n; q_1, \ldots, q_n; t)$ in the form

$$K = \tfrac{1}{2} \sum_{\alpha,\beta} M_{\alpha\beta}\, \dot{q}_\alpha\, \dot{q}_\beta + \sum_\alpha B_\alpha \dot{q}_\alpha + C, \tag{2.16}$$

where

$$M_{\alpha\beta} \equiv \sum_i m_i (\partial_{q_\alpha}\mathbf{x}_i) \cdot (\partial_{q_\beta}\mathbf{x}_i) = M_{\alpha\beta}(q_1, \ldots, q_n; t), \tag{2.17a}$$

$$B_\alpha \equiv \sum_i (\partial_{q_\alpha}\mathbf{x}_i) \cdot (\partial_t \mathbf{x}_i) = B_\alpha(q_1, \ldots, q_n; t), \tag{2.17b}$$

$$C \equiv \frac{1}{2} \sum_i (\partial_t \mathbf{x}_i)^2 = C(q_1, \ldots, q_n; t). \tag{2.17c}$$

The explicit time-dependence of $\mathbf{x}_i = \mathbf{x}_i(q_1, \ldots, q_n; t)$ results only from time dependent constraints. Consequently, for time independent constraints (or no constraints at all) the kinetic energy assumes the form

$$K = \frac{1}{2} \sum_{\alpha, \beta} M_{\alpha\beta}(q_1, \ldots, q_n)\dot{q}_\alpha \dot{q}_\beta, \tag{2.18}$$

where the $M_{\alpha\beta}$ are to be obtained from (2.17a).

By deriving Lagrange's equations, we have carried out once and for all the steps required to introduce any set of coordinates into Newton's equations and eliminate the holonomic forces of constraint. From now on we can avoid those steps by constructing Lagrange's equations straightaway. This is *Lagrange's method*.

Lagrange's Method can be summarized as a series of steps for attacking a given dynamical problem:

Step I

Express any holonomic constraints in parametric form by determining the particle positions $\mathbf{x}_i$ as explicit functions $\mathbf{x}_i(q_1, \ldots, q_n; t)$ of an appropriate set of *independent generalized coordinates*. Diagrams are often a valuable guide to the selection of coordinates. Sometimes it is best to begin with dependent coordinates and then eliminate some of them by applying constraints in the nonparametric form (2.3). The best set of coordinates is usually determined by symmetry properties of the potential energy function, but it may be difficult to find.

Step II

Express the Lagrangian and, if needed, the generalized force as explicit functions of the generalized coordinates q_α and their velocities $\dot{q}_\alpha$.

Step III

Solve Lagrange's equations. The equations may be quite complicated even for fairly simple problems. No single mathematical method suffices to handle all problems.

Example 2.1

Atwood's Machine consists of a pair of "weights" connected by an inextensible string passing over a pulley as shown in Figure 2.1. In this case the parametric equations for the positions of the weights are so simple that it is

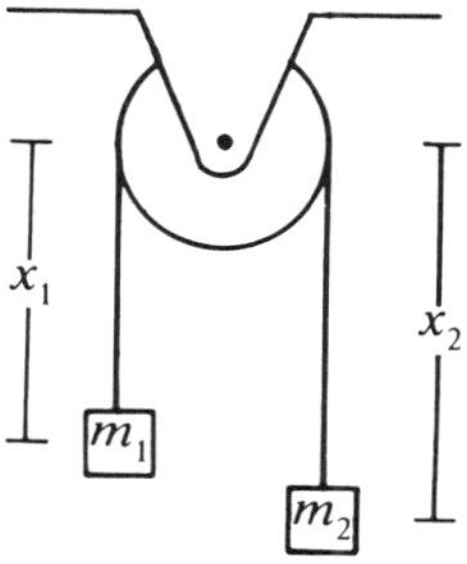

Fig. 2.1. Atwood's Machine

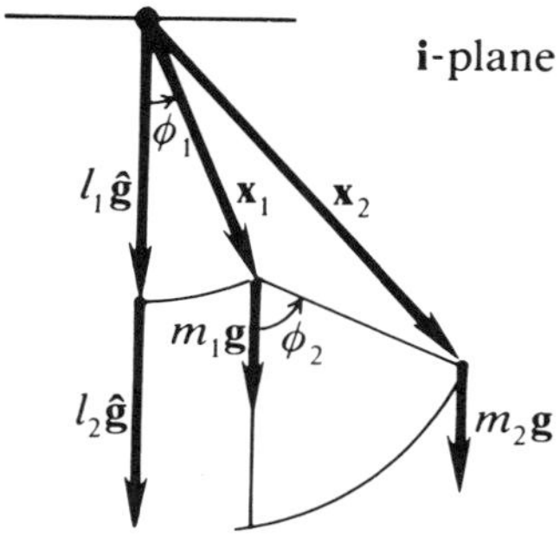

Fig. 2.2. The double pendulum.

unnecessary to write them down. We choose the vertical displacements x_1 and x_2 for generalized coordinates. So, neglecting friction and the masses of the string and pulley, the kinetic energy of the system is

$$K = \tfrac{1}{2}m_1\dot{x}_1^2 + \tfrac{1}{2}m_2\dot{x}_2^2,$$

while the potential energy is

$$V = -m_1gx_1 - m_2gx_2.$$

Since the string is inextensible, the coordinates are related by

$$x_1 + x_2 = C,$$

where C is a constant. Using this to eliminate one of the coordinates, we obtain the Lagrangian

$$L = \tfrac{1}{2}(m_1 + m_2)\dot{x}_1^2 + (m_1 - m_2)gx_1 + m_2gC.$$

Inserting this in Lagrange's equation

$$\frac{\mathrm{d}}{\mathrm{d}t}(\partial_{\dot{x}_1}L) - \partial_{x_1}L = 0,$$

we obtain the equation of motion

$$(m_1 + m_2)\ddot{x}_1 = (m_1 - m_2)g.$$

Note that the direction of the acceleration depends on the relative magnitudes of the masses.

Example 2.2

The *double pendulum* consists of one simple pendulum attached to the end of another. This is a two particle system subject to rigid constraints. Let the parameters of the system be specified by Figure 2.2. The plane of oscillation is specified algebraically by a unit bivector $\mathbf{i}$. With angles ϕ_1 and ϕ_2 as generalized coordinates, the parametric equations for the positions of the particles are

$$\mathbf{x}_1 = l_1\hat{\mathbf{g}}e^{\mathbf{i}\phi_1}$$

$$\mathbf{x}_2 = \hat{\mathbf{g}}(l_1e^{\mathbf{i}\phi_1} + l_2e^{\mathbf{i}\phi_2}).$$

Whence, the kinetic energy is put in the form

$$\begin{aligned} K &= \tfrac{1}{2}m_1\dot{\mathbf{x}}_1^2 + \tfrac{1}{2}m_2\dot{\mathbf{x}}_2^2 \\ &= \tfrac{1}{2}(m_1 + m_2)l_1^2\dot{\phi}_1^2 + \tfrac{1}{2}m_2(l_2^2\dot{\phi}_2^2 + 2l_1l_2\dot{\phi}_1\dot{\phi}_2\cos(\phi_2 - \phi_1)). \end{aligned}$$

The potential energy is given the form

$$\begin{aligned} V &= -m_1\mathbf{g}\cdot\mathbf{x}_1 - m_2\mathbf{g}\cdot\mathbf{x}_2 \\ &= -g\langle m_1l_1e^{\mathbf{i}\phi_1} + m_2(l_1e^{\mathbf{i}\phi_1} + l_2e^{\mathbf{i}\phi_2})\rangle_0 \\ &= -(m_1 + m_2)gl_1\cos\phi_1 - m_2gl_2\cos\phi_2. \end{aligned}$$

Forming the Lagrangian $L = K - V$ and substituting this into Lagrange's equation

$$\frac{\mathrm{d}}{\mathrm{d}t}(\partial_{\dot{\phi}_1} L) - \partial_{\phi_1} L = 0,$$

we obtain

$$\begin{aligned} &\frac{\mathrm{d}}{\mathrm{d}t}[(m_1 + m_2)l_1^2\dot{\phi}_2 + m_2l_1l_2\dot{\phi}_1\cos(\phi_2 - \phi_1)] \\ &\qquad + (m_1 + m_2)l_1g\sin\phi_1 + m_2l_1l_2\dot{\phi}_1\dot{\phi}_2\sin(\phi_2 - \phi_1) = 0. \end{aligned}$$

In a similar way we obtain

$$\begin{aligned} &\frac{\mathrm{d}}{\mathrm{d}t}[m_2l_2^2\dot{\phi}_2 + m_1l_1l_2\dot{\phi}_1\cos(\phi_2 - \phi_1)] \\ &\qquad + m_2l_2g\sin\phi_2 - m_2l_1l_2\dot{\phi}_1\dot{\phi}_2\sin(\phi_2 - \phi_1) = 0. \end{aligned}$$

For small oscillations, these equations reduce to

$$(m_1 + m_2)l_1^2\ddot{\phi}_1 + m_2l_1l_2\ddot{\phi}_2 + (m_1 + m_2)l_1g\phi_1 = 0,$$

$$m_2l_2^2\ddot{\phi}_2 + m_2l_1l_2\ddot{\phi}_1 + m_2l_2g\phi_2 = 0.$$

These are equations of motion for a pair of coupled harmonic oscillators. They can be solved by a change of variables that decouples the two equations. A systematic method for doing this will be developed in Section 6-4.

Now suppose we wish to account for the effect of air resistance on the motion of a double pendulum. The resistive force on each particle is proportional to its velocity, that is,

$$\mathbf{f}_1 = -\mu_1 \dot{\mathbf{x}}_1, \quad \mathbf{f}_2 = -\mu_2 \dot{\mathbf{x}}_2,$$

where μ_1 and μ_2 are positive constants. Therefore, the components of the generalized force (2.14) take the form

$$\begin{aligned} F_1 &= \mathbf{f}_1 \cdot (\partial_{\phi_1} \mathbf{x}_1) + \mathbf{f}_2 \cdot (\partial_{\phi_1} \mathbf{x}_2) \\ &= -\mu_1 \dot{\phi}_1 (\partial_{\phi_1} \mathbf{x}_1)^2 - \mu_2 (\dot{\phi}_1 \partial_{\phi_1} \mathbf{x}_2 + \dot{\phi}_2 \partial_{\phi_2} \mathbf{x}_2) \cdot (\partial_{\phi_1} \mathbf{x}_1) \\ &= -\mu l_1^2 \dot{\phi}_1 - \mu_2 (l_1^2 \dot{\phi}_1 + l_1 l_2 \dot{\phi}_2 \cos(\phi_2 - \phi_1)), \\ F_2 &= -(\dot{\phi}_1 \partial_{\phi_1} \mathbf{x}_2 + \dot{\phi}_2 \partial_{\phi_2} \mathbf{x}_2) \cdot (\partial_{\phi_2} \mathbf{x}_2) \\ &= -\mu_2 (\dot{\phi}_1 l_1 l_2 \cos(\phi_1 - \phi_2) + l_2^2 \dot{\phi}_2). \end{aligned}$$

Including these in the equations of motion, we get, in the small angle approximation,

$$(m_1 + m_2) l_1^2 \ddot{\phi}_1 + m_2 l_1 l_2 \ddot{\phi}_2 + (m_1 + m_2) l_1 g \phi_1 = -(\mu_1 + \mu_2) l_1^2 \dot{\phi}_1 - \mu_2 l_1 l_2 \dot{\phi}_2,$$

$$m_2 l_2^2 \ddot{\phi}_2 + m_2 l_1 l_2 \ddot{\phi}_1 + m_2 l_2 g \phi_2 = -\mu_2 l_1 l_2 \dot{\phi}_1 - \mu l_2^2 \dot{\phi}_2.$$

The solution of these equations is also discussed in Section 6-4.

Example 2.3

Consider a system of two particles connected by an inextensible, massless string of length a passing through a small hole in a table, as shown in Figure 2.3. Adopting polar coordinates r, θ in the plane of the table, the parametric equation for the position of the particle on the table is

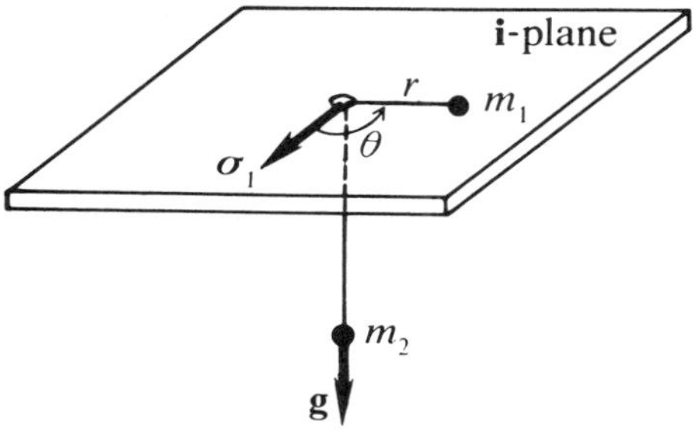

Fig. 2.3. A particle subject to a central force of constant magnitude.

$$\mathbf{x}_1 = \boldsymbol{\sigma}_1 r e^{\mathbf{i}\theta}.$$

Utilizing the constraint, the position of the suspended particle is specified by

$$\mathbf{x}_2 = \hat{\mathbf{g}}(a - r).$$

Consequently, the kinetic energy of the system is given the form

$$K = \tfrac{1}{2} m_1 (\dot{r}^2 + r^2 \dot{\theta}^2) + \tfrac{1}{2} m_2 \dot{r}^2,$$

and the potential energy is

$$V = -m_2 \mathbf{g} \cdot \mathbf{x}_2 = m_2 g (r - a)$$

so,

$$L = K - V = \tfrac{1}{2}(m_1 + m_2)\dot{r}^2 + \tfrac{1}{2} m_1 r^2 \dot{\theta}^2 - m_2 g (r - a).$$

Therefore, if friction is negligible, the equations of motion for the system are

$$\frac{d}{dt}(\partial_{\dot{r}}L) - \partial_r L = (m_1 + m_2)\ddot{r} - m_1 r\dot{\theta}^2 + m_2 g = 0,$$

$$\frac{d}{dt}(\partial_{\dot{\theta}}L) - \partial_\theta L = \frac{d}{dt}(m_1 r^2\dot{\theta}) = 0.$$

The last of these equations tells us that the angular momentum $l = m_1 r^2\dot{\theta}$ is a constant of the motion. Using this to eliminate $\dot{\theta}^2$ from the first equation, we obtain the radial equation of motion

$$(m_1 + m_2)\,\ddot{r} - \frac{l^2}{m_1 r^3} + m_2 g = 0.$$

It can be verified that $\dot{r}$ is an integrating factor for this equation by carrying out the differentiation in

$$\frac{d}{dt}\left\{\tfrac{1}{2}\,(m_1 + m_2)\dot{r}^2 + \frac{l^2}{2m_1 r^2} + mgr\right\} = 0.$$

Thus we obtain another constant of motion, the energy of the system, and we see that the orbit of the particle on the table is that of a particle subject to a conservative central force.

Ignorable Coordinates

The last example illustrates a valuable general principle. Lagrange's equation for the angle produced a constant of motion, the angular momentum, because the Lagrangian was independent of the angle. In general, for a conservative system, if

$$\partial_{q_\alpha} L = 0$$

for some coordinate q_α, then if follows trivially from Lagrange's equation (2.15) that

$$P_\alpha \equiv \partial_{\dot{q}_\alpha} L$$

is a constant of the motion. The coordinate q_α is then said to be *ignorable* or *cyclic*, because it has the following properties exhibited by the angle variable in the last example:

(1) The constant of motion P_α can be used to eliminate the q_α from the remaining equations of motion and so effectively reduce the number of variables in the problem.
(2) The coordinate q_α must be a periodic (i.e. *cyclic*) function of time.
(3) The condition $\partial_{q_\alpha} L = 0$ results from some symmetry property of the potential energy (such as its independence of angle in the example).

This suggests the possibility of a general method for choosing coordinates to simplify the equations of motion, but we shall not pursue it. In Section 6-4 we consider ways to use symmetry properties to help solve equations of motion. The systematic study of symmetries in equations of motion is a major topic in modern theoretical physics.

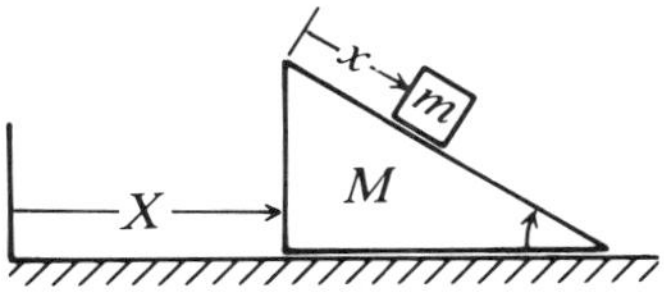

Fig. 2.4. Block sliding down a moveable inclined plane.

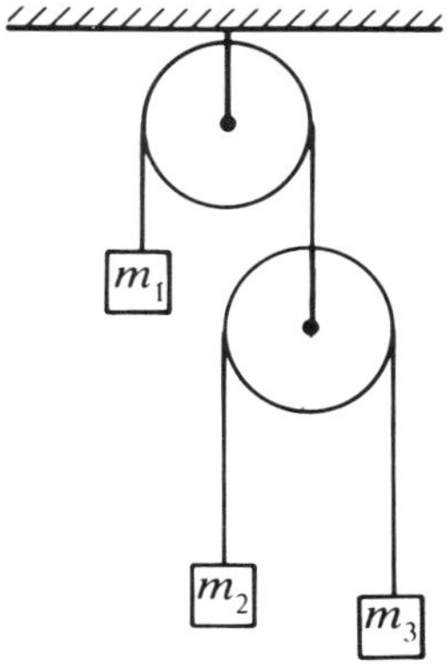

Fig. 2.5. A compound Atwood machine.

6-2. Exercises

(2.1) A block slides down the inclined plane surface of another block resting on horizontal plane, as in Figure 2.4. Assuming negligible friction, find the accelerations $\ddot{x}$ and $\ddot{X}$ of the coordinates indicated in the figure.

(2.2) Find the accelerations of the masses m_1, m_2, m_3 in the compound Atwood machine of Figure 2.5. Neglect friction and the masses of the pulleys.

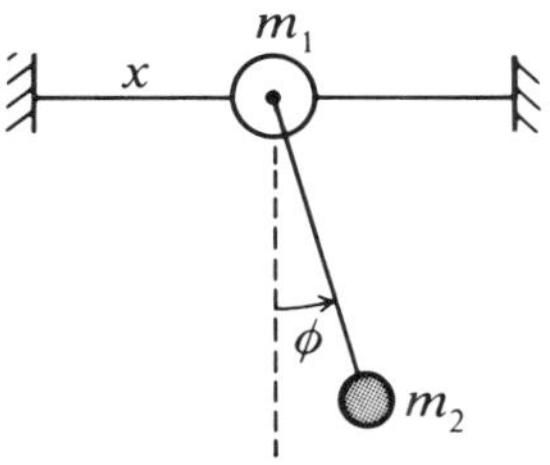

Fig. 2.6. A sliding pendulum.

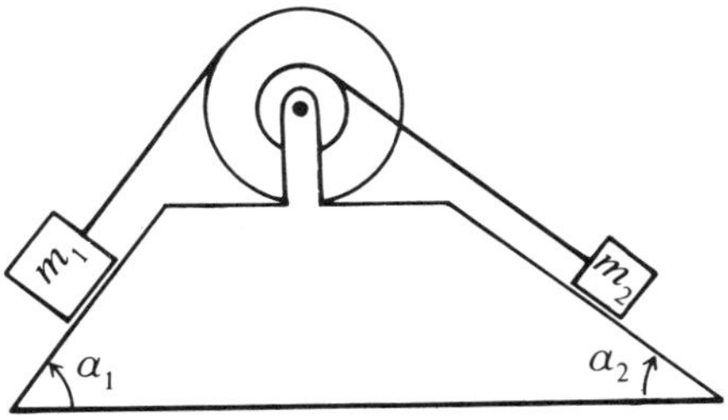

Fig. 2.7.

(2.3) A simple pendulum is suspended from a bead on a frictionless horizontal wire, as shown in Figure 2.6. Determine the equations of motion in terms of the coordinates x and ϕ.

(2.4) Two masses on smooth inclined planes are connected by string to a massless pulley consisting of two rigidly attached spools with diameters in ratio 2 to 1, as in Figure 2.7. Determine the motion of the system.

(2.5) Two simple pendulums are connected by a massless spring with stiffness constant k attached at a distance a from the supports as in Figure 2.8. The spring is unstretched when the pendulums are vertical. Determine Lagrange's equations of motion for the system.

(2.6) Two equal masses are suspended from identical massless pulleys as shown in Figure 2.9. Find their accelerations.

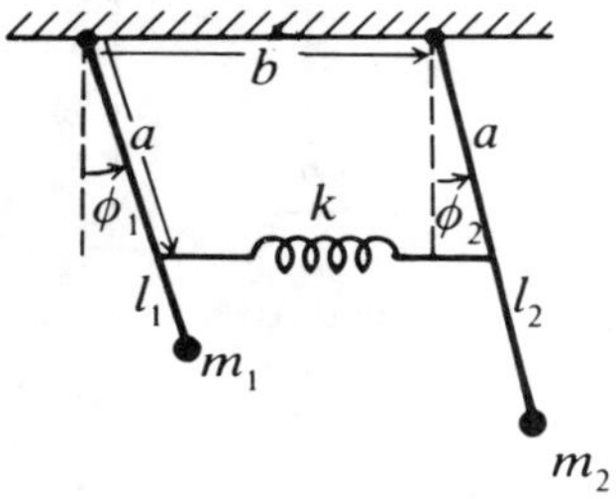

Fig. 2.8. Coupled pendulums.

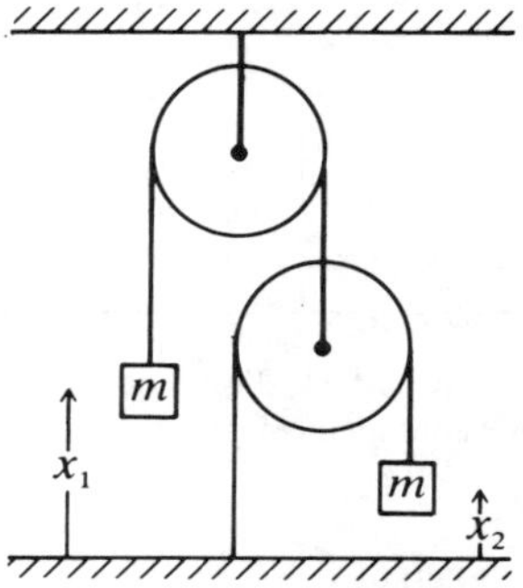

Fig. 2.9.

6-3. Coupled Oscillations and Waves

A rigid body model of a solid object consists of a system of particles with interparticle forces keeping the particles at fixed separations. A more realistic model accounts for deformations of a solid with internal forces allowing changes in interparticle separations. In an elastic solid the internal forces oppose small deformations from a stable equilibrium configuration. Therefore, by the general argument developed in Section 3-8, the interparticle restoring forces can be described by Hooke's law even without more specific knowledge about interparticle interactions. Thus, we arrive at a model of an elastic solid as a system of particles attached to their neighbors by (massless) springs, in other words, a system of coupled harmonic oscillators. The mathematical formulation of the model consists of a system of coupled linear second order differential equations. The theory of small oscillations is concerned with the analysis of such models. Before undertaking a systematic development of the theory, in this section we study the simplest examples to gain familiarity with the basic ideas.

Two Coupled Harmonic Oscillators

The main ideas in the theory of small oscillations appear in the simplest model, consisting of two identical harmonic oscillators with a linear coupling. Therefore, it will be profitable to study the ramifications of that model in detail. To have a specific physical realization of the abstract mathematical model in mind, consider an elastic string of two particles with fixed endpoints, as illustrated in Figure 3.1. The particles are connected to each other and the endpoints by (massless) springs.

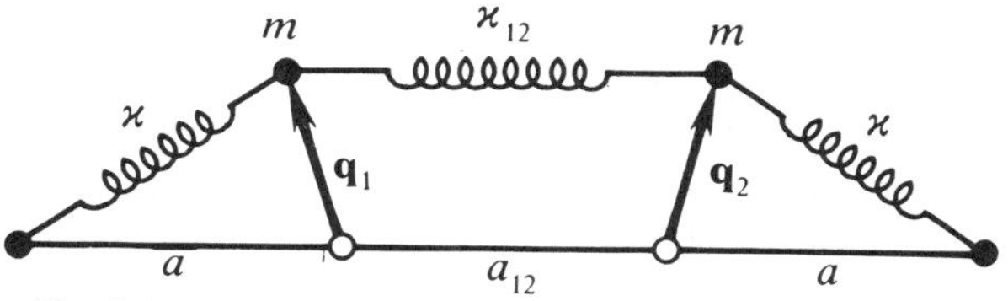

Fig. 3.1. A pair of coupled isotropic harmonic oscillators.

When the particles are at rest at the equilibrium points, the string has a uniform tension

$$T = \varkappa a = \varkappa_{12} a_{12}, \tag{3.1}$$

where $\varkappa$, $\varkappa_{12}$ are the force constants and a, a_{12} are the equilibrium lengths of the springs.

The forces on the particles are described by Hooke's law, so displacements $\mathbf{q}_1$ and $\mathbf{q}_2$ are governed by the equations of motion

$$m\ddot{\mathbf{q}}_1 = -\varkappa \mathbf{q}_1 - \varkappa_{12}(\mathbf{q}_1 - \mathbf{q}_2), \tag{3.2a}$$

$$m\ddot{\mathbf{q}}_2 = -\varkappa \mathbf{q}_2 - \varkappa_{12}(\mathbf{q}_2 - \mathbf{q}_1), \tag{3.2b}$$

We can decouple these equations by re-expressing them as equations for the "mean displacement" $\mathbf{Q}_+ = \frac{1}{2}(\mathbf{q}_1 + \mathbf{q}_2)$ and the "relative displacement" $2\mathbf{Q}_- = \mathbf{q}_1 - \mathbf{q}_2$. Thus, by adding the Equations (3.2a, b) we obtain

$$\ddot{\mathbf{Q}}_+ + \omega_+^2 \mathbf{Q}_+ = 0, \tag{3.3a}$$

where $\omega_+^2 = \varkappa/m$. The difference of Equations (3.2a, b) gives us

$$\ddot{\mathbf{Q}}_- + \omega_-^2 \mathbf{Q}_- = 0, \tag{3.3b}$$

where

$$\omega_-^2 = (\varkappa + 2\varkappa_{12})/m = \omega_+^2(1 + 2\varkappa_{12}/\varkappa). \tag{3.4}$$

We recognize (3.3a) and (3.3b) as equations for isotropic harmonic oscillators studied in Section 3-8, so we know that their solutions describe elliptical orbits and have the general mathematical form

$$\mathbf{Q}_+ = \mathbf{a}_+ \cos \omega_+ t + \mathbf{b}_+ \sin \omega_+ t, \tag{3.5a}$$

$$\mathbf{Q}_- = \mathbf{a}_- \cos \omega_- t + \mathbf{b}_- \sin \omega_- t, \tag{3.5b}$$

where $\mathbf{a}_\pm$ and $\mathbf{b}_\pm$ are constant vectors. Therefore, the displacements of the particles are superpositions

$$\mathbf{q}_1 = \mathbf{Q}_+ + \mathbf{Q}_-, \quad \mathbf{q}_2 = \mathbf{Q}_+ - \mathbf{Q}_- \tag{3.6}$$

of two harmonic oscillations with frequencies ω_+ and ω_-.

The particular combination of harmonic oscillations depends on the initial conditions. For example, initial conditions of the form $\mathbf{q}_1(0) = \mathbf{q}_2(0)$ and $\dot{\mathbf{q}}_1(0) = \dot{\mathbf{q}}_2(0)$ imply that $\mathbf{Q}_-(t) = 0$ for all t, so $\mathbf{q}_1(t) = \mathbf{Q}_+(t) = \mathbf{q}_2(t)$. Thus, the two particles oscillate *in phase* with a single frequency ω_+. On the other hand, initial conditions of the form $\mathbf{q}_1(0) = -\mathbf{q}_2(0)$ and $\dot{\mathbf{q}}_1(0) = -\dot{\mathbf{q}}_2(0)$ imply that $\mathbf{Q}_+(t) = 0$, so $\mathbf{q}_1(t) = \mathbf{Q}_-(t) = -\mathbf{q}_2(t)$. Then, the two particles oscillate *out of phase* with frequency ω_-.

Such collective oscillations with a single frequency of all particles in a system are called *normal modes* of the system. The frequencies ω_+ and ω_- of the normal modes are called *normal* (*natural* or *characteristic*) *frequencies*. The variables $\mathbf{Q}_+$ and $\mathbf{Q}_-$ for the normal modes may be called *normal coordinates*, though the term is usually reserved for scalar variables as done below. For a two particle system of coupled oscillators the normal modes are of two types, the *symmetrical mode* with coordinate $\mathbf{Q}_+$ and the *antisymmetrical mode* with coordinate $\mathbf{Q}_-$.

According to (3.4), the frequency ω_+ of the symmetrical mode is necessarily lower than the frequency ω_- of the antisymmetrical mode. This is the simplest case of a general result: In a system with any number of linearly coupled oscillators, *the mode with highest symmetry has the lowest frequency*. In a mode of lower symmetry, the springs work against each other, increasing the effective restoring force and thus producing a higher frequency.

The symmetrical and antisymmetrical modes are illustrated in Figure 3.2 for longitudinal oscillations and in Figure 3.3 for transverse oscillations. The symmetrical longitudinal and transverse modes may be regarded as different normal modes since they are linearly independent. However, in the present model, their normal frequencies are equal. Linearly independent modes with the same frequency are said to be *degenerate*. In the present case, the symmetrical normal mode is said to have a *3-fold degeneracy*, since the coordinate $\mathbf{Q}_+$ can be expressed as a linear combination of a longitudinal mode and two independent transverse modes. Similarly, the antisymmetric normal mode is *triply degenerate*.

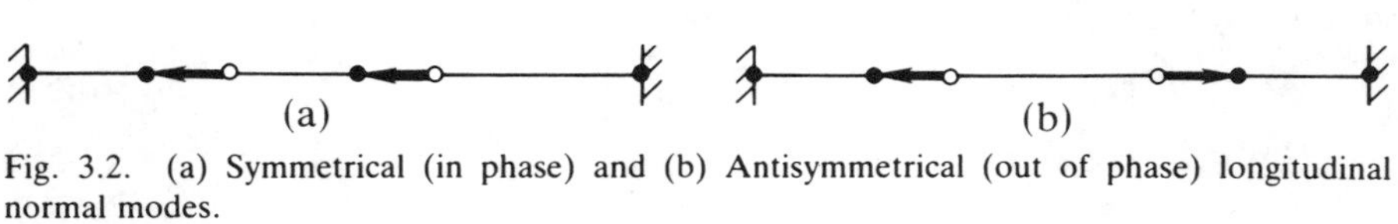

(a) (b)

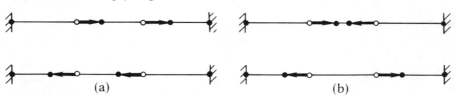

Fig. 3.2. (a) Symmetrical (in phase) and (b) Antisymmetrical (out of phase) longitudinal normal modes.

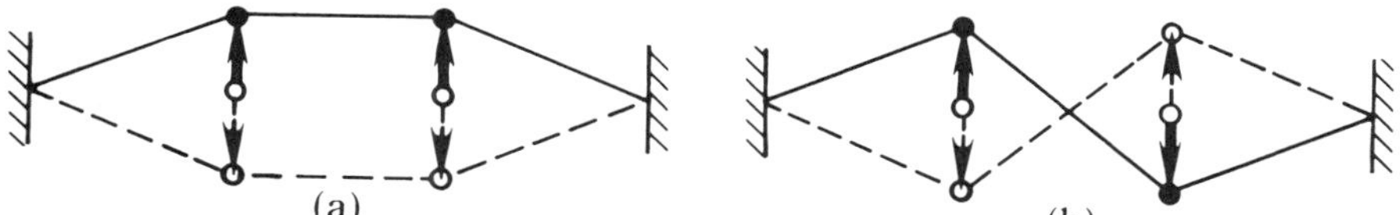

(a) (b)

Fig. 3.3. (a) Symmetrical (in phase) and (b) Antisymmetrical (out of phase) transverse normal modes.

Equation (3.5a) admits as a special case of circular solution of the form

$$\mathbf{Q}_+ = \mathbf{a}_+ e^{\mathbf{i}\omega_+ t}, \tag{3.7}$$

where $\mathbf{a}_+$ is a transverse vector in a plane with unit bivector $\mathbf{i}$. If the plane is transverse, then the mode is illustrated by Figure 3.3a, where the two particles circulate in phase about their equilibrium points. This normal mode can be expressed as a linear combination of two orthogonal transverse modes. On the other hand, if the plane contains the equilibrium points of the particles, then the mode can be expressed as a combination of a longitudinal and a transverse mode. Conversely, the linear longitudinal and transverse modes can be expressed as linear combinations of such circular modes. Of course, there are similar results for the antisymmetrical modes. Longitudinal circular normal modes are illustrated in Figure 3.4.

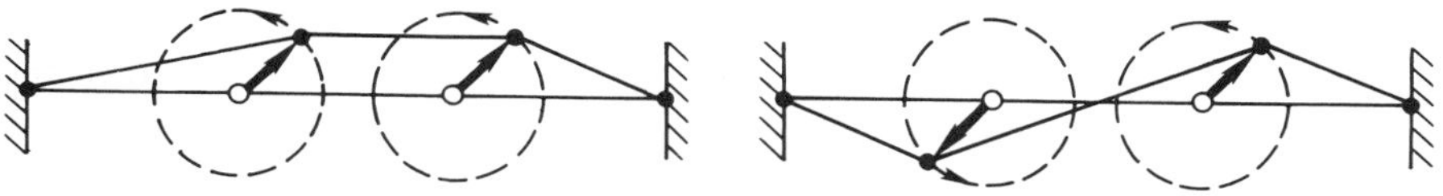

(a) (b)

Fig. 3.4. (a) Symmetrical and (b) Antisymmetrical circular normal modes in a longitudinal plane.

Taken together, the coupled equations (3.2a, b) are linear in the pair of variables $\mathbf{q}_1$ and $\mathbf{q}_2$; therefore, the *superposition principle* applies. This means that if we have two distinct solutions of the equations, then any linear combination of these solutions is also a solution. In particular, (3.6) tells us that any solution can be expressed as a linear combination of symmetrical and antisymmetrical normal modes which are themselves particular solutions. And each of these degenerate normal modes can be expressed as a combination of three linearly independent longitudinal, transverse and/or circular modes. Thus, we may select a set of six linearly independent normal modes and normalize them to unit amplitude (or energy). Let $\mathbf{a}_{nr}$ (for $n = 1, 2$, and $r = 1, 2, \ldots, 6$) be the normalized vectorial amplitude for the displacement of the nth particle in the rth normal mode. The pairs of vectors $\{\mathbf{a}_{1r}, \mathbf{a}_{2r}\}$ compose a *basis* for the six dimensional linear space of solutions to Equations (3.2a, b). Therefore, any solution of the equations can be written as the linear superposition

$$\mathbf{q}_1(t) = \sum_{r=1}^{6} \mathbf{a}_{1r}\, Q_r(t),$$

$$\mathbf{q}_2(t) = \sum_{r=1}^{6} \mathbf{a}_{2r}\, Q_r(t), \tag{3.8}$$

where the coefficients $Q_r(t)$ are scalar *normal coordinates*; they are harmonic functions which can be given the form $Q_r(t) = C_r \cos(\omega_r t + \delta_r)$. The expression (3.8) is called a *normal mode expansion*. It has the advantage of reducing apparently complex motions of the individual particles to the simple collective motions of the normal modes. A complete expansion into normal modes is not always appropriate. Often a partial expansion into degenerate modes with different frequencies is preferable. Thus, in the present case we prefer the partial expansion (3.6).

Energy Storage and Transfer by Coupled Oscillators

Since Hooke's Law is a conservative force, the total energy E of our coupled two particle system is conserved. In terms of particle coordinates

$$E = \tfrac{1}{2}m(\dot{\mathbf{q}}_1^2 + \dot{\mathbf{q}}_2^2) + \tfrac{1}{2}\varkappa(\mathbf{q}_1^2 + \mathbf{q}_2^2) + \tfrac{1}{2}\varkappa_{12}(\mathbf{q}_1 - \mathbf{q}_2)^2. \tag{3.9}$$

The last term is the "coupling energy", which may be regarded as residing in the connecting "spring", that is, in the "mutual bond" between the particles. In the absence of external interactions, the energy E remains stored in the system. The coupling between oscillators allows for a transfer of energy from one to the other, but the energy in each normal mode is separately conserved, as Equations (3.3a, b) imply. Thus, we can write

$$E_{\pm} = E_+ + E_-, \tag{3.10a}$$

where

$$E_{\pm} = \tfrac{1}{2}M(\dot{Q}_{\pm}^2 + \omega_{\pm}Q_{\pm}^2). \tag{3.10b}$$

Thus, energy can be stored independently in each normal mode.

In Section 3-9 we studied the motion of a harmonic oscillator driven by a periodic force without considering the possibility of a *feedback* effect of the oscillator on the driver. We saw that an oscillator can absorb and store energy supplied by the driver, so a feedback of energy should result from the action of the oscillator on the driver. The simplest example of such an effect is found in the present case of coupled oscillators. We can observe it by imparting all the energy initially to one of the oscillators. Thus, we strike the theoretical string by imposing the initial conditions $\dot{\mathbf{q}}_1(0) = \mathbf{v}_0$, $\dot{\mathbf{q}}_2(0) = 0$, $\mathbf{q}_1(0) = \mathbf{q}_2(0) = 0$. Then, from (3.5a, b) and (3.6) we get

$$\mathbf{Q}_{\pm}(t) = \frac{\mathbf{v}_0}{2\omega_{\pm}} \sin \omega_{\pm} t \tag{3.11}$$

and

$$\mathbf{q}_1 = \tfrac{1}{2}\mathbf{v}_0\left(\frac{\sin\omega_+ t}{\omega_+} + \frac{\sin\omega_- t}{\omega_-}\right) \tag{3.12a}$$

$$\mathbf{q}_2 = \tfrac{1}{2}\mathbf{v}_0\left(\frac{\sin\omega_+ t}{\omega_+} - \frac{\sin\omega_- t}{\omega_-}\right). \tag{3.12b}$$

To picture the motion described by these equations, we look at the limiting cases of weak and strong coupling.

For *weak coupling* ($\varkappa_{12} \ll \varkappa$), we can write the relation (3.4) in the approximate form

$$\omega_- = \omega_+(1 + 2\varepsilon), \tag{3.13}$$

where $\varepsilon = \varkappa_{12}/2\varkappa \ll 1$. In (3.12a, b), then, it is a good approximation to set $\omega_- = \omega_+$ in the denominators, but the small difference between ω_+ and ω_- cannot be neglected in the phases. Thus, we write

$$\begin{aligned}\mathbf{q}_1 &\approx \frac{\mathbf{v}_0}{2\omega_+}(\sin\omega_+ t + \sin\omega_- t)\\ &= \frac{\mathbf{v}_0}{\omega_+}\left[\cos\tfrac{1}{2}(\omega_+ - \omega_-)t\right]\sin\tfrac{1}{2}(\omega_+ + \omega_-)t\\ &\approx \frac{\mathbf{v}_0}{\omega_+}\left[\cos\varepsilon\omega_+ \mathrm{t}\right]\sin\omega_+ t,\end{aligned} \tag{3.14a}$$

and similarly,

$$\mathbf{q}_2 \approx \frac{-\mathbf{v}_0}{\omega_+}\left[\sin\varepsilon\omega_+ t\right]\cos\omega_+ t. \tag{3.14b}$$

The displacements are graphed in Figure 3.5 for $\varepsilon = 0.1$, showing the familiar phenomenon of *beats*, as the initial energy $E = \frac{1}{2}mv_0^2$ is passed back and forth between the oscillators. Each particle oscillates with frequency $\omega_+/2\pi$ while its amplitude is modulated with the lower frequency $\varepsilon\omega_+/2\pi$. The energy is transferred from one oscillator to the other in time $t = \pi/2\varepsilon\omega_+$. Complete transfer takes place even for very weak coupling.

For *strong coupling* ($\omega_- \gg \omega_+$), Equations (3.12a, b) are well approximated by

$$\mathbf{q}_1 = \frac{\mathbf{v}_0}{2\omega_+}\sin\omega_+ t = \mathbf{q}_2. \tag{3.15}$$

Thus, a blow delivered to either particle causes the two particles to move together as a single rigid body. Most of the energy delivered by the blow is stored as center of mass energy of the two particle system. Only a small fraction of it is stored in the antisymmetrical mode as "internal vibrational energy" of the system.

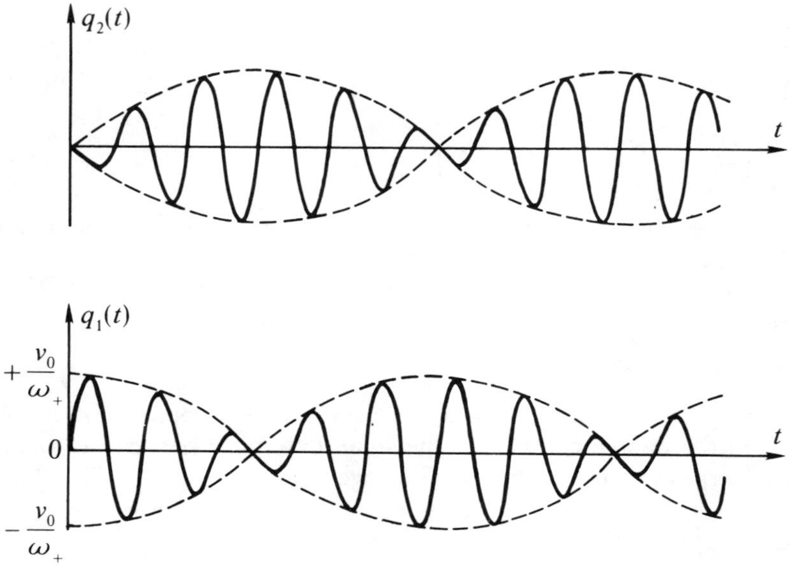

Fig. 3.5. Beats in coupled oscillations.

Note that the second oscillator can be treated as an "external agent" acting on the first oscillator by inserting the explicit expression (3.11) for $2\mathbf{Q}_- = \mathbf{q}_1 - \mathbf{q}_2$ into (3.2a) to get

$$m(\ddot{\mathbf{q}}_1 + \omega_+^2 \mathbf{q}_1) = \mathbf{a} \sin \omega_- t, \tag{3.16}$$

where $\mathbf{a}$ is a constant vector. This is the equation for an undamped, driven oscillator studied in Section 3-9, where we say that through the driving force on the right the external agent feeds energy into the oscillator. However, if the agent is another oscillator, we have seen that when its energy is depleted the direction of energy flow is reversed. Moreover, in this case resonance cannot occur since the "driving frequency" ω_- is necessarily greater than the oscillator frequency ω_+.

One-dimensional Lattice Vibrations

The simplest model of an elastic solid is a one-dimensional lattice (or string) of identical particles interacting linearly with nearest neighbors. This is a straightforward generalization of the two particle string we have just studied. Of course, the same model may be used to represent other physical systems, such as a string of macroscopic masses connected by springs or loaded on an elastic string. But it is of greatest interest in the theory of solids where, in spite of its simplicity, it has important physical implications.

We consider an N particle string with fixed end points (Figure 3.6). The equilibrium positions x_n of the particles are equally spaced with separation a

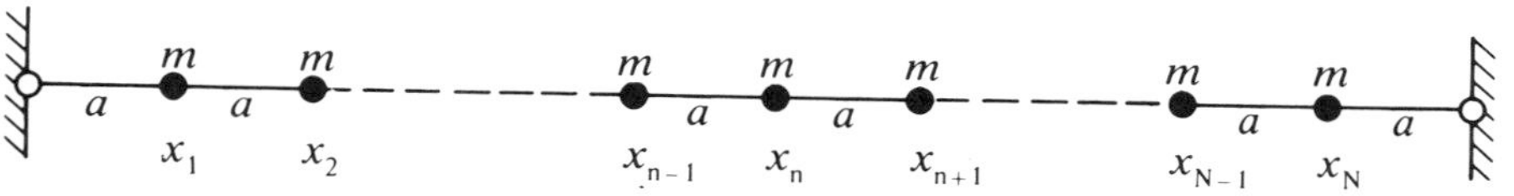

Fig. 3.6. Equilibrium positions for a 1-dimensional lattice of N identical particles.

called the *lattice constant*. The string has length $L = (N + 1)a$, and $x_n = na$ for $n = 1, 2, \ldots, N$. If we limit our considerations to transverse or longitudinal vibrations, the particle displacements from equilibrium can be represented by scalar variables $q_n = q_n(t) = q(x_n, t)$, where either the particle name n or the particle position x_n can be used to label the variables.

For particles of mass m, the equations of motion for small displacements are

$$m\ddot{q}_n = -\varkappa(q_n - q_{n-1}) - \varkappa(q_n - q_{n+1}), \tag{3.17}$$

If this is to describe a solid, then the force constant $\varkappa$ is a property of the material. If it is to describe a string under tension T, then $\varkappa$ is given by $\varkappa = T/a$. In any case, it is convenient to write the equations of motion in the form

$$\ddot{q}_n = \omega_0^2(q_{n-1} - 2q_n + q_{n+1}), \tag{3.18}$$

where $n = 1, 2, \ldots, N$, and $\omega_0^2 = \varkappa/m$. Boundary conditions at the ends of the string are imposed by writing

$$q_0 = q_{N+1} = 0. \tag{3.19}$$

In a normal mode all particles vibrate with the same frequency, so to find the normal modes we look for solutions of (3.18) with the form

$$q_n(t) = A_n e^{i\omega t}, \tag{3.20}$$

where $A_n = A(x_n)$ is constant. For the sake of algebraic convenience we follow the common practice of considering complex solutions and attributing physical significance only to their real parts. In that case, the unit imaginary i has no specified physical interpretation, and we are free to suppose that it is the unit pseudoscalar. When the complex solution has been determined, we get the physical solution by taking it real (= scalar) part, written

$$\mathcal{R}e\, q_n = \langle q_n \rangle_0 . \tag{3.21}$$

This trick of "complexifying" the solution works because the equations of motion are linear, so the superposition principle applies. It is well to remember, however, that there are situations where the entire complex solution has physical significance, as in the case of Equation (3.7), where the unit imaginary is a bivector for a plane in physical space. We will take advantage of this again later on.

Now, substituting the candidate solution (3.20) into the equations of motion (3.18), we get

$$-\omega^2 A_n = \omega_0^2 (A_{n-1} - 2A_n + A_{n+1}). \tag{3.22}$$

This system of N equations is most easily solved by allowing A_n to be complex and considering a trial solution of the form

$$A_n = Ae^{inak} = A(\cos nak + i \sin nak), \tag{3.23a}$$

or, indexed by position,

$$A_n = A(x_n) = Ae^{ikx_n} = A(\cos kx_n + i \sin kx_n), \tag{3.23b}$$

where A and k are scalars to be determined. Inserting this trial solution into (3.22), we obtain

$$\begin{aligned}\omega^2 &= -\omega_0^2 (e^{-ika} - 2 + e^{ika}) \\ &= 2\omega_0^2 (1 - \cos ka) \\ &= 4\omega_0^2 \sin^2\left(\frac{ka}{2}\right).\end{aligned} \tag{3.24}$$

Subject to this condition relating ω and k, the real and imaginary parts of A_n satisfy (3.22) separately. However from (3.23) we see that only the imaginary part satisfies the boundary condition (3.19) at $n = 0$. So we introduce a notation for the imaginary part by writing

$$a_n = \frac{1}{2i}(A_n - A_n^\dagger) = \frac{1}{2i}(A_n + A_{-n}). \tag{3.25}$$

Consequently,

$$a_n = A \sin kx_n = A \sin nak. \tag{3.26}$$

The constant k is now determined by imposing the boundary condition at the other end point:

$$a_{N+1} = A \sin[(N+1)ak] = 0. \tag{3.27}$$

For integer r, this has solutions $k = k_r$ of the form

$$k_r = \frac{r\pi}{(N+1)a} = \frac{r\pi}{L} \tag{3.28}$$

Equation (3.24) gives different values of ω for different k_r, so we rewrite it in the form

$$\omega_r = 2\omega_0 \sin\left(\frac{k_r a}{2}\right) = 2\omega_0 \sin\left(\frac{r\pi}{2(N+1)}\right). \tag{3.29}$$

Similarly, Equation (3.26) gives different values of a_n for different k_r, so we write

$$a_{nr} = A \sin k_r x_n = A \sin \left(\frac{rn\pi}{N+1} \right). \tag{3.30}$$

Equation (3.20) describes a different normal mode for each different normal frequency ω_r, so we write

$$q_{nr} = a_{nr} e^{i\omega_r t}. \tag{3.31}$$

The scalar part $q_{nr} = a_{nr} \cos \omega_r t$ is the displacement of the n-th particle in r-th normal mode. The real coefficient a_{nr} is the amplitude for the displacement of the n-th particle. We show below that there are exactly N distinct normal modes indexed by r in the range $r = 1, 2, \ldots, N$. It will be left as an exercise to show that the constant A in (3.30) has the value

$$A = \left(\frac{2}{N+1} \right)^{1/2} \tag{3.32}$$

if the normal modes are normalized by the condition

$$\sum_{r=1}^{N} a_{nr}^2 = 1. \tag{3.33a}$$

Since the energy of a particle in harmonic motion is proportional to the square of its amplitude, this amounts to a normalization of the total energy in a mode. The normalization condition (3.33a) is a special case of the "orthonormality relations"

$$\sum_{n=1}^{N} a_{nr} a_{ns} = \frac{2}{N+1} \sum_{n=1}^{N} \sin \left(\frac{rn\pi}{N+1} \right) \sin \left(\frac{sn\pi}{N+1} \right) = \delta_{rs}, \tag{3.33b}$$

where $[\delta_{rs}]$ is the $N \times N$ identity matrix. The right side of this expression vanishes when $r \neq s$, thus describing a kind of *orthogonality* of normal modes; this result follows easily from a general argument given in Section 6-4.

The above results from (3.28) to (3.31) completely characterize the normal modes of an N-particle string. For each normal mode, a *wave form* $a_r(x)$ with values for every x in the interval $0 \leqslant x \leqslant L$ is defined by

$$a_r(x) = A \sin k_r x = A \sin \frac{2\pi x}{\lambda_r}, \tag{3.34}$$

where

$$\lambda_r = \frac{2\pi}{k_r} = \frac{2L}{r} \tag{3.35}$$

is the *wavelength* and k_r is called the *wave number* of the mode. At the lattice points $x = x_n$, the wave form gives the particle amplitudes $a_{nr} = a_r(x_n)$. Normal modes for the case $N = 3$ are illustrated in Figure 3.7.

To prove that an N particle string has exactly N distinct normal modes, we examine the Equation (3.30) determining the particle amplitudes a_{nr}. First

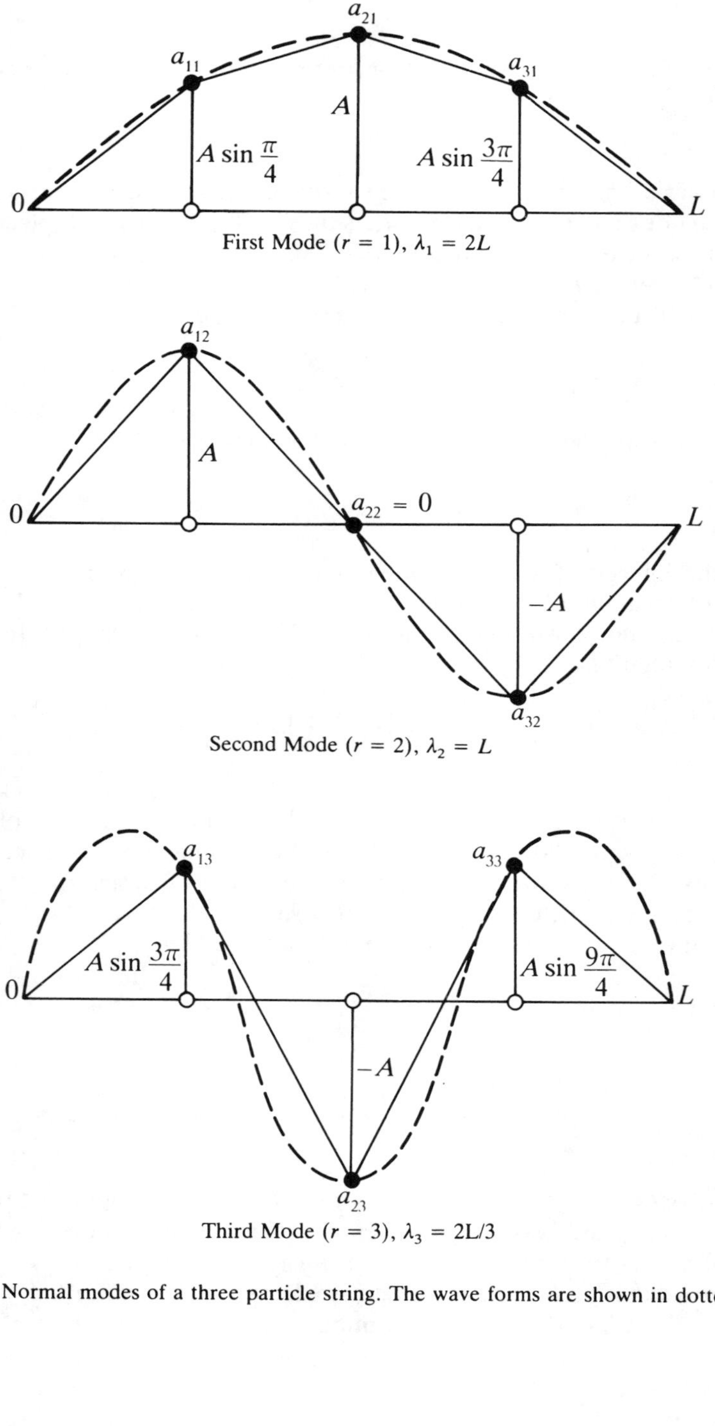

Fig. 3.7. Normal modes of a three particle string. The wave forms are shown in dotted lines.

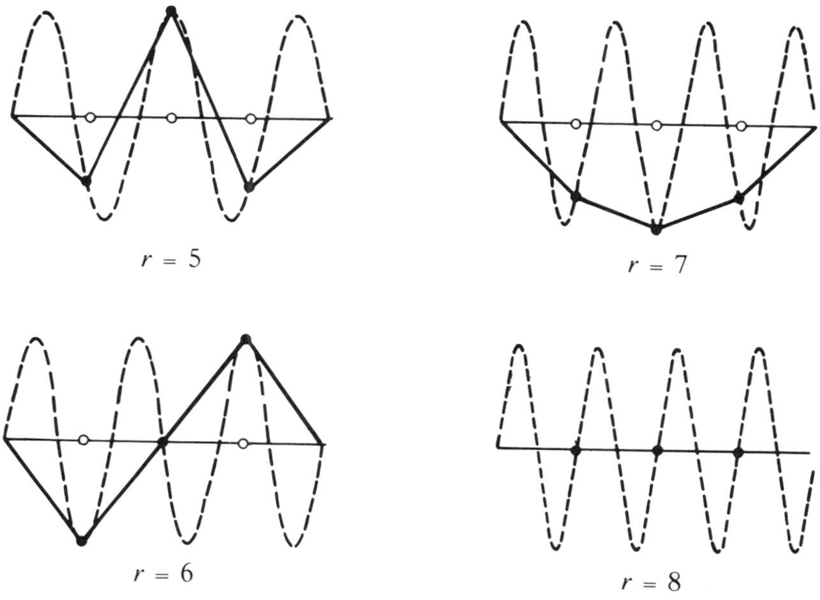

Fig. 3.8. Some unphysical wave forms for normal modes.

note that all the amplitudes vanish identically when $r = 0$ or $r = N + 1$, so these values for r describe a string at rest. Then note that for $r = N + 2$, $N + 3, \ldots, 2N + 1, 2N + 2$ the values of the a_{nr} are the same as for $r = 1$, $2, \ldots, N + 1$, except for a trivial reversal of order and sign. This is illustrated in Figure 3.8 for $N = 3$. Note that the wave forms in Figure 3.8 have twice as many oscillations as the corresponding forms in Figure 3.7, but these additional oscillations are physically meaningless, because they do not correspond to any difference in particle displacements. Thus, there is a smallest "physical wavelength" determined by the lattice constant, namely, $\lambda_N = 2L/N = 2a(N + 1)/N$, or more simply $\lambda_N \approx 2a$ for large N. By virture of (3.29), this smallest wavelength corresponds to a highest normal mode frequency ω_N, called the *cutoff frequency* of the system. Finally, to complete our proof we simply note that similar conclusions obtain for other integer values of r, positive or negative.

We have identified and characterized a complete set of normal modes for the equation of motion (3.18). These are special solutions of the equation of motion subject to the boundary conditions (3.19). The general solution is a superposition of the normal modes. To represent it compactly, it is convenient to introduce *complex normal coordinates* $Q_r = Q_r(t)$ defined by

$$Q_r = (C_r e^{i\delta_r}) e^{i\omega_r t} = C_r e^{i(\omega_r t + \delta_r)}, \tag{3.36}$$

where C_r and δ_r are scalar constants. As before, we attribute physical significance only to the real (i.e. scalar) part

$$\langle Q_r \rangle_0 = C_r \cos(\omega_r t + \delta_r). \tag{3.37}$$

Now the complex general solution $q_n = q_n(t)$ can be written in the form

$$q_n(t) = \sum_{r=1}^{N} a_{nr} Q_r(t), \tag{3.38}$$

where the amplitudes a_{nr} are given by (3.30) with A given by (3.32) or $A = 1$, as preferred.

There are exactly N normal coordinates, and, according to (3.36), each of these depends on two constants, so the general solution (3.38) depends on $2N$ constants, as we know it should from general theory. These constants can be determined from the initial conditions by inverting (3.38) to express the normal coordinates in the particle displacements. The orthogonality relations (3.33b) make this easy. Thus,

$$\sum_n a_{ns} q_n = \sum_n \sum_r a_{ns} a_{nr} Q_r = \sum_r \sum_n a_{ns} a_{nr} Q_r = \sum_r \delta_{sr} Q_r = Q_s,$$

proving that

$$Q_r(t) = \sum_{n=1}^{N} a_{nr} q_n(t). \tag{3.39}$$

The constants are consequently determined by the $2N$ equations

$$C_r e^{i\delta_r} = Q_r(0) = \sum_n a_{nr} q_n(0), \tag{3.40a}$$

$$i\omega_r C_r e^{i\delta_r} = \dot{Q}_r(0) = \sum_n a_{nr} \dot{q}_n(0). \tag{3.40b}$$

Traveling Waves

Each normal mode is a *standing wave* in the sense that its wave form (3.34) is time independent. It is also called a *harmonic wave*, because every particle in the mode oscillates harmonically with a single frequency. There are, however, other harmonic solutions of the equations of motion (3.18) which do not satisfy the boundary conditions (3.19) for standing waves. With minor changes in our analysis for normal modes, it is readily verified that, for arbitrary constants A and δ,

$$q_\omega(x, t) = (Ae^{i\delta})\, e^{i(\omega t - kx)} \tag{3.41}$$

is a harmonic solution of (3.18) provided k is related to ω by (3.29). As before, we attribute physical signficance only to the scalar part

$$\langle q_\omega(x, t) \rangle_0 = A \cos(\omega t - kx + \delta). \tag{3.42}$$

This describes particle displacements at the positions $x = x_n = na$.

The function $q_\omega(x, t)$ specified by (3.41) represents a *traveling harmonic wave* with velocity

$$v = \frac{\omega}{k} \cdot \tag{3.43}$$

To establish that, we write $x = x' + vt$ where x' is the position coordinate of a reference system moving with velocity v in the positive x-direction. Substituting this into (3.42) with (3.43) we obtain

$$\langle q_\omega(x, t)\rangle_0 = A \cos(kx' - \delta) . \tag{3.44}$$

As a function of x', this is a fixed wave form. Therefore, we may regard it as a fixed wave form moving with velocity v along the string. Thus, $q_\omega(-x, t)$ describes a similar wave traveling in the opposite direction. Now note that for $\delta = \pi/2$, (3.41) gives us

$$\tfrac{1}{2}[q_\omega(-x, t) - q_\omega(x, t)] = A(\sin kx)\, e^{i\omega t}, \tag{3.45}$$

which is identical to the expression for a complex standing wave. Therefore, every standing wave can be regarded as a superposition of two traveling waves moving in opposite directions.

When N is very large (as in the model of an elastic solid as a string of atoms), Equation (3.28) tells us that k is effectively a continuous variable. So we write the relation (3.29) between ω and k as a continuous function

$$\omega = \pm 2\omega_0 \sin\left(\frac{ka}{2}\right), \tag{3.46}$$

where the sign is chosen to make ω positive depending on the value of k. This equation is called a *dispersion relation* for the following reason. Using it to eliminate k from (3.43), we see that it implies that the velocity of a traveling harmonic wave depends on frequency, specifically

$$v = v(\omega) = \frac{a\omega}{2 \sin^{-1}\left(\dfrac{\omega}{2\omega_0}\right)} \tag{3.47}$$

Therefore, a "wave packet' composed of harmonic traveling waves with different frequencies will collapse, because the component waves traveling at different velocities will gradually separate, that is, *disperse*.

Dispersion relations are of the utmost importance in solid state physics, where they are used to describe many different characteristics of materials. The dispersion relation (3.46) is graphed in Figure 3.9. The two allowed signs for k correspond to wave propagation in opposite directions. The range of k is limited by the maximum value $k = \pi/a$ corresponding to the minimum wavelength $\lambda = 2\pi/k = 2a$ and the maximum (*cut off*) frequency $\omega = 2\omega_0$. As we

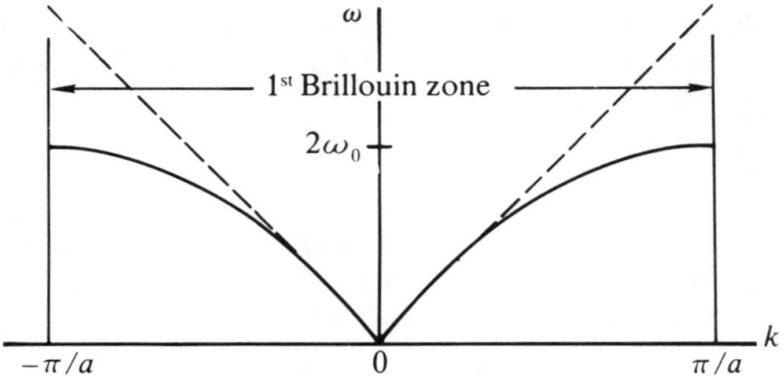

Fig. 3.9. Dispersion relation for a monatomic lattice.

determined earlier, waves of shorter wavelength (higher frequency) cannot be supported in a lattice. The limited allowed range for the wave number k is called the first *Brillouin zone* in solid state physics. The small slope of the dispersion curve near the boundary of the Brillouin zone implies that wave velocities vary rapidly with frequency thus producing large dispersion.

On the other hand, at low frequencies (the long-wavelength limit), the dispersion relation (3.46) reduces to

$$\omega \approx 2\omega_0 \left(\frac{ka}{2} \right) = \omega_0 ak . \tag{3.48}$$

Therefore, in this region harmonic waves with different frequencies have nearly the same velocity

$$v_0 = v(0) = \lim_{k \to 0} \frac{\omega}{k} = \omega_0 a . \tag{3.49}$$

The dotted line in Figure 3.9 is the dispersion curve that would obtain if all harmonic waves had this velocity. The expression (3.49) enables us to calculate the velocity of waves in a medium from measured elastic properties. Recall that $\omega_0^2 = \varkappa/m$, *and* $\varkappa = T/a$ for a string under tension. Whence (3.49) yields

$$v_0 = \left(\frac{T}{\varrho} \right)^{1/2} , \tag{3.50}$$

where $\varrho = m/a$ is the linear mass density of the string. Similarly, for an elastic solid the elastic modulus Y, defined as the ratio of applied force to elongation per unit length, is a measurable quantity. Whence, $\varkappa = Y/a$, and the velocity of a low frequency transverse wave is given by

$$v_0 = \left(\frac{Y}{\varrho} \right)^{1/2} . \tag{3.51}$$

It should be realized that we are talking about the velocity of waves with wavelengths much greater than the lattice constant. In this domain the waves are insensitive to the granular microstructure of the material, which may therefore be regarded as a continuous medium.

We can model a continuous string as the limit of a string of discrete particles as mass $m \to 0$ and particle separation $a \to 0$ but $\varrho = m/a$ remains finite. To get an equation of motion for the continuous string, we index particles by their positions and write (3.18) in the form

$$\frac{\partial^2}{\partial t^2} q(x, t) = \omega_0^2 a \left[\frac{q(x + a, t) - q(x, t)}{a} - \frac{q(x, t) - q(x - a, t)}{a} \right].$$

Inserting in this the Taylor expansion

$$q(x \pm a, t) = q(x, t) \pm a \frac{\partial q}{\partial x} + \tfrac{1}{2} a^2 \frac{\partial^2 q}{\partial x^2} \pm \ldots,$$

in the limit $a \to 0$ we obtain

$$\frac{\partial^2 q(x, t)}{\partial t^2} = v_0^2 \frac{\partial^2 q(x, t)}{\partial x^2}, \tag{3.52}$$

where $v_0 = \lim \omega_0 a$, in agreement with the long-wavelength result (3.49). Equation (3.52) is called the one-dimensional wave equation. The harmonic wave (3.41) is a solution of this equation, and its general solution is a superposition of such waves, all with the same constant velocity v_0.

As a final matter in our study of waves in strings, we note that all our results are easily generalized to the case where the particle coordinates are vectors representing displacement in three dimensions instead of scalars representing one-dimensional displacements, as we have already done for the two particle case. In particular, the expression (3.41) for a traveling harmonic wave generalizes to

$$\mathbf{q}_\omega(x, t) = \mathbf{a}_\omega e^{\mathbf{i}(\omega t - kx)}, \tag{3.53}$$

where now the unit imaginary $\mathbf{i}$ is a bivector and the constant vector amplitude $\mathbf{a}_\omega$ lies in the $\mathbf{i}$-plane. We can construct normal modes from this by superposition as in (3.45), yielding the expression (3.7) found previously in the two particle case.

In contrast to the scalar form (3.41), the imaginary part of (3.53) has a physical interpretation. This is easiest to picture when $\mathbf{i}$ is the bivector for a transverse plane. Then, (3.53) represents a *circularly polarized transverse harmonic* wave. To see that, note that for each fixed value of the particle coordinate x, (3.53) describes a displacement rotating with angular velocity ω about the equilibrium point. Alternatively, for fixed t and variable x, (3.53) represents a helical wave form as pictured in Figure 3.10. As time varies, the helical wave form can be pictured as rigidly rotating about the axis of

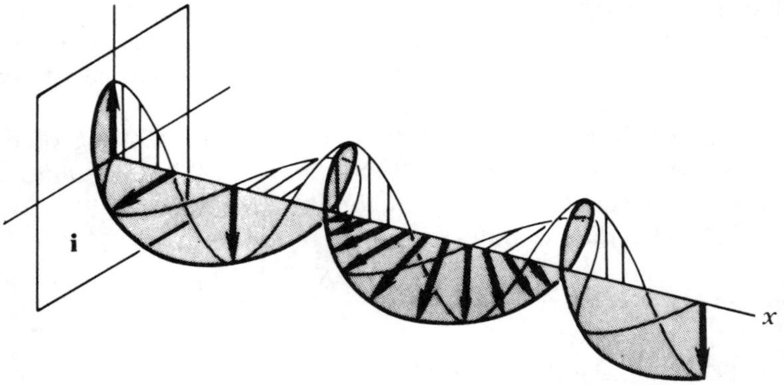

Fig. 3.10. A left-circularly polarized traveling wave has positive helicity.

equilibrium points, or moving rigidly without rotation along the axis with velocity $v = \omega/k$. If the wave form is a right-handed helix and the wave is said to have *positive helicity*. Unfortunately, in optics such a wave is said to be *left-circularly polarized*. Similarly, the function

$$\mathbf{q}_\omega(-x, -t) = \mathbf{a}_\omega e^{-\mathbf{i}(\omega t - kx)} \tag{3.54}$$

describes a harmonic wave with *negative helicity* (*right-circularly polarized*) moving in the positive x direction provided both ω and k are positive.

It is important to note that both cases (3.53) and (3.54) can be lumped into one described by (3.53) if we allow the frequency to have negative as well as positive values. Thus, *we can assign a physical interpretation to the sign of the frequency ω for a harmonic wave: it is the helicity of circular polarization*. The direction of wave motion is then determined by the sign of k relative to ω. As we have noted, it is in the positive (negative) x-direction if $v = \omega/k$ is positive (negative). All these considerations apply equally well to electromagnetic waves with $\mathbf{q}_\omega(x, t)$ replaced by the electric field vector $\mathbf{E}_\omega(x, t)$.

In solid state physics the one-dimensional lattice model studied here is generalized to three-dimensional lattices of various types composed of atoms of various kinds. To analyse more complicated models such as these efficiently, we need a systematic general theory of small oscillations, so we turn to that in the next section.

6-3. Exercises

(3.1) For a string of two identical particles coupled linearly, as in Figure 3.1, suppose the string is plucked with the initial conditions $\mathbf{q}_1(0) = \mathbf{A}$, $\mathbf{q}_2(0) = 0$, $\dot{\mathbf{q}}_1(0) = \dot{\mathbf{q}}_2(0) = 0$. Show that the displacements in time are given by

$$\mathbf{q}_1 = \mathbf{A}(\cos \omega_0 t) \cos(\varepsilon\omega_0 t)$$

$$\mathbf{q}_2 = \mathbf{A}(\sin \omega_0 t) \sin(\varepsilon\omega_0 t)$$

where $2\omega_0 = \omega_+ + \omega_-$ and $2\varepsilon\omega_0 = \omega_- - \omega_+$. Describe qualitative features of this solution in the weak and strong coupling limits. Why does the strong coupling result differ from the result in Equation (3.15)?

(3.2) For a string of three identical particles coupled linearly as in Figure 3.6, suppose the string is plucked by displacing the center particle longitudinally with initial condition $q_2(0) = A$, while $q_1(0) = q_3(0) = 0$ and $\dot{q}_1(0) = \dot{q}_2(0) = \dot{q}_3(0) = 0$. Determine the natural frequencies and the displacements $q_n(t)$ as explicit functions of time.

(3.3) For a string of five identical particles coupled linearly as in Figure 3.6, draw diagrams for the transverse normal modes from symmetry considerations alone and arrange them in order of increasing frequency. Then check your qualitative understanding by calculation.

(3.4) Derive the normalization factor (3.32) for normal modes by evaluating the sum-

$$\sum_{n=1}^{N} \sin^2\left(\frac{rn\pi}{N+1}\right) = \frac{N+1}{2} .$$

The following hints should be helpful. Write

$$4 \sin^2 n\theta = 2(1 - \cos 2n\theta) = 2 - (z^n + z^{-n}),$$

where z is a complex number obeying $z^{N+1} = 1$. The geometric series

$$\sum_{n=0}^{\infty} z^n = (1 - z)^{-1}$$

can be iterated to get

$$\sum_{n=0}^{\infty} z^n = 1 + z + z^2 + \ldots + z^k(1 - z)^{-1},$$

where k is any positive integer (consider $k = 1$ first). Then the finite sum

$$\sum_{n=1}^{N} z^n = \frac{z(1 - z^N)}{1 - z}$$

can be evaluated by expressing it as a difference of two infinite sums.

6-4. Theory of Small Oscillations

The study of specific examples in the preceding section has provided us with a nucleus of ideas from which the general theory of small oscillations can be developed. A systematic formulation of the general theory has several benefits. First, it clarifies the range of physical problems to which the theory applies. Second, it organizes techniques for efficient problem solving. Third, it provides a conceptual framework for thinking about many particle systems as a whole. With the examples and concepts from the preceding section in mind, we can procede rapidly to a concise formulation of the theory. We develop the theory in a form suitable for a wider range of applications than we can consider here. Unfortunately, we do not have space for applications of the mathematical theory of groups to the general theory, a major topic in modern theoretical physics. But that can be found in the specialist literature.

One should distinguish between mathematical and physical aspects of the theory of small oscillations. Mathematically, it belongs to the general linear systems theory, which is concerned with modeling the behavior of any system by systems of linear differential equations. Small oscillation theory deals with the case of deviations from a state of stable equilibrium. The analysis of any linear system can be carried out completely using techniques and results from the mathematical theory of linear differential equations. The well developed mathematical theory makes the analysis of any linear system a straightforward task; it may be computationally complex when many variables are involved, but this difficulty has been largely overcome by the development of modern computers and computer software.

The physical aspect of the theory of small oscillations concerns the physical interpretation of mathematical models. The same set of equations might be interpreted as a mathematical model for systems as diverse as an electrical network, a macroscopic system of springs and pendulums, or a microscopic system of atoms in a molecule or elastic solid. Our concern here will be with mechanical interpretations, specifically, with the mechanics of small displacements in a system of particles from a stable equilibrium configuration. Note that the term "small" refers to physical rather than mathematical aspects of the theory. It means that we are dealing with a linear approximations to nonlinear force laws, so the results have validity only for a range of states close to equilibrium. The equilibrium configuration is generally determined by nonlinear features of the force laws, so it must be taken for granted in a linear theory.

For "large" displacements from equilibrium, the restoring forces are nonlinear. The important subject of nonlinear oscillations has not yet been reduced to a theory with general results of wide applicability. It is currently an active field for research. Unfortunately, we don't have the space here for a suitable introduction to the exciting recent developments in this field.

Even the range of applications for the theory of small oscillations is too broad to survey here. Our objective will be to develop the main ideas and general results of the theory along with examples to show how they are applied. We consider only discrete systems of particles. The generalization to continuous systems is a major topic in continuum mechanics which we touched only briefly in the preceding section.

Harmonic Systems

We will employ the Lagrange formulation of mechanics developed in Section 6-2. We consider a conservative N-particle system with n degrees of freedom, so, with a suitable set of generalized coordinates $q_1, q_2, \ldots, q_n$, the interactions can be described by a potential energy function $V = V(q_1 \ldots, q_n)$. As explained in Section 6-2, this includes the possibility that the system is subject to time independent external constraints. We assume also that the system has a state of *stable static equilibrium* at $q_1 = q_2 = \ldots = q_n = 0$. This requires some explanation.

A system of particles is said to be in *static equilibrium* (in a given reference system) if all the particles remain at rest. This is possible only if the net force on each particle vanishes. In the Lagrange formulation, this *equilibrium condition* is expressed as a vanishing of the generalized force:

$$K_\alpha \equiv -\partial_{q_\alpha} V(q_1^0, q_2^0, \ldots, q_n^0) = 0\,, \tag{4.1}$$

where $\alpha = 1, 2, \ldots, n$. If the function $V(q_1, \ldots, q_n)$ is known, then this is a system of n equations which can be solved for the equilibrium values q_α^0. However, we shall see that in some physical problems the equilibrium values are known while the function V is not. In either case the equilibrium condition (4.1) is satisfied, and we are free to adjust our coordinate system so that $q_1^0 = \ldots = q_n^0 = 0$. Then the variables q_α directly describe departures from the static equilibrium state.

The effects of small departures from equilibrium can be described by approximating the potential with a Taylor expansion. Writing $V(q) = V(q_1, \ldots, q_n)$, we have

$$V(q) = V(0) + \sum_\alpha q_\alpha \partial_{q_\alpha} V(0) + \tfrac{1}{2} \sum_\alpha \sum_\beta q_\alpha q_\beta \partial_{q_\alpha} \partial_{q_\beta} V(0) + \ldots \tag{4.2}$$

We are free to choose $V(0) = 0$, and the equilibrium condition (4.1) implies that the second set of terms on the right vanish. So, to a first approximation the potential is given by the quadratic function

$$V(q) \approx \tfrac{1}{2} \sum_{\alpha,\beta} k_{\alpha\beta} q_\alpha q_\beta\,, \tag{4.3}$$

where

$$k_{\alpha\beta} = \frac{\partial^2 V(0)}{\partial q_\alpha \partial q_\beta} = k_{\beta\alpha}\,. \tag{4.4}$$

This is called the *harmonic* approximation and higher order terms in the Taylor expansion are said to be *anharmonic*. A departure from equilibrium is said to be "small" if the anharmonic contribution to the potential energy is negligible to the accuracy desired.

In the harmonic approximation, the potential (4.3) corresponds to a generalized force

$$K_\alpha(q) = -\,\partial_{q_\alpha} V_\alpha(q) = -\sum_\beta k_{\alpha\beta} q_\beta . \tag{4.5}$$

The coefficients $k_{\alpha\beta}$ are called *force constants*; they are measures of the coupling strength between different degrees of freedom.

The equilibrium point $\mathbf{q} = 0$ is said to be *stable* if the quadratic potential energy (4.3) is positive definite, that is, if

$$V(q) = \tfrac{1}{2} \sum_{\alpha,\beta} k_{\alpha\beta} q_\alpha q_\beta \geqslant 0 \tag{4.6}$$

for all values of the q_α. By setting all q_α but one to zero, we see that (4.6) implies $k_{\alpha\alpha} > 0$, so the corresponding generalized force $-k_{\alpha\alpha} q_\alpha$ draws the system back to equilibrium. Actually, (4.6) is merely a sufficient condition for stability; in Section 6.5 we shall see that stability is possible under more general conditions.

According to (2.18), in terms of the generalized coordinates the system kinetic energy K is a positive definite quadratic function of the $\dot{q}_\alpha$:

$$K = \tfrac{1}{2} \sum_{\alpha,\beta} m_{\alpha\beta} \dot{q}_\alpha \dot{q}_\beta \geqslant 0 . \tag{4.7}$$

In general, the mass coefficients $m_{\alpha\beta} = m_{\alpha\beta}(q)$ are functions of the coordinates as well as the masses of the particles. But we can expand them in a Taylor series,

$$m_{\alpha\beta}(q) = m_{\alpha\beta}(0) + \sum_\gamma q_\gamma \partial_{q_\gamma} m_{\alpha\beta}(0) + \ldots,$$

and, consistent with our approximation to the potential energy, we keep only the terms of lowest order. So we regard the mass coefficients as constants

$$m_{\alpha\beta} = m_{\alpha\beta}(0) = m_{\beta\alpha} . \tag{4.8}$$

From the fact that every particle has a mass it follows that $m_{\alpha\alpha} \neq 0$ for all α.

Now we form the Lagrangian $L = K - V$ for the system from (4.6) and (4.7), and from Lagrange's equation (2.15) we determine the n equations of motion for the system:

$$\sum_\beta (m_{\alpha\beta} \ddot{q}_\beta + k_{\alpha\beta} q_\beta) = F_\alpha , \tag{4.9}$$

where the F_α are components of the generalized force due to external agents.

To facilitate analysis, we write the system of Equations (4.9) as a single matrix equation

$$[m]|\ddot{q}) + [k]|q) = |F). \tag{4.10}$$

The mass matrix $[m]$ is defined by

$$[m] = \begin{bmatrix} m_{11} & m_{12} & \cdot & \cdot & \cdot & m_{1n} \\ m_{21} & m_{22} & \cdot & \cdot & \cdot & m_{2n} \\ \cdot & \cdot & & & & \cdot \\ \cdot & \cdot & & & & \cdot \\ \cdot & \cdot & & & & \cdot \\ m_{n1} & m_{n2} & & & & m_{nn} \end{bmatrix}, \tag{4.11}$$

and the matrix $[k]$ is defined similarly. The notation $|q)$ indicates the column of generalized coordinates defined by

$$|q) = \begin{bmatrix} q_1 \\ q_2 \\ \cdot \\ \cdot \\ \cdot \\ q_n \end{bmatrix} \tag{4.12}$$

Of course,

$$|\ddot{q}) = \begin{bmatrix} \ddot{q}_1 \\ \ddot{q}_2 \\ \cdot \\ \cdot \\ \cdot \\ \ddot{q}_n \end{bmatrix} \quad \text{and} \quad |F) = \begin{bmatrix} F_1 \\ F_2 \\ \cdot \\ \cdot \\ \cdot \\ F_n \end{bmatrix}.$$

It is convenient to introduce the notation

$$(q| = [q_1^\dagger, q_2^\dagger, \ldots q_n^\dagger] \tag{4.13}$$

for the row matrix corresponding to the column matrix $|q)$. The conjugation symbol is appropriate if we want to employ complex coordinates as defined by (3.20) and (3.21). Then

$$(q|q) = |q_1|^2 + |q_2|^2 + \ldots +, \tag{4.14}$$

where $|q_\alpha|^2 = q_\alpha^\dagger q_\alpha$, and $|q_\alpha|^2 = q_\alpha^2$ if q_α is real. Now the expressions (4.7) and (4.3) for kinetic and potential energies can be written

$$2K = (\dot{q}|[m]|\dot{q}), \tag{4.15}$$

$$2V = (q|[k]|q). \tag{4.16}$$

Any physical system modeled by a matrix equation of motion of the form (4.10), where the matrices $[m]$ and $[k]$ are positive definite and $[m]$ is nonsingular is called *harmonic system*, because it generalizes the single particle harmonic oscillator model treated in Sections 3.8 and 3.9. Our experience with the harmonic oscillator will serve as a valuable guide for the analysis of harmonic systems.

The matrix notation has conceptual as well as computational advantages. Just as we represent the position of a single particle by a vector in position space, so we can represent the configuration of a system of particles as a vector in the n-dimension configuration space. The notation $|q)$ provides us with a symbol for the configuration vector and so helps us think of the system as a whole rather than the collection of its parts. However, the symbol $|q)$ does not represent the configuration vector itself; it represents the matrix of components (or coordinates) of the configuration vector with respect to a particular basis in configuration space. The same configuration vector may be represented by a different matrix of coordinates $|Q)$ with respect to a different basis. Thus, in the matrix formulation we deal with multiple representations of the same physical configuration. The advantage of this is that different matrix representations $|q)$, $|Q)$, $|S)$, etc., of the configuration have different physical interpretations, and the notation helps keep track of this. It has the disadvantage of allowing ambiguities which can lead to confusion; for example, two different column matrices might be different sets of coordinates for the same configuration or coordinates for two different configurations.

If the matrix equation (4.10) is to amount to more than a mere abbreviation for the set of equations (4.9), we need a system of theorems from matrix algebra to facilitate computations. The theorems we need are all straightforward generalizations of results proved in Sections 5-1 and 5-2 for the 3-dimensional case, so we take them for granted here without further proof. Actually, for the purpose of illustration, we shall not consider calculations with matrices of dimension greater than 3×3, because algebraic labor becomes so great that it is best performed by computers.

Free Oscillations

In the absence of external forces, the matrix equation of motion (4.10) reduces to

$$[m]|\ddot{q}) + [k]|q) = 0. \tag{4.17}$$

Our first task is to find the general solution of this linear matrix equation. The most straightforward approach is to consider a trial solution of the form

$$|q) = |a)e^{i\omega t} \tag{4.18}$$

where ω and the matrix element of $|a)$ are real numbers. Inserting this into (4.17) we obtain the matrix equation

$$([k] - \omega^2[m])|a) = 0, \tag{4.19a}$$

which represents the system of scalar equations

$$\sum_\beta (k_{\alpha\beta} - \omega^2 m_{\alpha\beta})a_\beta = 0 \tag{4.19b}$$

for the constants a_β. The equations have a definite solution if and only if the *characteristic equation*

$$\det [k_{\alpha\beta} - \omega^2 m_{\alpha\beta}] = 0 \tag{4.20}$$

is satisfied. The determinant is a polynomial of degree n in the variable ω^2. Its n roots ω_α^2 are real and positive, because the matrices $[k]$ and $[m]$ are real, symmetric and positive definite, so a real, positive value for each ω_α is thereby determined. (This assertion can easily be proved as a byproduct of the alternative approach we take below.) Using the terminology introduced in the preceding section, we say that the roots ω_α are the *normal* or *natural frequencies* of the harmonic system. And if two distinct roots have the same value they are said to be *degenerate*. Each unique root is said to be *nondegenerate*.

For each root ω_α inserted into the matrix equation (4.19a), there is a solution $|a_\alpha)$ of the equation, which is unique up to a scale factor if the root is nondegenerate. We have already discussed the matter of degeneracy in Section 6-3, so it will be sufficient to confine our attention here to the nondegenerate case. It is convenient to fix the scale of the solution by normalizing it with the condition

$$(a_\alpha|[m]|a_\alpha) = 1\,. \tag{4.21}$$

Each $|a_\alpha)$ represents a *normal mode* with normal frequency ω_α.

The general solution $|q) = |q(t))$ of the matrix equation (4.17) can now be expressed as a superposition of normal modes $|a_\alpha)$ with normal coordinates $Q_\alpha = Q_\alpha(t)$; thus,

$$|q) = \sum_\alpha Q_\alpha\, |a_\alpha). \tag{4.22a}$$

Note that this can be put in the equivalent form

$$|q) = [a]|Q)\,, \tag{4.22b}$$

where $[a] = [\ |a_1)\ |a_2)\ .\ .\ .\ |a_n)]$ is a matrix with the $|a_\alpha)$ as columns, and $|Q)$ is the column matrix of normal coordinates. The normal coordinates can be given the explicit functional form

$$Q_\alpha(t) = C_\alpha \cos(\omega_\alpha t + \delta_\alpha)\,, \tag{4.23}$$

where C_α and δ_α are constants. Or if complex coordinates are preferred,

$$Q_\alpha(t) = (C_\alpha e^{i\delta_\alpha})e^{i\omega_\alpha t}\,. \tag{4.24}$$

Note that Equation (4.22b) can be regarded as a relation between two different sets of coordinates $|q)$ and $|Q)$ for the same system.

The method we have just outlined for solving the equation of motion (4.17) may be called the *brute force method*, since it does not take advantage of any special information that might be known about the matrices $[m]$ and $[k]$. There are alternative methods of solution which are simpler when certain kinds of information are available. For example, if the normal modes $|a_\alpha)$ can be determined first, then the normal frequencies can be most easily obtained from (4.19a), which yields

$$\omega_\alpha^2 = \frac{(a_\alpha|[k]|a_\alpha)}{(a_\alpha|[m]|a_\alpha)}. \tag{4.25}$$

Indeed, we used a variant of this approach in Section 6-3 to determine the normal frequencies of a one-dimensional lattice. As an example, it may be noted that the solution (3.38) for that system conforms to the general form for a solution given by (4.22a) and (4.23).

Whatever method we use to find the normal modes $|a_\alpha)$, we still have the problem of evaluating the constants in our expression (4.23) for the normal coordinates. That is most easily solved by using the relation

$$(a_\alpha|[m]|a_\beta) = \delta_{\alpha\beta}, \tag{4.26}$$

which combines the normalization condition (4.21) with the orthogonality relation

$$(a_\alpha|[m]|a_\beta) = 0 \quad \text{if} \quad \alpha \neq \beta. \tag{4.27}$$

This relation can be proved by multiplying (4.19a) in the form

$$\omega_\alpha^2[m]|a_\alpha) = [k]|a_\alpha)$$

by $(a_\beta\,|$ to get

$$\omega_\alpha^2(a_\beta|[m]|a_\alpha) = (a_\beta|[k]|a_\alpha)$$

Subtracting from this a similar equation with α and β interchanged and using the symmetry of $[m]$ and $[k]$, we obtain

$$(\omega_\alpha^2 - \omega_\beta^2)\,(a_\beta|[m]|a_\alpha) = 0.$$

Since $\omega_\alpha^2 \neq \omega_\beta^2$ when $\alpha \neq \beta$, this implies (4.27).

Now from (4.22a), (4.23) and (4.26) we obtain

$$\begin{aligned} C_\alpha \cos\delta_\alpha &= (a_\alpha|[m]|q(0)\,), \\ -\omega_\alpha C_\alpha \sin\delta_\alpha &= (a_\alpha|[m]|\dot{q}(0)\,), \end{aligned} \tag{4.28}$$

which determines the constants in terms of initial data.

Although the computational complexities of the brute force method have been largely overcome by computers, it is still worthwhile to consider alternative methods for the insight they provide. Let us first consider how the equation of motion (4.17) might be simplified.

Since $[m]$ is nonsingular it has an inverse $[m]^{-1}$, so we might consider simplifying (4.17) by multiplication by $[m]^{-1}$, to get

$$|\ddot{q}) + [K]|q) = 0,$$

where $[K]$ is a new matrix given by the matrix product $[K] = [m]^{-1}[k]$. The trouble with this procedure is that the product $[K]$of symmetric matrices $[m]^{-1}$ and $[k]$ is not necessarily symmetric, and we need to exploit the symmetry to solve the equation. Fortunately, the desired simplification can be achieved in a slightly different way.

Since $[m]$ is positive definite and symmetric, we know that it has a well-defined positive square root $[m]^{1/2}$. Indeed, if $[m]$ is diagonal with matrix elements $m_{\alpha\beta} = m_\alpha \delta_{\alpha\beta}$, then

$$[m]^{1/2} = \begin{bmatrix} m_1^{1/2} & 0 & \cdot & \cdot & \cdot & 0 \\ 0 & m_2^{1/2} & & & & 0 \\ \cdot & & \cdot & & & \cdot \\ \cdot & & & \cdot & & \cdot \\ \cdot & & & & \cdot & \cdot \\ 0 & 0 & \cdot & \cdot & \cdot & m_n^{1/2} \end{bmatrix}.$$

In any case, let us multiply (4.17) with the inverse $[m]^{-1/2}$ of $[m]^{1/2}$ to get it in the form

$$[m]^{1/2}|\ddot{q}) = [m]^{-1/2}[k][m]^{-1/2}[m]^{1/2}|q).$$

Thus, the equation of motion can be put in the simple form

$$|\ddot{q}') - [k']|q') = 0 \tag{4.29}$$

if we write

$$[k'] = [m]^{-1/2}[k][m]^{-1/2}, \tag{4.30}$$

and introduce a new set of coordinates q'_α for the system defined by

$$|q') = [m]^{1/2}|q). \tag{4.31}$$

The q'_α are sometimes called *mass-weighted coordinates.*

It follows from (4.19) that the matrix $[k']$ is real, symmetric and positive definite, since $[m]$ and $[k]$ have those properties. Therefore, it has n distinct eigenvectors $|b_\alpha)$ with corresponding eigenvalues ω_α^2, which, of course, will prove to be the normal frequencies. Thus, we have

$$[k']|b_\alpha) = \omega_\alpha^2|b_\alpha). \tag{4.32}$$

Defining a matrix $[b] = [\ |b_1)|b_2) \ . \ . \ . \ |b_n)]$, these n equations can be written as a single matrix equation

$$[k'][b] = [b][\omega^2], \tag{4.33}$$

where $[\omega^2]$ is the diagonal matrix

$$[\omega^2] = \begin{bmatrix} \omega_1^2 & 0 & . & . & . & 0 \\ 0 & \omega_2^2 & . & . & . & 0 \\ . & . & & & & \\ . & . & . & & & \\ . & . & & . & & \\ 0 & 0 & . & . & \vdots & \omega_n^2 \end{bmatrix}.$$

Since $[b]$ is nonsingular, we can put (4.33) in the form

$$[\omega^2] = [b]^{-1}[k'][b], \tag{4.34}$$

and we say that $[b]$ diagonalized $[k]$. We can use this to further simplify the equation of motion (4.29) by introducing new coordinates $|Q)$ defined by writing

$$|q') = [b]|Q), \tag{4.35}$$

Substituting this into (4.29) and using (4.34), we obtain

$$|\ddot{Q}) = [\omega^2]|Q). \tag{4.36}$$

Since $[\omega^2]$ is diagonal, this is equivalent to n uncoupled equations

$$\ddot{Q}_\alpha = \omega_\alpha^2 Q_\alpha, \tag{4.37}$$

which have the general solution found before.

These Q_α will be identical with the normal coordinates defined before provided we related $|b_\alpha)$ to $|a_\alpha)$ by

$$|b_\alpha) = [m]^{1/2}|a_\alpha), \tag{4.38}$$

which imposes on the $|b_\alpha)$ the normalization

$$(b_\alpha|b_\beta) = (a_\alpha|[m]|a_\beta) = \delta_{\alpha\beta}. \tag{4.39}$$

To establish the relation to our previous result explicitly, we simply invert (4.31) and substitute (4.36) to get $|q) = [a]\,|Q)$ with

$$[a] = [m]^{-1/2}[b] \tag{4.40}$$

Now we can identify $[a]$ as the matrix which transforms our original equation of motion (4.17) into the diagonal form (4.36) for which the solution is elementary.

Example 4.1

To illustrate the general method, let us apply it to the double pendulum. In Example 2.2 of Section 6-2 we determined the linearized equations of motions for the double pendulum, which can be put in the matrix form

$$[m]|\ddot{\phi}) + [k]|\phi) = 0,$$

where

$$[m] = \begin{bmatrix} m_{11} & m_{12} \\ m_{21} & m_{22} \end{bmatrix} = \begin{bmatrix} (m_1 + m_2)l_1^2 & m_2 l_1 l_2 \\ m_2 l_1 l_2 & m_2 l_2^2 \end{bmatrix},$$

$$[k] = \begin{bmatrix} k_{11} & 0 \\ 0 & k_{22} \end{bmatrix} = \begin{bmatrix} (m_1 + m_2)l_1 g & 0 \\ 0 & m_2 l_2 g \end{bmatrix},$$

$$|\phi) = \begin{bmatrix} \phi_1 \\ \phi_2 \end{bmatrix}.$$

Since $[k]$ is already diagonal in this case, it is algebraically simpler to introduce

$$|\phi) = [k]^{-1/2}|\phi')$$

to put the equation of motion in the form

$$|\ddot{\phi}') + [k']|\phi') = 0,$$

where

$$[k'] = [k]^{1/2}[m]^{-1}[k]^{1/2},$$

$$[k]^{1/2} = \begin{bmatrix} k_{11}^{1/2} & 0 \\ 0 & k_{22}^{1/2} \end{bmatrix}$$

and, by a result in Section 5-1.

$$[m]^{-1} = \frac{1}{\det [m]} \begin{bmatrix} m_{22} & -m_{12} \\ -m_{12} & m_{11} \end{bmatrix}$$

where $\det [m] = m_{11}m_{22} - m_{12}^2 = m_1 m_2 l_1^2 l_2^2$. This reduces the problem to solving the eigenvalue equation (4.32).

Vibrations of Triatomic Molecules

The frequencies at which molecules absorb electromagnetic radiation depend on their normal modes. So the determination of molecular normal modes is of great importance for molecular spectroscopy. Here we analyze the normal vibrations of bent triatomic molecules such as H_2O, SO_2 and Cl_2O. Our classical analysis is a necessary prelude to the more precise treatment with quantum mechanics.

The equilibrium configuration of a bent symmetric triatomic molecule is shown in Figure 4.1. For H_2O, the O-H bond length r and the valence angle θ have the values

$$r = 0.958 \times 10^{-8} \text{ cm}$$

$$\theta = 2\phi = 104.5°$$

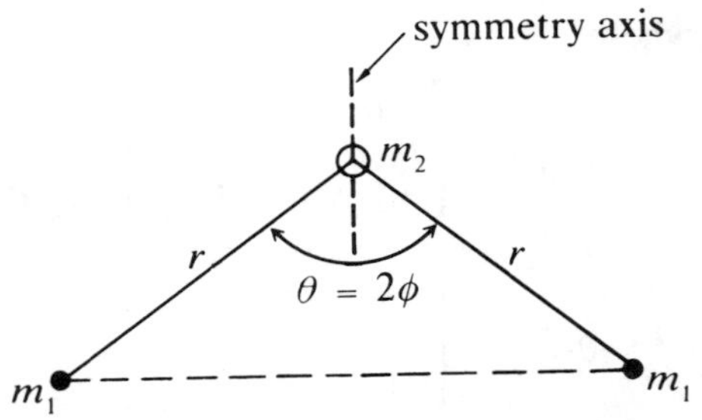

Fig. 4.1. Equilibrum configuration of a bent symmetric triatomic molecule.

Let $\mathbf{q}_1$, $\mathbf{q}_2$, $\mathbf{q}_3$ represent displacements of the atoms from their equilibrium position as indicated in Figure 4.2. In terms of these variables the kinetic energy K has the simple form

$$2K = m_1(\dot{\mathbf{q}}_1^2 + \dot{\mathbf{q}}_2^2) + m_2\dot{\mathbf{q}}_3^2. \tag{4.41}$$

However, we are interested here only in vibrations and not rotations and translations of the molecule as a whole. So the $3 \times 3 = 9$ degrees of freedom of the three displacement vectors must be restricted. First, an internal vibration cannot shift the center of mass, so we must require

$$m_1(\mathbf{q}_1 + \mathbf{q}_2) + m_2\mathbf{q}_3 = 0 \tag{4.42}$$

Second, the vibrations must lie in the plane of the molecule. And third, the molecule must not rotate in the plane. Thus, after subtracting the three rotational degrees of freedom, we have $9 - 2 \times 3 = 3$ independent vibrational degrees of freedom remaining. This leaves us with the problem of choosing an appropriate set of internal coordinates. We can't use normal coordinates until we have determined the normal modes.

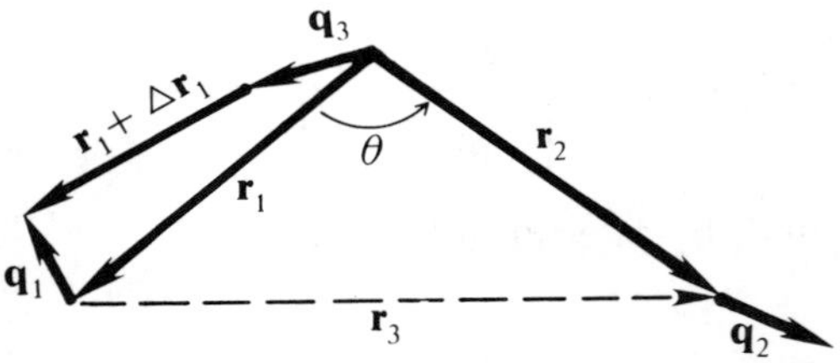

Fig. 4.2. Atomic displacements from equilibrium.

For direct physical description of the molecule, variations in the bond lengths, say R_1, R_2, R_3, and variations in the valence angle, say R_θ, are natural internal coordinates. Indeed, they provide the simplest expressions for the potential energy. The internal potential energy function is very difficult to calculate from first principles, and has been found for only a few simple molecules. So we must be content with estimating it from auxiliary assumptions. The first assumption which recommends itself is that when the system is in a near-equilibrium state, the forces of attraction and repulsion between atoms are *central*. In that case, the potential energy will depend only on the bond lengths, with the specific functional form

$$2V = k_1(R_1^2 + R_2^2) + k_3R_3^2. \tag{4.43}$$

The constants k_1 and k_3 are unknown but they can be evaluated from spectroscopic data after the normal frequencies have been calculated. Since there are only two unknown constants, one of the three normal frequencies can be computed from data for the other two, thus providing a specific prediction of the model. It turns out that the predictions for various molecules are accurate to about 25%. Therefore, the central force assumption is only a moderately accurate description of internal molecular forces.

Much better results have been obtained with a potential of the form

$$2V = k_1(R_1^2 + R_2^2) + k_\theta R_\theta^2. \tag{4.44}$$

the constant k_1 describes resistance of the main bonds to stretching while k_θ describes resistance to bending. The potential function (4.44) is also more reasonable than (4.43) because bending and stretching variations are orthogonal to one another, as we expect of uncoupled variables in the potential energy function. It has the additional advantage of applying to "straight" as well as "bent" molecules.

To use the potential energy function (4.44) along with the kinetic energy function (4.41), we need to relate the stretching and bending variables R_1, R_2, R_θ to the displacement vectors $\mathbf{q}_1$, $\mathbf{q}_2$, $\mathbf{q}_3$. To that end, it is convenient to represent the relative equilibrium position of the atoms by vectors $\mathbf{r}_1$, $\mathbf{r}_2$, $\mathbf{r}_3$ as in Figure 4.2. A variation in the bond length $R_1 = \Delta r_1$ is related to the relative atomic displacements $\Delta\mathbf{r}_1 = \mathbf{q}_1 - \mathbf{q}_3$ by differentiating the constraint $r_1^2 = \mathbf{r}_1^2$. Thus, $r_1\Delta r_1 = \mathbf{r}_1\cdot\Delta\mathbf{r}_1$, so

$$R_1 \equiv \Delta r_1 = \hat{\mathbf{r}}_1\cdot\Delta\mathbf{r}_1 = \hat{\mathbf{r}}_1\cdot(\mathbf{q}_1 - \mathbf{q}_3)\,, \tag{4.45a}$$

where $\hat{\mathbf{r}}_1 = \mathbf{r}_1/r_1$. Similarly,

$$R_2 \equiv \Delta r_2 = \hat{\mathbf{r}}_2\cdot\Delta\mathbf{r}_2 = \hat{\mathbf{r}}_2\cdot(\mathbf{q}_2 - \mathbf{q}_3)\,. \tag{4.45b}$$

To relate a variation in the bond angle $R_\theta = \Delta\theta$ to the atomic displacement we differentiate

$$\mathbf{r}_1\mathbf{r}_2 = r_1r_2e^{\mathbf{i}\theta}$$

where $\mathbf{i} = \sigma_1\sigma_2$ is the unit bivector for the plane of the molecules. Thus,

$$\Delta\mathbf{r}_1\mathbf{r}_2 + \mathbf{r}_1\Delta\mathbf{r}_2 = (r_2\Delta r_1 + r_1\Delta r_2 + r_1r_2\mathbf{i}\Delta\theta)e^{\mathbf{i}\theta}$$

Taking the scalar part of this expression and using (4.45a, b) we find

$$\Delta\theta = \frac{(\hat{\mathbf{r}}_1\cos\theta - \hat{\mathbf{r}}_2)\cdot\Delta\mathbf{r}_1}{r_1\sin\theta} + \frac{(\hat{\mathbf{r}}_2\cos\theta - \hat{\mathbf{r}}_1)\cdot\Delta\mathbf{r}_2}{r_2\sin\theta}\,. \tag{4.46}$$

This expression holds for the bond angle between any three atoms. For the symmetrical triatomic molecule we have $r_1 = r_2 = r$, and it will be convenient to employ the half angle $\phi = \theta/2$, so we write (4.46) in the form

$$R_\theta r \sin 2\phi = (\hat{\mathbf{r}}_1 \cos 2\phi - \hat{\mathbf{r}}_2)\cdot(\mathbf{q}_1 - \mathbf{q}_3) + (\hat{\mathbf{r}}_2 \cos 2\phi - \hat{\mathbf{r}}_1)\cdot(\mathbf{q}_2 - \mathbf{q}_3) \tag{4.47}$$

we could invert (4.45a), (4.45b), and (4.47) to express the q_k as functions of R_1, and R_2 and R_θ, but there is a better approach.

Experience has shown that internal vibrations are best described in terms of variables reflecting symmetries of the molecular structure. The water molecule is symmetrical about an axis through the equilibrium position of the oxygen atom, as shown in Figure 4.1. Accordingly, we introduce *symmetry coordinates* S_1, S_2, S_3 for three independent sets of atomic displacements, as indicated in Figure 4.3. In terms of these variables, the atomic displacements are given by

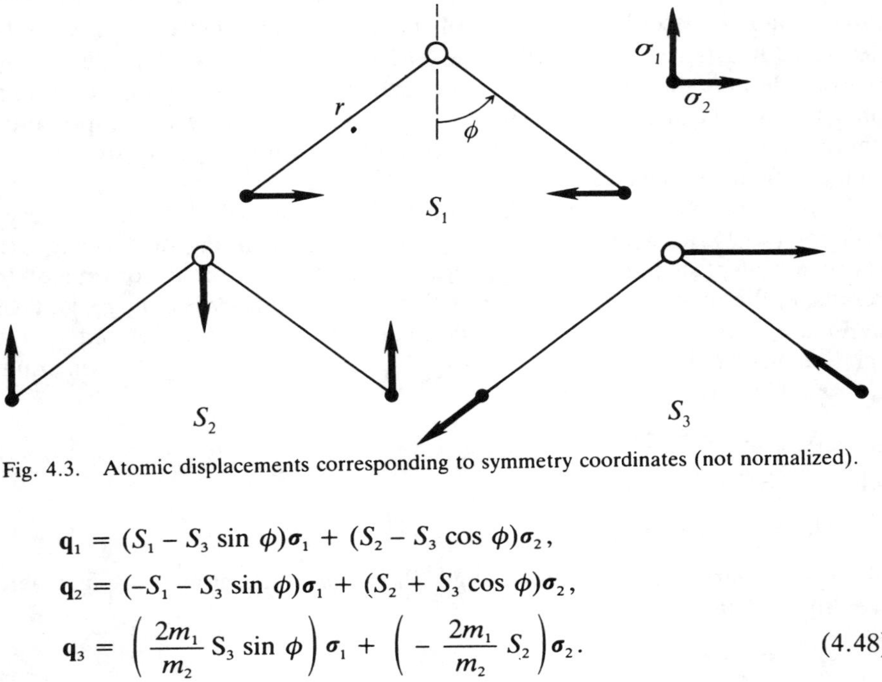

Fig. 4.3. Atomic displacements corresponding to symmetry coordinates (not normalized).

$$\begin{aligned}
\mathbf{q}_1 &= (S_1 - S_3 \sin\phi)\boldsymbol{\sigma}_1 + (S_2 - S_3 \cos\phi)\boldsymbol{\sigma}_2, \\
\mathbf{q}_2 &= (-S_1 - S_3 \sin\phi)\boldsymbol{\sigma}_1 + (S_2 + S_3 \cos\phi)\boldsymbol{\sigma}_2, \\
\mathbf{q}_3 &= \left(\frac{2m_1}{m_2} S_3 \sin\phi\right)\boldsymbol{\sigma}_1 + \left(-\frac{2m_1}{m_2} S_2\right)\boldsymbol{\sigma}_2.
\end{aligned} \tag{4.48}$$

These expressions were constructed by assigning unit displacements to particles 1 and 2 (for each variable) and then obtaining $\mathbf{q}_3$ from (4.42).

We express the kinetic energy in terms of symmetry coordinates by inserting (4.48) into (4.41);

$$2K = m_{11}\dot{S}_1^2 + m_{22}\dot{S}_2^2 + m_{33}\dot{S}_3^2, \tag{4.49a}$$

where

$$m_{11} = 2m_1, m_{22} = 2am_1, m_{33} = 2bm_1, \tag{4.49b}$$

with

$$a = 1 + 2\frac{m_1}{m_2}, \quad b = 1 + 2\frac{m_1}{m_2}\sin^2\phi. \tag{4.50}$$

The stretching and bending variables are expressed in terms of the symmetry coordinates by inserting (4.48) into (4.45a,b) and (4.47), with the results

$$R_1 = -S_1 \sin\phi - S_2\, a\cos\phi + bS_3, \tag{4.51a}$$

$$R_2 = -S_1 \sin\phi - S_2\, a\cos\phi - bS_3, \tag{4.51b}$$

$$rR_\theta = -2S_1\cos\phi + 2S_2\sin\phi. \tag{4.51c}$$

Inserting this into (4.44), we get the following expression for the potential energy in terms of symmetry coordinates:

$$2V = k_{11}S_1^2 + 2k_{12}S_1S_2 + k_{22}S_2^2 + k_{33}S_3^2, \tag{4.52}$$

where

$$\begin{aligned} k_{11} &= 2k_1\sin^2\phi + \frac{4k_\theta}{r^2}\cos^2\phi \\ k_{12} &= 2a\cos\phi\sin\phi\left(k_1 - \frac{2k_\theta}{r^2}\right) \\ k_{22} &= 2a^2\left(k_1\cos^2\phi + \frac{2k_\theta}{r^2}\sin^2\phi\right) \\ k_{33} &= 2b^2k_1. \end{aligned} \tag{4.53}$$

The introduction of symmetry coordinates has simplified the kinetic and potential energy functions so the normal frequencies can easily be calculated by the brute force method. In this case, the characteristic determinant (4.20) has the form

$$\begin{vmatrix} k_{11} - \omega^2 m_{11} & k_{12} & 0 \\ k_{12} & k_{22} - \omega^2 m_{22} & 0 \\ 0 & 0 & k_{33} - \omega^2 m_{33} \end{vmatrix} = 0. \tag{4.54}$$

This factors immediately into the two equations

$$k_{33} - \omega^2 m_{33} = 0, \tag{4.55}$$

$$m_{11}m_{22}\omega^4 - (m_{11}k_{22} + m_{22}k_{11})\omega^2 + k_{11}k_{22} - k_{12}^2 = 0. \tag{4.56}$$

The latter equation can be put in the factored form $(\omega^2 - \omega_1^2)(\omega^2 - \omega_2^2) = 0$ if ω_1^2 and ω_2^2 are its roots; so, by comparison of coefficients, we can express the roots of (4.56) by the equations

$$\omega_1^2 + \omega_2^2 = \frac{m_{11}k_{22} + m_{22}k_{11}}{m_{11}m_{22}}, \quad \omega_1^2\,\omega_2^2 = \frac{k_{11}k_{22} - k_{12}^2}{m_{11}m_{22}}. \tag{4.57}$$

Using (4.49b), (4.50) and (4.53) to evaluate the right sides of these equations in terms of molecular parameters, we obtain

$$\omega_1^2 + \omega_2^2 = \left(1 + 2\,\frac{m_1}{m_2}\cos^2\theta \right)\frac{k_1}{m_1} + \frac{2}{m_1}\left(1 + 2\,\frac{m_1}{m_2}\sin^2\phi \right)\frac{k_\theta}{r^2}\,, \tag{4.58}$$

$$\omega_1^2\omega_2^2 = 2\left(1 + 2\,\frac{m_1}{m_2} \right)\frac{k_1 k_\theta}{m_1^2 r^2}\,. \tag{4.59}$$

From (4.55) we get an independent expression for the other normal frequency

$$\omega_3^2 = \left(1 + 2\,\frac{m_1}{m_2}\sin^2\phi \right)^2 \frac{k_1}{m_1}\,. \tag{4.60}$$

Experimentally determined values for the normal frequencies of the water molecule are

$$\frac{\omega_1}{2\pi c} = 3654\ \text{cm}^{-1}, \qquad \frac{\omega_2}{2\pi c} = 1595\ \text{cm}^{-1}, \quad \frac{\omega_3}{2\pi c} = 3756\ \text{cm}^{-1},$$

where c is the speed of light. Substitution of these values into (4.59) and (4.60), yields the following values for the force constants:

$$k_1 = 7.76 \times 10^5\ \text{dyne cm}^{-1}, \quad \frac{k_\theta}{r^2} = 0.69 \times 10^5\ \text{dyne cm}^{-1}.$$

Comparison with the remaining equation (4.58) shows a two percent discrepancy, which is attributed to anharmonic forces. Note that the effective bending constant k_θ/r^2 is only about ten percent as large as the stretching constant k_1, indicating that bonds are easier to bend than stretch.

To complete the determination of normal modes, we need to find the matrix $[a]$ relating symmetry coordinates $|S)$ to normal coordinates $|Q)$ by $|S) = [a]|Q)$. From (4.54) it is evident that $[a]$ has the form

$$[a] = \begin{bmatrix} a_{11} & a_{12} & 0 \\ a_{22} & a_{22} & 0 \\ 0 & 0 & a_{33} \end{bmatrix}. \tag{4.61}$$

This shows that S_3 is already a normal coordinate differing from Q_3 only by a normalization factor, which we can read off directly from the expression (4.49) for kinetic energy;

$$a_{33} = m_{11}^{1/2} = (2bm_1)^{1/2}. \tag{4.62}$$

The brute force method gives us the ratios

$$\frac{a_{1\alpha}}{a_{2\alpha}} = \frac{\omega_\alpha^2 m_{22} - k_{22}}{k_{12}}$$

for $\alpha = 1, 2$. For the water molecule the ratios have the values

$$\frac{a_{11}}{a_{21}} = 1.2, \quad \frac{a_{12}}{a_{22}} = -0.875.$$

This information is sufficient for us to construct a diagrammatic representation of the displacements in the normal modes from Figure 4.3. The result is shown in Figure 4.4. The displacement vector for the oxygen atom in the figure should actually be reduced by a factor $2m_1/m_2 = 1/8$ to satisfy the center of mass constraint (4.42), and, of course, the displacements are exaggerated compared to the scale of interatomic distances.

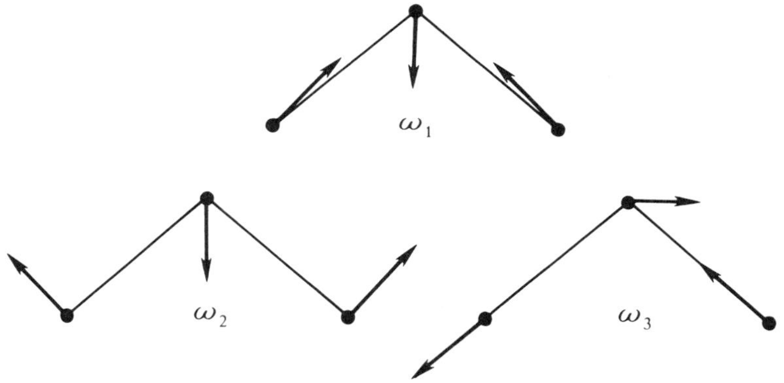

Fig. 4.4. Normal modes of the water molecule.

Damped Oscillations

For a harmonic system with linear damping, the equation of motion has the form

$$[m]|\ddot{q}) + [\mu]|\dot{q}) + [k]|q) = 0. \tag{4.64}$$

This is the matrix version of the system of coupled linear differential equations

$$\sum_\beta (m_{\alpha\beta}\ddot{q}_\beta + \mu_{\alpha\beta}\dot{q}_\beta + k_{\alpha\beta}q_\beta) = 0. \tag{4.65}$$

The matrix $[\mu]$ is symmetric and can be derived from the assumption of linear damping for each of the particles, as shown in Example 2.2 of Section 6-2. In general, the matrix elements $\mu_{\alpha\beta} = \mu_{\alpha\beta}(q)$ depend on the configuration, but in the linear approximation we use the constant values $\mu_{\alpha\beta} = \mu_{\alpha\beta}(0)$, just as we have done with the mass matrix elements $m_{\alpha\beta}$.

Equation (4.64) can be solved by the brute force method. One substitutes a trial solution of the form

$$|q) = |q)e^{i\Omega t}$$

into the equation to get

$$(-\Omega^2[m] + i\Omega[\mu] + [k])|a) = 0. \tag{4.66}$$

This has nontrivial solution only if

$$\det(-\Omega^2 m_{\alpha\beta} + i\Omega\mu_{\alpha\beta} + k_{\alpha\beta}) = 0. \tag{4.67}$$

This determinant is a polynomial of degree $2n$ in Ω, with $2n$ (complex) roots Ω_α. If the roots are all distinct, then (4.66) yields $2n$ corresponding solutions $|a_\alpha)$, and the general solution of (4.64) is a superposition of the $2n$ orthogonal solutions

$$|q_\alpha) = |a_\alpha)e^{i\Omega_\alpha t}. \tag{4.68}$$

If the characteristic determinant (4.67) is m-fold degenerate in a root Ω, then a trial solution of the form

$$|q) = (|a_1) + |a_2)t + \ldots + |a_m)t^{m-1})e^{i\Omega t}$$

will work. This generalizes the case of the critically damped harmonic oscillator discussed in Section 3-8.

The brute force method always works, but there are simpler methods exploiting special symmetries of the coefficient matricies. In some cases there is a change of variables which simultaneously diagonalizes all three matrices $[m]$, $[\mu]$ and $[k]$. Thus, if

$$[\mu] = \gamma[m], \tag{4.69}$$

then the change of variables $|q) = [a]|Q)$, which we found in the undamped case, puts the equation motion in the form

$$|\ddot{Q}) + \gamma|\dot{Q}) + [\omega^2]|Q) = 0, \tag{4.70}$$

where $[\omega^2]$ is the diagonal matrix (4.34). So each coordinate Q_α separately satisfies the equation for a damped harmonic oscillator:

$$\ddot{Q}_\alpha + \gamma\dot{Q}_\alpha + \omega_\alpha^2 Q_\alpha = 0. \tag{4.71}$$

This has the complex solution

$$Q_\alpha = (C_\alpha e^{i\delta_\alpha})e^{i\Omega_\alpha t}, \tag{4.72}$$

where

$$\Omega_\alpha = (\omega_\alpha^2 - \tfrac{1}{4}\gamma^2)^{1/2} + \tfrac{1}{4}i\gamma. \tag{4.73}$$

For light damping ($\omega_\alpha^2 \gg \gamma^2$); the "physical part" of the solution can therefore be written

$$\langle Q_\alpha \rangle_0 = C_\alpha e^{-\gamma t} \cos(\omega_\alpha t + \gamma). \tag{4.74}$$

This shows that all the normal modes are equally damped despite the differences in frequency.

Example 4.2

For the damped double pendulum discussed in Example 2.2, the results of Example 2.2 enable us to write the linearized equation of motion

$$[m]\ddot{\phi}) + [\mu]|\dot{\phi}) + [k]|\phi) = 0,$$

where

$$[\mu] = \begin{bmatrix} (\mu_1 + \mu_2)l_1^2 & \mu_2 l_1 l_2 \cdot \\ \mu_2 l_1 l_2 & \mu_2 l_2^2 \end{bmatrix}$$

and all the other matrices are as given in Example 4.1. Comparison of $[\mu]$ with $[m]$ in Example 4.1 shows that $[\mu] = \gamma[m]$ if $m_1 = m_1 = m$ and $\mu_1 = \mu_2 = \gamma m$. Then the method just discussed can be applied (Exercise 4.10).

Forced Oscillations

When the ith particle in a harmonic system is subject to an external sinusoidal force $\mathbf{f}_i \sin \omega t$, according to (2.14), the system is subject to a generalized force with components

$$F_\alpha \sin \omega t = \sum_i \mathbf{f}_i \cdot (\partial_{q_\alpha} \mathbf{x}_i) \sin \omega t.$$

Accordingly, Lagrange's equation yields the equations of motion

$$\sum_\beta (m_{\alpha\beta}\ddot{q}_\beta + \mu_{\alpha\beta}\dot{q}_\beta + k_{\alpha\beta}q_\beta) = F_\alpha e^{i\omega t}, \tag{4.75}$$

where the generalized force is taken to be complex for convenience, and the F_α are constant in the first order approximation. In matrix form,

$$[m]|\ddot{q}) + [\mu]|\dot{q}) + [k]|q) = |F)e^{i\omega t} \cdot \tag{4.76}$$

This generalizes the Equation (3–9.4) for a forced harmonic oscillator.

As in the harmonic oscillator case, the general solution of (4.76) consists of a particular solution plus a solution of the homogeneous equation (4.70) determined by the initial conditions. Since the homogeneous solution has been discussed, and it can be ignored in the presence of a steady driving force because it decays exponentially to zero, we can concentrate on the particular solution.

If the matrices $[m]$, $[\mu]$, $[k]$ can be diagonalized simultaneously by the change of variables $|q) = [a]|Q)$, we may choose $[a]$ in the form (4.40), to put the equation of motion (4.76) in the form

$$|\ddot{Q}) + [\gamma]|\dot{Q}) + [\omega^2]|Q) = |f)e^{i\omega t}, \tag{4.77}$$

where

$$|f) = [a]^{-1}|F). \tag{4.78}$$

The components of (4.77) obey the independent equations of motion

$$\ddot{Q}_\alpha + \gamma\dot{Q}_\alpha + \omega_\alpha^2 Q_\alpha = f_\alpha e^{i\omega t}. \tag{4.79}$$

This is the equation for the harmonic oscillator solved in Section 3-9. In the present case, the particular solution can be written

$$Q_\alpha(t) = \frac{f_\alpha e^{i(\omega t - \delta_\alpha)}}{[(\omega_\alpha^2 - \omega^2)^2 + \gamma_\alpha^2\omega^2]^{1/2}}, \tag{4.80a}$$

where

$$\tan\delta_\alpha = \frac{\omega\gamma_\alpha}{\omega_\alpha^2 - \omega^2}. \tag{4.80b}$$

Thus, the normal modes are excited independently of one another, and the excitation of each normal mode depends on the amplitude f_α of the generalized force as well as the driving frequency ω. Excitation is a maximum when the driving frequency is equal to one of the resonant frequencies

$$\omega_{\alpha R} = (\omega_\alpha^2 - \tfrac{1}{2}\gamma_\alpha^2)^{1/2}. \tag{4.81}$$

Molecules and crystals driven by electromagnetic waves absorb energy at such resonant frequencies.

The general case when the equation of motion (4.76) cannot be decomposed into independently excited normal modes will not be discussed here. But it should be mentioned that the general case is also characterized by multiple resonant frequencies.

6-4. Exercises

(4.1) Complete the solution of Example 4.1 to determine the normal modes $|a_\alpha)$ and normal frequencies ω_α when $m_2 = 2m_1 = 2m$, $l_1 = 3l$, $l_2 = 2l$. Determine the normal coordinates Q_1 and Q_2 as functions of the angles ϕ_1 and ϕ_2. Specificy initial conditions which will excite each of the normal modes.

(4.2) Show that in terms of normal coordinates the energy of a harmonic system is given by

$$2E = \sum_\alpha (\dot{Q}_\alpha^2 + \omega_\alpha^2 Q_\alpha^2) = (\dot{Q}|\dot{Q}) + (Q|[\omega^2]|Q).$$

(4.3) Three identical plane pendulums are suspended from a slightly yielding support, so their motions are coupled. (Figure 4.5). To simplify the mathematical description, adopt a system of units in which the masses, lengths, and weights of the pendulums are equal to unity. The linearized potential energy for the system has the form

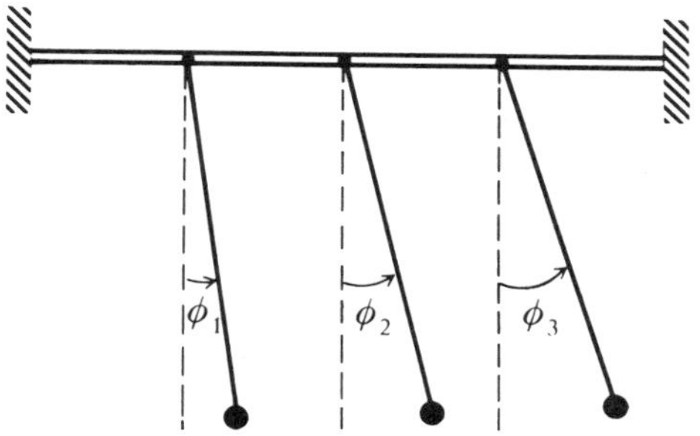

Fig. 4.5. Coupled pendulums.

$$2V = \dot{\phi}_1^2 + \dot{\phi}_2^2 + \dot{\phi}_3^2 - 2k\phi_1\phi_2 - 2k\phi_1\phi_3 - 2k\phi_2\phi_3 .$$

Determine the normal frequencies and modes of oscillation. The system has degenerate modes, but show that an orthogonal set of normal coordinates $|Q)$ can be chosen so that $|\phi) = [a]|Q)$, where

$$[a] = \frac{1}{\sqrt{6}} \begin{bmatrix} \sqrt{3} & 1 & \sqrt{2} \\ -\sqrt{3} & 1 & \sqrt{2} \\ 0 & -2 & \sqrt{2} \end{bmatrix} .$$

Illustrate the normal modes with diagrams. Find an alternative set of normal modes, and explain how they differ physically from the first set.

(4.4) Show that $R_3 = -2S_1$ and derive general expressions for the normal frequencies of a bent triatomic molecule assuming a potential energy of the form (4.44). Evaluate the consistency of the results with empirical data for the water molecule, and compare with the results in the text.

(4.5) Apply the results of the text to linear triatomic molecules by taking $\phi = 90°$. Sketch the normal modes. One of the modes is doubly degenerate, allowing circular displacements of the atoms without a net angular momentum of the atom as a whole. Sketch the circular modes.

(4.6) The experimentally determined normal frequencies of the CO_2 molecule are

$$\frac{\omega_1}{2\pi c} = 1337 \text{ cm}^{-1}, \quad \frac{\omega_2}{2\pi c} = 667 \text{ cm}^{-1}, \quad \frac{\omega_3}{2\pi c} = 2349 \text{ cm}^{-1}.$$

Use the results of the preceding exercise to evaluate the force constants k_1 and k_θ/r^2. Then check for self-consistency.

(4.7) Compute the normal frequencies for the two one-dimensional normal modes of symmetric linear triatomic molecules such as CO_2 and H_2S by assuming a potential energy of form

$$2V = k[(\mathbf{q}_1 - \mathbf{q}_3)^2 + (\mathbf{q}_2 - \mathbf{q}_3)^2].$$

Compare with the results of the preceding exercise for CO_2.

(4.8) Suppose the two particles in Figure 3.1 have different masses. Determine the normal modes and frequencies of the system as explicit functions of the physical parameters.

(4.9) For the two identical particles in Figure 3.1, determine the general solution to their equations of motion if they are subject to linear damping forces.

(4.10) Complete Example 4.2 by finding the general solution when $l_1 = l_2$.

(4.11) Determine the natural frequencies of the coupled pendulums in Exercise (2.5) in Section 6-2.

6-5. The Newtonian Many Body Problem

The *many body problem* in classical mechanics is this: For a system of N particles with known interactions, determine the evolution of the system from any given initial state. In the preceding section we studied one kind of many body problem appropriate for modeling elastic solids. In the *Newtonian many body problem*, the interactions are described by Newton's "Universal Law of Gravitation". Then the equations of motion for particles in the system have the specific form

$$m_i \ddot{\mathbf{x}}_i = -G \sum_{j \neq i} m_i m_j \frac{(\mathbf{x}_i - \mathbf{x}_j)}{|\mathbf{x}_i - \mathbf{x}_j|^3}, \tag{5.1}$$

for $i, j = 1, 2, \ldots, N$. The problem is to characterize all solutions of this system of coupled nonlinear differential equations. This is a mathematical problem of such importance and difficulty that it has engaged the best efforts of some of the greatest mathematicians. To this day, a general solution has not been found even for the case of three bodies, and much of the progress has been in understanding what actually constitutes a solution when the solution cannot be expressed in terms of known functions.

Up to the twentieth century, Newton's Law of Gravitation was the best candidate for an exact force law, so the most important reason for studying the many body problem was to work out detailed implications of the law which could be subjected to empirical test. Since the advent of Einstein's General Theory of Relativity it has been clear that Newton's law is not an exact description of the gravitational interaction in nature. The exact implications of Newton's law still provide baseline predictions from which to measure the small deviations predicted by Einstein's theory and possible alternatives. However, the most sensitive tests for distinguishing alternative theories are in situations free of many body complications.

Although the many body problem may no longer be so important as a test of fundamental gravitational theory, it is crucial to many applications in astronomy and spacecraft mechanics. Thus, an understanding of the many body problem is needed to trace the evolution of the solar system and answer such questions as, will any of the planets collide in the future? Have any collided in the past? Of course, the evolution of the solar system is greatly influenced by other factors such as the energy flux from the Sun and the dissipation of energy by tidal friction. But the many body dynamics is no less significant for that.

In Chapter 4 the gravitational two body problem was solved completely by finding four first integrals (or constants) of motion: the center of mass velocity and initial position, the angular momentum and the eccentricity vector. It is natural, therefore, to attack the general N body problem by looking for new first integrals. Unfortunately, that has turned out to be futile.

The system of equations (5.1) is of *order* $6N$, since it consists of N vector equations of order two. According to the theory of differential equations, then, its general solution depends on $6N$ scalar parameters or, equivalently $2N$ vector parameters, which amounts to specifying the initial position and velocity for each of the N particles. From the general many particle theory of Section 6-1, we know that, for an isolated system, the angular momentum and the center of mass momentum are constants of motion; from which it follows that the center of mass position is determined by its initial value. This gives us nine scalar (three vector) integrals of motion. In addition we know that the total energy is an integral of the motion, because the gravitational force is conservative. Generalizing work by Bruns and Poincaré, Painlevé (1897) proved that, besides these 10 integrals, the general Newtonian N body problem admits no other constants of motion which are algebraic functions of the positions and velocities of the particles or even integrals of such functions. There might be constants of motion expressible in terms of other variables, but none have been found. This leaves $6N-10$ variables to be determined by other means. For that no general method is known. That is where the many body problem gets difficult.

For a given initial conditions the N body problem can be solved by direct numerical integration of the equations of motion (5-1). More practical in most situations is the perturbation method, which generates N body solutions by calculating deviations from two body solutions. It will be developed in Chapter 8. Though these methods generate specific solutions, they give us little insight into qualitative features of solutions in general. They cannot tell us, for example, what specific initial conditions produce periodic or nearly periodic orbits, or when small changes in the initial conditions produce wildly different orbits, or what initial conditions allow one of the bodies to escape to infinity. To answer such qualitative questions about orbits, the great French mathematician Henri Poincaré developed new mathematical methods in the last quarter of the nineteenth century which proved to be seeds for whole new branches of mathematics blossoming in the twentieth century; including algebraic topology, global analysis and dynamical systems theory. Although here we cannot go deeply into the qualitative theory of differential equations incorporating such methods, we will survey what has been learned about the 3 body problem and add some observations about the generalization to N bodies.

The General Three Body Problem

The case of three bodies is not only the simplest unsolved many body problem, it is also the problem of greatest practical importance. Consequently, it is the most thoroughly studied and understood.

The three body problem has been attacked most successfully by analyzing the possibilities for reducing it to the two body problem. Absorbing the gravitational constant G into the definition of mass, for 3 bodies the Equations (5.1) can be written

$$\begin{aligned} \ddot{\mathbf{x}}_1 &= -m_2 \frac{\mathbf{x}_1 - \mathbf{x}_2}{|\mathbf{x}_1 - \mathbf{x}_2|^3} - m_3 \frac{\mathbf{x}_1 - \mathbf{x}_3}{|\mathbf{x}_1 - \mathbf{x}_3|^3}, \\ \ddot{\mathbf{x}}_2 &= -m_1 \frac{\mathbf{x}_2 - \mathbf{x}_1}{|\mathbf{x}_2 - \mathbf{x}_1|^3} - m_3 \frac{\mathbf{x}_2 - \mathbf{x}_3}{|\mathbf{x}_2 - \mathbf{x}_3|^3}, \\ \ddot{\mathbf{x}}_3 &= -m_1 \frac{\mathbf{x}_3 - \mathbf{x}_1}{|\mathbf{x}_3 - \mathbf{x}_1|^3} - m_3 \frac{\mathbf{x}_3 - \mathbf{x}_2}{|\mathbf{x}_3 - \mathbf{x}_2|^3}, \end{aligned} \tag{5.2}$$

With the center of mass as origin, the position vectors are related by

$$m_1\mathbf{x}_1 + m_2\mathbf{x}_2 + m_3\mathbf{x}_3 = 0. \tag{5.3}$$

This reduces the Equations (5.2) to a system or order $18 - 6 = 12$. By using the angular momentum and energy integrals, the order can be reduced to $12 - 4 = 8$. The order can be further reduced to 7 by eliminating time as a variable and then to 6 by a procedure called "the elimination of nodes". However, the actual reduction is messy and not conducive to insight, so we shall not carry it out. For the special case of orbits lying in a fixed plane, the order is reduced to $6 - 2 = 4$, but the problem is still formidable.

The three body equations of motion have their most symmetrical form when expressed in terms of the *relative position vectors* $\mathbf{s}_1$, $\mathbf{s}_2$, $\mathbf{s}_3$ defined by

$$\mathbf{s}_1 = \mathbf{x}_3 - \mathbf{x}_2$$
$$\mathbf{s}_2 = \mathbf{x}_1 - \mathbf{x}_3$$
$$\mathbf{s}_3 = \mathbf{x}_2 - \mathbf{x}_1 \,. \tag{5.4}$$

These variables are related by

$$\mathbf{s}_1 + \mathbf{s}_2 + \mathbf{s}_3 = 0\,. \tag{5.5}$$

(Figure 5.1). Solving (5.3) and (5.4) for the $\mathbf{x}_k$, we get

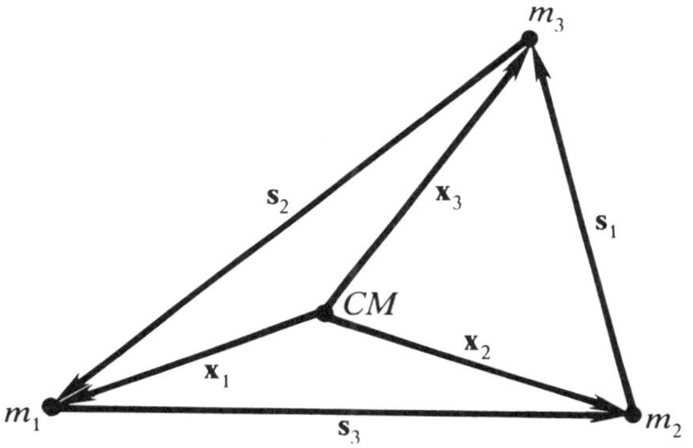

Fig. 5.1. Position vectors for the three body problem.

$$m\mathbf{x}_1 = m_3\mathbf{s}_2 - m_2\mathbf{s}_3$$
$$m\mathbf{x}_2 = m_1\mathbf{s}_3 - m_3\mathbf{s}_1$$
$$m\mathbf{x}_3 = m_2\mathbf{s}_1 - m_1\mathbf{s}_2 \,, \tag{5.6}$$

where

$$m = m_1 + m_2 + m_3\,. \tag{5.7}$$

We need (5.6) to relate a solution in terms of the symmetrical variables $\mathbf{s}_k$ to the fixed center of mass.

By substitution of (5.4) into (5.2), we get equations of motion in the symmetrical form

$$\ddot{\mathbf{s}}_1 = -m\frac{\mathbf{s}_1}{s_1^3} + m_1\mathbf{G}$$
$$\ddot{\mathbf{s}}_3 = -m\frac{\mathbf{s}_3}{s_2^3} + m_3\mathbf{G}$$
$$\ddot{\mathbf{s}}_2 = -m\frac{\mathbf{s}_2}{s_3^3} + m_2\mathbf{G} \tag{5.8}$$

where $s_k = |\mathbf{s}_k|$, and

$$\mathbf{G} = \frac{\mathbf{s}_1}{s_1^3} + \frac{\mathbf{s}_2}{s_2^3} + \frac{\mathbf{s}_3}{s_3^3}\,. \tag{5.9}$$

The noteworthy simplicity of this formulation was first pointed out by Broucke and Lass in 1973. Evidently it had been overlooked in two centuries of research on the three body problem. We see below that it provides a direct route to the known exact solutions of the three body problem. It has yet to be exploited in the analysis of more difficult questions.

The Triangular Solutions

Note that system of equations (5.8) decouples into a set of three similar two body equations if $\mathbf{G} = 0$. Comparing (5.5) and (5.9), we see that this will occur if

$$\mathbf{s}_1^2 = \mathbf{s}_2^2 = \mathbf{s}_3^2, \tag{5.10}$$

that is, if the particles are located at the vertices of an equilateral triangle. Then we can express two "sides" of the triangle in terms of the third by

$$\begin{aligned} \mathbf{s}_1 &= \mathbf{s}_3 e^{\mathbf{i}2\pi/3} = -\tfrac{1}{2}(1 + \mathbf{i}\sqrt{3})\mathbf{s}_3, \\ \mathbf{s}_2 &= \mathbf{s}_3 e^{-\mathbf{i}2\pi/3} = -\tfrac{1}{2}(1 - \mathbf{i}\sqrt{3})\mathbf{s}_3, \end{aligned} \tag{5.11}$$

where $\mathbf{i}$ is the unit bivector for the plane of the triangle. The triangular, relation will be maintained if $\mathbf{i}$ is constant, so $\mathbf{s}_1$ and $\mathbf{s}_2$ are determined by $\mathbf{s}_3$ at all times. This is consistent with $\mathbf{G} = 0$ in (5.8). Thus, we have found a family of solutions which reduce to solutions of the two body Kepler problem. As expressed by (5.11), the three particles remain at the vertices of an equilateral triangle, but the triangle may change its size and orientation in the plane as they move. The equilateral triangle solution was discovered by Lagrange. Note that it is completely independent of the particle masses.

To describe the particle orbits with respect to the center of mass, we substitute (5.11) into (5.6) and find

$$\begin{aligned} m\mathbf{x}_1 &= \tfrac{1}{2}[-2m_2 - m_3 + \mathbf{i}m_3\sqrt{3}]\mathbf{s}_3 \\ m\mathbf{x}_2 &= \tfrac{1}{2}[2m_1 + m_3 + \mathbf{i}m_3\sqrt{3}]\mathbf{s}_3 \\ m\mathbf{x}_3 &= \tfrac{1}{2}[m_1 - m_2 - \mathbf{i}(m_1 + m_2)\sqrt{3}]\mathbf{s}_3. \end{aligned} \tag{5.12}$$

This shows that the particles follow *similar* two body orbits differing only in size and orientation determined by the masses. It should be noted that the acceleration vector of each particle points towards the center of mass (Exercise 5.1). Orbits for elliptical motion are shown in Figure 5.2. Similar orbits for hyperbolic and parabolic motions (Section 4-3) are easily constructed.

The Collinear Solutions

Euler found another exact solution of the three body equations where all particles lie on a line separated by distances in fixed ratio. To ascertain the general conditions for such a solution, suppose that particle 2 lies between the other two particles. Then the condition (5.5) is satisfied by writing

$$\mathbf{s}_1 = \lambda\mathbf{s}_3, \quad \mathbf{s}_2 = -(1 + \lambda)\mathbf{s}_3 \tag{5.13}$$

where λ is a positive scalar to be determined. Now, we can eliminate $\mathbf{G}$ in the equations (5.8) to get

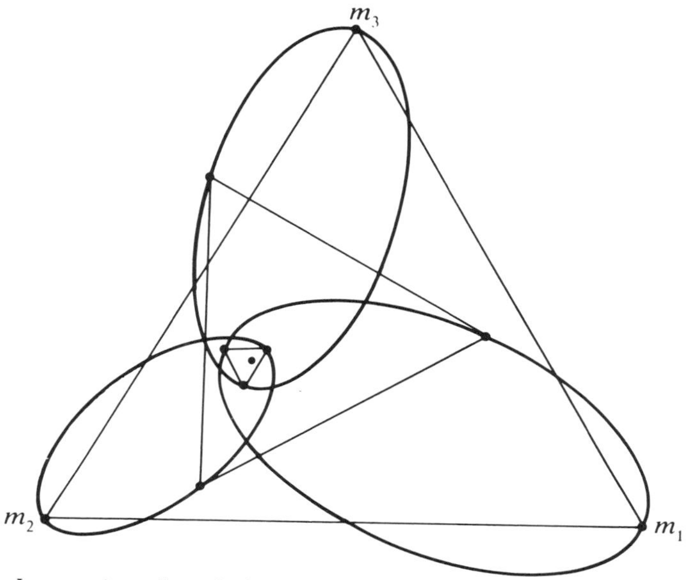

Fig. 5.2. Lagrange's equilateral triangle solution for masses in the ratio $m_1{:}m_2{:}m_3 = 1{:}2{:}3$.

$$\ddot{\mathbf{s}}_1 + m\frac{\mathbf{s}_1}{s_1^3} = \frac{m_1}{m_3}\left(\ddot{\mathbf{s}}_3 + m\frac{\mathbf{s}_3}{s_3^3}\right),$$

$$\ddot{\mathbf{s}}_2 + m\frac{\mathbf{s}_2}{s_2^3} = \frac{m_2}{m_3}\left(\ddot{\mathbf{s}}_3 + m\frac{\mathbf{s}_3}{s_3^3}\right) \tag{5.14}$$

Inserting (5.13) in these equations to eliminate $\mathbf{s}_1$ and $\mathbf{s}_2$, we obtain

$$(m_2 + m_3(1+\lambda))\ddot{\mathbf{s}}_3 = -(m_2 + m_3(1+\lambda)^{-2})\frac{m\mathbf{s}_3}{s_3^3}\,.$$

$$(m_1 - m_3\lambda)\ddot{\mathbf{s}}_3 = -(m_1 - m_3\lambda^{-2})\frac{m\mathbf{s}_3}{s_3^3}\,. \tag{5.15}$$

Consequently,

$$\frac{m_2 + m_3\,(1+\lambda)}{m_1 - m_3\lambda} = \frac{m_2 + m_3\,(1+\lambda)^{-2}}{m_1 - m_3\lambda^{-2}}\,.$$

Putting this in standard form, we see that it is a fifth degree polynomial:

$$(m_1 + m_2)\lambda^5 + (3m_1 + 2m_2)\lambda^4 + (3m_1 + m_2)\lambda^3 - \lambda^2(m_2 + 3m_3) -$$

$$-\lambda(2m_2 + 3m_3) - (m_2 + m_3) = 0\,. \tag{5.16}$$

The left side is negative when $\lambda = 0$ and positive as $\lambda \to \infty$; therefore the polynomial has a positive real root. By Descartes' rule of signs, it has no more than one positive real root. Therefore λ is a unique function of the masses, and the two body solutions of (5.15) determine a family of collinear three body solutions. Two other solutions are obtained by putting different particles between the others. Thus, *there are three distinct families of collinear three body solutions*. A solution for the elliptic case is shown in Figure 5.3.

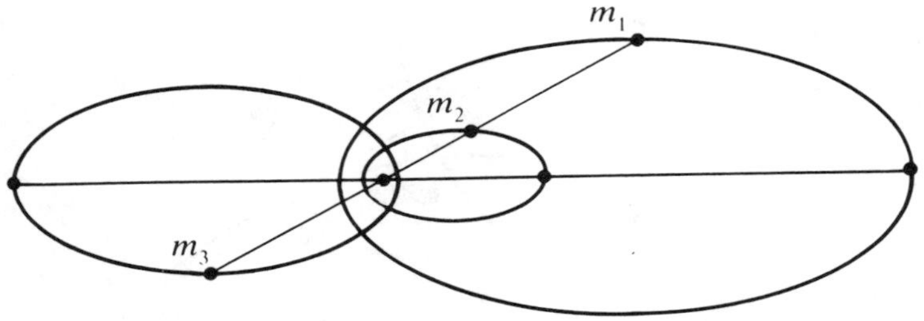

Fig. 5.3. Euler's collinear solution for masses in the ratio $m_1{:}m_2{:}m_3 = 1{:}2{:}3$.

Generalizations of the Lagrange and Euler solutions for systems with more than three particles have been found. As in the three particle case, in all these solutions the particles remain in a permanent configuration with accelerations directed towards the center of mass. The solutions are mainly of mathematical interest, since they require extremely special initial conditions to be realized physically.

The Euler and Lagrange solutions are the only known exact solutions of the three body problem. To get an understanding of the great range of other solutions, we turn to a qualitative analysis.

Classification of Solutions

A systematic classification and analysis of three body solutions based on energy and asymptotic behavior was initiated by Chazy in 1922 and refined by others since. Table 5.1 gives a current form of the classification using terminology suggested by Szebehely.

The three body system can be described as a pair of two body systems, one consisting of two particles, the second consisting of the third particle and the center of mass of the other two. The motion of each two body system can be described as elliptic, parabolic or hyperbolic, depending on the system's energy, and this description becomes exact asymptotically (that is, in the limit $t \to \infty$) if the separation of the third body from the others increases with time t when t is sufficiently large. The permissible asymptotic states of the two body system depend on the sign of the total energy of the three body system.

When the total energy E is positive the classification is easy, because the energy of at least one two body subsystem must be positive, so its asymptotic motion must be hyperbolic and the two body systems separate. There are three possibilities, as shown in Table 5.1. In hyperbolic explosion all three

TABLE 5.1. Classification of Three Body Solutions

$E > 0$		
	Explosion	hyperbolic
		hyperbolic-parabolic
	Escape	hyperbolic-elliptic
$E = 0$		
	Escape	hyperbolic-elliptic
	Explosion	parabolic
$E < 0$		
	Escape	hyperbolic-elliptic
		parabolic-elliptic
	Bounded motion	
		Interplay
		Ejection
		Revolution
		Equilibrium
		Periodic
	Oscillatory motion	

particles depart along hyperbolas. In hyperbolic-parabolic explosion two particles depart along parabolas while the third departs along a hyperbola. In the hyperbolic-elliptic case, one particle escapes along a hyperbola while the others form a *binary*, that is, a two particle system bound in elliptic motion.

The classification for zero total energy is similar to the positive energy case. Of course, in all cases the parabolic motions are unlikely to be realized physically, because they require a specific value for the energy.

When the total energy is negative, the classification is more complex. As in the positive energy case, the possibility exists that one particle escapes leaving a binary behind. However, there are also many types of bounded motion. The term *interplay* refers to motions with repeated close approaches among the particles. *Ejection* refers to motions where two particles form a binary, while the third is repeatedly ejected on large nearly elliptical orbits, much as comets are ejected from the solar system. Revolution is the case where the orbit of the third body surrounds a binary. We have discussed the *equilibrium* solutions of Lagrange and Euler already. As we show below in a special case, these solutions are unstable unless there are large differences in the masses. The *periodic* orbits can be of any type of bounded motion just mentioned.

The oscillatory motion listed in Table 5.1 was not discovered until 1960. It consists of a binary and a third particle which moves along a line perpendicular to the orbital plane of the binary and through its center of mass. The oscillating particle goes to infinity along this line while its velocity goes to zero in finite time, and this behavior is repeated as time goes to infinity. Obviously, this case is not of physical interest.

The classification gives us a general picture of the possible motions, but it is not specific enough to answer the most important questions, such as, which type of motion will ensue from given initial conditions with negative total energy. However, it has been determined that for arbitrary initial conditions hyperbolic-elliptic escape is the most likely. Interplay is a necessary prelude to ejection, and repeated ejections may lead to escape. Usually, the particle which escapes has the smallest mass.

The Restricted Three Body Problem

The three body problem can be viewed as a perturbation by the third particle of the two body motion of the other two particles, called *primaries*. For this purpose, it is convenient to employ *Jacobi coordinates* **x** and **r**. The vector **x** is the position vector of the third particle with respect to the center of mass of the primaries, located at

$$\frac{m_1\mathbf{x}_1 + m_2\mathbf{x}_2}{m_1 + m_2} = \frac{-m_3\mathbf{x}_3}{m_1 + m_2}.$$

Therefore,

$$\mathbf{x} = \mathbf{x}_3 + \frac{m_3\mathbf{x}_3}{m_1 + m_2} = \frac{m}{\mu}\mathbf{x}_3\,, \tag{5.17}$$

where $m = m_1 + m_2 + m_3$ and $\mu = m_1 + m_2$. The relative position vector for the primaries is

$$\mathbf{r} = \mathbf{x}_2 - \mathbf{x}_1\,. \tag{5.18}$$

In terms of the Jacobi coordinates, the relative positions of the third particle with respect to the primaries are

$$\mathbf{r}_1 = \mathbf{x}_3 - \mathbf{x}_1 = \mathbf{x} + m_2\mu^{-1}\mathbf{r}$$

$$\mathbf{r}_2 = \mathbf{x}_3 - \mathbf{x}_2 = \mathbf{x} - m_1\mu^{-1}\mathbf{r}\,. \tag{5.19}$$

In terms of the Jacobi coordinates, the three body equations of motion (5.2) take the form of an equation for the relative motion of the primaries

$$\ddot{\mathbf{r}} = -\mu\frac{\mathbf{r}}{r^3} + m_3\left(\frac{\mathbf{r}_2}{r_2^3} - \frac{\mathbf{r}_1}{r_1^3}\right), \tag{5.20}$$

coupled to an equation for the motion of the third body with respect to the primaries

$$\ddot{\mathbf{x}} = -\frac{m_1 m}{\mu}\frac{\mathbf{r}_1}{r_1^3} - \frac{m_2 m}{\mu}\frac{\mathbf{r}_2}{r_2^3}\,. \tag{5.21}$$

Here, $\mathbf{r}_1$ and $\mathbf{r}_2$ are to be regarded as auxiliary variables defined in terms of the Jacobi variables by (5.19).

Jacobi coordinates are most appropriate when the mass of the third body is much less than the mass of either primary. When the mass m_3 is so small that its influence on the primaries can be neglected, we can write $m = \mu = m_1 + m_2$ and Equations (5.20) and (5.21) reduce to

$$\ddot{\mathbf{r}} = -\mu \frac{\mathbf{r}}{r^3}, \tag{5.22}$$

$$\ddot{\mathbf{x}} = -m_1 \frac{\mathbf{r}_1}{r_1^3} - m_2 \frac{\mathbf{r}_2}{r_2^3}. \tag{5.23}$$

The problem of solving these equations is called *restricted three body problem*.

We already know the general solution of the two body equation (5.22) for the primaries. However, the general solution of (5.23) is still unknown, though many special solutions have been found. Thus, the restricted three body problem is still an open area for research.

Most three body research has concentrated on the *circular restricted* problem, which is restricted to circular solutions of the primary two body equation (5.22). Besides the helpful mathematical simplifications, this special case has important practical applications. It is a good model, for instance, of the Sun-Earth-Moon system, or of a spacecraft travelling between the Earth and Moon. So let us examine this case in more detail.

To that end, it is convenient to make a slight change of notation in (5.22) and (5.23), writing

$$\ddot{\mathbf{r}}' = -\mu \frac{\mathbf{r}'}{r^3} \tag{5.24}$$

$$\ddot{\mathbf{x}}' = -m_1 \frac{\mathbf{r}_1'}{r_1^3} - m_2 \frac{\mathbf{r}_2'}{r_2^3}. \tag{5.25}$$

The circular solution to (5.24) with angular frequency $\boldsymbol{\omega}$ is

$$\mathbf{r}' = R^\dagger \mathbf{r} R = \mathbf{r} R^2, \tag{5.26}$$

where

$$R = e^{1/2 i\boldsymbol{\omega} t} \tag{5.27}$$

and $\mathbf{r}$ is a fixed vector, the relative position vector of the primaries in the rotating system. The motion of the primaries is most easily accounted for in (5.25) by transforming it to the rotating system in which the primaries are at rest. Accordingly, we write

$$\mathbf{x}' = R^\dagger \mathbf{x} R, \quad \mathbf{r}_k' = R^\dagger \mathbf{r}_k R \tag{5.28}$$

and substitute into (5.25) to get the equation of motion in the rotating system:

$$\ddot{\mathbf{x}} + 2\boldsymbol{\omega} \times \dot{\mathbf{x}} = \mathbf{F}(\mathbf{x}) \tag{5.29}$$

where

$$\mathbf{F}(\mathbf{x}) = -\,\boldsymbol{\omega} \times (\boldsymbol{\omega} \times \mathbf{x}) - m_1 \frac{\mathbf{r}_1}{r_1^3} - m_2 \frac{\mathbf{r}_2}{r_2^3}\,, \tag{5.30}$$

is an "effective force" with $\mathbf{r}_1$ and $\mathbf{r}_2$ given by (5.19). Thus, the circular restricted problem has been reduced to solving this equation. Before looking for specific solutions, we ascertain some general characteristics of the equation.

We note that the "effective force" $F(\mathbf{x})$ is a conservative force with potential

$$U(\mathbf{x}) = -\,\tfrac{1}{2}(\boldsymbol{\omega} \times \mathbf{x})^2 - \frac{m_1}{r_1} - \frac{m_2}{r_2}\,, \tag{5.31}$$

that is,

$$\mathbf{F} = -\,\nabla U\,. \tag{5.32}$$

This can be proved by using properties of the gradient operator $\nabla = \nabla_{\mathbf{x}}$ established in Section 2-8 to carry out the differentiation of (5.31). Using $\nabla(\boldsymbol{\omega}\cdot\mathbf{x}) = \boldsymbol{\omega}$ and $\nabla \mathbf{x}^2 = 2\mathbf{x}$, we find that the gradient of the "centrifugal pseudopotential"

$$-\,\tfrac{1}{2}(\boldsymbol{\omega} \times \mathbf{x})^2 = \tfrac{1}{2}(\boldsymbol{\omega} \wedge \mathbf{x})^2 = \tfrac{1}{2}[\,(\boldsymbol{\omega}\cdot\mathbf{x})^2 - \boldsymbol{\omega}^2\mathbf{x}^2] \tag{5.33}$$

is the "centrifugal pseudoforce"

$$-\boldsymbol{\omega} \times (\boldsymbol{\omega} \times \mathbf{x}) = \boldsymbol{\omega}\cdot(\boldsymbol{\omega} \wedge \mathbf{x}) = \boldsymbol{\omega}^2\mathbf{x} - \boldsymbol{\omega}(\boldsymbol{\omega}\cdot\mathbf{x})\,. \tag{5.34}$$

The last two terms in (5.30) are obtained from the last two terms of (5.31) by using the chain rule and $\nabla r_k = \mathbf{r}_k/r_k$.

Multiplying the equation of motion (5.29) by $\dot{\mathbf{x}}$ and using (5.32), we easily prove that

$$C = \tfrac{1}{2}\dot{\mathbf{x}}^2 + U(\mathbf{x})\,, \tag{5.35}$$

known as *Jacobi's integral*, is a constant of the motion. Of course, this is just the energy integral for the three body system with the contributions of the primaries removed, as allowed by the approximation in the restricted problem.

Before studying particular solutions of the restricted problem, we can simplify computations by choosing a unit of length so that

$$\mathbf{r}^2 = 1\,, \tag{5.36a}$$

a unit of time so that

$$\boldsymbol{\omega}^2 = 1\,, \tag{5.36b}$$

and a unit of mass so that

$$\mu = m_1 + m_2 = 1\,. \tag{5.36c}$$

Defining a mass difference parameter γ by

$$m_1 - m_2 = \mu\gamma, \tag{5.36d}$$

we have

$$m_1 = \tfrac{1}{2}(1+\gamma), \quad m_2 = \tfrac{1}{2}(1-\gamma), \tag{5.36e}$$

where γ is positive for $m_1 > m_2$.

Equilibrium Points

A point $\mathbf{x}_0$ at which the force function $F(\mathbf{x})$ vanishes is called an *equilibrium point* (*stationary point* or *libration point*) of the differential equation (5.29). A particle initially at rest at an equilibrium point will remain at rest, because its acceleration vanishes. The equilibrium points are "critical points" of the potential $U(\mathbf{x})$, for, from (5.32) we see that

$$\mathbf{a}\cdot\nabla U(\mathbf{x}_0) = -\mathbf{a}\cdot\mathbf{F}(\mathbf{x}_0) = 0,$$

that is, at an equilibrium point $\mathbf{x}_0$ the directional derivative of the potential vanishes in every direction $\mathbf{a}$.

The equilibrium points are solutions of the equation

$$F(\mathbf{x}_0) = \mathbf{x}_0 - \boldsymbol{\omega}(\boldsymbol{\omega}\cdot\mathbf{x}_0) - m_1\frac{\mathbf{r}_1}{r_1^3} - m_1\frac{\mathbf{r}_2}{r_2^3} = 0, \tag{5.37}$$

where, by (5.19) and (5.36c),

$$\mathbf{r}_1 = \mathbf{x}_0 + m_2\mathbf{r}, \quad \mathbf{r}_2 = \mathbf{x}_0 - m_1\mathbf{r}. \tag{5.38}$$

Dotting (5.37) and (5.38) with $\boldsymbol{\omega}$, we determine that $\boldsymbol{\omega}\cdot\mathbf{r}_1 = \boldsymbol{\omega}\cdot\mathbf{r}_2 = \boldsymbol{\omega}\cdot\mathbf{x}_0 = 0$. Therefore, all equilibrium poins lie in the orbital plane of the primaries (called the *primary plane*).

The solutions of (5.37) are just restricted versions of the exact solutions of Lagrange and Euler which we found for the general three body problem. Therefore, since the primaries have a fixed separation, there are two eqiulibrium points L_4, L_5 at the vertices of equilateral triangles with $\mathbf{r}_1^2 = \mathbf{r}_2^2 = \mathbf{r}^2$, and three equilibrium

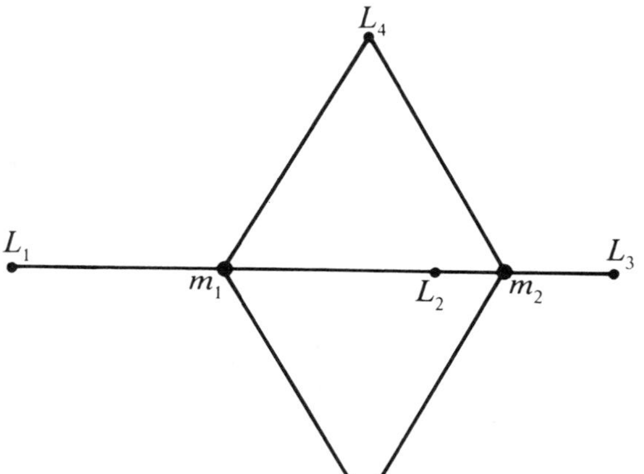

Fig. 5.4. The five Lagrange points of the restricted three body problem.

points L_1, L_2, L_3 collinear with the positions of the primaries (Figure 5.4). The five equilibrium points L_1, L_2, L_3, L_4, L_5 are called *Lagrange points*.

The Lagrange points are of more than mere academic interest. Groups of asteroids known as the "Trojans" are found near the points L_4 and L_5 of the Sun-Jupiter system. The Lagrange point L_4 of the Earth-Moon system has been suggested as a suitable place for a space colony. Reflected light from asteroids temporarily trapped near the point L_3 of the Sun-Earth system may be responsible for a faintly glowing spot in the night sky called the *gegenschein*.

Stability of the Lagrange Points

An equilibrium point is said to be *stable* if a particle stays near it when subjected to small disturbances. To investigate the stability at $\mathbf{x}_0$, we determine how the force function $\mathbf{F}(\mathbf{x})$ varies with small displacement $\boldsymbol{\varepsilon}$ from $\mathbf{x}_0$. A Taylor expansion (Section 2.8) gives us

$$\mathbf{F}(\mathbf{x}_0 + \boldsymbol{\varepsilon}) = \mathbf{F}(\mathbf{x}_0) + \boldsymbol{\varepsilon}\cdot\nabla\mathbf{F}(\mathbf{x}_0) + \ldots, \tag{5.39}$$

where higher order terms can be neglected. Introducing the notation $\mathbf{F}'(\boldsymbol{\varepsilon}) = \boldsymbol{\varepsilon}\cdot\nabla\mathbf{F}(\mathbf{x}_0)$ and differentiating (5.30) we obtain

$$\mathbf{F}'(\boldsymbol{\varepsilon}) = -\omega \times (\boldsymbol{\omega} \times \boldsymbol{\varepsilon}) - \left(\frac{m_1}{r_1^3} + \frac{m_2}{r_2^3}\right)\boldsymbol{\varepsilon} + 3\left(m_1\mathbf{r}_1\frac{\mathbf{r}_1\cdot\boldsymbol{\varepsilon}}{r_2^5} + m_2\mathbf{r}_2\frac{\mathbf{r}_2\cdot\boldsymbol{\varepsilon}}{r_2^5}\right), \tag{5.40}$$

where $\mathbf{r}_1$ and $\mathbf{r}_2$ are given by (5.38). With $\mathbf{F}(\mathbf{x}_0) = 0$, substitution of (5.39) and $\mathbf{x}(t) = \mathbf{x}_0 + \boldsymbol{\varepsilon}(t)$ into the equation of motion (5.29) yields the *variational equation*

$$\ddot{\boldsymbol{\varepsilon}} + 2\omega \times \dot{\boldsymbol{\varepsilon}} = \mathbf{F}'(\boldsymbol{\varepsilon}). \tag{5.41}$$

This is the linearized equation for motion near an equilibrium point. We say "linearized", because $\mathbf{F}'(\boldsymbol{\varepsilon})$ is the linear approximation to the force $\mathbf{F}(\mathbf{x}_0 + \boldsymbol{\varepsilon})$. The equation is called a *variational equation*, because it describes deviations (variations) from a reference orbit, namely the circular orbit of an equilibrium point in the primary plane.

The theory of differential equations tells us that stability (instability) of the linearized equation is necessary for stability (instability) of the nonlinear equation it approximates. To prove that stability of the linearized equation is also *sufficient* for stability of the nonlinear equation is a difficult mathematical problem that has been solved only for particular cases, too difficult to broach here. We must be content with the results of a linear analysis, a study of solutions to the linearized equation (5.41). Stability at the various Lagrange points must be examined separately. But first it is advisable to ascertain the qualitative features of the variational equation which contribute to stability.

For the displacement component $\varepsilon_3 = \boldsymbol{\omega}\cdot\boldsymbol{\varepsilon}$ along the direction $\hat{\boldsymbol{\omega}}$ normal to the primary plane, by dotting (5.40) and (5.41) with $\boldsymbol{\omega}$ we obtain the equation

$$\ddot{\varepsilon}_3 = -\left(\frac{m_1}{r_1^3} + \frac{m_2}{r_2^3}\right)\varepsilon_3 . \tag{5.42}$$

Since the coefficient is positive, $\boldsymbol{\varepsilon}_3$ is limited to small harmonic oscillations. Therefore, the equilibrium points are stable with respect to displacements normal to the primary plane. So, for the rest of our analysis we limit out attention to motions within the primary plane.

From (5.40) we see that $\mathbf{F}'(\boldsymbol{\varepsilon})$ is a symmetric linear function. Moreover, for displacements in the primary plane,

$$\boldsymbol{\varepsilon}\cdot\mathbf{F}'(\boldsymbol{\varepsilon}) = \left(1 - \frac{m_1}{r_1^3} - \frac{m_2}{r_2^3}\right)\boldsymbol{\varepsilon}^2 + 3\left(\frac{m_1}{r_1^5}(\mathbf{r}_1\cdot\boldsymbol{\varepsilon})\right)^2 + 3\left(\frac{m_2}{r_2^5}(\mathbf{r}_2\cdot\boldsymbol{\varepsilon})^2\right). \tag{5.43}$$

We shall see that the first coefficient vanishes when evaluated at the Lagrange points L_4 and L_5 so $\boldsymbol{\varepsilon}\cdot\mathbf{F}'(\boldsymbol{\varepsilon}) > 0$ for $\boldsymbol{\varepsilon} \neq 0$. This tells us that $\mathbf{F}'(\boldsymbol{\varepsilon})$ is a repulsive force increasing linearly with distance. We can express this in another way by inserting (5.32) into (5.43) to get

$$(\boldsymbol{\varepsilon}\cdot\nabla)^2\, U(\mathbf{x}_0) = -\boldsymbol{\varepsilon}\cdot\mathbf{F}'(\boldsymbol{\varepsilon}) < 0 , \tag{5.44}$$

which holds for any direction $\hat{\boldsymbol{\varepsilon}}$ in the primary plane. This tells us that the effective potential $U(\mathbf{x})$ is a maximum at the equilibrium points. In other words, the Lagrange points L_4 and L_5 are peaks of potential hills in the primary plane. Nevertheless, L_4 and L_5 are stable equilibrium points if a certain condition on the masses of the primaries is met.

To see how stability is possible on a potential hill, consider a particle at rest which has been nudged off the peak. The repulsive force accelerates it downhill. As its velocity increases, according to (5.41) the Coriolis force increases, and the particle is deflected to the right. If the deflection is sufficient, the particle starts back uphill and slows down until it starts to fall down again, repeating the process. Thus, the Coriolis force can bind a particle to a "pseudopotential" maximum, producing stability. The rightward deflection of the Coriolis force is opposite in sense to the rotation of the primaries, which we express by saying at the particle's motion is *retrograde*.

To see if the conditions for stability are met at the Lagrange points, we must examine the solutions of the variational equation (5.41) quantitatively. At the triangular Lagrange points L_4 and L_5, $r_1^2 = r_2^2 = r^2 = 1$, so (5.40) reduces to a strictly repulsive force

$$F'(\boldsymbol{\varepsilon}) = 3(m_1\mathbf{r}_1\mathbf{r}_1\cdot\boldsymbol{\varepsilon} + m_2\mathbf{r}_2\mathbf{r}_2\cdot\boldsymbol{\varepsilon}) = \tfrac{3}{2}(\boldsymbol{\varepsilon} + m_1\mathbf{r}_1\boldsymbol{\varepsilon}\mathbf{r}_1 + m_2\mathbf{r}_2\boldsymbol{\varepsilon}\mathbf{r}_2) . \tag{5.45}$$

At L_4, the sides of the triangle are related by

$$\begin{aligned}\mathbf{r}_1 &= \mathbf{r}e^{\mathbf{i}\pi/3} = \mathbf{r}\tfrac{1}{2}(1 + \mathbf{i}\sqrt{3}\,)\\ \mathbf{r}_2 &= \mathbf{r}e^{\mathbf{i}2\pi/3} = \mathbf{r}\tfrac{1}{2}(-1 + \mathbf{i}\sqrt{3}\,) ,\end{aligned} \tag{5.46}$$

where $\mathbf{i} = i\boldsymbol{\omega}$ is the unit bivector for the primary plane. Note that $\boldsymbol{\omega}\cdot\boldsymbol{\varepsilon} = 0$

implies $\mathbf{i}\boldsymbol{\varepsilon} = -\boldsymbol{\varepsilon}\mathbf{i}$, so substitution of (5.46) and (5.36) into (5.45) leads to the following explicit form for the linearized force:

$$\mathbf{F}'(\boldsymbol{\varepsilon}) = \tfrac{3}{2}\boldsymbol{\varepsilon} + \mathbf{r}\boldsymbol{\varepsilon}\mathbf{r}\,\tfrac{3}{4}(-1 + \mathbf{i}\gamma\sqrt{3}\,). \tag{5.47}$$

This can be simplified by expressing $\mathbf{F}'$ in terms of its eigenvectors. That is most easily done by writing

$$\mathbf{r} = \mathbf{e}_1 e^{\mathbf{i}\phi}, \tag{5.48}$$

where $\mathbf{e}_1$ is an eigenvector which can be obtained from $\mathbf{r}$ when ϕ is known. Now,

$$\mathbf{r}\boldsymbol{\varepsilon}\mathbf{r} = \mathbf{e}_1\boldsymbol{\varepsilon}\mathbf{e}_1 e^{\mathbf{i}2\phi},$$

and this will simplify the last term in (5.47) if

$$\tfrac{3}{4}(-1 + \mathbf{i}\gamma\sqrt{3}) = \beta e^{-\mathbf{i}2\phi}.$$

This implies that

$$\tan 2\phi = \gamma\sqrt{3}, \tag{5.49}$$

so we can put (5.47) in the general form

$$\mathbf{F}'(\boldsymbol{\varepsilon}) = \alpha\boldsymbol{\varepsilon} + \beta\mathbf{e}_1\boldsymbol{\varepsilon}\mathbf{e}_1, \tag{5.50}$$

with

$$\alpha = \tfrac{3}{2} \quad \text{and} \quad \beta = -\tfrac{3}{4}(1 + 3\gamma^2)^{1/2} \tag{5.51}$$

in this particular case. From (5.50)

$$\begin{aligned}\mathbf{F}'(\mathbf{e}_1) &= (\alpha + \beta)\mathbf{e}_1, \\ \mathbf{F}'(\mathbf{e}_2) &= (\alpha - \beta)\mathbf{e}_2,\end{aligned} \tag{5.52}$$

showing that $\mathbf{e}_1$ and $\mathbf{e}_2 = \mathbf{e}_1\mathbf{i}$ are eigenvectors of $\mathbf{F}'$ with eigenvalues $\alpha \pm \beta$. It is worth noting that we have illustrated here a new general method for solving the eigenvalue problem in two dimensions, which has some advantages over the methods developed in Section 5-2.

We use (5.50) to put (5.41) in the form

$$\ddot{\boldsymbol{\varepsilon}} + 2\dot{\boldsymbol{\varepsilon}}\mathbf{i} = \alpha\boldsymbol{\varepsilon} + \beta\mathbf{e}_1\boldsymbol{\varepsilon}\mathbf{e}_1. \tag{5.53}$$

It is convenient to reformulate this as an equation for the spinor

$$Z = \mathbf{e}_1\boldsymbol{\varepsilon} \tag{5.54}$$

which relates $\boldsymbol{\varepsilon}$ to $\mathbf{e}_1$ by $\boldsymbol{\varepsilon} = \mathbf{e}_1 Z$. Thus, we obtain

$$\ddot{Z} + 2\mathbf{i}\dot{Z} = \alpha Z + \beta Z^\dagger. \tag{5.55}$$

Our stability problem has been reduced to studying the solutions of this equation.

From our experience with linear differential equations in Chapter 3, we know that (5.55) has circular solutions of the form $Z = a\ \exp(\mathbf{i}\lambda t)$ if $\beta = 0$. However, when $\beta \neq 0$ the solutions cannot be of this form, because the conjugation $Z^\dagger$ changes the sign of the exponential. This suggests that we consider a trial solution of the form

$$Z = ae^{\mathbf{i}\lambda t} + be^{-\mathbf{i}\lambda t}, \tag{5.56}$$

where a and b are complex coefficients to be determined by the initial conditions. The parameter λ must be real (i.e. scalar) for a stable solution, for if it has a finite imaginary part one of the exponential factors will grow without bound.

Substituting (5.56) into (5.55) and separately equating coefficients of the different exponential factors, we obtain

$$\begin{aligned} a(\lambda^2 + 2\lambda + \alpha) &= -\beta b^\dagger, \\ b(\lambda^2 - 2\lambda + \alpha) &= -\beta a^\dagger. \end{aligned} \tag{5.57}$$

Assuming $\lambda = \lambda^\dagger$, we eliminate a and b from these equations to get

$$(\lambda^2 + \alpha)^2 - 4\lambda^2 = \beta^2.$$

This is a quadratic equation for λ^2 with the solution

$$\lambda_2 = (2-\alpha) \pm (\beta^2 - 4\alpha + 4)^{1/2}. \tag{5.58}$$

For the particular values of α and β in (5.51), this becomes

$$\lambda^2 = \tfrac{1}{2} \pm \tfrac{1}{4}(27\gamma^2 - 23)^{1/2}. \tag{5.59}$$

Both roots will be real if and only if

$$\gamma = \frac{m_1 - m_2}{m_1 + m_2} > \left(\frac{23}{27}\right)^{1/2} = 0.922958, \tag{5.60}$$

and both real roots will be positive since $\gamma^2 < 1$. Thus we have found a condition on the masses of the primaries necessary for stability at L_4.

For the Sun-Jupiter system, the mass ratio is about 1000:1, so $\gamma = 0.999$. For the Earth-Moon system, the mass ratio is about 81.4:1 so $\gamma = 0.977$. These values satisfy the inequality (5.46), so the Lagrange points L_4 and L_5 are stable for both systems.

For a real and positive root λ determined by (5.59), we know from our study of the harmonic oscillator that the solution (5.56) describes an ellipse. However, (5.56) is not the form for a general solution as it is in the harmonic oscillator case, for the coefficients a and b are not mutually independent, being related by (5.57). It is of some interest, therefore, to determine initial conditions which produce such a special solution. We choose initial time in (5.56) so that a is real and positive. Then (5.57) and (5.51) tell us that

$$\frac{b}{a} = \frac{\lambda^2 + 2\lambda + \alpha}{-\beta} = \frac{\lambda^2 + 2\lambda + \alpha}{[(\lambda^2 + \alpha)^2 - 4\lambda^2]^{1/2}} > 1. \tag{5.61}$$

Using (5.54) and (5.56), we can write the solution in the form

$$\boldsymbol{\varepsilon} = \mathbf{e}_1[(a + b)\cos\lambda t + \mathbf{i}\,(a - b)\sin\lambda t]. \tag{5.62}$$

This shows us that the major axis of the elliptical orbit must be aligned with the principal axis $\mathbf{e}_1$ of the force function $\mathbf{F}'$. Then (5.61) implies that the coefficient $(a - b)$ is negative, which tells us that the orbit is retrograde. For an initial position $\mathbf{x}_0$ on the principal axis, the constant $(a + b)$ is determined by

$$\boldsymbol{\varepsilon}_0 = \mathbf{e}_1(a + b). \tag{5.63a}$$

The initial velocity is then determined by

$$\dot{\boldsymbol{\varepsilon}}_0 = \mathbf{e}_1\mathbf{i}(a - b)\lambda = \mathbf{x}_0\mathbf{i}\left(\frac{a - b}{a + b}\right)\lambda, \tag{5.63b}$$

where the constant $(a - b)/(a + b)$ is determined by (5.61). More generally, Equation (5.62) determines a unique orbit, through any specified initial position.

According to (5.59), two unique positive values for λ are allowed, say λ_1 and λ_2 with $\lambda_1 > \lambda_2$. For each of these, there is a special solution of the form (5.62). So through any given point there pass exactly two retrograde elliptical orbits, an orbit with large angular frequency λ_1 and one with small frequency λ_2. Every allowed motion is a superposition of these two, that is to say, the general solution of the variational equation (5.53) has the form

$$\boldsymbol{\varepsilon} = \mathbf{e}_1(a_1 e^{\mathbf{i}\lambda_1 t} + b_1 e^{-\mathbf{i}\lambda_1 t} + a_2 e^{\mathbf{i}\lambda_2 t} + b_2 e^{-\mathbf{i}\lambda_2 t}), \tag{5.64}$$

where the a_i determine the b_i by (5.57), and a_1, a_2 can be regarded as two independent complex coefficients determined by the initial conditions.

Stability of the collinear Lagrange points L_1, L_2, L_3 can be investigated in the same way as L_4. The linearized force is found to have the same general form (5.50) with $\mathbf{e}_1 = \hat{\mathbf{r}}$, but the values of α and β differ from those in (5.51). They produce roots of opposite sign from (5.58). The positive root characterizes a retrograde elliptical orbit just as in the L_4 case. However, the negative root implies that λ is imaginary, and this leads to an exponentially divergent solution when inserted in (5.56). Thus, special initial conditions produce bounded elliptical motion at the collinear Lagrange points, but these equilibrium points must be regarded as unstable, because the general solution (of the form (5.58)) is divergent. In a real physical situation where an object is trapped in an elliptical orbit at one of the collinear points, external disturbances will eventually deflect it to an "unbounded orbit" so it escapes.

A global perspective on possible orbits with a given energy can be gained from a contour map of the effective potential, which has the form

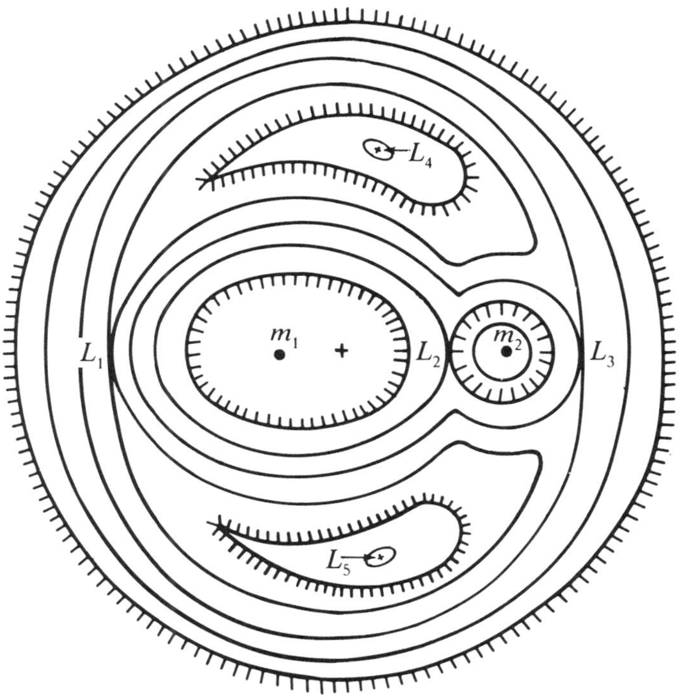

Fig. 5.5. Contour map of the Earth-Moon potential in the synodic (rotating) reference system.

$$U(\mathbf{x}) = -\tfrac{1}{2}\mathbf{x}^2 - \frac{m_1}{r_1} - \frac{m_2}{r_2} \tag{5.65}$$

in the primary plane. At large distances the first term dominates, so the potential decreases with $\mathbf{x}^2$ and the equipotential curves are nearly circular. Each mass is centered in a potential well, so nearby equipotentials are circles around it. The Lagrange points L_4 and L_5 are potential maxima, as we have shown. These are the critical features of the contour map in Figure 5.5. The map shows the three collinear Lagrange points as saddle points. This can be verified analytically by showing that, at the collinear points,

$$(\boldsymbol{\varepsilon}\cdot\nabla)^2 U = -\boldsymbol{\varepsilon}\cdot\mathbf{F}'(\boldsymbol{\varepsilon}) < 0 \tag{5.66a}$$

for $\boldsymbol{\varepsilon} = \mathbf{r}$, and

$$(\boldsymbol{\varepsilon}\cdot\nabla)^2 U > 0 \tag{5.66b}$$

for $\boldsymbol{\varepsilon} = \mathbf{ri}$.

The energy integral (5.35) implies that a particle with total energy C cannot cross a contour determined by the equation $U(\mathbf{x}) = C$; its motion is confined to regions where $U < C$. Thus, a particle trapped at L_4 will circle the peak but

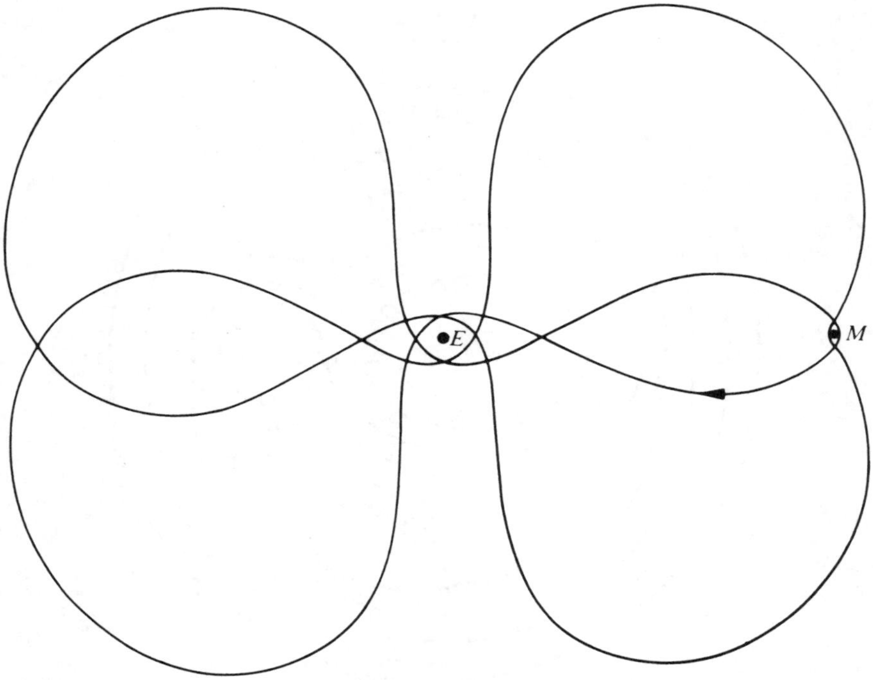

Fig. 5.6. Earth-Moon "bus route" in the synodic (rotating) reference system (found by Arenstorf (1963)).

never climb higher than $U = C$. It is worth noting that if the particle's kinetic energy is dissipated by collisions with gas, dust or small bodies, then it will slide down the peak, increasing the amplitude of its oscillations about L_4. This has been suggested as the origin of the large amplitude oscillations of the Trojan asteroids.

The regions excluded by energy conservation are called *Hill's regions* by astronomers, after G. W. Hill who pointed out that the stability of the Moon's orbit is assured by the fact that it lies within its bounding contour which encircles the Earth. The contour map is a helpful guide in the search for periodic solutions, to which we now turn.

Periodic Solutions

The systematic search for periodic solutions of the circular restricted three body problem was inaugurated by Poincaré in a masterful series of mathematical studies. He conjectured that every bounded solution is arbitrarily close to some periodic solution, possibly with very long period. This reduces the classification problem to classification of period solutions. Periodic solutions are more easily classified, because the behavior of a periodic function for

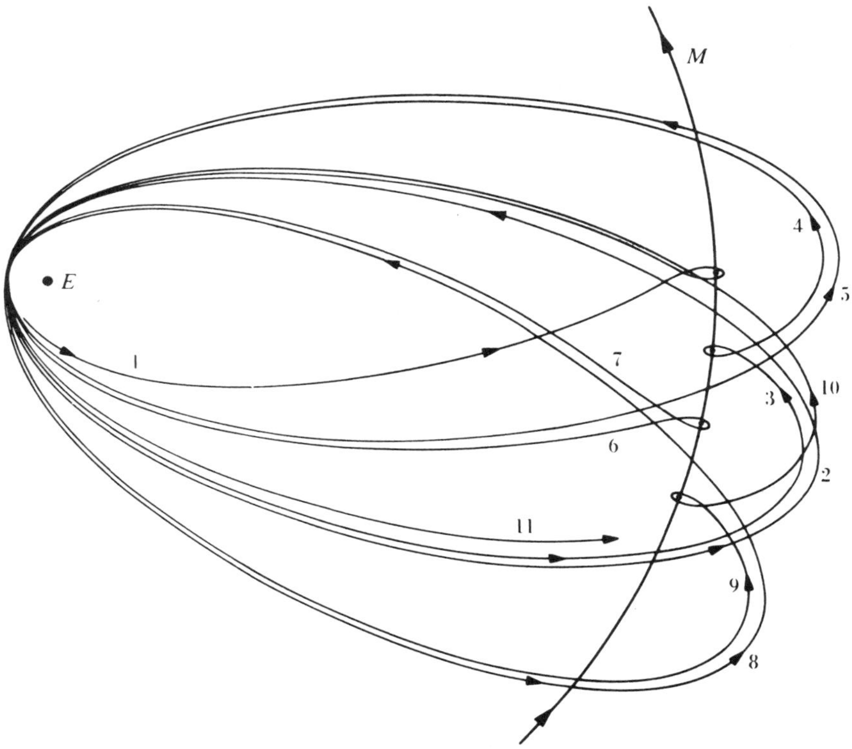

Fig. 5.7. Earth-Moon "bus route" in the sidereal (non-rotating) reference system (Arenstorf (1963)).

all time is known when its behavior for a finite time (the period) has been determined.

Another reason for studying periodic solutions is to use them as reference orbits for calculations which account for other physical effects by perturbation theory (Chapter 8). A prime example is the Hill-Brown lunar theory. With the Sun and Earth as primaries, Hill (1877) found a periodic solution within the Earth's potential well (Figure 5.5) known as Hill's *variational curve*. The curve is an oval, symmetrical about the primary axis and elongated perpendicular to the axis. With this as a reference curve, a description of the Moon's motion with high precision was developed by Hill and Brown. Since 1923, Brown's results have been used in preparing tables of lunar motion.

A solution $\mathbf{x} = \mathbf{x}(t; \gamma, C)$ is periodic with period T if, for any time t,

$$\mathbf{x}(t; \gamma, C) = \mathbf{x}(t + T; \gamma, C).$$

The solution depends parametrically on the mass and energy parameters γ and C. When a particular periodic solution has been found, a whole family of periodic solutions can be generated from it by varying the parameters. Thus,

Hill's variational curve generates a family of periodic orbits about the smaller primary. The shape of the curve varies as the parameters are changed; the oval develops unforeseeable cusps and loops; they can be found only by numerical calculation. Poincaré called this family of orbits *solutions of the first kind.*

Solutions of the second kind lie in the primary plane and loop around each primary. The existence of periodic orbits which pass arbitrarily close to each primary was first proved by Arenstorf (1963), and many such orbits have since been calculated. A particularly good candidate for a lunar bus route is shown in Figures 5.6 and 5.7. The buses would shuttle material and people between the Earth and Moon with minimal fuel consumption.

Before the invention of the modern computer, the computation of three body orbits was long and laborious. An enormous number of periodic solutions, many with bizarre shapes, have been found in the last 25 years. Interest in the 3-body problem has never been greater for both practical and mathematical reasons. It remains an open and active field for research. For more information the reader should consult the specialized literature. The most comprehensive account of 3-body research is by Szebehely (1967).

6-5. Exercises

(5.1) Use Equation (5.12) to show that the orbit $\mathbf{x}_1 = \mathbf{x}_1(t)$ of particle 1 solves the equation

$$\ddot{\mathbf{x}}_1 = -\frac{(m_2^2 + m_2 m_3 + m_3^2)^{3/2}}{m_2}\frac{\mathbf{x}_1}{x_1^{\,3}} \,.$$

(5.2) Solve Equation (5.16) and determine the collinear solutions explicitly when all three particles have identical masses.

(5.3) Verify the Jacobi equations of motion (5.20) and (5.41) and the following expressions for the total angular momentum $\mathbf{l}$, kinetic energy K and "central moment of inertia" $J = \sum_k m_k \mathbf{x}_k^{\,2}$

$$\mathbf{l} = g_1 \mathbf{r} \times \dot{\mathbf{r}} + g_2 \mathbf{x} \times \dot{\mathbf{x}},$$
$$2K = g_1 \dot{\mathbf{r}}^2 + g_2 \dot{\mathbf{x}}^2,$$
$$2J = g_1 \mathbf{r}^2 + g_2 \mathbf{x}^2,$$

where $g_1 = m_1 m_2/\mu$ and $g_2 = m_3 \mu/m$.

(5.4) Show that at the collinear Lagrange points L_1, L_2, L_3 the linearized equation for deviations in the primary plane has the form of equation (5.50), where $\alpha = 1 + \beta/3$ and

$$\beta = \frac{3}{4}\left(\frac{1+\gamma}{r_1^3} + \frac{1-\gamma}{r_2^3}\right).$$

Prove therefrom that the collinear Lagrange points are unstable.

Chapter 7

Rigid Body Mechanics

Rigid Body Mechanics is a subtheory of classical mechanics, with its own body of concepts and theorems. It is mainly concerned with working out the consequences of rigidity assumptions in models of solid bodies.

The formulation of rigid body theory in Section 7-1 is unique in its use of a spinor equation to describe rotational kinematics. This makes the whole spinor (quaternion), theory of rotations, with all its unique advantages, available for application to rigid body problems. Sections 7-3 and 7-4 present one of the most extensive mathematical treatments of spinning tops to be found anywhere, certainly the most extensive using spinor methods. Some of this material is likely to be difficult for the novice, but comparison with alternative approaches in the literature shows that it includes many simplifications. Since this is the first extensive spinor treatment of classical rotational dynamics to be published, it can probably be improved, and it is wide open for new applications. In a subsequent book (NFII), we shall see that this approach is closely related to the quantum mechanical theory of spinning particles.

The treatment of inertia tensors in Section 7-2 is intended to be complete and systematic enough to make it useful as a reference.

7-1. Rigid Body Modeling

This section is concerned with general principles and strategies for developing rigid body models of solid objects. We can distinguish three major stages in the development of a rigid body model. In the first stage, a suitable set of descriptive variables is determined to describe the structure and state of motion of the body as well as its interactions with other objects. In the second stage, the descriptive variables are combined with laws of motion and interaction to determine definite equations of motion for the body. In the final stage the equations of motion are solved and their consequences are analyzed. In this section we will be concerned with the first two stages only, but we will stop short of developing specific models. Specific rigid body models and their ramifications will be studied in subsequent sections.

A complete set of descriptive variables and laws of motion for an arbitrary rigid body is listed in Table 1.1. The dynamical laws of rigid body motion were derived from the laws of particle mechanics in Section 6.1. Now, however, when we do rigid body mechanics we take the laws of rigid body motion as axioms, so no further appeal to particle mechanics is necessary, unless one wants to consider alternatives to the working assumption of rigidity. To apply the laws of motion, we need to understand and control the descriptive variables, so we approach that first.

State Variables

We designate the *position* and *attitude* of a rigid body respectively by a vector $\mathbf{X}$ and a unitary spinor R. The vector $\mathbf{X}$ designates the center of mass of the body. The spinor R relates the *relative position* $\mathbf{r}$ of each particle in the body to a relative position $\mathbf{r}'$ in a fixed reference configuration; specifically, the spinor-valued function $R = R(t)$ determines the time dependent rotation

$$\mathbf{r} = R^{\dagger}\mathbf{r}'R \tag{1.1}$$

TABLE 1.1 Descriptive Variables and Laws of Motion for a Rigid Body

Translational Motion		Rotational Motion	
		Object and State Variables	
Mass	$m = \sum_i m_i$	Inertia Tensor	$\mathcal{I}\mathbf{u} = \sum_i m_i\mathbf{r}_i\mathbf{r}_i\wedge\mathbf{u}$ where $\mathbf{r}_i = \mathbf{x}_i - \mathbf{X}$
Position (vector)	$\mathbf{X} = \frac{1}{m}\sum_i m_i x_i$	Attitude (spinor)	R where $R^{\dagger}R = 1$
Velocity	$\dot{\mathbf{X}} = \frac{\mathrm{d}\mathbf{X}}{\mathrm{d}t}$	Rotational Velocity	$\boldsymbol{\Omega} = i\boldsymbol{\omega} = 2R^{\dagger}\dot{R}$
Momentum	$\mathbf{P} = m\dot{\mathbf{X}}$	Angular momentum	$\mathbf{l} = \mathcal{I}\boldsymbol{\omega}$ (vector) $\mathbf{L} = i\mathbf{l}$ (bivector)
Kinetic energy	$K_{\mathrm{tr}} = \frac{1}{2}\dot{\mathbf{X}}\cdot\mathbf{P}$ $= \frac{1}{2}m\dot{\mathbf{X}}^2$	Kinetic energy	$K_{\mathrm{rot}} = \frac{1}{2}\boldsymbol{\omega}\cdot\mathbf{l}$ $= \frac{1}{2}\boldsymbol{\omega}\cdot\mathcal{I}\boldsymbol{\omega}$
		Interaction Variables	
Force	$\mathbf{F} = \sum_i \mathbf{F}_i$	Torque	$\boldsymbol{\Gamma} = \sum_i \mathbf{r}_i\times\mathbf{F}_i$
		Laws of Motion	
Newton's law	$\dot{\mathbf{P}} = m\ddot{\mathbf{X}} = \mathbf{F}$	Euler's law	$\dot{\mathbf{l}} = \mathcal{I}\dot{\boldsymbol{\omega}} + \boldsymbol{\omega}\times\mathcal{I}\boldsymbol{\omega} = \boldsymbol{\Gamma}$

so that

$$\dot{\mathbf{r}} = \boldsymbol{\omega} \times \mathbf{r}, \tag{1.2}$$

where $\boldsymbol{\omega}$ is the rotational velocity of the body as specified in Table 1.1. The position $\mathbf{x}$ of a particle in the body is given by

$$\mathbf{x} = \mathbf{X} + \mathbf{r} = \mathbf{X} + R^{\dagger}\mathbf{r}'R, \tag{1.3}$$

It follows that the motion of a rigid body can be described completely by specifying its position $\mathbf{X} = \mathbf{X}(t)$ and its attitude $R = R(t)$ as functions of time, for the trajectory $\mathbf{x} = \mathbf{x}(t)$ of each particle in the body is then determined by (1.3). The translational and rotational trajectories $\mathbf{X} = \mathbf{X}(t)$ and $R = R(t)$ resulting from the action of specified forces are determined by the laws of motion in Table 1.1 together with specified initial values for $\dot{\mathbf{X}}$ and $\boldsymbol{\omega}$ as well as $\mathbf{X}$ and R. At any time t, the state of translational motion is described by the center of mass position $\mathbf{X}(t)$ and velocity $\dot{\mathbf{X}}(t)$, while the state of rotational motion is described by the attitude $R(t)$ and rotational velocity $\boldsymbol{\omega}(t)$. Accordingly, the descriptive variables $\mathbf{X}$, $\dot{\mathbf{X}}$, R, $\boldsymbol{\omega}$ are called *state variables* or *kinematic variables* for the rigid body.

Object Variables

Object variables describe intrinsic physical properties taking on particular fixed values for each particular object. Values of the object variables for a composite object depend on its structure. The object variables for a rigid body are the mass m, inertia tensor $\mathcal{I}$ and the location of the center of mass $\mathbf{X}$ with respect to the body. Size and shape are also intrinsic properties of a rigid body, but they play no role in rigid body kinematics, so they need not be represented by object variables. However, the geometrical properties of size and shape play an important role in dynamics, since they determine the points at which contact forces can be applied.

In modeling an object as a rigid body, the first problem is to determine the values of its object variables. Methods for solving this problem are discussed in Section 7-2. In our discussion here we take it for granted that the values of the object variables are known. We are interested here in general properties of the object variables.

To interpret, analyze and solve Euler's equation, we need to know the structure of the inertia tensor and its relation to the kinematic variables. We established in Section 6-1 that the general properties of the inertia tensor in Table 1.2 follow from its definition in Table 1.1. With these general properties in hand, we need not refer to the detailed definition of the inertia tensor in our analysis of Euler's equation.

Since the inertia tensor is linear and symmetric (Table 1.2), we know from Section 5-2 that it has three orthonormal *principal vectors* $\mathbf{e}_k$ ($k = 1, 2, 3$) satisfying the eigenvalue equation

TABLE 1.2 Some General Properties of the Inertia Tensor.
These properties hold for arbitrary vectors **u** and **v** and for an inertia tensor $\mathcal{I}$ relative to any base point.

Linear	$\mathcal{I}(\alpha\mathbf{u} + \beta\mathbf{v}) = \alpha\mathcal{I}\mathbf{u} + \beta\mathcal{I}\mathbf{v}$
Symmetric	$\mathbf{u}\cdot\mathcal{I}\mathbf{v} = \mathbf{v}\cdot\mathcal{I}\mathbf{u}$
Positive definite	$\mathbf{u}\cdot\mathcal{I}\mathbf{u} > 0$ if $\mathbf{u} \neq 0$
Kinematic	$\dot{\mathcal{I}}\mathbf{u} = \boldsymbol{\omega} \times \mathcal{I}\mathbf{u} + \mathcal{I}(\mathbf{u} \times \boldsymbol{\omega})$

$$\mathcal{I}\mathbf{e}_k = I_k\mathbf{e}_k \tag{1.4}$$

with *principal values* I_k ($k = 1, 2, 3$). From the positive definite property (Table 1.2), it follows that each principal value is a positive number. The determination of principal vectors and values for specific bodies is carried out in Section 7-2, but for the purpose of analyzing rotational motion, it is only necessary to know that they exist.

The principal vectors $\mathbf{e}_k$ specify directions in fixed relation to the rigid body. It is convenient to imagine that the $\mathbf{e}_k$ are rigidly attached to the body at the center of rotation (assumed to be the center of mass unless otherwise specified). The lines through the center of rotation with directions $\mathbf{e}_k$ are called *principal axes* of the body (Figure 1.1). Since the $\mathbf{e}_k$ rotate with the body, according to (1.2) they obey the equation of motion

$$\dot{\mathbf{e}}_k = \boldsymbol{\omega} \times \mathbf{e}_k . \tag{1.5}$$

According to (1.1), the solution to these equations can be given the form

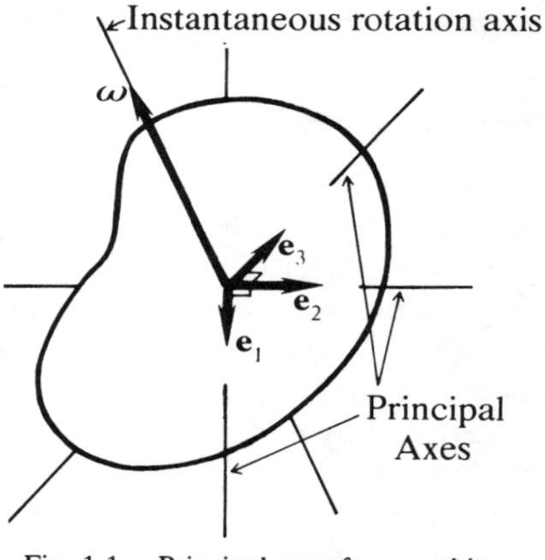

Fig. 1.1. Principal axes for an arbitrary body.

$$\mathbf{e}_k = R^\dagger\boldsymbol{\sigma}_k R, \tag{1.6}$$

where $\{\boldsymbol{\sigma}_k\}$ is any standard frame of constant vectors. Since $RR^\dagger = 1$, it follows from (1.6) that

$$\mathbf{e}_1\mathbf{e}_2\mathbf{e}_3 = \boldsymbol{\sigma}_1\boldsymbol{\sigma}_2\boldsymbol{\sigma}_3 = i,$$

where the body frame $\{\mathbf{e}_k\}$ has been chosen to be righthanded.

Euler's equation $\mathcal{I}\dot{\boldsymbol{\omega}} + \boldsymbol{\omega} \times \mathcal{I}\boldsymbol{\omega} = \boldsymbol{\Gamma}$ can be decomposed into its components with respect to the body frame, with the result

$$I_1\dot{\omega}_1 + (I_3 - I_2)\omega_2\omega_3 = \Gamma_1$$

$$I_2\dot{\omega}_2 + (I_1 - I_3)\omega_3\omega_1 = \Gamma_2 \tag{1.7}$$

$$I_3\dot{\omega}_3 + (I_2 - I_1)\omega_1\omega_2 = \Gamma_3$$

where $\omega_k = \boldsymbol{\omega}\cdot\mathbf{e}_k$ and $\Gamma_k = \boldsymbol{\Gamma}\cdot\mathbf{e}_k$. Most of the literature on rigid body motion deals with Euler's equation only in its component form (1.7). In contrast, we will develop techniques to handle Euler's equation without breaking it up into components. This makes it easier to interpret results and visualize the motion as a whole, and it has certain mathematical advantages. It should be noted, however, that (1.7) does not give the components of Euler's equation with respect to an arbitrary coordinate system. Rather (1.7) gives the components with respect to a special frame determined by the intrinsic structure of the body. So we should expect (1.7) to have some special advantages. Indeed, in Section 7-4 we shall see that (1.7) is most useful for treating the rotational motion of an asymmetric body.

The vector $\boldsymbol{\omega} = 2R^\dagger\dot{R}/i$ is the rotational velocity of the body relative to the nonrotating space frame. From the viewpoint of an observer on the spinning body, the body is at rest while the universe rotates around it with a rotational velocity $\boldsymbol{\omega}'$ which differs from $\boldsymbol{\omega}$ only in that it rotates with the body. More specifically, the rotational velocity with respect to the body frame is given by

$$\boldsymbol{\omega}' = R\boldsymbol{\omega}R^\dagger = -2i\dot{R}R^\dagger \tag{1.8}$$

note that

$$\omega_k = \boldsymbol{\omega}\cdot\mathbf{e}_k = \langle \boldsymbol{\omega}R^\dagger\boldsymbol{\sigma}_k R\rangle_0 = \boldsymbol{\omega}'\cdot\boldsymbol{\sigma}_k\,. \tag{1.9}$$

Consequently, Euler's equations (1.7) can be regarded as equations for either $\boldsymbol{\omega}$ or $\boldsymbol{\omega}'$. We will work mostly with $\boldsymbol{\omega}$, but $\boldsymbol{\omega}'$ will be needed when we wish to interpret observations with respect to a rotating frame such as the Earth.

Change of Base Point

The equations in Table 1.1 decompose the motion of a rigid body into translations and rotations which can be analyzed separately, although they may be coupled. This decomposition is achieved by choosing the center of mass $\mathbf{X}$ as a *base point* (or *center of rotation*) through which the rotation passes. Sometimes, however, the description of motion is simpler when referred to a different base point $\mathbf{Y}$ in the body. Let us examine this possibility. The chosen center of rotation $\mathbf{Y}$ can be designated by its directance

$$\mathbf{R} = \mathbf{X} - \mathbf{Y} \tag{1.10}$$

to the center of mass $\mathbf{X}$. The angular momentum about this point

$$\mathbf{l}_\mathbf{R} \equiv \sum_i m_i(\mathbf{x}_i - \mathbf{Y}) \times (\dot{\mathbf{x}}_i - \dot{\mathbf{Y}}) = \sum_i m_i\,(\mathbf{r}_i + \mathbf{R}) \times (\dot{\mathbf{r}}_i + \dot{\mathbf{R}}). \tag{1.11}$$

But $\sum_i m_i \mathbf{r}_i = \sum_i m_i(\mathbf{x}_i - \mathbf{X}) = 0$, so by expanding the right side of (1.11) we obtain

$$\mathbf{l}_\mathbf{R} = \mathbf{l} + m\mathbf{R} \times \dot{\mathbf{R}}. \tag{1.12}$$

Thus, $\mathbf{l}_\mathbf{R}$ is the sum of the body's "intrinsic angular momentum" $\mathbf{l}$ with respect to the center of mass and its "orbital angular momentum" $m\mathbf{R} \times \dot{\mathbf{R}}$ with respect to base point. Furthermore, from (1.2) it follows that $\dot{\mathbf{s}}_i = \boldsymbol{\omega} \times \mathbf{s}_i$ for $\mathbf{s}_i = \mathbf{x}_i - \mathbf{Y}$, so (1.11) yields

$$\mathbf{l}_\mathbf{R} = \mathcal{I}_\mathbf{R}\boldsymbol{\omega} \equiv \sum_i m_i \mathbf{s}_i \times (\boldsymbol{\omega} \times \mathbf{s}_i) = \sum_i m_i \mathbf{s}_i \mathbf{s}_i \wedge \boldsymbol{\omega} . \tag{1.13}$$

This defines the inertia tensor $\mathcal{I}_\mathbf{R}$ which determines the angular momentum $\mathbf{l}_\mathbf{R}$ as a function of the rotational velocity $\boldsymbol{\omega}$. Since $\dot{\mathbf{R}} = \boldsymbol{\omega} \times \mathbf{R}$, we can express (1.12) as a relation among inertia tensors

$$\mathcal{I}_\mathbf{R}\boldsymbol{\omega} = \mathcal{I}\boldsymbol{\omega} + m\mathbf{R}\mathbf{R}\wedge\boldsymbol{\omega}. \tag{1.14}$$

This important formula is called the *parallel axis theorem*, because it relates the rotational motion about an axis through the center of mass to the rotational motion about a parallel axis passing through another base point of the body (Figure 1.2).

We can get an equation for the rotational motion about $\mathbf{Y}$ by substituting (1.12) into the rotational equation of motion from Table 1.1; thus,

$$\dot{\mathbf{l}} = \dot{\mathbf{l}}_\mathbf{R} - m\mathbf{R} \times \ddot{\mathbf{R}} = \boldsymbol{\Gamma} = \boldsymbol{\Gamma}_\mathbf{R} - \mathbf{R} \times \mathbf{F}, \tag{1.15}$$

where

$$\boldsymbol{\Gamma}_\mathbf{R} \equiv \sum_i (\mathbf{x}_i - \mathbf{Y}) \times \mathbf{F}_i = \sum_i (\mathbf{r}_i + \mathbf{R}) \times \mathbf{F}_i \tag{1.16}$$

is the torque about the new base point. Inserting

$$\dot{\mathbf{P}} = m\ddot{\mathbf{Y}} + m\ddot{\mathbf{R}} = \mathbf{F} \tag{1.17}$$

into (1.15), we get the rotational equation of motion in the form

$$\dot{\mathbf{l}}_\mathbf{R} = \boldsymbol{\Gamma}_\mathbf{R} - m\mathbf{R} \times \ddot{\mathbf{Y}}. \tag{1.18}$$

The coupled equations (1.17) and (1.18) are useful in problems where $\mathbf{Y} = \mathbf{Y}(t)$ is a function specified by constraints, in other words, when the solution of the translational equation of motion is known. Most important is the case when $\mathbf{Y}$ is a fixed point so $\ddot{\mathbf{Y}} = 0$. Then (1.18) reduces to

$$\dot{\mathbf{l}}_\mathbf{R} = \mathcal{I}_\mathbf{R}\dot{\boldsymbol{\omega}} + \dot{\mathcal{I}}_\mathbf{R}\boldsymbol{\omega} = \boldsymbol{\Gamma}_\mathbf{R}. \tag{1.19}$$

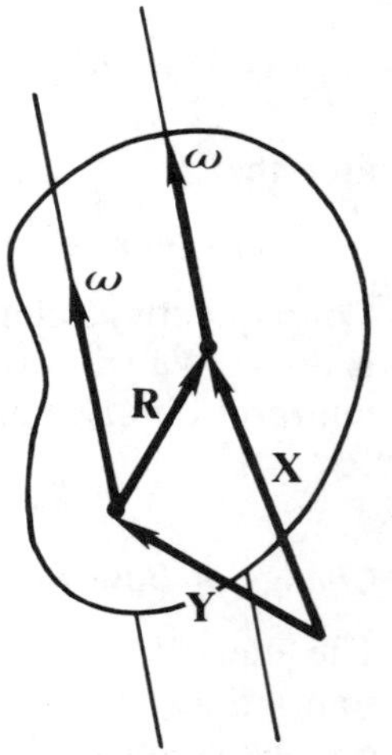

Fig. 1.2. Any point $\mathbf{Y}$ in a rigid body can be chosen as a center of rotation, and the rotational velocity has the same value $\boldsymbol{\omega}$ at all points of the body.

This is identical in form to Euler's equation (Table 1.1) for motion about the mass center. The only difference is in the choice of inertial tensor according to (1.13) or (1.14).

To sum up, a change of base point from **X** to **Y** can be regarded as a change of translational state variables. Although this does not affect the rotational state variables R and $\boldsymbol{\omega}$, it does induce changes in the inertia tensor and the equations of motion.

Constants of Motion

In rigid body mechanics, constants of motion arise from symmetries of the equations of motion, just as in particle mechanics. As in particle mechanics, constants of the translational motion derive from special properties of the applied forces. Similarly, constants of rotational motion derive from special properties of the applied torque. We derived the internal energy conservation law for a system of particles in Section 6-1. But if we are to carry out our program of developing rigid body mechanics independently of particle mechanics, we must see how energy conservation for rotational motion can be derived from Euler's equation.

Using the kinematic property of the inertia tensor in Table 1.2, we find that $\boldsymbol{\omega}\cdot\dot{\mathcal{I}}\boldsymbol{\omega} = 0$, and with the symmetry property we have

$$\boldsymbol{\omega}\cdot\dot{\mathbf{l}} = \boldsymbol{\omega}\cdot(\dot{\mathcal{I}}\boldsymbol{\omega} + \mathcal{I}\dot{\boldsymbol{\omega}}) = \dot{\boldsymbol{\omega}}\cdot\mathcal{I}\boldsymbol{\omega} = \frac{\mathrm{d}}{\mathrm{d}t}(\tfrac{1}{2}\boldsymbol{\omega}\cdot\mathcal{I}\boldsymbol{\omega}).$$

Therefore Euler's equation gives

$$\frac{\mathrm{d}}{\mathrm{d}t}(\tfrac{1}{2}\boldsymbol{\omega}\cdot\mathbf{l}) = \boldsymbol{\omega}\cdot\boldsymbol{\Gamma} \tag{1.20}$$

for the rate of change of the rotational kinetic energy $\frac{1}{2}\boldsymbol{\omega}\cdot\mathbf{l} = \frac{1}{2}\boldsymbol{\omega}\cdot\mathcal{I}\boldsymbol{\omega}$. This is the most we can say about rotational energy without some specific assumption about the torque.

In the important case of a single force **F** applied at a point **r** fixed in the rotating body, the torque is $\boldsymbol{\Gamma} = \mathbf{r}\times\mathbf{F}$, and

$$\boldsymbol{\omega}\cdot\boldsymbol{\Gamma} = (\boldsymbol{\omega}\times\mathbf{r})\cdot\mathbf{F} = \dot{\mathbf{r}}\cdot\mathbf{F}. \tag{1.21}$$

For a conservative force with potential $V = V(\mathbf{r})$, we know from Section 2-8 that

$$\dot{\mathbf{r}}\cdot\mathbf{F} = -\dot{\mathbf{r}}\cdot\nabla_{\mathbf{r}}V = -\frac{\mathrm{d}V}{\mathrm{d}t}. \tag{1.22}$$

Then, from (1.20) we get the rotational energy

$$E_{\mathrm{rot}} = \tfrac{1}{2}\boldsymbol{\omega}\cdot\mathbf{l} + V(\mathbf{r}) \tag{1.23}$$

as a constant of motion. For the important case of constant force, we use (1.21) directly to get

$$E_{\mathrm{rot}} = \tfrac{1}{2}\boldsymbol{\omega}\cdot\mathcal{I}\boldsymbol{\omega} - \mathbf{r}\cdot\mathbf{F}. \tag{1.24}$$

Of course, it is possible for translational and rotational energies to be conserved together even when they are not conserved separately.

Other constants of motion derive from the fact that the torque is a product of vectors. Thus, even for an arbitrary force we have

$$\mathbf{r}\cdot\dot{\mathbf{l}} = \mathbf{r}\cdot(\mathbf{r}\times\mathbf{F}) = 0.$$

If the force acts always on the same point in the body, then $\dot{\mathbf{r}} = \boldsymbol{\omega}\times\mathbf{r}$, and $\dot{\mathbf{r}}\cdot\mathbf{l} = (\boldsymbol{\omega}\times\mathbf{r})\cdot\mathbf{l} = \mathbf{r}\cdot(\mathbf{l}\times\boldsymbol{\omega})$. Obviously $\dot{\mathbf{r}}\cdot\mathbf{l}$ vanishes if $\mathbf{l} = I\boldsymbol{\omega}$; we shall see that it also vanishes for an axially symmetric body if $\mathbf{r}$ lies along the axis of symmetry. Therefore, in these cases $\mathbf{r}\cdot\dot{\mathbf{l}} = \mathrm{d}(\mathbf{r}\cdot\mathbf{l})/\mathrm{d}t = 0$, so $\mathbf{r}\cdot\mathbf{l}$ is a constant of the motion.

Similarly,

$$\mathbf{F}\cdot\dot{\mathbf{l}} = \mathbf{F}\cdot(\mathbf{r}\times\mathbf{F}) = 0.$$

Therefore, if $\mathbf{F}$ is constant, then $\mathbf{F}\cdot\mathbf{l}$ is a constant of the motion. Thus, for the case of a constant force acting at a fixed point on the body, the quantities E_{rot}, $\mathbf{r}\cdot\mathbf{l}$ and $\mathbf{F}\cdot\mathbf{l}$ are all constants of the rotational motion. By finding these three constants we have, in effect, integrated Euler's equation to determine the rotational velocity $\boldsymbol{\omega}$. This reduces the problem of determining the rotational motion to integrating $\dot{R} = \frac{1}{2}Ri\boldsymbol{\omega}$ for the attitude R. That is still a tricky problem, as we shall see in Sections 7-3 and 7-4.

A Compact Formulation of the Rigid Body Laws

The center of mass momentum $\mathbf{P}$ and "internal" angular momentum $\mathbf{l}$ of a rigid body can be combined into a single quantity P, defined by

$$P = \mathbf{P} + i\mathbf{l}. \tag{1.25}$$

Let's call this the *complex momentum* of the body. Similarly, a force $\mathbf{F}$ and torque Γ applied to a rigid body can be combined in a single quantity

$$W = \mathbf{F} + i\boldsymbol{\Gamma} \tag{1.26}$$

called a *complex force* or *wrench* on the body. With these definitions for P and W, the laws for rigid body translational and rotational motion in Table 1.1 can be combined in a single equation

$$\dot{P} = W, \tag{1.27}$$

which we might call the *complex law of motion* for a rigid body.

Now the question is whether the motion law (1.27) has any physical meaning or value beyond the fact that it is the most compact formulation of the rigid body motion laws which we could hope for. The first thing to note is that the complex combination of momentum and angular momentum in (1.25) is geometrically correct since, as we have noted before, the angular momentum is most properly represented geometrically as a bivector $\mathbf{L} = i\mathbf{l}$.

Thus, P is more than a mere formal combination of "real vectors" **P** and **l** with a "unit imaginary" i; it is a combination of physically distinct vector and bivector quantities. Similar remarks apply to the complex force, since the torque is most properly represented geometrically as a bivector $i\boldsymbol{\Gamma}$.

The "complex vector" P represents the state of rigid body motion in the 6-dimensional space of vectors plus bivectors. The dimension of this space is exactly right, because a rigid body has six degrees of freedom. Moreover, the partition of this space into the 3-dimensional subspaces of vectors and bivectors corresponds exactly to the partition of a rigid motion into translational and rotational motions. Thus, the description of rigid motion in this space by the law (1.27) makes sense physically. We shall see below that the "complex formulation" has a deeper physical meaning and some mathematical advantages. Since this compact formulation of rigid body theory with geometric algebra has not been published previously, the full extent of its usefulness remains to be determined. In the meantime, we always have the option of treating translational and rotational parts separately as usual.

Of course, our use of a complex momentum calls for a *complex velocity* V defined by

$$V = \dot{\mathbf{X}} + i\boldsymbol{\omega}. \tag{1.28}$$

In working with complex vectors, it is convenient to introduce a *scalar product* defined by

$$P * V \equiv \langle P^{\dagger} V \rangle_0 . \tag{1.29}$$

Using (1.25) and (1.28), then, we find that the *total kinetic energy* K of a rigid body is given by the simple expression

$$K = \tfrac{1}{2} P * V = \tfrac{1}{2} (\mathbf{P}\cdot\dot{\mathbf{X}} + \mathbf{l}\cdot\boldsymbol{\omega}) . \tag{1.30}$$

And from the motion law (1.27), we find that the change in kinetic energy is determined by the equation

$$\dot{K} = W * V = \mathbf{F}\cdot\dot{\mathbf{X}} + \boldsymbol{\Gamma}\cdot\boldsymbol{\omega}. \tag{1.31}$$

One would expect this to be most useful in problems where rotational and translational motions are coupled, as in rolling motion.

Equipollence and Reduction of Force Systems

A single force **F** applied to a rigid body at a point **r** in the body frame, exerts a torque $\boldsymbol{\Gamma} = \mathbf{r} \times \mathbf{F}$. Therefore, the full effect of the force on the body is represented by the "complex interaction variable"

$$W = \mathbf{F} + i\mathbf{r} \times \mathbf{F} = \mathbf{F} + \mathbf{r}\wedge\mathbf{F}. \tag{1.32a}$$

In Section 2-6 we saw that any vector **F** and its moment $\mathbf{r}\wedge\mathbf{F}$ determine a unique oriented line with *directance* $\mathbf{d} = (\mathbf{r}\wedge\mathbf{F})\mathbf{F}^{-1}$ from the origin (= center of mass here). For this reason, a wrench of the form (1.32a) is sometimes called a *line vector*, and it can be written in the equivalent form

$$W = \mathbf{F} + \mathbf{dF} = (1 + \mathbf{d})\mathbf{F}. \tag{1.32b}$$

The oriented line is called the *axode* and $\mathbf{d}$ is called the *moment arm* of $\mathbf{F}$ or W (Figure 1.3).

The equivalence of (1.32b) with (1.32a) means that $\mathbf{F}$ would have the same effect on the body if it were applied at the point $\mathbf{d}$ instead of at $\mathbf{r}$ or, indeed, at any other point on the axode. Forces applied at different points of a rigid body are said to be *equipollent* if and only if their line vectors (wrenches) are equal. This implies that two forces are equipollent if and only if they have the same axode and magnitude.

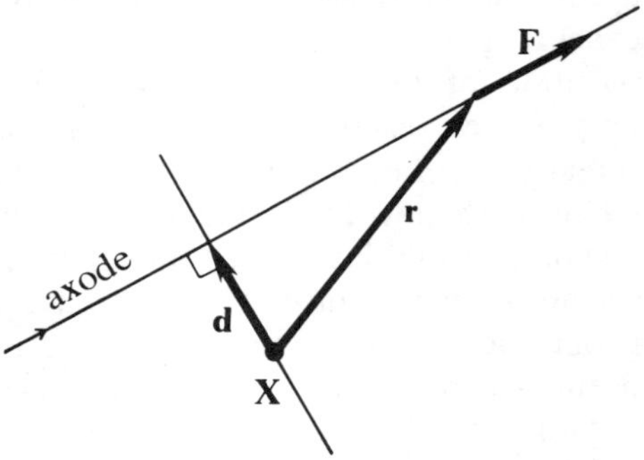

Fig. 1.3. The axode of a force $\mathbf{F}$.

The concept of equipollence is readily generalized to systems of forces. Suppose that forces $\mathbf{F}_1, \mathbf{F}_2, \ldots, \mathbf{F}_n$ are applied simultaneously to rigid body at points $\mathbf{r}_1, \mathbf{r}_2, \ldots, \mathbf{r}_n$ respectively. Each applied force $\mathbf{F}_i$ determines a wrench on the body

$$W_i = \mathbf{F}_i + i\mathbf{r}_i \times \mathbf{F}_i = \mathbf{F}_i + \mathbf{d}_i\mathbf{F}_i, \tag{1.33}$$

where $\mathbf{d}_i$ is the moment arm to the ith axode. The net effect on the body, of the entire system of forces, is determined by the superposition of wrenchs producing the *resultant wrench*

$$W = \sum_i W_i = \mathbf{F} + i\boldsymbol{\Gamma}, \tag{1.34}$$

where $\mathbf{F} = \sum_i \mathbf{F}_i$ is the resultant force and $\boldsymbol{\Gamma} = \sum_i \mathbf{r}_i \times \mathbf{F}_i$ is the resultant torque. This is the *superposition principle* for rigid body mechanics. According to the motion law (1.27), the entire effect of a system of forces on a rigid body is determined by its resultant wrench. Therefore, two different force systems will have the same effect on rigid body motion if and only if their resultant wrenches are equal. In that case, we say that the force systems are *equipollent*.

We can use the fact that equipollent force systems have identical effects on rigid body motion to simplify our models by replacing a given system of forces by a simpler system equipollent to it. This is called *reduction of forces* (or *wrenches*). To facilitate force reduction, we now develop a few general theorems.

First we note that *equipollence relations are independent of base point*. To prove it, we recall that a change in base point from the center of mass induces a change in the resultant torque $\boldsymbol{\Gamma}$ of a system of forces (given by (1.15)) while the resultant force **F** is unchanged. Therefore, a shift of base point changes a wrench $W = \mathbf{F} + i\Gamma$ to

$$W_{\mathbf{R}} = W - \mathbf{R}\wedge\mathbf{F}. \tag{1.35}$$

Since **F** is the same for equipollent force systems, their wrenches will be changed by the same amount and so remain equipollent.

The base point independence of equipollence relations is especially important in rigid body *statics*, the study of force systems that maintain mechanical equilibrium. A body is said to be in *mechanical equilibrium* if the resultant applied wrench vanishes, that is, if the system of applied forces is *equipollent to zero*. In a typical statics problem, one of the forces and/or its point of application is unknown and must be computed from the other known forces. If $W_1 = \mathbf{F}_1 + \mathbf{d}_1\mathbf{F}_1$ is the wrench of a unknown force $\mathbf{F}_1$ and W_2 is the resultant wrench of the known forces, then the equation for mechanical equilibrium can be written

$$W_1 + W_2 = 0, \tag{1.36}$$

which is trivially solved for F_1 and its moment arm $\mathbf{d}_1$. Evidently every statics problem is essentially a problem in wrench reduction. The main trick in such problems is to choose the base point to simplify the reduction. For example, the torque reduction,is often simplest if the origin is chosen at the intersection of concurrent axodes. Of course, the best choice of origin depends on the given information.

Parallel Forces

For a system of parallel forces $\mathbf{F}_i$, we can write $\mathbf{F}_i = F_i\mathbf{u}$, where $\mathbf{u}$ is a unit vector and $F_i = \mathbf{u}\cdot\mathbf{F}_i$. The resultant wrench of the system is then

$$W = \sum_i F_i\mathbf{u} + (\sum_i F_i\mathbf{d}_i)\mathbf{u}. \tag{1.37}$$

If the resultant force

$$\mathbf{F} = \sum_i \mathbf{F}_i = \sum_i F_i\mathbf{u}$$

does not vanish, then we can write (1.37) in the form

$$W = \mathbf{F} + \mathbf{d}\mathbf{F},$$

where

$$\mathbf{d} = \frac{\sum_i F_i\mathbf{d}_i}{\sum_i F_i}. \tag{1.38}$$

Thus, we have proved that *any system of parallel forces with nonvanishing resultant* **F** *is equipollent to a single force* **F** with moment arm **d** given by (1.38).

For a uniform gravitational field acting on the body, we have $\mathbf{F}_i = m_i\mathbf{g}$ where m_i is the mass of the ith particle, so (1.38) becomes

$$\mathbf{d} = \frac{\sum_i m_i \mathbf{d}_i}{\sum_i m_i} . \tag{1.39}$$

This is an expression for the directance of the center of mass from a line with direction **u** passing through the origin (= base point). Of course, **d** vanishes if the origin is the center of mass as we have been assuming. But our argument shows that the result (1.39) must hold for any chosen base point.

For the case of two parallel forces, (1.38) is simply

$$\mathbf{d} = \frac{F_1\mathbf{d}_1 + F_2\mathbf{d}_2}{F_1 + F_2}. \tag{1.40}$$

This should be recognized as the expression for a "point of division" discussed in Section 2-6. Many other geometrical results in that section are useful in the analysis of force systems. Note that (1.40) immediately gives us the elementary "law of the lever"; it tells us that the effects of parallel forces applied at $\mathbf{d}_1$ and $\mathbf{d}_2$ can be exactly cancelled by a single force applied to the point of division **d**.

Couples

A pair of equal and opposite forces applied to a rigid body is called an applied *couple*. The wrench C for a couple of forces $\mathbf{F}_1$ and $\mathbf{F}_2$ applied at points $\mathbf{r}_1$ and $\mathbf{r}_2$ is (Figure 1.4)

$$C = \mathbf{F}_1 + \mathbf{F}_2 + \mathbf{r}_1\wedge\mathbf{F}_1 + \mathbf{r}_2\wedge\mathbf{F}_2 = \mathbf{r}_1\wedge\mathbf{F}_1 - \mathbf{r}_2\wedge\mathbf{F}_1 .$$

Thus, the wrench of a couple can be written

$$C = (\mathbf{r}_1 - \mathbf{r}_2)\wedge\mathbf{F}_1 = \mathbf{d}\mathbf{F}_1, \tag{1.41}$$

where **d** is the directance between the axodes of the couple. Clearly, the resultant force of a couple is always zero and its resultant torque is zero if $\mathbf{d} = 0$.

Since $\mathbf{F} = \mathbf{F}_1 + \mathbf{F}_2 = 0$ for a couple, it follows from (1.35) that the torque exerted by a couple is independent of the base point. As (1.41) shows, the torque of a couple depends on the force $\mathbf{F}_1$ and the directance **d**. Therefore, equipollent couples can be obtained by

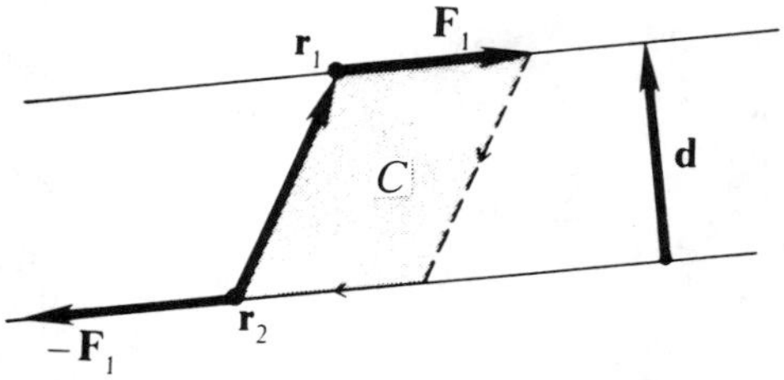

Fig. 1.4. A couple and its wrench C.

changing the magnitudes of $\mathbf{F}_1$ and $\mathbf{d}$ in inverse proportion or by displacing the point at which $\mathbf{F}_1$ is applied to any desired point as long as $\mathbf{d}$ is kept fixed.

Any system of applied forces with a vanishing resultant force can be reduced to a couple, that is, the system is equipollent to a couple. For, in general, such a system has a nonvanishing torque $\boldsymbol{\Gamma}$, so the wrench of the system has the form $W = i\boldsymbol{\Gamma}$. To find an equipollent couple, we pick any force $\mathbf{F}_1$ orthogonal to $\boldsymbol{\Gamma}$ and write

$$i\boldsymbol{\Gamma} = \mathbf{d}\mathbf{F}_1 .$$

This determines the directance of the couple:

$$\mathbf{d} = i\boldsymbol{\Gamma}\mathbf{F}_1^{-1} = \frac{i\boldsymbol{\Gamma}\wedge\mathbf{F}_1}{F_1^2} = \frac{\mathbf{F}_1 \times \boldsymbol{\Gamma}}{F_1^2} .$$

As a corollary to this result, we can conclude that *any system of couples is equipollent to a single couple.*

Reduction to a Force and Couple

We are now in position to ascertain the most general reduction theorem. The wrench of any system of forces has the form $W = \mathbf{F} + i\boldsymbol{\Gamma}$. A single force $\mathbf{F}$ acting at the base point produces no torque, and we have proved that a couple producing a given torque $\boldsymbol{\Gamma}$ can always be found. Therefore, *any system of forces can be reduced to a couple and a single force acting at the base point.* However, this reduction is not an equipollence relation, because it depends on the choice of base point.

To find the simplest equipollence reduction, we decompose $\boldsymbol{\Gamma}$ into components $\boldsymbol{\Gamma}_\perp$ and $\boldsymbol{\Gamma}_\parallel$, respectively orthogonal and parallel to $\mathbf{F}$. Define a moment arm $\mathbf{d}$ for the force $\mathbf{F}$ by

$$\mathbf{d} = i\boldsymbol{\Gamma}_\perp\mathbf{F}^{-1} = \frac{\mathbf{F} \times \boldsymbol{\Gamma}}{F^2} , \tag{1.42}$$

so that

$$\mathbf{F} + i\boldsymbol{\Gamma}_\perp = \mathbf{F} + \mathbf{d}\mathbf{F} . \tag{1.43}$$

This is a line vector describing the action of a single applied force. Now choose a couple with torque $\boldsymbol{\Gamma}_\parallel$ and note that we can write

$$\boldsymbol{\Gamma}_\parallel = h\mathbf{F} , \tag{1.44}$$

where

$$h = \frac{\mathbf{F}\cdot\boldsymbol{\Gamma}}{F^2} . \tag{1.45}$$

When the torque of a couple is collinear with a force $\mathbf{F}$ as expressed by (1.44) and (1.45), we say that *the couple is parallel to the force with pitch h*.

Now we add (1.43) and (1.44) to get

$$W = \mathbf{F} + i\boldsymbol{\Gamma} = \mathbf{F} + (\mathbf{d} + ih)\mathbf{F} = \mathbf{F} + i(\mathbf{d} \times \mathbf{F} + h\mathbf{F}). \tag{1.46}$$

The right side of this equation is the wrench for a single force $\mathbf{F}$ applied together with a parallel couple. A shift of base point will change the torque $\mathbf{d} \times \mathbf{F}$, but it will not change the body point at which the force $\mathbf{F}$ is applied or the torque of the couple. Since the left side of (1.46) may be regarded as the resultant wrench of an arbitrary system of forces, we have proved that *every system of forces is equipollent to a single force and parallel couple*. This includes the limiting cases $h = 0$ and $h = \infty$. When $h = 0$ the wrench reduces to a line vector, so the force system is equipollent to a single force without a couple. We may assume that the case $h = \infty$ describes a pure couple, obtained by taking the limit $h \to \infty$ as $|\mathbf{F}| \to 0$ in such a way that the product $h|\mathbf{F}|$ in (1.44) remains finite.

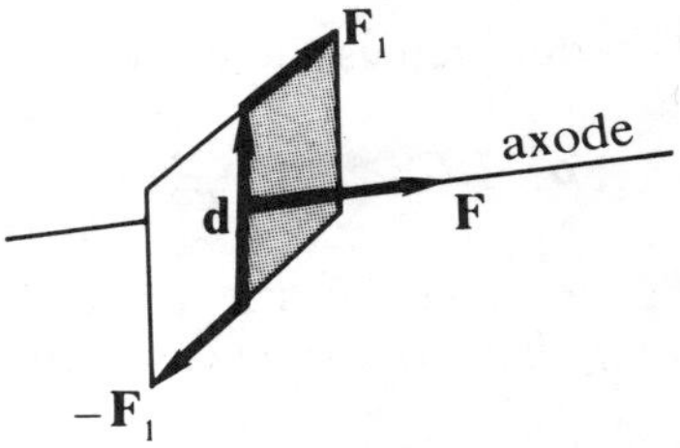

Fig. 1.5. A "wrench" generates a screw motion along its axode.

A force system consisting of a single applied force and a parallel couple is called a "wrench" in older literature on mechanics (Figure 1.5), whereas we have used the term wrench (without quotes) for the complex force of any force system. Our definition is more practical, since we then have frequent occasion to use the term, whereas an actual applied "wrench" is very rare physically. It is most convenient to use the same term wrench for either sense, since the two senses can be distinguished from the context, and they are intimately related. Then our major conclusion in the preceding paragraph can be expressed more succinctly as: *any system of forces can be reduced to a wrench*. For similar reasons, it is convenient to use the term *couple* for any applied torque which can be produced by a couple of forces, even if it is actually produced by a larger number of forces.

To appreciate the aptness of the term "wrench", consider a wrench applied to a rigid body at rest. The force $\mathbf{F}$ will produce an acceleration along its axode, whereas the couple $h\mathbf{F}$ will generate a rotation about the axode. The instantaneous composite motion is therefore an instantaneous screw displacement. It is analogous to the motion of a physical screw turned by a physical wrench.

Concurrent Forces

One other kind of force system is of general interest. A system of forces is said to be *concurrent* if the axodes of all the forces pass through a common point. Let $\mathbf{r}$ be the position vector of this point with respect to the body center of mass. Then the wrench of the system is

$$W = \sum_k \mathbf{F}_k + \sum_k \mathbf{r} \wedge \mathbf{F}_k .$$

Therefore,

$$W = \mathbf{F} + \mathbf{r} \wedge \mathbf{F} = \mathbf{F} + \mathbf{d}\mathbf{F} , \tag{1.47}$$

where **d** is the moment arm of the resultant force **F**. Thus, we have proved that *a system of concurrent forces is equipollent to a single force with axode passing through the intersection of their axodes*. The case of parallel forces with $\mathbf{F} \neq 0$ can be regarded as the limit of this case as $\mathbf{r} \to \infty$.

The example of two concurrent contact forces is illustrated in Figure 1.6. A more significant example is the case of a *central force field* acting on all the particles of the body, in particular, a gravitational field. According to (1.47), the effect of the entire field is equivalent to the effect of a single force **F** applied at the point **d**. In the gravitational case, this point is called the *center of gravity*. In general, the center of gravity differs from the center of mass, except in the limiting case of a uniform gravitational field. This means that a body in a nonuniform gravitational field experiences a torque about its center of mass. For example, the nonuniform gravitational fields of the Sun and the Moon exert a torque on the Earth. We shall investigate that in Chapter 8.

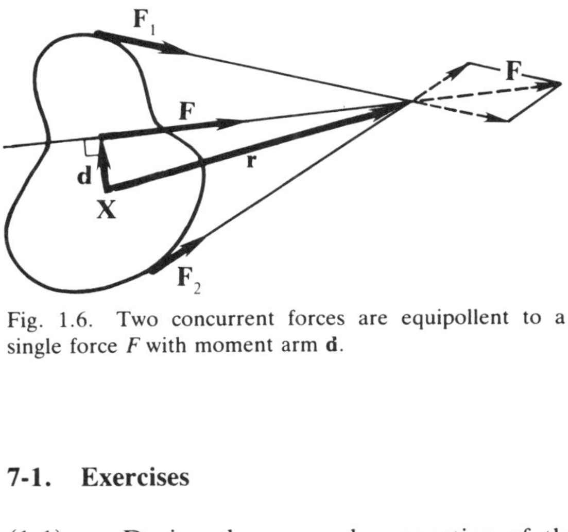

Fig. 1.6. Two concurrent forces are equipollent to a single force *F* with moment arm **d**.

7-1. Exercises

(1.1) Derive the general properties of the inertia tensor in Table 1.2 directly from its definition in Table 1.1.

(1.2) Derive the component form (1.7) for Euler's equation.

(1.3) Prove that the scalar product defined by (1.29) is symmetric and positive definite.

(1.4) Show that $W * V$ is invariant under a shift of base point, that is

$$\mathbf{F} \cdot \dot{\mathbf{X}} + \boldsymbol{\Gamma} \cdot \omega = \mathbf{F} \cdot \dot{\mathbf{Y}} + \boldsymbol{\Gamma}_{\mathbf{R}} \cdot \omega$$

Show that $P * V$ is not invariant in the same sense. What is the meaning of this?

7-2. Rigid Body Structure

The kinematics of rigid body motion depends on only a few intrinsic properties of the body, namely, its mass m, center of mass $\mathbf{X}$ and inertia tensor $\mathcal{I}$. In this section we develop methods for determining $\mathbf{X}$ and $\mathcal{I}$ from the distribution of mass in a given body.

So far we have regarded rigid bodies as composed of a finite number of point particles. However, our results can easily be generalized to describe continuous bodies by standard techniques of integral calculus. Thus, a continuous body can be subdivided into N parts, where the k-th part contains a point $\mathbf{x}_k$, has volume ΔV_k and mass Δm_k. The mass of the body is then

$$m = \sum_{k=1}^{N} \Delta m_k = \sum_{k=1}^{N} \left(\frac{\Delta m_k}{\Delta V_k}\right) \Delta V_k . \tag{2.1}$$

In the limit of infinite subdivision as $N \to \infty$ and $\Delta V_k \to 0$, the sum becomes an integral:

$$m = \int \mathrm{d}m = \int \varrho \, \mathrm{d}V , \tag{2.2}$$

where $\mathrm{d}V$ is the element of volume and $\varrho = \varrho(\mathbf{x})$ is the *mass density* at each point $\mathbf{x}$ of the body. Let it be understood that the integral in (2.2) is to be taken over all points of the body and further that the integral reduces to a sum for any part of the body composed of point particles.

Determining the center of mass

The distribution of mass in a body is described by the mass density $\varrho = \varrho(\mathbf{x})$. It determines the *center of mass* $\mathbf{X}$ defined by

$$\mathbf{X} = \frac{1}{m} \int \mathrm{d}m \mathbf{x} = \frac{1}{m} \int \mathrm{d}V \varrho \mathbf{x} . \tag{2.3}$$

This is the generalization of our earlier definition to continuous bodies.

The center of mass for a given body can always be calculated by performing the integral in (2.3). However, the calculation can often be simplified by using one of the general theorems which we now proceed to establish. In the first place, we are not much interested in the center of mass $\mathbf{X}$ relative to an arbitrary origin, because this can be given any value whatever merely by a shift of origin. Rather we are interested in the center of mass relative to some easily identifiable point $\mathbf{Y}$ of the body. Accordingly, let us write

$$\mathbf{r} = \mathbf{x} - \mathbf{Y}$$

for the position of a particle relative to $\mathbf{Y}$, so

$$m\mathbf{X} = \int \mathrm{d}m\mathbf{x} = \int \mathrm{d}m\mathbf{r} + \mathbf{Y}\int \mathrm{d}m.$$

Thus, the center of mass relative to our specially chosen origin **Y** is given by

$$\mathbf{R} \equiv \mathbf{X} - \mathbf{Y} = \frac{1}{m}\int \mathrm{d}m\mathbf{r} = \frac{1}{m}\int \mathrm{d}V\varrho\mathbf{r}, \tag{2.4}$$

where now we write $\varrho = \varrho(\mathbf{r})$. Equation (2.4) is mathematically identical to (2.3), but it differs in the physical assumption that the origin need not be a fixed point in an inertial frame. Thus, for the purpose of calculating the center of mass, we are free to choose any convenient point in the body as origin without considering motion of the body. Once the center of mass has been identified, its motion can be determined from the equation for translational motion.

Symmetry Principles for the Center of Mass

The center of mass can often be identified from symmetries of the body. There are three major types of symmetry: reflection, rotation and inversion. A body with "reflection symmetry" or "mirror symmetry" is symmetrical with respect to reflection in a plane. We can describe this with the mathematical formulation of reflections developed in Section 5-3. Let **n** be a unit normal to the *symmetry plane* and select some point in the plane as origin. The symmetry of the body is then described by the condition that the *mass density is invariant under the reflection*

$$\mathbf{r} \to \mathbf{r}' = -\mathbf{n}\mathbf{r}\mathbf{n}, \tag{2.5a}$$

that is,

$$\varrho(\mathbf{r}) = \varrho(-\mathbf{n}\mathbf{r}\mathbf{n}) = \varrho(\mathbf{r}'). \tag{2.5b}$$

This is illustrated in Figure 2.1. Now, from (2.4) and (2.5b) we obtain

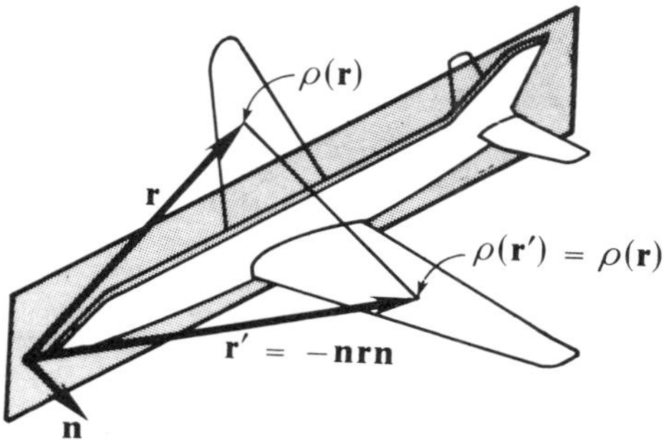

Fig. 2.1. A body with a single symmetry plane.

$$-\mathbf{n}\mathbf{R}\mathbf{n} = \frac{1}{m}\int \mathrm{d}m(\mathbf{r})(-\mathbf{n}\mathbf{r}\mathbf{n})$$

$$= \frac{1}{m}\int \mathrm{d}m(\mathbf{r}')\mathbf{r}' = \mathbf{R}.$$

Therefore,

$$\mathbf{n}\mathbf{R} + \mathbf{R}\mathbf{n} = 2\mathbf{n}\cdot\mathbf{R} = 0,$$

telling us that **R** lies in the symmetry plane. Thus we have proved the theorem: *If a body has a plane of symmetry, then the center of mass is located in that plane*. This theorem has some obvious corollaries: *If a body has two distinct symmetry planes, then the center of mass is located on their line of intersection. If a body has three symmetry planes which intersect at a single point, then that point is the center of mass.*

Rotational symmetry tells us more about the center of mass than a single reflection symmetry. A body is said to have a *symmetry axis* if it is symmetrical with respect to a *nontrivial* rotation about that axis. The adjective "nontrivial" here is meant to exclude the "trivial symmetries" under rotations by integer multiples of 2π which every body possesses. To describe a rotational symmetry mathematically, let us choose an origin on the axis of symmetry so a rotation $\mathcal{S}$ about this axis can be written

$$\mathbf{r} \to \mathbf{r}' = \mathcal{S}\mathbf{r} = S^\dagger \mathbf{r} S, \tag{2.6}$$

where S is a unitary spinor. This rotation is a symmetry of the body if it leaves the mass density invariant:

$$\varrho(\mathbf{r}) = \varrho(\mathcal{S}\mathbf{r}) = \varrho(\mathbf{r}'). \tag{2.7}$$

Applying this to the center of mass vector (2.4), we obtain

$$\mathcal{S}\mathbf{R} = \frac{1}{m}\int \mathrm{d}m(\mathbf{r})\mathcal{S}\mathbf{r} = \frac{1}{m}\int \mathrm{d}m(\mathbf{r}')\mathbf{r}' = \mathbf{R},$$

which tells us that the center of mass is invariant under $\mathcal{S}$. It follows (from 5–3.20) that the center of mass must lie on the axis of symmetry. Thus, we have proved the theorem: *If a body has an axis of symmetry, then its center of mass lies on that axis*. As an obvious corollary we have: *If a body has two distinct symmetry axes, then the axes intersect at a point which is the center of mass*.

A homogeneous plane lamina in the shape of a parallelogram, as shown in Figure 2.2, is invariant under a rotation by π about an axis along $\mathbf{n}$ through its center. It is obviously invariant under reflections in its plane as well. Therefore, its center of mass is located at the point

$$\mathbf{R} = \tfrac{1}{2}(\mathbf{a} + \mathbf{b})$$

if a corner is chosen as origin. This choice has the advantage of relating $\mathbf{R}$ to the vectors $\mathbf{a}$ and $\mathbf{b}$ in Figure 2.2 which characterize the shape of the body. If $\mathbf{a}\cdot\mathbf{b} \doteq 0$, the parallelogram reduces to a rectangle and the body has additional symmetries; specifically, it is symmetric with respect to reflections in the two planes passing through $\mathbf{R}$ with normals $\mathbf{a}$ and $\mathbf{b}$. However, the location of the center of mass is unaffected by this additional symmetry.

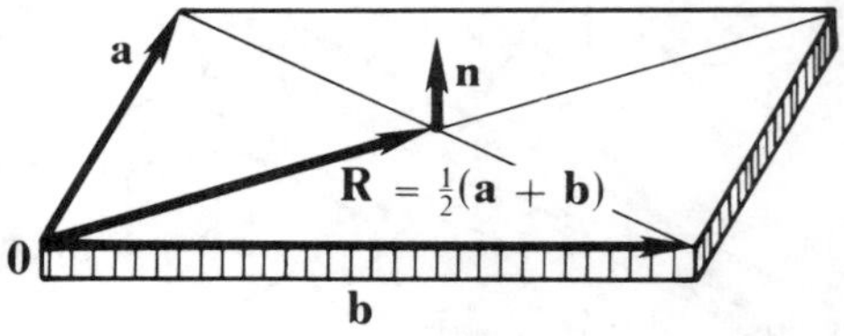

Fig. 2.2. The symmetry axis of a parallelogram passes through its center.

A body is said to have *inversion symmetry* if the origin can be chosen so that

$$\varrho(\mathbf{r}) = \varrho(-\mathbf{r}), \tag{2.8}$$

that is, the mass density is invariant under the inversion $\mathbf{r} \to -\mathbf{r}$. For a body with this kind of symmetry

$$m\mathbf{R} = \int \mathrm{d}m(\mathbf{r})\mathbf{r} = -\int \mathrm{d}m(-\mathbf{r})(-\mathbf{r}) = -m\mathbf{R},$$

so $\mathbf{R} = 0$. Thus, we have proved that *if a body has inversion symmetry, then the center of inversion is the center of mass.*

A homogeneous parallelopiped with intersecting edges which are not perpendicular provides an example of a body with inversion symmetry without any rotation or reflection symmetries.

It is a well-established convention that the phrase "a symmetry of the body" refers to invariance of the body under an orthogonal transformation of the three kinds (reflections, rotations, inversions) already discussed or any combination of them. But there is another common kind of symmetry which we have already mentioned without giving a proper definition, namely, the symmetry of a continuous body when all of its component particles are physically identical. A body with this kind of symmetry is said to be homogeneous. The mass density ϱ of a *homogeneous* body has the same value at each point of the body. It follows that

$$m = \varrho \int \mathrm{d}V = \varrho V, \tag{2.9}$$

that is, the mass m of the body is directly proportional to its volume V. Then from (2.4) it follows that

$$\mathbf{R} = \frac{1}{V}\int \mathrm{d}V\mathbf{r}, \tag{2.10}$$

so the location of the center of mass $\mathbf{R}$ is determined by the geometry of the body alone. In this case, the center of mass is called the *centroid.*

The Additivity Principle for the Center of Mass

Besides the symmetry principles just mentioned, the most useful general principle for determining the center of mass is *additivity*: If a body is composed of N bodies with known masses m_k and mass centers $\mathbf{R}_k$, then the mass center $\mathbf{R}$ of the composite body can be obtained by treating the parts as particles, that is,

$$m\mathbf{R} = m_1\mathbf{R}_1 + m_2\mathbf{R}_2 + \ldots + m_N\mathbf{R}_N, \tag{2.11}$$

where, of course, $m = m_1 + m_1 + \ldots + m_N$. This is an elementary consequence of the additivity of the integrals (2.4) and (2.7)

Example 2.1. Centroid of a Triangular Lamina

The additivity principle applies also to a continuous subdivision of a body. For example, let us calculate the centroid of the triangular lamina shown in Figure 2.3. We subdivide the triangle into narrow strips of width $d\lambda$ parallel to one side. By symmetry, the centroid of a strip is at its center

$$\mathbf{R}_\lambda = \frac{\lambda}{2}(\mathbf{a} + \mathbf{b}) = \lambda \mathbf{a}_+ , \tag{2.12}$$

where λ is a scalar parameter in the range $0 \leqslant \lambda \leqslant 1$. The directed area of the strip is

$$\mathrm{d}\mathbf{A}_\lambda = \tfrac{1}{2}[(\lambda + \mathrm{d}\lambda)\mathbf{a}]\wedge[(\lambda + \mathrm{d}\lambda)\mathbf{b}] - \tfrac{1}{2}(\lambda\mathbf{a})\wedge(\lambda\mathbf{b}) \approx 2\lambda \,\mathrm{d}\lambda\, \mathbf{A} , \tag{2.13}$$

where $\mathbf{A} = \frac{1}{2}\mathbf{a}\wedge\mathbf{b}$ is the directed area of the triangle. The mass of the strip is

$$\mathrm{d}m_\lambda = \left(\frac{m}{A}\right) \mathrm{d}A_\lambda = 2m\lambda \,\mathrm{d}\lambda , \tag{2.14}$$

where

$$\mathrm{d}A_\lambda = |\mathrm{d}\mathbf{A}_\lambda| \quad \text{and} \quad A = |\mathbf{A}| .$$

So the centroid of the entire triangle is

$$\mathbf{R} = \frac{1}{m}\int_0^1 (2m\lambda \,\mathrm{d}\lambda)(\lambda\mathbf{a}_+) = \frac{2}{3}\mathbf{a}_+ . \tag{2.15}$$

The vector $\mathbf{a}_+$ specifies a *median* of the triangle (the line segment from a vertex to the midpoint of the opposite side), so (2.15) says that the centroid is located on a median two thirds of the distance from a vertex. This must be true for all medians, so we can conclude that the *three medians of a triangle intersect at a common point, the centroid* of the triangle. This result is so simple that one suspects it can be obtained without integration. Indeed it can! Regard the triangle in Figure 2.3 as half a parallelogram with edges $\mathbf{a}$ and $\mathbf{b}$. We know by symmetry that the centroid of the parallelogram is $\mathbf{a}_+$. By additivity, $\mathbf{a}_+$ must lie on a line connecting the centroids of the two identical triangles. Therefore the centroid of each triangle must lie on

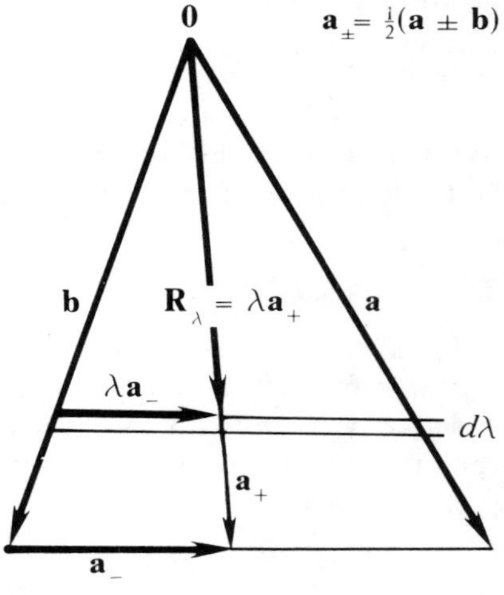

Fig. 2.3. Subdivision of a plane triangular lamina.

the median contained in the diagonal of the parallelogram. Since the centroid lies on one median, it must lie at the intersection of all three medians, as we found before by calculation. Unfortunately, our symmetry argument does not give the factor 2/3 in (2.15).

Example 2.2. Centroid of a Hemisphere

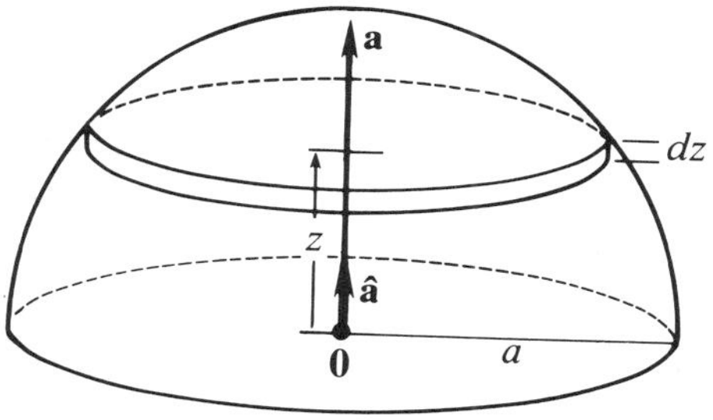

Fig. 2.4. Subdivision of a hemisphere into disks.

As another application of the additivity principle, let us calculate the centroid of a hemisphere. As indicated in Figure 2.4, we can subdivide the hemisphere into thin disks with centers on the axis of symmetry. A disk of thickness dz and centroid $z\hat{\mathbf{a}}$ has mass $\varrho\pi(a^2 - z^2)\,\mathrm{d}z$. So the mass of the hemisphere is

$$m = \varrho\pi \int_0^a (a^2 - z^2)\,\mathrm{d}z = \tfrac{2}{3}\pi a^3 \varrho ,$$

and the centroid of the hemisphere is

$$\mathbf{R} = \frac{1}{m}\int_0^a \varrho\pi(a^2 - z^2)\,\mathrm{d}z\, z\hat{\mathbf{a}} = \tfrac{3}{8}\,\mathbf{a}. \qquad (2.16)$$

The centroid of *any solid of revolution* can be found in a similar way. Such a body can always be subdivided into disks so its centroid is given by

$$\mathbf{R} = \frac{\displaystyle\int_0^a r^2(z)\,\mathrm{d}z\, z\hat{\mathbf{a}}}{\displaystyle\int_0^a r^2(z)\,\mathrm{d}z}, \qquad (2.17)$$

where $r^2(z)$ is the square of the radius of the disk with centroid $z\hat{\mathbf{a}}$ from an origin located at the intersection of the axis of symmetry with the base of the body.

Calculating the Inertia Tensor

The inertia tensor for a system of particles

$$\mathscr{I}\mathbf{u} = \sum_k m_k \mathbf{r}_k \mathbf{r}_k \wedge \mathbf{u}$$

generalizes to

$$\mathscr{I}\mathbf{u} = \int \mathrm{d}m\ \mathbf{r}\mathbf{r}\wedge\mathbf{u} \qquad (2.18)$$

for a continuous body. Although (2.18) holds for any chosen origin, it will be most convenient to assume that the origin for (2.18) is the center of mass unless an alternative is explicitly specified.

Example 2.3. Inertia tensor of a uniform rod.

The inertia tensor of a homogeneous rod can be calculated directly from (2.18) without difficulty. By symmetry the centroid of the rod is at its center, so we take that as the origin. Let the rod have directed length 2**a** and negligible thickness so we can regard it as a continuous line of mass points with direction specified by a vector **a**. Then $\mathbf{r} = \lambda\mathbf{a}$ designates the points of the rod when the values of λ are in the range $-1 \leqslant \lambda \leqslant 1$. In this case the volume integral (2.18) for the inertia tensor reduces to a line integral. Thus,

$$\mathcal{I}\mathbf{u} = \int \mathrm{d}m\, \mathbf{r}\mathbf{r}\wedge\mathbf{u} = \int_{-1}^{1} (a\, \mathrm{d}\lambda)\left(\frac{m}{2a}\right)(\lambda\mathbf{a})(\lambda\mathbf{a})\wedge\mathbf{u}$$

$$= \frac{m}{2}\,\mathbf{a}\mathbf{a}\wedge\mathbf{u} \int_{-1}^{1} \lambda^2\, \mathrm{d}\lambda .$$

Therefore

$$\mathcal{I}\mathbf{u} = \frac{m}{3}\mathbf{a}\mathbf{a}\wedge\mathbf{u} = \frac{m}{3}\,\mathbf{a}\times(\mathbf{u}\times\mathbf{a}) \tag{2.19}$$

is the inertia tensor for a homogeneous rod of length $2|\mathbf{a}|$. From the parallel axis theorem (1.14) we find that the inertia tensor $\mathcal{I}_{\mathbf{a}}$ with respect to either end of the rod is

$$\mathcal{I}_{\mathbf{a}}\mathbf{u} = \frac{m}{3}\,\mathbf{a}\mathbf{a}\wedge\mathbf{u} + m\mathbf{a}\mathbf{a}\wedge\mathbf{u} = \frac{4m}{3}\,\mathbf{a}\mathbf{a}\wedge\mathbf{u}. \tag{2.20}$$

Note that (2.19) is an explicit representation of the inertia tensor for a rod in terms of a vector **a** which directly represents the length and alignment of the rod. Similarly, for more complicated bodies we aim to evaluate the integral (2.18) to represent the inertia tensor for a body explicitly in terms of vectors describing prominent geometrical features of the body. But before attempting to integrate (2.18) for a complex body, it will be worthwhile to see how the problem can be simplified by exploiting the principles of symmetry and additivity which we used to simplify center of mass calculations.

Symmetry Principles for Inertia Tensors.

Let us first investigate what symmetries of a body tell about its inertia tensor. We have seen that a *symmetry* $\mathcal{S}$ of a body is best defined as an orthogonal transformation which takes each point **r** to a point $\mathbf{r}' = \mathcal{S}\mathbf{r}$ and leaves the mass density ϱ invariant, that is,

$$\varrho(\mathbf{r}') = \varrho(\mathcal{S}\mathbf{r}) = \varrho(\mathbf{r}). \tag{2.21a}$$

In Section 5-3 we proved that any orthogonal transformation can be written in the explicit form

$$\mathbf{r}' = \mathcal{S}\mathbf{r} = \pm\, S^\dagger \mathbf{r} S, \tag{2.21b}$$

where $S^\dagger S = 1$ and the plus (minus) sign is used if S is an even (odd) multivector. In particular, (2.21b) is identical to the reflection (2.5a) if $S = \mathbf{n}$ or to the inversion $\mathbf{r}' = -\mathbf{r}$ if $S = i$.

To determine the relation of $\mathcal{S}$ to the inertia tensor, we can use (2.21b) to obtain

$$\begin{aligned}\mathbf{r}'\mathbf{r}'\wedge\mathbf{u} &= \tfrac{1}{2} S^\dagger\mathbf{r}S(S^\dagger\mathbf{r}S\mathbf{u} - \mathbf{u}S^\dagger\mathbf{r}S)\\ &= \tfrac{1}{2} S^\dagger\mathbf{r}(\mathbf{r}S\mathbf{u}S^\dagger - S\mathbf{u}S^\dagger\mathbf{r})S = S^\dagger(\mathbf{r}\mathbf{r}\wedge(S\mathbf{u}S^\dagger))S.\end{aligned}$$

So, because of (2.21a).

$$\begin{aligned}\mathcal{I}\mathbf{u} &= \int \mathrm{d}m(\mathbf{r})\mathbf{r}\mathbf{r}\wedge\mathbf{u} = \int \mathrm{d}m(\mathbf{r}')\mathbf{r}'\mathbf{r}'\wedge\mathbf{u}\\ &= S^\dagger\left[\int \mathrm{d}m(\mathbf{r})\mathbf{r}\mathbf{r}\wedge(S\mathbf{u}S^\dagger)\right]S\\ &= S^\dagger\,(\mathcal{I}(S\mathbf{u}S^\dagger))S = \mathcal{S}\mathcal{I}\mathcal{S}^{-1}\mathbf{u},\end{aligned}$$

where $\mathcal{S}^{-1}\mathbf{u} = \pm\, S\mathbf{u}S^\dagger$. Thus we have the operator formula

$$\mathcal{S}\mathcal{I}\mathcal{S}^{-1} = \mathcal{I},$$

or, equivalently,

$$\mathcal{S}\mathcal{I} = \mathcal{I}\mathcal{S}. \tag{2.22}$$

This may be regarded as the precise mathematical formulation of the statement that "*the inertia tensor of a body is invariant under every symmetry of the body*". It is not true, however, that every orthogonal transformation which commutes with the inertia tensor is a symmetry of the body.

Equation (2.22) relates symmetries to principal vectors and principal values. If $\mathbf{a}$ is a principal vector of $\mathcal{I}$ with principal value A, then we write

$$\mathcal{I}\mathbf{a} = A\mathbf{a}. \tag{2.23}$$

From (2.22) it follows that

$$\mathcal{I}\mathcal{S}\mathbf{a} = \mathcal{S}\mathcal{I}\mathbf{a} = \mathcal{S}(A\mathbf{a}) = A\mathcal{S}\mathbf{a},$$

that is,

$$\mathcal{I}\mathbf{a}' = A\mathbf{a}' \quad \text{if} \quad \mathbf{a}' = \mathcal{S}\mathbf{a}.$$

Thus, *any symmetry-related vector* $\mathbf{a}' = \mathcal{S}\mathbf{a}$ *of a principle vector* $\mathbf{a}$ *is also a principle vector with the same principle value*. A number of corollaries follow easily for the various kinds of symmetry.

(1) *The normal of any symmetry plane is a principal vector.*
(2) *The axis of a nontrivial rotation symmetry is a principal axis.*
(3) *If there is a nontrivial rotation symmetry through an angle less than* π, *then all vectors orthogonal to the rotation axis are principal vectors with the same principal value.*

It should be noted that inversion symmetry tells us nothing about the inertia tensor, though it does tell us the location of the center of mass.

The Additivity Principle for Inertia Tensors

For direct calculation of an inertia tensor, additivity is the most important general principle. To formulate and derive this principle, it is convenient to designate a given body by the set $\mathscr{B}$ of mass points which make it up. Now if a body $\mathscr{B}$ is subdivided into N bodies $\mathscr{B}_1, \mathscr{B}_2, \ldots, \mathscr{B}_N$, it follows from the definition (2.18) that its inertial tensor is subject to the corresponding subdivision;

$$\int_{\mathscr{B}} \mathrm{d}m\, \mathbf{r}\mathbf{r}\wedge\mathbf{u} = \int_{\mathscr{B}_1} \mathrm{d}m\, \mathbf{r}\mathbf{r}\wedge\mathbf{u} + \int_{\mathscr{B}_2} \mathrm{d}m\, \mathbf{r}\mathbf{r}\wedge\mathbf{u} + \ldots + \int_{\mathscr{B}_N} \mathrm{d}m\, \mathbf{r}\mathbf{r}\wedge\mathbf{u},$$

where the integrals are over the indicated bodies (sets of particles). This relation can be expressed in operator form

$$\mathscr{I}' = \mathscr{I}'_1 + \mathscr{I}'_2 + \ldots + \mathscr{I}'_N, \tag{2.24}$$

describing the inertia tensor $\mathscr{I}'$ of a body as a sum of inertia tensors $\mathscr{I}'_k$ of its parts. The primes serve as a reminder that the inertia tensors in (2.24) are generally not referred to the mass center of the component bodies.

It is essential to realize that *the additivity relation (2.24) holds only for inertia tensors referred to a common origin*, so the parallel axis theorem (1.14) is needed to exploit it. Thus, if a body $\mathscr{B}$ is composed of N parts with known masses m_k, mass centers $\mathbf{R}_k$ and inertia tensors $\mathscr{I}_k$, then its inertia tensor can be found by the following steps. First, the inertia tensors of the parts must be referred to the origin by the parallel axis theorem (1.14), so we write

$$\mathscr{I}_{\mathbf{R}_k}\mathbf{u} = \mathscr{I}_k\mathbf{u} + m_k\mathbf{R}_k\mathbf{R}_k\wedge\mathbf{u}.$$

Then the inertia tensor $\mathscr{I}_{\mathbf{R}}$ of the body $\mathscr{B}$ is given by additivity;

$$\begin{aligned}\mathscr{I}_{\mathbf{R}}\mathbf{u} &= \mathscr{I}_{\mathbf{R}_1}\mathbf{u} + \mathscr{I}_{\mathbf{R}_2}\mathbf{u} + \ldots + \mathscr{I}_{\mathbf{R}_N}\mathbf{u}\\ &= \sum_k \mathscr{I}_k\mathbf{u} + \sum_k m_k\mathbf{R}_k\mathbf{R}_k\wedge\mathbf{u},\end{aligned} \tag{2.25}$$

where, of course,

$$\mathbf{R} = \frac{\sum_k m_k\mathbf{R}_k}{\sum_k m_k}$$

is the center of mass of the body. The inertia tensor $\mathscr{I}_{\mathbf{R}}$ is referred to the origin, so the inertia tensor $\mathscr{I}$ with respect to the center of mass is determined by the parallel axis theorem

$$\mathscr{I}_{\mathbf{R}}\mathbf{u} = \mathscr{I}\mathbf{u} + m\mathbf{R}\mathbf{R}\wedge\mathbf{u}. \tag{2.26}$$

It will be noted that the parallel axis theorem (2.26) can be interpreted as an additivity relation, for the last term in (2.26) is the inertia tensor with respect to the origin of a single particle located at the center of mass **R**.

The need to use (2.26) can be avoided by selecting the center of mass of the body $\mathscr{B}$ as origin before making the calculation, so (2.25) gives $\mathscr{I}$ directly; thus,

$$\mathscr{I}\mathbf{u} = \sum_{k=1}^{N} (\mathscr{I}_k\mathbf{u} + m_k\mathbf{R}_k\mathbf{R}_k\wedge\mathbf{u}), \tag{2.27}$$

where $\mathbf{R}_k$ is the directance from the *CM* of the entire body to the *CM* of its k-th part.

For a continuous subdivision of a body into parts the sum (2.27) goes over to an integral

$$\mathscr{I}\mathbf{u} = \int_{\alpha}^{\beta} \mathrm{d}m_\lambda\,(\mathscr{I}_\lambda\mathbf{u} + \mathbf{R}_\lambda\mathbf{R}_\lambda\wedge\mathbf{u}). \tag{2.28}$$

For each value of the parameter λ in the range $\alpha \leqslant \lambda \leqslant \beta$, $\mathscr{I}_\lambda$ is the inertia tensor per unit mass of the body part with mass $\mathrm{d}m_\lambda$ and center of mass $\mathbf{R}_\lambda$. Equation (2.28) is a powerful means for calculating inertia tensors. It enables us to calculate the tensors for 2-dimensional bodies from the tensors for 1-dimensional bodies, and then calculate the tensors for 3-dimensional bodies in from those of 2-dimensional bodies. This is best understood by working out some examples.

Example 2.4. Homogeneous triangular lamina

Let us see the additivity principle to calculate the inertia tensor of a homogeneous triangular lamina. We can subdivide the triangle into narrow strips indexed by a parameter λ as indicated in Figure 2.3. We have already determined in (2.14) that the mass of a strip is $\mathrm{d}m_\lambda = 2m\lambda\,\mathrm{d}\lambda$, and we know from (2.15) that the *CM* of the triangle is $\frac{2}{3}\,\mathbf{a}_+$, so the relative *CM* of the strip is $\mathbf{R}_\lambda = (\lambda - \frac{2}{3})\mathbf{a}_+$. From the expression (2.19) for the inertia tensor for a rod, we can write down the inertia tensor per unit mass for the strip

$$\mathscr{I}_\lambda\mathbf{u} = \tfrac{1}{3}\,(\lambda\mathbf{a}_-)\,(\lambda\mathbf{a}_-)\wedge\mathbf{u}.$$

Therefore (2.28) gives us

$$\mathscr{I}\mathbf{u} = \int_0^1 (2m\lambda\,\mathrm{d}\lambda)\left(\frac{\lambda^2}{3}\,\mathbf{a}_-\mathbf{a}_-\wedge\mathbf{u} + \left(\lambda - \frac{2}{3}\right)^2 \mathbf{a}_+\mathbf{a}_+\wedge\mathbf{u}\right).$$

Carrying out the elementary integrations, we obtain

$$\mathscr{I}\mathbf{u} = \frac{m}{6}(\mathbf{a}_-\mathbf{a}_-\wedge\mathbf{u} + \tfrac{1}{3}\,\mathbf{a}_+\mathbf{a}_+\wedge\mathbf{u}) \tag{2.29}$$

for the inertia tensor of a homogeneous triangular lamina.

The inertia tensor (2.29) can be expressed in forms that look more symmetrical as shown in the exercises, but (2.29) is prèferable for some purposes. For example, for an isocelles triangle we have $\mathbf{a}_+\cdot\mathbf{a}_- = 0$, and (2.29) shows immediately that $\mathbf{a}_+$ and $\mathbf{a}_-$ are the principal vectors of the triangle with principal values given by

$$\mathscr{I}\mathbf{a}_+ = \left(\frac{m\mathbf{a}_-^{\,2}}{6}\right)\mathbf{a}_+\,,$$

$$\mathscr{I}\mathbf{a}_- = \left(\frac{m\mathbf{a}_+^{\,2}}{18}\right)\mathbf{a}_-\,. \tag{2.30}$$

Of course we would recognize that $\mathbf{a}_+$ and $\mathbf{a}_-$ are principal vectors from symmetry principles, and that is sufficient reason to express the inertia tensor in terms of them.

Example 2.5. Elliptical lamina

For calculating the inertia tensor of a homogeneous elliptical lamina, the additivity formula (2.28) is not very helpful. It is easier to make a direct evaluation of the integral

$$\mathscr{I}\mathbf{u} = \int \frac{m\,\mathrm{d}A}{A}\,\mathbf{r}\mathbf{r}\wedge\mathbf{u}\,. \tag{2.31}$$

Let the semi-major and minor axes of the ellipse be specified by vectors $\mathbf{a}$ and $\mathbf{b}$. The points in the ellipse can be parametrized by the equation

$$\mathbf{r} = \mathbf{a}x + \mathbf{b}y, \tag{2.32a}$$

where

$$x^2 + y^2 = \lambda^2 \leqslant 1. \tag{2.32b}$$

This reduces the integration over the ellipse to integration over a circular disk, which is readily carried out by using polar coordinates λ and ϕ. Thus, the element of area of the ellipse is $\mathrm{d}A = (a\,\mathrm{d}x)\,(b\,\mathrm{d}y) = ab\lambda\,\mathrm{d}\lambda\,\mathrm{d}\phi$, so the area of the ellipse is

$$A = ab\int_0^1 \lambda\,\mathrm{d}\lambda \int_0^{2\pi} \mathrm{d}\phi = \pi ab,$$

in agreement with our calculation in Section 4-2 by a different method. Now, when (2.32a) is substituted into (2.31) we obtain

$$\mathscr{I}\mathbf{u} = \frac{m}{\pi}(\mathbf{a}\mathbf{a}\wedge\mathbf{u}\iint x^2\,\mathrm{d}x\,\mathrm{d}y + \mathbf{b}\mathbf{b}\wedge\mathbf{u}\iint y^2\,\mathrm{d}x\,\mathrm{d}y),$$

for it is obvious by symmetry that

$$\iint xy\,\mathrm{d}x\,\mathrm{d}y = 0.$$

Also by symmetry, we have

$$\iint x^2\,\mathrm{d}x\,\mathrm{d}y = \iint y^2\,\mathrm{d}x\,\mathrm{d}y = \tfrac{1}{2}\iint \lambda^2\,\mathrm{d}x\,\mathrm{d}y = \tfrac{1}{2}\int_0^1 \lambda^3\,\mathrm{d}\lambda \int_0^{2\pi}\mathrm{d}\phi = \tfrac{1}{4}\pi.$$

Thus, we obtain

$$\mathscr{I}\mathbf{u} = \frac{m}{4}(\mathbf{a}\mathbf{a}\wedge\mathbf{u} + \mathbf{b}\mathbf{b}\wedge\mathbf{u}) \tag{2.33}$$

for the inertia tensor of a homogeneous elliptical disk.

Note that before the last integral over λ was carried out we could have written the inertia tensor in the form

$$\mathscr{I}\mathbf{u} = \int_0^1 \mathrm{d}m_\lambda \frac{\lambda^2}{2}(\mathbf{a}\mathbf{a}\wedge\mathbf{u} + \mathbf{b}\mathbf{b}\wedge\mathbf{u}), \tag{2.34}$$

where

$$\mathrm{d}m_\lambda = \int_0^{2\pi} \frac{mab\lambda\,\mathrm{d}\lambda\,\mathrm{d}\phi}{\pi ab} = 2m\lambda\,\mathrm{d}\lambda.$$

So by using the additivity principle in reverse we can conclude that the inertia tensor for a homogeneous elliptical loop is

$$\mathscr{I}\mathbf{u} = \frac{m}{2}(\mathbf{a}\mathbf{a}\wedge\mathbf{u} + \mathbf{b}\mathbf{b}\wedge\mathbf{u}). \tag{2.35}$$

Note also that the inertia tensor for a flat elliptical ring can be calculated from (2.34) by raising the lower limit of integration.

Matrix Elements and Moments of Inertia

In most physics books the inertia tensor is calculated and used in matrix form only. We have developed techniques for dealing with the inertia tensor as a single entity, but we must consider its matrix elements to make contact with the literature.

Let $\{\boldsymbol{\sigma}_k\}$ be any righthanded basis of vectors. According to the definition of the inertia tensor (2.18),

$$\mathcal{I}\boldsymbol{\sigma}_k = \int \mathrm{d}m\mathbf{r}\mathbf{r}\wedge\boldsymbol{\sigma}_k = \int \mathrm{d}m(r^2\boldsymbol{\sigma}_k - r_k\mathbf{r}), \tag{2.36}$$

where $r_k = \mathbf{r}\cdot\boldsymbol{\sigma}_k$. From this we obtain the matrix elements

$$I_{jk} \equiv \boldsymbol{\sigma}_j\cdot(\mathcal{I}\boldsymbol{\sigma}_k) = \int \mathrm{d}m(\boldsymbol{\sigma}_j\wedge\mathbf{r})\cdot(\mathbf{r}\wedge\boldsymbol{\sigma}_k) = \int \mathrm{d}m(r^2\delta_{jk} - r_jr_k). \tag{2.37}$$

The conventional way to calculate an inertia tensor is by evaluating the integrals (2.37) to get its matrix elements. The diagonal elements of the matrix I_{jk} are called "moments of inertia". In particular,

$$I_{33} = \int \mathrm{d}m|\mathbf{r}\wedge\boldsymbol{\sigma}_3|^2 = \int \mathrm{d}m(r_1^2 + r_2^2) \tag{2.38}$$

is "the moment of inertia about $\boldsymbol{\sigma}_3$".

Of course, the numerical values of the matrix I_{jk} depend on the basis $\{\boldsymbol{\sigma}_k\}$ to which it is referred. The matrix takes its simplest form when referred to a basis of principal vectors $\mathbf{e}_k$, for then

$$\mathcal{I}\mathbf{e}_k = I_k\mathbf{e}_k, \tag{2.39}$$

which produces the diagonal matrix

$$\mathbf{e}_j\cdot(\mathcal{I}\mathbf{e}_k) = I_k\,\delta_{jk}, \tag{2.40}$$

where, according to (2.37) and (2.38),

$$I_k = \int \mathrm{d}m|\mathbf{r}\wedge\mathbf{e}_k|^2. \tag{2.41}$$

The *principal values* I_k of the inertia tensor are called *principal moments of inertia*.

The trace of the inertia tensor Tr $\mathcal{I}$ is defined as the sum of diagonal matrix elements. From (2.37) and (2.40)

$$\mathrm{Tr}\,\mathcal{I} = I_{11} + I_{22} + I_{33} = I_1 + I_2 + I_3 = 2\int \mathrm{d}mr^2. \tag{2.42}$$

The right hand side of (2.42) shows that the trace is independent of basis, so (2.42) relates any set of diagonal matrix elements to the principal moments of inertia. This relation is sometimes useful for calculating principal values.

Equation (2.38) shows that the moments of inertia I_{kk} must be positive numbers, but the special nature of the inertia tensor puts further restrictions on their relative values, as we shall now show. From (2.38) and (2.42) we have

$$I_{33} = \int \mathrm{d}m(r^2 - r_3^2) = \tfrac{1}{2}(I_{11} + I_{22} + I_{33}) - \int \mathrm{d}mr_3^2.$$

Hence,

$$I_{11} + I_{22} - I_{33} = 2 \int \mathrm{d}m r_3^2 \geq 0. \tag{2.43}$$

The integral here will vanish only if $r_3 = \mathbf{r}\cdot\boldsymbol{\sigma}_3 = 0$ for all points in the body, which can occur only if the points lie in a plane with normal $\boldsymbol{\sigma}_3$. Therefore, for a plane lamina with normal $\boldsymbol{\sigma}_3$, (2.43) reduces to

$$I_{11} + I_{22} - I_{33} = 0. \tag{2.44}$$

Of course, a plane lamina is only a convenient mathematical idealization justifiable for bodies of negligible thickness, so for a real body (2.44) can be only approximately true.

The relation (2.43) holds for any orthonormal basis, so it applies to the principle moments of inertia; thus,

$$I_1 + I_2 - I_3 \geq 0. \tag{2.45}$$

Furthermore, this relation holds for any permutation of the subscripts, so the sum of any two principal values can never be less than the other principal value.

Moment of Inertia and Radius of Gyration About a Line

So far we have discussed moments of inertia only as parts of a complete matrix of inertia. But when a body is rotating about a fixed axis with known direction $\mathbf{u}$, only the moment of inertia about that axis is of interest. The moment of inertia about (a line with direction) $\mathbf{u}$ is

$$I_{\mathbf{u}} \equiv \mathbf{u}\cdot(\mathcal{I}\mathbf{u}) = \int \mathrm{d}m\ |\mathbf{r}\wedge\mathbf{u}|^2. \tag{2.46}$$

Equation (2.46) implies that $I_{\mathbf{u}}$ is always a positive number for a real body. Vanishing moments of inertia occur only for mathematical idealizations such as a rod without thickness, as described by (2.19).

Note that $r_\perp \equiv |\mathbf{r}\wedge\mathbf{u}|$ is the perpendicular distance of the point $\mathbf{r}$ from the line with direction $\mathbf{u}$ passing through the origin. Therefore, the integral in (2.46) is a sum of the squared distances $r_\perp^2$ weighted by the mass of each point. Instead of the moment of inertia $I_{\mathbf{u}}$, it is sometimes convenient to use the radius of *gyration about* $\mathbf{u}$ defined by

$$I_{\mathbf{u}} = mk_{\mathbf{u}}^2. \tag{2.47}$$

The radius of gyration $k_{\mathbf{u}}$ can be interpreted as the distance from the axis of a single particle with mass m with the same moment of inertia as the body with tensor $\mathcal{I}$ in (2.46). This follows from the fact that the inertia tensor for a particular at $\mathbf{R}$ is $\mathcal{I}\mathbf{u} = m\mathbf{R}\mathbf{R}\wedge\mathbf{u}$, so

$$\mathbf{u}\cdot\mathcal{I}\mathbf{u} = m\ |\mathbf{R}\wedge\mathbf{u}|^2,$$

which agrees with (2.47) if $k_{\mathbf{u}} = |\mathbf{R}\wedge\mathbf{u}|$. Thus, we have a kind of equivalence in inertia of a whole body to a single particle. Our next task will be to investigate this kind of equivalence systematically and in complete generality.

Classification of Rigid Bodies

Two rigid bodies are said to be *equimomental* if they have the same mass and principal moments of inertia. A glance at the equations of motion for a rigid body in Section 7-1 shows that they are identical for equimomental bodies subject to equivalent forces. Thus, equimomental bodies are dynamically identical, though they may differ greatly in size and shape. The shape of a body is relevant to its motion only in the possibilities it gives for applying contact forces.

It is always possible to find a small set of particles equimomental to a given continuous body. For example, from the inertia tensor (2.19) for a rod we can see that the rod is equimomental to a system of three particles, one particle of mass $m/6$ at each of the points $\pm\mathbf{a}$ and one particle with mass $2m/3$ at the origin. The particles are symmetrically placed so their centroid will be at the origin, and the particle at the origin is needed so the system will have the correct total mass. It will be noted that other sets of particles will give the same result, for instance, two particles with mass $\frac{1}{2}m$ at the points $\pm\mathbf{a}/\sqrt{3}$.

Rigid bodies fall into three dynamically distinct classes, depending on the relative values of their three principle moments of inertia I_1, I_2, I_3. A body is said to be

(1) *centrosymmetric* if all its principle moments are equal ($I_1 = I_2 = I_3$),

(2) *axially symmetric* if it has exactly two distinct principal moments (e.g. $I_1 \neq I_2 = I_3$),

(3) *asymmetric* if it has three distinct principal moments (e.g. $I_1 > I_2 > I_3$).

This terminology necessarily differs from that in other books, since there is no standard terminology available for these three classes.

The inertia tensor of a centrosymmetric body is determined by a single number, its principal moment of inertia $I \equiv I_1 = I_2 = I_3$. Every line through the center of mass is a principal axis, so the dynamics of the body is completely independent of its attitude, that is, the dynamics would be unaffected by a finite rotation of the body about any axis through the center of mass. This rotational invariance of a centrosymmetric body is a dynamical symmetry and should not be confused with the geometrical symmetry of a body. The bodies listed in Table 2.1 are all centrosymmetric, so they have the same dynamical symmetries, but their geometrical symmetries are quite different. For example, the sphere is geometrically symmetrical under any rotation about its center, so its dynamical symmetry is identical to its geometrical symmetry. However, the cube is geometrically symmetrical only under particular finite rotations, such as a rotation by $2\pi/3$ about one of its

TABLE 2.1. The moment of inertia for some *centrosymmetric* homogeneous bodies

Body	Moment of inertia I
(Solid) sphere (radius a)	$\frac{2}{5}ma^2$
Hollow sphere	$\frac{2}{3}ma^2$
Hemisphere	$\frac{2}{5}ma^2$
Hemispherical shell	$\frac{2}{3}ma^2$
(Solid) cube (side a)	$\frac{1}{6}ma^2$
Regular tetrahedron (side a)	$\frac{ma^2}{24}$
Right circular cylinder of radius a and height $a\sqrt{3}$	$\frac{1}{2}ma^2$
Cone of radius a and height $\frac{1}{2}a$	$\frac{3ma^2}{10}$

diagonals. Nevertheless, the geometrical symmetries of the cube imply that all its principal moments of inertia are equal by theorems we have established above, so its dynamical symmetry is a consequence of its geometrical symmetry. For other bodies in Table 2.1, such as the hemisphere or the cylinder, geometrical symmetry alone is not sufficient to determine dynamical symmetry.

Inertia tensors of some axially symmetric bodies are expressed in "canonical form" in Table 2.2. These tensors are completely determined by a unit vector **e** along the "dynamical symmetry axis" and the two distinct principal moments of inertia I_1 and $I_2 = I_3$. The vector **e** is the principal vector of the inertia tensor $\mathscr{I}$ corresponding to the principal value I_1, so I_2 is the principal value of *any* vector orthogonal to **e**. To determine the effect of $\mathscr{I}$ on an arbitrary vector **u**, we break **u** into components parallel and perpendicular to **e**. Thus,

$$\mathscr{I}\mathbf{u} = \mathscr{I}(\mathbf{u}_\parallel + \mathbf{u}_\perp) = \mathscr{I}\mathbf{u}_\parallel + \mathscr{I}\mathbf{u}_\perp = I_1\mathbf{u}_\parallel + I_2\mathbf{u}_\perp = I_1\mathbf{e}\mathbf{e}\cdot\mathbf{u} + I_2\mathbf{e}\mathbf{e}\wedge\mathbf{u}.$$

The dependence on **e** can be simplified by using $\mathbf{e}\mathbf{e}\wedge\mathbf{u} = \mathbf{e}(\mathbf{e}\mathbf{u} - \mathbf{e}\cdot\mathbf{u}) = \mathbf{u} - \mathbf{e}\mathbf{e}\cdot\mathbf{u}$, so

$$\mathscr{I}\mathbf{u} = I_2\mathbf{u} + (I_1 - I_2)\mathbf{e}\mathbf{e}\cdot\mathbf{u}. \tag{2.48}$$

This is the "canonical form" adopted for the tensors in Table 2.2. There are other interesting forms for the inertia tensor. For example, substitution of $\mathbf{e}\cdot\mathbf{u} = \frac{1}{2}(\mathbf{e}\mathbf{u} + \mathbf{u}\mathbf{e})$ into (2.48) yields the "symmetrical" form

TABLE 2.2 Inertia tensors for some *axially symmetric* homogeneous bodies

Body		Inertia tensor $\mathcal{I}\mathbf{u}$
Flat circular ring	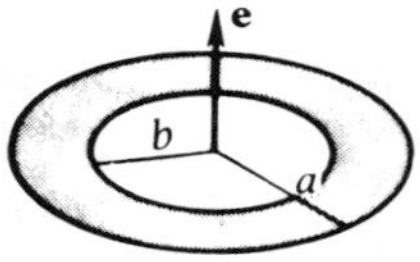	$\frac{m(a^2 + b^2)}{4}(\mathbf{u} + \mathbf{ee}\cdot\mathbf{u})$
Cylindrical Tube	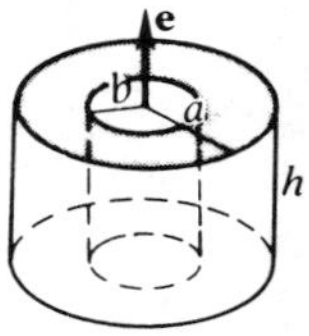	$\frac{m}{4}[(a^2 + b^2 + \frac{1}{3}h^2)\mathbf{u} + (a^2 + b^2 - \frac{1}{3}h^2)\mathbf{ee}\cdot\mathbf{u}]$
Ellipsoid of Revolution	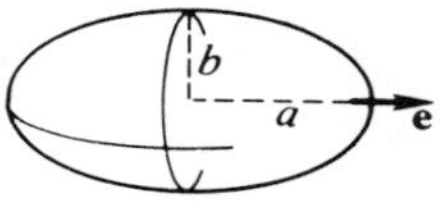	$\frac{m}{5}[(a^2 + b^2)\mathbf{u} + (a^2 - b^2)\mathbf{ee}\cdot\mathbf{u}]$
Circular Cone	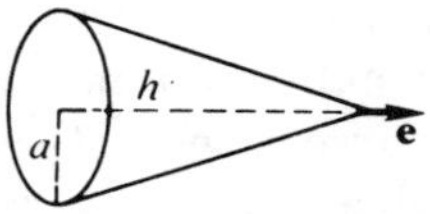	$\frac{3m}{20}[(a^2 + 4h^2)\mathbf{u} + (a^2 - 4h^2)\mathbf{ee}\cdot\mathbf{u}]$
Half Conical Shell	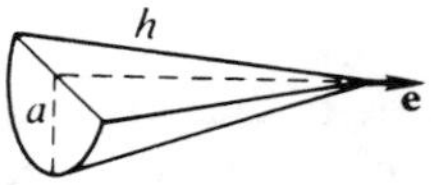	$\frac{m}{4}[(a^2 + 2h^2)\mathbf{u} + (a^2 - 2h^2)\mathbf{ee}\cdot\mathbf{u}]$
Half Cylindrical Shell	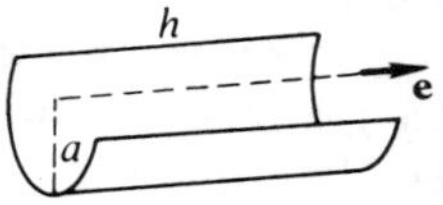	$m\left[\left(\frac{a^2}{2} + \frac{h^2}{3}\right)\mathbf{u} + \left(\frac{a^2}{4} - \frac{h^2}{3}\right)\mathbf{ee}\cdot\mathbf{u}\right]$
Solid Semicylinder	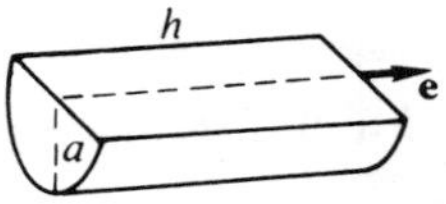	$m\left[\left(\frac{a^2}{4} + \frac{h^2}{3}\right)\mathbf{u} + \left(\frac{a^2}{4} - \frac{h^2}{3}\right)\mathbf{ee}\cdot\mathbf{u}\right]$
Half Torus	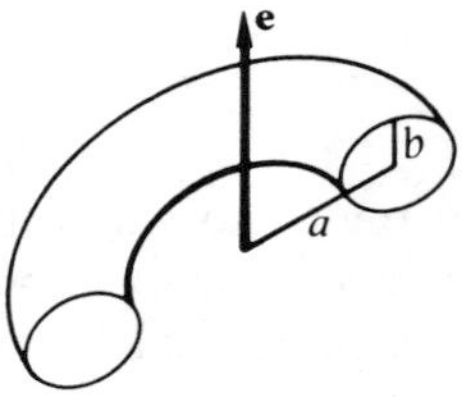	$\frac{m}{2}\left[\left(a^2 + \frac{5}{4}b^2\right)\mathbf{u} + \left(a^2 - \frac{1}{8}b^2\right)\mathbf{ee}\cdot\mathbf{u}\right]$

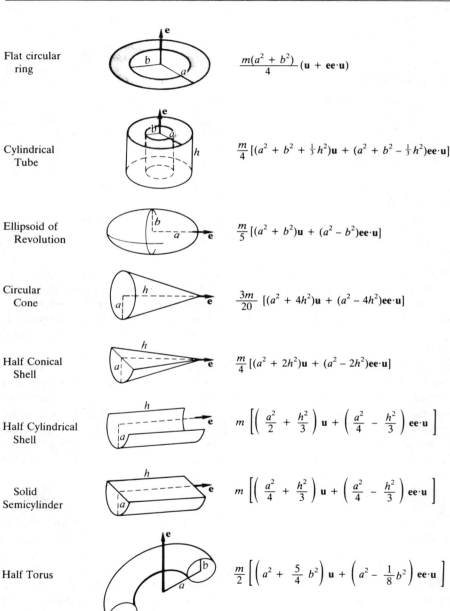

TABLE 2.3. Inertia tensors of some *asymmetric homogeneous bodies*

Body		Inertia tensor $\mathcal{I}\mathbf{u}$
Parallelepiped	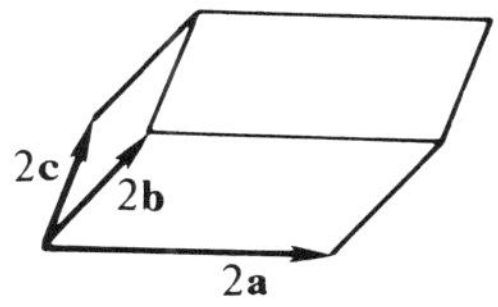	$\frac{m}{3}[\mathbf{aa}\wedge\mathbf{u} + \mathbf{bb}\wedge\mathbf{u} + \mathbf{cc}\wedge\mathbf{u}]$
Ellipsoid	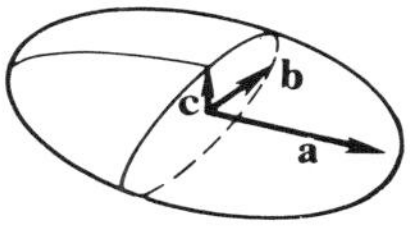	$\frac{m}{5}[\mathbf{aa}\wedge\mathbf{u} + \mathbf{bb}\wedge\mathbf{u} + \mathbf{cc}\wedge\mathbf{u}]$
Ellipsoidal Shell		$\frac{m}{3}[\mathbf{aa}\wedge\mathbf{u} + \mathbf{bb}\wedge\mathbf{u} + \mathbf{cc}\wedge\mathbf{u}]$
Elliptical Cylinder	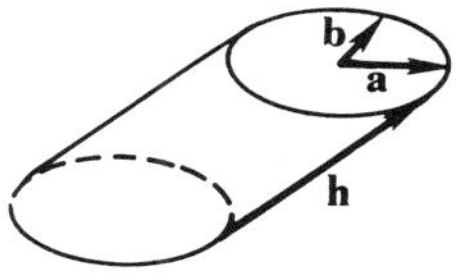	$\frac{m}{4}[\mathbf{aa}\wedge\mathbf{u} + \mathbf{bb}\wedge\mathbf{u} + \frac{1}{3}\mathbf{hh}\wedge\mathbf{u}]$
Elliptical Cone	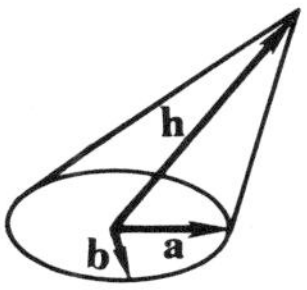	$\frac{3m}{20}[\mathbf{aa}\wedge\mathbf{u} + \mathbf{bb}\wedge\mathbf{u} + 4\mathbf{hh}\wedge\mathbf{u}]$
Hollow Elliptical Cone		$\frac{m}{4}[\mathbf{aa}\wedge\mathbf{u} + \mathbf{bb}\wedge\mathbf{u} + 2\mathbf{hh}\wedge\mathbf{u}]$
Triangular Lamina	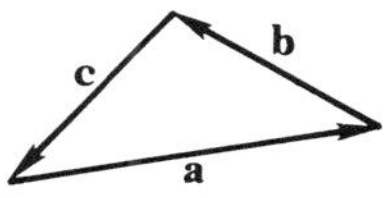	$\frac{m}{36}[\mathbf{aa}\wedge\mathbf{u} + \mathbf{bb}\wedge\mathbf{u} + \mathbf{cc}\wedge\mathbf{u}]$
Triangular Prism	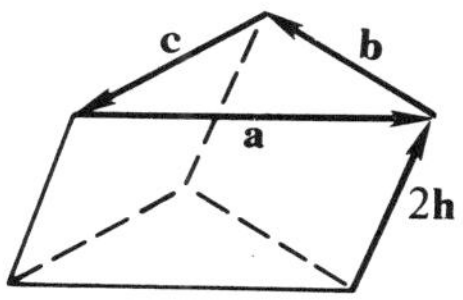	$\frac{m}{3}[\mathbf{aa}\wedge\mathbf{u} + \mathbf{bb}\wedge\mathbf{u} + \mathbf{cc}\wedge\mathbf{u} + \mathbf{hh}\wedge\mathbf{u}]$

$$\mathcal{I}\mathbf{u} = \left(\frac{I_1 + I_2}{2}\right)\mathbf{u} + \left(\frac{I_1 - I_2}{2}\right)\mathbf{eue}. \tag{2.49}$$

However, we shall see that (2.48) is most useful when we study dynamics.

The "canonical forms" for the inertia tensors of asymmetric bodies in Table 2.3 have been chosen to show their relation to the geometrical structure of the bodies. The principal axes and principal values are not evident from these forms, but must be calculated by the methods of Section 5-2. Some examples are given in the exercises. We shall see that it is important to know the principal axes when analyzing the motion of a body.

7-2. Exercises

(2.1) Calculate the centroid and moment of inertia of a hemispherical shell.

(2.2) Find the centroids of a solid circular cone, a hollow circular cone, and half conical shell (Origin as indicated in Table 2.2).

(2.3) Find the centroids of a solid semicylinder and a half cylindrical shell (Origin as indicated in Table 2.2).

(2.4) Find the centroid of a half torus (Origin as indicated in Table 2.2).

(2.5) Find the centroid of a spherical cap cut from a sphere of a radius a by a plane at a distance b from the center of the sphere.

(2.6) Determine the centroid and volume of the intersection of a sphere of radius a with a cone of vertex angle 2ϕ and vertex at the center of the sphere.

(2.7) A cylindrical hole of radius b is cut from a cube of width $2a$ with an axis at a distance d from the center of the cube and perpendicular to a face of the cube. Find the mass, centroid, and inertia tensor of the resulting body if it has uniform density ϱ.

(2.8) Calculate the moment of inertia of a homogeneous cube directly from Equation (2.18).

(2.9) Let three intersecting edges of a cube be designated by vectors $\mathbf{a}_1$, $\mathbf{a}_2$, $\mathbf{a}_3$, with $\mathbf{a}^2 = \mathbf{a}_1^2 = \mathbf{a}_2^2 = \mathbf{a}_3^2$. Calculate the inertia tensor about a corner of the cube. Determine its principle vectors and principle values and its matrix elements with respect to the edges of the cube.

(2.10) Find the moment of inertia of a rectangular parallelopiped with edges a, b, c about a diagonal.

(2.11) A particle of unit mass is located at each of the four points

$$\boldsymbol{\sigma}_1 - 5\boldsymbol{\sigma}_2 - \boldsymbol{\sigma}_3,\ 3(\boldsymbol{\sigma}_1 + \boldsymbol{\sigma}_2 + \boldsymbol{\sigma}_3),\ \boldsymbol{\sigma}_1 + \boldsymbol{\sigma}_2 + 5\boldsymbol{\sigma}_3,\ -5\boldsymbol{\sigma}_1 + \boldsymbol{\sigma}_2 - \boldsymbol{\sigma}_3.$$

Find the centroid and inertia tensor of the system.

(2.12) A homogeneous rod of mass m and length $2\mathbf{a}$ is rotating with angular velocity $\boldsymbol{\omega}$ about one end. What is its kinetic energy?

(2.13) Three particles of unit mass are located at the points $\mathbf{r}_1 = a\boldsymbol{\sigma}_1$, $\mathbf{r}_2 = a\boldsymbol{\sigma}_2 + 2a\boldsymbol{\sigma}_3$, $\mathbf{r}_3 = 2a\boldsymbol{\sigma}_2 + a\boldsymbol{\sigma}_3$. Find the principal moments of inertia about the origin and the corresponding principle values. Express the inertia tensor about the origin in terms of its principal vectors.

(2.14) Calculate the inertia tensor for an ellipsoid in the form given in Table 2.3 by generalizing the method of Example 2.5 to reduce integration over the ellipsoid to easy integrations over a sphere.

(2.15) For a given body, prove that if a line is a principal axis for one of its points, then it is a principal axis for all of its points and it passes through the centroid of the body, and conversely.
Prove also that a line with direction $\mathbf{u}$ and distance $\mathbf{d} \neq 0$ from the centroid can be a principal axis for one of its points if and only if

$$\mathbf{d} \wedge \mathbf{u} \wedge \mathcal{I}\mathbf{u} = 0.$$

(2.16) Derive the inertia tensor for a homogeneous triangle in the form shown in Table 2.3. Show that its principle values are

$$I_{\pm} = \frac{m}{36}\left(\frac{a^2 + b^2 + c^2}{2} \pm \sqrt{a^4 + b^4 + c^4 - b^2c^2 - c^2a^2 - a^2b^2}\right).$$

(2.17) If $\mathbf{r}_1, \mathbf{r}_2, \mathbf{r}_3$ locate the vertices with respect to the centroid, show that inertia tensor of a triangle of mass m can be put in the form

$$\mathcal{I}\mathbf{u} = \frac{m}{12} \sum_{k=1}^{3} \mathbf{r}_k \mathbf{r}_k \wedge \mathbf{u}.$$

(2.18) Let $\mathbf{a}_1, \mathbf{a}_2, \ldots, \mathbf{a}_6$ designate the edges of a homogeneous tetrahedron and let $\mathbf{r}_1, \mathbf{r}_2, \mathbf{r}_3, \mathbf{r}_4$ locate the vertices with respect to the centroid. Show that its inertia tensor is

$$\mathcal{I}\mathbf{u} = \frac{m}{80} \sum_{k=1}^{6} \mathbf{a}_k \mathbf{a}_k \wedge \mathbf{u} = \frac{m}{20} \sum_{k=1}^{4} \mathbf{r}_k \mathbf{r}_k \wedge \mathbf{u}.$$

(2.19) A homogeneous triangle of mass m is equimomental to a system of four particles with one at its centroid and three of the same mass either at its vertices or at the midpoints of its sides. Find the particle masses in each of the two cases.

(2.20) Show that a homogeneous tetrahedron of mass m is equimomental to a system of six particles with mass $m/10$ at the midpoints of its edges and one particle of mass $2m/5$ at its centroid.

(2.21) Prove that every body is equimomental to some system of four equal mass particles, and that every plane lamina is equimomental to three equal mass particles.

(2.22) *Legendre's Ellipsoid.* Prove that every body is equimomental to an ellipsoid.

(2.23) Inertia tensors $\mathcal{I}$ and $\mathcal{I}'$ are related by the displacement equation

$$\mathcal{I}'\mathbf{u} = \mathcal{I}\mathbf{u} + m\mathbf{R}\mathbf{R}\wedge\mathbf{u}.$$

Their principle vectors and principle values are specified by

$$\mathscr{I}\mathbf{e}_k = I_k\mathbf{e}_k, \quad \mathscr{I}'\mathbf{e}'_k = I'_k\mathbf{e}'_k.$$

For $\mathbf{R}\cdot\mathbf{e}_3 = 0$, prove that:
(a) $\mathbf{e}'_3 = \mathbf{e}_3$ and $I'_3 = I_3 + m\mathbf{R}^2$
(b) For $k = 1, 2$ and $\mathbf{i} = \mathbf{e}_1\mathbf{e}_2$,

$$\mathbf{e}'_k = \mathbf{e}_k e^{\mathbf{i}\phi}, \quad \text{where} \quad \tan 2\phi = \frac{2mR_1R_2}{I_1 - I_2},$$

(c) $$\begin{Bmatrix} I'_1 \\ I'_2 \end{Bmatrix} = \frac{I_1+I_2+mR^2}{2} \pm \tfrac{1}{2}\{(I_1 - I_2 + m(R_1^2 - R_2^2))^2 + 4m^2R_1^2R_2^2\}^{1/2}$$

where $R_k \equiv \mathbf{R}\cdot\mathbf{e}_k$.

(2.24) For the plane lamina in Figure 2.5, calculate the principal moments of inertia and the angle ϕ specifying the relative direction of the principal axes.

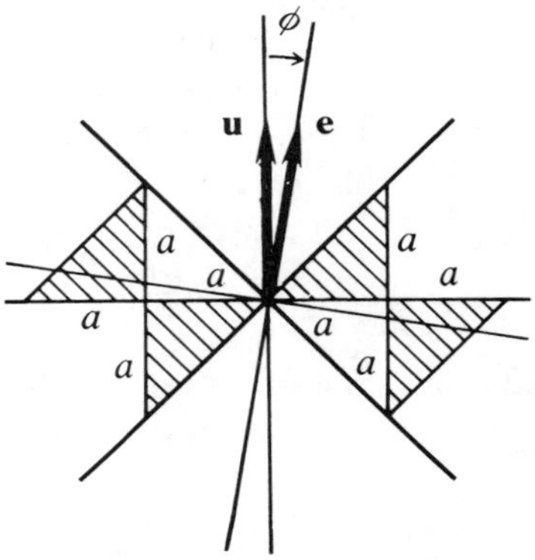

Fig. 2.5. Plane lamina indicated by shaded area.

7-3. The Symmetrical Top

In this section we study the rotational motion of an axially symmetric body, which might be referred to as a top, a pendulum or a gyroscope, depending on the nature of the motion. From Section 7-2, we know that the inertia tensor for an axially symmetric body can be put in the form

$$\mathbf{l} = \mathscr{I}\boldsymbol{\omega} = I\boldsymbol{\omega} + (I_3 - I)\boldsymbol{\omega}\cdot\mathbf{e}\,\mathbf{e}, \tag{3.1}$$

where I_3 and $I_1 = I_2 = I$ are the principle moments of inertia, and $\mathbf{e}$ is the direction of the symmetry axis. According to the parallel axis theorem, this form for the inertia tensor will be preserved by any shift of base point along the symmetry axis; the effect of such a shift is merely to change numerical values for the moments of inertia. Consequently, the results of this section apply to the motion of any symmetrical top with a fixed point somewhere on its symmetry axis.

Recall that the rotational equations of motion consists of a *dynamical equation* (Euler's equation)

$$\dot{\mathbf{l}} = \boldsymbol{\Gamma} \tag{3.2}$$

for the angular momentum $\mathbf{l}$ driven by an applied torque $\boldsymbol{\Gamma}$, and a *kinematical equation*

$$\dot{R} = \tfrac{1}{2}Ri\boldsymbol{\omega} \tag{3.3}$$

for the attitude spinor R depending on the rotational velocity $\boldsymbol{\omega}$. The spinor R determines the principle directions $\mathbf{e}_k$ of the body by

$$\mathbf{e}_k = R^\dagger \boldsymbol{\sigma}_k R \tag{3.4}$$

for fixed $\boldsymbol{\sigma}_k$. For an axially symmetric body we write $\mathbf{e} = \mathbf{e}_3$, so

$$\mathbf{e} = R^\dagger \boldsymbol{\sigma}_3 R\,. \tag{3.5}$$

This relation couples the two equations of motion (3.2) and (3.3) by virtue of (3.1), and further coupling derives from the fact that an applied torque usually depends on R as well. Such coupling makes the equations of motion difficult to solve. Consequently, closed solutions exist for only a few cases, and more general cases must be treated by approximations and numerical techniques. Fortunately, the simplest cases are the most common, and they provide a stepping stone for the analysis of more complex cases, so our primary aim will be to master the simplest cases of rotational motion completely.

Qualitative Features of Rotational Dynamics

Before getting involved in the details of solving Euler's equation, it is a good idea to identify the major qualitative features of rotational dynamics. Suppose a body is spinning rapidly with angular speed ω about a principal direction $\mathbf{e}$ with moment of inertia I_3. Then its rotational velocity is $\boldsymbol{\omega} = \omega\mathbf{e}$ and its angular momentum is $\mathbf{l} = \mathcal{I}\boldsymbol{\omega} = \omega\,\mathcal{I}\mathbf{e} = \omega I_3\mathbf{e}$. This state of affairs will persist as long as no forces act on the body, since $\dot{\mathbf{l}} = 0$. Now, if a small force $\mathbf{F}$ is applied to a point $\mathbf{r} = r\mathbf{e}$ on the axis of rotation, Euler's equation can be put in the approximate form

$$\dot{\mathbf{l}} = I_3\omega\dot{\mathbf{e}} = r\mathbf{e} \times \mathbf{F}\,,$$

or

$$\dot{\mathbf{e}} = \left(\frac{-r\mathbf{F}}{I_3\omega}\right) \times \mathbf{e}. \tag{3.6}$$

This equation displays two major features of gyroscopic motion. First, it tells us that a force applied to the axis of a rapidly spinning body causes the body to move in direction perpendicular to the force. This explains why leaning to one side on a moving bicycle makes it turn rather than fall over. Such behavior may seem paradoxical, because it is so different from the behavior of a nonspinning body. But it can be seen as a consequence ot elementary kinematics in the following way. Consider the effect of an impulse $\mathbf{F}\,\Delta t$ delivered in a short time to the spinning disk in Figure 3.1. Since the body is rigid, the torque about the center of the disk exerted by the force $\mathbf{F}$ is equivalent to the torque exerted by a force $\mathbf{F}'$ applied at a point on the rim as shown. But the effect of an impulse $\mathbf{F}'\,\Delta t$ is to alter the velocity $\mathbf{v}$ of the rim by, an amount $\Delta\mathbf{v}$, thus producing rotational motion perpendicular to $\mathbf{F}$ as asserted.

The second thing that Equation (3.6) tells us is that the larger the values of I and ω, the smaller the effect of $\mathbf{F}$. This is sometimes called *gyroscopic stiffness*, and it accounts for the great directional stability of a gyroscope. The dynamical properties of a slowly rotating body are quite different. Indeed, for validity of (3.6) it is necessary that $|rF/I_3\omega| \ll \omega$. This condition determines what is meant by a small force in gyroscopic problems.

Now let us turn to a detailed analysis of the equations of motion and their ramifications.

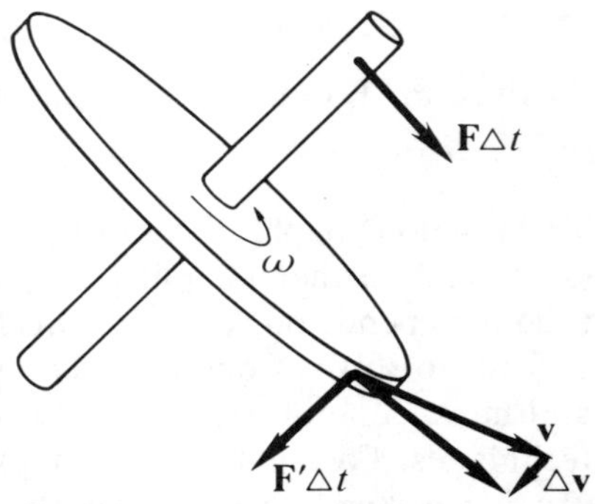

Fig. 3.1. Deflection of a spinning body by an impulse.

Free Precession

A spinning body is said to be *spinning freely* if the resultant torque $\boldsymbol{\Gamma}$ on it vanishes. According to Euler's equation $\dot{\mathbf{l}} = \boldsymbol{\Gamma}$, then, the angular momentum $\mathbf{l}$ of a freely spinning body is a constant of the motion. To obtain a complete description of motion, it is still necessary to determine the attitude R as a function of time by integrating the kinetic equation $\dot{R} = \frac{1}{2}Ri\boldsymbol{\omega}$. To do this, we need to know the rotational velocity $\boldsymbol{\omega}$, but that can be obtained from (3.1). Thus, we find

$$\boldsymbol{\omega} = \frac{\mathbf{l}}{I} + \frac{I - I_3}{I_3}\left(\frac{\mathbf{l}\cdot\mathbf{e}}{I}\right)\mathbf{e}. \tag{3.7}$$

Now, using $\mathbf{e} = R^{\dagger}\boldsymbol{\sigma}_3 R$ and $RR^{\dagger} = 1$, we can put the kinetic equation of motion in the form

$$\dot{R} = \tfrac{1}{2}Ri\boldsymbol{\omega} = \tfrac{1}{2}Ri\left[\frac{\mathbf{l}}{I} + \frac{(I - I_3)}{I_3}\frac{\mathbf{l}\cdot\mathbf{e}\;\mathbf{e}}{I}\right]$$
$$= \tfrac{1}{2}Ri\frac{\mathbf{l}}{I} + \tfrac{1}{2}\frac{(I - I_3)}{I_3}\frac{\mathbf{l}\cdot\mathbf{e}}{I}i\boldsymbol{\sigma}_3 R.$$

This is an equation of the form

$$\dot{R} = \tfrac{1}{2}i\boldsymbol{\omega}_1 R + \tfrac{1}{2}Ri\boldsymbol{\omega}_2, \tag{3.8a}$$

where

$$\boldsymbol{\omega}_2 = \frac{\mathbf{l}}{I} \quad \text{and} \quad \boldsymbol{\omega}_1 = \frac{(I - I_3)}{I_3}\,\mathbf{e}\cdot\boldsymbol{\omega}_2\boldsymbol{\sigma}_3. \tag{3.8b}$$

For a freely spinning body, both $\boldsymbol{\omega}_1$, and $\boldsymbol{\omega}_2$ are constant, so (3.8a) has the elementary solution

$$R = e^{(1/2)i\boldsymbol{\omega}_1 t}R_0 e^{(1/2)i\boldsymbol{\omega}_2 t}, \tag{3.9}$$

where R_0 describes the attitude at time $t = 0$.

From the solution (3.9), we find that the motion of the body's symmetry axis is given by

$$\mathbf{e} = R^{\dagger}\boldsymbol{\sigma}_3 R = e^{-(1/2)i\boldsymbol{\omega}_2 t}\mathbf{e}_0 e^{(1/2)i\boldsymbol{\omega}_2 t}, \tag{3.10}$$

where $\mathbf{e}_0 = R_0^{\dagger}\boldsymbol{\sigma}_3 R_0$ is the initial direction of the symmetry axis. Equation (3.10) tells us that the symmetry axis precesses about the angular momentum vector $\mathbf{l}$ with a constant angular velocity $\boldsymbol{\omega}_2 = \mathbf{l}/I$, as shown in Figure 3.2a, b. Therefore, by observing the free precession of the symmetry axis, the angular momentum can be determined.

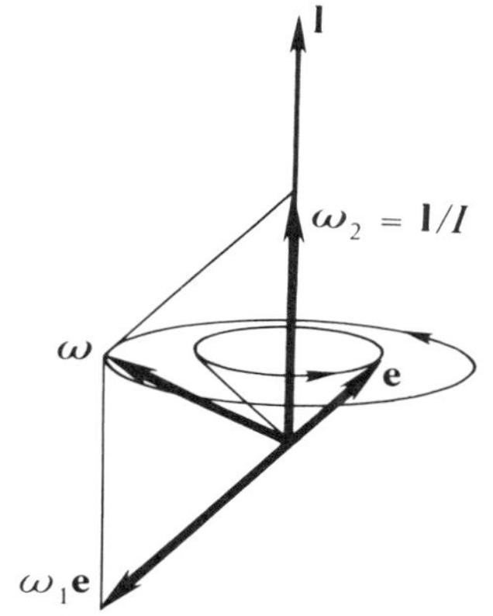

Fig. 3.2a. Free precession for an oblate body ($\omega_1 < 0$).

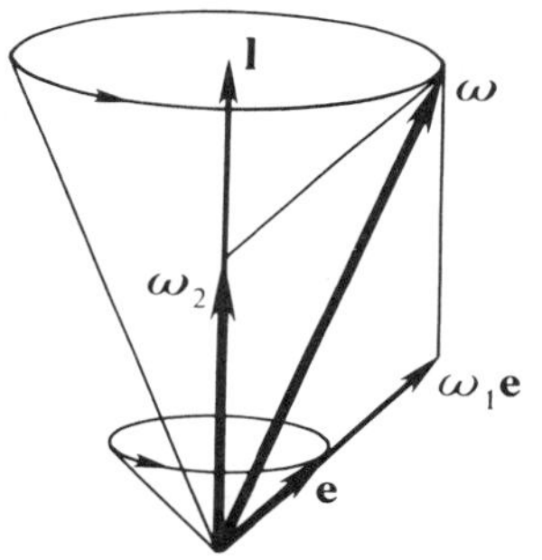

Fig. 3.2b. Free precession for a prolate body ($\omega_1 > 0$).

We can now supply the physical interpretation of the solution (3.9). The first factor in (3.9) describes a rotation of the body with constant angular speed ω_1 about its symmetry axis positioned along some arbitrary direction $\hat{\omega}_1 = \boldsymbol{\sigma}_3$. The second factor R_0 "tilts" the symmetry axis from $\boldsymbol{\sigma}_3$ to a specified direction $\mathbf{e}_0 = R_0^{\dagger}\boldsymbol{\sigma}_3 R_0$. Of course, we could take $R_0 = 1$ if we chose $\boldsymbol{\sigma}_3 = \mathbf{e}_0$. The third factor in (3.9) describes a precession of the body. Thus, the solution (3.9) describes a body spinning with angular speed ω_1 about its symmetry axis while it precesses with angular velocity $\boldsymbol{\omega}_2$. This motion is called *Eulerian free precession*.

The resultant rotational velocity $\boldsymbol{\omega}$ can be determined from (3.8a), (3.9) and (3.10); thus,

$$\begin{aligned}\boldsymbol{\omega} &= \frac{2}{i}R^{\dagger}\dot{R} = R^{\dagger}\omega_1 R + \boldsymbol{\omega}_2, = \omega_1\mathbf{e} + \boldsymbol{\omega}_2 \\ &= e^{-(1/2)i\boldsymbol{\omega}_2 t}(\omega_1\mathbf{e}_0 + \boldsymbol{\omega}_2)e^{(1/2)i\boldsymbol{\omega}_2 t}.\end{aligned} \tag{3.11}$$

This tells us that $\boldsymbol{\omega}$ precesses about $\mathbf{l} = I\boldsymbol{\omega}_2$ along with $\mathbf{e}$, with the three vectors $\boldsymbol{\omega}$, $\mathbf{l}$, $\mathbf{e}$ remaining in a common plane ($\boldsymbol{\omega}\wedge\mathbf{l}\wedge\mathbf{e} = 0$). The precession of **e** and $\boldsymbol{\omega}$ is illustrated in Figure 3.2a for an oblate body ($I < I_3$) such as the Earth of or a disk, and in Figure 3.2b for a prolate body ($I > I_3$) such as a football.

Free precession of the Earth produces an observable variation in latitude. From (3.9) and (3.11) we find that the resultant rotational velocity ω' in the body frame of the Earth is

$$\boldsymbol{\omega}' = R\boldsymbol{\omega}R^\dagger = \frac{2}{i}\dot{R}R^\dagger$$

$$= e^{(1/2)i\omega_1 t}R_0(\omega_1\mathbf{e}_0 + \boldsymbol{\omega}_2)R_0^\dagger e^{-(1/2)i\omega_1 t}. \tag{3.12}$$

This says that $\boldsymbol{\omega}'$ precesses with angular velocity $-\boldsymbol{\omega}_1$ about the Earth's symmetry axis $\hat{\boldsymbol{\omega}}_1 = \boldsymbol{\sigma}_3$. The vector $\hat{\boldsymbol{\omega}}'$ is the direction of the *celestial north pole*. On a short timelapse photograph with a camera pointing vertically upward the star trails are arcs of concentric circles centered on the celestial pole, so the direction of $\hat{\boldsymbol{\omega}}'$ relative to the earth can be measured with great accuracy. Successive measurements determine its motion relative to the Earth.

One complete rotation of the Earth about the celestial pole defines

$$1 \text{ siderial day} = \frac{2\pi}{\mathbf{e}\cdot\boldsymbol{\omega}} = \frac{2\pi}{\boldsymbol{\sigma}_3\cdot\boldsymbol{\omega}'} .$$

From an independent analysis of perturbations by the Moon (Chapter 8), it has been determined that $I/(I_3 - I) = 303$. Therefore, (3.12) and (3.11) predict that the celestial pole will precess about the Earth's axis with period

$$T = \frac{2\pi}{-\omega_1} = \frac{I_3}{I_3 - I}\,\frac{2\pi}{\mathbf{e}\cdot\omega_2} = \frac{I}{I_3 - I}\,\frac{2\pi}{\mathbf{e}\cdot\boldsymbol{\omega}} = 303 \text{ days}. \tag{3.13}$$

Empirically, it is found the celestial pole precesses irregularly about the Earth's axis, tracing in one year a roughly elliptical path on the Earth's surface with a mean radius of about four meters. Analysis of the data shows that the orbit has two distinct periods, one of 12-month and the other of 14-month duration. The 12-month period is accounted for by seasonal changes in the weather, primarily the formation of ice and snow in polar regions. The motion with a 14-month period is called the *Chandler wobble* after the man who discovered it. This is identified with the Eulerian wobble predicted above.

The large discrepancy between the observed 14-month period for the Chandler wobble and the prediction of a 10-month period in (3.13) deserves some explanation. It results from the fact that the Earth is not an ideal rigid body. Without developing a detailed theory, we can see qualitatively that the elasticity of the Earth will lengthen the period. Consider what would happen if the rotation of the Earth were to cease. If the Earth were a liquid body, it would clearly assume a spherical shape when the centrifugal force is turned off. But the Earth, or, at least, its shell, is an elastic body, so it will tend to retain its shape, though its *oblateness*, as defined by $(I_3 - I)/I$, will decrease. Similarly, the oblateness will be decreased, because the Earth's instantaneous axis of rotation is not along the polar axis, and (3.13) shows that this will lengthen the period of the wobble. A quantitative analysis of this effect is a complex problem in geophysics, requiring an analysis of fluctuations due to earthquakes, tidal effects, seasonal motions of air masses, etc. It is an active area for geophysical research today.

Reduction of the Symmetric Top

The analysis we have just completed has a significance that goes well beyond free precession. For it suggests a method of reducing the equations of motion for an axially symmetric body to the simpler equations of motion for a centrosymmetric body. Notice that Equations (3.8a, b) are valid for any motion of an axially symmetric body. Furthermore, we can separate (3.8a) into the two equations.

$$\dot{R}_1 = \tfrac{1}{2}\, i\boldsymbol{\omega}_1 R_1 = \tfrac{1}{2}\, R_1 i\boldsymbol{\omega}_1 \tag{3.14a}$$

and

$$\dot{R}_2 = \tfrac{1}{2}\, R_2 i\boldsymbol{\omega}_2 = \tfrac{1}{2}\, R_2 i\frac{\mathbf{l}}{I}\,, \tag{3.14b}$$

with

$$R = R_1 R_2 .$$

Now (3.14b) is the attitude equation for a centrosymmetric body with moment of inertia I, and it can be solved independently of (3.14a) using the equation of motion $\dot{\mathbf{l}} = \boldsymbol{\Gamma}$.

Then the solution to (3.14a) can be found from (3.14b); specifically, it has the form

$$R_1 = e^{(1/2)i\boldsymbol{\sigma}_3\phi_1}, \tag{3.15a}$$

where, by (3.8b)

$$\phi_1 = \int_0^t \omega_1\, \mathrm{d}t = \int_0^t \frac{(I - I_3)}{I_3}\, \mathbf{e}\cdot\boldsymbol{\omega}_2\, \mathrm{d}t\,, \tag{3.15b}$$

and

$$\mathbf{e} = R^\dagger \boldsymbol{\sigma}_3 R = R_2^\dagger \boldsymbol{\sigma}_3 R_2 . \tag{3.15c}$$

Note that the integral form of (3.15b) even allows time variations in I and I_3 to be taken into account.

Let us call the reduction of the equations of motion just described *the reduction theorem for a symmetric top*, and let us refer to the motion described by (3.15a) as the *Eulerian motion*, since it generalizes the Eulerian free precession. The reduction theorem tells us that the motion of an axially symmetric body can be obtained from the motion of a centrosymmetric body simply by superimposing the Eulerian motion. It also tells us that the Eulerian free precession is maintained even in the presence of a torque $\boldsymbol{\Gamma}$ if $\boldsymbol{\Gamma}\cdot\mathbf{e} = 0$; for then we know from Section 7-1 that $\mathbf{e}\cdot\boldsymbol{\omega}_2 = I^{-1}\,\mathbf{e}\cdot\mathbf{l}$ is constant and (3.15b) integrates to give us the same result as the one obtained for zero torque. This provides justification for ignoring torques when analyzing the Eulerian precession of the Earth.

The Spherical Top

From now on we restrict our analysis to the reduced equations of motion

$$\dot{\mathbf{l}} = I\dot{\boldsymbol{\omega}} = \mathbf{r} \times \mathbf{F}, \tag{3.16}$$

$$\dot{R} = \tfrac{1}{2}Ri\boldsymbol{\omega}. \tag{3.17}$$

These are general equations of motion for the attitude of a centrosymmetric body subject to a single force acting at a point $\mathbf{r}$ fixed in the body. We know from Section 7–2 that every centrosymmetric body is equimomental to a spherical body. So we refer to (3.16) as the dynamical equation of motion for a *spherical top*. The *reduction theorem* tells us that from the solution of the equations of motion for a spherical top, we get the solution for an axially symmetric body simply by multiplying by the Euler factor (3.15a). Actually, it won't be worth the trouble for us to include the Euler motion explicitly, because it affects only the rate of rotation about the symmetry axis. The motion of the symmetry axis is the most prominent feature of rotational motion, and is *completely* described by the reduced equations.

It should be noted that (3.16) and (3.17) are coupled equations, since the direction of $\mathbf{r}$ depends on the attitude R. Also, to relate the solution to the axially symmetric case, the direction of $\mathbf{r}$ must be that of the symmetry axis

$$\mathbf{e} = \mathbf{e}_3 = R^\dagger\boldsymbol{\sigma}_3 R,$$

so it will be unaffected by the Euler factor. To make this explicit and lump all the constants together, let us write $\mathbf{r} = r\mathbf{e}$ and write (3.16) in the form

$$\dot{\omega} = \mathbf{e} \times \mathbf{G}, \tag{3.18}$$

where the effective force $\mathbf{G}$ is defined by

$$\mathbf{G} = \frac{r\mathbf{F}}{I}. \tag{3.19}$$

Now a significant mathematical advantage of the reduction theorem can be seen. It enables us to combine the coupled equations (3.17) and (3.18) into a single spinor equation of motion for the body. We get the equation by differentiating (3.17) and using (3.18); thus,

$$\begin{aligned}\ddot{R} &= \tfrac{1}{2}R(i\dot{\omega} - \tfrac{1}{2}\omega^2) = \tfrac{1}{2}R(\mathbf{e}\wedge\mathbf{G} - \tfrac{1}{2}\omega^2)\\ &= \tfrac{1}{2}R(\mathbf{eG} - \mathbf{e}\cdot\mathbf{G} - \tfrac{1}{2}\omega^2)\\ &= \tfrac{1}{2}\boldsymbol{\sigma}_3 R\mathbf{G} - \tfrac{1}{2}R(\tfrac{1}{2}\omega^2 + \mathbf{e}\cdot\mathbf{G}).\end{aligned} \tag{3.20}$$

To make this a determinate equation, we need to express the last term as a definite function of R. This can be done in several different ways. We can eliminate either ω^2 or $\mathbf{G}\cdot\mathbf{e}$ in favor of the other by noting that for constant $\mathbf{G}$ Euler's Equation (3.18) admits the effective energy

$$E = \tfrac{1}{2}\omega^2 - \mathbf{e}\cdot\mathbf{G} \tag{3.21}$$

as a constant of the motion. Either ω^2 or $\mathbf{e}\cdot\mathbf{G}$ can be expressed in terms of R by using

$$\omega^2 = -4(R^\dagger\dot{R})^2, \tag{3.22}$$

or

$$\mathbf{e}\cdot\mathbf{G} = (R^\dagger\boldsymbol{\sigma}_3 R)\cdot\mathbf{G} = \langle R^\dagger\boldsymbol{\sigma}_3 R\mathbf{G}\rangle_0 . \tag{3.23}$$

To keep these alternatives in mind, let us write the *spinor equation of motion for a spherical top* in the *standard form*

$$\ddot{R} = \tfrac{1}{2}\boldsymbol{\sigma}_3 R\mathbf{G} - \tfrac{1}{2}LR, \tag{3.24}$$

where L can have the various forms

$$L = \tfrac{1}{2}\omega^2 + \mathbf{e}\cdot\mathbf{G} = E + 2\mathbf{e}\cdot\mathbf{G} = \omega^2 - E, \tag{3.25}$$

to be used in (3.24) along with (3.22) or (3.23).

The spinor equation of motion (3.24) can be put in many alternative forms by various parametrizations of the spinor R. For example, we can use the Euler parameters introduced in Section 5-3 by writing

$$R = \alpha + i\boldsymbol{\beta}. \tag{3.26}$$

This has the advantage of explicitly exhibiting the direction $\hat{\boldsymbol{\beta}}$ of the instantaneous axis of rotation. Substituting into the spinor equation (3.24) we obtain

$$\ddot{\alpha} + i\ddot{\boldsymbol{\beta}} = \tfrac{1}{2}(\alpha\boldsymbol{\sigma}_3\mathbf{G} + i\boldsymbol{\sigma}_3\boldsymbol{\beta}\mathbf{G}) - \tfrac{1}{2}L(\alpha + i\boldsymbol{\beta}).$$

Let us simplify this by choosing $\boldsymbol{\sigma}_3 = \hat{\mathbf{G}}$, which we are free to do. Then separating scalar and bivector parts and using $\hat{\mathbf{G}}\boldsymbol{\beta} = \boldsymbol{\beta}\hat{\mathbf{G}} + 2\hat{\mathbf{G}}\wedge\boldsymbol{\beta}$ we obtain a pair of coupled equations of motion for the Euler parameters

$$\ddot{\alpha} + \tfrac{1}{2}(L - G)\alpha = 0, \tag{3.27a}$$

$$\ddot{\boldsymbol{\beta}} + \boldsymbol{\beta}\wedge\hat{\mathbf{G}}\mathbf{G} + \tfrac{1}{2}(L - G)\boldsymbol{\beta} = 0. \tag{3.27b}$$

The Euler parameters are related by $R^\dagger R = \alpha^2 + \boldsymbol{\beta}^2 = 1$, but the Equations (3.27a, b) look more complicated if α is eliminated in favor of $\boldsymbol{\beta}$. As a function of the Euler parameters, the direction of the symmetry axis is

$$\mathbf{e} = (\alpha - i\boldsymbol{\beta})\hat{\mathbf{G}}(\alpha + i\boldsymbol{\beta}) = \alpha^2\hat{\mathbf{G}} + 2\alpha\boldsymbol{\beta}\times\hat{\mathbf{G}} + \boldsymbol{\beta}\hat{\mathbf{G}}\boldsymbol{\beta}$$
$$= \hat{\mathbf{G}} + 2\alpha\boldsymbol{\beta}\times\hat{\mathbf{G}} + 2\boldsymbol{\beta}\wedge\hat{\mathbf{G}}\boldsymbol{\beta}. \tag{3.28}$$

We shall find the Euler parameters useful for small angle approximations and motion in a vertical plane, but different parametrizations are better in other situations.

The spinor equation of motion (3.24) would be interesting to study for a variety of forces, but we shall use it only for the so-called *Lagrange problem*, in which case $\mathbf{G}$ and E are constants. To relate the equations to a concrete problem, let us consider the object in Figure 3.3, sometimes called a *gyroscopic pendulum*. The object is an axially symmetric body consisting of a disk attached to one end of a rigid rod along its axis. The other end of the rod is held fixed, but the object is free to spin about its axis and rotate in any way about the fixed point. Neglecting friction, the entire torque about the fixed point is due to the resultant gravitational force $\mathbf{F} = m\mathbf{g}$ acting at the center of mass at a directance $\mathbf{r}$ from the fixed point.

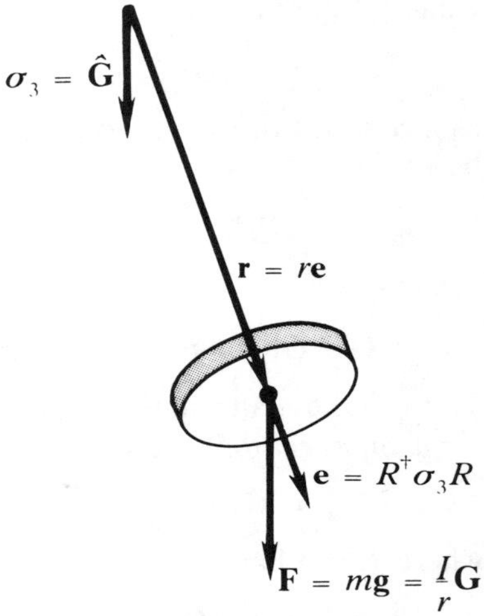

Fig. 3.3. The gyroscopic pendulum.

Now let us turn to the problem of finding solutions to the spinor equation of motion. We begin with the simplest special cases to gain insight.

Small Oscillations of a Pendulum

When we write $\mathbf{e} = R^\dagger\hat{\mathbf{G}}R$, the spinor $R = e^{(1/2)i\boldsymbol{\varepsilon}}$ measures the deviation of the symmetry axis $\mathbf{e}$ from the downward vertical $\hat{\mathbf{G}}$. For small deviations from the vertical, the angle $\boldsymbol{\varepsilon}$ is small, so we can use the power series expansion of the exponential function to get the approximate expression

$$R = e^{(1/2)i\boldsymbol{\varepsilon}} \approx 1 + \tfrac{1}{2}i\boldsymbol{\varepsilon}, \tag{3.29}$$

good to first order for $|\frac{1}{2}\boldsymbol{\varepsilon}| \ll 1$. Note that right side (3.29) is an expression for R in terms of Euler parameters $\alpha = 1$ and $\boldsymbol{\beta} = \frac{1}{2}\boldsymbol{\varepsilon}$, so we can insert these values into (3.27a, b) to obtain $L = G$ and the first order equation for $\boldsymbol{\varepsilon}$:

$$\ddot{\boldsymbol{\varepsilon}} + \boldsymbol{\varepsilon}\wedge\hat{\mathbf{G}}\mathbf{G} = 0. \tag{3.30}$$

To understand equation (3.30), notice that $\boldsymbol{\varepsilon}\wedge\hat{\mathbf{G}}\hat{\mathbf{G}}$ is merely the projection of $\boldsymbol{\varepsilon}$ onto the horizontal plane, so the vertical component of $\ddot{\boldsymbol{\varepsilon}}$ satisfies $\ddot{\boldsymbol{\varepsilon}}_{\parallel} = 0$.

The solution of this equation will be incompatible with the assumption that $\boldsymbol{\varepsilon}$ is small unless $\dot{\boldsymbol{\varepsilon}}_{\parallel} = 0$, so we may assume that $\boldsymbol{\varepsilon}_{\parallel} = 0$. Note, however, that if the object were hanging vertically and spinning about its symmetry axis with constant angular velocity $\boldsymbol{\omega}_{\parallel} = \omega_{\parallel}\hat{\mathbf{G}}$, then the rotation angle would be $\boldsymbol{\varepsilon}_{\parallel} = \boldsymbol{\omega}_{\parallel}t$. Therefore, the assumption that the angle $\boldsymbol{\varepsilon}$ is small implies that the object is not spinning about its axis, so its motion is that of a pendulum.

For horizontal $\boldsymbol{\varepsilon}$ we have $\boldsymbol{\varepsilon}\cdot\hat{\mathbf{G}} = 0$, so $\boldsymbol{\varepsilon}\wedge\hat{\mathbf{G}} = \boldsymbol{\varepsilon}\hat{\mathbf{G}}$, and (3.30) reduces to

$$\ddot{\boldsymbol{\varepsilon}} + G\boldsymbol{\varepsilon} = 0. \tag{3.31}$$

This will be recognized as the equation for a harmonic oscillator, so the solution is an ellipse in the horizontal plane with the parametric form

$$\boldsymbol{\varepsilon} = \mathbf{a}\cos G^{1/2}t + \mathbf{b}\sin G^{1/2}t. \tag{3.32}$$

For horizontal $\boldsymbol{\varepsilon}$, the Equation (3.28) for the axis reduces to

$$\mathbf{e} = \hat{\mathbf{G}} + \boldsymbol{\varepsilon}\times\hat{\mathbf{G}} - \tfrac{1}{2}\varepsilon^2\hat{\mathbf{G}} \tag{3.33}$$

This gives us the value of $\mathbf{e}$ to second order from the first order value of the angle $\boldsymbol{\varepsilon}$. The term $\boldsymbol{\varepsilon}\times\hat{\mathbf{G}} = \boldsymbol{\varepsilon}(-i\hat{\mathbf{G}})$ differs from $\boldsymbol{\varepsilon}$ only in being rotated by $\pi/2$ in the horizontal plane. Therefore, along with (3.32) the first two terms in (3.33) describe the orbit of the vector $\mathbf{e}$ to first order as an ellipse in a horizontal plane with its center at $\hat{\mathbf{G}}$. The second order term in (3.33) is directed entirely along the vertical; it has the effect of bending the elliptical orbit on the plane to fit it on the unit sphere and make it compatible with the condition $\mathbf{e}^2 = 1$.

The solution (3.32) tells us that the motion is periodic with period T given by

$$\left(\frac{2\pi}{T}\right)^2 = G = \frac{rmg}{I}, \tag{3.34}$$

where the value of G is taken from (3.19). According to the parallel axis theorem, $I = I_0 + mr^2 = mr_0^2 + mr^2$ where I_0 is the moment of inertia at the center of mass and r_0 is its radius of gyration. Therefore,

$$T = 2\pi\sqrt{\frac{r_0^2 + r^2}{rg}}\ . \tag{3.35}$$

For $r_0^2 \ll r^2$ this reduces to the formula for the period of a *simple pendulum*, for which the mass of the bob is supposed to be concentrated at a single point.

Steady Precession

Now let us return to the study of the exact spinor equation

$$\ddot{R} = \tfrac{1}{2}\boldsymbol{\sigma}_3 R\mathbf{G} - \tfrac{1}{2}LR,$$

with L given by (3.25). The terms on the right must be cancelled by the $\ddot{R}$ on the left, so $\boldsymbol{\sigma}_3$ and $\hat{\mathbf{G}}$ must be produced by differentiation. This suggests that we look for a solution of the form

$$R = R_1 R_2$$

where

$$\dot{R}_1 = \tfrac{1}{2} i\boldsymbol{\omega}_1 R_1 \quad \text{and} \quad \dot{R}_2 = \tfrac{1}{2} R_2 i\boldsymbol{\omega}_2 .$$

Then

$$\dot{R} = \tfrac{1}{2} i\boldsymbol{\omega}_1 R + \tfrac{1}{2} Ri\boldsymbol{\omega}_2 . \tag{3.36}$$

If a solution with constant $\boldsymbol{\omega}_1$ and $\boldsymbol{\omega}_2$ is possible, then

$$\ddot{R} = -\tfrac{1}{2}\boldsymbol{\omega}_1 R\boldsymbol{\omega}_2 - \tfrac{1}{4}(\omega_1^2 + \omega_2^2)R = \tfrac{1}{2}\boldsymbol{\sigma}_3 R\mathbf{G} - \tfrac{1}{2} LR.$$

From this we see that a solution is achieved if

$$\hat{\boldsymbol{\omega}}_1 = \boldsymbol{\sigma}_3, \qquad \hat{\boldsymbol{\omega}}_2 = \hat{\mathbf{G}}, \tag{3.37}$$

$$\omega_1 \omega_2 = -G, \tag{3.38}$$

$$\tfrac{1}{2}(\omega_1^2 + \omega_2^2) = L, \tag{3.39}$$

where $L = E + 2\mathbf{G}\cdot\mathbf{e}$ must be a constant, so $\mathbf{G}\cdot\mathbf{e}$ is constant. The last two equations can be solved to get ω_1 and ω_2 in terms of G and L but a different expression for ω_1 and ω_2 is more helpful.

Inserting (3.37) and (3.38) into (3.36), we obtain the total rotational velocity $\boldsymbol{\omega}$ of the body

$$\boldsymbol{\omega} = \frac{2}{i} R^\dagger \dot{R} = R^\dagger \boldsymbol{\omega}_1 R + \boldsymbol{\omega}_2 = \omega_1 \mathbf{e} + \omega_2 \hat{\mathbf{G}}, \tag{3.40}$$

where $\mathbf{e} = R^\dagger \boldsymbol{\sigma}_3 R$ as before. The total angular speed about the symmetry axis is

$$\boldsymbol{\omega}\cdot\mathbf{e} = \omega_1 + \omega_2\, \hat{\mathbf{G}}\cdot\mathbf{e}. \tag{3.41}$$

The right side of this equation shows that $\boldsymbol{\omega}\cdot\mathbf{e}$ is a constant of the motion. Now we solve (3.38) and (3.41) for ω_1 *and* ω_2, and we find two pairs of solutions

$$\omega_1 = \tfrac{1}{2}\left\{\boldsymbol{\omega}\cdot\mathbf{e} \mp [(\boldsymbol{\omega}\cdot\mathbf{e})^2 + 4\,\mathbf{G}\cdot\mathbf{e}]^{1/2}\right\} \tag{3.42}$$

$$\omega_2 = \frac{\boldsymbol{\omega}\cdot\mathbf{e} \pm [(\boldsymbol{\omega}\cdot\mathbf{e})^2 + 4\,\mathbf{G}\cdot\mathbf{e}]^{1/2}}{2\,\hat{\mathbf{G}}\cdot\mathbf{e}} . \tag{3.43}$$

The reader may verify that, if the expression (3.40) for ω is inserted into $L = \tfrac{1}{2}\omega^2 + \mathbf{e}\cdot\mathbf{G}$, then (3.39) is obtained as an identity when (3.38) is used. Therefore (3.39) does not give us any additional information about ω_2.

Having established that the conditions for a solution have been satisfied, we can write the solution explicitly:

$$R = R_1R_2 = e^{(1/2)i\boldsymbol{\omega}_1 t}R_0\, e^{(1/2)i\boldsymbol{\omega}_2 t}, \tag{3.44}$$

and we can take $R_0 = 1$ if we set $\boldsymbol{\sigma}_3 = \mathbf{e}_0$, the initial value for the symmetry axis. The solution (3.44) has exactly the same mathematical form as the free precession solution (3.9). Both solutions describe a body spinning with angular velocity ω_1 while it precesses *steadily* with angular velocity ω_2. However, their physical origins are totally different. The solution (3.44) describes a *forced steady precession*, because it results from an applied torque.

Of course, (3.44) is the reduced solution for an axially symmetric body, so to get the full solution we must multiply it by the Euler factor. According to (3.44), the angular speed of the Eulerian motion in this case is

$$\omega_0 = \left(\frac{I - I_3}{I_3}\right)\boldsymbol{\omega}\cdot\mathbf{e}. \tag{3.45}$$

The full solution for steady forced precession of an axially symmetric body is therefore

$$R = e^{(1/2)i\mathbf{e}_0(\omega_0 + \omega_1)t}\, e^{(1/2)i\boldsymbol{\omega}_2 t}. \tag{3.46}$$

The effect of the Eulerian motion is to shift the component of $\boldsymbol{\omega}$ along $\mathbf{e}$ to produce a new rotational velocity $\boldsymbol{\omega}'$. The amount of the shift is given by

$$\boldsymbol{\omega}'\cdot\mathbf{e} = \omega_0 + \boldsymbol{\omega}\cdot\mathbf{e} = \frac{I}{I_3}\boldsymbol{\omega}\cdot\mathbf{e}. \tag{3.47}$$

This relation makes it easy to convert results for centrosymmetric bodies to results for axially symmetric bodies. For example, using it in (3.43) we obtain

$$\omega_2 = \frac{I_3\boldsymbol{\omega}'\cdot\mathbf{e} \pm [I_3^2\,(\boldsymbol{\omega}'\cdot\mathbf{e})^2 + 4I^2\mathbf{G}\cdot\mathbf{e}]^{1/2}}{2I\hat{\mathbf{G}}\cdot\mathbf{e}}, \tag{3.48}$$

expressing the angular precession speed ω_2 in terms of the actual rotational velocity $\boldsymbol{\omega}'$ for an axially symmetry body. It will be noted that the conversion does not alter the functional form of the basic relations, so we might as well deal with the simpler relations in terms of $\boldsymbol{\omega}$ and make the conversion only at the end of calculations when numerical results are desired.

The expression (3.43) for the angular precession speed ω_2 is the item of greatest interest here, because it describes the motion of the symmetry axis. Let's see what it tells us about various special cases. For a rapidly spinning body the kinetic energy is much greater than the potential energy. Therefore, $(\boldsymbol{\omega}\cdot\mathbf{e})^2 \gg |4\,\mathbf{G}\cdot\mathbf{e}|$, and we can expand the square root in (3.43) to get

$$\omega_2 = \frac{\boldsymbol{\omega}\cdot\mathbf{e}}{2\,\hat{\mathbf{G}}\cdot\mathbf{e}}\left(1 \pm \left[1 + \frac{1}{2}\,\frac{4\,\mathbf{G}\cdot\mathbf{e}}{(\boldsymbol{\omega}\cdot\mathbf{e})^2} + \ldots\right]\right). \tag{3.49}$$

Thus, to a good approximation we have the two solutions

$$\omega_2 = \frac{\boldsymbol{\omega}\cdot\mathbf{e}}{\hat{\mathbf{G}}\cdot\mathbf{e}} \quad , \tag{3.50a}$$

and

$$\omega_2 = \frac{-G}{\boldsymbol{\omega}\cdot\mathbf{e}} \tag{3.50b}$$

The solution (3.50a) is said to describe a *fast top*, because the precession speed is large. It describes an *upright top* if $\hat{\mathbf{G}}\cdot\mathbf{e} < 0$ and a *hanging top* (or gyroscopic pendulum) if $\hat{\mathbf{G}}\cdot\mathbf{e} > 0$. The two possibilities are shown in Figure 3.4. The solution (3.50b) describes a *slow top* because the precession speed is small. The negative sign shows that the precession is *retrograde*, that is, in a sense opposite to the spin about the symmetry axis. The reciprocal relation $\omega_1\omega_2 = -G$ tells us that the spin ω_1, about the symmetry axis will be small for a fast top and large for a slow top.

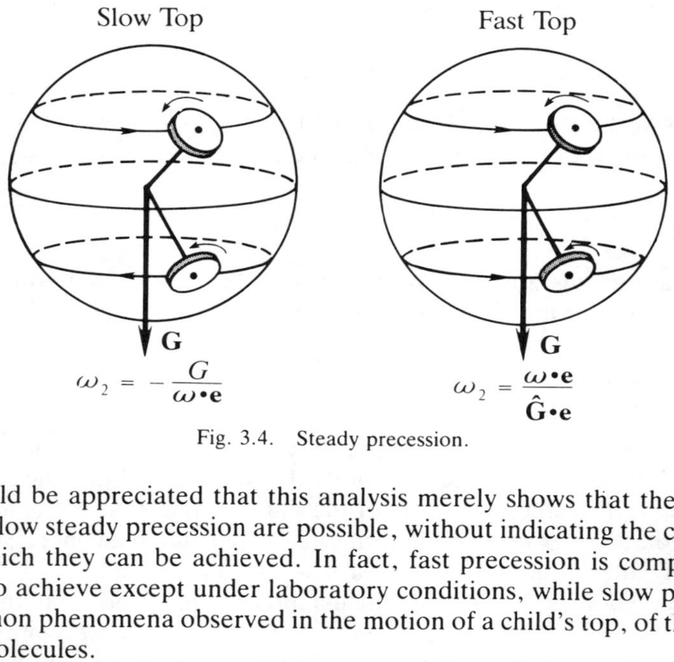

Fig. 3.4. Steady precession.

It should be appreciated that this analysis merely shows that the states of fast and slow steady precession are possible, without indicating the conditions under which they can be achieved. In fact, fast precession is comparatively difficult to achieve except under laboratory conditions, while slow precession is a common phenomena observed in the motion of a child's top, of the Earth, and of molecules.

The Sun and the Moon exert a torque on the Earth as a result of the Earth's oblateness, which is measured by the fractional difference $(I - I_3)/I_3$ of its moments of inertia. This produces a precession of the Earth's axis about the normal to the *ecliptic* (the plane of the Earth's orbit about the Sun). The intersections of the equator with the ecliptic are called equinoxes, so the effect is observed on Earth as a *precession of the equinoxes* on the celestial sphere, with different stars becoming the "pole star" at different times.

However, the torque is so weak that the precessional period is about 25,000 years. We shall see how to calculate this value in Chapter 8. Figure 3.5 shows the relation of the equinox precession to Eulerian precession.

Note that steady precession is possible only if $(\boldsymbol{\omega}\cdot\mathbf{e})^2 + 4\,\mathbf{G}\cdot\mathbf{e} > 0$, since otherwise the square root in (3.43) is not defined. This implies that a certain minimum energy is needed for precession of a top in an upright position, since then $\mathbf{G}\cdot\mathbf{e} < 0$. For an erect top, $\hat{\mathbf{G}}\cdot\mathbf{e} = -1$ and the condition that it will remain erect becomes strictly a condition its kinetic energy $\frac{1}{2}\omega^2 = \frac{1}{2}(\boldsymbol{\omega}\cdot\mathbf{e})^2 > 2G$. When this condition is met, precession is indistinguishable from spin about the symmetry axis, and the top is said to be *sleeping*.

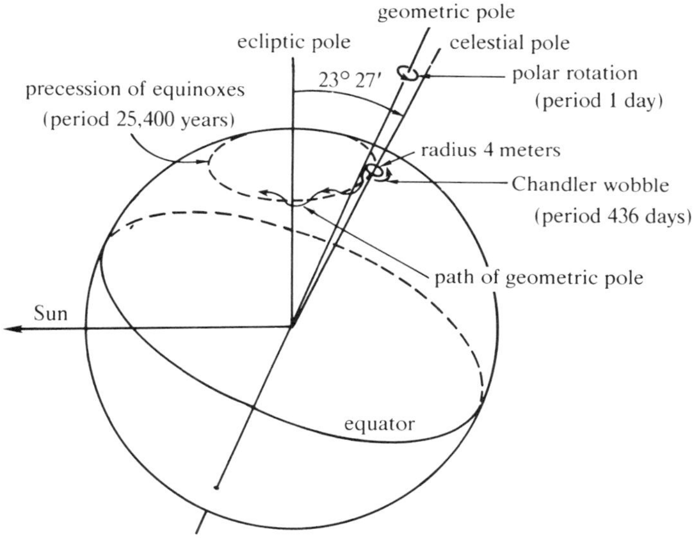

Fig. 3.5. Gyroscopic motion of the Earth; Eulerian motion and precession of the equinoxes.

For $\boldsymbol{\omega}\cdot\mathbf{e} = 0$, the expression (3.43) reduces to

$$\omega_2 = \pm\sqrt{\frac{G}{\hat{\mathbf{G}}\cdot\mathbf{e}}}\;, \tag{3.51}$$

and the motion is referred to as a *conical pendulum*. Since $\boldsymbol{\omega}\cdot\mathbf{e} = I_3\boldsymbol{\omega}'\cdot\mathbf{e}/I$ for an axially symmetric body, we will get $\boldsymbol{\omega}\cdot\mathbf{e} \approx 0$ when I_3 is relatively small. In any case, the conical pendulum requires that the kinetic energy of rotation about the symmetry axis be small.

Deviations from Steady Precession

The solution for steady precession which we have just examined is an exact solution of the equations of motion, but it is a special solution. Nevertheless, for any total energy there is always a solution with steady precession. It differs from other solutions with the same total energy in having the kinetic and potential energy as separate constants of motion. Therefore, we can describe any solution in terms of its deviation from steady precession. Accordingly, we write the solution in the form

$$R = R_1 U R_2 \tag{3.52}$$

where,

$$R_1 = e^{(1/2)i\boldsymbol{\omega}_1 t} \quad \text{with} \quad \boldsymbol{\omega}_1 = \omega_1 \mathbf{e}_0, \tag{3.53a}$$

$$R_2 = e^{(1/2)i\boldsymbol{\omega}_2 t} \quad \text{with} \quad \boldsymbol{\omega}_2 = \omega_2 \hat{\mathbf{G}}, \tag{3.53b}$$

$$\omega_1\omega_2 = -G, \tag{3.53c}$$

$$\tfrac{1}{2}(\omega_1^2 + \omega_2^2) = E + 2\mathbf{G}\cdot\mathbf{e}_0. \tag{3.53d}$$

The spinor U in (3.52) describes the deviation from steady precession. To obtain a differential equation for U, we substitute (3.52) into the equation of motion

$$\ddot{R} = \tfrac{1}{2}(\mathbf{e}_0 R\mathbf{G} - (E + 2\,\mathbf{G}\cdot\mathbf{e})R),$$

and use (3.53a, b, c, d). Thus,

$$\begin{aligned}\ddot{R} &= R_1\,[\ddot{U} + i\boldsymbol{\omega}_1\dot{U} + \dot{U}i\boldsymbol{\omega}_2 - \tfrac{1}{2}\boldsymbol{\omega}_1 U\boldsymbol{\omega}_2 - \tfrac{1}{4}(\omega_1^2 + \omega_2^2)U]R_2\\ &= \tfrac{1}{2}R_1[\mathbf{e}_0 U\mathbf{G} - (E + 2\,\mathbf{G}\cdot\mathbf{e})U]\,R_2.\end{aligned}$$

Hence,

$$\ddot{U} + i\boldsymbol{\omega}_1\dot{U} + \dot{U}i\boldsymbol{\omega}_2 + \mathbf{G}\cdot(\mathbf{e} - \mathbf{e}_0)U = 0. \tag{3.54}$$

Also, from (3.52)

$$\mathbf{e} = R^\dagger\mathbf{e}_0 R = R_2^\dagger\mathbf{e}_1 R_2, \tag{3.55a}$$

where

$$\mathbf{e}_1 = U^\dagger\mathbf{e}_0 U. \tag{3.55b}$$

Hence, from (3.53b),

$$\mathbf{G}\cdot\mathbf{e} = \mathbf{G}\cdot\mathbf{e}_1 = \langle\mathbf{G}U^\dagger\mathbf{e}_0 U\rangle_0. \tag{3.56}$$

This shows that the last term in (3.54) is a function of U only.

To study small deviations from steady precession, let us introduce Euler parameters by writing

$$U = \alpha + i\boldsymbol{\beta}. \tag{3.57}$$

Substituting this into (3.54) and separating bivector and scalar parts we obtain the two equations

$$\ddot{\boldsymbol{\beta}} + \dot{\boldsymbol{\beta}} \times \boldsymbol{\omega}_- + \dot{\alpha}\boldsymbol{\omega}_+ + \mathbf{G}\cdot(\mathbf{e} - \mathbf{e}_0)\boldsymbol{\beta} = 0 \tag{3.58a}$$

$$\ddot{\alpha} - \dot{\boldsymbol{\beta}}\cdot\boldsymbol{\omega}_+ + \mathbf{G}\cdot(\mathbf{e} - \mathbf{e}_0)\alpha = 0, \tag{3.58b}$$

where

$$\boldsymbol{\omega}_\pm = \boldsymbol{\omega}_1 \pm \boldsymbol{\omega}_2 . \tag{3.59}$$

From (3.55b) we obtain

$$\mathbf{e}_1 = U^\dagger \mathbf{e}_0 U = \mathbf{e}_0 + 2\alpha\boldsymbol{\beta} \times \mathbf{e}_0 + 2\boldsymbol{\beta}\wedge\mathbf{e}_0\boldsymbol{\beta}, \tag{3.60}$$

so from (3.56) we get

$$\mathbf{G}\cdot(\mathbf{e} - \mathbf{e}_0) = 2\alpha\boldsymbol{\beta}\cdot(\mathbf{e}_0 \times \mathbf{G}) + 2(\mathbf{e}_0 \times \boldsymbol{\beta})\cdot(\boldsymbol{\beta} \times \mathbf{G}). \tag{3.61}$$

This is to be used in (3.58a, b) to express the equations as functions of α and $\boldsymbol{\beta}$.

For the small angle approximation we have

$$U = e^{(1/2)i\boldsymbol{\varepsilon}} \approx 1 + \tfrac{1}{2} i\boldsymbol{\varepsilon}. \tag{3.62}$$

So using $\alpha = 1$ and $\boldsymbol{\beta} = \frac{1}{2}\boldsymbol{\varepsilon}$ in (3.58a, b) and (3.61), we obtain, to first order in $\boldsymbol{\varepsilon}$, the equations

$$\ddot{\boldsymbol{\varepsilon}} + \dot{\boldsymbol{\varepsilon}} \times \boldsymbol{\omega}_- = 0, \tag{3.63a}$$

$$-\dot{\boldsymbol{\varepsilon}}\cdot\boldsymbol{\omega}_+ + 2\,\boldsymbol{\varepsilon}\cdot(\mathbf{e}_0 \times \mathbf{G}) = 0. \tag{3.63b}$$

Equation (3.63a) integrates immediately to

$$\dot{\boldsymbol{\varepsilon}} = \boldsymbol{\omega}_- \times \boldsymbol{\varepsilon}, \tag{3.64}$$

where the integration constant has been set to zero so that (3.63b) is satisfied. From (3.59) and (3.53a, b, c) we have

$$\boldsymbol{\omega}_- \times \boldsymbol{\omega}_+ = 2\,\boldsymbol{\omega}_1 \times \boldsymbol{\omega}_2 = -2\,\mathbf{e}_0 \times \mathbf{G}, \tag{3.65}$$

which enables us to show that (3.63b) follows from (3.64).

The solution to (3.64) is the rotating vector

$$\boldsymbol{\varepsilon} = \boldsymbol{\varepsilon}_0\, e^{i\omega_- t} = \boldsymbol{\varepsilon}_0\, e^{i(\omega_1 - \omega_2)t} \tag{3.66}$$

where $\boldsymbol{\varepsilon}_0$ is a constant vector orthogonal to $\boldsymbol{\omega}_-$. An additive constant vector parallel to $\boldsymbol{\omega}_-$ has been omitted from (3.66), because its only effect would be to change initial conditions which are already taken care of in specifying the vector $\mathbf{e}_0$.

We now have a complete solution to the equation of motion, and we can exhibit it by writing the attitude spinor R in the form

$$R = e^{(1/2)i\omega_1 t}\, e^{(1/2)i\varepsilon t}\, e^{(1/2)i\omega_2 t}, \tag{3.67}$$

where $\boldsymbol{\varepsilon} = \boldsymbol{\varepsilon}(t)$ is the rotating vector given by (3.66). This shows explicitly the time dependence of the rotational motion and its decomposition into three simpler motions. As noted before, the first factor in (3.67) describes a rotation of the body about its symmetry axis, while the third factor describes a steady precession. The second factor describes an oscillation about steady precession called *nutation*. To visualize the motion, we consider the orbit $\mathbf{e} = \mathbf{e}(t)$ of the symmetry axis on the unit sphere.

To first order, substitution of (3.66) into (3.60) gives us

$$\mathbf{e}_1 = \mathbf{e}_0 + \boldsymbol{\varepsilon} \times \mathbf{e}_0 = \mathbf{e}_0 + (\boldsymbol{\varepsilon}_0 e^{i\omega_- t}) \times \mathbf{e}_0 . \tag{3.68}$$

Note that the term $\boldsymbol{\varepsilon} \times \mathbf{e}_0$ as a linear function which projects $\boldsymbol{\varepsilon}$ onto a plane with normal $\mathbf{e}_0$ and rotates it through a right angle. Therefore, it projects the circle $\boldsymbol{\varepsilon} = \boldsymbol{\varepsilon}(t)$ into an ellipse with its major axis in the plane containing $\mathbf{e}_0$ and the normal to the circle $\boldsymbol{\omega}_- = \omega_1\mathbf{e}_0 - \omega_2\hat{\mathbf{G}}$. Thus, (3.68) describes an ellipse $\mathbf{e}_1 = \mathbf{e}_1(t)$ centered at $\mathbf{e}_0$ and lying in the tangent plane to the unit sphere, as shown in Figure 3.6. The eccentricity of the ellipse depends on the angle between $\mathbf{e}_0$ and $\boldsymbol{\omega}_- = \omega_1\mathbf{e}_0 - \omega_2\hat{\mathbf{G}}$. For a slow top $\omega_1 > \omega_2$, so $\boldsymbol{\omega}_- \approx \omega_1\mathbf{e}_0$ and the ellipse is nearly circular.

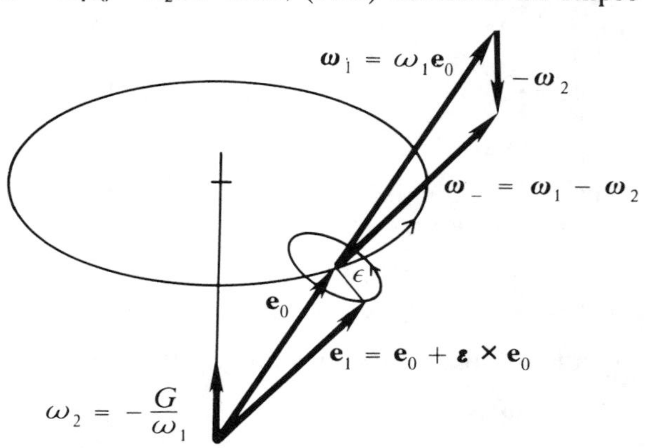

Fig. 3.6. First order orbit of the symmetry axis.

The orbit $\mathbf{e} = \mathbf{e}(t)$ of the symmetry axis is the composite of the elliptical motion (3.68) and the steady precession, as described by

$$\mathbf{e}(t) = R_2^\dagger \mathbf{e}_1 R_2 = \mathbf{e}^{-(1/2)i\boldsymbol{\omega}_2 t}\, \mathbf{e}_1(t) e^{(1/2)i\boldsymbol{\omega}_2 t} \tag{3.69}$$

The resulting curve oscillates with angular speed

$$\tfrac{1}{2}\omega_- = [\tfrac{1}{2}(E + \mathbf{e}_0\cdot\mathbf{G})]^{1/2} \tag{3.70}$$

between two circles on the unit sphere with angular separation $2\varepsilon = 2|\boldsymbol{\varepsilon}|$, as indicated in Figure 3.7. We use the term *nutation* to designate the elliptical oscillation about steady precession, though the term ordinarily refers only to the vertical 'nodding' component of this oscillation.

To determine the qualitative features of the orbit $\mathbf{e} = \mathbf{e}(t)$, we look at the velocity

$$\dot{\mathbf{e}} = R_2^\dagger[\mathbf{e}_0 \times (\boldsymbol{\omega}_- \times \boldsymbol{\varepsilon} + \boldsymbol{\omega}_2)]R_2. \tag{3.71}$$

The nutation velocity $\mathbf{e}_0 \times (\boldsymbol{\omega}_- \times \boldsymbol{\varepsilon})$ is exactly opposite to the precession velocity $\mathbf{e}_0 \times \boldsymbol{\omega}_2$ only when the orbit is tangent to the upper bounding circle in

Figure 3.6. Therefore, $|\dot{\mathbf{e}}|$ has its minimum values at such points, and $\dot{\mathbf{e}} = 0$ only on an orbit for which

$$\varepsilon|\boldsymbol{\omega}_1 - \boldsymbol{\omega}_2| = |\mathbf{e}_0 \times \boldsymbol{\omega}_2|. \tag{3.72}$$

This is the condition for the *cuspidal orbit* in Figure 3.7. A *looping orbit* occurs when $\varepsilon|\boldsymbol{\omega}_1 - \boldsymbol{\omega}_2| > |\mathbf{e}_0 \times \boldsymbol{\omega}_2|$, and a smooth orbit without loops occurs when $\varepsilon|\boldsymbol{\omega}_1 - \boldsymbol{\omega}_2| \ll |\mathbf{e}_0 \times \boldsymbol{\omega}_2|$. A cuspidal orbit can be achieved in practice by releasing the axis of a spinning top from an initial position at rest. Therefore, the two other kinds of orbits can be achieved with an initial impetus following or opposing the direction of precessional motion.

Equation (3.67) is an approximate solution of Euler's equation for the Lagrange problem with the same accuracy as the harmonic approximation for the motion of a pendulum. In Section 7-4 we will find the exact solution in terms of elliptic functions. Unfortunately, the exact solution is difficult to interpret and awkward to use. However, all its qualitative features are already displayed in a much more convenient form by the approximate solution (3.67). Moreover, in many applications the exact solution has little advantage, because of uncertainties about perturbing forces such as friction. As a rule, therefore, we expect the approximate solution to be more valuable than the exact solution.

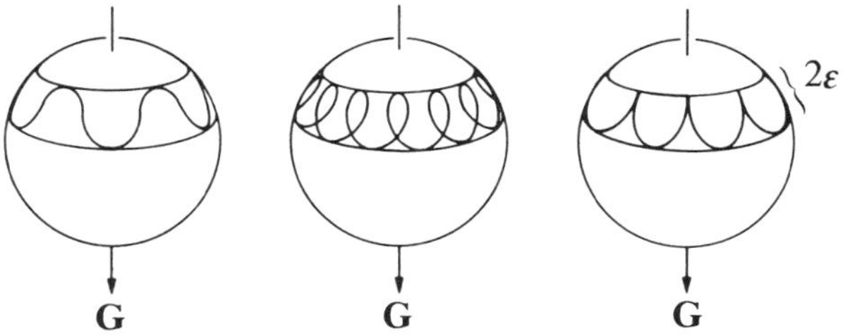

Fig. 3.7. Nutation of the symmetry axis of a precessing top.

Effects of Friction

Since friction is a contact force, the effect of friction on a spinning body depends on the distribution of frictional forces over the surface of the body. For a symmetric top spinning about its symmetry axis with its *CM* at rest, the forces of air friction are symmetrical about the axis. Consider the frictional forces $\mathbf{f}$ and $\mathbf{f}'$ at two symmetrically related points $\mathbf{r}$ and $\mathbf{r}'$ as shown in Figure 3.8. The frictional force is opposite to the velocity of the surface at the point of contact and $\mathbf{f} = -\mathbf{f}'$ by symmetry. Therefore, the two forces make up a couple with torque.

$$\mathbf{r} \times \mathbf{f} + \mathbf{r}' \times \mathbf{f}' = (\mathbf{r} - \mathbf{r}') \times \mathbf{f} \propto -\boldsymbol{\omega}_{\parallel} = -\boldsymbol{\omega}\cdot\mathbf{e}\mathbf{e}.$$

It is most important to note the direction of the torque is opposite to that of the angular velocity. Also, if the frictional force is proportional to the velocity at the point of contact, then it is also proportional to $\omega_{\parallel} = \boldsymbol{\omega}\cdot\mathbf{e}$. The same conclusions hold for all other pairs of symmetrically placed points. Therefore, the resultant torque due to air friction has the form

$$\boldsymbol{\Gamma}_f = -\lambda\boldsymbol{\omega}_{\parallel} = -\lambda\boldsymbol{\omega}\cdot\mathbf{ee}, \qquad (3.73)$$

where λ is a positive scalar depending on the shape of the body, the viscosity and density of the air and, to some extend, on $\boldsymbol{\omega}\cdot\mathbf{e}$ for reasons discussed in Section 3-5. From Euler's equation $\dot{\mathbf{l}} = \mathbf{I}\dot{\boldsymbol{\omega}} = \boldsymbol{\Gamma}_f$ we obtain

$$I\frac{\mathrm{d}}{\mathrm{d}t}(\mathbf{e}\cdot\boldsymbol{\omega}) = \mathbf{e}\cdot\boldsymbol{\Gamma}_f = -\lambda\mathbf{e}\cdot\boldsymbol{\omega}.$$

For a linear resistive force, λ is constant, so

$$\mathbf{e}\cdot\boldsymbol{\omega} = \mathbf{e}_0\cdot\boldsymbol{\omega}_0\, e^{-(\lambda/I)t}. \qquad (3.74)$$

Thus we have exponential decay of the spin $\boldsymbol{\omega}\cdot\mathbf{e}$.

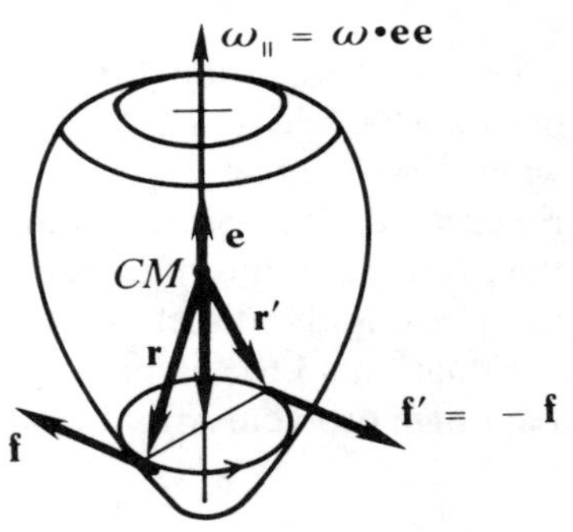

Fig. 3.8. Air friction on a spinning top.

For a slow top the effect of air resistance due to its precession will be small compared to the effect due to the spin about its symmetry axis. Therefore, the main effect of air resistance on a slow top will be simply to slow down its spin at a roughly exponential rate. Thus, for a sleeping top, the spin $\boldsymbol{\omega}\cdot\mathbf{e}$ is reduced by air friction as well as by friction at the point of support until the condition for stability no longer holds. Then it begins to fall so nutation and precession set in. As the spin continues to decrease, the amplitude of the nutation increases until it is so large that the top falls over.

When a rapidly spinning top is placed on a rough surface, as indicated in Figure 3.9, a force $\mathbf{f}$ of sliding friction is exerted at the point of contact. The torque exerted by the frictional force can be separated into two parts:

$$\mathbf{r}\times\mathbf{f} = \mathbf{r}_{\parallel}\times\mathbf{f} + \mathbf{r}_{\perp}\times\mathbf{f}, \qquad (3.75)$$

where $\mathbf{r}_{\parallel} = \mathbf{r}\cdot\mathbf{ee}$ and $\mathbf{r}_{\perp} = \mathbf{r}\wedge\mathbf{ee}$. The torque $\mathbf{r}_{\perp}\times\mathbf{f}$ simply reduces the spin $\boldsymbol{\omega}\cdot\mathbf{e}$ in the manner that has just been described, and it will be comparatively small for small $|\mathbf{r}_{\perp}|$. The torque $\mathbf{r}_{\parallel}\times\mathbf{f}$ has the form $\mathbf{e}\times\mathbf{G}$ which we have already studied, so we know that it will produce precession about $-\mathbf{f}$. Since $\mathbf{f}$ lies in a horizontal plane as shown in Figure 3.9, this torque will make the symmetry axis precess toward the vertical, and we speak of a

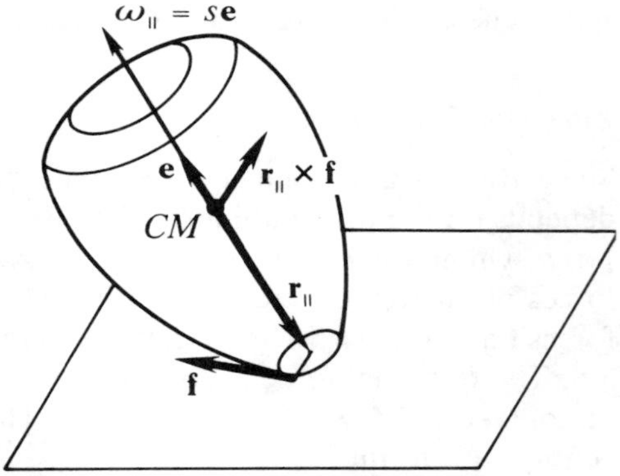

Fig. 3.9. The rising top.

rising top. As the top rises, its rotational speed $|\boldsymbol{\omega}|$ decreases, since kinetic energy is converted to potential energy. Once the top is erect it becomes a *sleeping top*, since $\mathbf{r}_{\|} \times \mathbf{f}$ vanishes. On the other hand, if the slipping ceases before the top is erect, rolling motion sets in.

Magnetic Spin Resonance

The spinor formulation of rotational dynamics has important applications to atomic physics, since atoms, electrons and nuclei have intrinsic angular momenta and magnetic moments. Consider an atom (electron or nucleus) with intrinsic angular momentum $\mathbf{l}$ and magnetic moment $\boldsymbol{\mu}$. According to electromagnetic theory, a magnetic field $\mathbf{B}$ will exert a torque $\boldsymbol{\mu} \times \mathbf{B}$, so the rotational equation of motion for the atom is

$$\dot{\mathbf{l}} = \boldsymbol{\mu} \times \mathbf{B}, \tag{3.76}$$

Atomic theory asserts $\mathbf{l}$ and $\boldsymbol{\mu}$ are related by the "constitutive equation"

$$\boldsymbol{\mu} = \gamma \mathbf{l}, \tag{3.77}$$

where γ is a scalar constant called the *gyromagnetic ratio*. Consequently, the equation of motion can be written

$$\dot{\mathbf{l}} = (-\gamma \mathbf{B}) \times \mathbf{l}. \tag{3.78}$$

This implies that $\mathbf{l}^2$ is a constant of motion, so the effect of $\mathbf{B}$ is to produce a time dependent rotation of $\mathbf{l}$. We know that such a rotation is most efficiently represented by the equation

$$\mathbf{l} = U^\dagger \mathbf{l}_0 U, \tag{3.79}$$

where $\mathbf{l}_0$ is the initial value of $\mathbf{l}$. Accordingly, we can replace (3.78) by the spinor equation of motion

$$\dot{U} = \tfrac{1}{2} U i(-\gamma \mathbf{B}) \tag{3.80}$$

subject to the initial condition $U(0) = 1$.

The spinor U in (3.80) should not be confused with the attitude spinor for a rigid body. According to current atomic theory, the attitude of an atom is not observable, so there is no attitude variable in the theory. The angular momentum and energy of an atom are observable, so there is a dynamical equation for rotational motion in atomic theory. But, in contrast with the macroscopic mechanics of rotating bodies, there is no kinematical equation for attitude.

Experimentalists wish to manipulate the magnetic moment $\boldsymbol{\mu}$ by applying suitable magnetic fields. To see how this can be done, we study the solution of (3.80) for particular applied fields. The dynamical spinor equation (3.80) is preferred over the vector equation (3.78), because it is easier to solve. For a static field $\mathbf{B} = \mathbf{B}_0$, the solution of (3.80) is simply

$$U = e^{-(1/2)i\gamma \mathbf{B}_0 t}. \tag{3.81}$$

This tells us that $\mathbf{l}$ and $\boldsymbol{\mu}$ precess about the magnetic field with an angular frequency $-\gamma\mathbf{B}_0$.

Now suppose we introduce a circularly polarized monochromatic plane wave propagating along the direction of the established static magnetic field $\mathbf{B}_0$. At the site of the atom, the magnetic field of such a wave is a rotating vector

$$\mathbf{b}(t) = \mathbf{b}_0 \mathbf{e}^{i\boldsymbol{\omega} t}, \tag{3.82}$$

where $\mathbf{b}_0$ and $\boldsymbol{\omega}$ are constant vectors for which $\boldsymbol{\omega}\mathbf{B}_0 = \mathbf{B}_0\boldsymbol{\omega}$ and $\mathbf{b}_0\boldsymbol{\omega} = -\boldsymbol{\omega}\mathbf{b}_0$. The resultant magnetic field acting on the atom is therefore

$$\mathbf{B} = \mathbf{B}_0 + \mathbf{b}_0 e^{i\boldsymbol{\omega} t} = S^\dagger(\mathbf{B}_0 + \mathbf{b}_0)S, \tag{3.83}$$

where

$$S = e^{(1/2)i\boldsymbol{\omega} t} \tag{3.84}$$

The form of (3.83) suggests that we should write U in the factored form

$$U = RS. \tag{3.85}$$

Then, the spinor equation of motion gives us

$$\dot{U} = -\tfrac{1}{2}Ui\gamma\mathbf{B} = -\tfrac{1}{2}Ri\gamma(\mathbf{B}_0 + \mathbf{b}_0)S = (\dot{R} + \tfrac{1}{2}Ri\boldsymbol{\omega})S.$$

Hence R obeys the equation

$$\dot{R} = -\tfrac{1}{2}Ri\gamma(\mathbf{B}_0 + \gamma^{-1}\boldsymbol{\omega} + \mathbf{b}_0). \tag{3.86}$$

This has the solution

$$R = e^{-(1/2)i\gamma\mathbf{B}'t}, \tag{3.87}$$

where

$$\mathbf{B}' = \mathbf{B}_0 + \gamma^{-1}\boldsymbol{\omega} + \mathbf{b}_0. \tag{3.88}$$

The motion of $\mathbf{l}$ is therefore completely described by the spinor

$$U = e^{-(1/2)i\gamma\mathbf{B}'t}e^{(1/2)i\boldsymbol{\omega} t} \tag{3.89}$$

This tells us that the motion is a composite of two precessions with constant angular velocities.

The experimentalist can tune the frequency $\boldsymbol{\omega}$ of the electromagnetic wave until the condition, $\boldsymbol{\omega} = -\gamma\mathbf{B}_0$, for magnetic resonance is achieved. Under this condition, (3.88) gives $\mathbf{B}' = \gamma\mathbf{b}_0$, which is orthogonal to $\boldsymbol{\omega}$ and $\mathbf{B}_0$. Then (3.89) tells us that $\boldsymbol{\mu} = \gamma\mathbf{l}$ is precessing with angular velocity $-\gamma\mathbf{b}_0$ orthogonal to $\mathbf{B}_0$ in a frame which is precessing about $\mathbf{B}_0$ with angular speed ω. Since $\omega = \gamma B_0 \gg \omega' = \gamma b_0$ typically, the composite motion will be a steady spiral motion, as illustrated in Figure 3.10. If $\mathbf{l}$ is initially aligned with $\mathbf{B}_0$ when the electromagnetic radiation is turned on, then its direction will be reversed in a time $T = 2\pi/\gamma b_0$. Consequently, a single "spin flip" can be produced by a pulse of duration T at resonance.

Since the gyromagnetic ratio γ has different values for different types of atoms, the radiation field can be tuned to detect specific atomic types by magnetic resonance even when they are buried in complex biological materials. For this reason, magnetic resonance is widely used for chemical analysis.

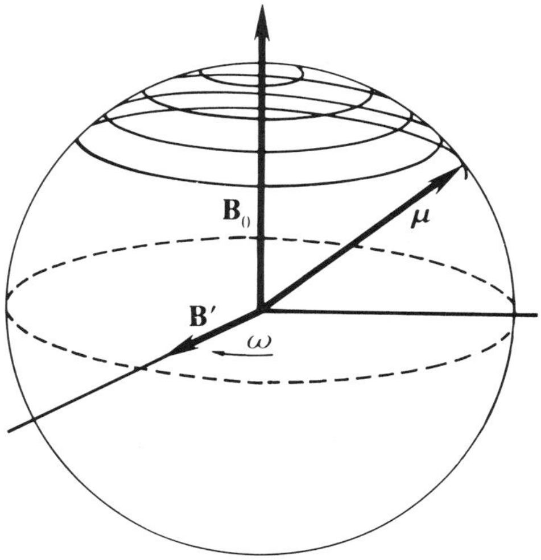

Fig. 3.10. Simultaneous precession of the magnetic moment $\boldsymbol{\mu}$ about the fields $\mathbf{B}'$ and $\mathbf{B}_0$ at resonance.

7-3. Exercises

(3.1) Let ϕ be the angle between the angular momentum $\mathbf{l}$ and the symmetry axis of a freely spinning axially symmetric satellite. Prove that

$$\sin^2 \phi = \frac{I}{I_3 - I}\left(\frac{2I_3 E}{l^2} - 1\right),$$

where $E = \frac{1}{2}\mathbf{l}\cdot\boldsymbol{\omega}$ is the kinetic energy.
The free Eulerian wobble of a satellite produces periodic internal stresses leading to energy dissipation without altering the angular momentum. Show, therefore, that the wobble tends to damp out if the satellite is oblate.

(3.2) A heavy homogeneous right circular cone spins with its vertex fixed. The axis of the cone is 10 cm. long, and the radius of its base is 5 cm. The cone processes steadily with a period of 4 sec. How many revolutions per second does the cone make about its own axis?

(3.3) A homogeneous circular disk of radius r spins on a smooth table about a vertical diameter. Prove that the motion is stable if the angular speed exceeds $2\sqrt{g/r}$.

(3.4) When $\boldsymbol{\omega}\cdot\mathbf{e} = 0$, Equation (3.69) describes the orbit of a so-called *spherical pendulum*. Sketch the orbit for appropriate choices of the free parameters. Show that the orbit of a spherical pendulum cannot have loops or cusps such as those in Figure 3.7.

(3.5) A spinning top is held fixed in an upright position and suddenly released. Describe the motion if the condition for stability of a sleeping top is not satisfied.

(3.6) The tippie-top is a child's toy consisting of a decapitated ball with a short stem as shown in Figure 3.11. If the tippie-top is set on a table, stem upward, with sufficient spin about its stem, it will do a complete flip-flop ending up in the steady precession standing on its stem. Similarly, if a hard boiled egg laying on a table is given sufficient spin about a vertical axis, it rises to steady precession standing on its narrow end. Provide a qualitative explanation for such behavior. Assuming that the ball is hollow, show that the "flip time" for the tippie-top is approximately $2\pi r\omega/3\mu g$. (The behavior of the tippie-top has been discussed in the *American Journal of Physics* on several occasions).

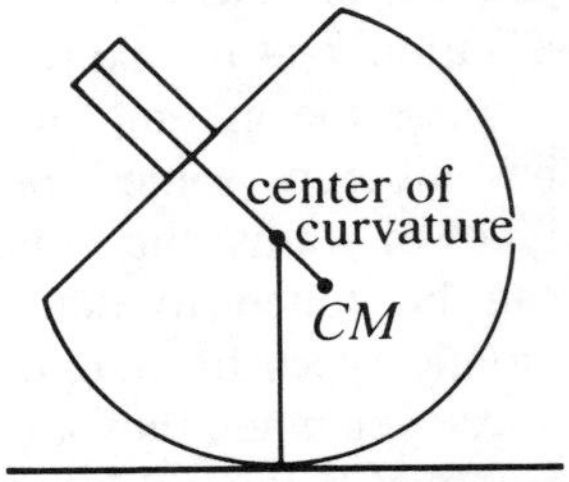

Fig. 3.11. The tippie-top.

7-4. Integrable Cases of Rotational Motion

When the general solution to a system of differential equations can be expressed in closed form, we say that the system is *integrable*. A solution in *closed form* is expressed in terms of a finite number of known functions (in contrast to an infinite series). Such a solution is sometimes said to be *exact*, as opposed to an *approximate* solution obtained by truncating an infinite series.

The rotational equations of motion are integrable in only a few simple cases, in particular, the plane pendulum, the symmetric top in a gravitational field, and a freely spinning asymmetric body. The general solutions in all these cases are expressible in terms of elliptic functions, and we shall find them in this section.

Constants of Motion for the Lagrange Problem

The problem of integrating the equations of motion for a symmetric top subject to a gravitational torque is called the *Lagrange problem*. In Section 7-3 we reduced the equations of motion for a symmetric top to those for a spherical top. So we begin here with the reduced equations.

$$\dot{R} = \tfrac{1}{2} R i \boldsymbol{\omega}, \tag{4.1}$$

$$\dot{\boldsymbol{\omega}} = \mathbf{e} \times \mathbf{G}, \tag{4.2}$$

where

$$\mathbf{e} = R^{\dagger} \boldsymbol{\sigma}_3 R \tag{4.3}$$

and

$$\mathbf{G} = \frac{\dot{r}\mathbf{F}}{I} \tag{4.4}$$

for a constant force (eg. $\mathbf{F} = m\mathbf{g}$) applied at the point $r\mathbf{e}$.

In Section 7-3 we combined (4.1) and (4.2) into a single second order differential equation, but to find the exact solution it is more efficient to exploit the constants of motion directly. As we have noted before, equation (4-2) admits three integrals of motion, namely,

$$\mathbf{e}\cdot\boldsymbol{\omega} = -2\langle i\sigma_3 \dot{R} R^\dagger\rangle_0 , \tag{4.5}$$

$$\mathbf{G}\cdot\boldsymbol{\omega} = -2\langle i\mathbf{G} R^\dagger \dot{R}\rangle_0 \tag{4.6}$$

and the energy integral

$$E = \tfrac{1}{2}\omega^2 - \mathbf{e}\cdot\mathbf{G} = 2|\dot{R}|^2 - \langle R^\dagger \boldsymbol{\sigma}_3 R\mathbf{G}\rangle_0 \tag{4.7}$$

By finding these three integrals of the motion, we have effectively integrated the dynamical equation (4.2), leaving us with three first order scalar equations to be integrated for R.

The Compound Pendulum

The motion of a pendulum is a special case of the motion of a symmetric top. A compound pendulum is a rigid body free to rotate about a fixed horizontal axis under the influence of gravity (Figure 4.1). If I is the moment of inertia for the axis, then the equations of motion for the pendulum are exactly as specified above.

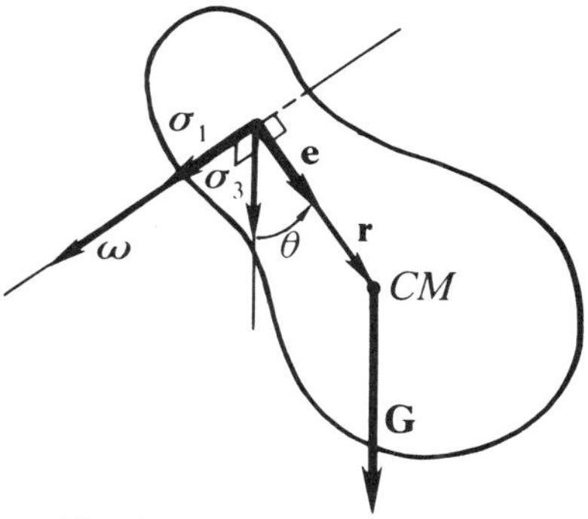

Fig. 4.1. A compound pendulum.

For a pendulum

$$\mathbf{e}\cdot\boldsymbol{\omega} = \mathbf{G}\cdot\boldsymbol{\omega} = 0 . \tag{4.8}$$

Therefore, the motion of the pendulum is completely determined by the energy integral (4.7). Equations (4.8) allow us to parametrize the attitude by

$$R = \alpha + i\boldsymbol{\sigma}_1\beta , \tag{4.9}$$

where α and β are scalars and $\boldsymbol{\sigma}_1$, is the direction of the fixed rotation axis. If R is to specify the deviation from the downward vertical direction $\boldsymbol{\sigma}_3 = \hat{\mathbf{G}}$, then it is related to the angle of deviation θ by

$$R = \cos\tfrac{1}{2}\theta + i\boldsymbol{\sigma}_1 \sin\tfrac{1}{2}\theta = e^{(1/2)i\boldsymbol{\sigma}_1\theta} , \tag{4.10}$$

However, the motion is more simply described in terms of the parameters α and β instead of θ, as we see below.

Substituting (4.9) into the energy integral (4.7) and using $\alpha^2 + \beta^2 = 1$ to eliminate α, we obtain

$$\dot{\beta}^2 = \tfrac{1}{2}(E + G - 2G\beta^2)(1 - \beta^2). \tag{4.11}$$

With a suitable identification of constants, this can be put in the standard form of Equation (B.2) in Appendix B; so the solution can be expressed in terms of elliptic functions. There are three cases to be considered, depending on the value of

$$E = \tfrac{1}{2}\omega_0^2 - G, \tag{4.12}$$

where $\frac{1}{2}\omega_0^2$ is the maximum kinetic energy of the pendulum.

(a) When $E < G$, let $k^2 = \omega_0^2/4G < 1$, so (4.11) assumes the form

$$\dot{\beta}^2 = G(k^2 - \beta^2)(1 - \beta^2).$$

With the change of variables $\beta = ky$ and $x = G^{1/2}t$, this equation becomes identical with (B.2), so it has the solution

$$\beta = k \operatorname{sn} x = k \operatorname{sn}(G^{1/2}t).$$

Since $\alpha^2 + \beta^2 = 1$, we have from (B.9)

$$\alpha = \operatorname{dn}(G^{1/2}t).$$

Also, we can write

$$E = G(2k^2 - 1) = -G\langle R_0^2\rangle_0 = -G(1 - 2\beta_0^2),$$

where $\beta_0 = \sin \frac{1}{2}\theta_0$ is the value of β at the angle θ_0 of greatest deflection from the vertical. Thus, the spinor solution to the equations of motion can be given the explicit form

$$R = \operatorname{dn}(G^{1/2}t) + i\boldsymbol{\sigma}_1\beta_0 \operatorname{sn}(G^{1/2}t), \tag{4.13}$$

where $\beta_0 = \sin \frac{1}{2}\theta_0 = k$ is the modulus of the elliptic function.

Comparing (4.13) with (B.7) we can conclude that the *period* of motion is

$$T = 4KG^{-1/2} \tag{4.14}$$

When $\theta_0 \approx 0$, then $k \approx 0$ and $K \approx \frac{1}{2}\pi$, so $T \approx T_0 = 2\pi G^{-1/2}$, the period we found in (3.34) for small oscillations. The exact value of the period for a pendulum depends on its amplitude, as shown in Table 4.1.

(b) When $E > G$ let $k^2 = 4G/\omega_0^2 < 1$, so (4.12) gives us $E + G = 2\,G/k^2$, and (3.34) becomes

$$\dot{\beta}^2 = \frac{G}{k^2}(1 - k^2\beta^2)(1 - \beta^2).$$

Hence,

$$\beta = \operatorname{sn}\left(\frac{G^{1/2}t}{k}\right) \quad \text{and} \quad \alpha = \operatorname{cn}\left(\frac{G^{1/2}t}{k}\right),$$

TABLE 4.1. Dependence of the period of a pendulum on its amplitude.

θ_0	$k = \beta_0 = \sin \frac{1}{2}\theta_0$	K	$T/T_0 = 2K/\pi$
30°	$\sin 15° = (\sqrt{3} - 1)/(2\sqrt{2})$	1.598	1.02
60°	$\sin 30° = \frac{1}{2}$	1.686	1.07
90°	$\sin 45° = 1/\sqrt{2}$	1.854	1.18
120°	$\sin 60° = \frac{1}{2}\sqrt{3}$	2.156	1.37
150°	$\sin 75° = (\sqrt{3} + 1)/(2\sqrt{2})$	2.768	1.76
180°	$\sin 90° = 1$	∞	∞

so

$$R = \text{cn}\left(\frac{G^{1/2}t}{k}\right) + i\boldsymbol{\sigma}_1 \,\text{sn}\left(\frac{G^{1/2}t}{k}\right). \tag{4.15}$$

This solution describes a pendulum which makes complete revolutions with period

$$T = \frac{2Kk}{G^{1/2}} = \frac{4K}{\omega_0}. \tag{4.16}$$

This is twice the period of R, because $\mathbf{e} = R^{\dagger}\hat{\mathbf{G}}R = \hat{\mathbf{G}}R^2$, so the period of the pendulum motion is the period of R^2. In the preceding case, the periods of R and R^2 were the same.

(c) When $E = G$, we have $k^2 \equiv \omega_0^2/4G = 1$, and (4.11) becomes

$$\dot{\beta}^2 = G(1-\beta^2)^2.$$

This has the solution $\beta = \tanh(G^{1/2}t)$, so

$$R = \text{sech}(G^{1/2}t) + i\boldsymbol{\sigma}_1 \tanh(G^{1/2}t). \tag{4.17}$$

The pendulum never quite reaches the upward vertical position.

Solution of the Lagrange Problem

We have reduced the Lagrange problem to solving the three integrals of motion (4.5), (4.6) and (4.7). The next step is to identify the parameters which provide the simplest description of the motion, as we did in the special case of the pendulum. To do that we note that the vector **G** introduces a preferred direction in the problem, the "downward vertical". For an erect top, we are interested in deviations of the top's symmetry axis from the upward vertical, so we specify $\boldsymbol{\sigma}_3 = -\hat{\mathbf{G}}$. For a hanging top, as in the case of the pendulum, $\boldsymbol{\sigma}_3 = \hat{\mathbf{G}}$ would be more appropriate. Both cases are taken care of, respectively, by writing

$$G \equiv -\mathbf{G}\cdot\boldsymbol{\sigma}_3 = \pm\,|\mathbf{G}|. \tag{4.18}$$

Now the factor $\boldsymbol{\sigma}_3 R\mathbf{G} = -G\boldsymbol{\sigma}_3 R\boldsymbol{\sigma}_3$ in (4.7) suggests that we might simplify the energy equation by writing

$$R = \alpha_+ - i\boldsymbol{\sigma}_2\alpha_- \tag{4.19}$$

where α_+ and α_- are quaternions which commute with $\boldsymbol{\sigma}_3$; for then

$$\boldsymbol{\sigma}_3 R \boldsymbol{\sigma}_3 = \alpha_+ + i\boldsymbol{\sigma}_2\alpha_- . \tag{4.20}$$

The commutivity with $\boldsymbol{\sigma}_3$ implies that we can write

$$\alpha_\pm = \lambda_\pm\, e^{i\boldsymbol{\sigma}_3\phi_\pm} , \tag{4.21}$$

where $\lambda_\pm$ and $\phi_\pm$ are scalars. Of course,

$$R^\dagger = \alpha_+^\dagger + \alpha_-^\dagger i\boldsymbol{\sigma}_2 , \tag{4.22}$$

so the parameters are related by

$$R^\dagger R = |\alpha_+|^2 + |\alpha_-|^2 = \lambda_+^2 + \lambda_-^2 = 1 . \tag{4.23}$$

The variables $\alpha_\pm$ are called the *Cayley-Klein parameters* in the scientific literature. However, our formulation identifies the imaginary unit in these parameters as the specific bivector $i\boldsymbol{\sigma}_3$. This enables us to see exactly when and why the Cayley-Klein parameters are useful, namely, in rotational problems where a preferred direction $\boldsymbol{\sigma}_3$ is specified. Readers who are familiar with advanced quantum mechanics will be interested to note that our decomposition of the spinor R into Cayley-Klein parameters corresponds exactly to the standard decomposition of an electron wave function into "spin up" and "spin down" amplitudes.

To express the integrals of motion in terms of the Cayley-Klein parameters, note that

$$\alpha_\pm\boldsymbol{\sigma}_2 = \boldsymbol{\sigma}_2\alpha_\pm^\dagger . \tag{4.24}$$

This helps us compute

$$\dot{R}R^\dagger = \alpha_+^\dagger\dot{\alpha}_+ + \alpha_-\dot{\alpha}_-^\dagger + i\boldsymbol{\sigma}_2(\dot{\alpha}_+^\dagger\alpha_- - \alpha_+^\dagger\dot{\alpha}_-)$$

$$R^\dagger\dot{R} = \alpha_+^\dagger\dot{\alpha}_+ + \alpha_-^\dagger\dot{\alpha}_- + i\boldsymbol{\sigma}_2(\dot{\alpha}_+\alpha_-^\dagger - \alpha_+\dot{\alpha}_-^\dagger)$$

Also we find

$$\alpha_+^\dagger\dot{\alpha}_+ = \lambda_+\dot{\lambda}_+ + i\boldsymbol{\sigma}_3\lambda_+^2\dot{\phi}_+ , \tag{4.25}$$

with a similar expression for $\alpha_-^\dagger\dot{\alpha}_+$. Now the integrals of motion (4.5) and (4.6) can be put in the form

$$\lambda_+^2\dot{\phi}_+ - \lambda_-^2\dot{\phi}_- = \tfrac{1}{2}\,\mathbf{e}\cdot\boldsymbol{\omega}$$

$$\lambda_+^2\dot{\phi}_+ + \lambda_-^2\dot{\phi}_- = \tfrac{1}{2}\,\hat{\mathbf{G}}\cdot\boldsymbol{\omega}$$

Or equivalently,

$$\lambda_+^2\dot{\phi}_+ = \tfrac{1}{2}\,\boldsymbol{\omega}\cdot(\hat{\mathbf{G}} + \mathbf{e}) \equiv \gamma_+ \tag{4.26a}$$

$$\lambda_-^2\dot{\phi}_- = \tfrac{1}{2}\,\boldsymbol{\omega}\cdot(\hat{\mathbf{G}} - \mathbf{e}) \equiv \gamma_- . \tag{4.26b}$$

It will be noted that these expressions are analogous to angular momentum conservation for central force motion. They enable us to calculate $\phi_\pm$ by straightforward integration when $\lambda_\pm$ is known.

To evaluate the energy integral we calculate

$$\boldsymbol{\sigma}_3\mathbf{e} = \boldsymbol{\sigma}_3 R^\dagger \boldsymbol{\sigma}_3 R = |\alpha_+|^2 - |\alpha_-|^2 - 2i\boldsymbol{\sigma}_2\alpha_+\alpha_- .$$

So with the help of (4.23), we find

$$\boldsymbol{\sigma}_3\cdot\mathbf{e} = \pm(2|\alpha_\pm|^2 - 1) = \pm(2\lambda_\pm^2 - 1). \tag{4.27}$$

For $\boldsymbol{\sigma}_3\cdot\mathbf{e} = \cos\theta$, this tells us that

$$\lambda_+ = \cos\tfrac{1}{2}\theta \quad \text{and} \quad \lambda_- = \sin\tfrac{1}{2}\theta. \tag{4.28}$$

Using (4.27) in the energy integral (4.7), we get

$$E = 2(|\dot\alpha_-|^2 + |\dot\alpha_-|^2) \pm G(2|\alpha_\pm|^2 - 1) \tag{4.29}$$

From (4.25) and (4.26a, b) we find

$$|\dot\alpha_\pm|^2 = \dot\lambda_\pm^2 + \gamma_\pm \lambda_\pm^{-2} . \tag{4.30}$$

Substituting this into (4.29) and using (4.23) to eliminate λ_-, we obtain

$$\lambda_+^2\dot\lambda_+^2 = G\lambda_+^6 - \tfrac{1}{2}(E + 3G)\lambda_+^4 + [\tfrac{1}{2}(E + G) + \boldsymbol{\omega}\cdot\mathbf{e}]\lambda_+^2 - \gamma_+ \tag{4.31}$$

This differential equation can be solved for λ_+, and then λ_- can be obtained from (4.23).

According to Appendix B, Equation (4.31) has a closed solution in terms of elliptic functions given by

$$\lambda_+^2 = a\,\mathrm{sn}^2\,\mu t + b\,, \tag{4.32}$$

where a and b are constants. This tells us that λ_+^2 is a periodic function of time with maximum and minimum values satisfying

$$0 \leqslant b \leqslant 1\,, \tag{4.33a}$$

$$0 \leqslant a + \mathrm{b} \leqslant 1\,. \tag{4.33b}$$

Thus, $\lambda_+ = \cos\frac{1}{2}\theta$ oscillates symmetrically about the value

$$\cos\tfrac{1}{2}\theta_0 = |b + \tfrac{1}{2}a|^{1/2}. \tag{4.34}$$

Comparing (4.31) with (B.12) and (B.16a, b, c, d) in Appendix B, we find that b is determined by the cubic equation

$$Gb^3 - \tfrac{1}{2}(E + 3G)b^2 + [\tfrac{1}{2}(E + G) + \boldsymbol{\omega}\cdot\mathbf{e}]b - \gamma_+ = 0 \tag{4.35}$$

and (4.33a) tells us how to identify the "physical root". Also, after b is known we can get a by solving the quadratic equation

$$2Ga^2 + [G(2b - 3) - E]a + E(1 - 2b) + G(6b^2 - 6b + 1) + 2\boldsymbol{\omega}\cdot\mathbf{e} = 0, \tag{4.36}$$

and (4.33a, b) tells us that the "physical root" must satisfy $|a| \leq 1$. Finally the "time constant" μ in the solution is obtained from a and b by using

$$\mu^2 = \tfrac{1}{2}(3G + E) - G(3b + a) \tag{4.37}$$

and the modulus k^2 of the elliptic function is given by

$$k^2 = \frac{aG}{\mu^2} . \tag{4.38}$$

Thus, we have determined all the constants in the solution (4.26a) for λ_+.

Using (4.32) we solve (4.26a) for the angle ϕ_+:

$$\phi_+(t) = \gamma_+ \int_0^t \frac{dt}{b + a\,\mathrm{sn}^2(\mu t)} .$$

This can be put in the standard form

$$\phi_+(t) = \frac{\gamma_+}{\mu b}\, \pi\!\left(\mu t, \frac{a}{b}\right), \tag{4.39}$$

where

$$\pi(\tau, n) = \int_0^\tau \frac{dt}{1 + n\,\mathrm{sn}^2\,\tau} \tag{4.40}$$

is a standard function known as the *incomplete elliptic integral of the third kind*. By the same method we find

$$\phi_-(t) = \frac{\gamma}{\mu(1-\mathrm{b})}\, \pi\left(\mu t, \frac{a}{b-1}\right) \tag{4.41}$$

$$\lambda_-^2 = 1 - b - a\,\mathrm{sn}^2\,\mu t. \tag{4.42}$$

Numerical values for the elliptic integrals as well as the elliptic functions can be found in standard tables, but nowadays it is much more convenient to get the results by computer calculation.

We now have a complete closed solution of the Lagrange problem. Unfortunately, our solution is not easy to interpret. A picture of the motion can be obtained by computer simulation, or, more laboriously, by further mathematical analysis of the solution. But we will not pursue the matter further, since we already have a clear picture from our approximate solution in Section 7-3.

Freely Spinning Asymmetric Body

We turn now to the problem of finding an analytic description for the motion of an arbitrary, freely spinning rigid body. The dynamical equation of motion for the body is

$$\dot{\mathbf{l}} = \mathcal{I}\dot{\boldsymbol{\omega}} + \boldsymbol{\omega} \times \mathbf{l} = 0. \tag{4.43}$$

Note that because of the vanishing torque, this equation is not coupled to the kinetic equation

$$\dot{R} = \tfrac{1}{2}Ri\boldsymbol{\omega}. \tag{4.44}$$

Therefore, it can be solved directly for $\boldsymbol{\omega} = \boldsymbol{\omega}(t)$, and the result can be inserted into (4.44) to determine the attitude spinor R. From (4.43) we can conclude immediately that the angular momentum $\mathbf{l} = \mathcal{I}\boldsymbol{\omega}$ and the energy $E = \frac{1}{2}\boldsymbol{\omega}\cdot\mathbf{l}$ are constants of motion. So we want to solve the differential equation (4.43) for $\boldsymbol{\omega}$ in terms of $\mathbf{l}$ and $\mathbf{E}$.

Before attempting a general solution, let us consider the possibility of steady rotational motion about a fixed axis. In that case $\dot{\boldsymbol{\omega}} = 0$ and (4.43) implies that $\boldsymbol{\omega} \times \mathbf{l} = 0$, so $\boldsymbol{\omega}$ must be collinear with $\mathbf{l}$. Since $\mathbf{l} = \mathcal{I}\boldsymbol{\omega}$, this is possible only if $\boldsymbol{\omega}$ is directed along one of the principal axes. Moreover, if $\boldsymbol{\omega} \times \mathbf{l} \neq 0$, then (4.43) implies that $\dot{\boldsymbol{\omega}} \neq 0$. Therefore, in the absence of an applied torque, *steady rotational motion is possible if and only if the axis of rotation coincides with a principal axis of the inertia tensor*. We will ascertain conditions for the stability of steady rotation later on.

Returning to the general problem, we follow (1.7) and put the equation of motion (4.43) in the form

$$I_1\dot{\omega}_1 = (I_2 - I_3)\omega_2\omega_3 \tag{4.45a}$$

$$I_2\dot{\omega}_2 = (I_3 - I_1)\omega_3\omega_1 \tag{4.45b}$$

$$I_3\dot{\omega}_3 = (I_1 - I_2)\omega_1\omega_2, \tag{4.45c}$$

where the

$$\omega_k = \boldsymbol{\omega}\cdot\mathbf{e}_k = I_k^{-1}\,\mathbf{l}\cdot\mathbf{e}_k \tag{4.46}$$

are components of the rotational velocity with respect to the principal axes. Since we have already analyzed the motion of a symmetric body, we assume that all the principle moments of inertia I_k have different values.

The three variables ω_1, ω_2, ω_3 are related by the integrals of motion

$$l^2 = \boldsymbol{\omega}\cdot\mathcal{I}^2\boldsymbol{\omega} = I_1^2\omega_1^2 + I_2^2\omega_2^2 + I_3^2\omega_3^2 \tag{4.47a}$$

$$2E = \boldsymbol{\omega}\cdot\mathcal{I}\boldsymbol{\omega} = I_1\omega_1^2 + I_2\omega_2^2 + I_3\omega_3^2. \tag{4.47b}$$

Therefore we should be able to eliminate two of the variables from (4.45a, b, c) to get an equation for the third alone. To this end, it is convenient to eliminate each variable in turn from (4.47a, b) to get

$$l^2 - 2EI_1 = I_2(I_2 - I_1)\omega_2^2 + I_3(I_3 - I_1)\omega_3^2 \tag{4.48a}$$

$$l^2 - 2EI_2 = I_1(I_1 - I_2)\omega_1^2 + I_3(I_3 - I_2)\omega_3^2 \tag{4.48b}$$

$$l^2 - 2EI_3 = I_1(I_1 - I_3)\omega_1^2 + I_2(I_2 - I_3)\omega_2^2. \tag{4.48c}$$

To obtain an equation for ω_3 alone, we square (4.45c) and use (4.48a, b) to eliminate ω_1^2 and ω_2^2; thus,

$$I_3^2\dot{\omega}_3^2 = (I_2 - I_1)^2\, \omega_1^2\omega_2^2$$
$$= (I_2 - I_1)^2\left[\frac{(2EI_2 - l^2) - I_3(I_2 - I_3)\omega_3^2}{I_1(I_2 - I_1)}\right]\left[\frac{(l^2 - 2EI_1) - I_3(I_3 - I_1)\omega_3^2}{I_2(I_2 - I_1)}\right]. \tag{4.49}$$

Comparing this with the equation

$$\dot{y}^2 = \mu^2\,(1 - y^2)\,(1 - k^2y^2), \tag{4.50}$$

we see that it will have a solution of the form

$$\omega_3 = a_3\,\mathrm{sn}\,\mu t \tag{4.51}$$

provided

$$I_1 < I_3 < I_2 \tag{4.52}$$

and

$$2EI_2 - l^2 > 0, \quad l^2 - 2EI_1 > 0. \tag{4.53}$$

Of course we are free to label the I_k so (4.52) holds, and we notice that the inequalities (4.53) are then consequences of (4.48a, b). Therefore, the conditions for the solution (4.51) are satisfied; and the constants can be evaluated by inserting (4.51) into (4.49). At the same time, by comparing the two lines of (4.49) we can determine ω_1 and ω_2. There are two cases, for which we find

Case (a):

$$\omega_1 = a_1\,\mathrm{dn}(\mu t), \quad \omega_2 = a_2\,\mathrm{cn}(\mu t). \tag{4.54a}$$

Case (b):

$$\omega_1 = a_1\,\mathrm{cn}(\mu t), \quad \omega_2 = a_2\,\mathrm{dn}(\mu t). \tag{4.54b}$$

For both cases we get

$$a_1^2 = \frac{2EI_2 - l^2}{I_1(I_2 - I_1)}, \quad a_2^2 = \frac{l^2 - 2EI_2}{I_2(I_2 - I_1)}. \tag{4.54}$$

For case (a),

$$a_3^2 = \frac{l^2 - 2EI_1}{I_3(I_3 - I_1)}, \quad \mu^2 = \frac{(I_3 - I_1)\,(2EI_2 - l^2)}{I_1I_2I_3},$$

$$k^2 = \left(\frac{I_3 - I_1}{I_2 - I_3}\right)\left(\frac{2EI_2 - l^2}{l^2 - 2EI_1}\right), \tag{4.56a}$$

and for case (b)

$$a_3^2 = \frac{2EI_2 - l^2}{I_3(I_2 - I_3)}, \quad \mu^2 = \frac{(I_2 - I_3)(l^2 - 2EI_1)}{I_1 I_2 I_3},$$

$$k^2 = \left(\frac{I_2 - I_3}{I_3 - I_1}\right)\left(\frac{l^2 - 2EI_1}{2EI_2 - l^2}\right). \tag{4.56b}$$

The quantities μ, k and a_3 can be taken to be positive. To determine the signs of a_1 and a_2, we substitute the solutions into (4.45a, b, c), and after carrying out the differentiation (see Appendix B), we obtain

$$a_1 a_2 a_3 = \frac{\mu^3 I_1 I_2 I_3}{(I_1 - I_2)(I_2 - I_3)(I_3 - I_1)} < 0. \tag{4.57}$$

Therefore, if we take $a_1 > 0$, then $a_2 < 0$. Finally, note that the two cases are distinguished by the requirement that the expression for k^2 in (4.56a) or (4.56b) must satisfy $k^2 < 1$. By substitution into (4.48a), we see that this requirement can be reexpressed as the condition that

$$l^2 - 2EI_3 > 0 \text{ for case (a)} \tag{4.58a}$$

$$l^2 - 2EI_3 < 0 \text{ for case (b)}. \tag{4.58b}$$

This completes our solution of the dynamical equation of motion.

The problem remains to determine the attitude spinor R from the known functions $\omega_k = \omega_k(t)$. We could proceed by integrating

$$\dot{R} = \tfrac{1}{2}Ri\boldsymbol{\omega} = \tfrac{1}{2}i(\omega_1\boldsymbol{\sigma}_1 + \omega_2\boldsymbol{\sigma}_2 + \omega_3\boldsymbol{\sigma}_3)R,$$

but there is a much simpler way which exploits the constants of motion and determines R almost completely by algebraic means. The *angular momentum direction cosines*

$$h_k = \hat{\mathbf{l}}\cdot\mathbf{e}_k = l^{-1}I_k\omega_k \tag{4.59}$$

are more convenient parameters than the ω_k, because then we can write

$$\hat{\mathbf{l}} = h_1\mathbf{e}_1 + h_2\mathbf{e}_2 + h_3\mathbf{e}_3 = R^\dagger(h_1\boldsymbol{\sigma}_1 + h_2\boldsymbol{\sigma}_2 + h_3\boldsymbol{\sigma}_3)R = \boldsymbol{\sigma}_3, \tag{4.60}$$

where we have used our prerogative to identify $\boldsymbol{\sigma}_3$ with the distinguished direction $\hat{\mathbf{l}}$ in our problem. Note that with this choice (4.51) gives

$$h_3 = \boldsymbol{\sigma}_3\cdot\mathbf{e}_3 = l^{-1}I_3\omega_3 = a_3 \operatorname{sn} \tau. \tag{4.61}$$

The question is now, what does (4.60) tell us about the functional form of R?

As before, the fact that $\boldsymbol{\sigma}_3$ is a distinguished direction suggests it may be convenient to express R in terms of Cayley-Klein parameters:

$$R = \alpha_+ - i\boldsymbol{\sigma}_2\alpha_-, \tag{4.62}$$

$$\alpha_\pm = \lambda_\pm e^{i\boldsymbol{\sigma}_3\phi_\pm}. \tag{4.63}$$

Using this parametrization for R, from (4.60) we obtain

$$h_1\boldsymbol{\sigma}_1 + h_2\boldsymbol{\sigma}_2 + h_3\boldsymbol{\sigma}_3 = R\boldsymbol{\sigma}_3R^\dagger = (\lambda_+^2 - \lambda_-^2)\boldsymbol{\sigma}_3 + 2\boldsymbol{\sigma}_1\alpha_+^\dagger\alpha_- . \tag{4.64}$$

Hence,

$$h_3 = \lambda_+^2 - \lambda_-^2 \tag{4.65}$$

and

$$h_1 + i\sigma_3 h_2 = 2\alpha_-\alpha_+^\dagger = 2\lambda_+\lambda_- e^{i\boldsymbol{\sigma}_3(\phi_- - \phi_+)}. \tag{4.66}$$

Since

$$R^\dagger R = \lambda_+^2 + \lambda_-^2 = 1,$$

from (4.65) we obtain

$$\lambda_\pm = \left(\frac{1 \pm h_3}{2}\right)^{1/2}. \tag{4.67}$$

And (4.66) give us

$$\phi_- - \phi_+ = \tan^{-1}\left(\frac{h_2}{h_1}\right), \tag{4.68}$$

Thus, we have determined $\lambda_\pm$ and $\phi_- - \phi_+$ as functions of the h_k, so we can complete our solution by determining $\phi_+ + \phi_-$. That requires an integration.

Instead of determining $\phi_+ + \phi_-$ directly, it is more convenient to determine the variable

$$\phi = \phi_+ + 2\phi_- - \pi, \tag{4.69}$$

which, as established in Exercise (4.6), is one of the Euler angles. Using

$$\boldsymbol{\omega} = -2iR^\dagger\dot{R},$$

we can express $\boldsymbol{\omega}$ in terms of the Cayley-Klein parameters or the Euler angles and their derivatives (Exercise 4.5). Whence we obtain

$$\dot{\phi} + h_3\dot{\phi}_+ = \boldsymbol{\omega}\cdot\boldsymbol{\sigma}_3 = \frac{2E}{l},$$

$$h_3\dot{\phi} + \dot{\phi}_+ = \boldsymbol{\omega}\cdot\mathbf{e}_3 = \frac{lh_3}{I_3}.$$

Eliminating $\dot{\phi}_+$, we get

$$\dot{\phi} = \dot{\phi}_+ + 2\dot{\phi}_- = \frac{l}{I_3} + \left(\frac{2EI_3 - l^2}{lI_3}\right)\frac{1}{1 - h_3^2}. \tag{4.70}$$

This integrates to

$$\phi(t) = \phi(t_0) + \frac{l}{I_3}(t - t_0) + \left(\frac{2EI_3 - l^2}{lI_3}\right)[\pi(\tau, a^2) - \pi(\tau_0, a^2)], \tag{4.71}$$

where $\pi(\tau, a^2)$ is the incomplete elliptic integral of the third kind defined by (4.40).

Although we now have a complete and exact analytic solution to our problem, the solution does not immediately provide a clear picture of the body's motion. For that purpose, we now look at the problem in a different way.

Poinsot's Construction

To develop a picture for the motion of a freely spinning asymmetric body, consider the restriction on the rotational velocity $\boldsymbol{\omega}$ due to the energy integral:

$$\tfrac{1}{2}\boldsymbol{\omega}\cdot(\mathcal{J}\boldsymbol{\omega}) = E. \tag{4.72}$$

This is the equation for an ellipsoid; call it the *energy ellipsoid*. Thus, for a given rotational kinetic energy E, $\boldsymbol{\omega}$ must be a point on this ellipsoid. Note that the principal axes of the energy ellipsoid are the same as those of the inertia tensor. So the attitude of the energy ellipsoid in space faithfully represents the attitude of the body itself.

The normal to the energy ellipsoid at a point $\boldsymbol{\omega}$ is given by the gradient

$$\nabla_{\boldsymbol{\omega}}(\tfrac{1}{2}\boldsymbol{\omega}\cdot\mathcal{J}\boldsymbol{\omega}) = \mathcal{J}\boldsymbol{\omega} = \mathbf{l}. \tag{4.73}$$

For a constant angular momentum $\mathbf{l}$, this puts a further restriction on the allowed values of $\boldsymbol{\omega}$. Indeed, for fixed $\mathbf{l}$ and variable $\boldsymbol{\omega}$, the energy integral

$$\boldsymbol{\omega}\cdot\mathbf{l} = 2E \tag{4.74}$$

is the equation for a plane with normal $\mathbf{l}$ and distance $2E/l$ from the origin. This plane is called the *invariable plane*. Therefore, for given E and $\mathbf{l}$, at any time t, *the invariable plane is tangent to the energy ellipsoid at the point* $\boldsymbol{\omega} = \boldsymbol{\omega}(t)$. (Figure 4.2) Moreover the energy ellipsoid can be said to roll on the invariable plane without slipping, since the point of contact is on the instantaneous rotation axis. This picture of the motion is due to Poinsot (1834). An apparatus that shows subtle features of the motion has been described by Harter and Kim (*Amer. J Phys.* **44**, 1080 (1976)).

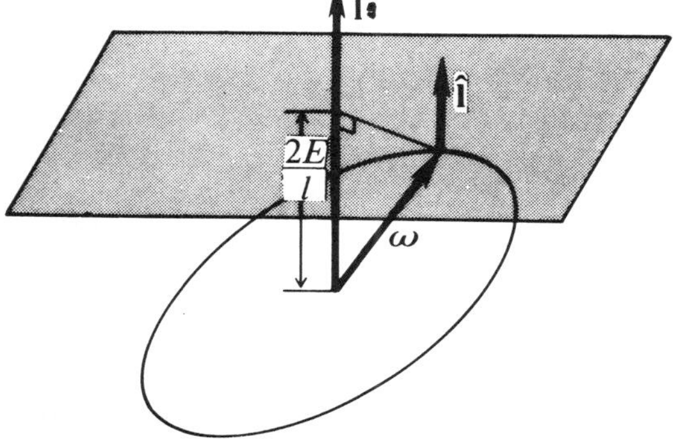

Fig. 4.2. The invariable plane is tangent to the energy ellipsoid at $\boldsymbol{\omega}(t)$.

As $\boldsymbol{\omega}(t)$ varies with time, it traces out a curve on the energy ellipsoid called the *polhode* and a curve on the invariable plane called the *herpolhode*. We have already found a parametric equation for the polhode, given by (4.51) and (4.54a, b). But a nonparametric representation makes it easier to picture the curve as a whole. This is easily found by noting that angular momentum conservation implies that $\boldsymbol{\omega}$ must lie on the ellipsoid

$$l^2 = \boldsymbol{\omega}\cdot(\mathcal{I}^2\boldsymbol{\omega}). \tag{4.75}$$

Therefore, the polhode is a curve of intersection of the two ellipsoids (4.72) and (4.75). This tells us at once that a polhode is a closed curve, so the motion is periodic. Polhodes for various initial conditions are shown in Figure 4.3. A polhode can be interpreted as the curve traced out by the tip of the $\boldsymbol{\omega}$-vector as "seen" by an observer rotating with the body.

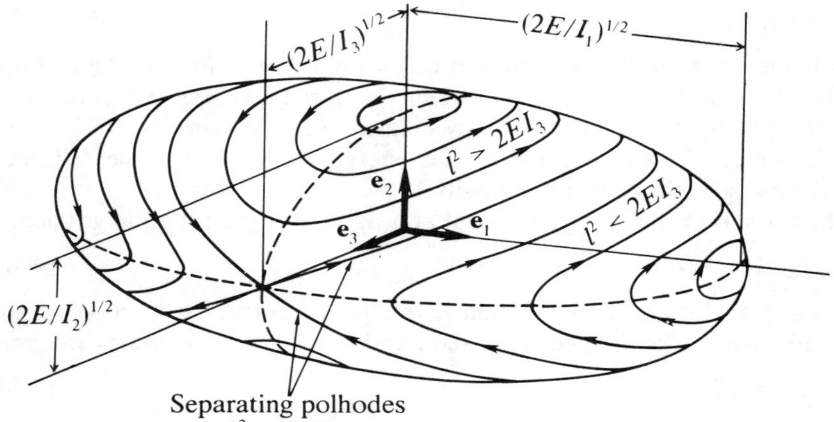

Fig. 4.3. The energy ellipsoid, showing polhodes for different initial conditions ($I_1 < I_3 < I_2$).

Questions about the stability of rotational motion are best answered by examining the family of polhodes with different initial conditions. We have already proved that steady rotation is possible only about a principal axis. To investigate the stability of steady rotation quantitatively, we consider a small departure from the steady motion by writing

$$\boldsymbol{\omega} = \boldsymbol{\omega}_0 + \boldsymbol{\varepsilon}, \tag{4.76a}$$

where

$$\mathcal{I}\boldsymbol{\omega}_0 = I_3\boldsymbol{\omega}_0 \tag{4.76b}$$

indicates that $\boldsymbol{\omega}_0$ is directed along the $\mathbf{e}_3$ principal axis, and

$$\boldsymbol{\varepsilon}\cdot\boldsymbol{\omega}_0 \approx 0 \tag{4.76c}$$

must be satisfied for a small deviation $\boldsymbol{\varepsilon}$. The plan now is to get an equation for $\boldsymbol{\varepsilon}$, so we can study its behavior. Substituting (4.76a) into (4.72) and (4.75) and using (4.76b), we get

$$2E = I_3\omega_0^2 + 2I_3\boldsymbol{\varepsilon}\cdot\boldsymbol{\omega}_0 + \boldsymbol{\varepsilon}\cdot\mathcal{I}\boldsymbol{\varepsilon}$$

$$l^2 = I_3{}^2\omega_0^2 + 2I_3^2\,\boldsymbol{\varepsilon}\cdot\boldsymbol{\omega}_0 + \boldsymbol{\varepsilon}\cdot\mathcal{I}^2\boldsymbol{\varepsilon}.$$

Using the approximation (4.76c) and eliminating ω_0^2 from these two equations, we get the desired equation for $\boldsymbol{\varepsilon}$

$$l^2 - 2I_3E = \boldsymbol{\varepsilon}\cdot(\mathcal{I}^2 - I_3\mathcal{I})\boldsymbol{\varepsilon}. \tag{4.77}$$

This is, in fact, an exact equation for the projection of the polhode onto a plane with normal $\mathbf{e}_3$. If we decompose $\boldsymbol{\varepsilon}$ into its components $\varepsilon_k = \boldsymbol{\varepsilon}\cdot\mathbf{e}_k$ with respect to the principal axes, this equation can be written

$$I_1(I_1 - I_3)\varepsilon_1^2 + I_2(I_2 - I_3)\varepsilon_2^2 = l^2 - 2I_3E. \tag{4.78}$$

This will be recognized as the equation for an ellipse if I_3 is greater than or less than I_1 and I_2. This means $\boldsymbol{\omega}$ will stay close to $\boldsymbol{\omega}_0$ during the entire motion. Therefore, *steady rotation about the two axes with the largest and smallest moment of inertia is stable.*

On the other hand, if the value of I_3 is between the values of I_1 and I_2, then (4.78) is the equation for a hyperbola. So if $\boldsymbol{\varepsilon}$ has any small value initially, it will increase with time, and $\boldsymbol{\omega}$ will wander away from $\boldsymbol{\omega}_0$. Therefore, steady rotation about the intermediate principal axis is unstable. This can be seen by examining the polhodes in Figure 4.3. And it can be empirically demonstrated by attempting to throw an asymmetric object like a tennis racket up in the air so that it spins about a principal axis.

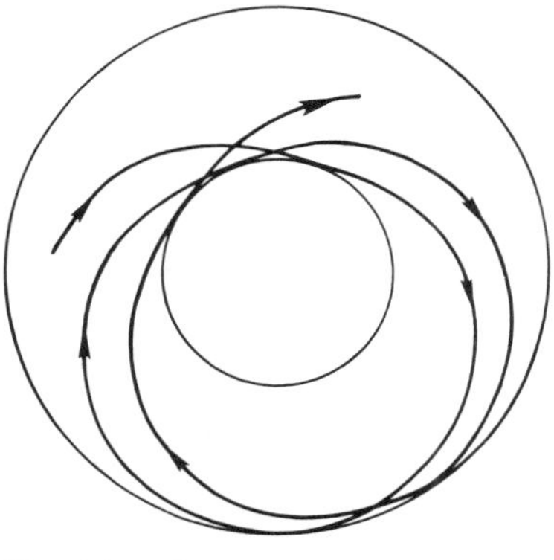

Fig. 4.4. The herpolhode is confined to an annulus in the invariable plane.

As the energy ellipsoid rolls on the invariable plane, the polhode rolls on the herpolhode. In contrast to the polhode, the herpolhode is not necessarily a closed curve, but as shown in Figure 4.4, it must oscillate between maximum and minimum values corresponding to maxima and minima of the polhode.

7-4. Exercises

(4.1) Verify the expressions (4.5), (4.6) and (4.7) for the constants of motion in terms of R and $\dot{R}$.

(4.2) For *compound pendulum*, show that the frequencies of oscillation about two different parallel axes will be the same if and only if $rr' = r_0^2$, where r and r' are the distances of the points from the CM, and r_0 is the CM radius of gyration. Show also that the oscillation frequency is that of a simple pendulum with length $r + r'$.

(4.3) Show that the *elliptic integral* (B.6) in Appendix B can be written in the form:

$$K = \int_0^{\pi/2} \frac{d\phi}{(1 - k^2 \sin^2 \phi)^{1/2}}$$

Expand the integrand in a series and perform term-by-term integration to get the following expression for the period of a plane pendulum:

$$T = \frac{2\pi}{G^{1/2}} \left[1 + \left(\frac{1}{2} \right)^2 k^2 + \left(\frac{1 \cdot 3}{2 \cdot 4} \right)^2 k^4 + \ldots \right] .$$

Thus, show that the first order correction for the period in the small angle approximation gives:

$$T = \frac{2\pi}{G^{1/2}} \left[1 + \frac{\theta_0^2}{16} \right] ,$$

where θ_0 is the angular amplitude.

(4.4) A plane pendulum beats seconds when swinging through an angle of 6°. If the angle is increased to 8°, show that it will lose approximately 10 beats a day.

(4.5) In the extensive literature on the Lagrange problem, the motion is usually parametrized by Euler angles. To compare the literature to the approach taken here, recall that the parametrization of a rotation by Euler angles ψ, θ, ϕ is given by:

$$R = e^{(1/2)i\boldsymbol{\sigma}_3\psi} e^{(1/2)i\boldsymbol{\sigma}_1\theta} e^{(1/2)i\boldsymbol{\sigma}_3\phi} ,$$

where $\boldsymbol{\sigma}_1$ and $\boldsymbol{\sigma}_3$ are orthogonal constant unit vectors. For a rotating body subject to a constant effective force $\mathbf{G}$, we take $\boldsymbol{\sigma}_3 = \hat{\mathbf{G}}$ and $\mathbf{e} = R^\dagger \boldsymbol{\sigma}_3 R = R^\dagger \hat{\mathbf{G}} R$ as the axis of symmetry. Then ϕ is called the *precession angle*, θ is called the *nutation angle*, and ψ is called the *phase angle*. Show that the rotational velocity in terms of Euler angles is given by:

$$\boldsymbol{\omega} = -2iR^\dagger \dot{R} = \hat{\mathbf{G}}\dot{\phi} + \mathbf{G} \hat{\times} \mathbf{e}\dot{\theta} + \mathbf{e}\dot{\psi} ,$$

where,

$$\mathbf{G} \hat{\times} \mathbf{e} \equiv \frac{\mathbf{G} \times \mathbf{e}}{|\mathbf{G} \times \mathbf{e}|} .$$

Show that the constants of motion $a = \boldsymbol{\omega} \cdot \mathbf{e}$, $b = \boldsymbol{\omega} \cdot \hat{\mathbf{G}}$ and $E = \frac{1}{2}\omega^2 - \mathbf{G} \cdot \mathbf{e}$ yield the equations

$$\dot{\psi} = \frac{a - b \cos \theta}{\sin^2 \theta} ,$$

$$\dot{\phi} = \frac{b - a\cos\theta}{\sin^2\theta} ,$$

$$\dot{\theta}^2 + \frac{(b - a\cos\theta)^2}{\sin^2\theta} + 2G\cos\theta = 2E - a^2 .$$

The solution of these equations is discussed by many authors. Use our results to express $\cos\theta$ in terms of elliptic functions and show that it is a solution to the last of these equations.

(4.6) Show that the Cayley-Klein parameters are related to the Euler angles by

$$\alpha_+ = \cos\tfrac{1}{2}\theta\, e^{(1/2)i\boldsymbol{\sigma}_3(\psi + \phi)} ,\ \alpha_- = i\boldsymbol{\sigma}_3 \sin\tfrac{1}{2}\theta\, e^{(1/2)i\boldsymbol{\sigma}_3(\phi - \psi)} ,$$

To establish the converse relations, show that

$$R = e^{(1/2)i\boldsymbol{\sigma}_3(\phi_+ - \phi_- + \pi/2)}(\lambda_+ + i\boldsymbol{\sigma}_1\lambda_-)\, e^{(1/2)i\boldsymbol{\sigma}_3(\phi_+ + \phi_- - \pi/2)} .$$

Whence,

$$\psi = \phi_+ - \phi_- + \pi/2$$
$$\phi = \phi_+ + \phi_- - \pi/2 .$$

(4.7) For the case of steady forced precession, compare the solution in terms of elliptic functions to the exact solution obtained in Section 7-3.

(4.8) Show that when the R is expressed in terms of Cayley-Klein parameters $\alpha_\pm$, the spinor equation (3.24) can be separated into two uncoupled second order equations:

$$\ddot{\alpha}_\pm + [\tfrac{1}{2}(E \pm 3G) \mp 2G\,|\alpha_\pm|^2]\alpha_\pm = 0$$

Of course, since G is a parameter which can have either sign, the two equations are essentially the same. To solve this equation directly, it is helpful to rewrite it by defining $\mathbf{r} = \boldsymbol{\sigma}_1\alpha_+$, which is a vector because of Equation (4.21). In terms of this variable the equation becomes

$$\ddot{\mathbf{r}} = [\tfrac{1}{2}(E + 3G) - 2Gr^2]\mathbf{r} = 0,$$

which we recognize as the equation for a particle in an unusual central force field. Use this fact to get Equations (4.26a) and (4.31) directly as first integrals of the motion with undetermined constants. Note that if we use Equation (4.32) we can put the above equation in the form

$$\ddot{\alpha} + [A + B\,\mathrm{sn}^2\,\mu t\,]\alpha = 0,$$

where A and B are scalar constants. This is called *Lamé's equation*. Although we have solved this equation for the case of the top, that does not end the matter, because the form of our solution is probably not optimal. We derived separate expressions for the modulus and angle of α, whereas there are alternative expressions for α as a unit which can probably be calculated more easily. This is a worthy issue for mathematical research. Evaluation of the solutions is discussed by E. T. Whitaker (*A Treatise on the Analytical Dynamics of Particles and Rigid Bodies*, Dover, N.Y., 4th Ed. (1944), especially p. 161). Lamé's equation is discussed by Whitaker and Watson (*Modern Analysis*, Cambridge U. Press (1952), Chapter 23). No doubt there are significant improvements yet to be made in the theory of the top, and we can expect new insights from bringing together spinor theory and the classical theory of elliptic functions.

7-5. Rolling Motion

The mathematical description of rolling motion requires both translational and rotational equations of motions coupled by a rolling constraint. Consider a centrosymmetric sphere of radius a, mass m, and moment of inertia $I = mk^2$ rolling on a rough surface with unit normal $\mathbf{n}$ at the point of contact. Let $\mathbf{f}$ denote the "reaction force" exerted by the constraining surface on the sphere (Figure 5.1).

The translational and rotational equations of motion for the sphere are

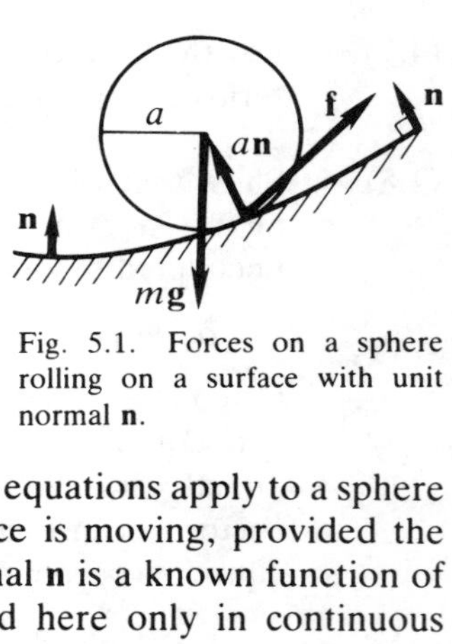

Fig. 5.1. Forces on a sphere rolling on a surface with unit normal $\mathbf{n}$.

$$m\dot{\mathbf{V}} = m\mathbf{g} + \mathbf{f}, \tag{5.1}$$

$$mk^2\dot{\boldsymbol{\omega}} = (-a\mathbf{n}) \times \mathbf{f}, \tag{5.2}$$

where $\mathbf{V} = \dot{\mathbf{X}}$ is the center of mass velocity. These equations apply to a sphere rolling on an arbitrary surface even if the surface is moving, provided the surface is mathematically prescribed, so the normal $\mathbf{n}$ is a known function of position and time. Of course, we are interested here only in continuous surfaces with a unique normal at every point.

The velocity $\mathbf{v}$ of the point on the sphere which is instantaneously in contact with the surface is determined by the kinematical relation

$$\mathbf{v} = \mathbf{V} + \boldsymbol{\omega} \times (-a\mathbf{n}). \tag{5.3}$$

Suppose the constraining surface is moving with a velocity $\mathbf{u}$ at the point of contact. The relative velocity $\mathbf{v} - \mathbf{u}$ must vanish if the sphere is not slipping. Therefore, the *equation of constraint for rolling contact* is

$$\mathbf{u} = \mathbf{V} - a\boldsymbol{\omega} \times \mathbf{n}. \tag{5.4}$$

The velocity $\mathbf{u}$ will be known if the motion of the constraining surface is prescribed.

Now we have sufficient equations to determine rolling motion of the sphere on a given surface. A general strategy for solving these equations is to eliminate the reaction force between (5.1) and (5.2) to get

$$k^2\dot{\boldsymbol{\omega}} = a\mathbf{n} \times (\mathbf{g} - \dot{\mathbf{V}}). \tag{5.5}$$

Then (5.4) can be used to get separate equations for $\boldsymbol{\omega}$ and $\mathbf{V}$. However, it must be remembered that (5.5) does not hold when $\mathbf{f} = 0$, that is, when the sphere loses contact with the constraining surface.

This is as far as we can go with the theory of rolling motion without assumptions about the constraining surface. So we turn now to consider special cases.

Rolling on an inclined plane

For a fixed inclined plane, the normal $\mathbf{n}$ is constant and the equation (5.4) for rolling contact reduces to

$$\mathbf{V} = a\boldsymbol{\omega} \times \mathbf{n} = -ia\, \boldsymbol{\omega}\wedge\mathbf{n} \tag{5.6}$$

From (5.2) we find that $\mathbf{n}\cdot\dot{\boldsymbol{\omega}} = 0$; hence the spin about the normal

$$s = \boldsymbol{\omega}\cdot\mathbf{n} \tag{5.7}$$

is a constant of the motion. We can combine (5.6) and (5.7) to solve for $\boldsymbol{\omega}$. Thus,

$$\mathbf{V} - ias = -ai\boldsymbol{\omega}\mathbf{n},$$

so

$$\boldsymbol{\omega} = \left(s + \frac{i\mathbf{V}}{a}\right)\mathbf{n} = s\mathbf{n} + \frac{\mathbf{n} \times \mathbf{V}}{a}. \tag{5.8}$$

Now we substitute this into (5.5) to get an equation for $\mathbf{V}$ alone:

$$\left(1 + \frac{k^2}{a^2}\right)\mathbf{n}\wedge\dot{\mathbf{V}} = \mathbf{n}\wedge\mathbf{g}. \tag{5.9}$$

The outer product is more convenient than the cross product form of this equation, because the condition $\mathbf{n}\cdot\mathbf{V} = 0$ from (5.6) tells us that we can divide by $\mathbf{n}$ to get

$$\left(1 + \frac{k^2}{a^2}\right)\dot{\mathbf{V}} = \mathbf{n}(\mathbf{n}\wedge\mathbf{g}) = \mathbf{g}_{\|}. \tag{5.10}$$

Thus, the sphere rolls in the plane with a constant acceleration, which has the value $(5/7)\mathbf{g}_{\|}$ if the sphere is a homogenous solid ($k^2 = 2a^2/5$). The trajectory is therefore a parabola

$$\mathbf{X} = \frac{5}{14}\mathbf{g}_{\|}t^2 + \mathbf{V}_0 t + \mathbf{X}_0 \tag{5.11}$$

for initial position $\mathbf{X}_0$ and velocity $\mathbf{V}_0$.

Substituting (5.11) into (5.8), we find the explicit time dependence of the rotational velocity

$$\omega = \left(\frac{5}{7a}\,\mathbf{n} \times \mathbf{g}\right)t + \left(s\mathbf{n} + \frac{\mathbf{n} \times \mathbf{V}_0}{a}\right) \tag{5.12}$$

The attitude $R = R(t)$ can be determined from this by integrating $\dot{R} = \frac{1}{2} Ri\omega$. However, the integration is not trivial unless the sphere starts from rest, since the direction of ω is not constant.

Rolling in a spherical bowl

For a sphere rolling inside a fixed spherical container of radius b, we can write $\mathbf{X} = (a - b)\mathbf{n}$ (Figure 5.2). Therefore the unit normal $\mathbf{n}$ is a natural position variable, and the rolling constraint can be written

$$\mathbf{V} = (a - b)\dot{\mathbf{n}} = a\omega \times \mathbf{n}. \tag{5.13}$$

This implies that $\omega \cdot \dot{\mathbf{n}} = 0$, and (5.2) implies that $\mathbf{n} \cdot \dot{\omega} = 0$. Therefore, in this case also

$$s = \omega \cdot \mathbf{n} \tag{5.14}$$

is a constant of motion.

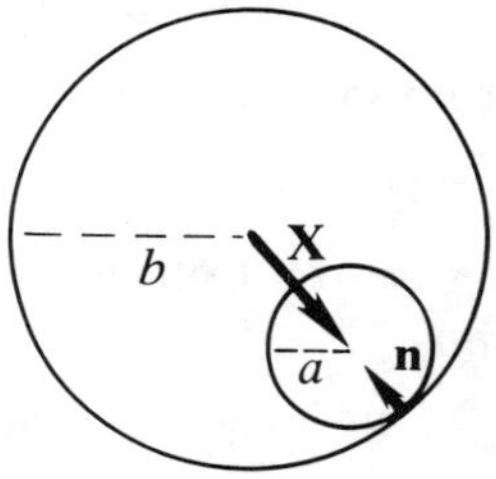

Fig. 5.2. Sphere rolling in a sphere.

Equation (5.5) subject to (5.13) is nearly the same as the equation of motion for a spherical top, so our experience with the top suggests that the best strategy is to look at once for constants of the motion. Using (5.13) to eliminate $\mathbf{V}$, we can put (5.5) in the form

$$\frac{\mathrm{d}}{\mathrm{d}t}[k^2\omega + a^2\mathbf{n} \times (\omega \times \mathbf{n})] = a\mathbf{n} \times \mathbf{g}. \tag{5.15}$$

The quantity is square brackets here can be identified as the angular momentum of the sphere (per unit mass) about the point of contact.

As in the case of the top, from (5.15) we find that

$$\mathbf{g} \cdot [k^2\omega + a^2\mathbf{n} \times (\omega \times \mathbf{n})] = (k^2 + a^2)\mathbf{g} \cdot \omega - sa^2\mathbf{g} \cdot \mathbf{n} \tag{5.16}$$

is a constant of motion. We can get one other constant of motion from (5.15), namely the total energy, which we can also write down from first principles. The kinetic energy is

$$\tfrac{1}{2} mV^2 + \tfrac{1}{2} I\omega^2 = \frac{m}{2}[(k^2 + a^2)(\omega \times \mathbf{n})^2 + k^2 s^2].$$

The potential energy is

$$-m\mathbf{X} \cdot \mathbf{g} = m(b - a)\mathbf{n} \cdot \mathbf{g}.$$

So the effective energy constant is

$$E = \tfrac{1}{2}(k^2 + a^2)(\boldsymbol{\omega} \times \mathbf{n})^2 + (b - a)\mathbf{n}\cdot\mathbf{g}. \tag{5.17}$$

The three integrals of motion (5.14), (5.16) and (5.17) can be expressed as equations for the attitude spinor R by writing

$$\mathbf{n} = -R^{\dagger}\hat{\mathbf{g}}R \tag{5.18}$$

and using $\dot{R} = \frac{1}{2}Ri\omega$. Clearly, a solution in terms of elliptic functions can be found by introducing Cayley-Klein parameters in the same way that we handled the top.

When $\mathbf{n}\cdot\mathbf{g} > 0$, it is possible for the sphere to lose contact with the container. For contact to be maintained, the normal component of the reaction force must be positive. Using (5.1), this condition is expressed by

$$\mathbf{n}\cdot\mathbf{f} = m\mathbf{n}\cdot(\dot{\mathbf{V}} - \mathbf{g}) \geqslant 0.$$

With (5.13), this *contact condition* can be put in the form

$$V^2 = a^2(\boldsymbol{\omega} \times \mathbf{n})^2 \geqslant (b - a)\mathbf{n}\cdot\mathbf{g}. \tag{5.19}$$

It can be further reduced by using the energy equation (5.17).

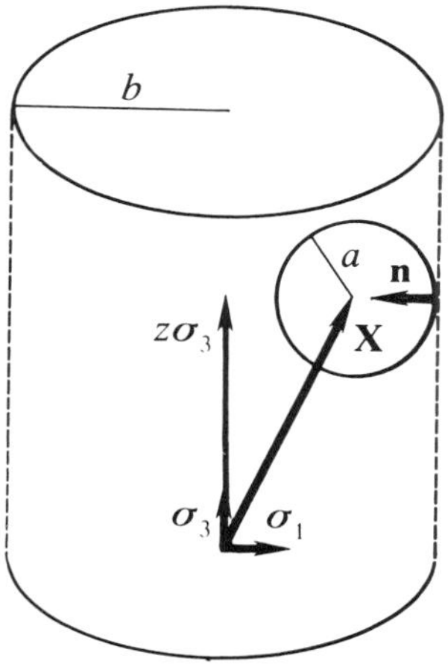

Fig. 5.3. Sphere rolling inside a vertical circular cylinder.

Rolling inside a cylinder

For a sphere rolling inside a fixed vertical cylinder of radius b, it is convenient to represent the upward vertical direction by $\boldsymbol{\sigma}_3 = -\hat{\mathbf{g}}$ and parametrize the inward normal $\mathbf{n}$ of the constraining cylinder in terms of an angle θ by writing

$$\mathbf{n} = \boldsymbol{\sigma}_1 e^{i\boldsymbol{\sigma}_3\theta} \tag{5.20}$$

Then an explicit parametrization $\mathbf{X} = \mathbf{X}(\theta, z)$ of the center of mass is given by

$$\mathbf{X} = (a - b)\mathbf{n} + z\boldsymbol{\sigma}_3 \tag{5.21}$$

(Figure 5.3).

To get suitable equations for the parameters, first note that (5.8) holds for rolling motion on any fixed surface even when $s = \boldsymbol{\omega}\cdot\mathbf{n}$ is not constant. Differentiating (5.8) we get

$$a\dot{\boldsymbol{\omega}} = \dot{\mathbf{n}} \times \mathbf{V} + \mathbf{n} \times \dot{\mathbf{V}} + as\dot{\mathbf{n}} + a\dot{s}\mathbf{n},$$

and substituting this into (5.5), we obtain

$$\left(\frac{k^2 + a^2}{a^2}\right)\mathbf{n} \times \dot{\mathbf{V}} + \frac{k^2}{a^2}\dot{\mathbf{n}} \times \mathbf{V} + \frac{k^2}{a}(s\dot{\mathbf{n}} + \dot{s}\mathbf{n}) - \mathbf{n} \times \mathbf{g} = 0. \tag{5.22}$$

Differentiation of (5.20) and (5.21) yields

$$\dot{\mathbf{n}} = i\mathbf{n}\boldsymbol{\sigma}_3\dot{\theta} = \dot{\theta}\boldsymbol{\sigma}_3 \times \mathbf{n}$$

$$\mathbf{V} = (a - b)\dot{\mathbf{n}} + \dot{z}\boldsymbol{\sigma}_3$$

$$\dot{\mathbf{V}} = (a - b)(i\mathbf{n}\boldsymbol{\sigma}_3\ddot{\theta} - \mathbf{n}\dot{\theta}^2) + \ddot{z}\boldsymbol{\sigma}_3 .$$

By substituting this into (5.22) we get

$$\frac{k^2 + a^2}{a^2}\left[(a - b)\,\ddot{\theta}\boldsymbol{\sigma}_3 + \ddot{z}\,\mathbf{n} \times \boldsymbol{\sigma}_3\right] + \frac{k^2}{a^2}\,\dot{\theta}\dot{z}\,\mathbf{n} +$$

$$+ \frac{k^2}{a}\,(s\dot{\theta}\boldsymbol{\sigma}_3 \times \mathbf{n} + \dot{s}\,\mathbf{n}) + g\mathbf{n} \times \boldsymbol{\sigma}_3 = 0$$

The coefficients of the orthogonal vectors $\boldsymbol{\sigma}_3$, $\mathbf{n}$ and $\mathbf{n} \times \boldsymbol{\sigma}_3$ in this equation can be equated separately to give us three scalar equations:

$$\ddot{\theta} = 0, \tag{5.23a}$$

$$a\dot{s} + \dot{\theta}\dot{z} = 0, \tag{5.23b}$$

$$\ddot{z} - \frac{ak^2 s\dot{\theta}}{k^2 + a^2} + \frac{a^2 g}{a^2 + k^2} = 0. \tag{5.23c}$$

From (5.23a) we find

$$\dot{\theta} = \alpha = \text{const.} \tag{5.24}$$

When this is inserted in (5.23b) we find

$$\alpha z + as = a\beta, \tag{5.25}$$

where β is a second constant of integration. Using these results in (5.23c), we obtain

$$\ddot{z} - \frac{k^2\alpha}{k^2 + a^2}(a\beta - \alpha z) + \frac{a^2 g}{a^2 + k^2} = 0. \tag{5.26}$$

From this equation we can conclude that the ball oscillates vertically with simple harmonic motion of period

$$T = 2\pi\left(\frac{k^2 + a^2}{k^2\alpha^2}\right)^{1/2} \tag{5.27}$$

about the horizontal level

$$z_0 = \frac{a}{\alpha}\left(\beta - \frac{ag}{\alpha k^2}\right). \tag{5.28}$$

This may explain why a golfball or basketball which appears to have been "sunk" sometimes rises up and out of the hole.

Rolling and Slipping

If the sphere is slipping as it rolls on a fixed surface, then (5.3) gives us the *slipping constraint*

$$\mathbf{v} = \mathbf{V} - a\boldsymbol{\omega} \times \mathbf{n} . \tag{5.29}$$

The appearance of this new variable $\mathbf{v}$ is offset by introducing an empirically based law for the reaction force of the form

$$\mathbf{f} = N(\mathbf{n} - \mu\hat{\mathbf{v}}) , \tag{5.30}$$

where $N = \mathbf{f}\cdot\mathbf{n} \geq 0$, and *the coefficient of friction* μ is a positive scalar constant characteristic of the surfaces in contact.

Let us consider the slipping motion for the simple case of a billiard ball on a horizontal table. In this case $\mathbf{g} = -\mathbf{n}g$ *and* $\mathbf{n}\cdot\mathbf{V} = 0$, so after (5.30) is inserted in (5.1), we can separate the vertical component of the equation

$$\mathbf{n}\cdot(\mathbf{f} + m\mathbf{g}) = N - mg = 0 \tag{5.31a}$$

from the horizontal component

$$m\dot{\mathbf{V}} = -\mu N\hat{\mathbf{v}} . \tag{5.31b}$$

By eliminating N between these two equations, the translational equation of motion is reduced to

$$\dot{\mathbf{V}} = -\mu g\hat{\mathbf{v}} . \tag{5.32}$$

Similarly, by substituting (5.30) into (5.2) and using (5.31a), the rotational equation of motion is reduced to

$$k^2\dot{\boldsymbol{\omega}} = a\mu g\mathbf{n} \times \hat{\mathbf{v}} . \tag{5.33}$$

By eliminating $\hat{\mathbf{v}}$ between these last two equations we get

$$k^2\dot{\boldsymbol{\omega}} = -a\mathbf{n} \times \dot{\mathbf{V}} = ai\mathbf{n}\dot{\mathbf{V}} , \tag{5.34}$$

where $\mathbf{n}\cdot\mathbf{V} = 0$ was used in the last step.

Next we differentiate the slipping equation (5.29) and eliminate $\boldsymbol{\omega}$ with (5.34) to get

$$\dot{\mathbf{v}} = \left(\frac{k^2 + a^2}{k^2}\right)\dot{\mathbf{V}} . \tag{5.35}$$

Using (5.32) to eliminate $\dot{\mathbf{V}}$, we get

$$\dot{\mathbf{v}} = -\mu g\left(\frac{k^2 + a^2}{k^2}\right)\hat{\mathbf{v}} . \tag{5.36}$$

This tells us that $\hat{\mathbf{v}}$ is constant and the speed $v = |\mathbf{v}|$ is determined by

$$\dot{v} = -\mu g\left(\frac{k^2 + a^2}{k^2}\right) .$$

So the speed decreases linearly:

$$v = v_0 - \mu g \left(\frac{k^2 + a^2}{k^2} \right) t. \tag{5.37}$$

And *slipping continues until* $v = 0$ at time

$$\tau = \frac{k^2 v_0}{\mu g (k^2 + a^2)}. \tag{5.38}$$

A billiard ball is a uniform solid, in which case $k^2 = 2a^2/5$ and $\tau = 2v_0/7\mu g$.

The trajectory of the ball during slipping is obtained by integrating (5.32); so

$$\mathbf{V} = \mathbf{V}_0 - \mu g t \hat{\mathbf{v}}, \tag{5.39}$$

$$\mathbf{X} = \mathbf{X}_0 + \mathbf{V}_0 t - \tfrac{1}{2}\mu g t^2 \hat{\mathbf{v}}. \tag{5.40}$$

This is a parabola if, $\mathbf{V}_0 \wedge \hat{\mathbf{v}} \neq 0$. This explains (in principle!) how a billiards trick shot artist can shoot around obstacles.

The rotational velocity is found by inserting (5.39) and (5.37) into (5.29) to get

$$a\boldsymbol{\omega} \times \mathbf{n} = \mathbf{V}_0 - \mathbf{v}_0 + \frac{\mu g a^2 t}{k^2} \hat{\mathbf{v}}.$$

Note also that (5.33) implies that $s = \boldsymbol{\omega} \cdot \mathbf{n}$ is a constant of the motion. Combining these results, we obtain

$$\boldsymbol{\omega} = \boldsymbol{\omega}_0 + \frac{\mu g a t}{k^2} \mathbf{n} \times \hat{\mathbf{v}}, \tag{5.41a}$$

where

$$\boldsymbol{\omega}_0 = \mathbf{n} s + a^{-1} \mathbf{n} \times (\mathbf{V}_0 - \mathbf{v}_0) \tag{5.41b}$$

is the initial angular velocity. After slipping ceases, the ball rolls with a constant angular velocity

$$\boldsymbol{\omega} = \boldsymbol{\omega}_0 + \frac{a \mathbf{n} \times \mathbf{v}_0}{k^2 + a^2}. \tag{5.42}$$

Rolling on a rotating surface

Consider a sphere rolling on a surface which is rotating with a constant angular velocity $\boldsymbol{\Omega}$. Then, if the origin is located on the rotation axis, the center of mass position vector $\mathbf{X}$ with respect to the rotating surface is related to the position vector $\mathbf{X}'$ in the "rest system" by

$$\mathbf{X}' = U^{\dagger} \mathbf{X} U, \tag{5.43}$$

where

$$U = e^{(1/2)i\boldsymbol{\Omega}t} \tag{5.44}$$

The kinematical variables in the two systems are therefore related by

$$\mathbf{V}' = U^\dagger(\mathbf{V} + \boldsymbol{\Omega} \times \mathbf{X})U$$

$$\dot{\mathbf{V}}' = U^\dagger(\dot{\mathbf{V}} + \boldsymbol{\Omega} \times \mathbf{V} + \boldsymbol{\Omega} \times (\boldsymbol{\Omega} \times \mathbf{X}))U$$

$$\omega' = U^\dagger \omega U$$

$$\dot{\omega}' = U^\dagger(\dot{\omega} + \boldsymbol{\Omega} \times \omega)U.$$

Therefore, the equations of motion in the rotating system are

$$\dot{\mathbf{V}} + \boldsymbol{\Omega} \times \mathbf{V} + \boldsymbol{\Omega} \times (\boldsymbol{\Omega} \times \mathbf{X}) = \mathbf{g} + m^{-1}\mathbf{f}, \tag{5.45}$$

$$mk^2(\dot{\omega} + \boldsymbol{\Omega} \times \omega) = -a\mathbf{n} \times \mathbf{f}. \tag{5.46}$$

The "pseudoforces" and "pseudotorque" due to the rotation are explicitly shown. Note that the "apparent gravitational force" $m\mathbf{g}$ is a rotating vector related to the *constant* gravitational force $m\mathbf{g}'$ in the rest system by

$$\mathbf{g} = U^\dagger \mathbf{g}' U. \tag{5.47}$$

The rolling constraint in the rotating system is, of course,

$$\mathbf{V} = a\omega \times \mathbf{n}. \tag{5.48}$$

By way of example, let us examine the rolling motion on a vertical plane rotating about a vertical axis (like an opening door). In this case $\boldsymbol{\Omega}\wedge\mathbf{g} = 0$, and $\mathbf{g} = \mathbf{g}'$ is a constant vector in the plane. Also, $\mathbf{n}\cdot\mathbf{g} = \mathbf{n}\cdot\boldsymbol{\Omega} = \mathbf{n}\cdot\mathbf{V} = 0$. It will be convenient to decompose $\mathbf{X}$ into a vertical component

$$\mathbf{X}_\parallel = \mathbf{X}\cdot\hat{\mathbf{g}}\hat{\mathbf{g}} \tag{5.49a}$$

and a horizontal component

$$\mathbf{X}_\perp = \hat{\mathbf{g}}\hat{\mathbf{g}}\wedge\mathbf{X}. \tag{5.49b}$$

Then $\boldsymbol{\Omega} \times (\boldsymbol{\Omega} \times \mathbf{X}) = -\Omega^2\mathbf{X}_\perp$, and when we eliminate the constraining force $\mathbf{f}$ between (5.45) and (5.46) we get

$$k^2(\dot{\omega} + \boldsymbol{\Omega} \times \omega) = -\mathbf{n} \times (\dot{\mathbf{V}} - \Omega^2\mathbf{X}_\perp - \mathbf{g}).$$

We use this to eliminate $\dot{\omega}$ from

$$\dot{\mathbf{V}} = a\dot{\omega} \times \mathbf{n}$$

to get

$$\left(\frac{k^2 + a^2}{a^2}\right)\dot{\mathbf{V}} = \frac{a^2}{h^2}\, a\mathbf{n}\cdot\omega\, \boldsymbol{\Omega} + \Omega^2\mathbf{X}_\perp + \mathbf{g}. \tag{5.50}$$

We can determine the time dependence of $\mathbf{n}\cdot\omega$ by using (5.46) and (5.48) to get

$$\mathbf{n}\cdot\dot{\omega} = -\mathbf{n}\cdot(\boldsymbol{\Omega} \times \omega) = -\boldsymbol{\Omega}\cdot(\omega \times \mathbf{n}) = \frac{-\boldsymbol{\Omega}\cdot\mathbf{V}}{a}.$$

Integrating this, we obtain

$$a\mathbf{n}\cdot\omega = -\boldsymbol{\Omega}\cdot(\mathbf{X} - \mathbf{X}_0) + a\mathbf{n}\cdot\omega_0. \tag{5.51}$$

Finally, by substituting this into (5.50) we get a determinate equation for the trajectory

$$\left(\frac{k^2 + a^2}{a^2}\right)\dot{\mathbf{V}} = \Omega^2(\mathbf{X}_\perp - \frac{k^2}{a^2}\mathbf{X}_\parallel) + \mathbf{g} + \frac{h^2}{a^2}(a\mathbf{n}\cdot\omega_0 + \boldsymbol{\Omega}\cdot\mathbf{X}_0)\boldsymbol{\Omega}. \tag{5.52}$$

This separates easily into uncoupled equations for horizontal and vertical displacements. The equations tell us that the ball recedes radially from the rotation axis with steadily increasing speed while it oscillates vertically with simple harmonic motion of period

$$T = \frac{2\pi}{\Omega}\left(\frac{k^2}{k^2 + a^2}\right)^{1/2}. \tag{5.53}$$

7-5. Exercises

(5.1) A homogeneous sphere of radius a rolls on the outer surface of a sphere of radius b. If it begins from rest at the highest point, at what point will the sphere lose contact? For what values of the coefficient of friction will slipping begin before contact is lost?

(5.2) A sphere rolls on the inner surface of a right circular cone at rest with a vertical axis. Compare its translational motion with that of a heavy particle constrained to move on the same surface. Show that the vertical component of the angular velocity is a constant of the motion.

(5.3) For a sphere rolling in a spherical bowl as described in the text, show that the condition for steady motion in a horizontal circle of radius r with constant angular speed Ω is

$$(b - a)^2 s^2 \geqslant 35 rg\Omega \cot\theta,$$

where $\mathbf{g}\cdot\mathbf{n} = -g\cos\theta$.

(5.4) Determine the orbit of a homogeneous sphere rolling on a horizontal turntable. Show that there are circular orbits with period completely determined by the period of the turntable. (K. Weltner, *Am. J. Phys.* **47**, 984 (1979)). Compare the motion with that of a charged particle moving in a magnetic field (J. Burns, *Am. J. Phys.* **49**, 56 (1981)).

(5.5) For a sphere rolling on an inclined plane which rotates with a constant angular velocity $\boldsymbol{\Omega}$, show that, with a proper choice of origin, its translational equation of motion in the rotating frame can be put in the form

$$\left(\frac{b^2}{a^2}+1\right)\ddot{\mathbf{r}}+\boldsymbol{\Omega}_\perp\times\dot{\mathbf{r}}=f(\mathbf{r})+\mathbf{g}_\parallel(t),$$

where $\boldsymbol{\Omega}_\perp$ is the component of $\boldsymbol{\Omega}$ perpendicular to the plane, and f is a linear vector function. Determine f, and discuss qualitative characteristics of the motion.

(5.6) For a sphere rolling on the inner surface of a cylinder rotating about its vertical axis with a constant angular velocity, show that the vertical components of the motion is simple harmonic and determine its period.

(5.7) Analyze the motion of a homogeneous sphere rolling on a horizontal plane subject to a central force specified by Hooke's law.

(5.8) Study the scattering of a sphere rolling on a $1/\varrho$ surface of revolution (C. Anderson and H. von Baeger, *Am. J. Phys.* **38**, 140 (1970)).

7-6. Impulsive Motion

An *impulsive force* $\mathbf{F}$ is very large during a short time interval $\Delta t = t - t_0$ and negligible outside that interval. The effect of an impulsive force on a particle is to produce a sudden change in velocity given, from $\mathbf{F} = m\dot{\mathbf{v}}$, by

$$m(\mathbf{v}-\mathbf{v}_0)=\mathbf{J}, \tag{6.1}$$

where

$$\mathbf{J}=\int_{t_0}^{t}\mathbf{F}\,\mathrm{d}t \tag{6.2}$$

is called the *impulse* of the force. In saying that the impulsive force is "large", we mean that during Δt the change in velocity is significant and the effect of other forces is negligible. As a rule, the time interval Δt can be taken to be so short that the change in velocity given by (6.1) can be regarded as instantaneous.

For any system of particles to which a system of impulsive forces $\mathbf{F}_i$ is applied during Δt, we can neglect all other forces and write

$$m\dot{\mathbf{V}}=\sum_i m_i\dot{\mathbf{v}}_i=\sum_i\mathbf{F}_i$$

during Δt. Whence the impulsive change in center of mass velocity $\mathbf{V}$ is given by

$$m(\mathbf{V}-\mathbf{V}_0)=\sum_i\mathbf{J}_i, \tag{6.3}$$

where $\mathbf{J}_i = m_i(\mathbf{v}_i - \mathbf{v}_{i0})$ is the impulse of $\mathbf{F}_i$ on the ith particle.

Similarly, during Δt the total angular momentum $\mathbf{l} = \mathcal{I}\boldsymbol{\omega}$ of the system of particles satisfies

$$\dot{\mathbf{l}} = \sum_i \mathbf{r}_i \times \mathbf{F}_i .$$

The $\mathbf{r}_i$ can be regarded as fixed during Δt. Therefore, the impulsive change in rotational velocity $\boldsymbol{\omega}$ is given by

$$\mathcal{I}(\boldsymbol{\omega} - \boldsymbol{\omega}_0) = \sum_i \mathbf{r}_i \times \mathbf{J}_i , \tag{6.4}$$

where the linearity of the inertia tensor has been used.

Though the impulsive forces produce only an infinitesimal displacement during Δt, they nevertheless do a finite amount of work. This produces a finite change in the kinetic energy which can be expressed in the form

$$K - K_0 = \tfrac{1}{2}\sum_i m_i(\mathbf{v}_i^2 - \mathbf{v}_{i0}^2) = \tfrac{1}{2}\sum_i m_i(\mathbf{v}_i + \mathbf{v}_{i0})\cdot(\mathbf{v}_i - \mathbf{v}_{i0})$$

or in the form

$$K - K_0 = \tfrac{1}{2} m(\mathbf{V} + \mathbf{V}_0)\cdot(\mathbf{V} - \mathbf{V}_0) + (\boldsymbol{\omega} + \boldsymbol{\omega}_0)\cdot\mathcal{I}(\boldsymbol{\omega} - \boldsymbol{\omega}_0) .$$

Therefore, by (6.3) and (6.4), *the impulsive change in kinetic energy K is given by*

$$K - K_0 = \tfrac{1}{2} \sum_i (\mathbf{v}_i + \mathbf{v}_{i0})\cdot\mathbf{J}_i \tag{6.5}$$

or

$$K - K_0 = \tfrac{1}{2}[(\mathbf{V} + \mathbf{V}_0)\cdot(\sum_i \mathbf{J}_i) + (\boldsymbol{\omega} + \boldsymbol{\omega}_0)\cdot(\sum_i \mathbf{r}_i \times \mathbf{J}_i)] . \tag{6.6}$$

The impulse equations (6.3) and (6.4) apply, of course, to a rigid body, and they suffice to determine the effect of given impulses on the body. Details of the impulse forces during Δt are unnecessary. In fact, such data are rarely available for actual impacts. The real circumstances where (6.3) and (6.4) apply are fairly limited. At the least, it is necessary that Δt be large compared to times for elastic waves to travel through the body, but small compared to the period of oscillation of the body as a physical pendulum.

From now on, we limit our considerations to the case of a single impulse $\mathbf{J}$ delivered to a rigid body at a point with position $\mathbf{r}$ in the center of mass system. Then the Equations (6.3) and (6.4) for translational and rotational impulse reduce to

$$m\Delta\mathbf{V} = \mathbf{J} , \tag{6.7}$$

$$\mathcal{I}(\Delta\boldsymbol{\omega}) = \mathbf{r} \times \mathbf{J} . \tag{6.8}$$

In addition, from kinematics we have

$$\mathbf{v} = \mathbf{V} + \boldsymbol{\omega} \times \mathbf{r} \tag{6.9}$$

for the velocity $\mathbf{v}$ of the point at which the impulse is applied. These three equations are linear in the seven vectors $\mathbf{V}$, $\mathbf{V}_0$, $\boldsymbol{\omega}$, $\boldsymbol{\omega}_0$, $\mathbf{J}$, $\mathbf{r}$, $\mathbf{v}$, so they can readily be solved for any three of the vectors in terms of the other four. Usually the initial values $\mathbf{V}_0$, $\boldsymbol{\omega}_0$ and $\mathbf{r}$ are given and the final values $\mathbf{V}$, $\boldsymbol{\omega}$ are to be determined. So we have two cases of particular interest, when either $\mathbf{J}$ or $\mathbf{v}$ is prescribed.

Motion initiated by an impulse

If a rigid body at rest is set in motion by a blow, then

$$m\mathbf{V} = \mathbf{J}, \tag{6.10}$$

Thus the center of mass moves in the direction of the blow, and for a prescribed impulse $\mathbf{J}$ the rotational velocity is given by

$$\boldsymbol{\omega} = \mathcal{I}^{-1}(\mathbf{r} \times \mathbf{J}). \tag{6.11}$$

Inserting this in (6.9) we find

$$\mathbf{v} = \mathbf{V} - m\mathbf{r} \times \mathcal{I}^{-1}(\mathbf{r} \times \mathbf{V}). \tag{6.12}$$

This tells us that the particle which receives the blow does not generally move in the direction of the blow.

Using (6.6) we find that the energy imparted to the object by the blow can be expressed in the form

$$K = \tfrac{1}{2}mV^2[1 + m(\mathbf{r} \times \hat{\mathbf{J}})\cdot\mathcal{I}^{-1}(\mathbf{r} \times \hat{\mathbf{J}})]. \tag{6.13}$$

This tells us how the energy imparted varies with the direction of the blow.

As an example, let us examine the effect of an impulse from a cue stick on a cue ball in billiards. Suppose that the cue stick is stroked in a horizontal direction in a vertical plane through the center of the cue ball. Then (6.10) and (6.11) give us the scalar relations

$$mV = J, \tag{6.14a}$$

$$I\omega = Jh, \tag{6.14b}$$

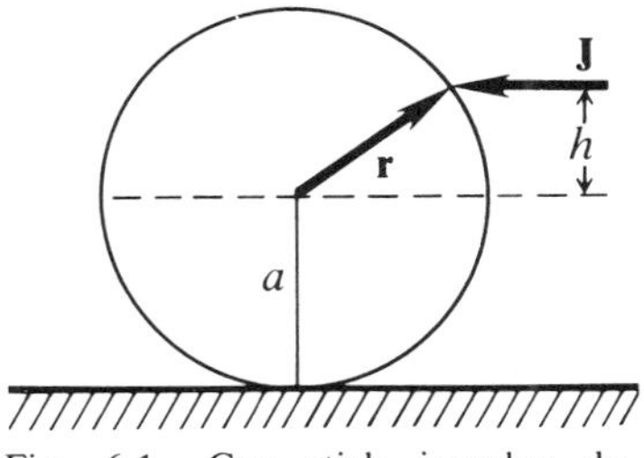

Fig. 6.1. Cue stick impulse delivered to a cue ball.

where $I = (2/5)ma^2$, and h is the height of the contact point above the center of the cue ball (Figure 6.1). If the cue ball is to roll immediately without slipping, then the rolling condition $V = \omega a$ must be satisfied and (6.14a, b) imply that

$$h = \tfrac{2}{5}a. \tag{6.15}$$

The cue ball will slip as it moves if it is "high struck" ($h > \frac{2}{5}a$) or "low struck" ($h < \frac{2}{5}a$). The ensuing motion is determined by results in Section 7-5 with (6.14a, b) as initial conditions. From (5.38) we find that the cue ball will slip for a time

$$\tau = \frac{2v_0}{7\mu g} = \frac{2}{7\mu g}\left|\frac{hJa}{I} - \frac{J}{m}\right| = \left|\frac{5h-2a}{7a}\right|\frac{J}{mg\mu} \tag{6.16}$$

after which, according to (5.39), it will have a speed $V = \mu g\tau$.

Billiard balls are quite smooth, so little angular momentum is transferred when they collide. Moreover, the collisions are nearly elastic. When the cue ball strikes an object ball "head on", conservation of momentum implies that all its velocity will be transferred to the object ball. However, if it is still slipping, the cue ball will then accelerate from rest and follow the object ball if it had been high struck or retreat back along its original path if it had been low struck. The first case is called a "follow shot", while the second is called a "draw shot".

The effect of striking the cue ball to the left or right of the median plane is to give it left or right English (= spin about the vertical axis) in addition to the rolling or slipping motions we have discussed. Ideally, English will be conserved during motion and collisions with the smooth balls, but not in collisions with the rough cushions on the billiard table.

The mechanics of billiards was developed by Coriolis (1835), and it is used in the design of equipment for the game. For example, the cushions on a billiard table are designed to make contact with a billiard ball at a height $h = (2/5)a$ above the ball's center, so that collision with the cushion does not impart to the ball any spin about a horizontal axis.

Constraint on the point of contact

An impulse may be known indirectly from its effect on the velocity of the particle to which it is applied. In that case, we eliminate **J** from (6.7) and (6.8) to get

$$\mathcal{I}(\Delta\boldsymbol{\omega}) = m\mathbf{r} \times (\Delta\mathbf{V}). \tag{6.17}$$

Then, since the velocity **v** in (6.9) is known, the two equations (6.9) and (6.17) can be solved for the unknowns **V** and $\boldsymbol{\omega}$. Eliminating **V** between these equations to solve for $\boldsymbol{\omega}$, we get

$$\mathcal{I}(\Delta\boldsymbol{\omega}) = m[\mathbf{r} \times (\mathbf{v} - \mathbf{V}_0) + \mathbf{r}\cdot\boldsymbol{\omega}\mathbf{r} - r^2\boldsymbol{\omega}]. \tag{6.18}$$

Note that (6.17) implies that

$$\mathbf{r}\cdot\mathcal{I}(\Delta\boldsymbol{\omega}) = \Delta\boldsymbol{\omega}\cdot(\mathcal{I}\mathbf{r}) = 0, \tag{6.19}$$

that is, the "radial component" of the angular momentum is conserved through the impact.

To complete our solution for $\boldsymbol{\omega}$, we need a specific form for the inertia tensor $\mathcal{I}$. Let us consider the important special case where $\mathcal{I}\boldsymbol{\omega} = I\boldsymbol{\omega}$ and $\mathcal{I}\boldsymbol{\omega}_0 = I\boldsymbol{\omega}_0$, and write $I = mk^2$. Then (6.19) implies

$$\mathbf{r}\cdot\boldsymbol{\omega} = \mathbf{r}\cdot\boldsymbol{\omega}_0\,, \tag{6.20}$$

and (6.18) yields

$$\boldsymbol{\omega} = \frac{k^2\boldsymbol{\omega}_0 + \mathbf{r}\times(\mathbf{v}-\mathbf{V}_0) + \mathbf{r}\mathbf{r}\cdot\boldsymbol{\omega}_0}{k^2+r^2}\,. \tag{6.21}$$

Finally, we can get **V** by substituting this into (6.9), and the impulse **J** which produces this result can be found from (6.7).

In the special case where the impulse brings the point of impact to rest, we have $\mathbf{v} = 0$. Consequently, the motion immediately after impact is a rotation about the point of impact. See Exercise (6.5) for an example.

Properties of impulsive forces

When a ball is bounced vertically off a fixed horizontal floor, it is found empirically to lose a fixed fraction of its translational kinetic energy in the bounce, irrespective of its initial velocity over a wide range. Thus, the kinetic energies before and after collision with the floor are related by

$$\tfrac{1}{2}mV^2 = e^2\left(\tfrac{1}{2}mV_0^2\right), \tag{6.22}$$

where e is a constant in the range $0 \leqslant e \leqslant 1$. When $e = 1$ energy is conserved, and the collision is said to be *elastic*. If $e = 0$ the collision is said to be *completely inelastic*. The constant e characterizes elastic properties of the objects in collision. It is called the *coefficient of restitution*, because it characterizes the fact that during collision the forces of compression deforming the ball are greater than the forces of restitution restoring its original shape. We can see this by decomposing the impulse **J** delivered by the ball to the wall into two parts,

$$\mathbf{J} = \mathbf{J}_C + \mathbf{J}_R\,. \tag{6.23a}$$

By definition, the compressive impulse $\mathbf{J}_C$ brings the ball to rest, so

$$-m\mathbf{V}_0 = \mathbf{J}_C\,. \tag{6.23b}$$

Then, in accordance with (6.7), the restitution impulse $\mathbf{J}_R$ propels the ball from rest to its final velocity;

$$m\mathbf{V} = \mathbf{J}_R\,. \tag{6.23c}$$

The relation between the forces of compression and restitution can be described by

$$\mathbf{J}_R = e\mathbf{J}_C\,. \tag{6.24}$$

It then follows immediately from (6.23b, c) that

$$\mathbf{V} = -e\mathbf{V}_0\,. \tag{6.25}$$

And this implies (6.22) as anticipated.

The value of this analysis lies in recognizing that the coefficient of restitution characterizes only the normal component of the impulse at the point of contact. The tangential component of the impulse derives from the frictional force, so it vanishes if the surfaces are ideally smooth. Experiments have shown that the empirical "laws of friction" are the same for impulsive forces as for the smaller forces between objects in continuous contact. Therefore, the empirical force law (5.30) implies a relation between the normal and tangential components of the impulse in a collision with slipping between surfaces. See Exercise (6.4) for an example.

To see how the coefficient of restitution is used to characterize a collision between moving bodies, let us consider a collision between two balls. Let the balls have masses m and M and center of mass velocities $\mathbf{V}$ and $\mathbf{U}$ respectively. According to (6.7), the impulse $\mathbf{J}$ applied by the second body on the first produces a change of velocity to

$$\mathbf{V} = \mathbf{V}_0 + \frac{1}{m}\mathbf{J}. \tag{6.26a}$$

By Newton's third law the impulse of the first on the second must be $-\mathbf{J}$. Therefore,

$$\mathbf{U} = \mathbf{U}_0 - \frac{1}{M}\mathbf{J}, \tag{6.26b}$$

and the impulsive change in the *relative velocity* of the spheres is given by

$$\mathbf{V} - \mathbf{U} = \mathbf{V}_0 - \mathbf{U}_0 + \frac{m + M}{mM}\mathbf{J}. \tag{6.27}$$

For application to the present problem, the relation expressed by (6.25) must be put in the more general form

$$\mathbf{n}\cdot(\mathbf{V} - \mathbf{U}) = -e\mathbf{n}\cdot(\mathbf{V}_0 - \mathbf{U}_0), \tag{6.28}$$

where $\mathbf{n}$ is a unit normal to the balls at the point of contact. Then from (6.27) we find that the normal component of the impulse has the value

$$\mathbf{n}\cdot\mathbf{J} = \frac{mM(1 + e)}{m + M}\mathbf{n}\cdot(\mathbf{V}_0 - \mathbf{U}_0). \tag{6.29}$$

If the balls are perfectly smooth, then this gives the entire impulse $\mathbf{J} = \mathbf{J}\cdot\mathbf{n}\mathbf{n}$, and velocities after impact are completely determined from (6.26a, b). In the limit $M \to \infty$, (6.29) gives the impulse for collision with a moving wall.

To see the impulsive effects of friction, let us consider a ball bouncing off a fixed, plane surface. Suppose that the coefficient of restitution is unity, and suppose that friction is sufficient to eliminate slipping during contact. A commercially produced ball that comes close to meeting these requirements of perfect elasticity and roughness is called a *Super-Ball*. From (6.28) we obtain immediately

$$\mathbf{n}\cdot\mathbf{V} = -\mathbf{n}\cdot\mathbf{V}_0, \tag{6.30}$$

where $\mathbf{n}$ is the unit normal to the plane (Figure 6.2). Thus, perfect elasticity implies that the normal components of the velocity is simply reversed by a bounce. Since there is no slipping during contact, the frictional force will not do work, so the total energy, will be conserved in a bounce.

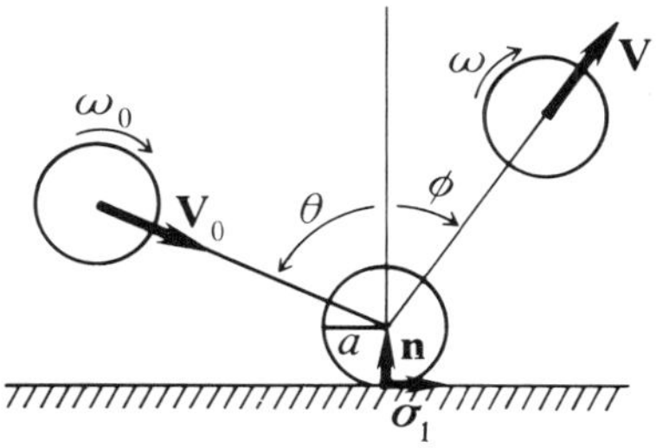

Fig. 6.2. A ball bouncing off a fixed surface.

For ball of radius a and moment of inertia $I = m\alpha a^2$, Equation (6.8) for the angular momentum impulse gives us

$$\alpha a \Delta\boldsymbol{\omega} = -\mathbf{n} \times \Delta\mathbf{V} \tag{6.31}$$

Multiplying this by $\mathbf{n}$, we note that the scalar part gives

$$\mathbf{n}\cdot\boldsymbol{\omega} = \mathbf{n}\cdot\boldsymbol{\omega}_0, \tag{6.32}$$

while the bivector part can be put in the form

$$\Delta\mathbf{V}_{\|} = \alpha\Delta\mathbf{u}, \tag{6.33}$$

where

$$\mathbf{V}_{\|} = \mathbf{n}(\mathbf{n}\wedge\mathbf{V}) = -\mathbf{n} \times (\mathbf{n} \times \mathbf{V})$$

is the tangential component of $\mathbf{V}$, and

$$\mathbf{u} \equiv \boldsymbol{\omega} \times (-a\mathbf{n}) \tag{6.34}$$

is the velocity of the point of contact with respect to the center of the mass.

Equation (6.33) describes the conversion of linear momentum $m\Delta\mathbf{V}_{\|}$ into angular momentum due to the action of the frictional force, while (6.32) tells us that the normal component of the angular momentum is conserved. Energy conservation puts an additional restriction on the kinematical variables.

To put energy conservation in its most useful form, we decompose the kinetic energy into normal and tangential parts by writing

$$K = \tfrac{1}{2}m[\mathbf{V}_{\|}^2 + (\mathbf{n}\cdot\mathbf{V})^2] + \tfrac{1}{2}m\alpha a^2[(\boldsymbol{\omega} \times \mathbf{n})^2 + (\boldsymbol{\omega}\cdot\mathbf{n})^2].$$

Since the normal components $(\mathbf{n}\cdot\mathbf{V})^2$ and $(\mathbf{n}\cdot\boldsymbol{\omega})^2$ are separately conserved in the collision, energy conservation reduces to a relation among the tangential components, which can be written

$$\mathbf{V}_{\|}^2 + \alpha\mathbf{u}^2 = \mathbf{V}_{0\|}^2 + \alpha\mathbf{u}_0^2\,. \tag{6.35}$$

Thus, we have reduced the description of frictional effects in a bounce to two equations (6.33) and (6.35). This is all we can learn from general dynamical principles. However, there is another property of the frictional force which we need to determine the direction of the tangential impulse.

During the bounce, the frictional force is opposite in direction to the velocity

$$\mathbf{v} = \mathbf{V}_{\|} + \boldsymbol{\omega} \times (-a\mathbf{n}) \tag{6.36}$$

of the ball at the point of contact. Therefore, if the initial angular velocity is orthogonal to the plane of incidence, as expressed by the equation

$$\boldsymbol{\omega}_0 \cdot (\mathbf{n} \wedge \mathbf{V}_0) = \boldsymbol{\omega}_0 \cdot \mathbf{n} \mathbf{V}_0 - \boldsymbol{\omega}_0 \cdot \mathbf{V}_0 \mathbf{n} = 0, \tag{6.37}$$

then the frictional force will lie within the plane. Consequently, the velocity impulse $\Delta\mathbf{V}$ and the trajectory after the bounce will lie in the incident plane. On the other hand, if $\boldsymbol{\omega}_0 \cdot \mathbf{V}_0 \neq 0$, then the ball will have a velocity component normal to the incident plane after bouncing, that is, the ball will bounce sideways.

Let us restrict our analysis to the case where the initial condition (6.37) is satisfied, as presumed in Figure 6.2. Then, if $\boldsymbol{\sigma}_1$ is a unit vector as indicated in the figure, we can write $\mathbf{V}_{\|} = V_{\|}\boldsymbol{\sigma}_1$ and $\mathbf{u} = u\boldsymbol{\sigma}_1$, so (6.33) reduces to the scalar equation

$$V_{\|} - V_{0\|} = \alpha(u - u_0). \tag{6.38}$$

Also (6.34) reduces to the scalar relation

$$u = \omega a. \tag{6.39}$$

Now the energy equation (6.35) can be put in the form

$$V_{\|}^2 - V_{0\|}^2 = -\alpha(u^2 - u_0^2)$$

which, with (6.38), can be reduced to the simpler condition

$$V_{\|} + V_{0\|} = -(u + u_0),$$

or

$$V_{\|} + u = -(V_{0\|} + u_0). \tag{6.40}$$

According to (6.36), this says that *the tangential velocity* $\mathbf{v}$ *of the contact point is exactly reversed by a bounce.*

Solving (6.38) and (6.40) for the final state variables, we get

$$a\omega = \left(\frac{\alpha - 1}{\alpha + 1}\right) a\omega_0 - \frac{2}{\alpha + 1} V_{0\|} \tag{6.41a}$$

$$V = \frac{-2\alpha a\omega_0}{\alpha + 1} - \left(\frac{\alpha - 1}{\alpha + 1}\right) V_{0\|}. \tag{6.41b}$$

For a Super-Ball $\alpha = 2/5$, and these equations become

$$\omega = -\frac{3}{7}\omega_0 - \frac{10}{7}\frac{V_{0\|}}{a} \tag{6.42a}$$

$$V_{\|} = -\frac{4}{7}a\omega_0 + \frac{3}{7}V_{0\|}. \tag{6.42b}$$

As a particular example, let $\omega_0 = 0$. Then recalling (6.30), we find that the angles of incidence and rebound (Figure 6.2) are related by

$$\tan\phi = \frac{3}{7}\tan\theta$$

and the final spin is

$$\omega = -\frac{10}{7}\frac{V_0}{a}\sin\theta.$$

On multiple bounces, the Super-Ball exhibits some surprising behavior (Exercise (6.7)).

7-6. Exercises

(6.1) Under what conditions can the motion of a free rigid body be arrested by a single impulsive force?

(6.2) An impulse **J** is applied at one end of a uniform bar of mass m and length $2a$ in a direction perpendicular to the bar. Find the velocity imparted to the other end of the bar if the bar is (a) free, or (b) fixed at the center of mass.

(6.3) A flat circular disk is held at its center and struck a blow on its edge in a direction perpendicular to the radius and inclined at 45° to the plane of the disk. About what axis will it begin to rotate? Describe its subsequent motion. How would the motion be altered if the disk were tossed into the air before being struck?

(6.4) A thin hoop of mass m and radius a slides on a frictionless horizontal table with its axis normal to the table and collides with a flat, rough, vertical wall. Initially, the hoop is not spinning and it is incident on the wall with speed V_0 at an angle of $\pi/4$. After momentarily sliding during contact (μ = coefficient of kinetic friction) the hoop rebounds. Assuming that the coefficient of restitution is unity, determine the angle of reflection ϕ and the angular velocity ω after collision.

(6.5) A hoop of mass m and radius a rolls on a horizontal floor with velocity V_0 towards an inelastic step of height $h(< \frac{1}{2}a)$, the plane of the hoop being vertical and perpendicular to the edge of the step. (Figure 6.3)

(a) Show that the angular velocity of the hoop just after colliding with the step is

$$\omega = \frac{V_0}{a}\left(1 - \frac{h}{2a}\right).$$

(b) Find the minimum initial velocity required for the hoop to mount the step if it does not slip.

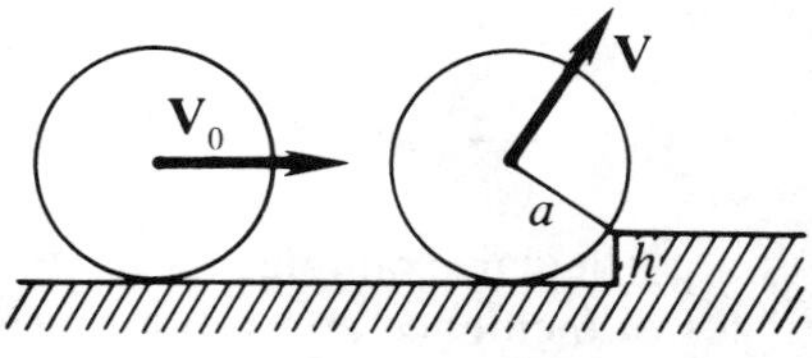

Fig. 6.3. Rolling hoop colliding with a step.

(c) Find the maximum initial velocity for which the hoop can mount the step without losing contact.

(6.6) A homogeneous solid cube is spinning freely about one of its long diagonals when suddenly an edge with one end on the rotation axis is held fixed. Show that the kinetic energy is reduced to one twelvth the original value.

(6.7) Consider a Super-Ball bouncing between two parallel planes, such as floor and the underside of a table. Show that with $\omega_0 = 0$, after three bounces

$$\omega_3 = -\frac{130}{343}\frac{V_{0\|}}{a},$$

$$V_3 = -\frac{333}{343}V_{0\|},$$

showing that the motion is almost exactly reversed as in (Figure 6.4a). What moment of inertia should a Super-Ball have if it is to return precisely along its original path, as in Figure 6.4b (R. Garwin, *Am. J. Physics* **37**, 88–92 (1969)).

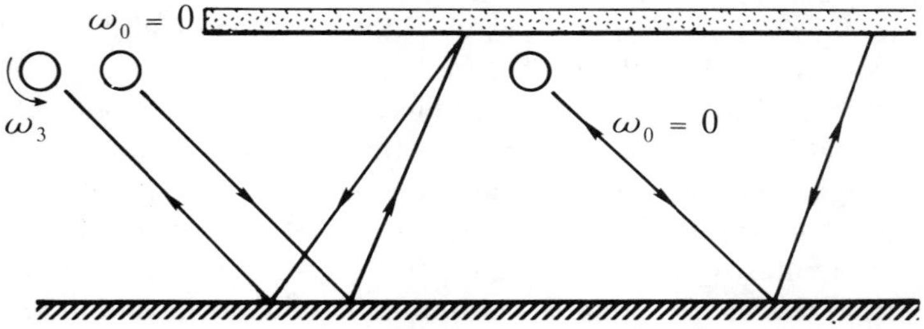

Fig. 6.4. A Super Ball thrown without spin will follow the path indicated in (a), bouncing from the floor to the underside of a table and back to the floor. The tangent of the angle of bounce is 3% greater than that of the angle of incidence. For comparison, the trajectory of a body which returns precisely along its original path is shown in (b).

(6.8) A ball strikes a plane surface at an angle of 45° and rebounds at an angle of 45°. Show that the coefficient of friction μ must have the value

$$\mu = \frac{1 - e}{1 + e} ,$$

where e is the coefficient of restitution.

Chapter 8

Celestial Mechanics

Celestial Mechanics is the crowning glory of Newtonian mechanics. It has revolutionized man's concept of the Cosmos and his place within it. Its spectacular successes in the 18th and 19th centuries established the unique power of mathematical theory for precise explanation and prediction. In the 20th century it has been overshadowed by exciting developments in other branches of physics. But the last three decades have seen a resurgence of interest in celestial mechanics, because it is a basic conceptual tool for the emerging *Space Age*.

The main concern of celestial mechanics (CM) is to account for the motion of celestial bodies (stars, planets, satellites, etc.). The same theory applies to the motion of artificial satellites and spacecraft, so the emerging science of space flight, *astromechanics*, can be regarded as an offspring of celestial mechanics. Space Age capabilities for precise measurements and management of vast amounts of data has made CM more relevant than ever. Celestial mechanics is used by observational astronomers for the prediction and explanation of occultation and eclipse phenomena, by astrophysicists to model the evolution of binary star systems, by cosmogonists to reconstruct the history of the Solar System, and by geophysicists to refine models of the Earth and explain geological data about the past. To cite one specific example, it has recently been established that major Ice Ages on Earth during the last million years have occurred regularly with a period of 100,000 years, and this can be explained with celestial mechanics as forced by oscillations in the Earth's eccentricity due to perturbations by other planets. Moreover, periodicities of minor Ice Ages can be explained as forced by precession and nutation of the Earth's axis due to perturbation by the Sun and Moon.

We have already covered a good bit of celestial mechanics in preceeding chapters — the one and two body Kepler problems in Chapter 4, and the Newtonian three body problem in Section 6-5. This chapter is concerned mainly with perturbation theory. The standard formulation of perturbation theory in general use and presented in recent texts is more than a hundred years old. It has the drawback of appearing unnecessarily complicated and difficult to interpret. This chapter presents new formulation of perturbation theory which exploits the advantages of geometric algebra, developing it to the point where calculations can be carried out efficiently. Its effectiveness is demonstrated by first order calculations for the principle perturbations in the Solar System. We stop just short of calculating the periodicities of the Ice Ages, which is a second order effect.

8-1. Gravitational Forces, Fields and Torques

The Newtonian theory of gravitation is based on Newton's law of gravitational attraction between two material particles, which can be put in the form,

$$\mathbf{f}_1 = -Gmm_1 \frac{\mathbf{x} - \mathbf{x}_1}{|\mathbf{x} - \mathbf{x}_1|^3}, \tag{1.1}$$

where G is the universal constant of gravitation, with the empirical value

$$G = 6.6732 \times 10^{-11} \frac{\mathrm{N{\cdot}m^2}}{\mathrm{kg^2}}. \tag{1.2}$$

The force law (1.1) specifies the force on a particle of mass m at $\mathbf{x}$ due to a particle of mass m_1 at $\mathbf{x}_1$ in an inertial system. As we have noted before, the potential energy of the 2-particle gravitational interaction is

$$V_1 = \frac{-Gmm_1}{|\mathbf{x} - \mathbf{x}_1|}, \tag{1.3}$$

and this determines the gravitational force by differentiation;

$$\mathbf{f}_1 = -\nabla V_1 . \tag{1.4}$$

It will be convenient to introduce an alternative formulation of gravitational interactions in terms of gravitational fields.

We define the gravitational field $\mathbf{g}_1(\mathbf{x}, t)$ of a single particle located at $\mathbf{x}_1 = \mathbf{x}_1(t)$ by

$$\mathbf{g}_1(\mathbf{x}, t) = -Gm_1 \frac{\mathbf{x} - \mathbf{x}_1(t)}{|\mathbf{x} - \mathbf{x}_1(t)|^3}. \tag{1.5}$$

The particle at $\mathbf{x}_1 = \mathbf{x}_1(t)$ is called the *source* of the field and the mass m_1 is the *source strength*. The field $\mathbf{g}_1$ is a function which assigns a definite vector $\mathbf{g}_1(\mathbf{x}, t)$ to every spatial point $\mathbf{x}$. The field at $\mathbf{x}$ may change with time t due only to motion of its source.

If a particle of mass m is placed "in the gravitation field" $\mathbf{g}_1$ at the point $\mathbf{x}$, we say that *the field exerts a force*

$$\mathbf{f}_1 = \mathbf{f}_1(\mathbf{x}, t) = m\mathbf{g}_1(\mathbf{x}, t). \tag{1.6}$$

Although this is mathematically identical to Newton's force law (1.1), the field concept provides a new view on the nature of physical reality which has evolved into a new branch of physics called *classical field theory*. From the Newtonian point of view, particles exert forces directly on one another in accordance with the force law (1.1), though they may be separated by large distances. From the viewpoint of field theory, however, particles interact indirectly through the intermediary of a field. Each material particle is the source of a gravitational field which, in turn, acts on other particles with a force depending on their masses, as specified by (1.6). The gravitational field is regarded as a real physical entity pervading all space surrounding its source and acting on any matter that happens to be present.

The development of gravitational field theory leads ultimately to the conclusion that the expression (1.5) for a Newtonian gravitational field must be modified, and *Einstein's Theory of General Relativity* proposes modifications which have been confirmed experimentally with increasing precision during the last two decades. Einstein's theory therefore sets definite limits on the validity of Newtonian theory, but it also tells us that the corrections to Newtonian theory are utterly negligible in most physical situations. So it will be worth our while to study the implications of Newtonian gravitation theory without getting involved in the deeper subtleties of field theory.

The concept of a gravitational field has a formal mathematical advantage even within the context of Newtonian theory. It enables us to separate gravitational interactions into two parts which can be analyzed separately, namely, (a) the production of gravitational fields by extended sources, and (b) the effect of a given gravitational field on given bodies. We study the production of fields first.

The one-particle Newtonian field (1.5), can be derived from the *gravitational potential*

$$\phi_1(\mathbf{x}, t) = \frac{-Gm_1}{|\mathbf{x} - \mathbf{x}_1(t)|}, \tag{1.7}$$

by differentiation; thus,

$$\mathbf{g}_1(\mathbf{x}, t) = -\nabla\phi_1(\mathbf{x}, t) \tag{1.8}$$

where $\nabla = \nabla_{\mathbf{x}}$ is the derivative (or gradient) with respect to $\mathbf{x}$. The gravitational potential energy (1.3) of a particle with mass m at $\mathbf{x}$ is given by $V_1(\mathbf{x}, t) = m\phi_1(\mathbf{x}, t)$. However, it is essential to distinguish clearly between the concepts of "potential" and "potential energy". The latter is shared energy of two interacting objects, while the former is characteristic of a single object, its source.

The gravitational field $\mathbf{g}(\mathbf{x}, t)$ of an N-particle system is given by the superposition of fields:

$$\mathbf{g}(\mathbf{x}, t) = \sum_{k=1}^{N} \mathbf{g}_k(\mathbf{x}, t) = -G\sum_{k=1}^{N} m_k \frac{\mathbf{x} - \mathbf{x}_k(t)}{|\mathbf{x} - \mathbf{x}_k(t)|^3}. \tag{1.9}$$

On a particle of mass m at $\mathbf{x}$, this field exerts a force

$$\mathbf{f} = m\mathbf{g} = \sum_k m\mathbf{g}_k = \sum_k \mathbf{f}_k\,, \tag{1.10}$$

as required by the superposition law for forces. This field can also be derived from a potential; thus,

$$\mathbf{g}(\mathbf{x}, t) = -\nabla\phi(\mathbf{x}, t)\,, \tag{1.11}$$

where

$$\phi(\mathbf{x}, t) = \sum_k \phi_k(\mathbf{x}, t) = -G\sum_k \frac{m_k}{|\mathbf{x} - \mathbf{x}_k(t)|}\,. \tag{1.12}$$

And the potential energy of a particle in the field is given by

$$V(\mathbf{x}, t) = m\phi(\mathbf{x}, t)\,. \tag{1.13}$$

Note that this does not include the potential energy of interaction between the particles producing the field. The internal energy of the N-particle system can be ignored as long as we are concerned only with the influence of the system on external objects.

The Gravitational Field of an Extended Object

The gravitational field of a continuous body is obtained from the field of a system of particles by the same limiting process used in Section 7–2 to define the center of mass and inertia tensor for continuous bodies. Thus, we subdivide the body into small parts which can be regarded as particulate, and in the limit of an infinitely small subdivision the sum (1.9) becomes the integral

$$\mathbf{g}(\mathbf{x}, t) = -G\int dm'\,\frac{(\mathbf{x} - \mathbf{x}')}{|\mathbf{x} - \mathbf{x}'|^3}\,, \tag{1.14}$$

where $dm' = dm(\mathbf{x}', t)$ is the mass of an "infinitesimal" corpuscle (small body) at the point $\mathbf{x}'$ at time t. Similarly, the limit of (1.12) gives us the gravitational potential of a continuous body:

$$\phi(\mathbf{x}, t) = -G\int \frac{dm'}{|\mathbf{x} - \mathbf{x}'|}\,. \tag{1.15}$$

Hereafter, we will not indicate the time dependence explicitly. The relation $\mathbf{g} = -\nabla\phi$ still applies here. This enables us to find $\mathbf{g}$ by differentiation after evaluating the integral for ϕ in (1.15).

For a spherically symmetric body the integral is easy to evaluate. We chose the origin at the body's center of mass and indicate this by writing $\mathbf{r}$ and $\mathbf{r}'$ instead of $\mathbf{x}$ and $\mathbf{x}'$ (Figure 1.1). The symmetry is expressed by writing the mass density as a function of radial distance only. Thus,

$$dm' = \varrho(r')r'^2 \, dr' \, d\Omega,$$

where $d\Omega = \sin\theta \, d\theta \, d\phi$ is the "element of solid angle", and

$$\phi(\mathbf{r}) = -G \int \frac{dm'}{|\mathbf{r} - \mathbf{r}'|} = -G \int_0^R \varrho' r'^2 \, dr' \oint \frac{d\Omega}{|\mathbf{r} - \mathbf{r}'|}$$

For $r = |\mathbf{r}| > r' = |\mathbf{r}'|$, we can easily evaluate the integral

$$\oint \frac{d\Omega}{|\mathbf{r} - \mathbf{r}'|} = 2\pi \int_0^\pi \frac{\sin\theta \, d\theta}{[r^2 + r'^2 - 2rr' \cos\theta]^{1/2}} = \frac{4\pi}{r} \tag{1.16}$$

And the remaining integral simply gives the total mass of the body

$$M = \int dm' = 4\pi \int_0^R \varrho(r')r'^2 \, dr'.$$

Therefore, *the external gravitational potential of a spherically symmetric body* is given by

$$\phi(\mathbf{r}) = -G \int \frac{dm'}{|\mathbf{r} - \mathbf{r}'|} = -\frac{GM}{r}. \tag{1.17}$$

This is identical to the potential of a point particle with the same mass located at the mass center, so the gravitational field $\mathbf{g} = -\nabla\phi$ of the body is also the same as for a particle. Since many celestial bodies are nearly spherically symmetric, this is an excellent first approximation to their gravitational fields, indeed, a sufficient approximation in many circumstances.

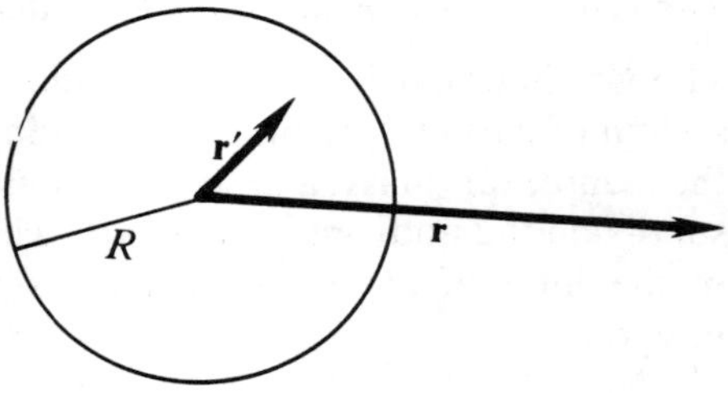

Fig. 1.1. Points inside the sphere are denoted by the primed variable; external points are denoted by the unprimed variable.

A more accurate description of gravitational fields is best achieved by evaluating the effects of deviations from spherical symmetry. We expand the potential of a given body in a Taylor series about its center of mass. Since $r > r'$, we have

$$\frac{1}{|\mathbf{r} - \mathbf{r}'|} = \frac{1}{r}\left\{1 + \frac{P_1(\mathbf{r}\,\mathbf{r}')}{r^2} + \frac{P_2(\mathbf{r}\,\mathbf{r}')}{r^4} + \ldots\right\}$$

$$= \frac{1}{r}\left\{1 + \sum_{n=1}^{\infty} \left(\frac{r'}{r}\right)^n P_n(\hat{\mathbf{r}}\cdot\hat{\mathbf{r}}')\right\}, \tag{1.18}$$

where the P_n are the Legendre polynomials (see Exercise (8.4) in Section 2-8). We will need explicit expressions only for

$$P_1(\mathbf{r}\mathbf{r}') = \mathbf{r}\cdot\mathbf{r}',$$
$$P_2(\mathbf{r}\mathbf{r}') = \tfrac{1}{2}[3(\mathbf{r}\cdot\mathbf{r}')^2 - r^2r'^2],$$
$$P_3(\mathbf{r}\mathbf{r}') = \tfrac{1}{2}[5(\mathbf{r}\cdot\mathbf{r}')^3 - 3r'^2r^2\mathbf{r}\cdot\mathbf{r}'] . \tag{1.19}$$

The last line of (1.18) shows that the magnitude of the nth term in the expansion is of the order $(r'/r)^n$, so the series converges rapidly at a distance r which is large compared to the dimensions of the body. The expansion (1.18) gives us an expansion of the potential,

$$\phi(\mathbf{r}) = -\frac{G}{r}\left\{M + \frac{1}{r^2}\int P_1(\mathbf{r}\mathbf{r}')\,\mathrm{d}m' + \frac{1}{r^4}\int P_2(\mathbf{r}\mathbf{r}')\,\mathrm{d}m' + \ldots\right\}.$$

By (1.19),

$$\int P_1(\mathbf{r}\mathbf{r}')\,\mathrm{d}m' = \mathbf{r}\cdot\left[\int \mathbf{r}'\,\mathrm{d}m'\right] = \mathbf{r}\cdot[0] = 0,$$

since the center of mass is located at the origin. Recall (from Section 7-2), that the inertia tensor $\mathscr{I}$ of the body is defined by

$$\mathscr{I}\mathbf{r} = \int \mathrm{d}m'\,\mathbf{r}'\,\mathbf{r}'\wedge\mathbf{r} = \int \mathrm{d}m'(r'^2\mathbf{r} - \mathbf{r}'\,\mathbf{r}'\cdot\mathbf{r}), \tag{1.20a}$$

and the trace of the inertia tensor is given by

$$\mathrm{Tr}\,\mathscr{I} = 2\int \mathrm{d}m' r'^2 = I_1 + I_2 + I_3, \tag{1.20b}$$

where I_1, I_2, I_3 are principal moments of inertia. Therefore,

$$\int \mathrm{d}m'\,P_2(\mathbf{r}\mathbf{r}') = \int \mathrm{d}m'\,\tfrac{1}{2}[3(\mathbf{r}\cdot\mathbf{r}')^2 - r^2r'^2]$$
$$= \tfrac{1}{2}[r^2\,\mathrm{Tr}\,\mathscr{I} - 3\mathbf{r}\cdot\mathscr{I}\mathbf{r}] = \tfrac{1}{2}\,\mathbf{r}\cdot\mathcal{Q}\mathbf{r}, \tag{1.21}$$

which defines a symmetric tensor

$$\mathcal{Q}\mathbf{r} = \mathbf{r}\,\mathrm{Tr}\,\mathscr{I} - 3\,\mathscr{I}\mathbf{r}. \tag{1.22}$$

Adopting well-established terminology from electromagnetic theory, we may refer to $\mathcal{Q}$ as the gravitational quadrupole tensor.

Now the expanded potential can be written

$$\phi(\mathbf{r}) = -\frac{G}{r}\left\{M + \frac{1}{2}\,\frac{\mathbf{r}\cdot\mathcal{Q}\mathbf{r}}{r^4} + \ldots\right\}. \tag{1.23}$$

This is called a *harmonic* (or *multipole*) *expansion* of the potential. The quadrupole term describes the first order deviation from the field of a spherically symmetric body. From this the gravitational field $\mathbf{g} = -\nabla\phi$ can be obtained with the help of

$$\nabla(\tfrac{1}{2}\,\mathbf{r}\cdot\mathcal{Q}\mathbf{r}) = \mathcal{Q}\mathbf{r},$$

$$\nabla r^n = n r^{n-1}\hat{\mathbf{r}}.$$

Thus,

$$\mathbf{g}(\mathbf{r}) = -\frac{G}{r^2}\left\{ M\hat{\mathbf{r}} - \frac{1}{r^2}\,(\mathcal{Q}\hat{\mathbf{r}} - \tfrac{5}{2}(\hat{\mathbf{r}}\cdot\mathcal{Q}\hat{\mathbf{r}})\hat{\mathbf{r}}) + \ldots \right\}. \tag{1.24}$$

This expression for the gravitational field holds for a body of arbitrary shape and density distribution.

From Section 7-2 we know that the inertia tensor for an axisymmetric body can be put in the form

$$\mathcal{I}\mathbf{r} = I_1\mathbf{r} + (I_3 - I_1)\mathbf{r}\cdot\mathbf{u}\mathbf{u}, \tag{1.25}$$

where $I_1 = I_2$ is the "equatorial moment of inertia", I_3 is "polar moment of inertia", and $\mathbf{u} = \hat{\mathbf{u}}$ is the direction of the symmetry axis. Then (1.22) and (1.20b) gives us

$$\mathcal{Q}\mathbf{r} = (I_3 - I_1)(\mathbf{r} - 3\mathbf{r}\cdot\mathbf{u}\mathbf{u}). \tag{1.26}$$

From (1.24), therefore, the gravitational field for an axisymmetric body is given by

$$\mathbf{g}(\mathbf{r}) = -\frac{MG}{r^2}\left\{ \hat{\mathbf{r}} + \tfrac{3}{2}J_2\left(\frac{R}{r}\right)^2 [(1 - 5(\hat{\mathbf{r}}\cdot\mathbf{u})^2)\hat{\mathbf{r}} + 2\mathbf{u}\cdot\hat{\mathbf{r}}\mathbf{u}] + \ldots \right\} \tag{1.27}$$

where R is the equatorial radius of the body and J_2 is defined by

$$J_2 = \frac{I_3 - I_1}{MR^2}. \tag{1.28}$$

The constant J_2 is a dimensionless measure of the oblateness of the body, and the factor $(R/r)^2$ in (1.27) measures the rate at which the oblateness effect falls of with distance.

For an axisymmetric body the effect of harmonics higher than the quadrupole are not difficult to find, because the series (1.18) integrates to a harmonic expansion for the potential with the form

$$\phi(\mathbf{r}) = -\frac{MG}{r}\left\{ 1 - \sum_{n=2}^{\infty} J_n\left(\frac{R}{r}\right)^n P_n(\hat{\mathbf{r}}\cdot\mathbf{u}) \right\}, \tag{1.29}$$

where the J_n are constant coefficients. As mentioned above, J_2 is a measure of oblateness and is related to the moments of inertia by (1.28). The constant J_3 measures the degree to which the body is "pearshaped" (i.e. southern hemisphere fatter than northern hemisphere). The advantage of (1.29) is that it can be immediately written down once axial symmetry has been assumed, and the J_n can be determined empirically, in particular, from data on orbiting satellites. For the Earth,

$$J_2 = 1.083 \times 10^{-3}$$
$$J_3 = -2.5 \times 10^{-6}$$
$$J_4 = -1.6 \times 10^{-6}$$
$$J_5 = -0.2 \times 10^{-6}. \tag{1.30}$$

Clearly, the quadrupole harmonic strongly dominates. Although for $n > 2$ the J_n are of the same order of magnitude, the contributions of the harmonics decrease with n because of the factor $(R/r)^n$ in (1.29). Since the J_n are independent of radius, comparison of the J_n for different planets is a meaningful quantitative way to compare shapes of planets.

Gravitational Force and Torque on an Extended Object

The total force exerted by a gravitational field on a system of particles is

$$\mathbf{f}(t) = \sum_k m_k \mathbf{g}(\mathbf{x}_k(t), t), \tag{1.31}$$

where $\mathbf{x}_k(t)$ is the position of the kth particle at time t. In the limit for a continuous body this becomes

$$\mathbf{f} = \int \mathrm{d}m(\mathbf{x})\,\mathbf{g}(\mathbf{x}), \tag{1.32}$$

where the time dependence has been suppressed.

For the force on an extended body due to a particle of mass M at the origin, (1.32) gives us

$$\mathbf{f} = -GM \int \mathrm{d}m\, \frac{\mathbf{x}}{x^3} = -GM \int \mathrm{d}m'\, \frac{\mathbf{r} + \mathbf{r}'}{|\mathbf{r} + \mathbf{r}'|^3}, \tag{1.33}$$

where $\mathbf{r}$ is the center of mass of the body and the variable of integration has been changed from $\mathbf{x}$ to $\mathbf{r}'$ (Figure 1.2). In accordance with Newton's third law, the expression (1.33) for the force of a particle on a body differs only in sign from the expression (1.14) gives for the force of a body on a particle. Consequently, the result of approximating the right side of (1.33) by expanding the denominator can be written down at once from previous approximation of the gravitational field. From (1.24) with (1.22) we get

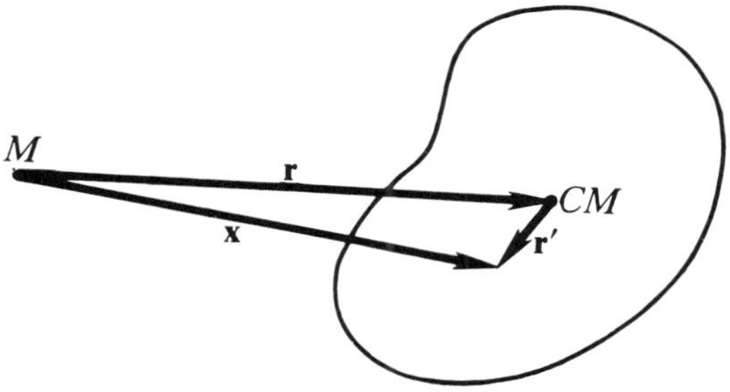

Fig. 1.2. Integration variable for the gravitational force of a point particle on an extended body.

$$\mathbf{f} = -\frac{GM}{r^3}\left\{ m\mathbf{r} + \tfrac{3}{2}(\mathrm{Tr}\,\mathcal{J} - 5\hat{\mathbf{r}}\cdot\mathcal{J}\hat{\mathbf{r}})\,\frac{\mathbf{r}}{r^2} + \frac{3}{r^2}\,\mathcal{J}\mathbf{r} \right\} \tag{1.34}$$

To second order, this is the gravitational force of a particle (or a spherically symmetric body) of mass M on an extended body with mass m and inertia tensor $\mathcal{J}$. This is quite a good approximation for many purposes in celestial mechanics.

The gravitational torque on a body (with center of mass at $\mathbf{r}$ as base point) is given by

$$\boldsymbol{\Gamma} = \int \mathrm{d}m'\,\mathbf{r}' \times \mathbf{g} = -\mathbf{r} \times \mathbf{f} + \int \mathrm{d}m\,\mathbf{x} \times \mathbf{g}, \tag{1.35}$$

where $\mathbf{r}' = \mathbf{x} - \mathbf{r}$, as in Figure 1.2. If the field is produced by a particle at the origin, then $\mathbf{x} \times \mathbf{g} = 0$. Therefore substitution of (1.34) into (1.35) gives us

$$\boldsymbol{\Gamma} = \frac{3GM}{r^5}\,\mathbf{r} \times \mathcal{J}\mathbf{r}. \tag{1.36}$$

This is a useful expression for the torque on a satellite. Note that it vanishes identically if the body is spherically symmetric.

Tidal Forces

In the preceding subsection we examined the gravitational force and torque on a body as a whole. Besides these effects, a nonuniform gravitational field produces internal stresses in a body called *tidal forces*. To have a specific example in mind, let us consider tidal forces on the Earth due to the Moon.

We aim to determine the tidal forces on the surface of a spherical Earth (Figure 1.3). For a particle of unit mass at rest at a point $\mathbf{R}$ on the surface of the Earth, we have the equation of motion

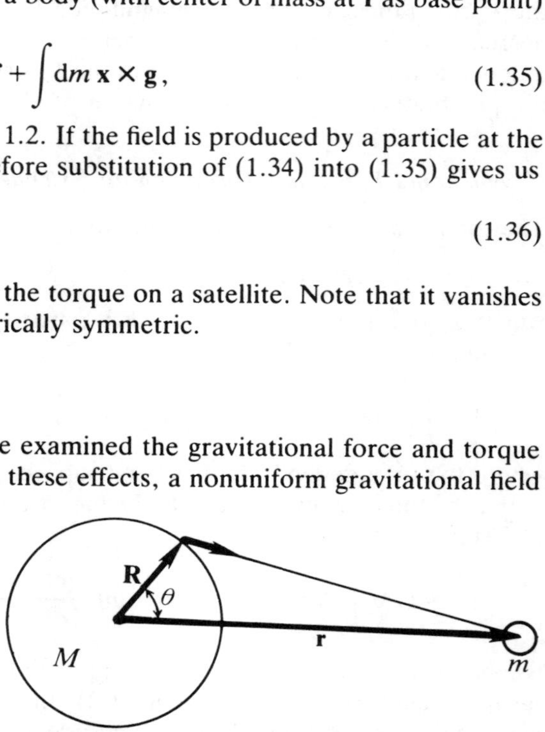

Fig. 1.3. Geocentric coordinates for the Earth and Moon.

$$-GM\frac{\mathbf{R}}{R^3} - Gm\frac{\mathbf{R} - \mathbf{r}}{|\mathbf{R} - \mathbf{r}|^3} + \mathbf{f} = \mathbf{a}. \tag{1.37}$$

The first two terms are the gravitational attractions of the Earth and Moon respectively, and $\mathbf{f}$ is a force of constraint due to the rigidity of the Earth. The term $\mathbf{a}$ on the right is the acceleration of the noninertial system in which the particle is at rest. Considering only the two body motion of the Earth and Moon about their common center of mass, we have

$$\mathbf{a} = Gm\frac{\mathbf{r}}{r^3}\,. \tag{1.38}$$

To this we could add the centripetal acceleration due to the rotation of the Earth, but we have already studied its effect in Section 5.6, and it is completely independent of tidal effects. The effect of the Earth's rotation on the tides is due entirely to the rotation of the position vector **R**, as we shall see.

Inserting (1.38) into (1.37), we find that the constraining force required to keep the particle at rest is given by

$$\mathbf{f} = GM\frac{\mathbf{R}}{R^3} + Gm\left[\frac{\mathbf{R}-\mathbf{r}}{|\mathbf{R}-\mathbf{r}|^3} + \frac{\mathbf{r}}{r^3}\right]. \tag{1.39}$$

The negative of the last term is the tidal force. Since the radius of the Earth $R = |\mathbf{R}|$ is far less than the Earth-Moon distance $r = |\mathbf{r}|$, we can employ the binomial expansion

$$\frac{1}{|\mathbf{R}-\mathbf{r}|^3} = \frac{1}{(r^2 - 2\mathbf{r}\cdot\mathbf{R} + R^2)^{3/2}} = \frac{1}{r^3}\left(1 + 3\frac{\mathbf{r}\cdot\mathbf{R}}{r^2} + \ldots\right).$$

So, as a first order approximation for the tidal force $\mathbf{g}_t = \mathbf{g}_t(\mathbf{R})$, we have

$$\mathbf{g}_t(\mathbf{R}) = -Gm\left[\frac{\mathbf{R}-\mathbf{r}}{|R-r|^3} + \frac{\mathbf{r}}{r^3}\right] \approx \frac{Gm}{r^3}(3\mathbf{R}\cdot\hat{\mathbf{r}}\hat{\mathbf{r}} - \mathbf{R})\,. \tag{1.40}$$

This expression applies to every point on the surface of the Earth; the distribution of forces is shown in Figure 1.4. Note that the Earth is under tension along the Earth-Moon axis and under compression perpendicular to the axis. The symmetry of the tidal field may be surprising, in particular, the fact that the tidal force at the point closest to the Moon has the same magnitude as at the furthest point. This symmetry disappears in higher order approximations, but they are negligible in the present case.

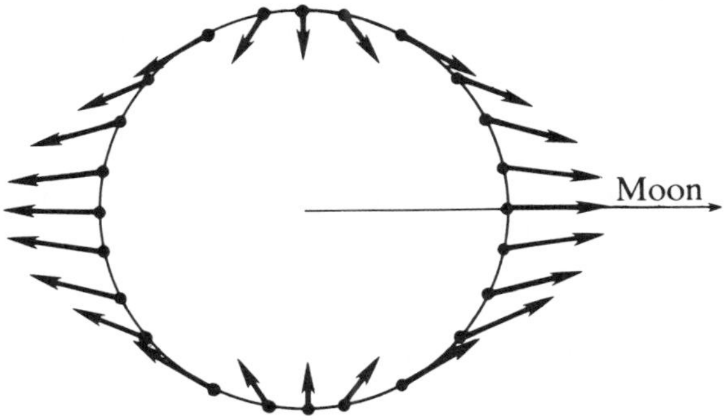

Fig.1.4. The tidal force field on the surface of the Earth.

In the Earth-Moon orbital plane, the tidal force (1.40) has a component tangent to the Earth's surface except at the four points $\theta = 0, \pi/2, \pi, 3\pi/2$. Consequently, water on the surface of the Earth piles up at these points into *two high* and *two low tides*. Since the Earth's axis is nearly perpendicular to the Earth-Moon orbital plane, the tidal bulges are swept over the surface of the Earth by the Earth's rotation, producing the *semi-diurnal tides*, that is, alternating high and low tides every twelve hours. A tidal bulge at one point on the Earth rotates "out from under the Moon" before it collapses, since frictional forces retard its collapse as well as its build-up. Consequently, an observer on the Earth will see a time lag between the appearance of the Moon overhead and the maximum tide (Figure 1.5).

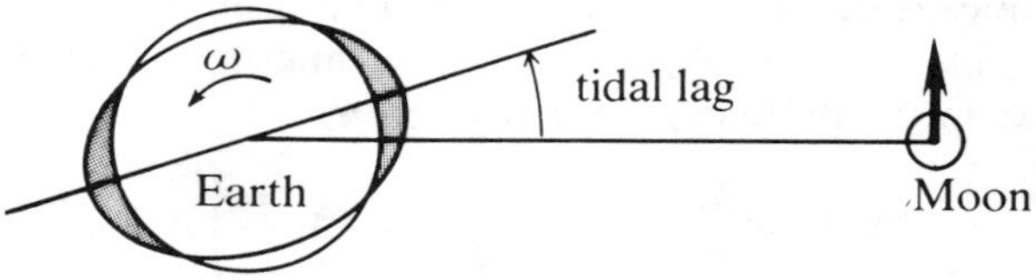

Fig. 1.5. Tidal lag.

Tidal friction dissipates the Earth's rotational energy, thus reducing the Earth's angular velocity and gradually increasing the length of a day. At the same time the tidal force reacts on the Moon to accelerate it. This increases the Moon's orbital energy, so it recedes from the Earth, and its orbital period gradually increases. Though energy is dissipated, the overall angular momentum of the Earth-Moon system is conserved under the action of tidal forces. So tidal forces drive a transfer of spin angular momentum to orbital angular momentum. This process will continue until the length of the day equals the length of the month, and the vanishing of tidal friction prevents further angular momentum exchange. However, solar tides will continue to slow down the Earth's rotation, so the Moon will begin to approach the Earth again with an orbital period locked in synchrony with the Earth's rotation.

Observational evidence leads to a value of 4.4 cm yr^{-1} for the rate of tidal recession of the Moon. By angular momentum conservation, one can infer from this a rate of change in the length of day of about two milliseconds per century. Techniques for measuring such small time changes have been developed only recently. The rate of tidal energy dissipation corresponding to such a change is on the order of 10^2 W.

A closer look at the tidal phenomena raises many questions which are subjects for active geophysical research today. By precisely what mechanism is tidal energy dissipated, and how accurately can it be estimated from geophysical models? What is the relative effectiveness of surface tides and body tides in dissipating energy? Has the rate of tidal energy dissipation been uniform over geologically long times? Then how close to the Earth was the Moon in the distant past? What bearing does this have on theories of the Moon's formation?

Of course, the tidal mechanism is operating throughout the solar system. In general, satellites in subsynchronous orbits spiral in towards the parent planet, while satellites in suprasynchronous orbits spiral outward. The moons of Mars, Phobos and Deimos, fall, respectively, into these two categories.

Now let us examine a tidal effect which has been ignored in our discussion so far. The inclination of the Moon's position with respect to the equator varies between 18° and 29° each month. To describe its effect on the distri-

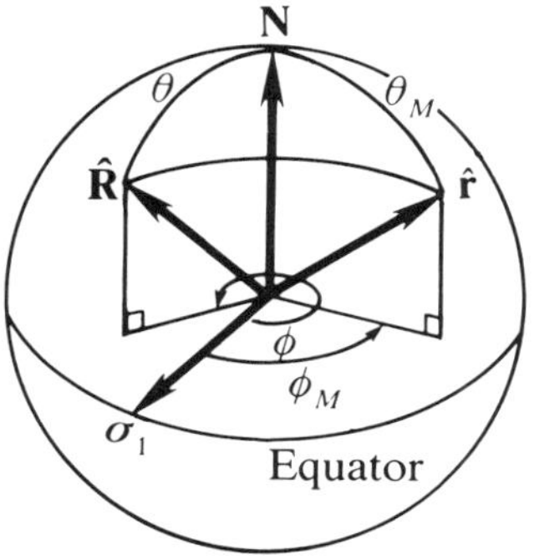

Fig. 1.6. Equitorial coordinates for the position of the Moon and a point on the Earth. **N** indicates north and $\boldsymbol{\sigma}_1$ designates the zero of longitude on the prime meridian.

bution and periodicity of the tides, we consider the radial component of the tidal force, which, by (1.40), has the value

$$\hat{\mathbf{R}}\cdot\mathbf{g}_t = \frac{GmR}{r^3}[3(\hat{\mathbf{R}}\cdot\hat{\mathbf{r}})^2 - 1]. \tag{1.41}$$

The tidal periodicity is completely determined by the term in brackets. We can analyze it by introducing equatorial coordinates as shown in Figure 1.6. Let θ be the colatitude and ϕ the longitude of a fixed point **R** on the Earth, and let θ_M and ϕ_M be the corresponding coordinates for the direction $\hat{\mathbf{r}}$ of the Moon. Applying the law of cosines from Appendix A to the spherical triangle in Figure 1.6, we obtain

$$\hat{\mathbf{R}}\cdot\hat{\mathbf{r}} = \cos\theta\cos\theta_M + \sin\theta\sin\theta_M\cos(\phi - \phi_M), \tag{1.42}$$

whence

$$\begin{aligned}[3(\hat{\mathbf{R}}\cdot\hat{\mathbf{r}})^2 - 1] &= \tfrac{3}{2}\sin^2\theta\sin^2\theta_M\cos 2(\phi - \phi_M) \\ &+ \tfrac{3}{2}\sin 2\theta\sin 2\theta_M\cos(\phi - \phi_M) \\ &+ \tfrac{1}{2}(3\cos^2\theta - 1)(3\cos^2\theta_M - 1).\end{aligned} \tag{1.43}$$

The three terms in this sum show three different periodic variations in the tidal force. With $\phi = \Omega t$, the first term shows the major effect of a semi-diurnal periodicity. The second term has diurnal periodicity, while the third oscillates twice a month due to the motion of the Moon.

The Sun produces tides on the Earth in the same way as the Moon. From (1.40), the relative magnitude of solar and lunar tidal forces is

$$\frac{M_M}{M_S}\left(\frac{r_S}{r_M}\right)^3 \approx 2.2. \tag{1.44}$$

The Sun and the Moon combine to produce high "spring" tides and low neap tides.

The Shape of the Earth

The shape of the Earth plays an important role in cartography and many geophysical phenomena, so we need a precise way to characterize it. A theory of the Earth's shape is developed by supposing that it originated from the cooling of a spinning molten mass to form a solid crust. Regarding the oscillating tides induced by other bodies as secondary effects to be considered separately, we model the Earth as a spinning fluid held together in steady state equilibrium by the gravitational field. In a geocentric frame spinning with the Earth the fluid will be at rest, with an effective gravitational potentional at its surface of the form

$$\Phi(\mathbf{r}) = \phi(\mathbf{r}) - \tfrac{1}{2}(\boldsymbol{\Omega} \times \mathbf{r})^2, \tag{1.45}$$

where $\phi(\mathbf{r})$ is the true gravitational potential and the last term is the centrifugal pseudopotential. The gravitational field

$$\mathbf{g} = -\nabla\Phi \tag{1.46}$$

must be normal to the surface, for if it had a tangential component that would force fluid to flow on the surface. That means the surface of the Earth is an equipotential surface defined by the equation

$$\Phi(\mathbf{r}) = K,$$

where K is a constant to be determined.

Since the spinning fluid is axisymmetrical, its gravitational potential ϕ can be described by the Legendre expansion (1.29), so, to second order, the shape of the Earth is described explicitly by the equation

$$\Phi(\mathbf{r}) = \frac{-GM}{r}\left\{1 - \tfrac{1}{2}J_2\left(\frac{a}{r}\right)^2[3(\hat{\mathbf{r}}\cdot\mathbf{u})^2 - 1]\right\} - \tfrac{1}{2}\Omega^2 r^2[1 - (\hat{\mathbf{r}}\cdot\mathbf{u})^2] = K, \tag{1.47}$$

where $\mathbf{u} = \hat{\boldsymbol{\Omega}}$ specifies the rotation axis and a is the *equatorial radius* of the Earth. The surface described by this equation is called the *geoid*. Its deviation from a sphere is characterized by a parameter ε called the *flattening* (or the *ellipticity*) and defined by

$$\varepsilon = \frac{a - c}{c}, \tag{1.48}$$

where c is the polar radius of the Earth. The constant K in (1.47) is evaluated by setting $r = c$ with $\mathbf{u}\cdot\hat{\mathbf{r}} = 1$, yielding

$$K = -\frac{GM}{c}\left\{1 - \frac{J_2 a^2}{c^2}\right\}. \tag{1.49}$$

The flattening ε can be expressed in terms of the other parameters by setting $r = a$ and $\hat{\mathbf{r}}\cdot\mathbf{u} = 0$ in (1.47). That gives

$$-\frac{GM}{a}\left\{1 + \tfrac{1}{2}J_2\right\} - \tfrac{1}{2}\Omega^2 a^2 = -\frac{GM}{a}\left\{1 - \frac{J_2 a^2}{c^2}\right\}(1 + \varepsilon).$$

Since ε and J_2 are known to be small quantities, it suffices to solve this equation to first order, and we get

$$\varepsilon = \tfrac{3}{2}J_2 + \tfrac{1}{2}\beta, \tag{1.50}$$

where

$$\beta \equiv \frac{\Omega^2 a^3}{GM} = \frac{\Omega^2 a}{GM/a^2} \tag{1.51}$$

is the ratio of centripetal force to gravitational force at the equator.

The expression for the *geopotential* can now be written in the form

$$\Phi(\mathbf{r}) = \frac{-GM}{r}\{1 + (\varepsilon - \tfrac{1}{2}\beta)\left(\frac{a}{r}\right)^3[\tfrac{1}{3} - (\hat{\mathbf{r}}\cdot\mathbf{u})^2] + \tfrac{1}{2}\beta\left(\frac{r}{a}\right)^3 [1 - (\hat{\mathbf{r}}\cdot\mathbf{u})^2]\}. \tag{1.52}$$

This can be used to determine the flattening from empirical data. For the gravitational acceleration at the pole g_p and at the equator g_e, it gives

$$g_p = \frac{GM}{a^2}(1 + \beta),$$

$$g_e = \frac{GM}{a^2}(1 + \varepsilon - \tfrac{3}{2}\beta). \tag{1.53}$$

Gravimetric measurements give the values

$$g_p = 983.217 \text{ cm/sec}^2,$$
$$g_e = 978.039 \text{ cm/sec}^2. \tag{1.54}$$

Using values for the a and Ω from Appendix C, from (1.53) we calculate

$$\varepsilon = 0.003376,$$
$$\beta = 0.003468. \tag{1.55}$$

As a check on the internal consistency of the theory, in (1.50) these numbers give a value for J_2 which agrees with the values (1.30) from satellite data to better than one percent.

The shape of the Earth as described by the geoid (1.47) agrees with measurements of sea level to within a few meters. However, radar ranging to measure the height of the ocean is accurate to a fraction of a meter. So geodocists are engaged in developing more refined models for the shape of the Earth. The main deviation from the geoid is an excessive bulge around the equator. This has been attributed to a retardation in the rotation of the Earth over past millions of years — one more clue among many to be fed into a conceptual reconstruction of the Earth's history.

8-1. Exercises

(1.1) For an axisymmetric body, a harmonic expansion of the gravitational field can be put in the form

$$\mathbf{g}(\mathbf{r}) = \frac{-MG}{r^3}\{\mathbf{r} + \sum_{n=2}^{\infty} \mathbf{g}_n(\mathbf{r})\}.$$

$\mathbf{g}_2(\mathbf{r})$ is given in Equation (1.27). From Equation (1.29), show that

$$\mathbf{g}_3(\mathbf{r}) = \tfrac{5}{2} J_3\left(\frac{R}{r}\right)^3 [(-7(\hat{\mathbf{r}}\cdot\mathbf{u})^3 + 3\hat{\mathbf{r}}\cdot\mathbf{u})\hat{\mathbf{r}} + (3(\hat{\mathbf{r}}\cdot\mathbf{u})^2 - \tfrac{3}{5})\mathbf{u}].$$

(1.2) Show that to second order the gravitational force of one extended body on another is given by

$$\mathbf{f} = \frac{-Gm_1m_2}{r^3}\mathbf{r} - m_1\mathbf{g}_2 - m_2\mathbf{g}_1,$$

where

$$\mathbf{g}_k = \frac{-G}{r^5}[3\,\mathcal{I}_k\mathbf{r} + \tfrac{3}{2}\,(\mathrm{Tr}\,\mathcal{I}_k - 5\hat{\mathbf{r}}\cdot\mathcal{I}_k\hat{\mathbf{r}})]$$

and $\mathcal{I}_k$ is the inertia tensor for body k.

(1.3) The variation Δh in the water level from low to high tides can be derived from Equation (1.40), assuming that the water is in equilibrium in the unperturbed field $-GMR/\mathbf{R}^3$. Show that variations at the equator are of the order

$$\Delta h = \tfrac{3}{2}\frac{m}{M}\frac{R^4}{r^3} \approx 0.5\ m.$$

This is comparable to the variations observed on small Pacific atolls which approximate open ocean conditions. How does Δh vary with latitude?

(1.4) The periodicity of the first term in Equation (1.43) is not quite diurnal, since the angle $2(\phi - \phi_M)$ depends on the motion of the Moon as well as the rotation of the Earth. Show that its actual periodicity is 12h and 26.5 min, so high tide is observed about 53 min later each day.

(1.5) Derive the Equations (1.53) and check the computation of (1.55).

(1.6) The geoid is nearly an oblate spheroid. That can be established by considering the equation for an oblate spheroid,

$$1 = \frac{(\mathbf{r}\cdot\mathbf{u})^2}{c^2} + \frac{(\mathbf{r}\times\mathbf{u})^2}{b^2} = \frac{r^2}{b^2}\left[1 - \left(\frac{b^2-c^2}{c^2}\right)(\hat{\mathbf{r}}\cdot\mathbf{u})^2\right].$$

For small $\varepsilon' = (b - c)/c$, this can be put in the approximate form

$$r \approx \frac{b}{[1 - 2\varepsilon'(\hat{\mathbf{r}}\cdot\mathbf{u})^2]^{1/2}} \approx b[1 + \varepsilon'(\hat{\mathbf{r}}\cdot\mathbf{u})^2].$$

Show that to first order in ε the Equation (1.52) for the geoid can be put in this form, and relate the parameters ε and a there to the parameters ε' and b here.

8-2. Perturbations of Kepler Motion

The Newtonian two-body problem is the only dynamical problem in celestial mechanics for which an exact general solution is known. We call the motion in that case *Kepler motion*. Approximate solutions to a large class of more difficult dynamical problems are best characterized as *perturbations* (or disturbances) of Kepler motion. To that end, we write the translational equation of motion for a celestial body (planet, satellite, spacecraft, etc.) in the form

$$\dot{\mathbf{v}} = -\frac{\mu\mathbf{r}}{r^2} + \mathbf{f}, \tag{2.1}$$

where $\mathbf{f}$ is referred to as the *perturbing force* (per unit mass). The perturbing force is said to be *small* if $|\mathbf{f}| << \mu/r^2$, that is, if it is much smaller in magnitude than the Newtonian force. If the primary body is large enough to be regarded at rest, then $\mu = MG$ in (2.1), where M is the mass of the primary. Otherwise, μ should include the two-body correction determined in Section 4-6. We can always insert the two-body correction at the end of our calculations if the degree of precision requires it.

Gravitational perturbation theory is concerned with general methods for solving Equation (2.1) for any specified perturbing force. Several methods have been widely employed for a long time. However, we shall develop here a new coordinate-free method exploiting the advantages of geometric algebra. A more sophisticated method will be developed in Section 8-4. Instead of attacking the equation directly, it is best to reformulate the problem by using our knowledge about Kepler motion to take the Newtonian force into account once and for all. Then we can analyze the effect of the perturbing force separately.

From our study of the Kepler problem in Section 4-3, we know that the instantaneous values of the position and velocity vectors $\mathbf{r}$ and $\mathbf{v}$ determine a unique *Kepler orbit*, which may be an ellipse, a hyperbola or a parabola. The orbit is completely characterized by an *angular momentum vector* (per unit mass)

$$\mathbf{h} = \mathbf{r} \times \mathbf{v} \tag{2.2}$$

and an *eccentricity vector* $\boldsymbol{\varepsilon}$ given by

$$\mu(\boldsymbol{\varepsilon} + \hat{\mathbf{r}}) = \mathbf{v} \times \mathbf{h} = i\mathbf{h}\mathbf{v}. \tag{2.3}$$

The vectors $\mathbf{h}$ and $\boldsymbol{\varepsilon}$ are called *orbital elements*. Although these two vectors characterize the orbit completely, we have seen that it is useful to classify orbits by the values of another orbital element, the energy (per unit mass) E, given by

$$E = \tfrac{1}{2}v^2 - \frac{\mu}{r} = \frac{\mu^2(\epsilon^2 - 1)}{2h^2} \tag{2.4}$$

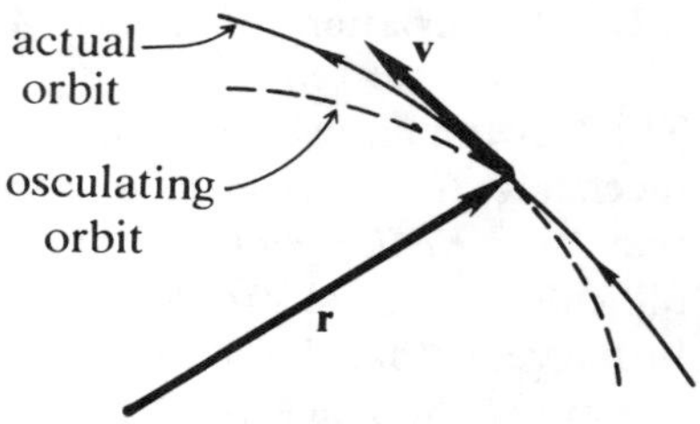

Fig. 2.1. The osculating (Kepler) orbit is instantaneously tangent to the true orbit at the point $\mathbf{r} = \mathbf{r}(t)$.

Unique values for the orbital elements $\mathbf{h}$ and $\boldsymbol{\varepsilon}$ are determined at each time t by (2.2) and (2.3), even in the presence of a perturbing force $\mathbf{f}$. They specify a Kepler orbit instantaneously tangent to the actual orbit at the point $\mathbf{r} = \mathbf{r}(t)$. This Kepler orbit is called the *osculating orbit* of the motion. Since $\mathbf{h}$ and $\boldsymbol{\varepsilon}$ are constants of the unperturbed motion, the osculating orbit will be identical with the actual orbit when $\mathbf{f} = 0$. When $\mathbf{f} \neq 0$, $\mathbf{h}$ and $\boldsymbol{\varepsilon}$ are no longer constant, so the osculating orbit must change continuously in time. We can picture the perturbed motion as motion of a particle on a Kepler orbit which is being continuously deformed by the perturbing force. This picture is of great value for understanding the effects of perturbations.

To describe the deformation of an osculating orbit analytically, we need equations of motion for $\boldsymbol{\varepsilon}$ and $\mathbf{h}$. These are easily found by differentiating (2.2) and (2.3) and eliminating derivatives of $\mathbf{v}$ and $\hat{\mathbf{r}}$ using (2.1) and the identity

$$\frac{\mathrm{d}\hat{\mathbf{r}}}{\mathrm{d}t} = \frac{i\hat{\mathbf{r}}\mathbf{h}}{r^2} = \frac{\mathbf{h} \times \mathbf{r}}{r^3}\,, \tag{2.5}$$

established in Section 4-3.

In this way, we obtain the coupled equations

$$\dot{\mathbf{h}} = \mathbf{r} \times \mathbf{f}, \tag{2.6}$$

$$\mu\dot{\boldsymbol{\varepsilon}} = i(\dot{\mathbf{h}}\mathbf{v} + \mathbf{h}\mathbf{f}) = \mathbf{v} \times \dot{\mathbf{h}} + \mathbf{f} \times \mathbf{h}. \tag{2.7}$$

Since $\mathbf{h}$ and $\boldsymbol{\varepsilon}$ are constants of the unperturbed motion, they will be slowly varying functions in the presence of a small perturbation, and approximate solutions to their equations of motion (2.5) and (2.6) will be easy to find. This is the main reason for considering perturbations of orbital elements rather than working directly with the Newtonian equation of motion (2.1). However, our formulation of perturbation theory can be further improved.

The main drawback of a perturbation theory for $\mathbf{h}$ and $\boldsymbol{\varepsilon}$ is the fact that these vectors are not independent of one another. Each vector alone is equivalent to three scalar elements, but together they are equivalent to only five because of the orthogonality condition $\mathbf{h}\cdot\boldsymbol{\varepsilon} = 0$. We can eliminate the redundancy due to this constraint by introducing a spinor R determined by the equations

$$\hat{\mathbf{h}} = R^{\dagger}\sigma_3 R = \mathbf{e}_3\,. \tag{2.8a}$$

$$\hat{\boldsymbol{\varepsilon}} = R^{\dagger}\sigma_1 R = \mathbf{e}_1\,. \tag{2.8b}$$

These equations determine a dextral frame

$$\mathbf{e}_k = R^\dagger \sigma_k R \qquad (k = 1, 2, 3), \tag{2.9}$$

which we call a *Kepler frame*, because it specifies the attitude in space of the osculating Kepler orbit. Given a fixed frame $\{\sigma_k\}$, the attitude of the osculating orbit is completely determined by the spinor R, so let us refer to R as the *attitude element*.

Instead of the redundant vector elements $\mathbf{h}$ and $\boldsymbol{\varepsilon}$, we can work with the independent elements R, $h = |\mathbf{h}|$ and $\varepsilon = |\boldsymbol{\varepsilon}|$. This choice has additional advantages of a direct geometrical meaning. While the attitude of the osculating orbit is described by the spinor R, its size and shape are described by h and ε. Actually, the orbit size is directly described by the Kepler energy element E defined by (2.4), while ε is the shape parameter. With (2.4) we can eliminate any one of the elements h, ε, E in favor of the other two, but it is best not to commit ourselves to a particular choice prematurely.

From our study of rotational kinematics, we know that the attitude element R obeys an equation of form

$$\dot{R} = \tfrac{1}{2} R i\omega, \tag{2.10}$$

so we need to determine ω from the perturbing force. Because of (2.8a) and (2.8b), the derivatives of $\mathbf{h}$ and $\boldsymbol{\varepsilon}$ can be put in the form

$$\dot{\mathbf{h}} = \boldsymbol{\omega} \times \mathbf{h} + \dot{h}\hat{\mathbf{h}}, \tag{2.11a}$$

$$\dot{\boldsymbol{\varepsilon}} = \boldsymbol{\omega} \times \boldsymbol{\varepsilon} + \dot{\varepsilon}\hat{\boldsymbol{\varepsilon}}. \tag{2.11b}$$

These equations can be solved for ω as follows: First eliminate the unwanted $\dot{h}$ and $\dot{\varepsilon}$ by the multiplications

$$\mathbf{h} \times \dot{\mathbf{h}} = \mathbf{h} \times (\boldsymbol{\omega} \times \mathbf{h}) = \mathbf{h}\cdot(\mathbf{h}\wedge\boldsymbol{\omega}) = h^2\boldsymbol{\omega} - \mathbf{h}\cdot\boldsymbol{\omega}\mathbf{h},$$

$$\boldsymbol{\varepsilon} \times \dot{\boldsymbol{\varepsilon}} = \boldsymbol{\varepsilon} \times (\boldsymbol{\omega} \times \boldsymbol{\varepsilon}) = \boldsymbol{\varepsilon}\cdot(\boldsymbol{\varepsilon}\wedge\boldsymbol{\omega}) = \varepsilon^2\boldsymbol{\omega} - \boldsymbol{\varepsilon}\cdot\boldsymbol{\omega}\boldsymbol{\varepsilon},$$

The first of these equations can be solved for ω if we get an independent relation for $\mathbf{h}\cdot\boldsymbol{\omega}$ from the second. Since $\boldsymbol{\varepsilon}\cdot\mathbf{h} = 0$, we get

$$(\boldsymbol{\varepsilon} \times \dot{\boldsymbol{\varepsilon}})\cdot\mathbf{h} = \varepsilon^2\boldsymbol{\omega}\cdot\mathbf{h}.$$

Therefore,

$$\boldsymbol{\omega} = \frac{\mathbf{h} \times \dot{\mathbf{h}}}{h^2} + \frac{(\boldsymbol{\varepsilon} \times \dot{\boldsymbol{\varepsilon}})\cdot\mathbf{h}}{\varepsilon^2 h^2}\mathbf{h}, \tag{2.12}$$

This is, of course, a general kinematical result giving the rotational velocity of the frame determined by *any two* time dependent orthogonal vectors.

To ascertain how ω depends on the perturbing force we insert the equation of motion for $\mathbf{h}$ and $\boldsymbol{\varepsilon}$ into the kinematic formula (2.12). Note that (2.6) give us

$$\mathbf{h} \times \dot{\mathbf{h}} = \mathbf{h} \times (\mathbf{r} \times \mathbf{f}) = \mathbf{h}\cdot\mathbf{f}\,\mathbf{r}, \tag{2.13}$$

since $\mathbf{h}\cdot\mathbf{r} = 0$. And (2.7) gives

$$\mu\boldsymbol{\varepsilon} \times \dot{\boldsymbol{\varepsilon}} = \boldsymbol{\varepsilon} \times (\mathbf{v} \times \dot{\mathbf{h}}) + \boldsymbol{\varepsilon} \times (\mathbf{f} \times \mathbf{h}) = \boldsymbol{\varepsilon}\cdot\dot{\mathbf{h}}\mathbf{v} - \boldsymbol{\varepsilon}\cdot\mathbf{v}\dot{\mathbf{h}} - \boldsymbol{\varepsilon}\cdot\mathbf{f}\mathbf{h}.$$

Hence,

$$\mu(\boldsymbol{\varepsilon} \times \dot{\boldsymbol{\varepsilon}})\cdot\mathbf{h} = -\boldsymbol{\varepsilon}\cdot\mathbf{v}(\mathbf{r} \times \mathbf{f})\cdot\mathbf{h} - h^2\boldsymbol{\varepsilon}\cdot\mathbf{f}, \tag{2.14}$$

Substituting (2.13) and (2.14) into (2.12), we get the desired expression

$$\boldsymbol{\omega} = \frac{\mathbf{h} \times (\mathbf{r} \times \mathbf{f})}{h^2} - \frac{1}{\mu\varepsilon^2 h^2}\,[\boldsymbol{\varepsilon}\cdot\mathbf{v}(\mathbf{r} \times \mathbf{f})\cdot\mathbf{h} + h^2\boldsymbol{\varepsilon}\cdot\mathbf{f}]\mathbf{h}. \tag{2.15}$$

If desired, we can eliminate **v** from this expression by using (2.3), which yields the relations

$$\boldsymbol{\varepsilon}\cdot\mathbf{v} = \frac{\mu}{h^2}\,\mathbf{h}\cdot(\hat{\mathbf{r}} \times \boldsymbol{\varepsilon}) = -\mathbf{v}\cdot\hat{\mathbf{r}} = -\dot{r}\,. \tag{2.16}$$

We can interpret (2.15) by regarding the entire osculating orbit as a rigid body slowly "spinning" in space with rotational velocity $\boldsymbol{\omega}$ and symmetry axis along **h**. The physical significance of the two terms on the right side of (2.15) can be identified at once. The first term $h^{-2}\,\mathbf{h} \times (\mathbf{r} \times \mathbf{f}) = h^{-2}\mathbf{h}\cdot\mathbf{f}\,\mathbf{r}$ describes an instantaneous rotation about the radius vector **r**. This can only tilt the orbital plane and the symmetry axis of the orbit, just as the axis of an axially symmetry spinning rigid body is tilted in precession and nutation. The last term describes an instantaneous rotation in the orbital plane about the symmetry axis along **h**, a motion called *pericenter* (perihelion or perigee) or *apse precession* by astronomers. Apse precession is most simply characterized mathematically as a change in direction of the eccentricity vector $\boldsymbol{\varepsilon}$ in the orbital plane. Note that the apse precession coefficient in (2.15) shows the contribution of angular momentum change in the term with $(\mathbf{r} \times \mathbf{f})\cdot\mathbf{h} = \dot{\mathbf{h}}\cdot\mathbf{h} = \dot{h}h$.

We get an equation of motion for the attitude of the osculating orbit by inserting the explicit expression (2.15) for ω into the spinor equation $\dot{R} = Ri\omega/2$. Also, we need independent equations for the orbital size and shape parameters. We can obtain such equations easily from (2.11a, b) by using $\mathbf{h}\cdot\dot{\mathbf{h}} = h\dot{h}$ and $\boldsymbol{\varepsilon}\cdot\dot{\boldsymbol{\varepsilon}} = \varepsilon\dot{\varepsilon}$. Thus, we obtain

$$h\dot{h} = (\mathbf{r} \times \mathbf{f})\cdot\mathbf{h} = \mathbf{f}\cdot(\mathbf{h} \times \mathbf{r}) \tag{2.17}$$

and

$$\begin{aligned}\mu\varepsilon\dot{\varepsilon} &= (\mathbf{r} \times \mathbf{f})\cdot(\varepsilon \times \mathbf{v}) + \mathbf{f}\cdot(\mathbf{h} \times \varepsilon)\\ &= \mathbf{f}\cdot[(\varepsilon \times \mathbf{v}) \times \mathbf{r} + \mathbf{h} \times \varepsilon].\end{aligned} \tag{2.18}$$

One of these equations can be replaced by the energy equation

$$\dot{E} = \mathbf{v}\cdot\mathbf{f}, \tag{2.19}$$

which is most easily derived from (2.1) in a manner we have noted before.

To complete this formulation of perturbation theory we need to add an equation for the effect of perturbations on the time of flight along the osculating orbit. However, we shall skip that, because it will not be needed, for the particular problems which we shall consider, and the method of Section 8-4 is probably better for that purpose anyway.

Orbital Averages

In many problems of satellite motion the variation of orbital elements is slow compared to the orbital period. In such problems we can simplify our perturbation equations considerably by averaging over an orbital period while holding the orbital elements fixed. This *time-smoothing* procedure eliminates oscillations in the orbital elements over a single orbital period. It eliminates **r** and **v** from the perturbation equations, reducing them to equations for the orbital elements R, h, ε and E alone. We shall refer to the resulting time-smoothed equations as *secular equations of motion*; since the changes in the orbital elements they describe are called *secular variations* by astronomers. Secular equations are most appropriate for investigating *long-term* perturbation effects.

The orbital average $\bar{f}$ of a physical quantity $f = f(t)$ is defined by

$$\bar{f} = \frac{1}{T}\int_0^T f(t)\,\mathrm{d}t, \tag{2.20}$$

where T is the period of orbital motion, and the orbital elements are held constant in the integration. To compute the average $\bar{f}$, therefore, f must be expressed as an explicit function of the orbital elements. After averaging, we release the time dependence of the orbital elements so $\bar{f}$ becomes a function of time.

Our secular equations of motion for the orbital elements can be written

$$\dot{R} = \tfrac{1}{2}Ri\bar{\omega}, \tag{2.21a}$$

where

$$\bar{\omega} = \frac{\mathbf{h}\times\overline{(\mathbf{r}\times\mathbf{f})}}{h^2} - \frac{\mathbf{h}}{\mu\varepsilon^2h^2}\left[\boldsymbol{\varepsilon}\cdot\overline{\mathbf{v}(\mathbf{r}\times\mathbf{f})}\cdot\mathbf{h} + h^2\boldsymbol{\varepsilon}\cdot\bar{\mathbf{f}}\right]. \tag{2.21b}$$

Also,

$$h\dot{h} = \mathbf{h}\cdot\overline{(\mathbf{r}\times\mathbf{f})}, \tag{2.22}$$

$$\mu\varepsilon\dot{\varepsilon} = \boldsymbol{\varepsilon}\cdot\overline{[\mathbf{v}\times(\mathbf{r}\times\mathbf{f})]} + (\mathbf{h}\times\boldsymbol{\varepsilon})\cdot\bar{\mathbf{f}}, \tag{2.23}$$

$$\dot{E} = \overline{\mathbf{v}\cdot\mathbf{f}}. \tag{2.24}$$

When a definite perturbing force function **f** is given, the indicated time averages can be performed, and these become definite differential equations to be solved for the time dependence of the orbital elements.

To facilitate applications of the theory, we collect relations needed to evaluate orbital averages efficiently and establish specific results which will be useful in calculations later on. Orbital averages are often easier to compute when the independent variable is an angle instead of time, because the parametric representation of the orbit is simpler. So we need explicit parametric representations for the dependent variables of interest and relations among the alternative parameters. The relations we need were derived in the first part of Chapter 4, primarily in Section 4-4, so we can just write them down here.

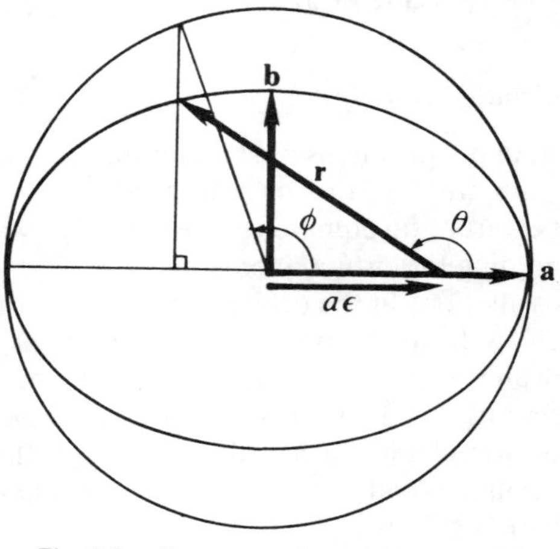

Fig. 2.2. Parameters for an elliptical orbit.

The angle variables of interest, θ and ϕ, relate the position vector to the orbital elements as shown in Figure 2.2. Astronomers call θ the *true anomaly* and ϕ the *eccentric anomaly*. The dimensionless time variable

$$M = \frac{2\pi}{T}(t - \tau) \tag{2.25a}$$

is called the *mean anomaly*, and τ is the initial *time of pericenter passage*. The orbital frequency parameter

$$n = \frac{2\pi}{T} = \left(\frac{\mu}{a^3}\right)^{1/2} \tag{2.25b}$$

is called the *mean motion* by astronomers. For parametric representations, the following relations are useful:

$$\mathbf{r} = \mathbf{a}(\cos\phi - \varepsilon) + \mathbf{b}\sin\phi, \tag{2.26}$$

$$\hat{\boldsymbol{\varepsilon}}\mathbf{r} = a(\cos\phi - \varepsilon) + \mathbf{i}b\sin\phi = re^{\mathbf{i}\theta}, \tag{2.27}$$

where $\mathbf{i} = i\hat{\mathbf{h}}$, $\hat{\mathbf{a}} = \hat{\boldsymbol{\varepsilon}}$, $\hat{\mathbf{b}} = \hat{\mathbf{a}}\mathbf{i} = \hat{\mathbf{h}} \times \hat{\boldsymbol{\varepsilon}}$.

$$r = \frac{b^2/a}{1 + \varepsilon\cos\theta} = a(1 - \varepsilon\cos\phi), \tag{2.28}$$

$$b^2 = a^2(1 - \varepsilon^2) = \frac{ah^2}{\mu}, \tag{2.29}$$

$$v^2 = \mu\left(\frac{2}{r} - \frac{1}{a}\right) = \dot{r}^2 + \frac{h^2}{r^2}. \tag{2.30}$$

The true anomaly is related to the time parameter by angular momentum conservation;

$$\dot{\theta} = \frac{h}{r^2} = \frac{2\pi ab}{r^2 T} . \tag{2.31}$$

By differentiating Kepler's Equation (4–4.4) and using (2.28) we find

$$\dot{\phi} = \frac{2\pi a}{rT} . \tag{2.32}$$

These last two equations can be reexpressed as the differential relations

$$\frac{dt}{T} = \frac{r\,d\phi}{2\pi a} = \frac{r^2\,d\theta}{2\pi ab} = \frac{dM}{2\pi} . \tag{2.33}$$

This enables us to express the time average in any of the equivalent forms

$$\bar{f} \equiv \frac{1}{T}\int_0^T f\,dt = \frac{1}{2\pi}\int_0^{2\pi} f\,dM = \frac{1}{2\pi a}\int_0^{2\pi} fr\,d\phi = \frac{1}{2\pi ab}\int_0^{2\pi} fr^2\,d\theta, \tag{2.34}$$

where r can be expressed in terms of θ or ϕ by (2.28) as needed.
To illustrate averaging calculations, we work out a few examples.

$$\begin{aligned}\bar{r} &= \frac{1}{2\pi a}\int_0^{2\pi} r^2\,d\phi = \frac{a}{2\pi}\int_0^{2\pi}(1-\varepsilon\cos\phi)^2\,d\phi \\ &= \frac{a}{2\pi}\left\{\int_0^{2\pi} d\phi - 2\varepsilon\int_0^{2\pi}\cos\phi\,d\phi + \varepsilon^2\int_0^{2\pi}\cos^2\phi\,d\phi\right\} \\ &= \frac{a}{2\pi}\{2\pi - 0 + \varepsilon^2\pi\} .\end{aligned}$$

Hence,

$$\bar{r} = a(1 + \tfrac{1}{2}\varepsilon^2) . \tag{2.35}$$

In general,

$$\begin{aligned}\overline{r^n} &= \frac{a^n}{2\pi}\int_0^{2\pi}(1-\varepsilon\cos\phi)^{n+1}\,d\phi \\ &= \overline{ar^{n-1}} - \frac{a^n\varepsilon}{2\pi}\int_0^{2\pi}\cos\phi(1-\varepsilon\cos\phi)^n\,d\phi .\end{aligned} \tag{2.36}$$

When the numerator is expanded here, odd powers in $\cos\phi$ can be discarded, since

$$\int_0^{2\pi}\cos^k\phi\,d\phi = 0$$

if k is an odd integer. From considering (2.28), it is evident that ϕ is a more convenient parameter than θ, in these cases, because the cosine appears in the numerator instead of the denominator. The opposite is true in the next example.

$$\overline{\left(\frac{1}{r}\right)^4} = \frac{1}{2\pi ab}\int_0^{2\pi} \frac{r^2\, d\theta}{r^4} = \frac{a}{2\pi b^5}\int_0^{2\pi} (1 - \varepsilon \cos \theta)^2\, d\theta,$$

Hence,

$$\overline{\left(\frac{1}{r}\right)^4} = \frac{a}{b^5}(1 + \tfrac{1}{2}\varepsilon^2) = \frac{\overline{r}}{b^5}\,. \tag{2.37}$$

Notice the reciprocity between θ and ϕ in the integrals for (2.35) and (2.37). Considering (2.28), it is evident that this reciprocity is a general relation, and it is not difficult to prove that

$$\overline{\left(\frac{b}{r}\right)^n} = \overline{\left(\frac{r}{b}\right)^{n-3}} \quad \text{for} \quad n \geqslant 3. \tag{2.38}$$

Results of computations by the above method are given in Table 2.1. This straightforward method is adequate for computing any desired orbital average, but we shall develop an elegant alternative approach.

TABLE 2.1. Orbital Averages

$\overline{r} = a(1 + \frac{1}{2}\varepsilon^2)$	$\overline{\left(\frac{1}{r}\right)} = \frac{1}{a}$
$\overline{r^2} = a^2(1 + \frac{3}{2}\varepsilon^2)$	$\overline{\left(\frac{1}{r^2}\right)} = \frac{1}{ab}$
$\overline{r^3} = a^3(1 + 3\varepsilon^3 + \frac{3}{8}\varepsilon^4)$	$\overline{\left(\frac{1}{r^3}\right)} = \frac{1}{b^3}$
$\overline{r^4} = a^4(1 + 5\varepsilon^2 + \frac{15}{8}\varepsilon^4)$	$\overline{\left(\frac{1}{r^n}\right)} = \frac{\overline{r^{n-3}}}{b^{2n-3}}$ for $n \geqslant 3$
$\overline{r^5} = a^5(1 + \frac{15}{8}\varepsilon^2 + \frac{45}{8}\varepsilon^4 + \frac{5}{8}\varepsilon^6)$	
$\overline{r^6} = a^6(1 + \frac{21}{2}\varepsilon^2 + \frac{105}{8}\varepsilon^4 + \frac{35}{16}\varepsilon^6)$	
$\overline{r^7} = a^7(1 + 14\varepsilon^2 + \frac{105}{8}\varepsilon^4 + \frac{35}{4}\varepsilon^6 + \frac{35}{128}\varepsilon^8)$	
$\overline{r^8} = a^8(1 + 18\varepsilon^2 + \frac{189}{4}\varepsilon^4 + \frac{105}{4}\varepsilon^6 + \frac{315}{128}\varepsilon^8)$	
$\overline{r^9} = a^9(1 + \frac{45}{2}\varepsilon^2 + \frac{315}{4}\varepsilon^4 + \frac{525}{8}\varepsilon^6 + \frac{1575}{128}\varepsilon^8 + \frac{63}{256}\varepsilon^{10})$	
$\overline{r^{10}} = a^{10}(1 + \frac{55}{2}\varepsilon^2 + \frac{495}{4}\varepsilon^4 + \frac{1165}{8}\varepsilon^6 + \frac{5775}{128}\varepsilon^8 + \frac{693}{256}\varepsilon^{10})$	

$$\overline{r^n\mathbf{r}} = \frac{\boldsymbol{\varepsilon}}{\varepsilon^2}\left[\frac{b^2}{a}\overline{r^n} - \overline{r^{n+1}}\right]$$

$$\overline{\left(\frac{\mathbf{r}}{r^n}\right)} = \frac{\boldsymbol{\varepsilon}}{\varepsilon^2 b^{2n-5}}\left[\frac{\overline{r^{n-3}}}{a} - \overline{r^{n-4}}\right], \quad \text{for } n \geqslant 4$$

For an arbitrary constant vector $\mathbf{u} = \mathbf{u}_1\mathbf{e}_1 + u_2\mathbf{e}_2 + u_3\mathbf{e}_3$ where $\mathbf{e}_1 = \hat{\boldsymbol{\varepsilon}}$, $\mathbf{e}_2 = \mathbf{e}_3 \times \mathbf{e}_1$, $\mathbf{e}_3 = \hat{\mathbf{h}}$.

$$\overline{\left(\frac{\mathbf{u}\cdot\mathbf{rr}}{r^n}\right)} = \frac{1}{\varepsilon^3}\left\{u_1\mathbf{e}_1\overline{\left(\frac{b^4}{a^2r^n} - \frac{2b^2}{ar^{n-1}} + \frac{1}{r^{n-2}}\right)} - u_2\mathbf{e}_2\overline{\left(\frac{b^4}{a^2r^n} - \frac{2b^2}{ar^{n-1}} + \frac{1-\varepsilon^2}{r^{n-2}}\right)}\right\}$$

or, for $n \geqslant 5$

$$\overline{\left(\frac{\mathbf{u}\cdot\mathbf{rr}}{r^n}\right)} = \frac{1}{\varepsilon^2 b^{2n-7}}\{u_1\mathbf{e}_1A_2 + u_2\mathbf{e}_2(A_2 - \varepsilon^2B_2)\},$$

where $A_2 = \dfrac{\overline{r^{n-3}}}{a^2} - 2\dfrac{\overline{r^{n-4}}}{a} + \overline{r^{n-5}}$, $B_2 = \overline{r^{n-5}}$, for $n \leqslant 5$

For arbitrary constant vectors $\mathbf{u}$ and $\mathbf{w}$,

$$\overline{\left(\frac{\mathbf{w}\cdot\mathbf{ru}\cdot\mathbf{rr}}{r^n}\right)} = \frac{1}{\varepsilon^2 b^{2n-9}}\{u_1w_1\mathbf{e}_1A_3 - [u_2w_2\mathbf{e}_1 + (u_1w_2 + u_2w_1)\mathbf{e}_2](A_3 - \varepsilon^2B_3)\},$$

where $A_3 = \dfrac{\overline{r^{n-3}}}{a^3} - 3\dfrac{\overline{r^{n-4}}}{a^3} + 3\dfrac{\overline{r^{n-5}}}{a} - \overline{r^{n-6}}$,

$B_3 = \dfrac{\overline{r^{n-5}}}{a} \quad \overline{r^{n-6}}$, for $n \leqslant 6$

$$\overline{\mathbf{u}\cdot\mathbf{rr}} = \frac{a^2}{2}\{u_1\mathbf{e}_1(1 + 4\varepsilon^2) + u_2\mathbf{e}_2(1 - \varepsilon^2)\} = \frac{a^2}{2}\{\mathbf{u}(1 - \varepsilon^2) + 5\boldsymbol{\varepsilon}\boldsymbol{\varepsilon}\cdot\mathbf{u}\}$$

$$\overline{\mathbf{w}\cdot\mathbf{ru}\cdot\mathbf{r}\hat{\mathbf{r}}} = -\frac{a^2\varepsilon}{2}\{u_1w_1\mathbf{e}_1(3 + 2\varepsilon^2) + [u_2w_2\mathbf{e}_1 + (u_1w_1 + \mathbf{u}_2w_2)\mathbf{e}_2](1 - \varepsilon^2)\}$$

Orbital averages can be systematically computed by exploiting the relation

$$i\mathbf{h}\mathbf{v} = \mu(\boldsymbol{\varepsilon} + \hat{\mathbf{r}}). \tag{2.39}$$

Since $\mathbf{v}$ is a time derivative its orbital average vanishes; explicitly,

$$\overline{\mathbf{v}} = \frac{1}{T}\int_0^T \frac{\mathrm{d}\mathbf{r}}{\mathrm{d}t}\,\mathrm{d}t = \frac{1}{T}\oint \mathrm{d}\mathbf{r} = 0.$$

Therefore, from (2.39) we immediately get

$$\overline{\hat{\mathbf{r}}} = -\boldsymbol{\varepsilon}. \tag{2.40}$$

Note that this implies the relations

$$\overline{\hat{\boldsymbol{\varepsilon}}\cdot\hat{\mathbf{r}}} = \overline{\cos\theta} = -\varepsilon,$$

$$\overline{\hat{\boldsymbol{\varepsilon}} \times \hat{\mathbf{r}}} = \hat{\mathbf{h}}\ \overline{\sin\theta} = 0 .$$

To transform (2.39) into a form for computing general orbital averages, we multiply it by **r** and separate scalar and bivector parts. The scalar part can be written

$$\mu(\boldsymbol{\varepsilon}\cdot\mathbf{r} + r) = i\mathbf{h}\mathbf{v}\wedge\mathbf{r} = h^2 = \frac{\mu b^2}{a} ,$$

or

$$\boldsymbol{\varepsilon}\cdot\mathbf{r} = \frac{b^2}{a} - r . \tag{2.41}$$

The bivector part gives us

$$\boldsymbol{\varepsilon}\wedge\mathbf{r} = i\mathbf{h}\frac{\mathbf{v}\cdot\mathbf{r}}{\mu} = i\,\mathbf{h}\,\frac{r\dot{r}}{\mu} . \tag{2.42}$$

Adding these two equations and solving for **r**, we obtain

$$\mathbf{r} = \boldsymbol{\varepsilon}^{-1}\left(\frac{b^2}{a} - r + i\mathbf{h}\frac{r\dot{r}}{\mu}\right) = \boldsymbol{\varepsilon}^{-1}\left(\frac{b^2}{a} - r\right) + \mathbf{h}\times\boldsymbol{\varepsilon}^{-1}\frac{r\dot{r}}{\mu} . \tag{2.43}$$

Since $2r\dot{r} = \mathrm{d}r^2/\mathrm{d}t$, this gives us the time average

$$\overline{\mathbf{r}} = \boldsymbol{\varepsilon}^{-1}\left(\frac{b^2}{a} - \overline{r}\right) = -\frac{3}{2}a\boldsymbol{\varepsilon} , \tag{2.44}$$

with the help of (2.29) and (2.35). More generally, (2.43) gives us

$$\frac{\mathbf{r}}{r^n} = \boldsymbol{\varepsilon}^{-1}\left(\frac{b^2}{ar^n} - \frac{1}{r^{n-1}}\right) + \frac{\mathbf{h}\times\boldsymbol{\varepsilon}^{-1}}{\mu}\left(\frac{\dot{r}}{r^{n-1}}\right) .$$

Since the last term here is again a total time derivative, it averages to zero, and we get

$$\overline{\left(\frac{\mathbf{r}}{r^n}\right)} = \boldsymbol{\varepsilon}^{-1}\left[\frac{b^2}{a}\overline{\left(\frac{1}{r^n}\right)} - \overline{\left(\frac{1}{r^{n-1}}\right)}\right] . \tag{2.45}$$

Using (2.38), for $n \geqslant 4$ this can put in the slightly simpler form given in Table 2.1.

Now let **u** be an arbitrary constant vector. With a Kepler frame (2.9) as basis, we can write **u** in the expanded form

$$\mathbf{u} = u_1\mathbf{e}_1 + u_2\mathbf{e}_2 + u_3\mathbf{e}_3 ,$$

where $u_k = \mathbf{u}\cdot\mathbf{e}_k$. Using (2.43) in the form

$$\mathbf{r} = \frac{1}{\varepsilon}\left[\mathbf{e}_1\left(\frac{b^2}{a} - r\right) + \mathbf{e}_2\frac{hr\dot{r}}{\mu}\right] , \tag{2.46}$$

we get

$$\mathbf{u}\cdot\mathbf{rr} = \frac{1}{\varepsilon^2}\left\{u_1\mathbf{e}_1\left(\frac{b^2}{a} - r\right)^2 + u_2\mathbf{e}_2\frac{b^2}{a}\frac{(r\dot{r})^2}{\mu}\right\}$$
$$+ \frac{1}{\varepsilon^2}(u_2\mathbf{e}_1 + u_1\mathbf{e}_2)\left(\frac{b^2}{a} - r\right)\frac{hr\dot{r}}{\mu}\,. \tag{2.47}$$

The last term will yield a vanishing orbital average even when multiplied by an arbitrary polynomial in r. Therefore, (2.47) gives us

$$\overline{\frac{(\mathbf{u}\cdot\mathbf{r})\mathbf{r}}{r^n}} = \frac{1}{\varepsilon^2}\left\{u_1\mathbf{e}_1\overline{\left(\frac{b^4}{a^2r^n} - \frac{2b^2r}{a\,r^n} + \frac{r^2}{r^n}\right)} + u_2\mathbf{e}_2\frac{b^2}{a\mu}\overline{\frac{(r\dot{r})^2}{r^n}}\right\} \tag{2.48}$$

The average of $(r\dot{r})^2$ can be computed from (2.30), which gives us

$$\frac{(r\dot{r})^2}{\mu} = -\left(\frac{b^2}{a} - 2r + \frac{r^2}{a}\right). \tag{2.49}$$

Then, for $n \geqslant 5$, (2.38) can be used to put (2.48) in the useful form given in Table 2.1

We now have all the techniques needed to reduce the orbital average of any homogeneous function of $\mathbf{r}$ to averages of powers of $r = |\mathbf{r}|$. Let us consider one more example, which will be needed for our calculations later on. Dotting (2.47) with an arbitrary constant vector $\mathbf{w}$ and multiplying by (2.46), we get

$$\mathbf{u}\cdot\mathbf{r}\;\mathbf{w}\cdot\mathbf{r}\,\mathbf{r} = \frac{1}{\varepsilon^3}\left\{u_1w_1\mathbf{e}_1\left(\frac{b^2}{a} - r\right)^3\right.$$
$$\left.+ [u_2w_2\mathbf{e}_1 + (u_1w_1 + u_2w_2)\mathbf{e}_2]\left(\frac{b^2}{a} - r\right)\frac{b^2}{a}\frac{(r\dot{r})^2}{\mu} + \frac{\mathrm{d}}{\mathrm{d}t}\cdots\right\} \tag{2.50}$$

Proceeding as before to compute the orbital average, we get the result in Table 2.1.

Astronomical Coordinates

Our specification of orbital elements and equations of motion is mathematically complete. However, we need to relate our attitude element to a conventional set of orbital elements to facilitate comparison of our results with observations and our theory with conventional perturbation theories.

To measure the attitude of a satellite orbit in space, a system of astronomical coordinates must be set up. These angular coordinates relating points on the *celestial sphere*, a unit sphere with points representing directions in physical space. Positions of the "fixed" stars are nearly constant on the celestial sphere, so they are good points of reference.

The first step in setting up a coordinate system is selection of a convenient orthonormal frame $\{\boldsymbol{\sigma}_k\}$ of fixed reference directions. A *pole vector* $\boldsymbol{\sigma}_3$ is chosen normal to a *reference plane*, which intersects the celestial sphere in a *reference equator*. The vector $\boldsymbol{\sigma}_1$ is an arbitrary direction in the reference plane, usually chosen as the direction of an easily identifiable star for observational convenience. Then, of course, $\boldsymbol{\sigma}_2$ is determined by $\boldsymbol{\sigma}_2 = \boldsymbol{\sigma}_3 \times \boldsymbol{\sigma}_1$.

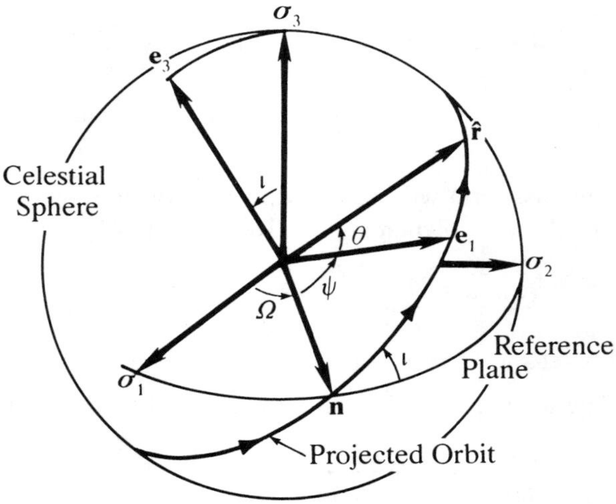

Fig. 2.3. Coordinates for the orbit of a satellite projected on the celestial sphere.

The osculating orbit of a satellite projects to a great circle on the celestial sphere. The Kepler frame $\{\mathbf{e}_k\}$ of the orbit is related to the reference frame $\{\boldsymbol{\sigma}_k\}$ by

$$\mathbf{e}_k = R^\dagger \boldsymbol{\sigma}_k R. \tag{2.51}$$

The attitude element R can be parametrized by a set of Euler angles ψ, ι, Ω, as indicated in Figure 2.3; from Section 5-3 we have

$$R = e^{(1/2)i\boldsymbol{\sigma}_3\psi}\, e^{(1/2)i\boldsymbol{\sigma}_1\iota}\, e^{(1/2)i\boldsymbol{\sigma}_3\Omega} = e^{(1/2)i\boldsymbol{\sigma}_3\Omega}\, e^{(1/2)i\mathbf{n}\iota}\, e^{(1/2)i\mathbf{e}_3\psi}. \tag{2.52}$$

Let us refer to these three angles as *Eulerian (orbital) elements*. To facilitate discourse, it is convenient to introduce some additional nomenclature from astronomy.

The angle ι is the *inclination* of the orbit ($0 \leq \iota \leq 180°$). The orbital motion is said to be *direct* if $0 \leq \iota \leq 90°$ and *retrograde* if $90° \leq \iota \leq 180°$. The vector

$$\mathbf{n} = \frac{\boldsymbol{\sigma}_3 \times \mathbf{e}_3}{|\boldsymbol{\sigma}_3 \times \mathbf{e}_3|} = \boldsymbol{\sigma}_1\, e^{i\boldsymbol{\sigma}_3\Omega} \tag{2.53}$$

is the direction of the *ascending node* where the orbit crosses the reference plane. The angle Ω is the *longitude of ascending node*. The angle ψ is the *argument of pericenter*.

For observational astronomy, the most practical coordinate system is the *equatorial system*, a geocentric system in which the *celestial north pole* $\boldsymbol{\sigma}_3$ is the point where the Earth's axis penetrates the celestial sphere. In this system the path of the sun is a great circle on the celestial sphere called the *ecliptic*. The ecliptic cuts the *celestial equator* at two points called the *equinoxes*. The *vernal equinox* is the ascending node of the ecliptic, and this is taken as the reference direction $\boldsymbol{\sigma}_1$. There is more special nomenclature for the equatorial system which need not concern us here.

Another widely used coordinate system is the *ecliptic system*, in which the north pole is the direction of angular momentum vector of the Earth's orbit about the sun and $\boldsymbol{\sigma}_1$ is also taken to be the direction of the Vernal equinox. The trouble with the equatorial and ecliptic systems is their reference directions are not truly constant, as we shall see.

Conventional perturbation theories use Eulerian elements to specify the attitude of an orbit, and so develop equations of motion for these variables. In contrast, use of the spinor element R enables us to develop the theory and applications without prior commitment to a particular set of angle variables. This has important practical as well as conceptual advantages: (1) A single set of Eulerian elements cannot be used for orbits of arbitrary eccentricity and inclination, because their equations of motion become ill-defined for some values of the parameters. (2) The variables which provide the simplest solution of the attitude equation of motion depend on the perturbing force, and they are not necessarily Euler angles. (3) Variables suitable for handling one perturbation may not be optimal when many perturbations are considered together.

Although we eschew the use of Eulerian elements in our perturbation theory, when the functional form of the attitude spinor R has been determined for a particular problem, we may wish to express the results in terms of Euler angles for practical reasons. That can be done by solving (2.52) for the Euler angles in terms of R. The derivatives of the Euler angles related to the rotational velocity of a Kepler frame by

$$\boldsymbol{\omega} = \frac{2}{i} R^{\dagger}\dot{R} = \boldsymbol{\sigma}_3\dot{\Omega} + \mathbf{n}\dot{\iota} + \mathbf{e}_3\dot{\psi}, \tag{2.54}$$

a result derived in Chapter 7. This can be solved to get the variations in the Eulerian elements from $\boldsymbol{\omega}$. Thus, multiplication of (2.54) gives us

$$\dot{\iota} = \mathbf{n}\cdot\boldsymbol{\omega} \tag{2.55a}$$

and

$$\boldsymbol{\omega}\times\mathbf{n} = \boldsymbol{\sigma}_3\times\mathbf{n}\,\dot{\Omega} + \mathbf{e}_3\times\mathbf{n}\,\dot{\psi},$$

whence,

$$\dot{\Omega} = \frac{\boldsymbol{\omega}\cdot(\mathbf{n}\times\mathbf{e}_3)}{\mathbf{n}\cdot(\mathbf{e}_3\times\boldsymbol{\sigma}_3)} = \frac{\boldsymbol{\omega}\cdot\mathbf{e}_3\cos\iota - \boldsymbol{\omega}\cdot\boldsymbol{\sigma}_3}{\sin^2\iota} \tag{2.55b}$$

$$\dot{\psi} = \frac{\boldsymbol{\omega}\cdot(\mathbf{n}\times\boldsymbol{\sigma}_3)}{\mathbf{n}\cdot(\boldsymbol{\sigma}_3\times\mathbf{e}_3)} = \frac{\boldsymbol{\omega}\cdot\mathbf{e}_3 - \boldsymbol{\omega}\cdot\boldsymbol{\sigma}_3\cos\iota}{\sin^2\iota}. \tag{2.55c}$$

Referring to Figure 2.2 for interpretation, a time variation of the inclination ι is called *nutation* of the orbit. A variation of Ω is called *precession* of the *nodes* or *longitudinal* precession. A variation of ψ is called, as before, *precession* of the *apses* or *major axis*. Either type of precession is said to be *direct* if the angle increases or *retrograde* if it decreases. The orbital tilt mentioned previously in the discussion of Equation (2.15) is a combination of nutation and longitudinal precession. Indeed, the decomposition of tilt into nutation and precession depends on the chosen reference direction.

Conventional equations of motion for the Euler elements can be derived from (2.55a, b, c) after inserting the expression for ω in terms of the perturbing force given by (2.15). But we have no need for those equations.

8-2. Exercises

(2.1) Begin with the general three body equations (6–5.2). Let particle 1 be the primary with particle 2 as its satellite and particle 3 a perturbing body. Derive the perturbed 2-body equation

$$\ddot{\mathbf{r}} = -\frac{\mu \mathbf{r}}{r^3} + \mathbf{f},$$

where $\mu = G(m_1 + m_2)$, $\mathbf{r} = \mathbf{x}_2 - \mathbf{x}_1$, $\mathbf{r}' = \mathbf{x}_3 - \mathbf{x}_1$, and the exact perturbing force is given by

$$\mathbf{f} = Gm_3 \left[\frac{\mathbf{r}' - \mathbf{r}}{|r - r'|^3} - \frac{\mathbf{r}}{r'^3} \right].$$

Note that $\mathbf{f}$ is a definite function $\mathbf{f}(\mathbf{r}, t)$ if $\mathbf{r}' = \mathbf{r}'(t)$ is a specified function of time. Show that $\mathbf{f}$ can be derived from the potential

$$\phi(\mathbf{r}, t) = Gm_3 \left[\frac{1}{|\mathbf{r} - \mathbf{r}'(t)|} - \frac{\mathbf{r} \cdot \mathbf{r}'(t)}{r(t)^3} \right].$$

(2.2) Evaluate A_2 in Table 2.1 for various values of n. Is there a pattern?

(2.3) Show that for a satellite subject to a central perturbing force $\mathbf{f} = -\nabla\phi(r)$,

$$C = \tfrac{1}{2}\varepsilon^2 + \frac{h^2}{\mu}\phi(r)$$

is a constant of the motion. Therefore, the eccentricity oscillates between maximum and minimum values depending on $\phi(r)$.

8-3. Peturbations in the Solar System

In this section we discuss the principal perturbations on satellites in the solar system and calculate their long-term effects to first order. The various perturbations can be classified into three main types:

(A) *Gravitational perturbations* following from Newton's law have the largest effects. Two subtypes are particularly important : (1) Quadrupole perturbations due to an asymmetric mass distribution in a nearby body, (2) Secular third body perturbations due to a distant orbiting body. These perturbations have a wide variety of effects on the orbital and rotational motions of planets and satellites.

(B) *Non-Newtonian perturbations* from Einstein's theory of Relativity have the most subtle effects, which can be identified only after the Newtonian effects have been accounted for with great precision.

(C) *Non-gravitational perturbations* such as atmospheric drag, the Solar wind and magnetic forces are negligible for the planets and larger satellites, but their effects on artificial satellites and the smaller asteroids are quite significant.

We will employ the secular perturbation theory developed in Section 8-2, so let us repeat the equations of motion for the orbital elements for easy reference:

$$\dot{R} = \tfrac{1}{2} Ri\overline{\omega} \,, \tag{3.1a}$$

where

$$\overline{\omega} = \frac{\mathbf{h} \times \overline{(\mathbf{r} \times \mathbf{f})}}{h^2} - \frac{\mathbf{h}}{\mu\varepsilon^2 h^2}\,[\overline{\boldsymbol{\varepsilon}\cdot\mathbf{v}(\mathbf{r} \times \mathbf{f})\cdot\mathbf{h}} + h^2\overline{\boldsymbol{\varepsilon}\cdot\mathbf{f}}\,], \tag{3.1b}$$

and

$$\mathbf{e}_k = R^\dagger \sigma_k R \tag{3.1c}$$

is a Kepler frame, with $\mathbf{e}_1 = \hat{\boldsymbol{\varepsilon}}$, $\mathbf{e}_2 = \mathbf{e}_3 \times \mathbf{e}_1$, $\mathbf{e}_3 = \hat{\mathbf{h}}$. And

$$h\dot{h} = \mathbf{h}\cdot\overline{(\mathbf{r} \times \mathbf{f})}, \tag{3.2}$$

$$\mu\varepsilon\dot{\varepsilon} = \boldsymbol{\varepsilon}\cdot[\overline{\mathbf{v} \times (\mathbf{r} \times \mathbf{f})}] + (\mathbf{h} \times \boldsymbol{\varepsilon})\cdot\overline{\mathbf{f}} \tag{3.3}$$

$$\dot{E} = \overline{\mathbf{v}\cdot\mathbf{f}} \,. \tag{3.4}$$

Also, we need the relations

$$E = \frac{\mu(\varepsilon^2 - 1)}{2h^2} = -\frac{\mu}{2a} = -\frac{2\pi^2 a^2}{T^2}, \tag{3.5}$$

where a is the semi-major axis and T is the period of the osculating elliptical orbit. The variable $\mathbf{v}$ can be eliminated in the orbital averages by using

$$\mathbf{v} = \mu(\boldsymbol{\varepsilon} + \hat{\mathbf{r}})i\mathbf{h}^{-1} = \frac{\mu}{h^2}(\mathbf{h} \times \boldsymbol{\varepsilon} + \mathbf{h} \times \hat{\mathbf{r}}). \tag{3.6}$$

In particular, in (3.1b) it is convenient to use

$$\boldsymbol{\varepsilon}\cdot\mathbf{v} = \frac{\mu}{h^2}(\boldsymbol{\varepsilon} \times \mathbf{h})\cdot\hat{\mathbf{r}} = -\frac{\mu\varepsilon}{h}\mathbf{e}_2\cdot\hat{\mathbf{r}}. \tag{3.7}$$

We will refer back to Section 8-2 for calculation of the averages in these equations. But note that if $\mathbf{f} = -\nabla V$ is a static conservative force, then

$$\dot{E} = \overline{\mathbf{v}\cdot\mathbf{f}} = \overline{\frac{-\mathrm{d}V}{\mathrm{d}t}} = 0. \tag{3.8}$$

Therefore, such a force can change the shape but not the size of the orbit.

Oblateness Perturbations

The Earth's oblateness has significant effects on the orbits of artificial satellites near the Earth, and an evaluation of the effect of the Sun's oblateness on the orbit of Mercury is needed for testing Einstein's theory of gravitation. To be more specific, the main effects of the Earth's oblateness on a near satellite are more than a million times greater than the effects of the Moon. Oblateness produces one of the two main perturbations on near satellites; the other is caused by the Earth's atmosphere. These two perturbations can be considered separately, because their effects are different in kind. Anyway, in first order perturbation theory, the effects of different perturbations are simply additive.

In Section 8-1, we found that oblateness of an axisymmetric planet produces a quadrupole gravitational force (per unit mass) on a satellite with the explicit form

$$\mathbf{f} = -\frac{k}{r^4}[(1 - 5(\mathbf{u}\cdot\hat{\mathbf{r}})^2)\hat{\mathbf{r}} + 2\hat{\mathbf{r}}\cdot\mathbf{u}\mathbf{u}], \tag{3.9}$$

with

$$k = \tfrac{3}{2}\mu J_2 r_p^2, \tag{3.10}$$

where r_p is the equatorial radius of the planet, and $\mathbf{u}$ is a unit vector along the planet's symmetry axis.

We derived the force (3.9) from a potential, so (3.8) tells us immediately that the Kepler energy E is a secular constant of the motion. To determine the secular equations for the other orbital elements, we need the following orbital averages which are easily calculated from Table 2.1 in Section 8.2.

- $$\overline{\mathbf{r}\times\mathbf{f}} = -2k\left(\overline{\frac{\mathbf{r}\cdot\mathbf{u}\mathbf{r}}{r^5}}\right)\times\mathbf{u} = -\frac{k}{b^3}u_3\,\mathbf{u}\times\mathbf{e}_3 \tag{3.11}$$

$$\overline{\mathbf{f}}\cdot\boldsymbol{\varepsilon} = -k\left\{\left(\overline{\frac{\mathbf{r}}{r^5}}\right) - 5\left(\overline{\frac{(\mathbf{u}\cdot\mathbf{r})^2\mathbf{r}}{r^7}}\right) + 2\left(\overline{\frac{\mathbf{r}}{r^5}}\right)\cdot\mathbf{uu}\right\}\cdot\boldsymbol{\varepsilon}$$

$$= -k\left\{\frac{a\varepsilon^2}{b^5} - \frac{5a\varepsilon^2}{4b^5}(3u_1^2 + u_2^2) + \frac{2a\varepsilon^2}{b^5}u_1^2\right\}$$

$$= -\frac{ka\varepsilon^2}{b^5}(1 - \tfrac{7}{4}u_1^2 - \tfrac{5}{4}u_2^2) \tag{3.12}$$

$$\left(\overline{\frac{\boldsymbol{\varepsilon}\cdot\dot{\mathbf{v}}(\mathbf{r}\times\mathbf{f})\cdot\mathbf{h}}{r^5}}\right) = 2k\varepsilon\mu\left(\overline{\frac{\mathbf{e}_2\cdot\mathbf{rr}\cdot\mathbf{ur}}{r^6}}\right)\cdot(\mathbf{u}\times\mathbf{e}_3) =$$

$$= \frac{2k\mu\varepsilon^2}{8b^3}(u_2^2 - u_1^2), \tag{3.13}$$

where $u_k = \mathbf{u}\cdot\mathbf{e}_k$.

Insertion of (3.11) into (3.2) tells us at once that h, like E, is a secular constant of the motion, so ε must be constant as well. Therefore the satellite orbit does not change size and shape under an oblateness perturbation; it only rotates rigidly in space with its focus at the center of the primary as a fixed point. In other words, the secular effects of oblateness are entirely determined by the secular rotational velocity $\bar{\boldsymbol{\omega}}$. Inserting the orbital averages (3.11), (3.12) and (3.13) into (3.1b), we get

$$\bar{\boldsymbol{\omega}} = \varkappa\{[2 - \tfrac{5}{2}(\mathbf{u}\times\mathbf{e}_3)^2]\mathbf{e}_3 - \mathbf{e}_3\cdot\mathbf{uu}\}, \tag{3.14}$$

where

$$\varkappa = \frac{k}{hb^3} = \frac{3}{2}\frac{\mu J_2 r_p^2}{b^4}\left(\frac{a}{\mu}\right)^{1/2} = \frac{3\pi J_2}{T(1-\varepsilon^2)^2}\left(\frac{r_p}{a}\right)^2 \tag{3.15}$$

The equation of motion for the attitude spinor can therefore be written

$$\dot{R} = \tfrac{1}{2}Ri\bar{\omega} = \tfrac{1}{2}i\omega_1 R + \tfrac{1}{2}Ri\omega_2 \tag{3.16a}$$

where

$$\omega_1 = \varkappa[2 - \frac{5}{2}(\mathbf{u}\times\mathbf{e}_3)^2]\boldsymbol{\sigma}_3 = \frac{\varkappa}{2}[5(\mathbf{u}\cdot\mathbf{e}_3)^2 - 1]\boldsymbol{\sigma}_3 \tag{3.16b}$$

and

$$\omega_2 = -\varkappa\mathbf{e}_3\cdot\mathbf{uu}. \tag{3.16c}$$

Since $\boldsymbol{\omega}_1$ and $\boldsymbol{\omega}_2$ are constant vectors, Equation (3.16a) integrates to

$$R = e^{(1/2)i\boldsymbol{\omega}_1 t}R_0 e^{(1/2)i\boldsymbol{\omega}_2 t} \tag{3.17}$$

where R_0 is the initial value of the attitude spinor R.

Equation (3.17) specifies the attitude of the orbit at any time. We have already met such an equation in our study of spinning bodies, and that experience is helpful in interpreting it. Since $\boldsymbol{\omega}_2$ is constant, the orbital angular momentum vector $\mathbf{h} = h\mathbf{e}_3$ undergoes steady gyroscopic precession about the symmetry axis $\hat{\omega}_2 = \mathbf{u}$ of the oblate primary. This is the same thing as longitudinal precession of the ascending and decending nodes, as shown in Figure 3.1 for the case of an artificial Earth satellite. Using (2.55b), we find that on one revolution the nodes precess through an angle

$$\Delta\Omega = T\dot{\Omega} = T\omega_2\cdot\mathbf{u}$$

$$= -\frac{3\pi J_2}{(1-\varepsilon^2)^2}\left(\frac{r_p}{a}\right)^2 \cos\iota. \qquad (3.18)$$

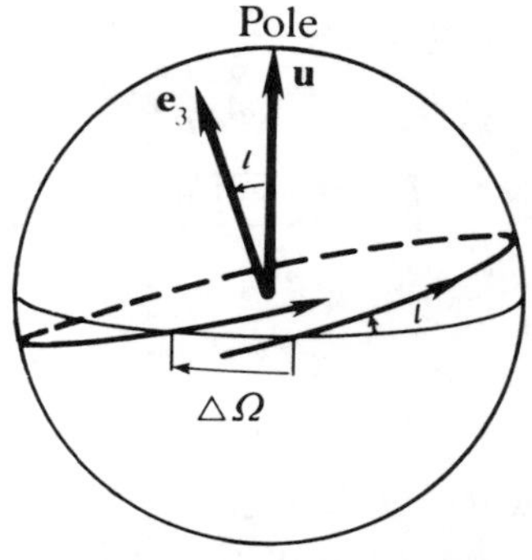

Fig. 3.1. Satellite orbit on the celestial sphere showing regression of the ascending node along the equator due to the Earth's oblateness.

The negative sign means that the precession is retrograde. The precession rate decreases with increasing inclination and vanishes for a polar orbit. Observations of this precession by artificial Earth satellites provide precise values for J_2 and higher order harmonic coefficients J_n when included in the perturbation theory.

The other vector ω_1 in (3.17) gives the precession rate of the apses in the orbital plane. In one revolution the axis turns through an angle $\Delta\psi = T\omega_1$, as shown in Figure 3.2. According to (3.16b), the precession rate depends on the inclination of the orbit and vanishes at the critical angle

$$\iota_c = \sin^{-1}(1/\sqrt{5}) = 63°.43. \qquad (3.19)$$

The precession is retrograde for larger inclinations and direct for smaller inclinations.

Note that the magnitude of the oblateness effects decrease with increasing range by a factor r_p^2/a^2, so they are quite negligible at the distance of the Moon.

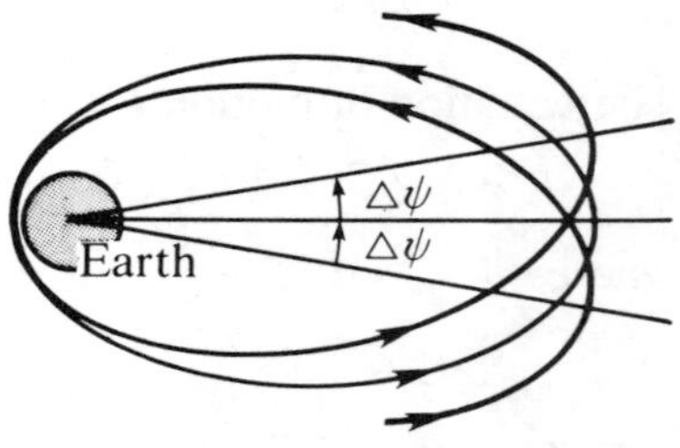

Fig. 3.2. Precession of the major axis (or perigee) due to oblateness.

Secular Third Body Forces

Consider two satellites orbiting a common primary. Suppose the primary is at rest at the origin and the orbit of the outer satellite is a given 2-body ellipse. The very long term effect of the outer satellite (the third body) on the inner satellite can be estimated by a method originally employed by Gauss. The time interval must be long enough for the third body to make many orbital revolutions. If the periods of the two satellites are incommensurable, there is no correlation between their positions on their respective orbits, and the net third body potential at any position **r** is given by the time average

$$\overline{\phi_p}(\mathbf{r}) = -\frac{Gm_p}{T_p}\int_0^{T_p}\frac{\mathrm{d}t}{|\mathbf{r}-\mathbf{r}_p(t)|} \tag{3.20}$$

This is the same as the potential of an elliptical ring formed by smearing the mass m_p of the perturbing body over its orbit with a density proportional to its transit time. We shall refer to it as a *secular* potential to remind us to the special conditions for its applicability.

We can estimate the time average in (3.20) by expanding the integrand in the Legendre series (1.18). Since $r_p = |\mathbf{r}_p| > r = |\mathbf{r}|$, we obtain

$$\begin{aligned}\overline{\phi_p} &= -Gm_p\left\{\left(\overline{\frac{1}{r_p}}\right) + \mathbf{r}\cdot\left(\overline{\frac{\mathbf{r}_p}{r_p^3}}\right) + \frac{3}{2}\left(\overline{\frac{(\mathbf{r}\cdot\mathbf{r}_p)^2}{r_p^5}}\right) - \frac{r^2}{2}\left(\overline{\frac{1}{r_p^3}}\right) + \ldots\right\}\\ &= -Gm_p\left\{\frac{1}{a_p} + 0 + \frac{3}{4}\frac{(\mathbf{r}\times\mathbf{u}_p)^2}{b_p^3} - \frac{r^2}{2b_p^3} + \ldots\right\}\\ &= -Gm_p\left\{\frac{1}{a_p} + \frac{r^2}{4b_p^3} - \frac{3}{4}\frac{(\mathbf{r}\cdot\mathbf{u}_p)^2}{b_p^3} + \ldots\right\},\end{aligned} \tag{3.21}$$

where the subscripts indicate parameters of the third body orbit, and $\mathbf{u}_p$ is the unit normal of the orbital plane. To lowest order, the *secular* gravitational field of the perturbing body is

$$\overline{\mathbf{g}_p}(\mathbf{r}) = -\nabla\overline{\phi_p} = \frac{Gm_p}{2b_p^3}[\mathbf{r} - 3(\mathbf{r}\cdot\mathbf{u}_p)\mathbf{u}_p]. \tag{3.22}$$

It may be surprising to get an axially symmetric field. The field is independent of the alignment of the major axis for two reasons: the origin is at a focus rather than the center of the ring, and the mass density on the ring increases with distance from the origin at a rate just sufficient to cancel the $1/r$ fall off in potential. Note that the field strength for an elliptical ring is greater than that for a circular ring with the same mass and major axis, since $b_p^2 = a_p^2(1 - \varepsilon_p^2)$.

At points which are closer to the ring than the origin, the Legendre expansion converges so slowly that a great many terms are needed to approximate the secular potential accurately. A more efficient method is available when the ring is circular. In that case the potential can be evaluated explicitly in terms of elliptic integrals, but a more elementary approach will suffice here. We suppose that the satellite orbits are coplanar, so we need to evaluate the secular field only in that plane. It will be convenient to calculate the field directly instead of indirectly by calculating the potential first. The secular field at **r** is given by

$$\overline{\mathbf{g}_p}(\mathbf{r}) = \frac{Gm_p}{2\pi}\int_0^{2\pi} d\theta \, \frac{\mathbf{r}_p - \mathbf{r}}{|\mathbf{r}_p - \mathbf{r}|^3}\,, \tag{3.23}$$

where the true anomaly has been substituted for the time variable. The circular orbit can be parameterized by $a_p = |\,\mathbf{r}_p\,|$ and

$$\hat{\mathbf{r}}_p = \hat{\mathbf{r}}e^{\mathbf{i}\theta}, \tag{3.24}$$

where $\mathbf{i}$ is the unit bivector for the orbital plane. On the other hand,

$$\mathbf{r}_p - \mathbf{r} = R\hat{\mathbf{r}}e^{\mathbf{i}\psi} \tag{3.25}$$

is the simplest parameterization of the relative position variable (Figure 3.3.), enabling us to put (3.23) in the form

$$\overline{\mathbf{g}_p}(\mathbf{r}) = \frac{Gm_p}{2\pi}\,\hat{\mathbf{r}}\int_0^{2\pi} d\theta \, \frac{e^{\mathbf{i}\psi}}{R^2}\,. \tag{3.26}$$

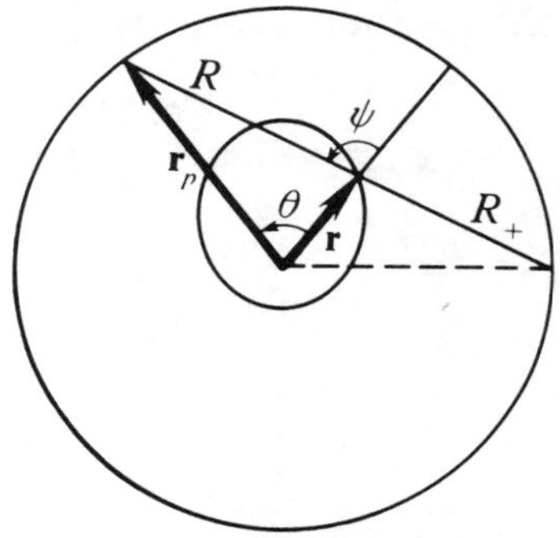

Fig. 3.3. Coplanar orbits with a common primary. The outer orbit is circular.

This suggests that ψ would be the most appropriate integration variable. The variables are related by

$$a_p e^{\mathbf{i}\theta} = r + Re^{\mathbf{i}\psi}, \tag{3.27}$$

which we obtain from (3.24) and (3.25). To express R as a function of ψ, we eliminate θ from (3.27); thus,

$$a_p^2 = r^2 + R^2 + 2rR\cos\psi.$$

The positive root of this equation gives us

$$R = R(\psi) = [a_p^2 - r^2\sin^2\psi]^{1/2} - r\cos\psi. \tag{3.28}$$

It will be convenient to define

$$R_+ \equiv R(\psi + \pi) = [a_p^2 - r^2\sin^2\psi]^{1/2}.$$

Note that

$$\begin{aligned} RR_+ &= a_p^2 - r^2, \\ R_+ - R &= 2r\cos\psi, \\ R_+ + R &= 2[a_p^2 - r^2\sin^2\psi]^{1/2}. \end{aligned} \tag{3.29}$$

The differential for the change in variables from θ to ψ is obtained by differentiating (3.27). After some algebra, we get

$$\frac{d\theta}{R} = \frac{d\psi}{[a_p^2 - r^2\sin^2\psi]^{1/2}} = \frac{2\,d\psi}{R + R_+}\,. \tag{3.30}$$

Now we are prepared to derive an explicit formulation of the integral in (3.26); thus,

$$\int_0^{2\pi} d\theta \, \frac{e^{i\psi}}{R} = \int_0^{2\pi} \frac{2\,d\psi}{R + R_+} \, \frac{e^{i\psi}}{R} = 2\int_0^{\pi} \frac{d\psi \, e^{i\psi}}{R + R_+} \left(\frac{1}{R} - \frac{1}{R_+} \right)$$

$$= \frac{2r}{a_p^2 - r^2} \int_0^{\pi} \frac{d\psi \cos^2 \psi}{[a_p^2 + r^2 \sin^2 \psi]^{1/2}} \,.$$

Therefore, our expression for the secular field can be written

$$\mathbf{g}_p(\mathbf{r}) = \frac{Gm_p}{\pi a_p} \, \frac{\mathbf{r}}{a_p^2 - r^2} \int_0^{\pi} \frac{d\psi \cos^2 \psi}{[1 + (r/a_p)^2 \sin^2 \psi]^{1/2}} . \tag{3.31}$$

The integral can be evaluated in terms of elliptic integrals, but we will be satisfied with a binomial expansion of the denominator to get

$$I(r) \equiv \int_0^{\pi} \frac{d\psi \cos^2 \psi}{[1 + (r/a)^2 \sin^2 \psi]^{1/2}}$$

$$= \int_0^{\pi} d\psi \cos^2 \psi \left[1 + \tfrac{1}{2}\left(\frac{r}{a}\right)^2 \sin^2 \psi + \tfrac{3}{8}\left(\frac{r}{a}\right)^4 \sin^4 \psi + \ldots \right]$$

$$= \frac{\pi}{2}\left[1 + \tfrac{1}{8}\left(\frac{r}{a}\right)^2 + \tfrac{3}{64}\left(\frac{r}{a}\right)^4 + \ldots . \right]. \tag{3.32}$$

This should be compared with the expansion

$$\frac{1}{a^2 - r^2} = \frac{1}{a^2}\left[1 - \left(\frac{r}{a}\right)^2 + \left(\frac{r}{a}\right)^4 - \ldots \right].$$

That shows that the second order term in (3.32) is essential for second order accuracy better than ten percent. Inserting these expansions into (3.31), we get

$$\overline{\mathbf{g}}_p(\mathbf{r}) = \frac{Gm_p}{2a_p^3}\,\mathbf{r}\left[1 - \tfrac{7}{8}\left(\frac{r}{a_p}\right)^2 + \tfrac{59}{64}\left(\frac{r}{a_p}\right)^4 + \ldots \right]. \tag{3.33}$$

To first order this agrees with the result (3.22) from the Legendre expansion, except for a factor $(1 - \varepsilon_p^2)^{-3/2} \approx 1 + \frac{3}{2}\varepsilon_p^2$ which we could include in (3.33) to account for a small eccentricity in the orbit.

Solar and Lunar Perturbations of an Earth Satellite

To investigate the influence of the Sun on the orbit of an Earth satellite, we adopt a geocentric reference system. In this system the Sun orbits the Earth, and thus generates a secular gravitational field which, according to (3.22), is given in the vicinity of the Earth by

$$\mathbf{g}_S = \frac{GM_S}{2b_S^3}[\mathbf{r} - 3(\mathbf{r}\cdot\mathbf{u})\mathbf{u}], \tag{3.34}$$

where M_S is the mass of the Sun, $\mathbf{u}$ is the pole vector of the ecliptic, and b_S is the minor axis of the Sun's orbit about the Earth, so $b_S^2 = a_S^2(1 - \varepsilon_S^2)$. This is the average field of the Sun over a period of one year, so it is appropriate for calculating secular perturbations over one or many years.

The secular rotational velocity of the orbit $\bar{\boldsymbol{\omega}}$ is calculated from (3.1b), with the help of Table 2.1 to compute the averages. Writing

$$\bar{\boldsymbol{\omega}} = \omega_1\mathbf{e}_3 + \boldsymbol{\omega}_2 \tag{3.35}$$

we find

$$\omega_1 = \frac{1}{h\varepsilon}\overline{\mathbf{e}_2\cdot\hat{\mathbf{r}}(\mathbf{r}\times\mathbf{g}_S)}\cdot\mathbf{e}_3 - \frac{h}{\mu\varepsilon}\,\mathbf{e}_1\cdot\overline{\mathbf{g}}_S$$

$$= \frac{GM_S}{2b_S^3 h\varepsilon}\{-3\overline{\mathbf{e}_2\cdot\hat{\mathbf{r}}\;\mathbf{u}\cdot\mathbf{rr}}\cdot(\mathbf{u}\times\mathbf{e}_3) - \frac{h^2}{\mu}\mathbf{e}_1\cdot[\overline{\mathbf{r} - 3\mathbf{r}\cdot\mathbf{uu}}]\}$$

$$= \frac{GM_S}{2b_S^3 h}\{\tfrac{3}{2}\,a^2\varepsilon(1-\varepsilon^2)[(\mathbf{u}\times\mathbf{e}_3)^2 - 2(\mathbf{u}\cdot\mathbf{e}_1)^2]$$

$$- a(1-\varepsilon^2)[-\tfrac{3}{2}\,a\varepsilon(1 - 3(\mathbf{u}\cdot\mathbf{e}_1)^2]\},$$

$$= \frac{3}{4}\frac{GM_S b^2}{hb_S^3}[1 + (\mathbf{u}\times\mathbf{e}_3)^2 - 5(\mathbf{u}\cdot\mathbf{e}_1)^2].$$

Using $h^2 = GMa(1-\varepsilon^2)$ and $b^2 = a^2(1-\varepsilon^2)$, it will be convenient to write this in the form

$$\omega_1 = \varkappa[1 + (\mathbf{u}\times\mathbf{e}_3)^2 - 5(\mathbf{u}\cdot\mathbf{e}_1)^2], \tag{3.36}$$

where $\varkappa$ is defined by

$$\left(\frac{1-\varepsilon_S^2}{1-\varepsilon^2}\right)^{3/2}\varkappa = \frac{3}{4}\left(\frac{GM_S}{a_S^3}\right)\left(\frac{a^3}{GM}\right)^{1/2} = \frac{3}{4}\left(\frac{2\pi}{T_S}\right)^2\left(\frac{T}{2\pi}\right) = \frac{3}{4}\frac{n_S^2}{n}, \tag{3.37}$$

where T_S is the orbital period of the Sun and T is the orbital period of the satellite. Also, we calculate

$$\boldsymbol{\omega}_2 = h^{-1}\mathbf{e}_3\times\overline{(\mathbf{r}\times\mathbf{g}_S)} = h^{-1}\overline{\mathbf{e}_3\cdot\mathbf{g}_S\mathbf{r}} = -\frac{3}{2}\frac{GM_S}{hb_S^3}\overline{\mathbf{u}\cdot\mathbf{rr}}\;\mathbf{u}\cdot\mathbf{e}_3$$

$$= -\frac{3}{4}\frac{GM_S a^2}{hb_S^3}[\mathbf{u}(1-\varepsilon^2) + 5\mathbf{u}\cdot\boldsymbol{\varepsilon\varepsilon}]\mathbf{u}\cdot\mathbf{e}_3$$

which we can write as

$$\boldsymbol{\omega}_2 = \omega_p \mathbf{u} + \omega_N \mathbf{e}_1 , \tag{3.38a}$$

where

$$\omega_p = -\varkappa \mathbf{u}\cdot\mathbf{e}_3 \tag{3.38b}$$

and

$$\omega_N = -\left(\frac{5\varepsilon^2\varkappa}{1-\varepsilon^2}\right)\mathbf{u}\cdot\mathbf{e}_1\mathbf{u}\cdot\mathbf{e}_3 . \tag{3.38c}$$

Inserting (3.38a) into (3.35) we have

$$\bar{\boldsymbol{\omega}} = \omega_1\mathbf{e}_3 + \omega_N\mathbf{e}_1 + \omega_p\mathbf{u}. \tag{3.39}$$

Note the special form of this vector; $\mathbf{u}$ is constant and $\mathbf{e}_k = R^\dagger \boldsymbol{\sigma}_k R$. This observation and our experience with Euler angles enables us to formally integrate the attitude equation

$$\dot{R} = \tfrac{1}{2} i R\bar{\boldsymbol{\omega}},$$

with the result

$$R = e^{(1/2)i\boldsymbol{\sigma}_3\psi}\, e^{(1/2)i\boldsymbol{\sigma}_1\theta}\, e^{(1/2)i\mathbf{u}\phi} , \tag{3.41}$$

where

$$\phi - \phi_0 = \int_0^t \omega_p \,\mathrm{d}t$$

$$\theta - \theta_0 = \int_0^t \omega_N \,\mathrm{d}t$$

$$\psi - \psi_0 = \int_0^t \omega_1 \,\mathrm{d}t. \tag{3.42}$$

We can interpret (3.41) as follows: The third factor on the right side describes precession of the orbital plane about the pole vector of the ecliptic $\mathbf{u}$ with an angular velocity $\omega_p = \dot{\phi}$; so ϕ is the angle of precession. The second factor describes the inclination of the orbital plane and nutation about an average inclination if ω_N is periodic, as it happens to be in (3.38c). The third factor describes precession of the apses in the orbital plane at the rate $\omega_1 = \dot{\psi}$.

The precession of the nodes which astronomers observe is actually a combination of the second and third factors, as can be seen by comparing (3.39) with (2.54). The parametrization used here is better than the conventional one, because it simplifies integrations.

Let us see what these results tell us about the motion of Earth's most important satellite, the Moon. The expression (3.38b) for ω_p tells us that the precession is retrograde and constant if the eccentricity in (3.37) is constant. Using the data $\varepsilon_\mathrm{S} = 0.0167$ and $\varepsilon = 0.0549$ for the orbital eccentricities of Sun and Moon respectively, we find

$$\left(\frac{1-\varepsilon_S^2}{1-\varepsilon^2}\right)^{3/2} \approx 1 + \tfrac{3}{2}(\varepsilon^2 - \varepsilon_S^2) = 1.0041. \tag{3.43}$$

Then, using (3.37) and the empirical value $\theta_0 = 5.14°$ for the average inclination, we estimate that the period of precession of the Moon's orbit is

$$\frac{2\pi}{-\omega_p} = \frac{2\pi}{\varkappa \cos\theta_0} = \frac{4}{3}\left(\frac{365.25 \text{ yr}}{27.32}\right)\left(\frac{1.0041}{0.996}\right) = 17.93 \text{ yr}. \tag{3.44}$$

This is reasonably close to the observed value of 18.61 yr. For we can estimate the accuracy of our approximation by recalling that, in the Legendre expansion used to calculate the secular perturbing force, the neglected third order terms are smaller than the second order terms by the ratio of Earth-Moon distance to Earth-Sun distance, that is, by

$$\frac{a}{a_S} = \frac{3.844 \times 10^5 \text{ km}}{1.495 \times 10^8 \text{ km}} = 0.00257. \tag{3.45}$$

This is of the same order of magnitude as the eccentricity correction (3.43), so third order terms would be needed for an accurate calculation.

The inclination of the Moon's orbit with respect to the ecliptic is nearly constant with a value of 5°.14. So, in the equatorial system of an observer on Earth, the steady precession of the Moon's orbit will appear as an oscillation in the inclination of the Moon's orbit over a range of 10°.28 with a period of 18.61 yr.

To estimate ω_1, we use (3.41) to write (3.36) in the form

$$\omega_1 = \varkappa[1 + \sin^2\theta - 5\sin^2\theta\sin^2\psi]. \tag{3.46}$$

Estimating θ by its average value $\theta_0 = 5.14°$ and using the average value $\sin^2\psi = \frac{1}{2}$, this gives us

$$\omega_1 \approx \varkappa[1 - \tfrac{3}{2}\sin^2\theta_0] = \varkappa(0.9880). \tag{3.47}$$

So, for the period of precession of the Moon's perigee, (3.44) yields the estimate

$$\frac{2\pi}{\omega_1} \approx 18.08 \text{ yr}. \tag{3.48}$$

This is about twice the observed value of 8.85 yr. In view of our estimate (3.45), this discrepancy appears to be too large to be accounted for by a third order correction. It suggests, rather, that a factor of 2 has been overlooked in the calculation. The matter deserves to be examined more carefully.

To estimate the amplitude and period of the secular nutation of the Moon's orbit, we use (3.41) to write (3.38c) in the form

$$\omega_N = \dot\theta = \dot\psi\frac{d\theta}{d\psi} = -\left(\frac{5\varepsilon^2\varkappa}{1-\varepsilon^2}\right)\frac{\sin 2\theta}{2}\sin\psi. \tag{3.49}$$

Because of the small coefficient on the right, it is permissible to replace θ by its average value and assume that $\psi = \omega_1$ is constant, so (3.49) is easily integrated to get

$$\theta - \theta_0 = \frac{5}{2}\left(\frac{\varepsilon^2 \sin 2\theta}{1 - \varepsilon^2}\right)\frac{\varkappa}{\omega_1} \cos \psi. \tag{3.50}$$

This tells us that the nutation has period $2\pi/\omega_1$ and if we use empirical value $\omega_1 \approx 2\varkappa$, for the amplitude of the nutation we obtain the estimate

$$\tfrac{5}{4}\,\varepsilon^2 \sin \theta_0 = \tfrac{5}{4}\,(0.0030)\,(0.1785) = 0.0007 \text{ rad}, \tag{3.51}$$

or 2.4 sec of arc. Of course, this effect is much too small to be observed in the approximation we are working with here. But it gives us a preview of an effect to be expected in the next approximation.

It should be remembered that we have been considering only long-term perturbations of the Moon's motion. Perturbation by the Sun also produces significant periodic effects of shorter term, including nutation with amplitude of nine minutes, and a 40% oscillation in eccentricity.

For long-term solar effects on a near-Earth artificial satellite our results will be much more accurate than for the Moon and especially significant since they apply for arbitrary inclination and eccentricity. However, the satellite will be perturbed by the Moon as well as the Sun. In fact, the lunar effect is more than twice as great as the solar effect, for an estimate of the ratio of their effects from (3.37) gives

$$\frac{\varkappa_M}{\varkappa_S} = \frac{M_M}{M_S}\left(\frac{a_S}{a_M}\right)^3\left(\frac{1 - \varepsilon_S^2}{1 - \varepsilon_M^2}\right)^{3/2} = 2.194. \tag{3.52}$$

This is the same as our earlier estimate for the ratio of lunar/solar tidal forces on Earth.

To calculate the lunar-solar effects on a satellite, we simply add separate versions of (3.39) for Sun and Moon; thus,

$$\bar{\boldsymbol{\omega}} = (\omega_{1M} + \omega_{1S})\mathbf{e}_3 + (\omega_{NM} + \omega_{NS})\mathbf{e}_1 + (\omega_{PM}\mathbf{u}_M + \omega_{PS}\mathbf{u}_S). \tag{3.53}$$

Most notable here is the fact that the axis of precession specified by the last term is determined by vector addition, and, of course, it will change slowly with time since $\mathbf{u}_M$ precesses about $\mathbf{u}_S$ with an 18 yr period, as we learned in our study of the Moon. It should be noted that a second order calculation of lunar effects should not be expected to give better than 10% accuracy, for a bound on the accuracy is given by the ratio.

$$\frac{\text{Earth radius}}{\text{Earth-Moon distance}} \approx \frac{1}{81} = 0.012. \tag{3.54}$$

Luni-Solar Precession and Nutation

Since the Earth is oblate, the nonuniform gravitational fields of the Sun and the Moon exert torques on the Earth, driving changes in the direction of

Earth's axis known as Luni-Solar precession and nutation.

To calculate the long-term effects of the Sun on the Earth's rotation we use the secular field (3.34). This field exerts a torque on the Earth

$$\boldsymbol{\Gamma}_S = \int \mathrm{d}m\ \mathbf{r} \times \mathbf{g}_S = -\frac{3}{2}\frac{GM_S}{b_S^3}\left[\int \mathrm{d}m\ \mathbf{rr}\cdot\mathbf{u}_S\right] \times \mathbf{u}_S .$$

Using the expression (1.20a) for the inertia tensor $\mathscr{I}$, this can be written

$$\boldsymbol{\Gamma}_S = \frac{3}{2}\frac{GM_S}{b_S^3}\left(\mathscr{I}\mathbf{u}_S\right) \times \mathbf{u}_S . \tag{3.55}$$

Then, using the special form (1.25) for the inertia tensor of an axisymmetric body, we get

$$\boldsymbol{\Gamma}_S = -\frac{3}{2}\frac{GM_S}{b_S^3}(I_1 - I_3)\mathbf{u}_S\cdot\mathbf{ee} \times \mathbf{u}_S , \tag{3.56}$$

where $\mathbf{e}$ designates the symmetry axis of the Earth. To this we must add a completely analogous expression for the Moon to get the total Luni-Solar torque $\boldsymbol{\Gamma}_S + \boldsymbol{\Gamma}_M$.

Now, by the method developed in Section 7-3, we can reduce the equation of motion for the Earth's angular velocity $\boldsymbol{\Omega}$ to the form

$$I_1\dot{\boldsymbol{\Omega}} = \boldsymbol{\Gamma}_S + \boldsymbol{\Gamma}_M$$

or

$$\dot{\boldsymbol{\Omega}} = \mathbf{e} \times \mathbf{F} \tag{3.57}$$

where $\mathbf{F}$ is defined by

$$\mathbf{F} = -\frac{3}{2}\left(\frac{I_1 - I_3}{I_2}\right)\left[\left(\frac{GM_S}{b_S^3}\right)\mathbf{u}\cdot\mathbf{eu}_S + \left(\frac{GM_M}{b_S^3}\right)\mathbf{u}_M\cdot\mathbf{eu}_M\right] . \tag{3.58}$$

We know from our studies in Sections 7-3 and 7-4 that the solution of (3.57) can be quite complicated even when $\mathbf{F}$ is constant, but it is sensitive to initial conditions. Fortunately, any large nutation that the Earth may have had as a result of initial conditions has long since been damped out by the time dependence of $\mathbf{F}$.

To extract the main effects from (3.58) we assume that $\mathbf{u}_S$ is constant, though it actually precesses slowly due to perturbations by the other planets. We know from (3.44) that $\mathbf{u}_M$ precesses about $\mathbf{u}_S$ with a period of 18.6 yr. Therefore, the direction of $\mathbf{F}$ oscillates and this drives a nutation of the Earth's axis with the same period. This nutation with a period of 18.6 yr is called *Luni-Solar nutation*. The nutation is an oscillation about steady precession superimposed on the Chandler wobble shown in Figure 3.5 of Section 7-3.

To study the steady precession, we separate it from the nutation by replacing $\mathbf{u}_M$ in (3.58) by its mean value over its period $\mathbf{u}_M = \alpha \mathbf{u}_S$, where

$$\alpha = \mathbf{u}_M \cdot \mathbf{u}_S = \cos(5.14°) = 0.996.$$

Introducing empirical values for the other parameters, we get

$$|\mathbf{F}| = \frac{3}{2}\left(\frac{I_1 - I_3}{I_3}\right)\mathbf{u}_S\cdot\mathbf{e}\left[1 + \alpha^2\left(\frac{M_M}{M_S}\right)\left(\frac{b_S}{b_M}\right)^3\right]\frac{GM_M}{b_M^3}$$

$$= \frac{3}{2}\left(\frac{1}{305.3}\right)\cos\,(23.5°)\,[1 + (0.996)^2\,(2.194)]\left(\frac{2\pi}{\text{yr}}\right)^2$$

$$= \frac{1}{69.87}\left(\frac{2\pi}{\text{yr}}\right)^2.$$

By Equation (3.50b) of Section 7-3, the period of precession for slow top is given by

$$T_P = \frac{2\pi\boldsymbol{\Omega}\cdot\mathbf{e}}{|\mathbf{F}|}\ . \tag{3.59}$$

Here $\boldsymbol{\Omega}\cdot\mathbf{e} = 2\pi$ day. Hence, for the period of *Luni-Solar precession* we get the estimate

$$T_P = (69.87)(365.25)\text{yr} = 25\,521\text{ yr}. \tag{3.60}$$

This is reasonably close to the observed value of 25400 yr. The Luni-Solar precession is directly observable as a *precession of the equinoxes*. Note that the minus sign in (3.58) means that the precession is retrograde.

Satellite Attitude Stability

In Section 8-1 we found that an asymmetric body with inertia $\mathcal{I}$ in the field of a centrosymmetric body with mass M is subjected to a gravitational torque

$$\boldsymbol{\Gamma} = \frac{3GM}{r^3}\,\hat{\mathbf{r}} \times \mathcal{I}\hat{\mathbf{r}}. \tag{3.61}$$

This is appropriate for investigating short-term effects of gravitational torques, in contrast to (3.55), which applies only to long-term effects.

Note that the torque (3.61) vanishes if one of the principle axes is aligned with $\mathbf{r}$. It is of interest to investigate the stability of such an alignment for orbiting satellites. Such knowledge can be used, for example, to keep communication satellites facing the Earth. Fortunately, the simplest alignment is also the most significant, so we shall limit our attention to that case.

We suppose that one principal vector $\mathbf{e}_3$ is perpendicular to the orbital plane, so it is not affected by the orbital motion. Then we can write

$$\mathcal{I}\hat{\mathbf{r}} = I_1\mathbf{e}_1 \cos\theta - I_2\mathbf{e}_2 \sin\theta,$$

where θ measures the deviation of $\mathbf{e}_1$ from $\hat{\mathbf{r}}$, so $\hat{\mathbf{r}}\cdot\mathbf{e}_1 = \cos\theta$ and $\hat{\mathbf{r}}\cdot\mathbf{e}_2 = -\sin\theta$. It follows that

$$\hat{\mathbf{r}} \times \mathcal{I}\hat{\mathbf{r}} = \mathbf{e}_3(I_1 - I_2)\cos\theta \sin\theta,$$

Also, if Ω is the angular velocity of the satellite, then $\mathcal{I}\boldsymbol{\Omega} = I_3\boldsymbol{\Omega} = I_3\dot{\theta}\mathbf{e}_3$. Therefore, Euler's equation for the satellite reduces to

$$I_3\ddot{\theta} = \frac{3}{2}\frac{GM}{r^3}(I_1 - I_2)\sin 2\theta. \tag{3.62}$$

For a circular orbit, we can write $GM/r^3 = n^2$, where n is the constant orbital angular velocity. Then in the small angle approximation $\sin 2\theta \approx 2\theta$, it is evident that solutions to (3.62) are stable about $\theta = 0$ if $I_1 < I_2$ and unstable if $I_1 > I_2$.

Assuming $I_1 < I_2$ and $n^2 = GM/r^3$ constant, we recognize (3.62) as the equation for a pendulum, for which we found the general solution in terms of elliptic functions in Section 7-4. For small displacements from equilibrium, the satellite oscillates harmonically with natural period

$$T_0 = \frac{2\pi}{n}\sqrt{\frac{I_3}{3(I_2 - I_1)}}\ . \tag{3.63}$$

This result is pertinent to the Moon. Its mean rotational period is equal to its orbital period, so it orbits with the same face (centered at one end of its longest principal axis) directed towards the Earth. A small perturbation would be sufficient to set the face in harmonic oscillation about this equilibrium motion. In principle, one could measure the period of this oscillation and obtain a value for $(I_2 - I_1)/I_3$ from (3.63). But the actual oscillations are so small and the period is so long that such a measurement was not feasible until quite recently. However,the same quantity can be determined from the observation the forced oscillation which synchronizes the orbital and rotational motions. This is due to the fact that the Moon's orbit is actually elliptical, so the factor r^{-3} undergoes small oscillations over an orbital period which force oscillations of the angle θ according to (3.62).

The equivalence of lunar spin and orbital periods is an example of *spin-orbit resonance* which exists throughout the solar system. It is believed to come about in the following way: The Moon loses an arbitrary initial spin gradually by tidal effects until it falls into a stable resonant state, where the rotational motion is sustained by parametric orbital forcing through the coefficient r^{-3} in the torque. Mercury is in a 3:2 resonant state. Other spin-orbit resonances are found among the moons of the major planets, and some moons are coupled in orbit-orbit resonances as well.

The Advance of Mercury's Perihelion

We turn now to secular perturbations by the planets. As a specific example, we evaluate the secular perturbations of Mercury. We shall see that the main effect is a forward precession of Mercury's perihelion. Historically, difficulty in accounting for this effect provided the first evidence for limitations of Newtonian theory, and the resolution of the difficulty was one of the first triumphs of Einstein's General Theory of Relativity. More accurate data from the space program will undoubtedly make this a more stringent test of gravitational theories in the future.

We aim to calculate the main effect on Mercury to an accuracy of a few percent. From the table of planetary data in Appendix C, we conclude that, to this accuracy, we can neglect the inclinations and eccentricities of the perturbing planets; for our experience with Solar perturbations has taught us that the relative effect of inclination is on the order of the sine of the angle, and eccentricity corrections are on the order of eccentricity squared. Accordingly, we can use (3.31) and (3.32) in a heliocentric reference system to describe the secular force $\mathbf{f}_p$ of the pth planet on Mercury. The force is a central force

$$\mathbf{f}_p = \hat{\mathbf{r}} f_p(r) \tag{3.64a}$$

where $f_p(r)$ has the explicit form

$$f_p(r) = \frac{Gm_p}{2a_p}\left(\frac{r}{a_p^2 - r^2}\right)\left[1 + \frac{1}{8}\left(\frac{r}{a_p}\right)^2 + \frac{3}{64}\left(\frac{r}{a_p}\right)^4 + \ldots\right]. \tag{3.64b}$$

Our perturbation equations (3.1) to (3.4) tell us that the *only* secular effects of a central force $\mathbf{f}$ is a precession of the apses at a rate

$$\bar{\omega}_p = -\frac{h}{\mu\varepsilon}\,\hat{\boldsymbol{\varepsilon}}\cdot\overline{\mathbf{f}_p}\,. \tag{3.65}$$

Computation of the orbital average $\mathbf{f}_p$ can be simplified by exploiting the fact that the radius r oscillates about Mercury's semi-major axis a. To second order, a Taylor expansion gives us

$$f_p(r) = f_p(a) + (r-a)\,f_p'(a) + \frac{(r-a)^2}{2}\,f_p''(a)\,. \tag{3.66}$$

Inserting this into (3.64b) and computing the orbital averages

$$\bar{\hat{\mathbf{r}}} = -\boldsymbol{\varepsilon}, \quad \overline{(r-a)\hat{\mathbf{r}}} = -\tfrac{1}{2}a\boldsymbol{\varepsilon}, \quad \overline{(r-a)^2\hat{\mathbf{r}}} = -\tfrac{1}{4}a^2\varepsilon^2\boldsymbol{\varepsilon},$$

we obtain

$$\bar{\mathbf{f}}_p = -\boldsymbol{\varepsilon}\{f_p(\mathrm{a}) + \tfrac{1}{2}af_p'(a) + \tfrac{1}{4}\varepsilon^2a^2 f_p''(a)\}\,.$$

Although Mercury has a large eccentricity $\varepsilon = 0.2056$, the coefficient $\varepsilon^2/4 = 0.0106$ shows that the last term can be neglected in our approximation.

Inserting (3.67) into (3.65) and using (3.64b) as well as the identity

$$\frac{h}{\mu} = \frac{2\pi}{T} \frac{a^2(1-\varepsilon^2)^{1/2}}{GM_S},$$

we get the secular precession rate in a form which is convenient for computation:

$$\overline{\omega_p} = (1-\varepsilon^2)^{1/2} \frac{\pi}{T} \frac{a^2}{M_S} \frac{2}{G} [f_p(a) + \frac{1}{2} a f_p'(a)] \tag{3.68}$$

Here T is Mercury's orbital period and M_S is the mass of the sun. In units with the Earth mass and Earth-Sun distance set equal to 1, the coefficient in (3.68) has the value

$$\frac{(1-\varepsilon^2)^{1/2}\pi}{T} \frac{a^2}{M_S} = \frac{(0.97864)\pi}{87.969 \text{ day}} \frac{(0.381)^2}{3.329 \times 10^5} \left(365.25 \times 10^2 \frac{\text{day}}{\text{cent}} \right)$$

$$\times \left(20.628 \times 10^4 \frac{\text{arc sec}}{\text{rad}} \right) = 118.53 \frac{\text{arc sec}}{\text{cent}}, \tag{3.69}$$

where the units of radians per day have been converted to seconds of arc per century. To evaluate the remaining factor in (3.68), it is convenient to express $2G^{-1}f_p(r)$ as the product $A_p(r)B_p(r)$ of a dominant term

$$A_p(r) = \frac{m_p}{a_p} \left(\frac{r}{a_p^2 - r^2} \right) \tag{3.70a}$$

and a correction factor

$$B_p(r) = 1 + \frac{1}{8} \left(\frac{r}{a_p} \right)^2 + \frac{3}{64} \left(\frac{r}{a_p} \right)^4. \tag{3.70b}$$

Then, we compute the derivatives

$$aA_p'(a) = A_p \left(\frac{a_p^2 + a^2}{a_p^2 - a^2} \right), \tag{3.70c}$$

$$aB_p'(a) = \frac{1}{4} \left(\frac{a}{a_p} \right)^2 + \frac{12}{64} \left(\frac{a}{a_p} \right)^4. \tag{3.70d}$$

And we use these results in

$$C_p \equiv \frac{2}{G} [f_p + \tfrac{1}{2} a f_p'] = A_p(B_p + \tfrac{1}{2} aB_p') + \tfrac{1}{2} aA_p' B_p. \tag{3.71}$$

This quantity is evaluated numerically for each of the perturbing planets in Table 3.1, which also displays values for the various factors so their contributions can be assessed. Values for the three most distant planets Uranus,

TABLE 3.1. Data and Calculations

Planet	m_p	a_p	a/a_p	A_p	$\frac{1}{2}aA_p'$	B_p	$\frac{1}{2}aB_p'$	C_p	$\overline{\omega}_p$
1. Mercury	0.0554	0.38710	1.0						
2. Venus	0.815	0.72333	0.53516	1.168	0.90135	1.0396	1.0831	2.3600	279.73
3. Earth	1.000	1.00000	0.38710	0.45533	0.67625	1.0198	1.0406	0.7878	93.38
4. Mars	0.1075	1.52369	0.25405	0.01258	0.56900	1.0060	1.0167	0.01999	2.37
5. Jupiter	317.83	5.20280	0.07440	0.87845	0.50539	1.0007	1.0013	1.3271	157.30
6. Saturn	95.147	9.53884	0.05481	0.04798	0.50147	1.0002	1.0003	0.0720	8.53
7. Uranus	14.54	19.1819							
8. Neptune	17.23	30.0578							
9. Pluto	0.17	39.44						Total	541.31
Sun	3.329×10^5								

Neptune and Pluto have been omitted, since it is evident that, by comparison with the small contribution of the larger and closer planet Saturn, their contributions are negligible. Figures in the last column are the calculated precession rates $\overline{\omega}_p$ due to each of the five closest planets. For the total *secular precession rate of Mercury's perihelion due to planetary perturbations*, we obtain

$$\overline{\omega} = \sum_{p=2}^{6} \overline{\omega}_p = 541''.3/\text{cent}$$

This is within 2% of the accepted value 531''.5/cent, about as close as we should expect.

For comparison with our results, results of the most precise and carefully checked calculations are displayed in Table 3.2 for Earth as well as Mercury. In the case of Mercury, the difference of 43''/cent between the observed advance and the calculated planetary effects was recognized as a serious problem from the time of the first accurate calculations by Leverrier in 1859. Many theories were proposed to account for it. These theories are of two general types: (1) those in which Newton's law of gravitation is retained, but the existence of an unobserved planet or ring of material particles inside Mercury's orbit is postulated; (2) proposals to modify Newton's law. More recently, the possibility that some or all of the discrepancy may be due to the oblateness of the Sun has been considered seriously. Einstein's theory of General Relativity, proposed in 1915, accounted for the discrepancy in spectacular fashion. His theory provides an explanation of the second type, but it is distinguished from alternatives by (a) its lack of arbitrariness (no adjustable constants are introduced), (b) its accurate prediction of other phenomena such as gravitational deflection of light passing near the Sun, (c) the fact that it is a derived consequence of deep revisions in the foundations of mechanics. Observations have been sufficient to completely rule out theories

TABLE 3.2. Contributions to the Perihelion Advance of Mercury and Earth

Cause	Precession rate (arc sec/century)	
	Mercury	*Earth*
Mercury		−13″.75 ± 2″.3
Venus	277″.856 ± 0″.68	345.49 ± 0.8
Earth	90.038 ± 0.08	
Mars	2.536 ± 0.00	97.69 ± 0.1
Jupiter	153.584 ± 0.00	696.85 ± 0.0
Saturn	7.302 ± 0.01	18.74 ± 0.0
Uranus	0.141 ± 0.00	0.57 ± 0.0
Neptune	0.042 ± 0.00	0.18 ± 0.0
Moon		7.68 ± 0.0
Sum	531.50 ± 0.85	1153.45 ± 2.7
Observed	574.09 ± 0.41	1158.05 ± 0.5
Difference	42.56 ± 0.94	4.6 ± 2.7
Relativity effect	43.03 ± 0.03	3.8 ± 0.0

Reference: G. M. Clemence, *Reviews of Modern Physics* **19**, 361 (1947).

of the first type only recently, but the exact magnitude of the Sun's oblateness effect is still uncertain.

Although we cannot go into Einstein's theory here, we can evaluate its implications for planetary motion. According to Einstein's theory, the Newtonian gravitational force on a planet should be modified by adding the terms

$$\mathbf{f}_{Rel} = \frac{1}{c^2}\,[-2\nabla\phi^2 + 4\mathbf{v}(\mathbf{v}\cdot\nabla)\phi - v^2\nabla\phi], \tag{3.73}$$

where ϕ is the gravitational potential, and c is the speed of light in a vacuum. For a spherically symmetric Sun, $\phi = -\mu/r$, so

$$\begin{aligned}\mathbf{f}_{Rel} &= \frac{-\mu}{c^2}\left[2\mu\nabla\frac{1}{r^2} + 4\mathbf{v}\left(\mathbf{v}\cdot\nabla\frac{1}{r}\right) - v^2\nabla\frac{1}{r}\right]\\ &= \frac{\mu}{c^2}\left[4\mu\frac{\hat{\mathbf{r}}}{r^3} + 4\mathbf{v}\frac{(\mathbf{v}\cdot\hat{\mathbf{r}})}{r^2} - v^2\frac{\hat{\mathbf{r}}}{r^2}\right].\end{aligned} \tag{3.74}$$

To evaluate its secular effects on the motion of a planet, we first note that its secular torque vanishes:

$$\overline{\mathbf{r}\wedge\mathbf{f}_{Rel}} = -4\mu\mathbf{h}\left(\overline{\mathbf{v}\cdot\nabla\frac{1}{r}}\right) = 0. \tag{3.75}$$

Therefore, like a central force, it will not contribute to secular precession or nutation of planetary orbits, and its effect on apse precession is completely determined by its secular average $\overline{\mathbf{f}}_{Rel}$.

Before computing the average, it is convenient to use $v^2 = 2\mu/r + 2E$ to write (3.74) in the form

$$\mathbf{f}_{Rel} = \frac{\mu}{c^2}\left[2\mu\frac{\mathbf{r}}{r^4} + 4\mathbf{v}\frac{(\mathbf{v}\cdot\mathbf{r})}{r^3} - 2E\frac{\mathbf{r}}{r^3}\right].$$

We can easily compute the average of each term with the help of Table 2.1. Since the orbital element E is to be regarded as constant when computing averages, the last term does not contribute, because

$$\overline{\left(\frac{\mathbf{r}}{r^3}\right)} = 0.$$

For the first term, we compute

$$\overline{\left(\frac{\mathbf{r}}{r^4}\right)} = \frac{1}{2b^3}\,\boldsymbol{\varepsilon}.$$

To evaluate the second term, we use $\mathbf{v} = \mu h^{-2}(\mathbf{h} \times \varepsilon + \mathbf{h} \times \hat{\mathbf{r}})$, which also implies $\mathbf{v}\cdot\hat{\mathbf{r}} = \mu h^{-2}(\mathbf{h} \times \boldsymbol{\varepsilon})\cdot\hat{\mathbf{r}}$. Therefore

$$\overline{\mathbf{v}\frac{\mathbf{v}\cdot\mathbf{r}}{r^3}} = \frac{\mu^2}{h^2}\left\{\mathbf{h} \times \boldsymbol{\varepsilon}(\mathbf{h} \times \boldsymbol{\varepsilon})\cdot\overline{\left(\frac{\mathbf{r}}{r^3}\right)} + \mathbf{h} \times \left[\overline{\frac{(\mathbf{h} \times \boldsymbol{\varepsilon})\cdot\mathbf{r}\mathbf{r}}{r^4}}\right]\right\}.$$

The first term on the right vanishes, and to evaluate the second term we compute

$$\overline{\frac{(\mathbf{h} \times \boldsymbol{\varepsilon})\cdot\mathbf{r}\mathbf{r}}{r^4}} = \frac{\mathbf{h} \times \boldsymbol{\varepsilon}}{2ab}.$$

Then, since $\mu/h^2 = a/b^2$, we have

$$\overline{\mathbf{v}\frac{\mathbf{v}\cdot\mathbf{r}}{r^3}} = \frac{\mu}{h^2}\frac{a}{b^2}\frac{\mathbf{h} \times (\mathbf{h} \times \boldsymbol{\varepsilon})}{2ab} = \frac{-\mu}{b^3}\,\boldsymbol{\varepsilon}.$$

Thus, for the secular value of (3.76), we find

$$\overline{\mathbf{f}}_{Rel} = \frac{\mu}{c^2}\left[2\mu\left(\frac{\boldsymbol{\varepsilon}}{2b^3}\right) + 4\left(\frac{-\mu\boldsymbol{\varepsilon}}{b^3}\right)\right] = \frac{-3\mu^2\boldsymbol{\varepsilon}}{c^2b^3}. \tag{3.77}$$

Then, for the relativistic contribution to perihelion precession, we find

$$\begin{aligned}\overline{\omega_{Rel}} &= -\frac{h}{\mu\varepsilon}\,\hat{\boldsymbol{\varepsilon}}\cdot\overline{\mathbf{f}}_{Rel} = -\frac{2\pi}{T}\frac{ab}{\mu\varepsilon}\left(\frac{-3\mu^2\varepsilon}{c^2b^3}\right) \\ &= \frac{6\pi}{T}\frac{a\mu}{b^2c^2} = \frac{6\pi}{T}\frac{GM_s}{a(1-\varepsilon^2)c^2}.\end{aligned} \tag{3.78}$$

When empirical values are inserted, this gives 43″/cent for Mercury and 3.8″/cent for Earth, as reported in Table 3.2.

From observation of the Sun's surface shape and rotation rate, the Sun's quadrupole moment J_2 is believed to be comparatively small, contributing less than one arc second per century to the advance of Mercury's perihelion. A much larger J_2 would be inconsistent with Einstein's explanation of the advance. Skeptics have pointed out that the Sun may be rotating more rapidly beneath its visible surface, thus producing a larger J_2. However, it is not difficult to show that a J_2 large enough to contribute more than 8″/cent to Mercury's advance would be inconsistent with empirical data on precession of nodes and change of inclination for orbits of the inner planets. For we have seen that J_2 contributes to these effects while, because of (3.75), relativity does not. This issue will be set to rest when NASA completes its goal of determining the Sun's J_2 accurately with observations on artificial satellites orbiting close to the Sun.

8-3. Exercises

(3.1) The gravitational force of the Sun on the Moon is much larger than that of the Earth. How is it, then, that we can ignore the Sun in a first approximation when calculating the orbit of the Moon? Support your explanation with order of magnitude estimates.

(3.2) Use Equation (3.18) to calculate the rate of regression of the lunar nodes. Compare with the observed result of 6.6″/yr for regression along the ecliptic. Take into account the fact that the inclination of the Moon's orbit to the equator varies by about 10°.

(3.3) The greatest known oblateness effect in the Solar System occurs for Jupiter's fifth satellite which is so close to the highly oblate (1/15.4) planet that the nodes regress more than $2\frac{1}{2}$ complete revolutions per year. The inclination of the orbit is 0.4°, and $a = 181 \times 10^3$ km for the planet. Use this information to estimate the quadrupole moment J_2 of Jupiter.

(3.4) The artificial Earth satellite Vanguard I launched in 1958 had the following orbital elements:

Semi-major axis	$a = 1.3603\ R_E$
Eccentricity	$\varepsilon = 0.1896$
Inclination	$\iota = 34.26°$
Mean motion	$n = 3867.3$ deg/day
Period	$T = \frac{2\pi}{n} = 134.05$ min

For Vanguard I and a synchronous (24 hr) satellite, calculate the oblateness and lunar-solar perturbations and compare with the results of L. Blitzer in the following table:

Secular Perturbations

		Oblateness (deg/day)	Moon (deg/day)	Sun (deg/day)
Vanguard I	$\dot{\Omega}$	−3.02	−0.00028	−0.00013
	$\dot{\psi}$	+4.41	+0.00039	+0.00018
24-Hr Orbit	$\dot{\Omega}$	−0.057	−0.0030	−0.0014
	$\dot{\psi}$	+0.084	+0.0042	+0.0019

The estimated lifetime of Vanguard I is 200 yr. How much has its orbit been changed by the above perturbations since it was launched?

(3.5) At what distance from the Earth will Lunar-Solar and Oblateness effects on an Earth satellite be of the same order of magnitude?

(3.6) The Sun produces radial oscillations in the orbit of an Earth satellite which average to zero over the orbital period. To investigate the magnitude of this effect, suppose that the unperturbed orbit is a circle of radius r and the Sun is in the orbital plane. Derive a first order expression for the radial variation δr as a function of the time. Show that it is harmonic with period $T/2$ and a maximum displacement

$$(\delta r)_{\max} = \frac{n_S^2}{n^2}\, r,$$

where $n = 2\pi/T = (GM/a^3)^{1/2}$ is the "mean motion" of the satellite about the Earth and n_S is the mean motion of the Earth about the Sun. Show that δr ranges from less than one meter for a near Earth satellite to 2500 km for the Moon. Note the similarity of this effect to the tides.

(3.7) To account for the 43″ discrepancy in the advance of Mercury's perihelion, Seeliger proposed the "screened Kepler potential"

$$\phi(r) = -\frac{GM_S}{r}\, e^{-r/d}$$

for the gravitational potential of the Sun. Assuming $d \ll r$, determine the value of d required. (Reference: N. T. Roseveare, *Mercury's Perihelion from LeVerrier to Einstein*, Clarendon, Oxford, 1982).

(3.8) Calculate the precession rate of Earth's perihelion due to Mars, Jupiter and Saturn and compare with the results in Table 3.2.

(3.9) Equations (3.31) and (3.32) for the secular field of a planet apply only inside the planet's orbit. Derive corresponding equations for the secular field outside the planet's orbit. Use these equations to calculate the precession rate of Earth's perihelion due to Mercury and Venus. Compare your results with those in Table 3.2.

(3.10) For a charged particle bound by a Coulomb force, determine the secular effects of perturbing constant magnetic field **B**, including the Larmor precession frequency.

(3.11) For a charged particle bound by a Coulomb force determine the secular effects of a perturbing constant electric field **E**.

(3.12) The Newtonian equation of motion for a planet under the influence of the Sun alone can be written

$$\dot{\mathbf{p}} = -GM_S m \frac{\mathbf{r}}{r^3}$$

where $\mathbf{p} = m\mathbf{v}$. Einstein's Theory of Special Relativity simply changes the expression for the momentum to $\mathbf{p} = m\gamma\mathbf{v}$, where $\gamma = (1 - v^2/c^2)^{-1/2}$ and c is the speed of light. Adopting this change, show that the equation of motion can be written in the form of a perturbed Newtonian equation

$$\dot{\mathbf{v}} = -GM_S \frac{\mathbf{r}}{r^3} + \mathbf{f}_{sp}$$

where, since $v^2 \ll c^2$ the perturbing force is given by

$$\mathbf{f}_{sp} = -\dot{\gamma}\mathbf{v} + (1 - \gamma)\dot{\mathbf{v}} \approx \frac{GM_S}{c^2}\left[\frac{\mathbf{r}\cdot\mathbf{v}}{r^3}\mathbf{v} + \frac{1}{2}v^2\frac{\mathbf{r}}{r^3}\right].$$

How much does this contribute to the advance of Mercury's perihelion?

(3.13) Show that for a constant perturbing force the orbital elements satisfy the secular equations of motion

$$\dot{E} = 0, \quad \dot{\mathbf{h}} = -\frac{3a}{2}\boldsymbol{\varepsilon} \times \mathbf{f}, \quad \mu\dot{\boldsymbol{\varepsilon}} = -\frac{3}{2}\mathbf{h} \times \mathbf{f}.$$

Study the consequent changes in the shape and attitude of the osculating orbit.

The averaging can be simplified by noting that

$$\mathbf{f} \times \mathbf{h} = \mathbf{f} \times (\mathbf{r} \times \mathbf{v}) = \mathbf{f}\cdot\mathbf{v}\mathbf{r} - \mathbf{f}\cdot\mathbf{r}\mathbf{v}$$

and

$$\overline{\mathbf{f}\cdot\mathbf{r}\mathbf{v}} + \overline{\mathbf{f}\cdot\mathbf{v}\mathbf{r}} = 0,$$

so

$$\overline{\mathbf{f}\cdot\mathbf{v}\mathbf{r}} = \tfrac{1}{2}\mathbf{f} \times \mathbf{h}.$$

(3.14) *Solar Wind.* The intensity of Solar radiation at the mean Earth-distance from the Sun (the solar constant) is

$$I = 1.36 \times 10^{3} \text{ J m}^{-2} \text{ s}^{-1}.$$

On an absorbing body, this produces a radiation pressure normal to the incident beam of magnitude

$$P = \frac{I}{c} = 4.5 \times 10^{-6} \text{ N m}^{-2}$$

where c is the speed of light.

On the Earth satellites at altitudes above 800 km, the effect of radiation pressure is greater than atmosphere drag. The greatest effect has been observed on the 30 km ECHO balloon in its nearly circular orbit at an altitude of 1600 km. Assuming that the balloon is perfectly reflecting with area/mass = 102 $\text{cm}^2 \text{ gm}^{-1}$, estimate its daily variation in perigee due to radiation pressure. Thus, show that radiation pressure can have a substantial effect on a satellite's lifetime. Note that the Solar radiation force can be regarded as constant, so the results of the preceding exercise can be used.

(3.15) According to the drag paradox, atmospheric drag increases the speed of a satellite as it spirals inwards. Prove this statement using only very general assumptions about the drag force.

(3.16) From Section 3-4, we know that the atmospheric drag force on a satellite is of the form

$$\mathbf{F}_D = F_D \hat{\mathbf{v}}, \quad \text{with} \quad F_D = \tfrac{1}{2} C_D \varrho A V^2,$$

where drag coefficient $C_D \approx 2$, ϱ is the atmospheric density, A is the effective cross-section area of the satellite, and V is the satellite velocity with respect to the local atmosphere. The simplest model atmosphere is spherically symmetric with a density distribution $\varrho = \varrho(r)$ which falls off exponentially with distance, so

$$\varrho(r) = \varrho_0 \, e^{-(r - r_0)/\sigma},$$

where ϱ_0 is the density at some chosen level $r = r_0$ and σ is a scale factor.

Derive secular equations of motion for perturbation by drag alone. Averages over the exponential density can be expressed as Bessel functions, but a simpler approach, sufficient for rough analysis, is to expand $\varrho(r)$ about $r = a$ before averaging over a period of the osculating orbit.

Show that drag does not alter the orbital plane. Derive an expression for the secular decay in size of the orbit, and show that the orbit tends to become increasingly circular as it decays. (For a detailed treatment of atmospheric drag, see D. G. King-Hele, *Theory of Satellite Orbits in an Atmosphere*, Butterworth, London, 1964).

8-4. Spinor Mechanics and Perturbation Theory

This section develops a new spinor formulation of classical mechanics which has not yet been widely applied, so it is a promising starting point for new research. The spinor formulation of perturbation theory in celestial mechanics has clear advantages over alternative formulations, so we will concentrate on that. But the approach is not without interest in atomic physics as well, for one cannot help asking if the classical spinor variables have some definite relation to the spinor wave functions in quantum mechanics. The question has not yet been studied in any depth. Nor are the purely classical applications of *spinor mechanics* sufficiently well worked out to present here. So we will be content with a formulation of the general theory without applications.

Position Vector and Spinor

From our study of linear transformations in Chapter 5, we know that geometric algebra enables us to write any rotation-dilation of Euclidean 3-space in the canonical form

$$\mathbf{x}' = U^\dagger \mathbf{x} U, \tag{4.1}$$

where $\mathbf{x}$ and $\mathbf{x}'$ are vectors and U is a nonzero quaternion (or spinor) with conjugate $U^\dagger$. This equation describes the rotation and dilation of any given vector $\mathbf{x}$ into a unique vector $\mathbf{x}'$. The modulus $|U|$ of the spinor U is a positive scalar determined by

$$|U|^2 = U^\dagger U = U U^\dagger .$$

Consequently, the spinor U, like any nonzero quarternion, has an inverse

$$U^{-1} = |U|^{-2} U^\dagger .$$

Equation (4.1) can now be written in the form

$$\mathbf{x}' = |U|^2 (U^{-1}\mathbf{x}U) .$$

This exhibits the transformation as the composite of a rotation $U^{-1}\mathbf{x}U$ and a dilation by a scale $|U|^2$.

We can use this result to represent the position of a particle by a *position spinor* U instead of a position vector $\mathbf{r}$. We simply choose an arbitrary fixed unit vector $\boldsymbol{\sigma}_1$ and write

$$\mathbf{r} = U^\dagger \boldsymbol{\sigma}_1 U . \tag{4.2}$$

This is just Equation (4.1) applied to a single vector rather than regarded as a linear transformation of the whole vector space. Squaring (4.2), we get $\mathbf{r}^2 = |U|^4$, so

$$r = |\mathbf{r}| = |U|^2. \tag{4.3}$$

Thus, the radial distance r is represented as the scale factor $|U|^2$ of a rotation-dilation.

Although the position vector $\mathbf{r}$ is uniquely determined by the position spinor U according to (4.2), the converse is not true. Indeed, if S is a spinor such that

$$S^\dagger \boldsymbol{\sigma}_1 S = \boldsymbol{\sigma}_1, \tag{4.4}$$

then (4.2) gives us

$$\mathbf{r} = U^\dagger \boldsymbol{\sigma}_1 U = V^\dagger \boldsymbol{\sigma}_1 V, \tag{4.5}$$

where

$$V = SU, \tag{4.6}$$

and S is arbitrary except for the condition (4.4). The condition (4.4) simply states that $\boldsymbol{\sigma}_1$ is an eigenvector of the rotation $S^\dagger \mathbf{x} S$. In other words, S may be any spinor describing a rotation about the σ_1 axis, so it can be written in the parametric form

$$S = e^{(1/2)i\boldsymbol{\sigma}_1\phi}, \tag{4.7}$$

where ϕ is the scalar angle of rotation and i is the unit pseudoscalar.

Let us refer to the transformation (4.6) of U into V as a *gauge transformtion*, because it is similar to the gauge transformation of a spinor state function in quantum theory. We say then that Equation (4.2) is invariant under the one-parameter group of gauge transformations specified by (4.6) and (4.4) or (4.7). If Equation (4.2) is regarded as a linear transformation of the vector $\boldsymbol{\sigma}_1$ into $\mathbf{r}$, the *gauge invariance* simply means that this transformation is invariant under a rotation about the radial axis. We suppose that $\boldsymbol{\sigma}_1$ is some definite unit vector, though the choice is arbitrary. Given $\boldsymbol{\sigma}_1$, by Equation (4.2) a spinor U determines a unique vector $\mathbf{r}$, but the vector $\mathbf{r}$, determines U only up to a gauge transformation. This nonunique correspondence between spinors and vectors is to be expected, of course, because it takes four scalar parameters to specify the quaternion U but only 3 parameters to specify the vector $\mathbf{r}$. To associate a unique spinor U with the vector $\mathbf{r}$, we must impose some *gauge condition* consistent with (4.2) to *fix* the *gauge* uniquely. A natural gauge condition appears when we consider kinematics.

Velocity Vector and Spinor

Let $\mathbf{r} = \mathbf{r}(t)$ be the orbit of a particle in *position space*, so $\mathbf{h} = \mathbf{r} \times \dot{\mathbf{r}}$ is the angular momentum (per unit mass), and

$$\mathbf{r}\dot{\mathbf{r}} = \mathbf{r}\cdot\dot{\mathbf{r}} + i(\mathbf{r} \times \dot{\mathbf{r}}) = r\dot{r} + i\mathbf{h}. \tag{4.8}$$

The kinematic significance of this quantity will become apparent in the following.

Equation (4.2) relates an orbit $U = U(t)$ in *spinor space* to an orbit $\mathbf{r} = \mathbf{r}(t)$ in *position space*. We still need to relate the velocity $\dot{U}$ in spinor space to the velocity $\dot{\mathbf{r}}$ in position space. Differentiating $r = |U|^2$, we obtain

$$\dot{r} = \dot{U}U^\dagger + U\dot{U}^\dagger = 2\langle \dot{U}U^\dagger\rangle_0. \tag{4.9}$$

Next, it will be convenient to introduce a quaternion W defined by

$$W = 2U^{-1}\dot{U} = 2r^{-1}U^\dagger\dot{U} = r^{-1}\dot{r} + i\boldsymbol{\omega}, \tag{4.10}$$

where $\boldsymbol{\omega}$ is a vector and (4.9) has been used to determine that $\langle W\rangle_0 = r^{-1}\dot{r}$.

We can put (4.10) in the form

$$\dot{U} = \tfrac{1}{2}UW, \tag{4.11}$$

from which we obtain $\dot{U}^\dagger = \frac{1}{2}W^\dagger U^\dagger$. If we insert these expressions into the equation

$$\dot{\mathbf{r}} = \dot{U}^\dagger\boldsymbol{\sigma}_1 U + U^\dagger\boldsymbol{\sigma}_1\dot{U}$$

obtained by differentiating (4.2), we get

$$\dot{\mathbf{r}} = \tfrac{1}{2}(W^\dagger\mathbf{r} + \mathbf{r}W). \tag{4.12}$$

Using (4.10) this can be written

$$\dot{\mathbf{r}} = \dot{r}\hat{\mathbf{r}} + \tfrac{1}{2}i(\mathbf{r}\boldsymbol{\omega} - \boldsymbol{\omega}\mathbf{r}).$$

Then using $\mathbf{r}\boldsymbol{\omega} = \mathbf{r}\cdot\boldsymbol{\omega} + i(\mathbf{r} \times \boldsymbol{\omega})$, we obtain

$$\dot{\mathbf{r}} = \dot{r}\hat{\mathbf{r}} + \boldsymbol{\omega} \times \mathbf{r}. \tag{4.13}$$

Thus, we identify $\boldsymbol{\omega}$ as the *angular velocity* of the orbit $\mathbf{r} = \mathbf{r}(t)$.

According to (4.13), the radial component of $\boldsymbol{\omega}$ is irrelevant to $\dot{\mathbf{r}}$. Hence, we are free to eliminate it by introducing the *subsidiary condition*

$$\boldsymbol{\omega}\cdot\mathbf{r} = \langle\boldsymbol{\omega}\mathbf{r}\rangle_0 = 0. \tag{4.14}$$

This condition can be written in several equivalent ways; thus,

$$\boldsymbol{\omega}\mathbf{r} = -\mathbf{r}\boldsymbol{\omega}$$

or,

$$W^\dagger\mathbf{r} = \mathbf{r}W, \tag{4.15}$$

which after inserting (4.10) and (4.2), gives us

$$\dot{U}^\dagger\boldsymbol{\sigma}_1 U = U^\dagger\boldsymbol{\sigma}_1\dot{U}. \tag{4.16}$$

This is equivalent to the scalar condition

$$\langle iU^\dagger \boldsymbol{\sigma}_1 \dot{U}\rangle_0 = \langle i\boldsymbol{\sigma}_1 \dot{U}U^\dagger\rangle_0 = 0 . \tag{4.17}$$

Thus, we have expressed the subsidiary condition as a relation between U and its derivative $\dot{U}$.

Now, using (4.15) in (4.12) we obtain

$$\dot{\mathbf{r}} = \mathbf{r}W . \tag{4.18}$$

Solving for W and using (4.10), we get the fundamental result

$$W = 2U^{-1}\dot{U} = \mathbf{r}^{-1}\dot{\mathbf{r}} = r^{-1}\dot{r} + i\boldsymbol{\omega} . \tag{4.19}$$

Comparison with (4.8) shows us that the angular velocity is related to the angular momentum by

$$\boldsymbol{\omega} = r^{-2}\mathbf{h} .$$

This is a consequence of or, if you prefer, an alternative form of the subsidiary condition (4.14).

Equation (4.19) specifies completely our desired relation between $\dot{U}$ and $\dot{\mathbf{r}}$. Various special relations between $\dot{U}$ and $\dot{\mathbf{r}}$ are easily derived from it. For example,

$$|W|^2 = WW^\dagger = \frac{4}{r}\,|\dot{U}|^2 = \frac{\dot{\mathbf{r}}^2}{r^2} . \tag{4.20}$$

Useful alternative forms of (4.19) are obtained by multiplying it by $\mathbf{r}$ and using (4.12). Thus, we obtain

$$\dot{\mathbf{r}} = 2U^\dagger \boldsymbol{\sigma}_1 \dot{U} , \tag{4.21}$$

or, equivalently,

$$2r\dot{U} = \boldsymbol{\sigma}_1 U \dot{\mathbf{r}} . \tag{4.22}$$

The subsidiary condition (4.17) is a gauge condition. To see how it determines the gauge, consider an arbitrary time dependent gauge transformation $V = SU$. We wish to relate $\dot{V}$ to $\dot{U}$ to determine the effect of the gauge transformation. Differentiating (4.7) with $\phi = \phi(t)$, we have

$$\dot{S} = \tfrac{1}{2} i\boldsymbol{\sigma}_1 \dot{\phi} S = S \tfrac{1}{2} i\boldsymbol{\sigma}_1 \dot{\phi} . \tag{4.23}$$

So, using (4.11), we have

$$\dot{V} = \dot{S}U + S\dot{U} = \tfrac{1}{2}(i\boldsymbol{\sigma}_1 \dot{\phi} V + VW) .$$

With the help of (4.5) we can put this in the form

$$\dot{V} = \tfrac{1}{2} V(i\hat{\mathbf{r}}\dot{\phi} + W) . \tag{4.24}$$

This is a completely general relation showing how W can be altered by a gauge transformation. Using the specific form (4.19) for W, we obtain

$$2V^{-1}\dot{V} = 2U^{-1}\dot{U} + i\hat{\mathbf{r}}\dot{\phi} = r^{-1}\dot{r} + i(r^{-2}\mathbf{h} + \hat{\mathbf{r}}\dot{\phi}). \tag{4.25}$$

This shows explicitly that the gauge transformation adds a radial component $\hat{\mathbf{r}}\dot{\phi}$ to the angular velocity. We can solve (4.25) for $\dot{\phi}$, with the result

$$\dot{\phi} = -\langle i\hat{\mathbf{r}}2V^{-1}\dot{V}\rangle_0 = -\frac{2}{r}\langle iV^{\dagger}\boldsymbol{\sigma}_1\dot{V}\rangle_0. \tag{4.26}$$

This reduces to the subsidiary condition (4.17) if and only if $\dot{\phi} = 0$. Thus, the subsidiary condition fixes the gauge to a constant value. In other words, the gauge can be choosen freely at one time, but its value for all other times is then fixed by the subsidiary condition.

We have proved that any alternative to our gauge condition will have, in general, an angular velocity with a nonvanishing radial component. Equation (4.13) shows that a radial component of the angular velocity will not affect the velocity $\dot{\mathbf{r}}$ in position space, so we are free to adopt alternative gauge conditions. A physically significant alternative will be discussed later.

The Spinor Equation of Motion

The spinor acceleration corresponding to the acceleration vector in position space is most easily found by differentiating (4.22). Thus,

$$2\frac{d}{dt}(r\dot{U}) = \boldsymbol{\sigma}_1 U\ddot{\mathbf{r}} + \boldsymbol{\sigma}_1\dot{U}\dot{\mathbf{r}}$$

$$= UU^{-1}\boldsymbol{\sigma}_1 U\ddot{\mathbf{r}} + \boldsymbol{\sigma}_1\left(\frac{\boldsymbol{\sigma}_1 U\dot{\mathbf{r}}}{2r}\right)\dot{\mathbf{r}}.$$

Hence,

$$2\frac{d^2U}{ds^2} = U(\mathbf{r}\ddot{\mathbf{r}} + \tfrac{1}{2}\dot{\mathbf{r}}^2), \tag{4.27}$$

where $d/ds = r\,d/dt$. Thus, in spinor space it is natural to introduce a new time variable s related to inertial time t by

$$\frac{dt}{ds} = r = |U|^2. \tag{4.28}$$

Now we have a complete system of equations relating position, velocity, acceleration and time variables in position space to corresponding variables in spinor space. These are general kinematic results, enabling us to transform any problem or relation from position space to spinor space or vice-versa.

Given the vector equation of motion in position space

$$\ddot{\mathbf{r}} = -\mu\frac{\mathbf{r}}{r^3} + \mathbf{f}, \tag{4.29}$$

where **f** is an arbitrary perturbing force (per unit mass), the spinor equation of motion is obtained by substitution into (4.27). Thus, we obtain

$$2\frac{\mathrm{d}^2U}{\mathrm{d}s^2} - EU = U\mathbf{rf}(= r\boldsymbol{\sigma}_1 U\mathbf{f}), \tag{4.30}$$

where E is the Kepler energy

$$E = \tfrac{1}{2}\dot{\mathbf{r}}^2 - \frac{\mu}{r} = |U|^{-2}\left(2\left|\frac{\mathrm{d}U}{\mathrm{d}s}\right|^2 - \mu\right). \tag{4.31}$$

The spinor equation of motion (4.30) becomes a determinate equation in spinor space when **f** is given as an explicit function of **r** and $\dot{\mathbf{r}}$ so **rf** can be expressed as a function of U and $\dot{U}$ by using (4.2) and (4.21). It can be solved subject to the subsidiary condition in the form (4.16) or (4.17). The subsidiary condition can be shown to be a constant of motion, so if it is imposed initially, it is automatically maintained for all subsequent times. Note that the perturbation factor $\mathbf{rf} = \mathbf{r}\cdot\mathbf{f} + \mathrm{i}(\mathbf{r}\times\mathbf{f})$ in (4.30) decomposes naturally into a radial part $\mathbf{r}\cdot\mathbf{f}$ which can alter the size and shape of the osculating Kepler orbit and a torque $\mathrm{i}(\mathbf{r}\times\mathbf{f})$ which can alter the attitude of the orbit in space. This is closely related to the alternative gauge condition discussed below.

The spinor equation of motion (4.30) was first derived in a more complicated form by P. Kustaanheimo in 1964. Kustaanheimo and E. Stiefel recast it in a matrix form, which is now known as the KS equation. Geometric algebra has enabled us to further simplify the derivation and formulation of the equation as well as clarify its interpretation. For example, it helped us see the elementary kinematic meaning of the subsidiary condition (4.17), which was never recognized in the matrix formulation.

Stiefel and Schiefele (1971) have shown that solving the perturbed Kepler problem by integrating the KS equation is numerically more efficient and accurate than standard methods for integrating the Newtonian equation of motion. Therefore, we can confidently expect no lesser advantage from developing the theory for integrating our spinor equation (4.30). We could, of course, simply translate the integration methods of Stiefel and Scheifele into our language. But we could probably do better by developing new methods which exploit the special advantages of geometric algebra. That is a task for the future.

To see what has been gained in the transformation from vector to spinor equation of motion, we briefly examine solutions of the unperturbed spinor equation, which we can write in the form

$$U'' - \frac{E}{2}U = 0, \tag{4.32}$$

where the primes represent differentiation with respect to s. For the case $E < 0$, this has the mathematical form of the equation for a harmonic oscillator with natural frequency

$$\omega_0 = (-E/2)^{1/2}. \tag{4.33}$$

So it has the general solution

$$U = U_0 \cos \omega_0 s + \frac{U'_0}{\omega_0} \sin \omega_0 s. \tag{4.34}$$

where U_0 and U_0' are the initial spinor position and velocity. We can evaluate U_0 and U_0' in terms of the initial position and velocity vectors $\mathbf{r}_0$ and $\dot{\mathbf{r}}_0$ using (4.2) and (4.21). The evaluation is simplified if we use our prior knowledge that motion lies in a plane. We are free to choose $\boldsymbol{\sigma}_1 = \hat{\mathbf{r}}_0$; then the rotation $\hat{\mathbf{r}} = U^{-1}\boldsymbol{\sigma}_1 U$ is confined to the orbital plane, and (4.2) can be put in the form

$$\mathbf{r} = \boldsymbol{\sigma}_1 U^2. \tag{4.35}$$

Therefore,

$$U = (\boldsymbol{\sigma}_1 \mathbf{r})^{1/2}, \tag{4.36}$$

and, in particular,

$$U_0 = (\hat{\mathbf{r}}_0 \mathbf{r}_0)^{1/2} = r_0^{1/2}. \tag{4.37}$$

Similarly, (4.22) and (4.28) give us

$$U' = \tfrac{1}{2} \boldsymbol{\sigma}_1 U \dot{\mathbf{r}}, \tag{4.38}$$

and, in particular,

$$U_0' = \tfrac{1}{2} r_0^{1/2} \hat{\mathbf{r}}_0 \dot{\mathbf{r}}_0. \tag{4.39}$$

Therefore, the solution (4.34) can be put in the form

$$U = r_0^{1/2} \left(\cos \omega_0 s + \frac{1}{2\omega_0} \hat{\mathbf{r}}_0 \dot{\mathbf{r}}_0 \sin \omega_0 s \right). \tag{4.40}$$

Of course, from (4.33) the value of ω_0 is determined by

$$\omega_0 = \frac{1}{2} \left(\frac{2\mu}{r_0} - \dot{\mathbf{r}}_0^2 \right)^{1/2}. \tag{4.41}$$

Now that we have $U = U(s)$ as an explicit function of s determined by the initial conditions, our solution will be completed by integrating $\mathrm{d}t/\mathrm{d}s = |U|^2$ to get s as a function of t. We shall see that, in fact, this last step is equivalent to solving Kepler's equation.

Additional insight into the spinor solution is gained by writing it in the alternative form

$$U = U_+ e^{\mathbf{i}\omega_0 s} + U_- e^{\mathbf{i}\omega_0 s} \tag{4.42}$$

where $\mathbf{i} = i\boldsymbol{\sigma}_3$ is the unit bivector for the orbital plane. From (4.35) it follows that the bivector part of U must be proportional to $\mathbf{i}$. Inserting (4.42) into (4.35) we get

$$\mathbf{r} = \boldsymbol{\sigma}_1[U_+^2 e^{\mathrm{i}2\omega_0 s} + U_-^2 e^{-\mathrm{i}2\omega_0 s} + 2U_+U_-]. \tag{4.43}$$

This should be compared with the parametrization of **r** with respect to the eccentric anomaly ϕ:

$$\mathbf{r} = a\hat{\boldsymbol{\varepsilon}}[(\cos\phi - \varepsilon) + \mathbf{i}(1-\varepsilon^2)^{1/2}\sin\phi]. \tag{4.44}$$

If we choose $\boldsymbol{\sigma}_1 = \hat{\boldsymbol{\varepsilon}}$, then the comparison tells us that

$$U_\pm = \pm\left(\frac{a}{2}\right)^{1/2}[1 \pm (1-\varepsilon^2)^{1/2}]^{1/2} \tag{4.45}$$

and

$$\phi = 2\omega_0 s. \tag{4.46}$$

Equation (4.45) tells us how $U_\pm$ are related to the standard orbital elements. Of course U_+ and U_- are alternative orbital elements appropriate in the spinor theory.

Equation (4.46) tells us that the parameter s differs from the eccentric anomaly only by a scale factor $2\omega_0 = (-2E)^{1/2}$, which is itself an important orbital element. Thus, the eccentric anomaly appears naturally in the spinor theory, in contrast to its rather *ad hoc* introduction in the vectorial theory through Kepler's equation. Since the parameter s is equivalent to the eccentric anomaly when $E < 0$, we can be sure that the Equation (4.28), $\mathrm{d}t/\mathrm{d}s = 2\omega_0\,\mathrm{d}t/\mathrm{d}\phi = |U|^2$, integrates to Kepler's equation, so we need not discuss its solution here. However, Kepler's equation applies only when $E < 0$, whereas (4.28) applies also when $E = 0$ or $E > 0$. Thus, s is a *universal* parameter, generalizing the eccentric anomaly to apply to all cases.

It is readily verified that the solutions (4.40) and (4.42) apply also when $E > 0$, provided one understands that the imaginary root of $(-E/2)^{1/2}$ is a bivector, namely, $\omega_0 = (-E/2)^{1/2} = \mathbf{i}(E/2)^{1/2}$. Stiefel and Scheifele show that the solution can be cast in a form which applies also when $E = 0$. Then we have a *universal* solution of the spinor equation which applies for any energy. This is important, for perturbations can change the sign of the energy, so one does not want solutions which break down when that happens.

The striking thing about the unperturbed spinor equation $2U'' - EU = 0$ is the fact that it is a linear differential equation. Thus, the change in variables from vectors to spinors has *linearized* the Newtonian equation $\ddot{\mathbf{r}} + \mu\mathbf{r}/r^3 = 0$. Moreover, it has eliminated the *singularity* at $r = 0$, where r^{-2} becomes infinite. The elimination of a singularity in this way is called *regularization*. Regularization has real practical value, for it eliminates the instabilities (errors) in numerical integration that occur near a singularity. This is computationally important in close encounters between celestial bodies, such as a comet grazing the Sun.

To sum up, the *universality*, *linearity*, and *regularity* of the spinor formulation are three major reasons for its computational superiority over the standard vectorial formulation of the general two body problem, and this becomes more significant when perturbations are included.

An Alternative Gauge Condition

We have seen that the spinor state function U is related to any acceptable alternative state function V by a gauge transformation $V = SU$. According to (4.5), U and V determine the same orbit $\mathbf{r} = \mathbf{r}(t)$. As a geometrically significant alternative to the gauge condition (4.16), consider

$$V^{-1}\boldsymbol{\sigma}_3 V = \hat{\mathbf{h}} = h^{-1}\mathbf{h}\,, \tag{4.47}$$

where $\boldsymbol{\sigma}_3$ is an arbitrarily chosen fixed unit vector orthogonal to $\boldsymbol{\sigma}_1$. Equation (4.47) is consistent with (4.5) since $\mathbf{h}\cdot\mathbf{r} = 0$. Therefore it is acceptable as a gauge condition.

The condition (4.47) has a number of advantages. To begin with, it assures that V has a direct geometrical interpretation. The spinor V determines both the position $\mathbf{r}$ by (4.5) and the plane of motion in position space by (4.47). Conversely, given the position $\mathbf{r}$ and the plane of motion specified by $\mathbf{h}$, then V is determined uniquely (except for sign) by Equations (4.5) and (4.47). Thus, V provides a unique and direct description of the position and plane of motion at every time.

A further advantage of using V appears when we relate it to the spinor R which determines the *Kepler frame*

$$\mathbf{e}_k = R^\dagger \boldsymbol{\sigma}_k R \tag{4.48}$$

(k = 1, 2, 3). This frame, with $\boldsymbol{\sigma}_2 = \boldsymbol{\sigma}_3 \times \boldsymbol{\sigma}_1$, is specified by the physical conditions

$$\mathbf{e}_3 = R^\dagger \boldsymbol{\sigma}_3 R = \hat{\mathbf{h}} \tag{4.49}$$

and

$$\mathbf{e}_1 = R^\dagger \boldsymbol{\sigma}_1 R = \hat{\boldsymbol{\varepsilon}}\,, \tag{4.50}$$

where $\boldsymbol{\varepsilon}$ is the eccentricity vector pointing towards periapse of the osculating orbit.

Equations (4.47), (4.49), and (4.50) determine a unique factorization of the *spinor state function* V into

$$V = ZR\,, \tag{4.51}$$

where Z and R can be regarded as 'internal' and 'external' state functions respectively. Consistency of (4.47) with (4.49) implies that

$$Z^\dagger \boldsymbol{\sigma}_3 Z = r\boldsymbol{\sigma}_3\,.$$

Hence, we can write Z in the form

$$Z = (re^{i\boldsymbol{\sigma}_3\theta})^{1/2}. \tag{4.52}$$

Then, using (4.47) and (4.50) we obtain

$$\mathbf{r} = V^\dagger \boldsymbol{\sigma}_1 V = R^\dagger \boldsymbol{\sigma}_1 Z^2 R = r\hat{\boldsymbol{\varepsilon}} e^{i\hat{\mathbf{h}}\theta}. \tag{4.53}$$

This exhibits θ as the *true anomaly* of the osculating orbit.

The *internal state function* Z describes the size and shape of the osculating orbit as well as location on the orbit. If we take the eccentricity $\varepsilon = |\boldsymbol{\varepsilon}|$, the angular momentum $h = |\mathbf{h}|$, and the true anomaly as internal state variables, then Z is a determinate function $Z = Z(\varepsilon, h, \theta)$ of these variables. Actually, we can identify Z with U in the unperturbed case, so, according to (4.42) and (4.45), it is better to choose ε, a, s as internal state variables, so $Z = Z(\varepsilon, h, s)$.

Although the fixed reference frame $\{\boldsymbol{\sigma}_k\}$ can be chosen arbitrarily, it will most often be convenient to associate it with an initial osculating orbit of the particle. For Kepler motion the best choice is

$$\boldsymbol{\sigma}_1 = \hat{\boldsymbol{\varepsilon}}_0 \quad \text{and} \quad \boldsymbol{\sigma}_3 = \hat{\mathbf{h}}_0, \tag{4.54}$$

where $\mathbf{h}_0$ is the initial angular momentum and $\boldsymbol{\varepsilon}_0$ is the initial eccentricity vector. The initial value of the spinor V is then

$$V_0 = Z_0 = (\hat{\boldsymbol{\varepsilon}}_0 \mathbf{r}_0)^{1/2} = r_0^{1/2} e^{(1/2)i\boldsymbol{\sigma}_3\theta_0}, \tag{4.55}$$

where θ_0 is the initial true anomaly.

The *external state function* R determines the attitude of the osculating orbit in position space. Of course, R is exactly the attitude spinor used in Sections 8-2 and 8-3.

The factorization $V = ZR$ should be of value in perturbation theory, because it admits a systematic separation of perturbation effects determined by the geometry of the orbital elements. Unfortunately, the spinor equation (4.30) loses its simplicity when translated into an equation for V instead of U, although, of course V can be identified with U in the absence of perturbations. On the other hand, if the factorization $V = ZR$ is used, it might be best to work with a pair of weakly coupled equations for R and Z, but we cannot pursue that theme here.

Chapter 9

Foundations of Mechanics

Now that we have become familiar with the content and applications of mechanics, we are prepared to examine its conceptual foundations systematically. This calls for an explicit formulation and analysis of all presuppositions of the theory. It goes beyond a mere statement of Newton's laws to an analysis of the status of laws in a theory and nature of scientific theories in general. This kind of study belongs to the philosophy of science, but it is no mere academic exercise. The profound revolutions in physics due to Newton and Einstein were changes in the conceptual foundations resulting from careful analysis. So it takes a study of foundations to fully understand the evolution of physics, or, if the facts demand, to instigate a new revolution. Improvements in the foundations are truly revolutionary, because they are so rare and their repercussions are so extensive, bearing on every application of the theory.

Newton's original formulation of mechanics nearly 300 years ago is followed with little change in most mechanics books even today. Nevertheless, it is not entirely satisfactory for several reasons. First, it is incomplete in the sense that not all major assumptions of the theory are explicitly spelled out. Second, in the last century Newtonian theory has undergone profound modifications and extensions which should be taken into account. To begin with, Einstein's Theory of Relativity has revolutionized the scientific concepts of space and time. We now know that any adequate formulation of space and time has empirical content with testable consequences. So a clear and explicit formulation of these concepts is scientifically as essential to Newtonian mechanics as it is to relativity theory. Pedagogically, it is needed to help students distinguish between their own vague intuitions of space and time and an objective scientific formulation of these concepts. Fortunately, the formulation can be designed so a small change in the concept of simultaneity generates a smooth transition from Newtonian mechanics to relativistic mechanics.

Another big change in mechanics since Newton has been brought about by the development of the field concept. Even introductory physics courses move rapidly from interactions between particles to interactions of particles with electric and magnetic fields. We need a formulation of Newton's laws which readily accomodates this profound theoretical change. We need to provide for a smooth transition from pure particle mechanics to the classical theory of fields and particles.

A modern formulation of mechanics should also incorporate profound changes in the concept of a theory which have evolved since Newton. Today it is widely recognized that physics is concerned with constructing and testing mathematical models of physical systems. Thus, the concept of a mathematical model is central to the modern conception of a scientific theory. Yet physics textbooks scarcely mention models, let alone explain that mathematical modeling is the essential core of the scientific method.

9-1. Models and Theories

> Philosophy is written in that great book which ever lies before our eyes – I mean the Universe – but we cannot understand it if we do not first learn the language and grasp the symbols in which it is written. This book is written in the mathematical language, and the symbols are triangles, circles, and other geometrical figures, without whose help it is impossible to comprehend a single word of it; without which one wanders in vain through a dark labyrinth.*
>
> Galileo Galilei

This magnificent passage is the capstone of Galileo's great intellectual achievements. It is the first incisive formulation of a philosophical viewpoint which played a crucial role in the development of modern science. This viewpoint has been so thoroughly assimilated into modern science that most scientists take it for granted without recognizing that a profound issue is involved. On the other hand, it is still debated endlessly in philosophical circles, where it is called *scientific realism*. The importance that Galileo himself attached to the above passage is clear from his order that it be placed at the head of his collected works.

Scientific realism must be distinguished from the *naive realism* of common sense. The presumption common to all forms of realism is that a "real world" of things exists independently of any person to observe them. According to common sense, things in the real world are just as we see them; they are known to us directly through experience, provided the senses are operating properly so the view is not distorted. But, as Galileo puts it, scientific realism holds that the real world is known only indirectly; it is merely posed to us through the senses as a cipher, so to know real things we must decode the

*Translation from p. 67 of E. A. Burtt, *The Metaphysical Foundations of Modern Science*, Routledge and Kegen Paul LTD, London (1932). Burtt gives a historical account of the origins of scientific realism.

messages of experience. Moreover, the code can be broken only by recognizing that geometrical properties of things are primary, and we can know them only conceptually by representing them mathematically.

Galileo's profound scientific realism evolved from long contemplation and a variety of astute observations. Throughout his writings Galileo was occupied with an analysis of experience to distinguish the "primary properties" essential to real objects from "secondary properties" which depend on the mode of human sensation. The analysis was continued by Descartes and Boyle among others, and it was a crucial preliminary to Newton's definitive formulation of mechanics in the *Principia*, from which all reference to secondary properties was banished. This decisive step severed psychology cleanly from physics, enabling physics to progress without being distracted by the complexities of subjective experience. It is the basis today for such distinctions as between the perceived color of light (a secondary property) and the frequency of light (a primary property), or the pitch of a tone and its frequency. The properties ascribed to objects by physics, such as mass, velocity, force and frequency, are very different from the directly perceived properties of things. Physical properties are primary properties which can be represented as quantities. Thus, the distinction between primary and secondary properties was a crucial preliminary to developing a mathematical theory of the real world.

In this chapter we adopt a modern version of scientific realism, which holds that objective knowledge about the real world is obtained by developing validated mathematical models to represent real objects. Scientific realism maintains a sharp distinction between a physical thing and its model, between the real world of physical things and the mental world of concepts. One should realize, however, that *this dualism is only methodological*. It by no means requires that the physical and mental worlds exist independently of one another. It is entirely compatible with an explanation of mental phenomena in terms of physical brain states. Indeed, the distinction between primary and secondary properties opens the possibility of explaining secondary properties in terms of primary properties. But this is an issue for neuropsychology to investigate. What matters here is that scientific realism holds that a clear distinction between physical things and their models can be made and must be maintained against the contrary tendencies of natural language which is infected with naive realism.

Scientific realism has been vigorously challenged recently by physicists and philosophers who hold that it is incompatible with quantum mechanics. They claim that quantum mechanics does not allow a sharp separation between the state of a real object and an observer's knowledge of that state. We cannot get involved in that debate here. Suffice it to say that the issue has not been resolved to the satisfaction of all concerned physicists. Without further apology, in this chapter we strive for a sharply formulated theory of scientific knowledge from the viewpoint of scientific realism.

Models

The term "model" is often used in the scientific literature with only a vague meaning. To sharpen the concept of model, we need terminology which expresses clear distinctions and specifications. We assume that a *model* is a conceptual representation of a real object. The represented object is said to be a *referent* of the model. A model may have more than one referent. For example, a model of the hydrogen atom has all hydrogen atoms for referents, while a model of the solar system has a single referent. The set of all referents of a model is called its *reference class*. If its reference class is empty, a model is said to be *fictitious*. An assignment of a particular referent or reference class to a given model is called a *factual interpretation* of the model, or a *physical interpretation* if the model belongs to physics. A single model may be given many different factual interpretations, especially in a mature science like physics. For example, the one-dimensional harmonic oscillator may be interpreted as a model for such diverse objects as an elastic solid, a pendulum, a diatomic molecule or an atom.

We are concerned here with mathematical models, though much of our discussion applies more generally. A **mathematical model** has four components:

(1) A set of **names** for the object and agents that interact with it, as well as for any parts of the object represented in the model.
(2) A set of **descriptive variables** (or **descriptors**) representing properties of the object.
(3) **Equations** *of the model*, describing its structure and time evolution.
(4) An **interpretation** relating the descriptive variables to properties of objects in the reference class of the model.

Each of these components needs some explication.

Numerals are often used as object names; thus, we may speak of "particle 1" and "particle 2". Descriptive variables are functions of the object names, since each descriptor represents a property of a particular object. For example, the velocity descriptor $\mathbf{v}_k$ for the kth object in a system is an explicit function of the object name k. Often, however, the dependence of descriptors on object name is tacitly understood, as when we write $\mathbf{v}$ for the velocity of some object.

There are **three types of descriptors**: object variables, state variables and interaction variables.

Object variables represent intrinsic properties of the object. For example, mass and charge are object variables for a material particle, while moment of inertia and specifications of size and shape are object variables for a rigid body. The object variables have fixed values for a particular object, but they have different values for different objects, so they are indeed variables from a general modeling perspective.

State variables represent intrinsic properties with values which may vary with time. For example, position and velocity are state variables for a particle. A descriptor regarded as a state variable in one model may be regarded as an object variable in another model. Mass, for example, is a state variable in a model that allows it to change, though it is usually constant in particle models. Thus, object variables can be regarded as state variables with constant values.

An **interaction variable** represents the interaction of some external object (called an *agent*) with the object being modeled. The basic interaction variable in mechanics is the force vector; work, potential energy and torque are alternative interaction variables.

Different kinds of property can be distinguished by characteristics of their representations as descriptive variables. A property is said to be *quantitative* if it can be represented by mathematical *quantities*, such as elements of the Geometric Algebra. Otherwise it is said to be *qualitative*. Physics is concerned with a particular set of quantitative properties called *physical properties*. The corresponding descriptors are called *physical variables* or *physical quantities*.

The equations of a mathematical model describe relations among quantitative properties. Equations determining the time evolution of the state variables are called *dynamical equations*, or *equations of motion* in mechanics. In a mature scientific theory, the equations are derived from laws of the theory. Otherwise, they must be assumed as hypotheses subject to verification.

It is common practice in the literature to say that a particular dynamical equation constitutes a mathematical model. This should be recognized as a loose use of language, for an equation represents nothing unless its variables are given factual interpretations.

The **interpretation** of a model is specified by a set of attribute functions for its properties. The set of objects with a given property is called the *scope* or *reference class* of that property. The *attribute function* for a property assigns particular *values* of the *descriptive variable* to objects in its reference class. When specific numerical values are assigned to certain variables, these variables are said to be *instantiated*. As examples of *instantiation* in particle mechanics, we have the assignment of a particular mass to a particle or particular initial conditions for its trajectory.

When, for specific instantiations, the equations of a model are sufficient to determine specific values for all its descriptors, the model is said to be a *specific model*. A specific model can thus describe a particular object under particular circumstances.

Theory

Evidently we have tacitly employed a theory of some sort in specifying the general characteristics of a model. A vaguely defined theory of this sort is frequently called *Systems Theory* in the scientific literature; although it is seldom formulated in the generality we need here. We may regard Systems Theory as a theory of theories, or more specifically, a general theory of mathematical models. Thus, Systems Theory specifies the characteristics of models common to all scientific theories. Consider, for example, the distinction between state variables and interaction variables in a model. That distinction was first sharply drawn in mechanics. But, as other theories developed, many people noticed that the distinction has a wider significance if the concepts of state and interaction are suitably generalized. Based on this distinction, Systems Theory goes on to describe how complex objects can be modeled as systems of interacting parts. Thus, it provides a general theory of structure and composition of objects of any kind. This too is a generalization of concepts developed in physics. A complete development of Systems Theory will not be attempted here. However, the general characterization of a scientific theory, to which we now turn, may be regarded as part of Systems Theory.

A **scientific theory can be regarded as a system of design principles for modeling real objects**. The theory consists of:

I. A **framework** of generic and specific laws characterizing the descriptive variables of the theory.
II. A **semantic base** of correspondence rules relating the descriptive variables to properties of real objects.
III. A **superstructure** of definitions, conventions and theorems to facilitate modeling in a variety of situations.

The mathematical language used to formulate a theory is usually taken for granted. However, it should be recognized that most of the mathematics used in physics was developed to meet the theoretical needs of physics. In Chapter 1, we saw that this is true of the real number system and its generalization to Geometric Algebra. Moreover, differential equations were first invented to formulate dynamical laws of physics. The moral is that the symbolic calculus (mathematics) employed by a scientific theory should be tailored to the theory, not the other way around.

The key concept in a scientific theory is the concept of a scientific law, so it should be explicated carefully. A *scientific law* is a relation or system of relations among descriptive variables presumed to represent an objective relation or pattern among the corresponding properties. If the relation is among physical variables, it is called a *physical law*. Most physical laws are formulated as mathematical equations. Scientific realism maintains that it is important to distinguish between a law and the objective pattern it represents, because the latter is an unchanging property of the real world while the former may be changed when we understand the world better. Moreover, a law may be true or false or approximately true, but the property pattern it is presumed to represent just "is". To qualify as a law, a relation among descriptive variables must represent a property which is *universal* in the sense that its scope is not limited to a finite number of objects, and it must be *corroborated* in some empirical domain by scientific methods. A proposed law which has not been experimentally tested and confirmed is called a *hypothesis*. Thus, a law is a corroborated hypothesis.

There are several types of law. The *generic laws* of a theory define the basic descriptive variables of the theory. The generic laws of Classical Mechanics fall into two groups: (a) The *Zeroth Law*, which defines the concepts of position, motion and composition of bodies, and (b) *Dynamical Laws* (Newton's Laws), which implicitly define the concepts of mass and force. The Zeroth Law is so general that it belongs to every physical theory; indeed, it is presumed (tacitly at least) in *every* scientific theory. The Dynamical Laws apply only to material objects. In Sections 9-2 and 9-3, these laws will be formulated and discussed in detail.

The specific laws of a theory specify relations among the descriptive variables defined by the generic laws. As a rule they apply only to special circumstances, whereas the generic laws are presumed to hold in everv application of the theory. The specific laws of *Classical Mechanics* are *interaction laws* such as Coulomb's Law, Newton's Law of Gravitation and Stokes' Law of fluid friction.

Taking the Zeroth Law for granted, the other basic laws of any scientific theory can be classified into *dynamical laws*, which determine the time evolution of state variables, and *interaction laws*, which interrelate the state variables of different objects.

The *basic laws* of a theory are included in the theory by assumption. The superstructure of the theory also contains *derived laws*, such Galileo's law of falling bodies. As a rule, the scope of a basic law is much wider than the scope of a derived law.

We must be clear about what it means to say that concepts like motion and mass are *defined* by generic laws. All sorts of unnecessary difficulties are caused by a sloppy or inadequate concept of definition, so it will be worth our while to explicate the concept. The purpose of a definition is to establish the meaning of a concept (or the term (symbol) which designates it) by specifying its relation to other concepts (terms). When this has been done, we say that the concept (term) is well-defined. There are two ways to do it, yielding *two kinds of definition*: *explicit and implicit*.

A concept is *defined explicitly* by expressing it in terms of other concepts. This is the conventional notion of definition, used, for example, in defining the kinetic energy K by the equation $K = mv^2/2$.

A concept (term) is *defined implicitly* by a set of *axioms* which relates to it other concepts (terms). Thus, the concept of "point" is defined by the axioms of geometry which specify its relations to other points, lines and planes. Similarly the concept of "vector" is defined implicitly by specifying how to add and multiply vectors. In each case axioms define concepts by specifying relations. Axioms are set apart from other statements or equations by accepting them as definitions, so they need not be proved. Nevertheless, terms like "point" and "vector" introduced by axioms are commonly said to be "undefined terms". This is a misleading expression that ought to be discarded. Novices often interpret it in the sense of "ill-defined" or "obscure". At least they find it unnecessarily mysterious. Evidently it conflicts with established usage of the term "well-defined". It would be better to say that "some terms in a theory must be defined implicitly" rather than "some terms must be undefined".

Generic laws are axioms defining basic descriptive variables. Our definition of model might have given the impression that descriptive variables can be defined independently of any laws. But why are descriptive variables scalar- or vector-valued, that is what makes them quantitative? It will be seen that this is a consequence of the Zeroth Law, which introduces geometrical attributes into every physical theory. The generic laws of space and time are usually taken for granted, so they are seldom mentioned in the formulation of a model. They are essential, nevertheless. A variable which is undefined by laws is completely nondescript; it is no more than a name. To be definite concepts, descriptive variables must be well-defined by laws.

Newton's Laws are sometimes called axioms. That invites confusion between the purely mathematical concept of an axiom and the factual concept of a law. A law is an axiom, but the converse is not true. *A physical law is an axiom with a physical interpretation.*

The correspondence rules of a theory determine factual interpretations for its descriptive variables and laws, and so for models designed with it. They include operational procedures for *measurement*, that is, the assignment of particular values for the descriptors of particular objects. Thus, they determine attribute functions relating descriptors to the properties they represent. The correspondence rules are not independent of physical laws; rather they are specified in accordance with the laws. For example, any operational procedure for measuring length must be consistent with the Euclidean properties of physical space, as specified by the Zeroth Law. Moreover, the laws of physics often enable us to measure the same physical quantity in many different ways, so the results of measurement must be independent of the particular procedure employed.

A correspondence rule for measuring a physical quantity is often called an "operational definition". But this is an abuse of language, confusing the concepts of definition and measurement. A definition, whether explicit or implicit, relates concepts to concepts, not concepts to things. Mario Bunge has suggested that the term "operational definition" be replaced by "operational referition", since it is concerned with the semantic concept of *reference*; it relates a descriptor (a concept) to its referents (things).

The set of real objects which can be modeled with a theory is called the *reference class* of the theory. The reference class of Classical Mechanics is enormous, the set of all material bodies. Yet the generic laws of Mechanics model a very small number of properties. The theory asserts that these are properties that all bodies have in common, so we call them *basic properties*. The fact that the generic laws describe only basic properties does not mean that other properties cannot be described by the theory. A composite body has new properties not possessed by its parts which emerge when it is assembled. They are called *emergent properties*. This challenges theory to explain emergent properties in terms of basic properties. Indeed, it challenges physicists to explain all physical properties of matter — geometrical, mechanical, electrical, thermodynamic, optical — in terms of a small number of basic properties. This grand challenge has long been a major motivation for research.

The emergent geometrical properties of size and shape can be explained in terms of basic properties by the Zeroth Law, which incorporates the physical content of Greek geometry. Geometry can be regarded as the theory of size and shape. This may be obvious, but it is far from trivial, as witnessed by the whole field of architectural design. The Kinetic Theory of Gases is a subtheory of Mechanics which explains temperature as an emergent property. The problem of explaining all thermodynamic properties as emergent from physical properties of molecules is so complex that a separate theory, Statistical Mechanics, has been developed to handle it. More specialized theories like Plasma Physics, Solid State Physics and Theoretical Chemistry are also concerned with explaining emergent properties. All these theories are founded on Classical Mechanics as well as Quantum Mechanics.

Having discussed the general features of models and theories, let us turn now to a formulation of generic laws for Classical Mechanics.

9-2. The Zeroth Law of Physics

Everyone has well-developed notions of space and time abstracted from personal experience. Perceptual categories of space and time are essential for sorting out sensory data. However, perceptual space and time must sharply be distinguished from the concepts of physical space and time. The former is a *modus operandi* of the human brain – the proper study of psychology, psychophysics and neuroscience. It provides an intuitive base for the physical concepts. But the concepts of physical space and time are objective rather than intuitive. Intuitive concepts are subjective, which is to say that they vary from person to person; whereas objective concepts are the same for everyone. Objectivity is achieved in science by providing concepts with explicit mathematical definitions and factual interpretations in terms of rules which might be applied by anyone, or by a computer for that matter. Of course, everyone's conception of space and time combines intuitive and objective components. But only the objective component will concern us here.

Objective concepts evolve with changes in their definitions and interpretations. Since Newton's day two major improvements in the concepts of space and time have evolved which should be incorporated into the foundations of mechanics. First, we have learned to distinguish between mathematical and physical geometries. Scientific realism regards physical geometry as a feature of the real world which we model with a mathematical geometry. Thus, our model geometries should be subjected to empirical tests. In Newton's day no one had conceived of an alternative to Euclidean geometry or the idea of testing it, though, of course, it had been subjected to many crude informal tests when employed in architectural design and construction. Alternatives to Euclidean geometry were first conceived by mathematicians in the nineteenth century, but none was incorporated into a viable physical theory until Einstein's General Theory of Relativity in the twentieth century. We shall formulate a Euclidean model of physical geometry, since that is appropriate for classical mechanics. But we aim to do it in a way which makes its "physical content" explicit, and allows for easy generalization to "relativistic theories."

The second major improvement in concepts of space and time is due mainly to Einstein. He recognized that the concept of *distant simultaneity* is an essential part of the time concept which had not previously been explicitly defined in classical physics. Rather, physicists had unwittingly adopted an implicit concept of simultaneity which was inconsistent with ideas of causality and experimental fact. By supplying an appropriate definition of distant simultaneity and analyzing its consequences, Einstein created his Special Theory of Relativity. Thus, the Special Theory is best regarded as a completion of classical physics with a full elucidation of the time concept.

The change instituted by Einstein in the classical time concept appears to be comparatively small, but its consequences are immense. It implies that space and time are relative concepts which cannot be defined independently of one another and do not correspond to unique features of the real world. It implies that the real physical geometry is a non-Euclidean geometry of a 4-dimensional entity space-time, with respect to which the separate concepts of space and time only describe the viewpoint of a particular observer. Thus, a small change in the time concept has profoundly altered the physicists' conception of reality.

The Special Theory of Relativity will be discussed in a sequel to this book, NF II. Here we will be content with preparing the way for a smooth transition to the modern space-time concept by elucidating the classical concepts of space and time. We begin with the concept of space.

The problem of providing the concept of space with a precise mathematical formulation has been solved to nearly everyone's satisfaction. But physicists are still far from agreement on the physical status of space. Is space a thing or a property of things? Or is it a property of the human mind, a "category of the understanding," as the philosopher Immanual Kant proposed? Every kind of answer can be found in the literature. This attests to widespread confusion about the conceptual foundations of physics. Confusion is perpetuated by an outmoded concept of space which infects our natural language. Thus, we speak of physical objects *in* space as if space were a container with an existence independent of its contents. The literature shows that physicists are not immune to this infection, but a cure can be achieved by a careful conceptual analysis. The source of the infection is easy to identify. The natural language was developed to describe features of perceptual experience, which it can do with remarkable fidelity. The brain does, indeed, contain a *sensorium*, a carrier of perceptions which exists independently of its contents. This is reflected in perceptual experience and so in the natural language. Thus, a cure for confusion about the nature of space begins with a clear distinction between the perceptual space of subjective experience and the objective concept of physical space. The complete cure requires a rigorous formulation of the physical concept in perfect accord with experimental practice.

To ascertain a suitable physical interpretation for the concept of space, we must examine the role of geometry in experimental practice. We note that every measurement of distance determines a relation between two objects. Every measurement of position determines a relation between one object and some other object or system of objects. In accordance with the standpoint of scientific realism, we regard such measured relations as representations of real properties of real objects. These are mutual (or shared) properties relating one object to another. We call them *geometrical properties*. We are now prepared for an explicit formulation of physical space as a system of relations among physical objects.

To begin with, we recognize two kinds of objects, *particles* and *bodies* which are composed of particles. Given a body $\mathcal{R}$ called a *reference frame*, each particle has a geometrical property called its *position with respect to* $\mathcal{R}$. We characterize this property indirectly by introducing the concept of *Position Space*, or *Relative Space*, if you prefer. **For each reference frame $\mathcal{R}$, a position space $\mathcal{P}$ is defined by the following postulates:**

A. **$\mathcal{P}$ is a 3-dimensional Euclidean space**.
B. **The position (with respect to $\mathcal{R}$) of any particle can be represented as a point in $\mathcal{P}$.**

The first postulate specifies the mathematical structure of a position space while the second postulate supplies it with a physical interpretation. Thus, the postulates define a physical law, for the mathematical structure implies geometrical relations among the positions of distinct particles. Let us call it the **Law of Spatial Order**.

Notice that this law asserts that every particle has a property called position and it specifies properties of this property. But it does not tell us how to measure position. Measurement is a separate matter, since it entails correspondence rules as well as laws. In actual practice the reference frame is often fictitious, though it is related indirectly to a physical body. Our discussion is simplified by feigning that the reference frame is always a real body.

We turn now to the problem of formulating the scientific concept of time. We begin with the idea that time is a measure of motion, and *motion is a change of position with respect to a given reference frame*. The concept of time embraces two distinct relations: temporal order and distant simultaneity. To keep this clear we introduce each relation with a separate postulate.

First we formulate the **Law of Temporal Order**:

> **The motion of any particle with respect to a given reference frame can be represented as an orbit in position space.**

This postulate has a semantic component as well as a mathematical one. It presumes that each particle has a property called motion and attributes a mathematical structure to that property by associating it with an orbit in position space. Recall that an *orbit* is a continuous, oriented curve. Thus, a particle's orbit in position space represents an ordered sequence of positions. We call this order a *temporal order*, so we have attributed a distinct temporal order to the motion of each particle.

To define a physical time scale as a measure of motion, we select a *moving* particle which we call a *particle clock*. We refer to each successive position of this particle as an *instant*. We define the *time interval* Δt *between two instants* by

$$\Delta s = c\Delta t,$$

where c is a positive numerical constant and Δs is the arclength of the clock's orbit between the two instants. Our measure of time is thus related to the measure of distance in position space.

To use this time scale as a measure for the motions of other particles, we need to relate the motions of particles at different places. The necessary relation can be introduced by postulating the **Law of Simultaneity**:

> **At every instant, each particle has a unique position.**

This postulate determines a correspondence between the points on the orbit of any particle and points on the orbit of a clock. Therefore, every particle orbit can be parametrized by a time parameter defined on the orbit of a particle clock.

Note that this postulate does not tell us how to determine the position of a given particle at any instant. That is a problem for the theory of measurement.

So far our laws permit orbits which are nondifferentiable at isolated points or even at every point. These possibilities will be eliminated by Newton's laws which require differentiable orbits. We include in the class of allowable orbits, orbits which consist of a single point during some interval. A particle with such an orbit is said to be *at rest* with respect to the given reference frame during that interval. Of course, we require that the particles composing the reference frame itself be at rest with respect to each other, so the reference frame can be regarded as a rigid body.

Note that the speed of a particle is just a comparison of the particle's displacement to the displacement of a particle clock. The speed of the particle clock has the constant value $c = \Delta s/\Delta t$, so the clock moves uniformly by definition. In principle, we can use any moving particle as a clock, but the dynamical laws we introduce later suggest a preferred choice. It is sometimes asserted that a periodic process is needed to define a clock. But any moving particle automatically defines a periodic process, because it moves successively over spatial intervals of equal length. It should be evident that any real clock can be accurately modeled as a particle clock. By regarding the particle clock as the fundamental kind of clock, we make clear in the foundations of physics that the scientific concept of time is based on an objective comparison of motions.

We now have definite formulations of space and time, so we can define a *reference system* as a representation $\mathbf{x}$ for the possible position of any particle at each time t in some time interval. Each reference system presumes the selection of a particular reference frame and particle clock, so $\mathbf{x}$ is to be interpreted as a point in the position space of that frame. Also, a reference system presumes the selection of a particular origin for time and space and particular choices for the units of distance and time, so each position and time is assigned a definite numerical value. The term "reference system" is sometimes construed as a system of procedures for constructing a numerical representation of space and time.

After we have formulated our dynamical laws, it will be clear that certain reference systems called *inertial systems* have a special status. Then it will be necessary to supplement our *Law of Simultaneity* with a postulate that relates simultaneous events in different inertial systems. That is the critical postulate that distinguishes Newtonian theory from Special Relativity, but we defer discussion of it until we are prepared to handle it completely. It is mentioned now, because our formulation of space and time will not be complete until such a postulate is made.

It is convenient to summarize and generalize our postulates with a single law statement, the **Zeroth** (or *Spatiotemporal*) **Law of Physics**:

Every real object has a continuous history in space and time.

To explicate this law, we assert that it has four major components:
1. *The Law of Spatial Order.*
2. *The Law of Temporal Order.*
3. *The Law of Simultaneity.*
4. *The Generic Law of Composition.*

The **Generic Law of Composition** asserts:

The properties of any real object can be represented mathematically by the values of a state function defined on the position and time variables of a given reference system.

A *Specific Law of Composition* imposes some condition on the nature of the *state function* for a particular object or class of objects. The successive values of the state function as a function of time describes the *history* of the object. The Zeroth Law does not specify the history of any object; dynamical laws determining the state function are needed for that. But it does assert that every object has a history.

In classical physics, *every model of a real object is one of three kinds: particle, body or field*. Each model is distinguished by a particular state function. We have already specified the state function for a particle, namely, the function $\mathbf{x} = \mathbf{x}(t)$ for its orbit in position space. A *material particle* also has a property called mass, so a complete state function must specify any time variation of the mass.

A *body* is an *extended object*, which is to say that more than one point in Position Space is required to specify its *location*. We have modeled bodies as systems of particles. In this case, the *location* of a body is the set of positions of its particles, and its *history* is the set of particle histories. Alternatively, a *material body* (or *material medium*) can be modeled as a spatially continuous object which does not have a unique decomposition into particles.

A *field* is also an extended object, but its state function as well as its physical interpretation is quite different from that of a body. The *Classical Theory of Fields* will be discussed in NF II. By way of illustration, let us only note here that the theory asserts the existence of real objects called electric fields, each of which can be represented by a vector-valued state function $\mathbf{E}(\mathbf{x}, t)$.

The Zeroth Law applies to Quantum Mechanics as well as classical physics, but the state functions for particles are different. The quantum mechanical state function for an electron will be discussed in NF II.

The Zeroth Law is the most universal of all scientific laws. It asserts that every real thing that ever existed or will exist has definite spatiotemporal properties, that is, definite spatiotemporal relations to every other real thing. Some aspect of the Zeroth Law is presumed in every scientific theory and investigation.

Other scientists can take the Zeroth Law for granted, but physicists are responsible for refining its formulation and testing its consequences. The present formulation has been designed for compatibility with the Special Theory of Relativity, and it can be directly generalized to the "curved spacetime" of the General Theory of Relativity, but that must be left for another book. The mathematical structure attributed to space and time in our formulation is widely accepted by physicists, but the physical interpretation is controversial. We have adopted a *relational view*, interpreting space and time as a system of relations among real objects. But some physicists prefer a *material view*, interpreting spacetime as a primal material out of which all things are composed, so that objects can be regarded as local variations in the properties of spacetime. Thus, the material view interchanges the objects and properties of the relational view. Is there a definite empirical distinction between these two interpretations so we can decide on one over the other? That is a profound question which will not be easily answered. At least both interpretations are consistent with scientific realism.

9-3. Generic Laws and Principles of Particle Mechanics

The spatiotemporal properties of real objects are described by the Zeroth Law. To produce a complete physical theory, the Zeroth Law must be supplemented by a set of dynamical laws which describe the nature and effect of interactions between objects. In Particle Mechanics the *interaction property* is represented by force functions. A set of *generic laws* implicitly define the concepts of mass and force and assign them a physical interpretation. To produce a specific model of interacting particles, the generic laws must be supplemented by *specific force laws* which specify definite force functions.

Our formulation of the general theory consists of four generic laws, one hypothesis and three generic principles. Let us present them all at once, and then comment on each one separately. Of course, our formulation presumes the Zeroth Law, so the notions of particle, time, position, velocity and acceleration are all well-defined. In addition, the formulation is presumed to hold only for a certain kind of reference system called an *inertial system*, which is implicitly defined by the First Law. Now we are ready.

First Law (*Law of Inertia*):
In an inertial system, every free particle has a constant velocity. A particle is said to be *free* if the total force on it vanishes.

Second Law (*Law of Causality*):
The total force exerted on a particle by other objects at any specified time can be represented by a vector f such that

$$\mathbf{f} = m\mathbf{a},$$

where $\mathbf{a}$ is the particle's acceleration and *m is a positive scalar constant* called the *mass* of the particle.

Third Law (*Law of Reciprocity*):
To the force exerted by any object on a particle there corresponds an equal and opposite force exerted by the particle on that object.

Fourth Law (*Superposition Law*):
The total force f due to several objects acting simultaneously on a particle is equal to the vector sum of forces $\mathbf{f}_k$ due to each object acting independently, that is,

$$\mathbf{f} = \sum_k \mathbf{f}_k .$$

To relate formulations of the laws in different inertial systems, we adopt the

Hypothesis of Absolute Simultaneity:

Local events which are simultaneous in one inertial system are simultaneous in every inertial system.

A *local event* is defined as a change in the position or velocity of a particle.

Specific force laws need not be regarded as part of the general theory. However, they are restricted in form by generic principles. The principles function as laws when force functions are unknown. In particular, they sharpen the general concept of force defined by the generic laws. However, when we have specific force laws that satisfy the principles, the principles are superfluous. For this reason we do not call them laws. In Section 3-1 we introduced *The Principle of Analyticity*. Two other principles are important:

The Principle of Local Interaction:

The force on a particle at any time is a unique function of particle position and position time derivatives; it is independent of the particle's past or future history.

The Principle of Relativity:

The laws of mechanics have the same functional form in all inertial systems.

Comments on the First Law

The First Law implicitly defines a time scale for an inertial system. For it requires that displacements of the system's particle clock are proportional to displacements of a free particle. This amounts to requiring that equal intervals of time be *defined* by equal displacements of a free particle. Thus, the motion of a free particle determines the time scale for an inertial system up to a multiplicative constant. This fundamental kind of scale is called an *inertial time scale*.

Besides determining a time scale, the First Law associates straight lines in an inertial system with free particle motion. So within an inertial system, the deviation of any particle from uniform motion in a straight line can be attributed to the action of other objects in accordance with the Second Law.

The First Law defines an inertial system implicitly by specifying a physically-grounded criterion which distinguishes it from noninertial reference systems. It tells us that an inertial frame can be identified in principle by examining the motion of free particles. In practice, such a procedure is usually impossible. For inertial frames do not occur naturally, so most measurements are done with respect to an accelerated reference frame. And free particles are not usually available for experiments either. Such practical difficulties should not be construed as casting doubt on the utility of the First

Law. Rather, they pose the experimentalist with the problem of distinguishing real forces, due to the interaction of objects, from pseudoforces, due to the acceleration of his reference frame. He needs the First Law to make such a distinction, but he needs the other Laws as well.

The First Law is evidently not independent of the other Laws, because they are needed to define what is meant by "free particles".

Comments on the Second Law

The formula $\mathbf{f} = m\mathbf{a}$ by itself is frequently presented as a complete statement of the Second Law. Although such a formula is an acceptable mathematical axiom, a law statement should include a physical interpretation of the mathematical terms it employs. Of course, an interpretation for $\mathbf{f}$ could be supplied by a separate postulate, but it is best included in the Second Law since that is where $\mathbf{f}$ first appears in the theory.

Thus, our formulation asserts that $\mathbf{f}$ represents a physical property called "force" which a particle shares with other objects. The vector $\mathbf{f}$ itself is commonly referred to as "the force on a particle", so $\mathbf{f}$ serves also as a name for the property it represents. The term "objects" in our law statement is presumed to be defined previously by the Laws of Composition as part of the Zeroth Law. As said before, we recognize three kinds of objects: particles, bodies and fields, and this reduces to two basic kinds if bodies are modeled as systems of particles. To produce a pure particle theory, one need only replace the word object with the word particle in all our law statements. But our formulation generalizes particle theory by allowing interactions with fields. Indeed, in a pure field theory particles never interact directly, but only through the intermediary of a field. In that case the term "object" in our law statements should always be interpreted as "field," and we need additional laws to fully describe the properties of fields.

It is often claimed that $\mathbf{f} = m\mathbf{a}$ is a definition of force. On the contrary, an explicit definition of force is impossible. Rather, the complete set of generic laws is required to define $\mathbf{f}$ implicitly by specifying the common characteristics of all forces. The equation $\mathbf{f} = m\mathbf{a}$ represents only one characteristic of force. It relates the general property of interaction to the general spatiotemporal property of motion. The $\mathbf{f}$ represents the action of the universe on a particle while the $m\mathbf{a}$ represents the particle's response with a change in its state of motion. This provides us with a physical interpretation of mass as a measure of the strength of a particle's response to a given force. No other definition or interpretation of mass is needed in the theory.

Comments on the Third Law

For two interacting particles, the Third Law can be written

$$\mathbf{f}_{12} = -\mathbf{f}_{21},$$

where $\mathbf{f}_{12}$ is the force of particle 2 on particle 1. This was called the *weak form* on the Third law in Section 6-1. One readily verifies that this relation is satisfied by Newton's gravitational force law and the similar Law of Coulomb. However, it fails for direct magnetic interactions between charged particles (see Exercise 5).

This failure of the Third Law for a force law of such great physical importance raises a serious problem of determining precisely under what conditions the Third Law can be expected to hold and what is responsible for its failure. The problem is best addressed by considering the Third Law in a different form. For a 2-particle system, the Second Law gives us

$$\frac{d\mathbf{p}_1}{dt} = \mathbf{f}_{12} \quad \text{and} \quad \frac{d\mathbf{p}_2}{dt} = \mathbf{f}_{21},$$

where $\mathbf{p}_1$ and $\mathbf{p}_2$ are momenta of the particles. So the Third Law can be written

$$\frac{d\mathbf{p}_1}{dt} = -\frac{d\mathbf{p}_2}{dt} .$$

Thus, the Third Law can be interpreted as a *Law of Momentum Exchange*. Hence a failure of the Third Law would be a failure of momentum conservation. Today, physicists regard the *Law of Momentum Conservation* as more fundamental than Newton's Laws because it holds in Quantum Mechanics as well as Classical Mechanics with no known exception. Any apparent violation of momentum conservation prompts the question: "What happened to the missing momentum?" On several occasions attempts to answer this question have led to the discovery of new physical objects, of which the elementary particle called the *neutrino* is a spectacular example.

Classical Field Theory accounts for the apparent failure of magnetic interactions to satisfy momentum conservation by attributing momentum to the electromagnetic field. We are not prepared for a quantitative discussion of this matter using Field Theory, so we must be content with qualitative remarks. Electromagnetic Field Theory allows a particle to interact only with fields at the position of the particle. This extends our stated *Principle of Local Interaction* to include field variables. It precludes the possibility of instantaneous interparticle interactions except as an approximation. Rather, the interaction between particles is indirect with the field as intermediary. It proceeds by a transfer of momentum from one particle to the field; then the field transports some of the momentum at the speed of light to the position of the second particle where it can be transferred from field to particle, while the rest of the momentum may travel freely as electromagnetic radiation.

The point to be made here is that the Third Law is completely consistent with Field Theory if we extend the Principle of Local Interaction and interpret the "object" in the law statement as a field. Physicists do not ordinarily speak of "a force exerted by a particle on a field" as in the law statement. But this just means "rate of momentum transfer from particle to field", which is a conventional expression.

For a unified view of physics, particle mechanics should be regarded as an approximation to Classical Field Theory. In this approximation, then, the Third Law can be applied to particles acting instantaneously at a distance, as in Newton's theory of gravitation.

Comments on the Fourth Law

This Law is sometimes regarded as part of the Second Law, but it deserves an independent formulation to emphasize its importance. It helps us "divide and conquer" in mechanics by allowing us to decompose complex forces into simpler parts for separate analysis, just as the Law of Composition allows us to decompose extended bodies into particles. Conversely, it allows us to lump a great many forces into a single force to be analyzed as a unit. In a word, the Third and Fourth Laws are the main mathematical tools for assembling and disassembling interactions.

Comments on Absolute Simultaneity

The Hypothesis of Absolute Simultaneity is best regarded as a supplement to the First Law. It implies that an inertial time scale set up in one inertial system can be employed in any other inertial system, so one time scale suffices for all inertial systems. This is equivalent to Newton's assumption that there exists a unique absolute time variable that can be employed in any reference system.

Absolute simultaneity is called a hypothesis rather than a law here, because it is now known to be empirically false, though it is approximately true in a large empirical domain. Explicit formulation of this hypothesis, which is implicit in Newtonian theory, shows us exactly where Relativity differs from Classical Mechanics. Einstein replaced absolute simultaneity with the

Law of Light Propagation:
The speed of light is constant with all inertial systems.

With this law we can use an idealized light pulse or photon to construct a model particle clock, a *photon clock*. The photon clock establishes an inertial time scale which is the same for all inertial systems and uniquely relates the time scale to the distance scale. Moreover, the Law of Simultaneity which we introduced as part of the Zeroth Law can now be reduced to a mere *definition of simultaneity*. All this leads to a conceptual fusion of space and time into a unified concept of *spacetime*. The mathematical formulation and analysis of these ideas using geometric algebra will be developed in the following volume NF II. It should be mentioned here that the Light Propagation Law requires a small but significant alteration of the Second Law because it modifies the concept of time. But no other changes in the laws are needed to give us relativistic mechanics.

Comments on the Local Interaction Principle

The Principle of Local Interaction is implicit in every treatment of mechanics, yet it has not been singled out peviously as a postulate of the general theory. It is essential if we are to conclude from the generic laws that specific forces determine definite differential equations for particle orbits. And our aim is to formulate the general presumptions of mechanics as explicitly and completely as possible.

Our formulation of Local Interaction allows the force to be a function of time derivatives of the position vector to any order. As a rule, the velocity is the only time derivative to appear in a specific force law. But there is an exception of great theoretical importance, namely, the *radiative reaction force* due to the reaction of electromagnetic radiation on a particle emitting it. This force law depends on the third time derivative of position. However, this is not the place to study it.

We have already noted the closed relation of Local Interaction to the Third Law. In a pure particle theory we can combine these postulates to draw conclusions about the functional form of the two particle force. Thus, for a force that depends only on position and velocity we find

$$\mathbf{f}_{12} = \mathbf{f}[\mathbf{x}_1(t), \mathbf{v}_1(t), \mathbf{x}_2(t), \mathbf{v}_2(t)] = -\mathbf{f}[\mathbf{x}_2(t), \mathbf{v}_2(t), \mathbf{x}_1(t), \mathbf{v}_1(t)]. \tag{4.1}$$

The Relativity Principle restricts the function form still further. This significantly restricts the force laws to be considered in a pure particle theory.

Comments on the Relativity Principle

The effect of a change in reference system on the equations of motion for a particle has already been discussed in Section 5-5. Here we will merely comment on its general theoretical implications in accordance with the Relativity Principle. We saw that the most general transformation of position vectors relating one inertial system to another has the form

$$\mathbf{x} \to \mathbf{x}' = R^\dagger(\mathbf{x} + \mathbf{a} + \mathbf{u}(t + t_0))R, \tag{4.2}$$

Where R is a unitary spinor and R, $\mathbf{a}$, $\mathbf{u}$ and t_0 are constants. This transformation is a composite of a space translation, a Galilean transformation, a time translation and a rigid rotation. It maps a particle orbit $\mathbf{x} = \mathbf{x}(t)$ onto an orbit $\mathbf{x}' = \mathbf{x}'(t') = \mathbf{x}'(t + t_0)$. Differentiation therefore gives the general *velocity addition theorem*

$$\dot{\mathbf{x}} \to \dot{\mathbf{x}}' = R^\dagger(\dot{\mathbf{x}} + \mathbf{u})R. \tag{4.3}$$

Another differentiation gives us the transformation of the Causality Law,

$$m\ddot{\mathbf{x}} = \mathbf{f} \to m\ddot{\mathbf{x}}' = \mathbf{f}', \tag{4.4}$$

showing that its form is unchanged, as required by the Relativity Principle, provided the force undergoes the induced transformation

$$\mathbf{f} \to \mathbf{f}' = R^\dagger \mathbf{f} R. \tag{4.5}$$

The Relativity Principle requires more, however. It requires that the functional form of the force law must be preserved by the transformation (4.2). In particular, the two particle force law (4.1) must be of the more restricted form

$$\mathbf{f}_{12} = \mathbf{f}[\mathbf{x}_1 - \mathbf{x}_2, \dot{\mathbf{x}}_1 - \dot{\mathbf{x}}_2]; \tag{4.6}$$

it must depend only on the relative position $\mathbf{x}_1 - \mathbf{x}_2$ to be form invariant under translations, and on the relative velocity $\dot{\mathbf{x}}_1 - \dot{\mathbf{x}}_2$ to be invariant under Galilean transformations. Moreover, invariance under time translations implies that $\mathbf{f}$ cannot be an explicit function of time. To be even more specific, if $\mathbf{f}_{12}$ is an algebraic function of $\mathbf{x}_1 - \mathbf{x}_2$ and $\dot{\mathbf{x}}_1 - \dot{\mathbf{x}}_2$, then (4.5) is automatically a consequence of (4.2) and (4.3). This is, indeed, characteristic of the most fundamental force laws we have considered.

Clearly the Relativity Principle is an important modeling principle. It tells us that our models should be independent of our chosen (inertial) reference system, so interactions should be functions only of *relative* positions and velocities. We have interpreted the transformation (4.2) as a *passive* change in descriptive variables without altering the state of motion of any object. Alternatively, for $t_0 = 0$, we can regard (4.2) as a change in description due to an *active* rigid displacement and boost in velocity of a single reference body (or frame). If our models are to be unaffected by such a shift of the reference body, as required by the Relativity Principle, we conclude that the reference body must not be interacting with real objects. In other words, the reference body must be regarded theoretically as fictitious. Of course, real objects are needed as reference bodies in experiments. So the Relativity Principle serves as a guide to the idealizations required for a theoretical description of experiments.

Another profound implication of the Relativity Principle is found by interpreting (4.2) as a transformation with respect to a single reference system. The transformation (4.2) maps any orbit $\mathbf{x} = \mathbf{x}(t)$ onto an orbit $\mathbf{x}' = \mathbf{x}'(t + t_0)$ at a different time and place. According to the Relativity Principle, these orbits describe physically equivalent (or congruent) processes. Thus, *the Relativity Principle can be regarded as a general congruence law*, providing a precise criterion for the equivalence of different physical processes at different places and times. This makes it possible to compare results of different experiments performed at different places and times. Thus, the Relativity Principle provides a theoretical basis for the *reproducibility* and *predictability* of physical results.

It should be noted that the Relativity Principle is a semantic principle, because it is concerned with the interpretation of descriptive variables, that is, with the relation of models to their referents. It is appropriate to regard the Relativity Principle as a "congruence law", because it describes an equivalence relation under rigid transformations in space and time, so it generalizes the notion of congruence from elementary geometry. This geometrical character of the Relativity Principle shows that it should be grouped together with the Zeroth and First Laws. These three laws together determine the model of space and time used in classical mechanics, and they must all be modified to characterize the model of space-time proposed in Einstein's General Theory of Relativity.

Comments on the Theoretical Structure of Mechanics

We have completed our formulation of the generic laws and principles of Particle Mechanics. These laws and principles compose an axiom system from which all results of the theory can, in principle, be derived as theorems. We say "in principle" because no one has bothered to develop the theory as an orderly system of theorems and proofs based on well-defined axioms. A major reason for this has been the lack of a complete and appropriate set of axioms. Under the influence of recent mathematical fashion, some authors have developed axiomatic formulations of mechanics using set theory. But set theory is not the right mathematical tool, because it is too general. Consequently, theorems and proofs in this approach are inordinately unwieldy. Geometric algebra is a better tool, because it was designed for the geometrical job. And our formulation of the axioms conforms well to physical practice.

Of course, we have already derived the results of major interest in mechanics in an informal way, so there is no point to embarking on a formal development here. However, it is worth pointing out that formalization of mechanics should have some advantages. It can be expected to clarify the structure of the theory, eliminate unnecessary redundancy and make results more accessible for applications. On the other hand, it must be recognized that the organization of mechanics should be dictated by physical rather than mathematical considerations. For the purpose of theory is to make specific models.

9-4. Modeling Processes

Scientific knowledge is of two kinds, factual and procedural. The *factual knowledge* consists of theories, models, and empirical data interpreted (to some degree) by models in accordance with theory. A theory is to be regarded as factual, rather than hypothetical, because the laws of the theory have been corroborated, though theories differ in range of application and corroboration. The *procedural knowledge* of science consists of strategies, tactics and techniques for developing, validating, and utilizing factual knowledge. It is commonly referred to under the rubric of *scientific method*.

The structure of factual knowledge has been explicated in our general discussion of models and theories and our detailed analysis of classical mechanics. Our aim in this section is to explicate the structure of procedural scientific knowledge. The subject is complex, so we cannot hope to produce much more than an outline. We will do well to identify organizing principles which give the subject some coherence.

The key to an explication of scientific method is recognizing that the central activity of scientists is the development and validation of mathematical models. Thus we need to analyze the processes of mathematical modeling. We can distinguish *two types of modeling process*: model development and model deployment. The first is concerned primarily with theoretical aspects of modeling, while the second is concerned with empirical aspects. Theoretical and empirical aspects are often interrelated, so the distinction between development and deployment is a matter of emphasis rather than sharp separation. Let us proceed to a discussion of each process in turn.

Model Development

A model is a surrogate object; it depicts or portrays a real object by representing its properties. The properties of a real object are known only through their representation in a model; they are never experienced directly. Moreover, our knowledge of any real object is always incomplete. Every model is an *idealization* or *partial representation* of its referent, which is to say that some but not all properties are represented in the model. Nevertheless, physicists strive to construct complete models of the most elementary constituents of matter, such as electrons. (These are the only objects that might be simple enough to model in all detail – but that is pure speculation).

Deliberate idealization is a method of simplification. A model which fails to represent known properties of its referent is often useful when those properties are regarded as irrelevant or uninteresting. Thus, we model the Earth as a particle when concerned with its motion in the solar system.

The method of deliberate idealization generalizes to the *method of successive refinements*, which is one of the major modeling strategies in science. Beginning with a simple model, a sequence of increasingly complex models is constructed by successively incorporating additional attributes to represent the object with increasing detail. Thus, the simple particle model of the Earth is refined by modeling it as a rigid body to describe its rotation, further refined by modeling it as an elastic solid to account for the effects of tidal forces; then it may be assigned a model atmosphere and molten core to account for its thermal properties. The modeling is never finished, as any geophysicist or climatologist can attest.

Figure 4.1

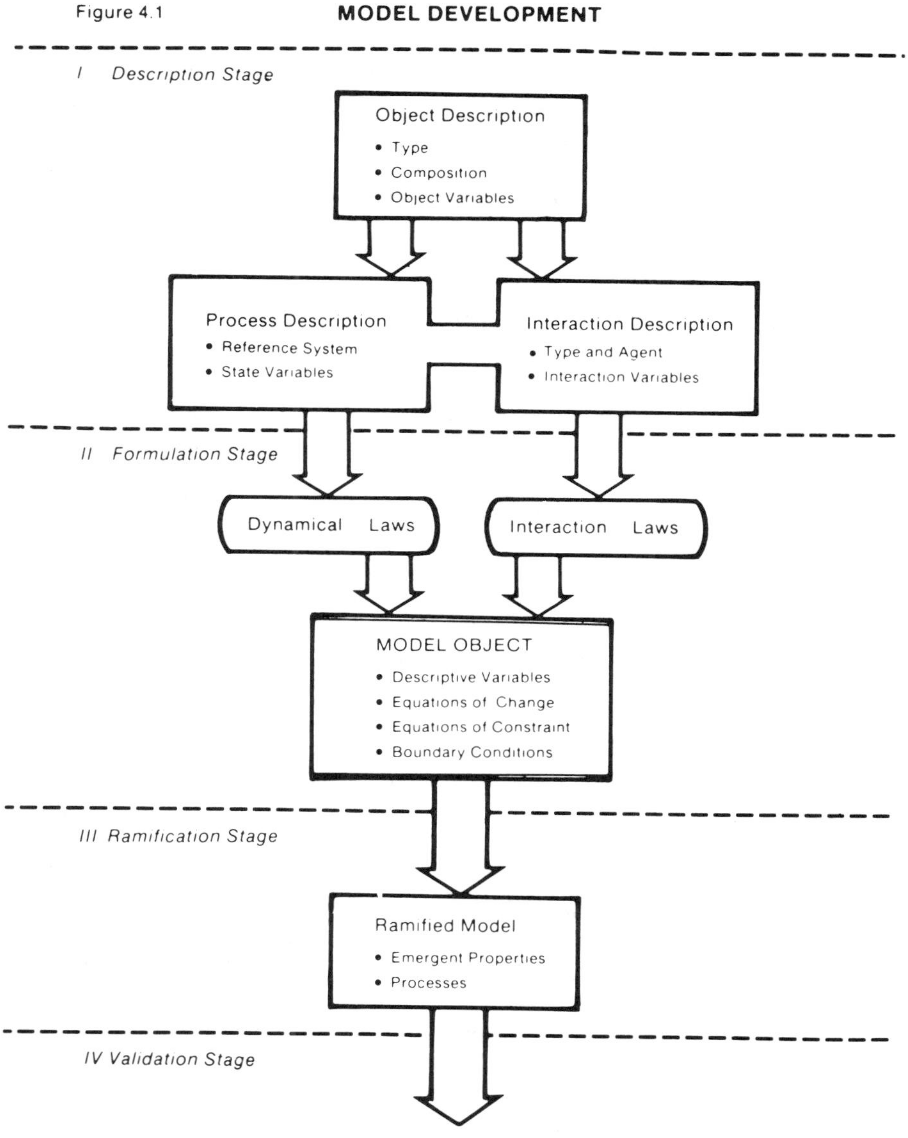

The process of developing a mathematical model can be analyzed into four *essential stages*: (1) *Description*, (II) *Formulation*, (III) *Ramification*, and (IV) *Validation*. The stages are implemented consecutively, though back-tracking to revise the results of an earlier stage is not uncommon. The entire model development process is outlined schematically in Figure 4.1 to indicate the kind of information processing in each stage. The figure can be regarded as the outline of a modeling strategy as well as a description of the modeling process. Moreover, it can be regarded as a problem solving strategy, since, by and large, physics problems are solved by developing models.

The modeling strategy outlined in Figure 4.1 is sufficiently general to apply to any branch of physics, indeed, to any branch of science. Therefore it can be regarded as a *general scientific method*. However, the implementation of each stage in a particular model is theory-specific, that is, the tactical details in modeling vary from theory to theory. To understand how the strategy applies to mechanics, we need to elaborate on the details of each modeling stage.

(I) The Description Stage begins with a choice of objects and properties to be modeled. The theory to be used in modeling depends on the kinds of property to be modeled – physical, chemical or biological, for example. When an appropriate theory has been chosen, the theory provides a system of principles which constrain and direct the modeling process.

Object description is the first step in modeling. The object description begins with a decision on the type of model to be developed. For example, a given solid object could be modeled either as a material particle, or as a rigid body. Mechanics provides subtheories to facilitate the modeling of objects of each type. Complex objects are modeled as composite systems of interacting parts, for example, a system of particles or rigid bodies. In that case, the object description must specify the composition of the system and the model type of each part. Each part can then be modeled separately, and the model for the whole system is determined by the way the interacting parts are assembled.

In a process description the state variables of the model are specified. The state variables may be either basic or derived. *Basic variables* are defined implicitly by the generic laws (including the Zeroth Law). *Derived variables* must be defined explicitly in terms of basic variables. In mechanics, position and velocity are basic variables, while momentum, kinetic energy and angular momentum are derived variables. A process description necessarily employs the Zeroth Law, so some reference system must be adopted, even if it is not mentioned explicitly.

A *process* is defined as the time evolution of some set of state variables. Motion is the basic process in mechanics. The energy conservation law makes it convenient, sometimes, to consider the process of energy flow independently of the objects processing the energy. In such a case, one is modeling a process rather than an object. A *process model* omits reference to objects underlying the process.

Graphical or diagrammatic methods are often useful in a process description. As a rule, only a qualitative graph of the process can be made in the description stage of modeling; although a few points, such as initial and final states may be specified completely. A quantitative graph is usually possible in the ramification stage of modeling.

An interaction description specifies the interaction type and agent for all interactions in the model, along with appropriate interaction variables, basic or derived. This includes internal interactions among the parts of a composite system, as well as interactions with external agents. The interaction description must be coordinated with the process description; a consistent set of variables must be chosen, and any changes in interactions between different stages of the process must be indicated. In mechanics, for example, use of kinetic energy as a state variable calls for use of potential energy and work as interaction variables. And the description of interactions differs in the processes of projectile motion and collision.

To sum up, the descriptive stage produces complete lists of object names and descriptive variables for the model and supplies the model with a *physical interpretation* by providing referential meanings for the variables.

(II) In **The Formulation Stage**, the laws of dynamics and interaction are applied to get definite equations of change for the state variables. Within a given theory, the appropriate choice of laws depends on the type of model and descriptive variables, as is clear in examples from mechanics. In a particle model, Newton's Second Law is the dynamical law relating basic descriptive variables, but conservation laws for energy, momentum and angular momentum may be more appropriate when derived variables are used. For a system of particles with interactions described by equations of constraint, we have seen that Lagrange's equation is the most convenient dynamical law. In a rigid body model, we employ separate dynamical laws for translational and rotational motion of the body. These laws belong to the superstructure of mechanics, being derived from the basic laws and the definition of a rigid body. The derivation is a special exercise in model formulation which can be carried out once and for all. The results can then be applied directly to the formulation of any rigid body model.

Besides equations of change, a model may include equations of constraint (as indicated in Figure 4.1). The *equations of constraint* in a model are functional relations among descriptive variables (rather than differential equations). There are many different kinds, including the so-called *Constitutive Relations* or *Equations of State*, such as the ideal gas law ($PV = nRT$) in thermodynamics and fluid mechanics.

Implementation of the formulation stage produces an *abstract model object* consisting of the set of descriptive variables and equations of change and constraint sufficient to determine values of the state variables. The adjective "abstract" signifies that in an abstract model the descriptive variables are detached from the referential meanings determined in the descriptive stage. Thus, the descriptive variables in an abstract model describe nothing in particular. The adjective "descriptive" remains appropriate, however, because in principle a descriptive variable can always be interpreted by associating it with a referent.

An abstract model does not represent a particular object. A model of a particular object consists of an abstract model together with an interpretation of its descriptive variables; in brief, *a concrete model is an interpreted abstract model*. The detachment of an abstract model from any physical interpretation is a step of major scientific and psychological importance. For the abstract model takes on a theoretical life of its own which can be studied apart from the complexities of a real physical situation. This process of *model abstraction* is crucial to scientific understanding. Paradoxically, physical insight into a given physical situation is achieved by sharply separating the perceived situation from its conceptual representation,that is, by constructing an abstract model. The physicist uses the same abstract model of a particle subject to a constant force to represent many different physical situations, such as a falling body or a body sliding on a rough surface. Thus, the model abstraction process enables the physicist to recognize common elements in different physical situations. Undoubtedly, it plays a role in the discovery of general physical laws from *ad hoc* models constructed wthout the help of general laws.

(III) In **The Ramification Stage** the special properties and implications of the abstract model are worked out. The equations of change are solved to determine trajectories of the state variables with various initial conditions; the time dependence of significant derived descriptors, such as energy, is determined; results may be represented graphically as well as analytically to facilitate analysis. Let us refer to a model object together with one or more of its main ramifications as a *ramified model.*

The ramification process is largely mathematical, but the analysis of results is just as important. Especially important is the identification of *emergent properties* in composite systems, such as resonances, stabilities and instabilities.

A large part of this book has been devoted to ramifications. We were able to work out ramifications of the gravitational two body problem at length, because the equations of motion can be solved exactly. On the other hand, we found that the ramifications of the gravitational three body problem are only partially known.

(IV) The Validation Stage is concerned with evaluating the ramified model by comparing it with some real object-in-situation which it is supposed to describe. This may range from a simple check on the reasonableness of numerical results to a full-blown experiment test. Validation is a model deployment process, so we will see it in perspective as we analyze model deployment.

Model Deployment

Model deployment is the process of matching a ramified model to a specific empirical situation. The result is a *concrete model* that *represents* objects and/or processes in that situation. We say, then, that the situation has been *modeled* by the scientific theory from which the model was developed. The *match* of the model to the situation is a correspondence between the values of descriptive variables in the model and properties of objects in the situation. The correspondence is established by measurement procedures, including the so-called operational referitions for the variables measured. Measurement involves error and uncertainty, so the match of model to situation must be characterized by some measure for "goodness of fit" and criteria for an adequate match must be set up. These issues are handled by a theory of measurement, which can be regarded as one part of a general theory of model deployment. We mention measurement here only to indicate how it fits into modeling theory.

Different kinds of model deployment can be classified according to different purposes they subserve. A model may be deployed for the purposes of scientific explanation, prediction or design. Indeed, we say that an empirical phenomenon can be *explained* scientifically if and only if it can be adequately modeled by a scientific theory. Scientific predictions are generated by process models which relate the values of property variables at different times. Scientific design involves the development of models to be deployed as plans for the construction of physical systems with specified properties.

The assertion that scientific explanation is a kind of model deployment deserves some further comment, since scientific explanation is not ordinarily characterized that way. There are two common kinds of scientific explanation: causal and inferential. A *causal explanation* of an event *A* is supplied by identifying its cause, consisting of agents and conditions sufficient to produce A. An *inferential explanation* of *A* is supplied by identifying a mechanism (or law) which accounts for *A*. Each kind of explanation employs one of the essential ingredients of a model. Thus they employ partial models and should be regarded as partial explanations only. A complete explanation requires a complete model.

Empirical tests of a scientific theory are variants of the three major kinds of model deployment we have just discussed. A theory can be tested only indirectly by testing for the empirical adequacy of models developed from the theory. A particular hypothesis can be tested only as part of a theory which is sufficient for the design of testable models and only against an alternative hypothesis which is a candidate to replace it. A test is made by comparing the adequacies of models generated with the alternative hypotheses.

This discussion of model deployment was necessarily brief, because a systematic theory of deployment processes is yet to be developed. The subject is complex, but the concept of model deployment appears to be the thread needed to tie up a lot of loose ends in the methodology of science.

Exercises for Chapter 9

1. Does the Zeroth Law imply the existence of a unique physical entity which we might identify as *physical space*?

2. Are space and time objectively real in the sense that they exist independently of any human mind?
3. Develop an explicit formulation of a *Law of Molecular Composition*, providing suitable definitions for the key terms, and carefully distinguishing between mathematical structure and physical interpretation. Discuss the scope and validity of the law.
4. Design a thought experiment for determining if a given reference system is an inertial system.
5. According to the *Biot-Savart Law*, a moving particle with charge q_1 produces a magnetic field

$$\mathbf{B}(\mathbf{x}, t) = \frac{q_1}{c} \frac{\mathbf{v}_1 \times (\mathbf{x} - \mathbf{x}_1)}{|\mathbf{x} - \mathbf{x}_1|^3},$$

where c is a constant, $\mathbf{x}_1 = \mathbf{x}_1(t)$, and $\mathbf{v}_1 = \mathbf{v}_1(t)$ is the velocity of the particle. Examine the magnetic interaction between two charged particles and show that the Law of Reciprocity is not satisfied. How is this result affected by including electric interactions? Evaluate the rate at which this two particle system transfers momentum to the electromagnetic field. What if one particle is initially at rest?
6. Suppose that during all of recorded history the earth was surrounded by a dense cloud cover so that the sun, moon and stars could not be seen. Suppose also that Newtonian mechanics had developed in spite of this handicap. Explain how earthbound physicists could nevertheless detect the rotation of the earth and the orbital motion about the sun and thus separate the associated pseudoforces from real forces.
7. Examine the change in form of the Second Law induced by changing to a time variable which is an arbitrary monotonic function of inertial time.
8. Discuss the change in form of equations of motion when transformed from an inertial system to an accelerated reference system.
9. Discuss the following assertion by J. L. Synge:

   "It is futile to ask whether nature is ultimately discrete or continuous, for "discrete" and "continuous" are categories of the understanding, not properties of nature".

10. Make a thorough critique of Eisenbud's influential article on mechanics (below), comparing it in detail with the formulation of mechanics in this chapter. Note how the concept of definition is used. Carefully distinguish between explicit and implicit definitions, interpretations, correspondence rules and measurements.

    L. Eisenbud, 'On the Classical Laws of Motion', *Am. J. Phys.* **26**, 144–159 (1958).

Appendix A

Spherical Trigonometry

In Section 2-4 we saw how efficiently geometric algebra describes relations among directions and angles in a plane. Here we turn to the study of such relations in 3-dimensional space. Our aim is to see how the traditional subjects of solid geometry and spherical trigonometry can best be handled with geometric algebra. Spherical trigonometry is useful in subjects as diverse as crystalography and celestial navigation.

First, let us see how to determine the angle that a line makes with a plane from the directions of the line and the plane. The direction of a given line is represented by a unit vector $\hat{\mathbf{a}}$, while that of a plane is represented by a unit bivector $\hat{\mathbf{A}}$. The angle $\boldsymbol{\alpha}$ between $\hat{\mathbf{a}}$ and $\hat{\mathbf{A}}$ is defined by the product

$$\hat{\mathbf{a}}\hat{\mathbf{A}} = \hat{\boldsymbol{\alpha}} \cos \alpha + i \sin \alpha = \hat{\boldsymbol{\alpha}} e^{i\boldsymbol{\alpha}} \tag{A.1}$$

where i is the unit righthanded trivector, $\hat{\boldsymbol{\alpha}}$ is a unit vector and $\boldsymbol{\alpha} = \alpha\hat{\boldsymbol{\alpha}}$. The vector and trivector parts of (A.1) are

$$\hat{\mathbf{a}} \cdot \hat{\mathbf{A}} = \hat{\boldsymbol{\alpha}} \cos \alpha, \tag{A.2a}$$

$$\hat{\mathbf{a}} \wedge \hat{\mathbf{A}} = i \sin \alpha. \tag{A.2b}$$

Our assumption that i is the unit righthanded trivector fixes the sign of $\sin \alpha$, so, by (A.2b), $\sin \alpha$ is positive (negative) when $\mathbf{a} \wedge \hat{\mathbf{A}}$ is righthanded (lefthanded). The angle α is uniquely determined by (A.1) if it is restricted to the range $0 \leqslant \alpha \leqslant 2\pi$.

Equations (A.2a, b) are perfectly consistent with the conventional interpretation of $\cos \alpha$ and $\sin \alpha$ as components of projection and rejection, as shown in Figure A.1. Thus, according to the definition of projection by Equation (4.5b) of Section 2-4, we have

$$P_{\hat{\mathbf{A}}}(\hat{\mathbf{a}})\hat{\mathbf{A}} = \hat{\mathbf{a}} \cdot \hat{\mathbf{A}} = \hat{\boldsymbol{\alpha}} \cos \alpha.$$

We can interpret this equation as follows. First the unit vector $\hat{\mathbf{a}}$ is projected into a vector $P_{\hat{\mathbf{A}}}(\hat{\mathbf{a}})$ with magnitude $\cos \alpha$. Right multiplication by $\hat{\mathbf{A}}$ then rotates the projected vector through a right angle into the vector $\hat{\mathbf{a}} \cdot \hat{\mathbf{A}}$, which can be expressed as a unit vector $\hat{\boldsymbol{\alpha}}$ times its magnitude $\cos \alpha$. Thus, the product $\hat{\mathbf{a}}\cdot\hat{\mathbf{A}}$ is equivalent to a projection of $\hat{\mathbf{a}}$ into the $\hat{\mathbf{A}}$-plane followed by a rotation by $\frac{1}{2}\pi$.

To interpret the exponential $e^{i\boldsymbol{\alpha}}$ in (A.1), multiply (A.1) by $\hat{\mathbf{a}}$ to get

$$\hat{\mathbf{A}} = (\hat{\mathbf{a}}e^{i\boldsymbol{\alpha}})\hat{\boldsymbol{\alpha}}\,.$$

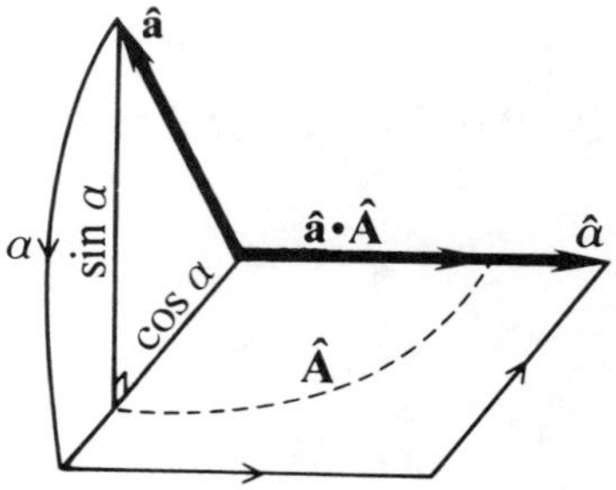

Fig. A.1. The angle between a vector and a bivector.

This expresses the unit bivector $\hat{\mathbf{A}}$ as the product of orthogonal unit vectors $\hat{\mathbf{a}}e^{i\boldsymbol{\alpha}}$ and $\hat{\boldsymbol{\alpha}}$. The factor $e^{i\boldsymbol{\alpha}}$ has rotated $\hat{\mathbf{a}}$ into the $\hat{\mathbf{A}}$-plane. We recognize $e^{i\boldsymbol{\alpha}}$ as a spinor which rotates vectors in the $(\iota\hat{\alpha})$-plane through an angle α. The unit vector $\hat{\boldsymbol{\alpha}}$ specifies the *axis* of rotation.

Now let us consider how to represent the angle between two planes algebraically. The directions of two given planes are represented by unit bivectors $\hat{\mathbf{A}}$ and $\hat{\mathbf{B}}$. We define the *dihedral angle* $\mathbf{c}$ between $\hat{\mathbf{A}}$ and $\hat{\mathbf{B}}$ by the equation

$$\hat{\mathbf{B}}\hat{\mathbf{A}} = e^{i\mathbf{c}}. \tag{A.3}$$

Both the magnitude of the angle $c = |\mathbf{c}|$ (with $0 \leqslant c \leqslant 2\pi$) and the direction of $i\hat{\mathbf{c}}$ of its plane are determined by (A.3). Separating (A.3) into scalar and bivector parts, we have

$$\langle\hat{\mathbf{B}}\,\hat{\mathbf{A}}\rangle_0 = \hat{\mathbf{B}}\cdot\hat{\mathbf{A}} = \cos c \tag{A.4a}$$

$$\langle\hat{\mathbf{B}}\hat{\mathbf{A}}\rangle_2 = i\hat{\mathbf{c}}\sin c\,. \tag{A.4b}$$

Note that for c = 0, Equation (A.3) becomes $\hat{\mathbf{B}}\hat{\mathbf{A}} = 1$, so $\hat{\mathbf{B}} = -\hat{\mathbf{A}}$, since $\hat{\mathbf{B}}^2 = \hat{\mathbf{A}}^2 = -1$.

The geometrical interpretation of (A.3) is indicated in Figure A.2, which shows the two planes intersecting in a line with direction $\hat{\mathbf{c}}$. The vector $\hat{\mathbf{c}}$ is therefore a common factor of the bivectors $\hat{\mathbf{A}}$ and $\hat{\mathbf{B}}$. Consequently there exist unit vectors $\hat{\mathbf{a}}$ and $\hat{\mathbf{b}}$ orthogonal to $\hat{\mathbf{c}}$ such that $\hat{\mathbf{A}}$ and $\hat{\mathbf{B}}$ have the factorizations

$$\hat{\mathbf{A}} = \hat{\mathbf{a}}\hat{\mathbf{c}} = -\hat{\mathbf{c}}\hat{\mathbf{a}}$$

$$\hat{\mathbf{B}} = \hat{\mathbf{c}}\hat{\mathbf{b}} = -\hat{\mathbf{b}}\hat{\mathbf{c}}\,.$$

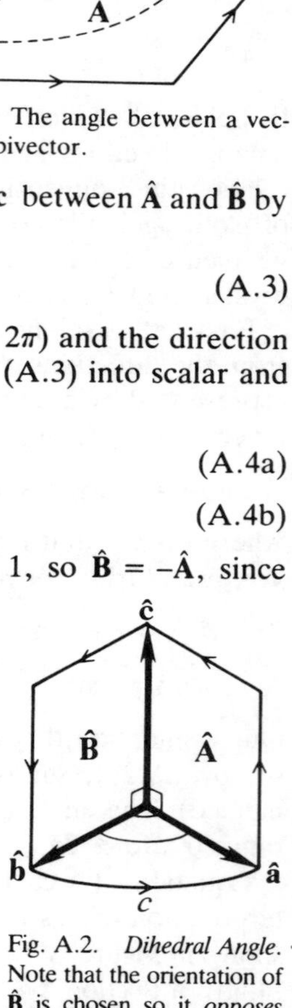

Fig. A.2. *Dihedral Angle.* Note that the orientation of $\hat{\mathbf{B}}$ is chosen so it *opposes* that of $\hat{\mathbf{A}}$ if $\hat{\mathbf{B}}$ is brought into coincidence with $\hat{\mathbf{A}}$ by rotating it through the angle *C*.

Note that the order of factors corresponds to the orientations assigned in Figure A.2. The common factor $\hat{\mathbf{c}}$ vanishes when $\hat{\mathbf{B}}$ and $\hat{\mathbf{A}}$ are multiplied;

$$\hat{\mathbf{B}}\hat{\mathbf{A}} = (-\hat{\mathbf{b}}\hat{\mathbf{c}})\,(-\hat{\mathbf{c}}\hat{\mathbf{a}}) = \hat{\mathbf{b}}\hat{\mathbf{a}} = e^{i\hat{\mathbf{c}}c}.$$

Note that this last equality has the same form as Equation (4.9) with the unit bivector of the ($\hat{\mathbf{b}}\wedge\hat{\mathbf{a}}$)-plane expressed as the dual $i\hat{\mathbf{c}}$ of the unit vector $\hat{\mathbf{c}}$. It should be clear now that (A.4b) expresses the fact that $\langle\hat{\mathbf{B}}\hat{\mathbf{A}}\rangle_2$ is a bivector determining a plane perpendicular to the line of intersection of the $\hat{\mathbf{B}}$- and $\hat{\mathbf{A}}$-planes and so intersecting these planes at right angles.

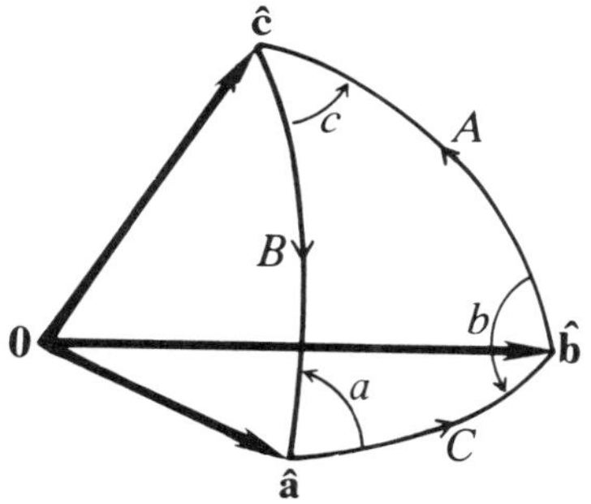

Fig. A.3. A spherical triangle.

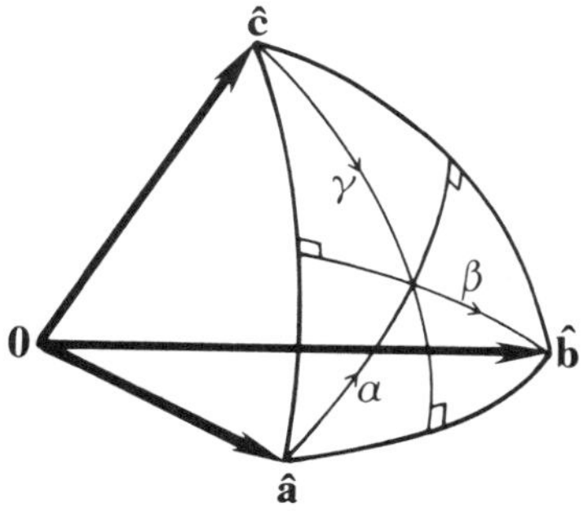

Fig. A.4. Altitudes of a spherical triangle.

Now we are prepared to analyze the relations among three distinct directions. Three unit vectors $\hat{\mathbf{a}}$, $\hat{\mathbf{b}}$, $\hat{\mathbf{c}}$ can be regarded as vertices of a *spherical triangle* on a unit sphere, as shown in Figure A.3. The sides of the triangle are arcs of great circles determined by the intersection of the circle with the planes determined by each pair of vectors. The sides of the triangle have lengths A, B, C which are equal to the angles between the vectors $\hat{\mathbf{a}}$, $\hat{\mathbf{b}}$, $\hat{\mathbf{c}}$. This relation is completely described by the equations

$$\hat{\mathbf{a}}\hat{\mathbf{b}} = e^{\mathbf{C}} = \cos C + \hat{\mathbf{C}} \sin C, \tag{A.5a}$$

$$\hat{\mathbf{b}}\hat{\mathbf{c}} = e^{\mathbf{A}} = \cos A + \hat{\mathbf{A}} \sin A, \tag{A.5b}$$

$$\hat{\mathbf{c}}\hat{\mathbf{a}} = e^{\mathbf{B}} = \cos B + \hat{\mathbf{B}} \sin B, \tag{A.5c}$$

where $\mathbf{A} = \hat{\mathbf{A}}A$, $\mathbf{B} = \hat{\mathbf{B}}B$, $\mathbf{C} = \hat{\mathbf{C}}C$. The unit bivectors $\hat{\mathbf{A}}$, $\hat{\mathbf{B}}$, $\hat{\mathbf{C}}$ are directions for the planes determining the sides of the spherical triangle by intersection with the sphere. Of course, we have the relations

$$\hat{\mathbf{A}}^2 = \hat{\mathbf{B}}^2 = \hat{\mathbf{C}}^2 = -1, \tag{A.6a}$$

as well as

$$\hat{\mathbf{a}}^2 = \hat{\mathbf{b}}^2 = \hat{\mathbf{c}}^2 = 1. \tag{A.6b}$$

The angles a, b, c of the spherical triangle in Figure A.3 are dihedral angles between planes, so they are determined by equations of the form (A.3), namely,

$$\hat{\mathbf{B}}\hat{\mathbf{A}} = e^{i\mathbf{c}} = \cos c + i\hat{\mathbf{c}} \sin c, \tag{A.7a}$$

$$\hat{\mathbf{C}}\hat{\mathbf{B}} = e^{i\mathbf{a}} = \cos a + i\hat{\mathbf{a}} \sin a, \tag{A.7b}$$

$$\hat{\mathbf{A}}\hat{\mathbf{C}} = e^{i\mathbf{b}} = \cos b + i\hat{\mathbf{b}} \sin b, \tag{A.7c}$$

where $\mathbf{a} = \hat{\mathbf{a}}a$, $\mathbf{b} = \hat{\mathbf{b}}b$, $\mathbf{c} = \hat{\mathbf{c}}c$.

Equations (A.5a, b, c) and (A.7a, b, c) with (A.6a, b) understood determine all the relations among the sides and angles of a spherical triangle. Let us see how these equations can be used to derive the fundamental equations of spherical trigonometry. Taking the product of Equations (A.5a, b, c) and noticing that

$$(\hat{\mathbf{a}}\hat{\mathbf{b}})\,(\hat{\mathbf{b}}\hat{\mathbf{c}})\,(\hat{\mathbf{c}}\hat{\mathbf{a}}) = 1\,,$$

we get

$$e^{\mathbf{C}}e^{\mathbf{A}}e^{\mathbf{B}} = 1\,. \tag{A.8}$$

This equation can be solved for any one of the angles **A**, **B**, **C** in terms of the other two. To solve for **C**, multiply (A.8) by $(e^{\mathbf{C}})^{-1} = e^{-\mathbf{C}}$ to get

$$e^{-\mathbf{C}} = e^{\mathbf{A}}e^{\mathbf{B}}. \tag{A.9}$$

If the exponentials are expanded into scalar and bivector parts and (A.7a) is used in the form $\hat{\mathbf{A}}\hat{\mathbf{B}} = (\hat{\mathbf{B}}\hat{\mathbf{A}})^{\dagger} = e^{-i\mathbf{c}}$, then (A.9) assumes the expanded form

$$\begin{aligned}\cos C - \hat{\mathbf{C}}\sin C &= (\cos A + \hat{\mathbf{A}}\sin A)(\cos B + \hat{\mathbf{B}}\sin B)\\ &= \cos A\cos B + \hat{\mathbf{A}}\sin A\cos B + \hat{\mathbf{B}}\sin B\cos A\\ &\quad + (\cos c - i\hat{\mathbf{c}}\sin c)\sin A\sin B\,.\end{aligned}$$

Separating this into scalar and bivector parts, we get

$$\cos C = \cos A\cos B + \sin A\sin B\cos c\,, \tag{A.10}$$

$$\begin{aligned}-\hat{\mathbf{C}}\sin C &= \hat{\mathbf{A}}\sin A\cos B + \hat{\mathbf{B}}\sin B\cos A\\ &\quad - i\hat{\mathbf{c}}\sin A\sin B\,.\end{aligned} \tag{A.11}$$

Equation (A.10) is called the *cosine law for sides* in spherical trigonometry. It relates three sides and an angle of a spherical triangle and determines any one of these quantities when the other two are known.

Since the value of C can be determined from **A** and **B** by (A.10), the direction $\hat{\mathbf{C}}$ is then determined by (A.11). Thus equations (A.10) and (A.11) together determine **C** from **A** and **B**. Nothing resembling Equation (A.11) appears in traditional spherical trigonometry, because it relates directions $\hat{\mathbf{A}}$, $\hat{\mathbf{B}}$, $\hat{\mathbf{C}}$, whereas the traditional theory is concerned only with scalar relations. Of course, all sorts of scalar relations can be generated from (A.11) by multiplying by any one of the available bivectors, but they are only of marginal interest. The great value of (A.11) is evident in our study of rotations in 3 dimensions in Section 5-3.

We can analyze consequences of (A.7a, b, c) in the same way we analyzed (A.5a, b, c). Observing that

$$(\hat{\mathbf{B}}\hat{\mathbf{A}})\,(\hat{\mathbf{A}}\hat{\mathbf{C}})\,(\hat{\mathbf{C}}\hat{\mathbf{B}}) = -1\,,$$

we obtain from the product of Equations (A.7a, c, b)

$$e^{i\mathbf{c}}e^{i\mathbf{b}}e^{i\mathbf{a}} = -1\,. \tag{A.12}$$

This should be compared with (A.8). From (A.9) we get

$$e^{-i\mathbf{c}} = -e^{i\mathbf{b}}e^{i\mathbf{a}}. \tag{A.13}$$

The scalar part of this equation gives us

$$\cos c = -\cos a \cos b + \sin a \sin b \cos C\,. \tag{A.14}$$

This is called the *cosine law for angles* in spherical trigonometry. Obviously, it determines the relation among three angles and a side of the spherical triangle.

The cosine law was derived by considering products of vectors in pairs, so we may expect to find a different "law" by considering the product $\hat{\mathbf{a}}\wedge\hat{\mathbf{b}}\wedge\hat{\mathbf{c}}$. Inserting $\hat{\mathbf{b}}\wedge\hat{\mathbf{c}} = \hat{\mathbf{A}} \sin A$ from (A.5b) and the corresponding relations from (A.5c, a) into $\hat{\mathbf{a}}\wedge\hat{\mathbf{b}}\wedge\hat{\mathbf{c}}$, we get

$$\hat{\mathbf{a}}\wedge\hat{\mathbf{b}}\wedge\hat{\mathbf{c}} = \hat{\mathbf{a}}\wedge\hat{\mathbf{A}} \sin A = \hat{\mathbf{b}}\wedge\hat{\mathbf{B}} \sin B = \hat{\mathbf{c}}\wedge\hat{\mathbf{C}} \sin C\,. \tag{A.15}$$

We can find an analogous relation from the product $\hat{\mathbf{A}}\hat{\mathbf{B}}\hat{\mathbf{C}}$. Using (A.7b), we ascertain that

$$\langle\hat{\mathbf{A}}\hat{\mathbf{B}}\hat{\mathbf{C}}\rangle_0 = -\langle\hat{\mathbf{A}}i\hat{\mathbf{a}}\rangle_0 \sin a = -i\hat{\mathbf{a}}\wedge\hat{\mathbf{A}} \sin a\,.$$

Obtaining the corresponding relations from (A.7a, b), we get

$$i\langle\hat{\mathbf{A}}\hat{\mathbf{B}}\hat{\mathbf{C}}\rangle_0 = \hat{\mathbf{a}}\wedge\hat{\mathbf{A}} \sin a = \hat{\mathbf{b}}\wedge\hat{\mathbf{B}} \sin b = \hat{\mathbf{c}}\wedge\hat{\mathbf{C}} \sin c\,. \tag{A.16}$$

The ratio of (A.15) to (A.16) gives us

$$\frac{\hat{\mathbf{a}}\wedge\hat{\mathbf{b}}\wedge\hat{\mathbf{c}}}{i\langle\hat{\mathbf{A}}\hat{\mathbf{B}}\hat{\mathbf{C}}\rangle_0} = \frac{\sin A}{\sin a} = \frac{\sin B}{\sin b} = \frac{\sin C}{\sin c}\,. \tag{A.17}$$

This is called the *sine law* in spherical trigonometry. Obviously, it relates any two sides of a spherical triangle to the two opposing angles.

We get further information about the spherical triangle by considering the product of each vector with the bivector of the opposing side. Each product has the form of Equation (A.1) which corresponds to Figure A.1. Thus, we have the equation

$$\hat{\mathbf{a}}\hat{\mathbf{A}} = \hat{\boldsymbol{\alpha}}e^{i\boldsymbol{\alpha}} = \hat{\boldsymbol{\alpha}} \cos \alpha + i \sin \alpha\,, \tag{A.18a}$$

$$\hat{\mathbf{b}}\hat{\mathbf{B}} = \hat{\boldsymbol{\beta}}e^{i\boldsymbol{\beta}} = \hat{\boldsymbol{\beta}} \cos \beta + i \sin \beta\,, \tag{A.18b}$$

$$\hat{\mathbf{c}}\hat{\mathbf{C}} = \hat{\boldsymbol{\gamma}}e^{i\boldsymbol{\gamma}} = \hat{\boldsymbol{\gamma}} \cos \gamma + i \sin \gamma. \tag{A.18c}$$

The angles $i\boldsymbol{\alpha}$, $i\boldsymbol{\beta}$, $i\gamma$ are "*altitudes*" of the spherical triangle with lengths α, β, γ, as shown by Figure A.4. If the trivector parts of (A.18a, b, c) are substituted into (A.15) and (A.16) we get the corresponding equations of traditional spherical trigonometry:

$$\frac{\hat{\mathbf{a}}\wedge\hat{\mathbf{b}}\wedge\hat{\mathbf{c}}}{i} = \sin\alpha\sin A = \sin\beta\sin B = \sin\gamma\sin C\,, \tag{A.19}$$

$$\langle\hat{\mathbf{A}}\hat{\mathbf{B}}\hat{\mathbf{C}}\rangle_0 = \sin\alpha\sin a = \sin\beta\sin b = \sin\gamma\sin c\,. \tag{A.20}$$

This completes our algebraic analysis of basic relations in spherical trigonometry.

Exercises

(A.1) Prove that Equation (A.8) is equivalent to the equations

$$e^{\mathbf{A}}e^{\mathbf{B}}e^{\mathbf{C}} = 1 \quad \text{and} \quad e^{-\mathbf{C}}e^{-\mathbf{B}}e^{-\mathbf{A}} = 1\,.$$

(A.2) The spherical triangle ($\hat{\mathbf{a}}$, $\hat{\mathbf{b}}$, $\hat{\mathbf{c}}$) satisfying Equations (A.5) and (A.7) determines another spherical triangle ($\hat{\mathbf{a}}'$, $\hat{\mathbf{b}}'$, $\hat{\mathbf{c}}'$) by the duality relations

$$\hat{\mathbf{A}} = i\hat{\mathbf{a}}', \quad \hat{\mathbf{B}} = i\hat{\mathbf{b}}', \quad \hat{\mathbf{C}} = i\hat{\mathbf{c}}'\,.$$

The triangle ($\hat{\mathbf{a}}'$, $\hat{\mathbf{b}}'$, $\hat{\mathbf{c}}'$) is called the *polar triangle* of the triangle ($\hat{\mathbf{a}}$, $\hat{\mathbf{b}}$, $\hat{\mathbf{c}}$), because its sides are arcs of great circles with $\hat{\mathbf{a}}$, $\hat{\mathbf{b}}$, and $\hat{\mathbf{c}}$ as poles.

Prove that the sides A', B', C' of the polar triangle are equal to the exterior angles supplementary to the interior angles α, β, γ of the primary triangle, and, conversely, that the sides A, B, C of the primary triangle are equal to the exterior angles supplementary to the interior angles α', β', γ' of the polar triangle.

From Equations (A.18a, b, c) prove that corresponding altitudes of the two triangles lie on the same great circle and that the distance along the great circle between $\hat{\mathbf{a}}$ and $\hat{\mathbf{a}}'$ is $|\alpha - \frac{1}{2}\pi|$.

(A.3) For the *right* spherical triangle with $c = \pi/2$, prove that

$$\sin a = \frac{\sin A}{\sin C}\,, \quad \sin b = \frac{\sin B}{\sin C}\,,$$

$$\cos A = \frac{\cos a}{\sin b}\,, \quad \cos B = \frac{\cos b}{\sin a}\,,$$

$$\cos C = \cos A\cos B = \cot a\cot b\,.$$

(A.4) Prove that an equilateral spherical triangle is equiangular with angle a related to side A by

$$\cos a - \cos A + \cos a\cos A = 0\,.$$

(A.5) Assuming (A.5a, b, c), prove that

$$2\hat{\mathbf{a}}\wedge\hat{\mathbf{b}}\wedge\hat{\mathbf{c}}\hat{\mathbf{c}} = e^{\mathbf{B}}e^{\mathbf{A}} - e^{\mathbf{A}}e^{\mathbf{B}} = 2i\hat{\mathbf{c}}\sin c\sin A\sin B\,.$$

Hence,

$$\frac{\hat{\mathbf{a}}\wedge\hat{\mathbf{b}}\wedge\hat{\mathbf{c}}}{i} = \sin A \sin B \sin c = \sin A \sin b \sin C$$

$$= \sin a \sin B \sin C.$$

Use this to derive the sine law, and, from (A.19), expressions for the altitudes of a spherical triangle.

(A.6) Prove that

$$|\langle\mathbf{abc}\rangle_1|^2 = (\mathbf{a}\cdot\mathbf{b})^2 + (\mathbf{b}\cdot\mathbf{c})^2 + (\mathbf{c}\cdot\mathbf{a})^2 - 2\mathbf{a}\cdot\mathbf{b}\;\mathbf{b}\cdot\mathbf{c}\;\mathbf{c}\cdot\mathbf{a}.$$

Assuming (A.5a, b, c) use the identity

$$\mathbf{abc} = \mathbf{a}\wedge\mathbf{b}\wedge\mathbf{c} + \langle\mathbf{abc}\rangle_1$$

to prove that

$$|\hat{\mathbf{a}}\wedge\hat{\mathbf{b}}\wedge\hat{\mathbf{c}}|^2 = 1 - \cos^2 A - \cos^2 B - \cos^2 C + 2\cos A \cos B \cos C.$$

Note that this can be used with (A.19) to find the altitudes of a spherical triangle from the sides.

(A.7) Find the surface area and volume of a parallelopiped with edges of lengths a, b, c and face angles A, B, $C \leqslant \frac{1}{2}\pi$.

(A.8) Establish the identity

$$(\mathbf{a}\wedge\mathbf{b})\cdot(\mathbf{c}\wedge\mathbf{d}) + (\mathbf{b}\wedge\mathbf{c})\cdot(\mathbf{a}\wedge\mathbf{d}) + (\mathbf{c}\wedge\mathbf{a})\cdot(\mathbf{b}\wedge\mathbf{d}) = 0.$$

Note that the three terms differ only by a cyclic permutation of the first three vectors. Use this identity to prove that the altitudes of a spherical triangle intersect in a point (compare with Exercise 2–4.11a).

(A.9) Prove that on a unit sphere the area Δ of a spherical triangle with interior angles a, b, c (Figure A.3) is given by the formula

$$\Delta = a + b + c - \pi.$$

Since Δ is given by the difference between the sum of interior angles for a spherical triangle and a plane triangle, it is often called the *spherical excess*.
Hint: The triangle is determined by the intersection of great circles which divide the sphere into several regions with area Δ, Δ_a, Δ_b or Δ_c as shown in Figure A.5. What relation exists between the angle a and the area $\Delta + \Delta_a$?

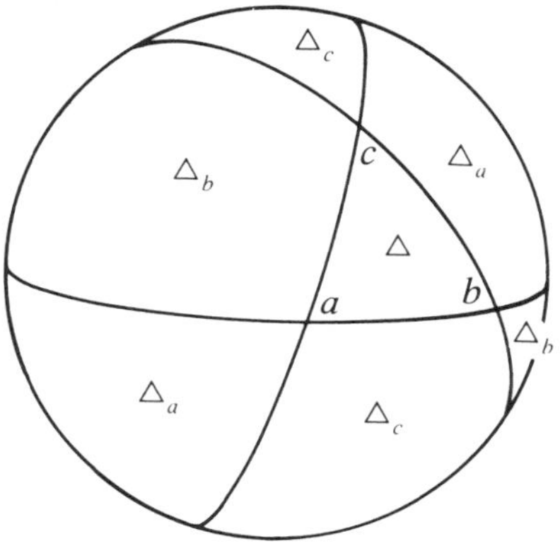

Fig. A.5. Spherical excess.

Appendix B

Elliptic Functions

Elliptic functions provide general solutions of differential equations with the form

$$\left(\frac{dy}{dx}\right)^2 = f(y), \tag{B.1}$$

where $f(y)$ is a polynomial in $y = y(x)$. Such equations are very common in physics, arising frequently from energy integrals where the left side of the equation comes from a kinetic energy term.

Since different polynomials can be related by such devices as factoring and change of variables, it turns out that the general problem of solving (B.1) for a large class of polynomials can be reduced to solving a differential equation of the standard form

$$\left(\frac{dy}{dx}\right)^2 = (1-y^2)(1-k^2y^2), \tag{B.2}$$

for $0 \leqslant k \leqslant 1$ and $-1 \leqslant y \leqslant 1$. The solution of this equation for the conditions

$$y = 0, \quad \frac{dy}{dx} > 0 \quad \text{when } x = 0, \tag{B.3}$$

is denoted by

$$y = \operatorname{sn} x \tag{B.4}$$

(Pronounced "ess-en-ex"). Of course, this function depends on the value of the parameter k, which is called the *modulus*.

Direct integration of (B.2) produces the inverse function

$$x = \operatorname{sn}^{-1} y = \int_0^y \frac{dy}{[(1-y^2)(1-k^2y^2)]^{1/2}} \,. \tag{B.5}$$

It is an odd function of y, which increases steadily from 0 to

$$K = \int_0^1 \frac{\mathrm{d}y}{[(1-y^2)(1-k^2y^2)]^{1/2}} \quad . \tag{B.6}$$

as y increases from 0 to 1. Consequently, $y = \operatorname{sn} x$ is an odd function of x, and it has period $4K$; that is

$$\operatorname{sn}(x+4K) = \operatorname{sn} x. \tag{B.7}$$

The integral (B.6) is called a *complete elliptic integral of the first kind*. We can evaluate it by a change of variables and a series expansion:

$$K(k) = \int_0^{\pi/2} \frac{\mathrm{d}u}{(1-k^2\sin^2 u)^{1/2}} = \int_0^{\pi/2} \left[1 + \frac{k^2}{2}\sin^2 u + \ldots\right] \mathrm{d}u$$

$$= \frac{\pi}{2}\left\{1 + \sum_{n=1}^{\infty}\left[\frac{1\cdot 3\ldots(2n-1)}{2.4\ldots 2n}\right]^2 k^{2n}\right\} \tag{B.8}$$

The function $K(k)$ is graphed in Figure B.2.

Two other functions cn x and dn x can be defined by the equations

$$\operatorname{cn}^2 x = 1 - \operatorname{sn}^2 x, \quad \operatorname{cn} 0 = 1,$$

$$\operatorname{dn}^2 x = 1 - k^2 \operatorname{sn}^2 x, \quad \operatorname{dn} 0 = 1, \tag{B.9}$$

along with the condition that their derivatives be continuous to determine the sign of the square root. Since $k \leqslant 1$, dn x is always positive with period $2K$, while cn x has period $4K$.

The three functions sn x, cn x and dn x are called *Jacobian elliptic functions*, or just elliptic functions. They may be regarded as generalizations or distortions of the familiar trigonometric functions. Indeed, from the above relations it is readily verified that for $k = 0$,

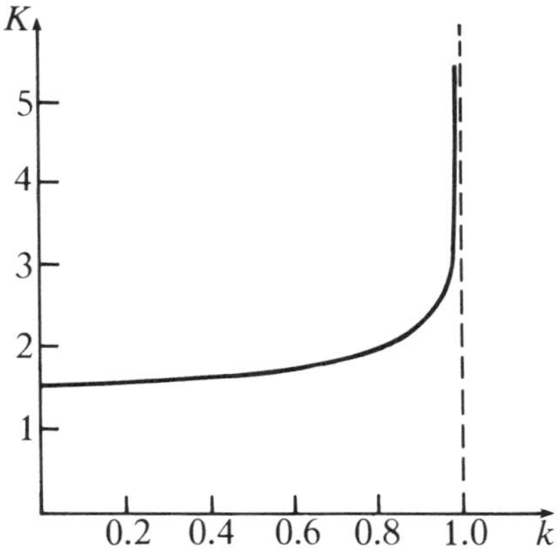

Fig. B.1. Graph of the period $4K$ as a function of the modulus k.

$$\operatorname{sn} x = \sin x, \quad \operatorname{cn} x = \cos x, \quad K = \tfrac{1}{2}\pi, \tag{B.10}$$

and for $k = 1$,

$$\operatorname{sn} x = \tanh x, \quad \operatorname{cn} x = \operatorname{dn} x = \operatorname{sech} x, \quad K = \infty. \tag{B.11}$$

Traditionally, the nomenclature of elliptic functions is used only when k is in range $0 < k < 1$.

Graphs of the elliptic functions are shown in Figure B.2. Tables of elliptic functions can be found in standard references such as Jahnke and Emde

(1945), but programs to evaluate elliptic functions on a computer are not difficult to write, and some are available commercially.

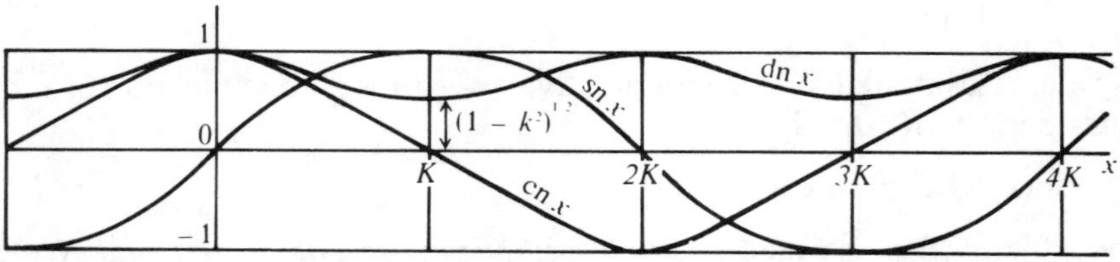

Fig. B.2. Graphs of the elliptic functions sn x, cn x, dn x for $k^2 = 0.7$.

For applications we need some systematic procedures for reducing equations to the standard form (B.2). Consider the equation

$$\left(\frac{\mathrm{d}r}{\mathrm{d}x}\right)^2 = Ar^4 + Br^2 + C + Dr^{-2}, \tag{B.12}$$

where A, B, C and D are given scalar constants. This can be reduced to standard form by the change of variables

$$r^2 = ay^2 + b, \quad \text{where} \quad y = \mathrm{sn}(\mu x) \tag{B.13}$$

and a, b, μ, are constants. To perform the reduction and determine the constants, we differentiate (B.13) to get

$$r^2\left(\frac{\mathrm{d}r}{\mathrm{d}x}\right)^2 = a^2y^2\left(\frac{\mathrm{d}y}{\mathrm{d}x}\right)^2 . \tag{B.14}$$

The left side of this equation can be expressed in terms of y^2 by substituting (B.12) and (B.13), while the right side can be expressed in terms of y^2 by using

$$\left(\frac{\mathrm{d}y}{\mathrm{d}x}\right)^2 = \mu^2(1-y^2)(1-k^2y^2). \tag{B.15}$$

Then, by equating coefficients of like powers in y, we obtain

$$Ab^3 + Bb^2 + Cb + D = 0$$

$$3b^2A + 2bB + C = a\mu^2$$

$$3bA + B = -\mu^2(1 + k^2)$$

$$aA = \mu^2k^2.$$

When these four equations are solved for the four unknowns b, a, μ^2 and k^2, the solution to (B.12) is given explicitly by (B.13). To prepare for this, we eliminate k^2 from the third equation and μ^2 from the second, putting the equations in the form

$$Ab^3 + Bb^2 + bC + D = 0 \tag{B.16a}$$

$$Aa^2 + (3bA + B)a + (3b^2A + 2bB + C) = 0 \tag{B.16b}$$

$$\mu^2 = -A(3b + a) - B > 0 \tag{B.16c}$$

$$k^2 = \frac{aA}{\mu^2} \ . \tag{B.16d}$$

After the cubic equation (B.16a) has been solved for b, the quadratic equation (B.16b) can be solved for a. Then the values of a and b can be used to evaluate μ^2 and k^2.

The theory of elliptic functions is rich and complex, a powerful tool for mathematical physics. We have discussed only some simpler aspects of theory needed for applications in the text.

Exercises

(B.1) Establish the derivatives

$$\frac{\mathrm{d}}{\mathrm{d}x}\mathrm{sn}\ x = \mathrm{cn}\ x\ \mathrm{dn}\ x$$

$$\frac{\mathrm{d}}{\mathrm{d}x}\mathrm{cn}\ x = -\mathrm{sn}\ x\ \mathrm{dn}\ x$$

$$\frac{\mathrm{d}}{\mathrm{d}x}\mathrm{dn}\ x = -k^2\ \mathrm{sn}\ x\ \mathrm{cn}\ x.$$

(B.2) Show that $y = \mathrm{cn}(\mu t)$ and $r = \mathrm{dn}(\mu t)$ are solutions of the differential equations

$$\dot{y}^2 = \mu^2 (1 - y^2)(k'^2 + k^2y^2)$$

$$\dot{r}^2 = \mu^2 (1 - r^2)(r^2 - k'^2)$$

where $k'^2 = 1 - k^2$.

(B.3) Find a change of variables that transforms

$$\left(\frac{\mathrm{d}w}{\mathrm{d}x}\right)^2 = Aw + Bw^2 + Cw^3 + Dw^4$$

into an equation of the form (B.12).

Appendix C

Tables of Units, Constants and Data

C-1. Units and Conversion Factors

Length

1 kilometer (km)	$= 10^6$ meter (m)
1 angström (Å)	$= 10^{-10}$ m
1 fermi	$= 10^{-12}$ m
1 light-year	$= 9.460 \times 10^{12}$ m
1 astronomical unit (AU)	$= 1.49596 \times 10^{11}$ m

Time, frequency

1 sidereal day	$= 24 \times 60 \times 60$ sidereal seconds (s)
	= 0.997 269 57 mean solar days (d)
	$= 23^h\ 56^m\ 4^s.091$ mean solar time
1 mean solar day	= 1.002 737 91 sidereal days
1 sidereal year	= 365.256 36 d
1 sidereal month	= 27.321 661 0 d
1 Hertz	= 1 vibration (or cycle) per second

Force, Energy, Power

1 newton (N)	$= 1$ kg·m/s^2 $= 10^5$ dynes
1 joule (J)	$= 1$ Nm $= 10^7$ ergs
	$= 6.242 \times 10^{18}$ electron-volts (eV)
1 MeV	$= 10^6$ eV
1 watt	= 1 J/s

Magnetic Field

1 tesla	$= 1$ Weber/m^2 $= 10^4$ gauss

C-2. Physical Constants

Gravitational constant	G	$= 6.668 \times 10^{-11}$ Nm^2/kg^2
Speed of light	c	$= 2.99791 \times 10^8$ m/s
Electron mass	m_e	$= 9.1096 \times 10^{-31}$ kg $= 0.511$ MeV/c^2
Proton mass	m_p	$= 1.6725 \times 10^{-27}$ kg $= 938.3$ MeV/c^2

Neutron mass	$m_n = 1.6748 \times 10^{-27}$ kg $= 939.6$ MeV/c^2
Electron charge	$e = 1.602 \times 10^{-19}$ Coulombs

C-3. The Earth

Mass	$M_{\oplus} = 5.976(4) \times 10^{24}$ kg
Equatorial radius	$a = 6.37816(4) \times 10^{6}$ m
Polar radius	$c = 6.35677(9) \times 10^{6}$ m
Flattening	$(a-c)/a = 1/298.25 \times 3.2529 + 10^{3}$
Principal moments of inertia	
Polar	$I_3 = 0.3306\ M_{\oplus}a^2$
Equitorial	$I_1 = 0.3295\ M_{\oplus}a^2$
	$(I_3 - I_1)/I_3 = 1/305.3 = 3.276 \times 10^{3}$
Inclination of equator	$= 23° 27'$
Length of year (Julian)	$= 365.25$ days

C-4. The Sun

Solar mass	$M_{\odot} = 1.989(2) \times 10^{30}$ kg
Solar radius	$R_{\odot} = 6.9598(7) \times 10^{8}$ m
Solar luminosity	$L_{\odot} = 3.90(4) \times 10^{26}$ Joule-sec
Mean Earth-Sun distance	$= 1$ AU $= 1.49596 \times 10^{11}$ m

C-5. The Moon

Lunar mass	$M_{☾} = M_{\oplus}/81.301 = 7.350 \times 10^{22}$ kg
Lunar radius	$R_{☾} = 1.738 \times 10^{6}$ m
Inclination of lunar equator	
to ecliptic	$= 1°32'.5$
to orbit	$= 6°41'$
Mean Earth-Moon distance	$r_{☾} = 3.84 \times 10^{8}$ m
Eccentricity of orbit	$= 0.05490$
Sidereal period	$= 27.321\ 661$ d

C-6. The Planets

Planet	Mass	Sidereal Period	Semi-major axis AU	Semi-major axis (10^6 km)	Eccentricity of orbit	Inclination to ecliptic
Mercury	0.0554	87.97d	0.38710	57.9	0.2056	7.00°
Venus	0.815	224.70	0.72333	108.2	0.0068	3.39
Earth	1.000	365.256	1.00000	149.6	0.0167	–
Mars	0.1075	686.98	1.52369	227.9	0.0934	1.85
Jupiter	317.83	4332.59	5.20280	778.3	0.0485	1.31
Saturn	95.147	10759.22	9.53884	1427.0	0.0557	2.49
Uranus	14.54	30685.4	19.1819	2869.6	0.0472	0.77
Neptune	17.23	60189.0	30.0578	4496.6	0.0086	1.77
Pluto	0.17	90465.0	39.44	5900.0	0.250	17.16

Hints and Solutions for Selected Exercises

"An expert is someone who has made all the mistakes"

H. Bethe

"Therefore we should strive to make mistakes as fast as possible"

J. Wheeler

Section 1-7.

(7.1c)

$$\begin{aligned} A + B &= A + C && \text{(Given)} \\ (-A) + (A + B) &= (-A) + (A + C) && \text{(Addition Property)} \\ (-A) + A + B &= (-A) + A + C && \text{(Associativity)} \\ 0 + B &= 0 + C && \text{(Additive Inverse)} \\ B &= C && \text{(Additive Identity)} \end{aligned}$$

(7.1d) $AB = BC$, $A^{-1}(AB) = A^{-1}(AC)$, $(A^{-1}A)B = (A^{-1}A)C$, $1B = 1C$, $B = C$.

(7.2a) $$(\alpha + \mathbf{a})^{-1} = \frac{1}{\alpha + \mathbf{a}} = \frac{1}{\alpha + \mathbf{a}} \frac{\alpha - \mathbf{a}}{\alpha - \mathbf{a}} = \frac{\alpha - \mathbf{a}}{\alpha^2 - \mathbf{a}^2} .$$

This is undefined if $\alpha^2 = \mathbf{a}^2$.

(7.2b) If $\alpha^2 = \mathbf{a}^2$, then $\dfrac{\alpha + \mathbf{a}}{2\alpha} = \frac{1}{2}(1 + \hat{\mathbf{a}})$ is idempotent.

(7.2c) If $R = AN$, then $AR = R$. So if $RR^{-1} = 1$, then $ARR^{-1} = RR^{-1} = A$ which is a contradiction.

(7.2d) If $P^2 = P$ and $PP^{-1} = 1$, then $P^2P^{-1} = PP^{-1} = P$. Hence $P = 1$.

Section 2-1.

(1.1) $$\begin{aligned} (\mathbf{a}\wedge\mathbf{b})\cdot(\mathbf{c}\wedge\mathbf{d}) &= \mathbf{a}\cdot(\mathbf{b}\cdot(\mathbf{c}\wedge\mathbf{d})) = \mathbf{a}\cdot(\mathbf{b}\cdot\mathbf{c}\,\mathbf{d} - \mathbf{c}\,\mathbf{b}\cdot\mathbf{d}) = \mathbf{b}\cdot\mathbf{c}\,\mathbf{a}\cdot\mathbf{d} - \mathbf{a}\cdot\mathbf{c}\,\mathbf{b}\cdot\mathbf{d} \\ &= \langle\mathbf{a}\wedge\mathbf{b}\,\mathbf{c}\wedge\mathbf{d}\rangle_0 = \langle\mathbf{abc}\wedge\mathbf{d}\rangle_0 - \langle\mathbf{bc}\wedge\mathbf{ba}\rangle_0 \\ &= \mathbf{b}\cdot[(\mathbf{c}\wedge\mathbf{d})\cdot\mathbf{a}] = [\mathbf{b}\cdot(\mathbf{c}\wedge\mathbf{d})]\cdot\mathbf{a}. \end{aligned}$$

Note that the ambiguity in writing $\mathbf{b}\cdot(\mathbf{c}\wedge\mathbf{d})\cdot\mathbf{a}$ is inconsequential.

(1.2) If $\alpha\mathbf{a} + \beta\mathbf{b} + \gamma\mathbf{c} = 0$ and $\alpha \neq 0$, then $\mathbf{b}\wedge\mathbf{c}\wedge(\alpha\mathbf{a} + \beta\mathbf{b} + \gamma\mathbf{c}) =$

$\alpha\mathbf{b}\wedge\mathbf{c}\wedge\mathbf{a} = 0$, so $\mathbf{a}\wedge\mathbf{b}\wedge\mathbf{c} = 0$.
If $\mathbf{a}\wedge\mathbf{b}\wedge\mathbf{c} = 0$ and $\mathbf{a}\wedge\mathbf{b} \neq 0$, then, from Exercise (1.1),
$(\mathbf{a}\wedge\mathbf{b})^2\mathbf{c} - (\mathbf{a}\wedge\mathbf{b})\cdot(\mathbf{a}\wedge\mathbf{c})\mathbf{b} + (\mathbf{a}\wedge\mathbf{b})\cdot(\mathbf{b}\wedge\mathbf{c})\mathbf{a} = 0$.

(1.3) $\mathbf{x} = \dfrac{\mathbf{c}}{\alpha} - \dfrac{\mathbf{c}\cdot\mathbf{b}\mathbf{a}}{\alpha(\alpha + \mathbf{a}\cdot\mathbf{b})}$ provided the denominator does not vanish.

(1.4) First show that $\alpha\mathbf{x}\wedge\mathbf{B} = \mathbf{a}\wedge\mathbf{B}$. Then use $\mathbf{x}\mathbf{B} = \mathbf{x}\cdot\mathbf{B} + \mathbf{x}\wedge\mathbf{B}$ to get

$$\mathbf{x} = (\mathbf{a} + \alpha^{-1}\mathbf{a}\wedge\mathbf{B})\,(\alpha + \mathbf{B})^{-1} = \frac{\alpha^2\mathbf{a} + \alpha\mathbf{B}\cdot\mathbf{a} - \mathbf{a}\wedge\mathbf{B}\mathbf{B}}{\alpha(\alpha^2 + |\,\mathbf{B}\,|^2)}$$

(1.5) $\mathbf{x} = \dfrac{\alpha\mathbf{a} + \mathbf{c}\cdot\mathbf{B}}{\mathbf{c}\cdot\mathbf{a}}$.

(1.6) $\mathbf{a}\mathbf{b}^2\mathbf{c} = \mathbf{b}^2(\mathbf{a}\cdot\mathbf{c} + \mathbf{a}\wedge\mathbf{c}) = (\mathbf{a}\cdot\mathbf{b} + \mathbf{a}\wedge\mathbf{b})\,(\mathbf{b}\cdot\mathbf{c} + \mathbf{b}\wedge\mathbf{c})$.
Separate scalar and bivector parts.

(1.7)
$$\begin{aligned}(\mathbf{a}\wedge\mathbf{b}\wedge\mathbf{c})\cdot(\mathbf{u}\wedge\mathbf{v}\wedge\mathbf{w}) &= (\mathbf{a}\wedge\mathbf{b})\cdot[\mathbf{c}\cdot(\mathbf{u}\wedge\mathbf{v}\wedge\mathbf{w})]\\ &= \mathbf{c}\cdot\mathbf{u}(\mathbf{a}\wedge\mathbf{b})\cdot(\mathbf{v}\wedge\mathbf{w}) - \mathbf{c}\cdot\mathbf{v}(\mathbf{a}\wedge\mathbf{b})\cdot(\mathbf{u}\wedge\mathbf{w}) +\\ &\quad\ \mathbf{c}\cdot\mathbf{w}(\mathbf{a}\wedge\mathbf{b})\cdot(\mathbf{u}\wedge\mathbf{v}).\end{aligned}$$
$\langle\mathbf{abuv}\rangle_0 = \mathbf{a}\cdot\mathbf{b}\mathbf{u}\cdot\mathbf{v} + (\mathbf{a}\wedge\mathbf{b})\cdot(\mathbf{u}\wedge\mathbf{v})$.

(1.8) $\langle\mathbf{abcuvw}\rangle_0 = 2\mathbf{a}\cdot\mathbf{b}\langle\mathbf{cuvw}\rangle_0 - \langle\mathbf{bacuvw}\rangle_0$.
Repeat the operation until $\mathbf{a}$ has been moved to the far right within the bracket. Then use $\langle\mathbf{abcuvw}\rangle_0 = \langle\mathbf{bcuvwa}\rangle_0$, which follows from Exercise (2.4).

(1.11) One proof uses equations (1.8) and (1.14).

(1.12) $(\mathbf{a}_1\mathbf{a}_2\cdots\mathbf{a}_r)^{-1} = \mathbf{a}_r^{-1}\cdots\mathbf{a}_2^{-1}\mathbf{a}_1^{-1} = \dfrac{\mathbf{a}_r\cdots\mathbf{a}_2\mathbf{a}_1}{\mathbf{a}_r^2\cdots\mathbf{a}_2^2\mathbf{a}_1^2}$

(1.13)
$$\begin{aligned}\langle A\rangle_r^\dagger &= \mathbf{a}_r\cdots\mathbf{a}_2\mathbf{a}_1 = -\mathbf{a}_r\cdots\mathbf{a}_3\mathbf{a}_1\mathbf{a}_2 = (-1)^2\mathbf{a}_r\cdots\mathbf{a}_4\mathbf{a}_1\mathbf{a}_3\mathbf{a}_2\\ &= \cdots = (-1)^{r-1}\mathbf{a}_1\mathbf{a}_r\cdots\mathbf{a}_2 = (-1)^{r-1}\,(-1)^{r-2}\mathbf{a}_1\mathbf{a}_2\mathbf{a}_r\cdots\mathbf{a}_3\\ &= (-1)^{(1/2)r(r-1)}\mathbf{a}_1\mathbf{a}_2\cdots\mathbf{a}_r.\end{aligned}$$

(1.14) Use (1.24a, b) and Exercise (1.13) as follows,
$\langle AB\rangle_0 = \langle B^\dagger A^\dagger\rangle_0 = \sum_r\langle\langle B^\dagger\rangle_r\langle A^\dagger\rangle_r\rangle_0 = \sum_r\langle\langle B\rangle_r\,\langle A\rangle_r\rangle_0 = \langle BA\rangle_0$.

(1.15) If $\mathbf{a}\wedge\mathbf{b}\wedge\mathbf{c}\wedge\mathbf{d} = 0$, then there exist scalars α, β, γ such that $\mathbf{d} = \alpha\mathbf{a} + \beta\mathbf{b} + \gamma\mathbf{c}$. Hence $\mathbf{a}\wedge\mathbf{b} + \mathbf{c}\wedge\mathbf{d} = \mathbf{a}\wedge\mathbf{b} + \alpha\mathbf{c}\wedge\mathbf{a} + \beta\mathbf{c}\wedge\mathbf{b} = (\mathbf{a} + \beta\mathbf{c})\wedge(\mathbf{b} - \alpha\mathbf{c}) = \mathbf{u}\wedge\mathbf{v}$. If $\mathbf{a}\wedge\mathbf{b} + \mathbf{c}\wedge\mathbf{d} = \mathbf{u}\wedge\mathbf{v}$, then
$(\mathbf{a}\wedge\mathbf{b} + \mathbf{c}\wedge\mathbf{d})\wedge(\mathbf{a}\wedge\mathbf{b} + \mathbf{c}\wedge\mathbf{d}) = 2\mathbf{a}\wedge\mathbf{b}\wedge\mathbf{c}\wedge\mathbf{d} = \mathbf{u}\wedge\mathbf{v}\wedge\mathbf{u}\wedge\mathbf{v} = 0$.

(1.17) $A\cdot B \equiv \sum_r\sum_s\langle A\rangle_r\cdot\langle B\rangle_s$, $A\wedge B \equiv \sum_r\sum_s\langle A\rangle_r\wedge\langle B\rangle_s$.
If (1.16) and (1.12) were allowed to apply to the case $r = 1$ and $s = 0$, we would have $\mathbf{a}\cdot\lambda = \langle\mathbf{a}\lambda\rangle_{1-0} = \lambda\mathbf{a}$ and $\mathbf{a}\wedge\lambda = \langle\mathbf{a}\lambda\rangle_{1+0} = \lambda\mathbf{a}$, which is inconsistent with $\mathbf{a}\lambda = \mathbf{a}\cdot\lambda + \mathbf{a}\wedge\lambda$.

Section 2-3.

(3.1a) By (1.5), $\mathbf{x}\wedge\mathbf{B} = \frac{1}{2}(\mathbf{xB} + \mathbf{Bx}) = 0$,
or by (1.4), $\mathbf{xB} = \mathbf{x}\cdot\mathbf{B} = -\mathbf{B}\cdot\mathbf{x} = -\mathbf{Bx}$.

(3.1b) $\mathbf{x}' = \mathbf{xB} = \mathbf{x}\cdot\mathbf{B} = \langle\mathbf{xB}\rangle_1$, a vector.
$|\mathbf{x}'|^2 = \mathbf{x}'^2 = \mathbf{xBxB} = \mathbf{x}(-\mathbf{xB})\mathbf{B} = \mathbf{x}^2\mathbf{B}^\dagger\mathbf{B} = |\mathbf{x}|^2|\mathbf{B}|^2$,
$\mathbf{xx}' = \mathbf{x}\cdot\mathbf{x}' + \mathbf{x}\wedge\mathbf{x}' = \mathbf{x}^2\mathbf{B}$
implies $\mathbf{x}\cdot\mathbf{x}' = 0$,
and also $\mathbf{x}\wedge\mathbf{x}' = \mathbf{x}^2\mathbf{iB}$.

(3.2) $$\frac{\mathbf{a}\wedge\mathbf{b}}{\mathbf{i}} = -\mathbf{ia}\wedge\mathbf{b} = \boldsymbol{\sigma}_2\boldsymbol{\sigma}_1\,\mathbf{a}\wedge\mathbf{b} = (\boldsymbol{\sigma}_2\wedge\boldsymbol{\sigma}_1)\cdot(\mathbf{a}\wedge\mathbf{b})$$
$$= \boldsymbol{\sigma}_1\cdot\mathbf{a}\boldsymbol{\sigma}_2\cdot\mathbf{b} - \boldsymbol{\sigma}_1\cdot\mathbf{b}\boldsymbol{\sigma}_2\cdot\mathbf{a}$$

(3.3) $i(\mathbf{ab}) = i\mathbf{ab} = \mathbf{a}(i\mathbf{b})$,
$i(\mathbf{a}\cdot\mathbf{b} + i\mathbf{a}\times\mathbf{b}) = \mathbf{a}\cdot(i\mathbf{b}) + \mathbf{a}\wedge(i\mathbf{b})$.
Equate vector and trivector parts separately to get the first two identities.
$$\mathbf{ab}\wedge\mathbf{c} = \mathbf{a}\cdot(\mathbf{b}\wedge\mathbf{c}) + \mathbf{a}\wedge(\mathbf{b}\wedge\mathbf{c}) = \mathbf{a}i(\mathbf{b}\times\mathbf{c})$$
$$= i[\mathbf{a}\cdot(\mathbf{b}\times\mathbf{c}) + \mathbf{a}\wedge(\mathbf{b}\times\mathbf{c})]$$
$$= i\mathbf{a}\cdot(\mathbf{b}\times\mathbf{c}) - \mathbf{a}\times(\mathbf{b}\times\mathbf{c}).$$
Equate vector and trivector parts separately.
$$\mathbf{abc} - \mathbf{cba} = \mathbf{a}(\mathbf{b}\wedge\mathbf{c}) + (\mathbf{b}\wedge\mathbf{c})\mathbf{a} = 2\mathbf{a}\wedge(\mathbf{b}\wedge\mathbf{c}).$$

(3.4) $$(\mathbf{a}\wedge\mathbf{b})\cdot(\mathbf{c}\wedge\mathbf{d}) = \langle\mathbf{a}\wedge\mathbf{bc}\wedge\mathbf{d}\rangle_0 = \langle i\mathbf{a}\times\mathbf{b}i\mathbf{c}\times\mathbf{d}\rangle$$
$$= -\langle\mathbf{a}\times\mathbf{bc}\times\mathbf{d}\rangle_0 = -(\mathbf{a}\times\mathbf{b})\cdot(\mathbf{c}\times\mathbf{d}).$$
$$\mathbf{ab}\cdot(\mathbf{c}\times\mathbf{d}) = \mathbf{a}\cdot\mathbf{bc}\times\mathbf{d} - \mathbf{a}\cdot\mathbf{cb}\times\mathbf{d} + \mathbf{a}\cdot\mathbf{db}\times\mathbf{c}$$
$$(\mathbf{u}\times\mathbf{v})\mathbf{a}\cdot(\mathbf{b}\times\mathbf{c}) = (\mathbf{u}\times\mathbf{v})\cdot(\mathbf{a}\times\mathbf{b})\mathbf{c} - (\mathbf{u}\times\mathbf{v})\cdot(\mathbf{a}\times\mathbf{c})\mathbf{b} + (\mathbf{u}\times\mathbf{v})\cdot(\mathbf{b}\times\mathbf{c})\mathbf{a}.$$

(3.5) $$\mathbf{a}\times\mathbf{b} = (\mathbf{a}\wedge\mathbf{b})i^\dagger = (\mathbf{a}\wedge\mathbf{b})\cdot(\boldsymbol{\sigma}_3\wedge\boldsymbol{\sigma}_2\wedge\boldsymbol{\sigma}_1)$$
$$= (\mathbf{a}\wedge\mathbf{b})\cdot(\boldsymbol{\sigma}_3\wedge\boldsymbol{\sigma}_2)\boldsymbol{\sigma}_1 - (\mathbf{a}\wedge\mathbf{b})\cdot(\boldsymbol{\sigma}_3\wedge\boldsymbol{\sigma}_1)\boldsymbol{\sigma}_2 + (\mathbf{a}\wedge\mathbf{b})\cdot(\boldsymbol{\sigma}_2\wedge\boldsymbol{\sigma}_1)\boldsymbol{\sigma}_3.$$

(3.7) Make the identifications $\mathbf{i} = -\mathbf{i}_1$, $\mathbf{j} = -\mathbf{i}_2$, $\mathbf{k} = -\mathbf{i}_3$. Hamilton chose a lefthanded basis, in contrast to our choice of a righthanded basis.

(3.8) $$B_{ij} = (\boldsymbol{\sigma}_j\wedge\boldsymbol{\sigma}_i)\cdot(i\mathbf{b}) = -\langle\boldsymbol{\sigma}_i\wedge\boldsymbol{\sigma}_j i\boldsymbol{\sigma}_k\rangle_0 b_k$$
$$= -\langle i\boldsymbol{\sigma}_i\wedge\boldsymbol{\sigma}_j\boldsymbol{\sigma}_k\rangle_0 b_k = -i\boldsymbol{\sigma}_i\wedge\boldsymbol{\sigma}_j\wedge\boldsymbol{\sigma}_k b_k$$
$$\mathbf{a}\times\mathbf{b} = -i\mathbf{a}\wedge\mathbf{b} = -i\mathbf{a}\wedge\mathbf{b}\wedge\boldsymbol{\sigma}_k\boldsymbol{\sigma}_k.$$
The last step follows from
$$\mathbf{a}\wedge\mathbf{b}\wedge\boldsymbol{\sigma}_k\boldsymbol{\sigma}_k = (\mathbf{a}\wedge\mathbf{b}\wedge\boldsymbol{\sigma}_k)\cdot\boldsymbol{\sigma}_k$$
$$= \boldsymbol{\sigma}_k\cdot\boldsymbol{\sigma}_k\mathbf{a}\wedge\mathbf{b} - \boldsymbol{\sigma}_k\cdot\mathbf{ba}\wedge\boldsymbol{\sigma}_k + \boldsymbol{\sigma}_k\cdot\mathbf{ab}\wedge\boldsymbol{\sigma}_k$$
$$= 3\mathbf{a}\wedge\mathbf{b} - \mathbf{a}\wedge\mathbf{b} + \mathbf{b}\wedge\mathbf{a} = \mathbf{a}\wedge\mathbf{b}.$$

(3.9) Since $\mathbf{a}\cdot\boldsymbol{\sigma}_k = \frac{1}{2}(\mathbf{a}\boldsymbol{\sigma}_k + \boldsymbol{\sigma}_k\mathbf{a})$,
$$\boldsymbol{\sigma}_k\mathbf{a}\boldsymbol{\sigma}_k = \boldsymbol{\sigma}_k(-\boldsymbol{\sigma}_k\mathbf{a} + 2\mathbf{a}\cdot\boldsymbol{\sigma}_k)$$
$$= -3\mathbf{a} + 2\mathbf{a} = -\mathbf{a}.$$
$$\boldsymbol{\sigma}_k\mathbf{a}\wedge\mathbf{b}\boldsymbol{\sigma}_k = \boldsymbol{\sigma}_k i\mathbf{a}\times\mathbf{b}\boldsymbol{\sigma}_k = i\boldsymbol{\sigma}_k\mathbf{a}\times\mathbf{b}\boldsymbol{\sigma}_k$$
$$= -i\mathbf{a}\times\mathbf{b}.$$
$$\boldsymbol{\sigma}_k i\boldsymbol{\sigma}_k = i\boldsymbol{\sigma}_k\boldsymbol{\sigma}_k = 3i.$$

(3.10) This problem is the same as Exercise (1.4) ,

$$\mathbf{x} = \frac{\alpha^2\mathbf{a} + \alpha\mathbf{a} \times \mathbf{b} + \mathbf{b}\mathbf{a}\cdot\mathbf{b}}{\alpha(\alpha^2 + \mathbf{b}^2)} .$$

Section 2-4.

(4.5b) $(\mathbf{a} - \mathbf{b})^2 = (\mathbf{a} - \mathbf{b})\mathbf{a}^2(\mathbf{a} - \mathbf{b}) = (1 - \mathbf{ba})(1 - \mathbf{ab})$
$= (1 - e^{\mathrm{i}\theta})(1 - e^{-\mathrm{i}\theta}) = -(e^{(1/2)\mathrm{i}\theta} - e^{-(1/2)\mathrm{i}\theta})^2.$

(4.7) The figure is a regular hexagon with external angle $\theta = 2\pi/6 = \pi/3$; there are two other

(4.8a) $\mathbf{a}^2\mathbf{b}^2 = \mathbf{abba} = (\mathbf{a}\cdot\mathbf{b} + \mathbf{a}\wedge\mathbf{b})(\mathbf{a}\cdot\mathbf{b} - \mathbf{a}\wedge\mathbf{b}) = (\mathbf{a}\cdot\mathbf{b})^2 - (\mathbf{a}\wedge\mathbf{b})^2$
For $\mathbf{ab} = e^{\mathrm{i}\theta}$, this reduces to $\cos^2\theta + \sin^2\theta = 1$.

(4.8b, c) These identities were proved in Exercise 1.1. Note that the second identity admits the simplification $(\mathbf{a}\wedge\mathbf{b})\cdot(\mathbf{b}\wedge\mathbf{c}) = \mathbf{a}\wedge\mathbf{b}\mathbf{b}\wedge\mathbf{c}$ if $\mathbf{a}\wedge\mathbf{b}\wedge\mathbf{c} = 0$.

For $\mathbf{ab} = e^{\mathrm{i}\theta}$ and $\mathbf{bc} = e^{\mathrm{i}\phi}$, the identities reduce to

$$\cos\theta \sin\phi - \sin(\theta + \phi) + \cos\phi \sin\theta = 0,$$
$$-\sin\theta \sin\phi = \cos(\theta + \phi) - \cos\theta \cos\phi.$$

(4.8d, e) Note that $\mathbf{bcb} = \langle\mathbf{bcb}\rangle_1$ because $\mathbf{b}\wedge\mathbf{c}\wedge\mathbf{b} = 0$. Thus,

$$\mathbf{a}\cdot(\mathbf{bcb}) = \langle\mathbf{abcb}\rangle_0 = \mathbf{a}\cdot\mathbf{b}\mathbf{c}\cdot\mathbf{b} + (\mathbf{a}\wedge\mathbf{b})\cdot(\mathbf{c}\wedge\mathbf{b}).$$

The desired identities are obtained by adding and subtracting this from the identity

$$\mathbf{b}^2\mathbf{a}\cdot\mathbf{c} = \langle\mathbf{abbc}\rangle_0 = \mathbf{a}\cdot\mathbf{b}\mathbf{b}\cdot\mathbf{c} - (\mathbf{a}\wedge\mathbf{b})\cdot(\mathbf{c}\wedge\mathbf{b})$$

For $\mathbf{abbc} = e^{\mathrm{i}(\theta + \phi)}$ and $\mathbf{abcb} = e^{\mathrm{i}(\theta - \phi)}$, these identities reduce to

$$2\sin\theta \sin\phi = \cos(\theta + \phi) - \cos(\theta - \phi)$$
$$2\cos\theta \cos\phi = \cos(\theta + \phi) + \cos(\theta - \phi).$$

(4.9b) Eliminate $\mathbf{a}\cdot\mathbf{b}$ **from** $c^2 = a^2 + b^2 + 2\mathbf{a}\cdot\mathbf{b}$
and $4|A|^2 = -(\mathbf{a}\wedge\mathbf{b})^2 = a^2b^2 - (\mathbf{a}\cdot\mathbf{b})^2$
$= (ab + \mathbf{a}\cdot\mathbf{b})(ab - \mathbf{a}\cdot\mathbf{b})$

(4.10) $(\mathbf{b} - \mathbf{a})\cdot(\mathbf{b} - \mathbf{c}) = (\mathbf{b} - \mathbf{a})\cdot(\mathbf{b} + \mathbf{a}) = b^2 - a^2 = 0.$

(4.11a) Establish and interpret the identity

$$(\mathbf{a} - \mathbf{b})\cdot(\mathbf{p} - \mathbf{c}) + (\mathbf{b} - \mathbf{c})\cdot(\mathbf{p} - \mathbf{a}) + (\mathbf{c} - \mathbf{a})\cdot(\mathbf{p} - \mathbf{b}) = 0.$$

Note that if any two terms in the identity vanish, the third vanishes also. Note that this is an instance of logical transitivity, and that the transitivity breaks the symmetry of the relation.

(4.11b) Establish, interpret and use the identity

$$(\mathbf{a} - \mathbf{b})\cdot\left(\mathbf{q} - \frac{\mathbf{a} + \mathbf{b}}{2}\right) + (\mathbf{b} - \mathbf{c})\cdot\left(\mathbf{q} - \frac{\mathbf{b} + \mathbf{c}}{2}\right) +$$
$$+ (\mathbf{c} - \mathbf{a})\cdot\left(\mathbf{q} - \frac{\mathbf{c} + \mathbf{a}}{2}\right) = 0,$$

Alternatively, one can argue that $(\mathbf{q}-\mathbf{a})^2 = (\mathbf{q}-\mathbf{b})^2$ and $(\mathbf{q}-\mathbf{b})^2 = (\mathbf{q}-\mathbf{c})^2$ implies $(\mathbf{q}-\mathbf{a})^2 = (\mathbf{q}-\mathbf{c})^2$. Here the transitivity in the argument is quite explicit.

(4.11c) Use the facts that $(\mathbf{a}-\mathbf{b})\cdot(\mathbf{p}-\mathbf{c}) = 0$, and $(\mathbf{a}-\mathbf{b})\cdot(2\mathbf{q}-\mathbf{a}-\mathbf{b}) = 0$. Whence, $(\mathbf{a}-\mathbf{b})\cdot(\mathbf{p}+2\mathbf{q}-3\mathbf{r}) = 0$.
What more is needed to conclude that $\mathbf{p}+2\mathbf{q}-3\mathbf{r} = 0$?

Section 2-5.

(5.8) $\mathbf{a}-\mathbf{b} = \mathbf{a}(1-\mathbf{a}^{-1}\mathbf{b})$.

$$\frac{1}{\mathbf{a}-\mathbf{b}} = (\mathbf{a}-\mathbf{b})^{-1} = (1-\mathbf{a}^{-1}\mathbf{b})^{-1}\mathbf{a}^{-1}.$$

The quantity $z = \mathbf{a}^{-1}\mathbf{b}$ has the form of a complex number and, as can be verified by long division, submits to the binomial expansion

$$\frac{1}{1-z} = 1 + z + z^2 + \ldots,$$

which converges for $|z|^2 = |\mathbf{b}|^2/|\mathbf{a}|^2 < 1$.

Section 2-6.

(6.1) $[(\mathbf{x}-\mathbf{a})\wedge\mathbf{u}]\cdot(\sigma_2\wedge\sigma_1) = u_2(x_1-a_1) - u_1(x_2-a_2)$, etc.

(6.2) (a) Equation (6.2) implies $(\mathbf{x}-\mathbf{a})\mathbf{u} = (\mathbf{x}-\mathbf{a})\cdot\mathbf{u} \equiv \lambda$.

(6.2) (b) $\{\mathbf{x}\}$ = half line with the direction $\mathbf{u}$ and endpoint $\mathbf{a}$.

(6.3) (a) $\mathbf{d} = (\mathbf{a}\wedge\mathbf{b})(\mathbf{b}-\mathbf{a})^{-1} = \dfrac{(\mathbf{a}\wedge\mathbf{b})\cdot(\mathbf{b}-\mathbf{a})}{|\mathbf{b}-\mathbf{a}|^2}$

(b) $\mathbf{d} = \dfrac{(\mathbf{a}-\mathbf{c})\wedge(\mathbf{b}-\mathbf{a})\cdot(\mathbf{b}-\mathbf{a})}{|\mathbf{b}-\mathbf{a}|^2} = P_{\mathbf{b}-\mathbf{a}}(\mathbf{a}-\mathbf{c})$

(6.4) $(\mathbf{x}-\mathbf{a})\wedge(\mathbf{b}-\mathbf{a})\wedge(\mathbf{c}-\mathbf{a}) = 0$.

(6.5) The solution set is the line of intersection of the **A**-plane with the **B**-plane.

(6.6) $\mathbf{p} = \dfrac{(\mathbf{b}-\mathbf{a})\wedge\mathbf{B}}{\mathbf{u}\wedge\mathbf{B}}\mathbf{u} + \mathbf{a}$.

(6.7) $\mathbf{x}\wedge\mathbf{u} = \mathbf{a}\wedge\mathbf{u} \Rightarrow \mathbf{x}\wedge\mathbf{u}\wedge\mathbf{v} = \mathbf{a}\wedge\mathbf{u}\wedge\mathbf{v}$,
$\mathbf{y}\wedge\mathbf{v} = \mathbf{b}\wedge\mathbf{v} \Rightarrow \mathbf{y}\wedge\mathbf{v}\wedge\mathbf{u} = \mathbf{b}\wedge\mathbf{v}\wedge\mathbf{u}$.
Hence,

$$(\mathbf{b}-\mathbf{a})\wedge\mathbf{u}\wedge\mathbf{v} = (\mathbf{y}-\mathbf{x})\wedge\mathbf{u}\wedge\mathbf{v} = \mathbf{d}\wedge\mathbf{u}\wedge\mathbf{v}.$$

But $\mathbf{d}\cdot(\mathbf{u}\wedge\mathbf{v}) = 0$, so

$$\mathbf{d} = (\mathbf{b} - \mathbf{a})\wedge\mathbf{u}\wedge\mathbf{v}(\mathbf{u}\wedge\mathbf{v})^{-1}$$
$$= \mathbf{b} - \mathbf{a} + \frac{[(\mathbf{b} - \mathbf{a})\wedge\mathbf{u}]\cdot(\mathbf{v}\wedge\mathbf{u})}{|\mathbf{u}\wedge\mathbf{v}|^2}\mathbf{v} + $$
$$+ \frac{[(\mathbf{a} - \mathbf{b})\wedge\mathbf{v}]\cdot(\mathbf{v}\wedge\mathbf{u})}{|\mathbf{u}\wedge\mathbf{v}|^2}\mathbf{u}.$$

(6.8) $\mathbf{d} = [(\mathbf{a} - \mathbf{b})\wedge\mathrm{U}]\mathrm{U}^{-1}$

(6.10) Comparing Figure 6.2 with Figure 6.4b, we see that we can use (6.17) to get

$$\frac{A_1}{A_2} = \frac{C_1 + C_2}{B_1 + B_2}, \quad \frac{B_1}{B_2} = \frac{A_1 + A_2}{C_1 + C_2}, \quad \frac{C_1}{C_2} = \frac{B_1 + B_2}{A_1 + A_2}.$$

(6.11) The area of the quadrilateral **0**, **a**, **b**, **c** is divided into four parts by the diagonals. The theorem can be proved by expressing the division ratios in terms of these four area.

(6.12) An immediate consequence of Equation (6.13).

(6.13) By Equation (6.12), we may write

$$\alpha\mathbf{a} + \alpha'\mathbf{a}' = \beta\mathbf{b} + \beta'\mathbf{b}' = \gamma\mathbf{c} + \gamma'\mathbf{c}' = \mathbf{s},$$
$$\alpha + \alpha' = \beta + \beta' = \gamma + \gamma' = 1.$$

Whence,

$$\alpha\mathbf{a} - \beta\mathbf{b} = -\alpha'\mathbf{a}' + \beta'\mathbf{b}',$$
$$\alpha - \beta = -(\alpha' - \beta')$$

So, if $\alpha - \beta \neq 0$,

$$\mathbf{r} = \frac{\alpha\mathbf{a} - \beta\mathbf{b}}{\alpha - \beta} = \frac{\alpha'\mathbf{a} - \beta'\mathbf{b}}{\alpha' - \beta'}.$$

If $\alpha - \beta = 0$, the lines are parallel and may be regarded as intersecting at ∞. After deriving similar expressions for **p** and **q** we can show that

$$(\beta - \gamma)\mathbf{p} + (\gamma - \alpha)\mathbf{q} + (\alpha - \beta)\mathbf{r} = 0,$$

and Exercise (6.12) can be used again.

(6.14) $\mathrm{U} = i\mathbf{u}$ implies $(\mathbf{x} - \mathbf{a})\wedge\mathrm{U} = i(\mathbf{x} - \mathbf{a})\cdot\mathbf{u}$.

(6.15) $(\mathbf{b} - \mathbf{a})\wedge(\mathbf{c} - \mathbf{a})\wedge(\mathbf{x} - \mathbf{a}) = 0$, or, to use the special form of Exercise (6.14), $(\mathbf{x} - \mathbf{a})\cdot(\mathbf{b} - \mathbf{a}) \times (\mathbf{c} - \mathbf{a}) = 0$.

(6.16) Expand $(\mathbf{a}\wedge\mathbf{b}\wedge\mathbf{c})^2 = 0$.

(6.17) At the points of intersection $r^2 = (\mathbf{a} + \lambda\mathbf{u})^2$.

(6.18) The equation for the line tangent to the circle at the point $\mathbf{d} = r\hat{\mathbf{d}}$ can be written $\mathbf{x} = \mathbf{d} + \lambda\mathbf{u} = (1 + \mathbf{i}\lambda/r)\mathbf{d}$. Its square is $\mathbf{x}^2 = r^2 + \lambda^2$. Evaluate at $\mathbf{x} = \mathbf{a}$ and solve for **d**.

(6.19) $r = \dfrac{|\mathbf{a}-\mathbf{b}|}{2\sin\phi}$, $\mathbf{c} = \mathbf{a} + \dfrac{(\mathbf{b}-\mathbf{a})e^{\mathbf{i}(\pi/2-\phi)}}{2\sin\phi}$.

(6.21a) $\mathbf{a}_0 = \mathbf{a}+\mathbf{c}$, $\mathbf{a}_1 = 2\mathbf{b}$, $\mathbf{a}_2 = \mathbf{c}-\mathbf{a}$, $\cos\phi = \dfrac{1-\lambda^2}{1+\lambda^2}$, $\sin\phi = \dfrac{2\lambda}{1+\lambda^2}$.

(6.22) $\mathbf{a}_0 = \mathbf{b}$, $\mathbf{a}_1 = -\mathbf{a}e^{\mathbf{i}\phi} - \mathbf{b}e^{-\mathbf{i}\phi} = -(\mathbf{a}+\mathbf{b})\cos\phi + (\mathbf{b}-\mathbf{a})\cdot\mathbf{i}\sin\phi$ (since $(\mathbf{b}-\mathbf{a})\wedge\mathbf{i} = 0$), $\mathbf{a}_2 = \mathbf{a}$, $\alpha_0 = \alpha_2 = 1$, $\alpha_1 = -2\cos\phi$.

(6.23) (a) Write $z = x + \mathbf{i}y = (1+\mathbf{i}\lambda)^{1/2}$. Then $z^2 = x^2 - y^2 + 2xy\mathbf{i} = 1 + \mathbf{i}\lambda$. Hence, $x^2 - y^2 = 1$ and $2xy = \lambda$, which describe a hyperbola in terms of rectangular coordinates.

(b) A circle with radius $\frac{1}{2}$ and center at $\frac{1}{2}\boldsymbol{\sigma}_1$.

(c) The evolvents of the unit circle, i.e. the path traced by a point on a taut string being unwound from around the unit circle.

(d) A lemniscate. Note that it can be obtained from the hyperbola in (a) by the inversion $\mathbf{x} \to \mathbf{x}^{-1}$.

(6.24) (a) For $|\mathbf{x}| \neq 0$, $\hat{\mathbf{a}}\cdot\hat{\mathbf{x}} = \pm a^{-1}$; this describes a cone with vertex at the origin and vertex angle α given by $\cos\frac{1}{2}\alpha = a^{-1} \leqslant 1$; it reduces to a line when $a = 1$ and a plane when $a = \infty$. Only zero is a solution when $a < 1$.

(b) Interior of a half cone for $\mathbf{a}^2 > 1$.

(c) Cone with vertex angle $\frac{1}{2}\pi$; symmetry axis and plane of the cone.

(d) $\langle(\mathbf{ax})^2\rangle_0 = (\mathbf{a}\cdot\mathbf{x})^2 - |\mathbf{a}\wedge\mathbf{x}|^2 = 1$ describes a hyperboloid asymptotic to the cone in (c).

(e) Paraboloid.

(f) For $\mathbf{a}^2 > 1$, $b^2 = 1$ and $c \neq 0$; circle if $\mathbf{a}\wedge\mathbf{b} = 0$, ellipse if $(\mathbf{a}\cdot\mathbf{b})^2 > 1$, parabola if $(\mathbf{a}\cdot\mathbf{b})^2 = 1$, hyperbola if $(\mathbf{a}\cdot\mathbf{b})^2 < 1$.

Section 2-7.

(7.1) $\dfrac{\mathrm{d}}{\mathrm{d}t}\left(\dfrac{\mathbf{u}}{u}\right) = \dfrac{\dot{\mathbf{u}}}{u} - \dfrac{\dot{u}}{u^2}\mathbf{u} = \dfrac{u^2\dot{\mathbf{u}} - u\dot{u}\mathbf{u}}{u^3} = \dfrac{\mathbf{u}\cdot\mathbf{u}\dot{\mathbf{u}} - \mathbf{u}\cdot\dot{\mathbf{u}}\mathbf{u}}{u^3} = \dfrac{\mathbf{u}(\mathbf{u}\wedge\dot{\mathbf{u}})}{u^3}$.

(7.2b) Differentiate $F^{-1}F = 1$.

(7.3) Generate the exponential series by a Taylor expansion about $t = 0$, and write $F(0) = \mathrm{B}$. Conversely, differentiate the exponential series to get $\dot{F}$.

(7.4) (a) Use the fact that the square of a k-blade is a scalar.

(b) Consider $F(t) = e^{At}$ where A is a constant bivector.

(7.5) Separate $\mathrm{d}/\mathrm{d}t\,(\mathbf{rp}) = \dot{\mathbf{r}}\mathbf{p} + \mathbf{r}\dot{\mathbf{p}}$ into scalar and bivector parts.

$$\frac{\mathrm{d}}{\mathrm{d}t}(\mathbf{p}\wedge\mathbf{q}\wedge\mathbf{r}) = \dot{\mathbf{p}}\wedge\mathbf{q}\wedge\mathbf{r} + \mathbf{p}\wedge\frac{\mathrm{d}}{\mathrm{d}t}(\mathbf{q}\wedge\mathbf{r})$$

$$= \dot{\mathbf{p}}\wedge\mathbf{q}\wedge\mathbf{r} + \mathbf{p}\wedge\dot{\mathbf{q}}\wedge\mathbf{r} + \mathbf{p}\wedge\mathbf{q}\wedge\dot{\mathbf{r}}$$

(7.7) $\mathbf{v} = v\hat{\mathbf{v}}$; $\mathrm{d}/\mathrm{d}t \log \mathbf{v}^2 = 2\dot{v}/v$, and according to Equation (7.11), $\mathrm{d}\hat{\mathbf{v}}/\mathrm{d}t = \hat{\mathbf{v}}\cdot\boldsymbol{\Omega}$.

Section 2-8.

(8.2a)
$$r^2 = (x - \mathbf{x}')\cdot(\mathbf{x} - \mathbf{x}')$$
$$\mathbf{a}\cdot\nabla r^2 = \mathbf{a}\cdot(\mathbf{x} - \mathbf{x}') + (\mathbf{x} - \mathbf{x}')\cdot\mathbf{a} = 2\mathbf{a}\cdot\mathbf{r} = 2r\mathbf{a}\cdot\nabla r$$
$$\mathbf{a}\cdot\nabla\mathbf{r} = \mathbf{a}\cdot\nabla(\mathbf{x} - \mathbf{x}') = \mathbf{a}\cdot\nabla\mathbf{x} = \mathbf{a}.$$

(8.2b)
$$\mathrm{a}\cdot\nabla\left(\frac{\mathbf{r}}{r}\right) = \frac{1}{r}(\mathbf{a}\cdot\nabla\mathbf{r}) + \mathbf{r}\mathbf{a}\cdot\nabla\left(\frac{1}{r}\right)$$
$$= \frac{1}{r}\,\mathbf{a} - \mathbf{r}\,\frac{\mathbf{a}\cdot\hat{\mathbf{r}}}{r^2} = \frac{r^2\mathbf{a} - \mathbf{r}\mathbf{a}\cdot\mathbf{r}}{r^3} = \frac{\mathbf{r}\cdot(\mathbf{r}\wedge\mathbf{a})}{r^3}.$$

(8.2f)
$$\mathbf{a}\cdot\nabla\left(\frac{1}{\mathbf{r}}\right) = \mathbf{a}\cdot\nabla\left(\frac{\mathbf{r}}{r^2}\right) = \frac{1}{r^2}\,\mathbf{a}\cdot\nabla\mathbf{r} + \mathbf{r}\mathbf{a}\cdot\nabla\left(\frac{1}{r^2}\right)$$

(8.5) Write $\mathbf{a}^{-1}\mathbf{b} = \dfrac{b}{a}\,e^{\mathbf{i}\theta}$ and $z = \mathbf{a}^{-1}\mathbf{x} = \dfrac{x}{a}\,e^{\mathbf{i}\phi}$.

Then $z^{-1} = \mathbf{x}^{-1}\mathbf{a} = \dfrac{a}{x}\,e^{-\mathbf{i}\phi}$,

and $\mathrm{d}z = \mathbf{a}^{-1}\,\mathrm{d}\mathbf{x} = \dfrac{e^{\mathbf{i}\phi}}{a}\,(\mathrm{d}x + x\mathbf{i}\,\mathrm{d}\phi)$.

So $\mathbf{x}^{-1}\,\mathrm{d}\mathbf{x} = z^{-1}\,\mathrm{d}z = \dfrac{\mathrm{d}x}{x} + \mathbf{i}\,\mathrm{d}\phi$,

and
$$\int_{\mathbf{a}}^{\mathbf{b}} \mathbf{x}^{-1}\,\mathrm{d}\mathbf{x} = \int_{\mathbf{a}}^{\mathbf{b}} \frac{\mathrm{d}x}{x} + \mathbf{i}\int_0^\theta \mathrm{d}\phi = \log\frac{b}{a} + \mathbf{i}\theta.$$

This has the scalar part
$$\int_{\mathbf{a}}^{\mathbf{b}} \frac{\mathbf{x}\cdot\mathrm{d}\mathbf{x}}{\mathbf{x}^2} = \int_{\mathbf{a}}^{\mathbf{b}} \frac{\mathrm{d}x}{x} = \log\frac{b}{a},$$

and the bivector part
$$\int_{\mathbf{a}}^{\mathbf{b}} \frac{\mathbf{x}\wedge\mathrm{d}\mathbf{x}}{\mathbf{x}^2} = \int_{\hat{\mathbf{a}}}^{\hat{\mathbf{b}}} \hat{\mathbf{x}}\wedge\mathrm{d}\hat{\mathbf{x}} = \mathbf{i}\int_0^\theta \mathrm{d}\phi = \mathbf{i}\theta.$$

The principal value is obtained by integrating along the straight line from **a** to **b** or along any curve in the plane which can be continuously deformed into that line without passing through the origin. If the straight line itself passes through the origin, the bivector part of the principal part can be assigned either of the values $\pm$ $\mathbf{i}\pi$. If the curve winds about the origin k times, the value of the integral differs from the principal part by the amount $\pm\, 2\pi k\mathbf{i}$, with the positive (negative) sign for counterclockwise (clockwise winding.

Section 3-2.

(2.2) $t_{\pm} = -\mathbf{v}_0 \cdot \mathbf{g}^{-1} \pm \{(\mathbf{v}_0 \cdot \mathbf{g}^{-1})^2 + 2\mathbf{r} \cdot \mathbf{g}^{-1}\}^{1/2}$. This expression has the drawback that $\mathbf{r}$ and $\mathbf{v}_0$ are not independent variables. For given $\mathbf{v}_0$ and $\mathbf{r}$, the two roots $t_{\pm}$ are times of flight to the same point by different paths. For given $\mathbf{v}_0$, they are times of flight to two distinct points on the same path equidistant from the vertical maximum.

(2.4) $\mathbf{g}\wedge\mathbf{r} = xg\mathbf{i} = \mathbf{v}\wedge\mathbf{v}_0 = \hat{\mathbf{v}}\wedge\hat{\mathbf{v}}_0 \;\; v_0(v_0^2 - 2gy)^{1/2}$: Therefore, horizontal range x is a maximum for fixed v_0 and y when $\hat{\mathbf{v}}\wedge\hat{\mathbf{v}}_0 = i$.

(2.6) Suggestion: Use the Jacobi identity for $\mathbf{g}$, $\mathbf{v}$, $\mathbf{r}$ and the fact that the vectors are coplanar.

(2.7) $$\mathbf{A}(t) = \tfrac{1}{2}\int \mathbf{r}\wedge\dot{\mathbf{r}}\, dt = \frac{1}{12}\, \mathbf{v}_0\wedge\mathbf{g}t^3 .$$

Section 3-5.

(5.1a) $v = 63$ m/s $= 141$ mph.
b) $v = 89$ m/s $= 198$ mph.
c) $v = 4.8 \times 10^6$ m/s $= 1 \times 10^7$ mph.
(5.2) 11 m
(5.3) $v = 6.6$ m/s $= 15$ mph $= 22$ ft/s; 120 m/s.
(5.5) The heavy ball beats the light one by 2.2 m and 1/20 sec.

Section 3-7.

(7.2) $$t = 2\pi/\omega,\ \mathbf{r} = \mathbf{c}t = \frac{2\pi c}{\omega}\ \mathbf{E} \times \mathbf{B}^{-1} .$$

Section 3-8.

(8.5) $\ddot{z} + \omega_0^2 z = 0$,
$\ddot{\mathbf{x}} - 2\dot{\mathbf{x}}\mathbf{i}\omega_L + \omega_0^2\mathbf{x} = 0$, where $\mathbf{i} = i\hat{\mathbf{B}}$,
and the so-called *Larmor frequency* ω_L is defined by $\omega_L \equiv \dfrac{q|\mathbf{B}|}{2mc}$.

general solution

$$\mathbf{r} = (\mathbf{a}_+ e^{i\Omega t} + \mathbf{a}_- e^{-i\Omega t})e^{-i\omega_L t} + \mathbf{c}\cos(\omega_0 t + \phi_0),$$

where

$$\Omega = (\omega_0^2 + \omega_L)^{1/2}, \quad \mathbf{B}\cdot\mathbf{a} = 0, \quad \mathbf{c}\wedge\mathbf{B} = 0.$$

The solution can be interpreted as an ellipse with period $2\pi/\Omega$, precessing (retrograde) with angular velocity ω_L while it vibrates along the **B** direction with frequency $\omega_0/2\pi$ and amplitude $|\,\mathbf{c}\,|$.

Section 4-2.

(2.4) $T^2 = \dfrac{4\pi^2(60)^3 R_E}{g}$

(2.5) $v = (gR_E)^{1/2} = 7.91$ km/s $= 18\ 000$ mph; $T = 84.4$ min.

(2.6) $T^2 = \dfrac{4\pi^2(R_E+h)^3}{gR_E^2}$; $h = 3.58 \times 10^4$ km.

(2.7) $m_s/m_E = 3.38 \times 10^5$

(2.9) $(\mathbf{r} - \mathbf{a})^2 = \mathbf{a}^2, \quad m\ddot{r} = \dfrac{L^2}{mr^3} - \dfrac{8L^2a^2}{mr^5}$.

Section 4-3.

(3.5) $v = (2gR_E)^{1/2} = 11.2$ km/s. $= 25\ 000$ mph.

(3.6) $v_0 = \left|\dfrac{2GM_S r_M}{r_E(r_M+r_E)}\right|^{1/2} - v_E = 2.8$ km/s $= 1.2 \times 10^4$ ft/s;

$$\tau = \frac{\pi[\frac{1}{2}(r_E+r_M)]^{3/2}}{(GM_S)^{1/2}} \cdot = 250 \text{ days}$$

(3.9) $\delta = 30°$.

(3.10) $h = R_E\varepsilon(1 - \cos\frac{1}{2}\beta)/(1 - \varepsilon)$; $\sin^2\alpha_0 = (2 - v_0^2/gR_E)^{-1}$ provided $v_0^2/gR_E < 1$.

(3.11) For $\hat{\mathbf{r}} = \hat{\boldsymbol{\varepsilon}}$, $\mathbf{L}\mathbf{v} = m\hat{\boldsymbol{\varepsilon}}v^2 a = k(\varepsilon+1)\hat{\boldsymbol{\varepsilon}}$, and Exercise (3.8) gives $\Delta\boldsymbol{\varepsilon} = -2\alpha a v^2\hat{\boldsymbol{\varepsilon}}$; $\varepsilon/2\alpha(\varepsilon + 1) \approx 24$ revolutions

Section 4-5.

(5.1) $\dot{r}^2 > 0$ for all values of r implies

$$r^2V(r) < -\frac{L^2}{2m} + Er^2 \xrightarrow[r \to 0]{} -\frac{L^2}{2m} .$$

(5.2) In each case the integral can be simplified by changing to the variable $u = r^{-1}$ or to $v = r^{-2}$.

(5.3) Investigate derivatives of the effective potential higher than second order.

Section 4-7.

(7.4) $$\frac{m_1}{m_2} = 1 + 2\frac{\sin\phi}{\sin\theta}\cos(\theta + \phi).$$

(7.5) (a) $E_1 = m_2Q/m$.
(b) $K_1 = m^{-1}(m_1K + m_2Q \pm 2\sqrt{m_1m_2QK})$.

(7.6) (a) $\Delta t_0 \approx 16$ yr, $v = 16$ km/s, $\cos\alpha = 1/2$.
(b) $\Delta v = 11.7$ km/s, $\theta = 109°$.
(c) $d = 0.9\ R_J$
(d) $\varepsilon = 3.5$, $\Delta t_2 = 3.7$ yr, $\Delta t_1 = 1.7$ yr.

Section 5-1.

(1.1) $\mathbf{x}\wedge\mathbf{y} = \alpha\mathbf{x}\wedge\mathbf{z}$, so $\underline{f}(\mathbf{x}\wedge\mathbf{y}) = \alpha\underline{f}(\mathbf{x}\wedge\mathbf{z}) = \alpha f(\mathbf{x})\wedge f(\mathbf{z})$.

(1.2) $\underline{f}(\alpha\mathbf{x}_1\wedge \ldots \wedge\mathbf{x}_k) = f(\alpha\mathbf{x}_1)\wedge f(\mathbf{x}_2)\wedge \ldots \wedge f(\mathbf{x}_k)$
$= \alpha f(\mathbf{x}_1)\wedge \ldots \wedge f(\mathbf{x}_k)$.

(1.3) $$\begin{vmatrix} \mathbf{v}\cdot f(\mathbf{x}) & \mathbf{v}\cdot f(\mathbf{y}) \\ \mathbf{u}\cdot f(\mathbf{x}) & \mathbf{v}\cdot f(\mathbf{y}) \end{vmatrix} = \begin{vmatrix} \overline{f}(\mathbf{v})\cdot\mathbf{x} & \overline{f}(\mathbf{v})\cdot\mathbf{y} \\ \overline{f}(\mathbf{u})\cdot\mathbf{x} & \overline{f}(\mathbf{u})\cdot\mathbf{y} \end{vmatrix} .$$

(1.4) (a) $f(\mathbf{x}) = \mathrm{x}_i f(\boldsymbol{\sigma}_i) = 0$ only if all x_i vanish provided $\underline{f}(\boldsymbol{\sigma}_1\wedge\boldsymbol{\sigma}_2\wedge\boldsymbol{\sigma}_3) = 0$.
(b) $\mathbf{x} = f^{-1}(\mathbf{y})$.

(1.5) $\underline{gf}(i) = \underline{g}\,\underline{f}(i) = (\det g)\underline{f}\,(i) = (\det g)(\det f)i$.

(1.6) Solution from Exercise (2–1.3)

$$f^{-1}\,\mathbf{x} = \frac{\mathbf{x}}{\alpha} - \frac{\mathbf{x}\cdot\mathbf{ba}}{\alpha(\alpha+\mathbf{a}\cdot\mathbf{b})}$$

(1.7) Solution from Exercise (2–1.4)

$$g^{-1}\mathbf{x} = \frac{\alpha^2\mathbf{x} + \alpha\mathbf{x}\cdot\mathbf{B} - \mathbf{B}\,\mathbf{x}\wedge\mathbf{B}}{\alpha(\alpha^2 - \mathbf{B}^2)} = (\alpha + \mathbf{B})^{-1}\,(\mathbf{x} + \alpha^{-1}\mathbf{x}\wedge\mathbf{B}).$$

(1.8) Use Equations (2–1.16) and (2–1.18).

(1.10) $(\alpha_1\mathbf{a}_1 + \ldots + \alpha_n\mathbf{a}_n)\wedge(\mathbf{a}_1\wedge \ldots \check{\mathbf{a}}_k \ldots \wedge\mathbf{a}_n) = \alpha_k\mathbf{a}_k\wedge\mathbf{a}_1 \ldots$

$\check{\mathbf{a}}_k \ldots \wedge \mathbf{a}_a = \mathbf{c}\wedge\mathbf{a}_1 \ldots \check{\mathbf{a}}_k \ldots \wedge\mathbf{a}_n$.
$B_n A_n = B_n{\cdot}A_n = A_n^{\cdot}B_n$, hence

$$\alpha_k = \frac{\mathbf{a}_1\wedge \ldots (\mathbf{c})_k \ldots \wedge \mathbf{a}_n}{A_n} = \frac{B_n^\dagger}{B_n^\dagger}\ \frac{\mathbf{a}_1\wedge \ldots (\mathbf{c})_k \ldots \wedge \mathbf{a}_n}{A_n}$$

$$= \frac{B_n^\dagger{\cdot}(\mathbf{a}_1\wedge \ldots (\mathbf{c})_k \ldots \wedge \mathbf{a}_n)}{B_n^\dagger{\cdot}A_n}$$

(1.12) $f_{jk}^{-1} = \boldsymbol{\sigma}_j{\cdot}(f^{-1}\boldsymbol{\sigma}_k) = \boldsymbol{\sigma}_j\wedge \underline{f}(\boldsymbol{\sigma}_k i)/\underline{f}(i) = \sum_k(-1)^{k+1}\boldsymbol{\sigma}_j\wedge\underline{f}(\boldsymbol{\sigma}_1 \ldots \check{\boldsymbol{\sigma}}_k \ldots \boldsymbol{\sigma}_n)/\underline{f}(\boldsymbol{\sigma}_1 \ldots \boldsymbol{\sigma}_n)$.

Section 5-2.

(2.1) $\overline{f}\mathbf{x} = \alpha\mathbf{x} + \mathbf{ba}{\cdot}\mathbf{x} + \mathbf{A}{\cdot}\mathbf{x}$,
$f_+\mathbf{x} = \alpha\mathbf{x} + \frac{1}{2}(\mathbf{ab}{\cdot}\mathbf{x} + \mathbf{ba}{\cdot}\mathbf{x})$,
$f_-\mathbf{x} = \mathbf{x}{\cdot}(\mathbf{A} + \mathbf{b}\wedge\mathbf{a})$.

(2.3b) 4, 10, –8; $\sqrt{3}\ \boldsymbol{\sigma}_1 \mp \sqrt{2}\ \boldsymbol{\sigma}_2 \pm \boldsymbol{\sigma}_3$, $\boldsymbol{\sigma}_2 + \sqrt{2}\ \boldsymbol{\sigma}_3$

(2.5) $\lambda_1 = 3$, $\mathbf{e}_1 = \frac{1}{3}(-\boldsymbol{\sigma}_1 + 2\boldsymbol{\sigma}_2 + 2\boldsymbol{\sigma}_3)$;
$\lambda_2 = 6$, $\mathbf{e}_2 = \frac{1}{3}(2\boldsymbol{\sigma}_1 - \boldsymbol{\sigma}_2 + 2\boldsymbol{\sigma}_3)$;
$\lambda_3 = 9$, $\mathbf{e}_3 = \frac{1}{3}(2\boldsymbol{\sigma}_1 + 2\boldsymbol{\sigma}_2 - \boldsymbol{\sigma}_3)$.

(2.6b) $\lambda_1 = a^{-2}$, $\lambda_2 = b^{-2}$, $\lambda_3 = -c^{-2}$.

(2.7) A quadric surface centered at the point $\mathbf{a}$, as described in Exercise (2.6) with $\mathcal{S} = \overline{f}f$. No solution if all eigenvalues of $\mathcal{S}$ are negative.

Section 5-3.

(3.1) $R = e^{i\mathbf{u}\pi/2}$.

(3.2) $\mathcal{U}^{-1} = \mathcal{U}$.

(3.3) $(-1)^3\ \boldsymbol{\sigma}_3\boldsymbol{\sigma}_2\boldsymbol{\sigma}_1\mathbf{x}\boldsymbol{\sigma}_1\boldsymbol{\sigma}_2\boldsymbol{\sigma}_3 = -\ i^\dagger\mathbf{x}i = -\mathbf{x}$.

(3.5) $R\mathbf{x}' = \mathbf{x}R$; hence $(1 + i\boldsymbol{\gamma})\mathbf{x}' = \mathbf{x}(1 + i\boldsymbol{\gamma})$.

(3.6) Use the relations $RR^\dagger = \alpha^2 + \boldsymbol{\beta}^2 = 1$ and $e^{i\mathbf{a}} = R^2 = \alpha^2 - \boldsymbol{\beta}^2 + 2i\alpha\boldsymbol{\beta}$.

(3.7) From Exercise (3.6)
$e_{jk} = \delta_{jk} + i\hat{\mathbf{a}}\wedge\boldsymbol{\sigma}_j\wedge\boldsymbol{\sigma}_k \sin a + (\hat{\mathbf{a}}\wedge\boldsymbol{\sigma}_j){\cdot}(\hat{\mathbf{a}}\wedge\boldsymbol{\sigma}_k)(1 - \cos a)$.

(a) $$\begin{bmatrix} \cos a & -\sin a & 0 \\ \sin a & \cos a & 0 \\ 0 & 0 & 1 \end{bmatrix}.$$

(3.12) $\mathrm{Tr}\ \mathcal{R} = \sum_k\langle\boldsymbol{\sigma}_k R^\dagger\boldsymbol{\sigma}_k R\rangle_0 = 4\langle R\rangle_0^2 - 1 = 4\cos^2\frac{1}{2}\mathbf{a} - 1 = 1 + 2\cos\mathbf{a}$.

(3.13) Consider the products of a symmetric operator with reflections by its eigenvectors.

(3.15) $\mathcal{S}\mathbf{x} = \mathbf{A}\mathbf{x}{\cdot}\hat{\mathbf{A}}$; $R = e^{\hat{\mathbf{A}}\pi/4}$.

(3.16) Eigenvector $\boldsymbol{\sigma}_1 = f\boldsymbol{\sigma}_1$; Principal values $\lambda_\pm = (1+\alpha^2)^{1/2} \pm \alpha$,
Principle vectors $\mathbf{e}_\pm = \boldsymbol{\sigma}_1 \pm \lambda_\pm \boldsymbol{\sigma}_2$;

$$R^2 = \lambda_\pm^{-1}\mathbf{e}^{-1}(f\mathbf{e}_\pm) = \frac{\lambda_\pm - \alpha\lambda_\pm\,\boldsymbol{\sigma}_1\boldsymbol{\sigma}_2}{1 \pm \alpha\lambda_\pm} = e^{\boldsymbol{\sigma}\,{}_{12}\theta},$$

whence $\tan\theta = -\alpha$.

Section 5-4.

(4.2) $\{S \mid \mathbf{b}\}\{R \mid \mathbf{a}\}\mathbf{x} = \{S \mid \mathbf{b}\}(\tilde{R}\mathbf{x}R + \mathbf{a}) = \tilde{S}(\tilde{R}\mathbf{x}R + \mathbf{a})S + \mathbf{b}$
$= (RS)^{\sim}\,\mathbf{x}(RS) + \tilde{S}\mathbf{a}S + \mathbf{b} = \{RS \mid \tilde{S}\mathbf{a}S + \mathbf{b}\}\mathbf{x}$.

(4.6) $\mathbf{c} = \mathbf{a} - \mathbf{b} + R^\dagger\mathbf{b}R$.

(4.7) $\{R \mid \mathbf{a}\}^{-1}\{1 \mid \mathbf{b}\}\{R \mid \mathbf{a}\} = \{1 \mid R\mathbf{b}\tilde{R}\}$

(4.8) $\mathcal{S}_\mathbf{a}\mathbf{x} = -\mathbf{a}^{-1}(\mathbf{x} - \mathbf{a})\mathbf{a} + \mathbf{a} = -\mathbf{a}^{-1}\mathbf{x}\mathbf{a} + 2\mathbf{a}$.

Section 5-5.

(5.3) Using Equations (3.42a, b, c),
$2\dot{R} = \dot{\psi} i\sigma_3 R + R_\psi Q_\theta i\sigma_1\dot{\theta}R_\psi + Ri\sigma_3\dot{\psi}$
$= i(\sigma_3\dot{\psi} + R_\psi Q_\theta\sigma_1 Q_\theta^\dagger R_\psi^\dagger\dot{\theta} + R\sigma_3 R^\dagger\dot{\phi})R$

(5.7) (a) $\boldsymbol{\omega}_1 = \dfrac{v}{b}\mathbf{a}_0,\ \boldsymbol{\omega}_2 = -\dfrac{v}{a}\mathbf{b},\ \boldsymbol{\omega} = \dfrac{v}{ab}(\mathbf{a} - \mathbf{b})$.

(b) $\dot{\mathbf{x}} = \dot{\mathbf{r}} + \dot{\mathbf{a}} = \dfrac{v}{ab}(\mathbf{a} - \mathbf{b}) \times \mathbf{r} + \mathbf{a} \times \mathbf{b}$,

$$\ddot{\mathbf{r}} = \frac{v^2}{a^2b^2}\,\mathbf{a} \times (\mathbf{a} \times \mathbf{r}) + \mathbf{b} \times (\mathbf{b} \times \mathbf{r}) + \mathbf{b} \times (\mathbf{r} \times \mathbf{a}) + \mathbf{a} \times (\mathbf{r} \times \mathbf{b}),$$

$$\ddot{\mathbf{a}} = -\frac{v^2}{a^2}\mathbf{a}.$$

(c) At $\mathbf{r} = \mathbf{b}$, $\dot{\mathbf{x}} = \dfrac{2v}{ab}\mathbf{a} \times \mathbf{b},\ \ddot{\mathbf{x}} = -v^2\,\mathbf{b}^{-1}$;

At $\mathbf{r} = -\mathbf{b}$, $\dot{\mathbf{x}} = \mathbf{0},\ \ddot{\mathbf{x}} = v^2(\mathbf{b}^{-1} - 2\mathbf{a}^{-1})$.

Section 5-6.

(6.1) $\alpha \approx \dfrac{|\mathbf{g}\wedge\boldsymbol{\omega}|\mathbf{r}\cdot\boldsymbol{\omega}}{g^2} = \dfrac{\omega^2 r}{g}\cos\theta\sin\theta$;

$\alpha_{\max} = 0°6'$ at $\theta = 45°$.

(6.2) (a) $\Delta\mathbf{r} = \frac{1}{3}t^3\,\mathbf{g} \times \boldsymbol{\omega} = (1.55\text{ cm})$ East.
(b) In an inertial frame, the Easternly velocity of the ball when it is released is greater than that of the Earth's surface.
(c) $(2.5 \times 10^{-4}\text{ cm})$ South.

(6.4) $r_S\omega_S^2/r_E\omega_E^2 = 0.2\%$.

Section 6-1.

(1.1) Consider the universe.

(1.3) Energy dissipated = $mga/6$.

(1.6) (a) $v = v_0/(1 + m/M)$.
(b) $v = v_0 e^{-m/M}$.

Section 6-2.

(2.1) $$\ddot{x} = g \sin\alpha \left(1 - \frac{m\cos^2\alpha}{m+M}\right)^{-1},$$

$$\ddot{X} = -g\sin\alpha\cos\alpha\left(\frac{m+M}{m} - \cos^2\alpha\right)^{-1}.$$

(2.3) $(m_1 + m_2)\ddot{x} + m_2(l\ddot{\phi}\cos\phi - l\dot{\phi}^2\sin\phi) = 0$,
$l\ddot{\phi} + \ddot{x}\cos\phi + g\sin\phi = 0$.

(2.6) $\ddot{x}_1 = \frac{1}{5}g = -\frac{1}{2}\ddot{x}_2$.

Section 6-3.

(3.2) $\omega_r = 2(\varkappa/m)^{1/2}\sin(r\pi/8)$ for $r = 1, 2, 3$.
$q_1(t) = (1/8)^{1/2}A(\cos\omega_1 t - \cos\omega_3 t) = q_3(t)$,
$q_2(t) = \frac{1}{2}A(\cos\omega_1 t - \cos\omega_3 t)$.

Section 6-4.

(4.1) $\omega_1 = 2\omega_2 = (g/l)^{1/2}$, with unnormalized $|\,a_1) = \left[\begin{smallmatrix}1\\-3\end{smallmatrix}\right]$, $|\,a_2) = \left[\begin{smallmatrix}2\\3\end{smallmatrix}\right]$. $Q_1 = \alpha_1(3\phi_1 - 2\phi_2)$, $Q_2 = \alpha_2(3\phi_1 + \phi_2)$, where α_1 and α_2 are normalization factors.

(4.3) $\omega_1 = \omega_2 = (1 + k)^{1/2}$, $\omega_3 = (1 - 2k)^{1/2}$.

Section 7-2.

(2.1) $\frac{1}{2}\mathbf{a}$.

(2.2) $\frac{1}{4}\mathbf{h},\ \frac{1}{3}\mathbf{h},\ \frac{1}{3}\mathbf{h} + \frac{2\mathbf{a}}{3\pi}$.

(2.3) $\frac{4\mathbf{a}}{3\pi}$, $\frac{2\mathbf{a}}{\pi}$.

(2.4) $\left(\frac{b^2 + 4a^2}{2\pi a}\right)\hat{\mathbf{a}}$.

(2.5) $\frac{3(a+b)^2}{4(2a+b)}$ from the center of the sphere.

(2.6) $\frac{3}{8}\,\mathbf{a}(1 + \cos\phi),\quad \frac{2\pi}{3}a^3(1 - \cos\phi)$.

(2.9) $$\mathcal{I}\mathbf{u} = \frac{m}{4}\left[\frac{2a^2}{3}\mathbf{u} - (\mathbf{a}_1 + \mathbf{a}_2 + \mathbf{a}_2)(\mathbf{a}_1 + \mathbf{a}_2 + \mathbf{a}_3)\cdot\mathbf{u}\right]$$

$$(\hat{\mathbf{a}}_j\cdot\mathcal{I}\hat{\mathbf{a}}_k) = ma^2\begin{pmatrix} \frac{2}{3} & -\frac{1}{4} & -\frac{1}{4} \\ -\frac{1}{4} & \frac{2}{3} & -\frac{1}{4} \\ -\frac{1}{4} & -\frac{1}{4} & \frac{2}{3} \end{pmatrix},\ I_1 = \tfrac{1}{6}\,ma^2,\ I_2 = I_3 = \tfrac{11}{12}\,ma^2$$

(2.10) $\mathrm{I} = \frac{m}{6}\,\frac{b^2c^2 + c^2a^2 + a^2b^2}{a^2 + b^2 + c^2}$.

(2.11) $\mathcal{I}\mathbf{u} = 72\mathbf{u}$

(2.12) $\frac{2}{3}\,(\mathbf{a} \times \boldsymbol{\omega})^2$

(2.13) $\mathbf{e}_1 = \mathbf{r}_1,\ \mathbf{e}_2 = \mathbf{r}_2 + \mathbf{r}_3,\ \mathbf{e}_3 = \mathbf{r}_2 - \mathbf{r}_3$

$$\mathcal{I}\mathbf{u} = a^2[\hat{\mathbf{e}}_1\hat{\mathbf{e}}_1\wedge\mathbf{u} + 9\hat{\mathbf{e}}_2\hat{\mathbf{e}}_2\wedge\mathbf{u} + \hat{\mathbf{e}}_3\hat{\mathbf{e}}_3\wedge\mathbf{u}].$$

(2.23) Use the method at the end of Section 5–2.

(2.24) $I = \frac{ma^2}{12}(8 + \sqrt{37}),\ \phi = 4.7°$

Section 7-6.

(6.4) $\tan\phi = 1 - 2\mu,\ \omega = \frac{\sqrt{2}\,\mu V_0}{a}$.

Appendix A

(A.2) $\hat{\mathbf{a}}\hat{\mathbf{a}}' = -i\hat{\mathbf{a}}\hat{\mathbf{A}} = -i\boldsymbol{\alpha}e^{i\boldsymbol{\alpha}} = e^{i\boldsymbol{\alpha}(\alpha - \pi/2)}$.

(A.7) Area $= 2|\mathbf{a}\wedge\mathbf{b}| + 2|\mathbf{b}\wedge\mathbf{c}| + 2|\mathbf{c}\wedge\mathbf{a}|$

$$= 2ab \sin C + 2bc \sin A + 2ca \sin A\,.$$

From Exercise (A.6),
Volume $= |\mathbf{a}\wedge\mathbf{b}\wedge\mathbf{c}|$
$= abc[1 - \cos^2 A - \cos^2 B - \cos^2 C + 2 \cos A \cos B \cos C]^{1/2}$.

(A.9) $4\pi = 2\Delta + 2(2a - \Delta) + 2(2b - \Delta) + 2(2c - \Delta)$.

References

There are many fine textbooks on classical mechanics, but only a couple are mentioned below as supplements to the present text. Most students spend too much time studying textbooks. They should begin to familiarize themselves with the wider scientific literature as soon as possible. The sooner a student penetrates the specialist literature on topics that interest him, the more rapidly he will approach the research frontier. He should not be afraid to tackle advanced monographs, for he will find that they often contain more licid treatments of the basics than introductory texts,and the difficult parts will alert him to specifics in his background that need to be filled in. Rather than read aimlessly in a broad field, he should focus on specific topics, search out the relevant literature, and determine what is required to master them. Above all, he should learn to see the scientific literature as a vast lode of exciting ideas which he can mine at will by himself.

Most of the references below are intended as entries to the literature on offshoots and applications of mechanics. Many are classics in their fields, and some are advanced monographs, but all of them will yield rich rewards to the dedicated student. This is just a sampling of the literature with no attempt at completeness on any topic.

Supplementary Texts

The books by French and Feynman are notable for their rich physical insight communicated with a minimum of mathematics. Although both are introductory textbooks, they can be read with profit by advanced students. French's book is especially valuable for its historical information, which should be part of every physicists education.

A. P. French, *Newtonian Mechanics*, Norton, N.Y. (1971).

R. P. Feynman, *Lectures on Physics*, Vol. I, Addison-Wesley, Reading (1963).

References on Geometric Algebra

The present book (NFI) is the only available introduction to geometric algebra and its applications to mechanics. A sequel (NFII) in preparation will considerably broaden the range of applications.

The only other published books on geometric algebra are advanced monographs, *Space-Time Algebra* and *Geometric Calculus*. The first deals tersely with applications to relativity. The second is devoted exclusively to mathematical developments of the calculus. It is advisable to become thoroughly familiar with NFI before addressing either of these books.

D. Hestenes and G. Sobczyk, *Clifford Algebra to Geometric Calculus, a Unified Language for*

Mathematics and Physics, D. Reidel, Dordrecht (1984). [Referred to as *Geometric Calculus* in the text]
D. Hestenes, *Space-Time Algebra*, Gordon and Breach, N.Y. (1966).
D. Hestenes, *New Foundations for Mathematical Physics*, D. Reidel, Dordrecht (estimated publication date: Spring 1989). [Referred to as NFII in the text]

Chapter 1

A satisfactory history of geometric algebra has not yet been written. But Kline traces the interplay between geometry and algebra, mathematics and physics in their historical development. The scholarly work by Van der Waarden shows clearly the common historical origins of geometry and algebra. Clifford's book is one of the best popular expositions ever written on the role of mathematics in science.

M. Kline, *Mathematical Thought from Ancient to Modern Times*, Oxford U. Press, N.Y. (1972).
B. L. Van der Waarden, *Science Awakening*, Wiley, N.Y. (1963).
W. K. Clifford, *Common Sense of the Exact Sciences* (1978), reprinted by Dover, N.Y. (1946).

Section 2-6

Zwikker gives an extensive treatment of plane analytic geometry, using complex numbers in a manner closely related to the techniques of Geometric Algebra.

C. Zwikker, *The Advanced Geometry of Plane Curves and Their Applications*, Dover, N.Y. (1963).

Section 3-1.

This magnificently edited and annotated collection of Newton's papers provides valuable insight into Newton's genius. One can see, for instance, the extensive mathematical preparation in analytic geometry that preceded his great work in mechanics.

D. T. Whiteside (ed.), *The Mathematical Papers of Isaac Newton*, Cambridge U. Press, Cambridge (1967–81), 8 Vol.

Section 3-5.

The forces of fluids on moving objects are extensively analyzed theoretically and empirically in Batchelor's classic.

G. K. Batchelor, *An Introduction to Fluid Dynamics*, Cambridge U. Press, N.Y. (1967).

Section 3-6 and 3-7.

Feynman gives a good introduction to electromagnetic fields and forces.

R. P. Feynman, *The Feynman Lectures on Physics*, Vol. II, Addison-Wesley, Reading (1964).

Section 6-3.

Brillouin's classic is an object lesson in how much can be accomplished with a minimum of mathematics. He discusses electrical-mechanical analogies as well as waves in crystals.

L. Brillouin, *Wave Propagation in Periodic Structures*, McGraw-Hill, N.Y. 1946 (Dover, N.Y. 1953).

Physics students will do well to sample the vast engineering and applied mathematics literature on linear systems theory.

Section 6-4.

Herzberg is still one of the most important references on molecular vibrations. Califano gives a more up-to-date treatment of group theoretic methods to account for molecular symmetry. Further improvement in these methods may be expected from employment of geometric algebra.

G. Herzberg, *Molecular Spectra and Molecular Structure*, II. *Infrared and Raman Spectra of Polyatomic Molecules*, D. Von Nostrand Co., London, (1945).

S. Califano, *Vibrational States*, Wiley, N.Y. (1976).

Section 6-5.

The most extensive survey of work on the restricted three body problem is,

V. Szebehely, *Theory of Orbits*, Academic Press, N.Y. (1967).

Section 7-4.

This one of the standard advanced references on the theory of spinning bodies, as well as the three body problem.

E. T. Whittaker, *A Treatise on the Analytical Dynamics of Particles and Rigid Bodies*, Cambridge, 4th Ed. (1937).

Chapter 8

Stacey and Kaula present fine introductions to the rich field of geophysics and its generalization to planetary physics, showing connections to celestial mechanics. The book by Munk and MacDonald is a classic on the Earth's rotation.

Roy gives an up-to-date introduction to celestial mechanics and astromechanics combined. Kaplan gives a more complete treatment of spacecraft physics.

F. D. Stacey, *Physics of the Earth*, Wiley, N.Y. (1969).

W. M. Kaula, *An Introduction to Planetary Physics*, Wiley, N.Y. (1968).

W. Munk and G. MacDonald, *The Rotation of the Earth*, Cambridge U. Press, London (1960).

A. E. Roy, *Orbital Motion*, Adam Hilger, Bristol, 2nd Ed. (1982).

M. H. Kaplan, *Modern Spacecraft Dynamics and Control*, Wiley, N.Y. (1976).

Section 8-4.

Steifel and Schiefele is the sole reference on applications of the KS equation. The book provides many important insights into computational theory and technique.

E. L. Stiefel and G. Scheifele, *Linear and Regular Celestial Mechanics*, Springer-Verlag, N.Y. (1971).

Chapter 9. References on the Philosophy of Science

The philosophy of science is in a disorderly state. Contradictory viewpoints and ill-considered opinions abound in the scientific and philosophical literature. An assessment of the situation in accord with the viewpoint in this book has been made by philosopher-physicist Mario Bunge:

M. Bunge (1973), *Philosophy of Physics*, Dordrecht: D. Reidel Publ. Co.

In his mature years, Bunge has undertaken a systematic formulation of the philosophy of science from the viewpoint of scientific realism and systems theory. His results will appear in an ambitious seven volume treatise of which several volumes have been published so far:

M. Bunge (1974–). *Treatise on Basic Philosophy*, 7 vol., Dordrecht: D. Reidel Publ. Co.

This sophisticated treatise is not likely to appeal to readers with only a superficial interest in philosophy. However, Bunge's systematic and careful analysis of basic concepts was invaluable in the preparation of the Chapter 9 here.

Supplementing our discussion in Chapter 9, Rosen discusses the fundamental role of symmetry principles in the foundations of scientific theory.

J. Rosen (1983), *A Symmetry Primer for Scientists*, Wiley, N.Y.

As a worthy sample of writings on the foundations of physics, we have the following books by eminent mathematicians as well as physicists and philosophers: But beware of conflicting viewpoints and tacit assumptions!

Campbell (1957). *Foundations of Science*, New York: Dover Publ., Inc., (formerly, *Physics: The Elements*, Cambridge, 1920).

H. Jeffries (1973). *Scientific Interference*, Cambridge U. Press, Cambridge.

J. C. Maxwell (1977). *Matter and Motion*, New York: Dover Publ., Inc.

E. Nagel (1961). *The Structure of Science*, Harcourt, Brace & World, New York.

R. Nevalinna (1964). *Space Time and Relativity*, Addison-Wesley, New York.

E. T. Whittaker (1949). *From Euclid to Eddington: A Study of Conceptions of the External World*, Cambridge U. Press, Cambridge.

Appendix B

Jahnke and Emde is a standard reference on elliptic functions and elliptic integrals.

E. Jahnke and F. Emde, *Tables of Functions*, Dover, N.Y. (1945).

Index

acceleration, 98, 312
  centripetal, 312
  Coriolis, 312
ambient velocity, 146
amplitude of an oscillation, 168
analyticity principle, 122
angle, 66
  radian measure of, 219
angular momentum, 195ff
  base point, 423
  bivector, 196
  change of, 423
  conservation, 196, 338
  induced, 330
  internal, 337
  intrinsic, 424
  orbital, 337
  total, 337
  vector, 196
Angular Momentum Theorem, 338
anharmonic oscillator, 165
anomaly,
  eccentric, 532
  mean, 532
  true, 532, 573
apocenter, 213
apse (see turning point),
area,
  directed, 70
  integral, 112ff, 196
associative rule, 27, 32, 35
astromechanics, 512
asymptotic region, 210, 236
attitude, 420
  element, 529, 549
  spinor, 420
Atwood's machine, 354
axode, 428

ballistic trajectory, 215
barycentric coordinates, 82
basis, 53
  of a linear space, 53, 363
  multivector, 53
  vectorial, 49, 260
beats, 365
billiards, 498, 503
bivector (2-vector), 21
  basis of, 56
  codirectional, 24
  interpretations of, 49
blade, 34
Brillouin zone, 374

Cayley-Klein parameters, 480, 485, 495
celestial mechanics, 512ff
celestial pole, 458, 538
celestial sphere, 466, 537
center of gravity, 433
center of mass, 230, 336
  additivity principles for, 437
  of continuous body, 434
  symmetry principles for, 435ff
  (see centroid)
center of mass theorem, 336
centroid, 438
chain rule, 100, 105, 108
Chasles' theorem, 305
Chandler wobble, 458
characteristic equation, 166, 171, 383
chord, 79
circle, equations for, 87ff
Classical Field Theory, 514, 591
Clifford, 59
clocks,
  particle, 585
  photon, 592

coefficient of restitution, 505, 511
collision,
  elastic, 236, 505
  inelastic, 346, 505
commutative rule, 15, 35
commutator, 44
configuration space, 351, 382
congruence, 3, 303
conicoid, 91
conic (section), 90ff, 207
constants of motion,
  for Lagrange problem, 476
  for rigid motion, 425
  for three body problems, 399
constraint, 181, 599
  bilateral, 188
  for rolling contact, 492
  for slipping, 497
  holonomic, 185, 351, 354
  unilateral, 188
continuity, 97
coordinates,
  complex, 371
  ecliptic, 238ff
  equitorial, 238ff
  generalized, 350, 381
  ignorable (cyclic), 358
  Jacobi, 406
  mass-weighted, 385
  normal (characteristic), 362, 364
  polar, 132, 194
  rectangular, 132
  symmetry, 388
correspondence rules, 579, 581
couple, 430
Cramer's rule, 254
cross product, 60

degrees of freedom, 351
derivative,
  by a vector, 117
  convective, 109
  directional, 105, 107
  of a spinor, 307
  partial, 108
  scalar, 98
  total, 109
Descartes, 5
descriptors, 577
definition,
  explicit, 580
  implicit, 580
determinant, 62, 255
  of a frame, 261
  of a linear operator, 255, 260
  of a matrix, 258, 260
differential, 107
  exact, 116
differential equation, 125
dihedral angle, 604
dilation, 13, 52
dimension, 34, 54
Diophantes, 9
directance, 82, 87, 93, 427
direction, 11
  of a line, 48
  of a plane, 49
dispersion relation, 373
displacement,
  rigid, 303, 305
  screw, 305
distance, 79
distributive rule, 18, 25, 31, 35
drag, 146
  atmospheric, 215, 563
  pressure, 149
  viscous, 149
  (see force law)
drag coefficient C_D, 147
drift velocity, 159
dual, 56, 63
dynamical equations, 454, 578
  (see equations of motion),
dynamics, 198

eccentricity, 90, 205
eccentricity vector, 91, 205, 527
ecliptic, 466, 539
eigenvalue problem, 264ff
  brute force method, 384
eigenvalues, 264ff
  degenerate, 266
eigenvectors, 264ff, 272
Einstein, 574
elastic modulus, 374
elastic solid, 360
electromagnetic wave, 174
ellipsoid, 276
ellipse, 91, 96, 173, 174, 199, 203, 208
  semi-major axis of, 212
elliptic functions, 222, 478, 481, 610ff
  modulus of, 610
elliptic integral, 482, 490, 545, 547,
  B-611ff

energy,
  conservation, 170
  Coriolis, 342
  diagrams, 223, 229
  dissipation, 177, 241
  ellipsoid, 487
  internal, 342ff
  kinetic, 182, 337
    internal, 341
    rotational, 338
    translational, 337
  potential, 182
  storage, 177, 364
  total, 182, 206, 528
  transfer, 238, 344, 364
  vibrational, 342
epicycle, 201
epitrochoid, 201, 204
equality, 12, 37
equations of motion, 125
  rotational (see spinor equations), 340, 420
  secular, 531
    for orbital elements, 531
  translational, 335, 420
equiangular spiral, 155
equilibrium, 379
  mechanical, 429
  point, 409
equimomental rigid bodies, 448
equinoxes, 539
equipotential surface, 116, 185
escape velocity, 214
Euclid, 29
Euclidean spaces
  2-dimensional, 54
  3-dimensional, 54
  n-dimensional, 80
Euler, 121
Euler angles, 289, 294, 486, 490, 538
Euler's Law (equation), 340, 420, 454
  components of, 422ff
Euler parameters, 382, 315
exponential function, 66, 73ff, 281

factorization, 45
Faraday effect, 179
field, 104
first law of thermodynamics, 344
fluid resistance, 146ff
force, 121
  binding, 164
  body, 125
  centrifugal, 318, 332
  conservative, 181, 219
  contact, 125
  Coriolis, 319, 322, 324, 328
  fictitious (see force law), 317
  generalized, 353
  impulsive, 214, 501
  perturbing, 143, 165, 527
  superposition, 122
  tidal, 520
force constants, 380
force field, 184
  central, 219
  conservative, 184, 219
force law, 122
  conservative, 181ff
  constant, 126
  Coulomb, 205,
    with cutoff, 251
  electromagnetic, 123, 155
  frictional, 192, 471
  gravitational, 123, 200, 205, 513
  Hooke's, 122, 361, 364
  inverse square, 200
  magnetic, 151
  phenomenological, 195
  resistive (see drag), 146
    linear, 134, 154
    quadratic, 140
forces on a rigid body,
  concurrent, 433
  equipollent, 428
  parallel, 429
  reduction of, 428
frame (see basis), 261
  body, 339
  Kepler, 529
  reciprocal, 262
frequency,
  cutoff, 371
  cyclotron, 154
  Larmor, 328
  normal (characteristics), 362
    degenerate, 362
  oscillator, 168
  resonant, 177
    multiple, 397

Galileo, 575
geoid, 524, 526
geometric algebra, 53, 55, 80

geometric product, 31, 39
geometry,
  analytic, 78ff
  coordinate, 78
  Euclidean, 79
  non-Euclidean, 79
geopotential, 525
Gibbs, 60
golden ratio, 226
grade, 22, 30, 34
gradient, 116
Grassman, 12, 14, 28
gravitational field, 513
  force exerted by, 513, 520
  of an axisymmetric body, 518
  of an extended object, 515ff
  source, 513
  superposition, 514
gravitational potential, 514
  harmonic (multipole) expansion of, 517
  of a spherically symmetric body, 516
gravitational quadrupole tensor, 517, 542
gravity assist, 239, 242
group,
  abstract, 296
  continuous, 298
  dirotation, 296
  Euclidean, 301ff
  Galilean, 313
  orthogonal, 299
  representation, 297
  rotation, 296ff
  subgroup of, 299, 306
  transformation, 295
  translation, 300ff
guiding center, 158
gyroscope, 454
gyroscopic stiffness, 455ff

Hall effect, 160
Halley's comet, 214
Hamilton, 59, 286
Hamilton's theorem, 295
harmonic approximation, 380
harmonic oscillator, 165
  anisotropic, 168
  coupled, 361
  damped, 170
  forced, 174
  in a uniform field, 173, 202, 325
  isotropic, 165
heat transfer, 345
helix, 154
Hill's regions, 416
hodograph, 127, 204
Hooke's law, 122, 166
hyperbola, 91, 96, 208
  branches, 213
hyperbolic functions, 74
hypothesis, 580
hypotrochoid, 202, 204

idempotent, 38
impact parameter, 211, 245
impulse, 501
impulsive motion, 501
inertia tensor, 253, 339, 421, 439
  additivity principles for, 442
  calculation of, 439
  canonical form for, 451
  derivative of, 340
  matrix elements of, 445
  of a plane lamina, 274
  principle axes of, 422
  principle values of, 422
  symmetries of, 448
initial conditions, 125
initial value problem, 208
integrating factor, 134, 139, 152, 173
interaction, 121
  gravitational, 513
inner product, 16ff, 33, 36, 39
  of blades, 43
inverse, 35, 37
inversion, 293, 437

Jacobi identity, 47, 83
Jacobi's integral, 408

Kepler, 200
Kepler motion, 527
Kepler problem, 204
  2-body effects on, 233
Kepler's equation, 216ff, 533
Kepler's Laws,
  first, 198
  second, 196, 198
  third, 197, 198, 200, 203
    modification of, 233
kinematical equation,
  for rotational motion, 454
kinematics, 198
KS equation, 569

Lagrange points, 409
Lagrangian, 353
Lagrange's equation, 190, 353, 380
Lagrange's method, 354
Lame's equation, 491
Laplace expansion, 43, 261
Laplace vector (see eccentricity vector),
Larmor's theorem, 328
lattice constant, 367
law of composition,
  generic, 586
  specific, 587
law of cosines, 19, 69
  spherical, 523, 606, 607
law of sines, 26, 70
  spherical, 607
law of tangents, 294
lemniscate, 204
lever, law of, 430
line,
  equations for, 48, 81ff
  moment of, 82
line integral, 109ff, 115
line vector, 428
linear algebra, 254
linear dependence, 47
linear functions, 107, 252ff
  (see linear operators)
linear independence, 53
linear operators, 253ff
  adjoint (transpose), 254
  canonical forms, 263, 270, 282
  derivative of, 316
  determinant of, 255, 260
  inverse, 260
    matrix element, 262
  matrix element, 257
  matrix representation of, 257
  nonsingular, 256
  orthogonal, 277
    improper, 278
    proper, 278
  polar decomposition, 291
  product, 253
  secular equation for, 265
    complex roots, 268
    degenerate, 266
  shear, 295
  skewsymmetric, 263
  symmetric (self-adjoint), 263, 269ff
    spectral form, 270
    square root, 271
  trace, 295
linear space, 53
  dimension of, 54
linear transformation (see linear operator),
Lissajous figure, 169
local interaction principle, 588, 591, 593
logarithms, 75ff
Lorentz electron theory, 179
Lorentz force, 123

Mach number, 149
magnetic spin resonance, 473ff
magnetron, 202
magnitude, 3, 6
  of a bivector, 24
  of a multivector, 46
  of a vector, 12
many body problem, 398
  constants of motion, 399
mass, 230
  density, 434
  reduced, 230
  total, 336, 434
matrix, 257
  determinant of, 259, 260
  equation, 258
  identity, 258
  product, 258
  sum, 258
mean motion, 532
measurement, 2, 581, 600
model, 378, 577ff
  abstract, 599
  concrete, 599
  deployment, 596, 600
  development, 596
  process, 598
  ramified, 600
modeling, 596ff
  stages, 596
modulus,
  of a complex number, 51
  of an elliptic function, 610
  of a multivector, 46
Mohr's algorithm, 273
moment arm, 428
momentum, 236
  conservation, 236, 336, 591
  flux, 347
  transfer, 238, 240, 591
motion, 121
  in rotating systems, 317ff

rigid, 306ff
translational, 335
(see rotational motion, periodic motion)
multivector, 34
even, 41
homogeneous, 41, 42
k-vector part of, 34, 39
odd, 41
reverse, 45

natural frequency, 168
Newton, 1, 120, 124, 574
Newton's Law of Gravitation, 398
universality of (see force law), 201, 203
Newton's Laws of Motion,
first, 588
second, 41, 588, 590
third, 588
strong form, 335
weak form, 335, 591
nodes,
ascending, 538
line of, 290
precession of, 540
normal (to a surface), 116
normal modes, 362
degenerate, 383
expansion, 364
nondegenerate, 383
normalization, 377
orthogonality, 369
wave form, 369
number, 3, 5
complex, 57
directed, 11, 12, 34
imaginary, 51
real, 10, 11, 12
nutation,
luni-solar, 551
of a Kepler orbit, 540
of a top, 470
of Moon's orbit, 550

oblateness,
constant J_2, 518
of Earth, 459, 467
perturbation, 542, 560
Ohm's law, 137
orbit, 121, 585
orbital averages, 253ff
orbital elements, 527
Eulerian, 538
secular equations for, 531
orbital transfer, 214
orientation, 16, 23, 51
origin, 79
operators (see linear operators), 50
operational referition, 581
oscillations,
damped, 393
forced, 395
free, 382
phase of, 168
small (see vibrations), 378
osculating orbit, 528
outer product, 20, 23, 36, 39
of blades, 43
outermorphism, 255

parabola, 91, 96, 126, 207
Parallel Axis Theorem, 424
parenthesis, 42
preference convention, 42
for sets, 48
particle, 121
unstable, 242
pendulum,
compound, 463, 477, 489
conical, 467
double, 355, 386
damped, 395
Foucault, 223
gyroscopic, 462
simple, 191, 463
small oscillations of, 462
spherical, 475
pericenter, 91, 213
perigee, 213
perihelion, 213
period,
of an oscillator, 168
of central force motion, 221
of the Moon, 203
periodic motion, 168, 478
perturbation,
oblateness, 542, 560
theory, 141, 320
gravitational, 527, 541
third body, 541
physical space, 80, 583
plane,
equations for, 86ff
Poincaré, 399, 416

Poinsot's construction, 487
point of division, 84, 430
polygonal approximation, 143
position, 80, 121, 420, 584ff
  spinor, 564
position space, 314, 584
potential, 116
  attractive, 229
  barrier, 228
  central, 220
  centrifugal, 224
  effective, 220, 408, 414
  gravitational, 514, 516
  screened Coulomb, 224
  secular, 545
  Yukawa, 224
precession, 222
  luni-solar, 552
  of Mercury's perihelion, 542ff
  of pericenter, 530
  of the equinoxes, 468, 553
  relativistic,
    General, 559
    Special, 562
precession of a rigid body,
  Eulerian free, 456, 467, 475
  steady, 463, 483
    deviations from, 467
principle moments of inertia, 446
principle values, 269, 292
principle vectors, 269, 292
  of inertia tensor, 422
projectile,
  Coriolis deflection, 321ff
  range, 127, 136
  terminal velocity, 135
  time of flight, 130, 132
projection, 16, 65, 270, 603
properties,
  emergent, 582, 600
  physical, 576
  primary, 576
  qualitative, 578
  quantitative, 578
  secondary, 576
pseudoscalar,
  dextral (right handed), 55
  of a plane, 49, 53
  of 3-space, 54, 57

quantity, 34
quaternion, 58, 62
  theory of rotations, 286

radius of gyration, 447, 463
reference class,
  of a model, 577
  of a property, 578
  of a theory, 581
reference frame (body), 314, 584
reference system, 317
  geocentric, 317
  heliocentric, 317
  inertial, 311, 586, 588, 589, 593
  topocentric, 317
    motion of, 327
reflection, 278ff
  law of, 280
rejection, 65
regularization, 571
Relativity,
  General Theory, 514, 542, 557, 574, 583
  Principle, 589, 593
  Special Theory, 562, 583
relaxation time, 135
resonance, 176
  cyclotron, 162
  electromagnetic, 175
  magnetic, 331, 473ff
  multiple, 396
  spin-orbit, 554
reversion, 45
Reynolds Number, 147
rigid body classification, 448
  asymmetric, 448
  axially symmetric, 448, 454
  centrosymmetric, 448
rocket propulsion, 348
Rodrigues' formula, 293
rolling motion, 492ff
rotation, 50, 278, 280ff
  axis, 304
  canonical form, 282, 288
  composition, 283
  group, 295ff
    matrix representation, 296
    spin representation, 296
  matrix elements, 286, with Euler angles, 294
  oriented, 283
  parametric form, 282
  physical, 297
  right hand rule, 282
  spinor theory of, 286
  rotational motion, 317
  integrable cases (see spinning top), 476
  of a particle system, 338
  of asymmetric body, 482

of the Earth, 327, 551
stability of, 488

satellite,
orbital precession, 544
perturbation of, 547
synchronous orbit, 203
scalar, 12
scalar integration, 100
scalar multiplication, 12, 24, 31, 35
scattering,
angle, 210, 245
in CM system, 237, 242
in LAB system, 239, 242
Coulomb, 247, 250
cross section, 243ff
LAB and CM, 248ff
Rutherford, 247
elastic, 236
for inverse square force, 210ff
hard sphere, 246
scientific explanation, 601
scientific knowledge, 596
scientific law, 579
basic, 580
derived, 580
dynamical, 580
generic, 580, 581
interaction, 580, 588
physical, 579
scientific method, 595, 597
scientific realism, 575
scientific theory, 579
semi-latus rectum, 91, 212
sense (or orientation), 51
siderial day, 458
simultaneity, 583, 585, 592
hypothesis of, 589, 592
solar wind, 563
solid angle, 244
sphere, equations for, 87ff
spherical excess, 609
spinning top,
fast, 466
hanging (see precession, rotational motion), 466
Lagrange problem for, 462, 479, 490
rising, 473
sleeping, 473
slow, 466
spherical, 460
symmetrical, 454ff
Eulerian motion of, 460
reduction of, 459
spinor, 51, 52, 67
derivative of, 307
Eulerian form, 284
improper, 300
mechanics, 564
parametrizations, 286
unitary (unimodular), 280
spinor equation of motion,
for a spherical top, 461
for a particle, 569
stability, 165, 227, 380
of circular orbits, 228
of Lagrange points, 410ff
of rotational motion, 488
of satellite attitude, 553
state function, 587
state variables, 126
Stokes' Law, 147
summation convention, 63
super-ball, 507, 510
superposition principle,
for fields, 514
for forces, 122, 588
for vibrations, 363
symmetry of a body, 435ff, 441
system,
2-particle, 230ff
closed, 346
configuration of, 350
Earth-Moon, 234
harmonic, 382
isolated, 232, 336
many-particle, 334ff
open, 346
systems theory, 578
linear, 378

Taylor expansion, 102, 107, 164
temperature, 346
tensor, 253
three body problem, 400
circular restricted, 407
periodic solutions, 416
classification of solutions, 404
collinear solutions, 402
restricted, 406
triangular solutions, 402
tidal friction, 522
tides, 522
time scale, 589
tippie-top, 476
torque, 338

base point, 424
gravitational, 520
moment arm, 428
translation, 300
(see group)
trivector, 26
trochoid, 159, 217
turning points, 213, 221, 227
trigonometric functions, 74, 281
trigonometry, 20, 68
identities, 71, 294
spherical, 603

units, 614

variables,
descriptive, 577
instantiation of, 585
interaction, 334, 578
kinematic, 421
macroscopic, 345
object, 421, 577
position, 334, 584
state, 420, 577
vector, 12
addition, 15
axial, 61
collinear (codirectional), 16, 64
identities, 62
negative, 15
orthogonal, 49, 64
orthonormal, 55
polar, 61
rectangular components, 49, 56
square, 35
units, 13
vector field, 184
vector space, 49, 53
velocity, 98
additional theorem, 314, 593
angular, 309
complex, 427
rotational with Euler angles, 308, 423, 315, 490
spinor, 564
translational, 309
velocity filter, 160
vibrations,
of H_2O, 392
lattice, 366
molecular, 341, 387ff
small, 341, 378
Vieta, 9

wave,
harmonic, 372
polarized, 375
standing, 372
traveling, 372
wavelength, 369
wave number, 369
weight,
apparent, 318
true, 318
work, 183, 342ff
microscopic, 345
Work-Energy Theorem, 343
wrench, 426
reduction of, 431
superposition principle, 428

Zeeman effect, 332
zero, 14
Zeroth Law, 80, 580, 582ff, 586